CBS PROBLEMS & SOLUTIONS SERIES

Problems & Solutions of Engineering Electromagnetics

AF673217

CBS PROBLEMS & SOLUTIONS SERIES

Problems & Solutions of Engineering Electromagnetics

By Experienced Teachers

CBS

CBS PUBLISHERS & DISTRIBUTORS PVT. LTD.

New Delhi • Bengaluru • Chennai • Kochi • Mumbai • Pune

ISBN : 81-239-0559-9

First Edition : 1998
Reprint : 1999, 2002, 2003, 2004, 2005, 2006,
2007, 2009, 2010, 2012, 2014

Copyright © by Author

All rights reserved. No part of this book may be reproduced or transmitted in any form or by any means, electronic or mechanical, including photocopying, recording, or any information storage and retrieval system without permission, in writing, from the publisher.

Published by Satish Kumar Jain and produced by V.K. Jain for
CBS Publishers & Distributors Pvt. Ltd.,
CBS Plaza, 4819/XI Prahlad Street, 24 Ansari Road, Daryaganj,
New Delhi - 110002, India. • Website: www.cbspd.com
e-mail: delhi@cbspd.com, cbspubs@airtelmail.in
Ph.: 23289259, 23266861, 23266867 • Fax: 011-23243014

Branches:

- ***Bengaluru:*** Seema House, 2975, 17th Cross, K.R. Road,
  Bansankari 2nd Stage, Bengaluru - 560070
  • Ph.: +91-80-26771678/79 • Fax: +91-80-26771680
  • E-mail: cbsbng@gmail.com, bangalore@cbspd.com
- ***Pune:*** Bhuruk Prestige, Sr. No. 52/12/2+1+3/2,
  Narhe, Haveli (Near Katraj-Dehu Road By-pass), Pune - 411041
  • Ph.: +91-20-64704058/59, 020-32392277 • E-mail: pune@cbspd.com
- ***Kochi:*** 36/14, Kalluvilakam, Lissie Hospital Road,
  Kochi - 682018, Kerala • Ph.: +91-484-4059061-65
  • Fax: +91-484-4059065 • E-mail: cochin@cbspd.com
- ***Chennai:*** 20, West Park Road, Shenoy Nagar, Chennai - 600030
  Ph.: +91-44-26260666, 26208620 • Fax: +91-44-42032115
  • E-mail: chennai@cbspd.com
- ***Mumbai:*** 83-C, Ist Floor, Dr. E. Moses Road, Worli, Mumbai-400 018,
  Maharashtra Ph.: +91-9833017933, 022-24902340/24902341
  • E-mail: mumbai@cbspd.com

Printed at :
India Binding House, Noida (UP)

Preface

The Engineer of the coming decades, in which the world will be more like a global village connected through *infobahns*, is supposed to be more diversified in his knowledge than the conventional engineers of the past used to be. He is expected to be highly versatile in his compatibility, whether it be in the field of *financial engineering* or in the making of *Jurassic Park*.

With this thought in our mind, we have made this book to which we don't claim as solution to something but an approach to make engineering easy to the students, so as to provide them with enough time and scope to broaden their horizons.

Publishers

Contents

CHAPTER 1

Review Of Vector Analysis

1.1 CARTESIAN COORDINATE SYSTEM :

1.1.1 Coordinates :

In cartesian coordinate system a point is represented by the coordinates (x, y, z). All the three coordinates represent planes. A right handed cartesian coordinate system is that one, in which a rotation (through the smaller angle) of the x axis into the y axis would cause a right handed screw to progress in the direction of z axis. Fig. 1.1 (a), (b), (c) shows possible orientations of right handed coordinate system. Origin is denoted by the letter O and the coordinates are (0, 0, 0).

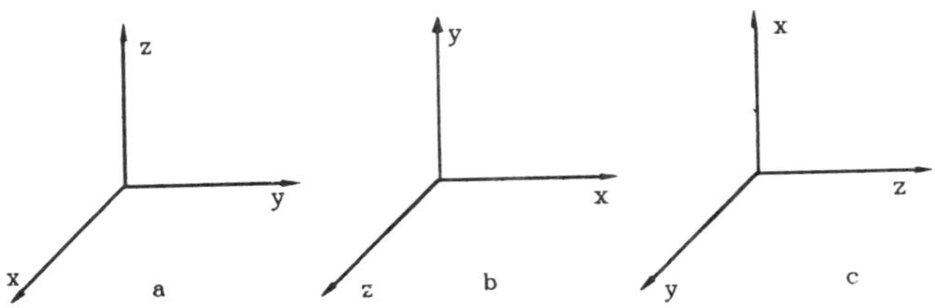

(a) (b) (c)

FIG. 1.1

1.1.2 Planes :

In xy plane, z coordinate of any point is zero. Hence equation of xy plane is $z = 0$. Similarly the equation of yz plane is $x = 0$ and the equation of xz plane is $y = 0$.

The $x = 0, y = 0$, and $z = 0$ are basic planes. Any other value of x, y, and z coordinate represents a plane parallel to respective basic planes, e.g. $z = 5$ represents a plane parallel to $z = 0$ ($x\,y$) plane.

In $x\,y$ plane, z coordinate is zero. Hence, general point in $z = 0$ plane is $(x, y, 0)$. Similarly general point in $y = 0$ plane is $(x, 0, z)$ and general point in $x = 0$ plane is $(0, y, z)$.

To write the general point for any plane whose equation is given by x, y, or z coordinate, include the equation of the plane as one out of the three coordinates where as the remaining two coordinates are variables. For example, for the plane $z = 5$, the general point will be $(x, y, 5)$. The three planes intersect each other to form a point as shown in Fig. 1.2.

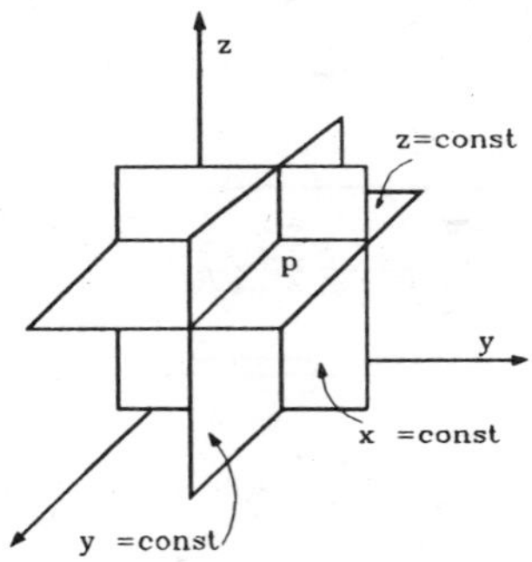

FIG.1.2

1.1.3 Axes :

A line is an intersection of two planes. As seen from Fig. 1.3, x axis is formed by the intersection of $z = 0$ (x, y) and $y = 0$ (x, z) planes. Hence, the equations of x axis are $y = 0, z = 0$. Similarly the equations of y axis are $x = 0, z = 0$ and that of z axis are $x = 0, y = 0$. Further the general point on the x axis is $(x, y = 0, z = 0)$, on the y axis is $(x = 0, y, z = 0)$ and on the z axis is $(x = 0, y = 0, z)$.

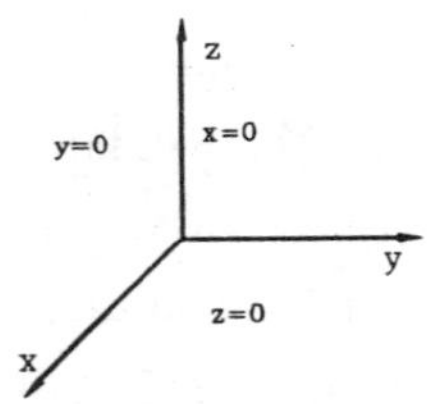

FIG. 1.3

1.1.4 Vectors :

Vectors are those quantities which can be represented by magnitude and direction both.

Unit vector :

It is a vector with magnitude equal to unity. It is denoted by the letter $\bar{a}$. A unit vector is obtained by dividing a vector by its magnitude. It is used to indicate the direction of a vector quantity.

In cartesian coordinate system the three unit vectors are $\pm\ \bar{a}_x$, $\pm\ \bar{a}_y$ and

$\pm\ \bar{a}_z$. All the three unit vectors are mutually perpendicular to each other.

The unit vectors $\pm\ \bar{a}_x$ are perpendicular where as the unit vectors $\pm\ \bar{a}_y$ and $\pm\ \bar{a}_z$ are tangent to the plane represented by $x =$ constant.

The unit vectors $\pm\ \bar{a}_y$ are perpendicular where as the unit vectors $\pm\ \bar{a}_x$ and $\pm\ \bar{a}_z$ are tangent to the plane represented by $y =$ constant.

The unit vectors $\pm\ \bar{a}_z$ are perpendicular whereas the unit vectors $\pm\ \bar{a}_x$ and $\pm\ \bar{a}_y$ are tangent to the plane represented by $z =$ constant.

Position vector :

A vector directed from origin towards any point is known as position vector (p.v.) of that point. It is denoted by the letter $\bar{r}$. If point A has the coordinates (x, y, z) , then position vector of point A is equal to

$$\bar{r}_A = x\,\bar{a}_x + y\,\bar{a}_y + z\,\bar{a}_z$$

A vector directed from point A (starting or tail point) towards point B (end or head point) can be obtained by applying triangle law of vector addition to Fig. 1.4.

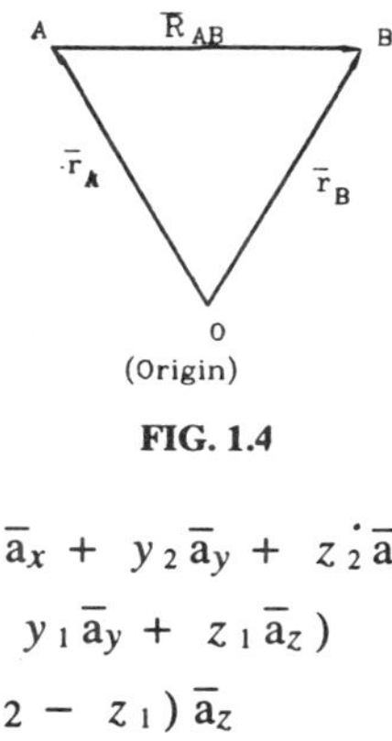

FIG. 1.4

$$\bar{r}_A + \bar{R}_{AB} = \bar{r}_B$$

or $\bar{R}_{AB} = \bar{r}_B - \bar{r}_A$

where $\bar{r}_B =$ p . v. of point B

$\bar{r}_A =$ p . v. of point A

Suppose coordinates of point A are (x_1, y_1, z_1) and that of point B are (x_2 , y_2 , z_2) . Then

$$\bar{r}_A = x_1\,\bar{a}_x + y_1\,\bar{a}_y + z_1\,\bar{a}_z \quad \text{and} \quad \bar{r}_B = x_2\,\bar{a}_x + y_2\,\bar{a}_y + z_2\,\bar{a}_z$$

$$\therefore\ \bar{R}_{AB} = (x_2\,\bar{a}_x + y_2\,\bar{a}_y + z_2\,\bar{a}_z) - (x_1\,\bar{a}_x + y_1\,\bar{a}_y + z_1\,\bar{a}_z)$$

$$= (x_2 - x_1)\,\bar{a}_x + (y_2 - y_1)\,\bar{a}_y + (z_2 - z_1)\,\bar{a}_z$$

A vector directed from point A towards point B is written as $\bar{R}_{AB}$. The first subscript (or letter A) represents the starting point where as the second subscript (or letter B) represents the end point.

Magnitude of a vector :

As far as $\bar{R}_{AB}$ is concerned, the magnitude of $\bar{R}_{AB}$ represents the distance between the points A and B . It is written as $|\ \bar{R}_{AB}\ |$. The magnitude of a vector is always positive.

To calculate the magnitude of a vector, square the coefficients of unit vectors, add them and then take the square root.

$$\therefore |\overline{R}_{AB}| = \sqrt{(x_2 - x_1)^2 + (y_2 - y_1)^2 + (z_2 - z_1)^2}$$

Hence unit vector in the direction of $\overline{R}_{AB}$ is

$$\overline{a}_{AB} = \frac{\overline{R}_{AB}}{|\overline{R}_{AB}|} = \frac{(x_2 - x_1)\overline{a}_x + (y_2 - y_1)\overline{a}_y + (z_2 - z_1)\overline{a}_z}{\sqrt{(x_2 - x_1)^2 + (y_2 - y_1)^2 + (z_2 - z_1)^2}}$$

Problem 1.1 : Given the two vectors, $\overline{r}_A = -\overline{a}_x - 3\overline{a}_y - 4\overline{a}_z$ and $\overline{r}_B = 2\overline{a}_x + 2\overline{a}_y + 2\overline{a}_z$, and point $C(1, 3, 4)$, find : (a) $\overline{R}_{AB}$; (b) $|\overline{r}_A|$; (c) $\overline{a}_A$; (d) $\overline{a}_{AB}$; (e) a unit vector directed from C toward A .

Solution :

(a) $\overline{R}_{AB} = \overline{r}_B - \overline{r}_A = (2\overline{a}_x + 2\overline{a}_y + 2\overline{a}_z) - (-\overline{a}_x - 3\overline{a}_y - 4\overline{a}_z)$

$= 3\overline{a}_x + 5\overline{a}_y + 6\overline{a}_z$

(b) $|\overline{r}_A| = \sqrt{(-1)^2 + (-3)^2 + (-4)^2} = 5.099$

(c) $$\overline{a}_A = \frac{\overline{r}_A}{|\overline{r}_A|} = \frac{-\overline{a}_x - 3\overline{a}_y - 4\overline{a}_z}{5.099}$$

$= -0.038\overline{a}_x - 0.588\overline{a}_y - 0.784\overline{a}_z$

(d) $$\overline{a}_{AB} = \frac{\overline{R}_{AB}}{|\overline{R}_{AB}|} = \frac{3\overline{a}_x + 5\overline{a}_y + 6\overline{a}_z}{\sqrt{(3)^2 + (5)^2 + (6)^2}}$$

$= 0.359\overline{a}_x + 0.598\overline{a}_y + 0.717\overline{a}_z$

(e) $\overline{r}_C = 1\overline{a}_x + 3\overline{a}_y + 4\overline{a}_z$

$\overline{R}_{CA} = \overline{r}_A - \overline{r}_C = (-\overline{a}_x - 3\overline{a}_y - 4\overline{a}_z) - (1\overline{a}_x + 3\overline{a}_y + 4\overline{a}_z)$

$= -2\overline{a}_x - 6\overline{a}_y - 8\overline{a}_z$

$$\overline{a}_{CA} = \frac{\overline{R}_{CA}}{|\overline{R}_{CA}|} = \frac{-2\overline{a}_x - 6\overline{a}_y - 8\overline{a}_z}{\sqrt{(-2)^2 + (-6)^2 + (-8)^2}}$$

$= -0.196\overline{a}_x - 0.588\overline{a}_y - 0.784\overline{a}_z$

Problem 1.2 : Given the vector field, $\overline{F} = 0.4(y - 2x)\overline{a}_x - [200/(x^2 + y^2 + z^2)]\overline{a}_z$: (a) evaluate $|\overline{F}|$ at $P(-4, 3, 5)$: (b) find a

unit vector specifying the direction of $\overline{F}$ at P. Describe the locus of all points for which : (c) $F_x = 1$; (d) $|F_z| = 2$.

Solution :

$$(a)\overline{F} \text{ at } P = 0.4\,[3 - 2(-4)]\,\overline{a}_x - \left[\frac{200}{(-4)^2 + (3)^2 + (5)^2}\right]\overline{a}_z$$

$$= 4.4\,\overline{a}_x - 4\,\overline{a}_z$$

$$\therefore |\overline{F}| \text{ at } P(-4, 3, 5) = \sqrt{(4.4)^2 + (-4)^2} = 5..946$$

$$(b)\ \overline{a} = \frac{\overline{F} \text{ at } P}{|\overline{F}| \text{ at } P} = \frac{4.4\,\overline{a}_x - 4\,\overline{a}_z}{5.946} = 0.740\,\overline{a}_x - 0.673\,\overline{a}_z$$

(c) $\overline{F} = F_x\,\overline{a}_x + F_z\,\overline{a}_z$. Hence $F_x = 0.4(y - 2x) = 1$

$\therefore y = 2x + 2.5$. This equation represents a plane.

$$(d)\ \text{Now } F_z = -\frac{200}{x^2 + y^2 + z^2}$$

$$\therefore |F_z| = \frac{200}{x^2 + y^2 + z^2} = 2 \quad \text{or} \quad x^2 + y^2 + z^2 = 100$$

It represents a sphere of radius 10 and centered at origin.

1.1.5 The Dot Product :

The result of dot product is a scalar quantity. Hence it is also known as scalar product. The dot product of two vectors, $\overline{A}$ and $\overline{B}$ is defined as

$$\overline{\mathbf{A}} \cdot \overline{\mathbf{B}} = |\overline{\mathbf{A}}|\ |\overline{\mathbf{B}}| \cos\theta \qquad (\text{ read" } \overline{A} \text{ dot } \overline{B}\text{" })$$

where θ is the smaller angle between $\overline{A}$ and $\overline{B}$. The commutative law holds good for the dot product, i.e., $\overline{A} \cdot \overline{B} = \overline{B} \cdot \overline{A}$

Important applications of the dot product :

(1) From the defination of dot product,

$$\cos\theta = \frac{\overline{A} \cdot \overline{B}}{|\overline{A}|\ |\overline{B}|} \quad \text{or} \quad \theta = \cos^{-1}\left(\frac{\overline{A} \cdot \overline{B}}{|\overline{A}|\ |\overline{B}|}\right)$$

where θ is the smaller angle between the vectors $\overline{A}$ and $\overline{B}$

(2) Scalar projection of $\overline{A}$ in the direction of $\overline{B}$. Fig. 1.5 shows two vectors, $\overline{A}$ and $\overline{B}$. A perpendicular is dropped from head of $\overline{A}$ on $\overline{B}$. The foot of perpendicular is Q. Then,

PQ = Scalar component or projection of $\overline{A}$ in the direction of $\overline{B}$.

$= |\overline{A}| \cos\theta$

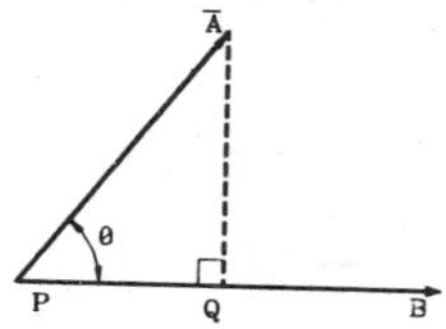

FIG.1.5

Now $\overline{A} \cdot \overline{a}_B = |\overline{A}|\ |\overline{a}_B| \cos\theta = |\overline{A}| \cos\theta$

where $\overline{a}_B$ is a unit vector in the direction of $\overline{B}$. Hence angle between $\overline{A}$ and $\overline{B}$ remains same for $\overline{A}$ and $\overline{a}_B$. Hence $PQ = \overline{A} \cdot \overline{a}_B$

Scalar component or projection of $\overline{A}$ in the direction of $\overline{B}$ = $\overline{A} \cdot \overline{a}_B$

(3) Vector projection of $\overline{A}$ in the direction of $\overline{B}$. If PQ is expressed as a vector, then it is known as vector projection of $\overline{A}$ in the direction of $\overline{B}$. The direction of $\overline{PQ}$ is same as that of $\overline{B}$, i.e., $\overline{a}_B$. Hence the required vector projection is equal to $\overline{PQ} = PQ\ \overline{a}_B = (\overline{A} \cdot \overline{a}_B)\ \overline{a}_B$ or

Vector component or projection of $\overline{A}$ in the direction of $\overline{B}$ = $(\overline{A} \cdot \overline{a}_B)\ \overline{a}_B$

Expression for dot product :

As the angle between a vector and itself is zero, we have

$$\overline{a}_x \cdot \overline{a}_x = \overline{a}_y \cdot \overline{a}_y = \overline{a}_z \cdot \overline{a}_z = 1$$

As all the three unit vectors are mutually perpendicular to each other, we have

$$\overline{a}_x \cdot \overline{a}_y = \overline{a}_y \cdot \overline{a}_x = \overline{a}_y \cdot \overline{a}_z = \overline{a}_z \cdot \overline{a}_y = \overline{a}_z \cdot \overline{a}_x = \overline{a}_x \cdot \overline{a}_z = 0$$

Let $\overline{A} = A_x \overline{a}_x + A_y \overline{a}_y + A_z \overline{a}_z$ and $\overline{B} = B_x \overline{a}_x + B_y \overline{a}_y + B_z \overline{a}_z$.

Then, $\overline{A} \cdot \overline{B} = A_x \overline{a}_x \cdot B_x \overline{a}_x + A_y \overline{a}_y \cdot B_x \overline{a}_x + A_z \overline{a}_z \cdot B_x \overline{a}_x +$

$A_x \overline{a}_x \cdot B_y \overline{a}_y + A_y \overline{a}_y \cdot B_y \overline{a}_y + A_z \overline{a}_z \cdot B_y \overline{a}_y +$

$A_x \overline{a}_x \cdot B_z \overline{a}_z + A_y \overline{a}_y \cdot B_z \overline{a}_z + A_z \overline{a}_z \cdot B_z \overline{a}_z$

$= A_x B_x + A_y B_y + A_z B_z$

Hence we have

$$\overline{\mathbf{A}} \cdot \overline{\mathbf{B}} = A_x B_x + A_y B_y + A_z B_z$$

1.1.6 General representation of a vector :

Any vector in cartesian coordinate system may be written as

$\overline{A} = A_x \overline{a}_x + A_y \overline{a}_y + A_z \overline{a}_z$ where

A_x = Component of $\overline{A}$ in the direction of $\overline{a}_x = \overline{A} \cdot \overline{a}_x$

A_y = Component of $\overline{A}$ in the direction of $\overline{a}_y = \overline{A} \cdot \overline{a}_y$

A_z = Component of $\overline{A}$ in the direction of $\overline{a}_z = \overline{A} \cdot \overline{a}_z$

Also $|\overline{A}| = \sqrt{A_x^2 + A_y^2 + A_z^2}$

and $\overline{a}_A = \dfrac{\overline{A}}{|\overline{A}|} = \dfrac{A_x \overline{a}_x + A_y \overline{a}_y + A_z \overline{a}_z}{\sqrt{A_x^2 + A_y^2 + A_z^2}}$

Problem 1.3 : Given points A (2, 5, –1), B (3, –2, 4), and C (–2, 3, 1), find: (a) $\overline{R}_{AB} \cdot \overline{R}_{AC}$; (b) the angle between $\overline{R}_{AB}$ and $\overline{R}_{AC}$; (c) the length of the projection of $\overline{R}_{AB}$ on $\overline{R}_{AC}$; (d) the vector projection of $\overline{R}_{AB}$ on $\overline{R}_{AC}$.

Solution :

(a) $\overline{R}_{AB} = \overline{r}_B - \overline{r}_A = (3\overline{a}_x - 2\overline{a}_y + 4\overline{a}_z) - (2\overline{a}_x + 5\overline{a}_y - \overline{a}_z)$

$= \overline{a}_x - 7\overline{a}_y + 5\overline{a}_z$

$\overline{R}_{AC} = \overline{r}_C - \overline{r}_A = (-2\overline{a}_x + 3\overline{a}_y + \overline{a}_z) - (2\overline{a}_x + 5\overline{a}_y - \overline{a}_z)$

$= -4\overline{a}_x - 2\overline{a}_y + 2\overline{a}_z$

$\therefore \overline{R}_{AB} \cdot \overline{R}_{AC} = (1)(-4) + (-7)(-2) + (5)(2) = 20$

(b) $\cos\theta = \dfrac{\overline{R}_{AB} \cdot \overline{R}_{AC}}{|\overline{R}_{AB}|\ |\overline{R}_{AC}|}$

$= \dfrac{20}{\sqrt{(1)^2 + (-7)^2 + (5)^2}\ \sqrt{(-4)^2 + (-2)^2 + (2)^2}}$

$= 0.4714$ or $\theta = \cos^{-1}(0.4714) = 61.874°$

(c) AP = Scalar projection of $\overline{R}_{AB}$ on $\overline{R}_{AC}$ (see Fig. P1.3)

$= \overline{R}_{AB} \cdot \overline{a}_{AC}$

$= \overline{R}_{AB} \cdot \dfrac{\overline{R}_{AC}}{|\overline{R}_{AC}|}$

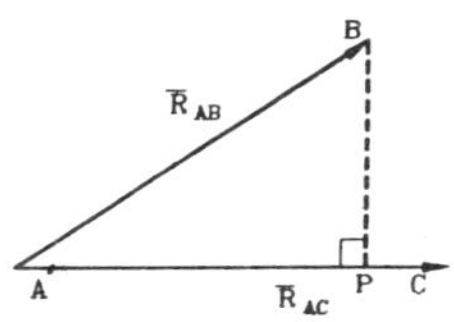

FIG. P 1.3

$$= \frac{20}{\sqrt{(-4)^2 + (-2)^2 + (2)^2}}$$

$$= 4.082$$

Hence length of projection $= 4.082$

(d) $\overline{R}_{AP} =$ Vector projection of $\overline{R}_{AB}$ on $\overline{R}_{AC} = (\overline{R}_{AB} \cdot \overline{a}_{AC})\,\overline{a}_{AC}$

$$= 4.082 \left[\frac{-4\,\overline{a}_x - 2\,\overline{a}_y + 2\,\overline{a}_z}{\sqrt{(-4)^2 + (-2)^2 + (2)^2}} \right]$$

$$= -3.333\,\overline{a}_x - 1.667\,\overline{a}_y + 1.667\,\overline{a}_z$$

1.1.7 The Cross Product :

The result of cross product is a vector quantity. Hence it is also known as the vector product.The cross product of two vectors, $\overline{A}$ and $\overline{B}$ is defined as

$$\overline{A} \times \overline{B} = |\overline{A}|\,|\overline{B}|\,\sin\theta\,\overline{a}_n$$

(read "$\overline{A}$ cross $\overline{B}$")

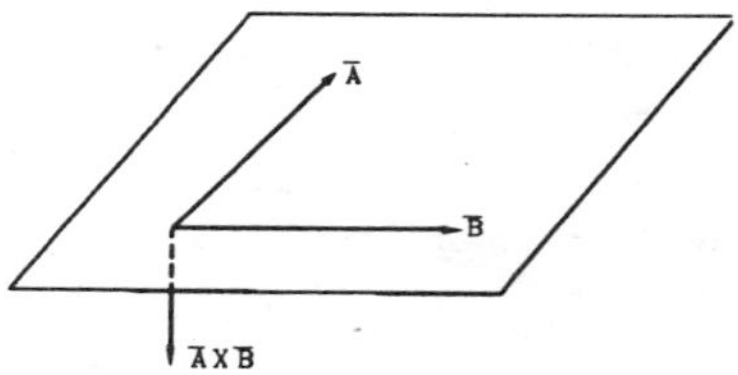

FIG.1.6

where θ is the smaller angle between $\overline{A}$ and $\overline{B}$ and $\overline{a}_n$ is a unit vector perpendicular to both $\overline{A}$ and $\overline{B}$ and is in the direction of advance of a right handed screw as $\overline{A}$ is turned into $\overline{B}$. This direction is shown in Fig.1.6 . Because of this direction requirement, the commutative law does not apply to the cross product , instead ,$\overline{A} \times \overline{B} = -\overline{B} \times \overline{A}$

Important applications of cross product :

(1) From the defination of cross product, we have

$$|\overline{A} \times \overline{B}| = |\overline{A}|\,|\overline{B}|\,\sin\theta$$

$$\therefore \sin\theta = \frac{|\overline{A} \times \overline{B}|}{|\overline{A}|\,|\overline{B}|} \quad \text{or} \quad \theta = \sin^{-1}\left(\frac{|\overline{A} \times \overline{B}|}{|\overline{A}|\,|\overline{B}|}\right)$$

(2) Possible unit vectors perpendicular to the vectors, $\overline{A}$ and $\overline{B}$.
From the defination of cross product, we have

$$\bar{a}_n = \frac{\bar{A} \times \bar{B}}{|\bar{A}|\ |\bar{B}|\ \sin\theta} = \frac{\bar{A} \times \bar{B}}{|\bar{A} \times \bar{B}|}$$

As there are two possible unit vectors, we can write

$$\bar{a}_n = \pm \frac{\bar{A} \times \bar{B}}{|\bar{A} \times \bar{B}|}$$

(3) Area of a parallelogram.

If $\bar{A}$ and $\bar{B}$ represents two adjacent sides of a parallelogram, then

$$\text{Area of parallelogram} = |\bar{A} \times \bar{B}|$$

(4) Area of a triangle .

If $\bar{A}$ and $\bar{B}$ represents two adjacent sides of a triangle, then

$$\text{Area of triangle} = \frac{1}{2}|\bar{A} \times \bar{B}|$$

Expression for the Cross product :

As the angle between a vector and itself is zero, we have

$$\bar{a}_x \times \bar{a}_x = \bar{a}_y \times \bar{a}_y = \bar{a}_z \times \bar{a}_z = 0$$

As all the three unit vectors are mutually perpendicular to each other, we have

$$\bar{a}_x \times \bar{a}_y = \bar{a}_z \ ;\ \bar{a}_y \times \bar{a}_z = \bar{a}_x \ ;\ \bar{a}_z \times \bar{a}_x = \bar{a}_y$$

$$\bar{a}_y \times \bar{a}_x = -\bar{a}_z \ ;\ \bar{a}_z \times \bar{a}_y = -\bar{a}_x \ ;\ \bar{a}_x \times \bar{a}_z = -\bar{a}_y$$

let $\bar{A} = A_x\bar{a}_x + A_y\bar{a}_y + A_z\bar{a}_z$ and $\bar{B} = B_x\bar{a}_x + B_y\bar{a}_y + B_z\bar{a}_z$

$$\begin{aligned}\therefore\ \bar{A} \times \bar{B} &= A_x\bar{a}_x \times B_x\bar{a}_x + A_y\bar{a}_y \times B_x\bar{a}_x + A_z\bar{a}_z \times B_x\bar{a}_x + \\ &\quad A_x\bar{a}_x \times B_y\bar{a}_y + A_y\bar{a}_y \times B_y\bar{a}_y + A_z\bar{a}_z \times B_y\bar{a}_y + \\ &\quad A_x\bar{a}_x \times B_z\bar{a}_z + A_y\bar{a}_y \times B_z\bar{a}_z + A_z\bar{a}_z \times B_z\bar{a}_z \\ &= (A_yB_z - A_zB_y)\,\bar{a}_x - (A_xB_z - A_zB_x)\,\bar{a}_y + \\ &\quad (A_xB_y - A_yB_x)\,\bar{a}_z\end{aligned}$$

which can be written as

$$\bar{A} \times \bar{B} = \begin{vmatrix} \bar{a}_x & \bar{a}_y & \bar{a}_z \\ A_x & A_y & A_z \\ B_x & B_y & B_z \end{vmatrix}$$

Problem 1.4 : A triangle is defined by the three points *A* (2, –5, 1),*B* (–3, 2, 4), and *C* (0, 3, 1). Find : (a) $\bar{R}_{BC} \times \bar{R}_{BA}$; (b) the area of the triangle ; (c) a unit vector perpendicular to the plane in which the triangle is located.

Solution :

(a) $\overline{R}_{BC} = \overline{r}_C - \overline{r}_B = (3\,\overline{a}_y + \overline{a}_z) - (-3\,\overline{a}_x + 2\,\overline{a}_y + 4\,\overline{a}_z)$

$= 3\,\overline{a}_x + \overline{a}_y - 3\,\overline{a}_z$

$\overline{R}_{BA} = \overline{r}_A - \overline{r}_B = (2\,\overline{a}_x - 5\,\overline{a}_y + \overline{a}_z) - (-3\,\overline{a}_x + 2\,\overline{a}_y + 4\,\overline{a}_z)$

$= 5\,\overline{a}_x - 7\,\overline{a}_y - 3\,\overline{a}_z$

$$\overline{R}_{BC} \times \overline{R}_{BA} = \begin{vmatrix} \overline{a}_x & \overline{a}_y & \overline{a}_z \\ 3 & 1 & -3 \\ 5 & -7 & -3 \end{vmatrix} = -24\,\overline{a}_x - 6\,\overline{a}_y - 26\,\overline{a}_z$$

(b) Area of triangle $ABC = 0.5 \left| \overline{R}_{BC} \times \overline{R}_{BA} \right|$

$= 0.5\sqrt{(-24)^2 + (-6)^2 + (-26)^2}$

$= 17.944$

(c) $$\overline{a}_n = \pm \frac{\overline{R}_{BC} \times \overline{R}_{BA}}{\left| \overline{R}_{BC} \times \overline{R}_{BA} \right|} = \pm \frac{-24\,\overline{a}_x - 6\,\overline{a}_y - 26\,\overline{a}_z}{\sqrt{(-24)^2 + (-6)^2 + (-26)^2}}$$

$= \pm(-0.669\,\overline{a}_x - 0.167\,\overline{a}_y - 0.724\,\overline{a}_z)$

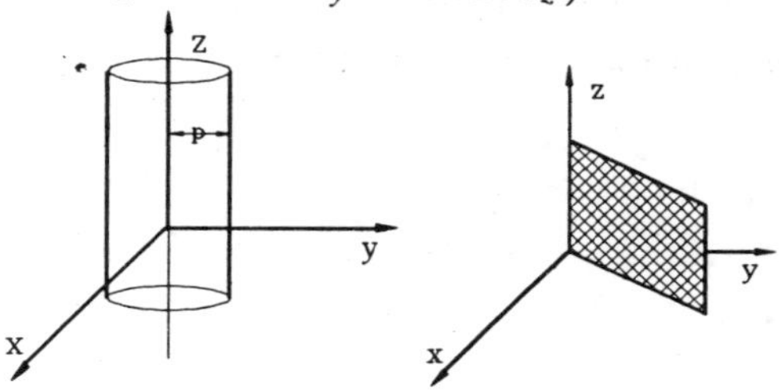

(a) (b)

FIG. 1.7

1.2 CYLINDRICAL COORDINATE SYSTEM :

1.2.1 Coordinates :

In cylindrical coordinate system a point is represented by the coordinates (ρ, ϕ, z). The ρ coordinate represents a cylinder of radius ρ with z axis as its axis as shown in Fig.1.7 (a). The ϕ coordinate represents the angle made by a semi infinite plane with positive x

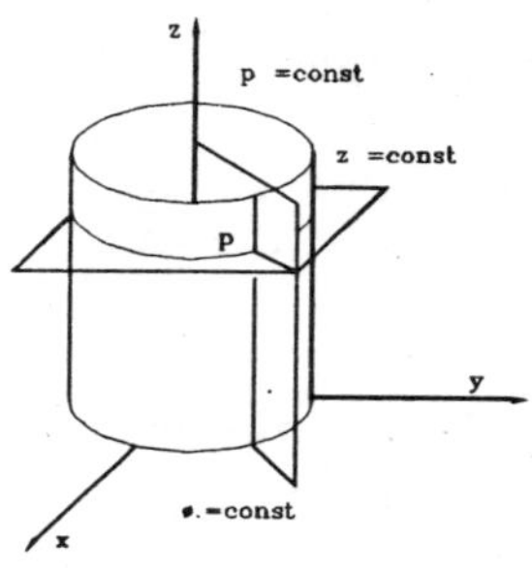

FIG.1.8

axis. It is shown in Fig.1.7 (b) The z coordinate is same as that of cartesian coordinate system. These three surfaces intersect each other orthogonally to form a point as shown in Fig.1.8.

Transformation of cylindrical coordinates into cartesian coordinates :

The formule are

$$x = \rho \cos \phi$$
$$y = \rho \sin \phi$$
$$z = z$$

Transformation of cartesian coordinates into cylindrical coordinates :

The formulae are

$$\rho = \sqrt{x^2 + y^2}$$
$$\phi = \tan^{-1}\left(\frac{y}{x}\right)$$
$$z = z$$

Important Note :

While using the formula for ϕ coordinate it is important to see in which quadrant the point lies by neglecting the z coordinate Otherwise wrong values are obtained.

Problem 1.5 : Given points A ($x = 2, y = 3, z = -1$) and B ($\rho = 4, \phi = -50^\circ, z = 2$), find the distance from : (a) A to the origin ; (b) B to the origin ; (c) A to B.

Solution :

(a) $d(AO) = \sqrt{(2-0)^2 + (3-0)^2 + (-1-0)^2} = 3.742$

(b) Cartesian coordinates of point B are $x = 4\cos(-50^\circ) = 2.571$; $y = 4\sin(-50^\circ) = -3.064$; $z = 2$

$\therefore\ d(BO) = \sqrt{(2.571-0)^2 + (-3.064-0)^2 + (2-0)^2} = 4.472$

(c) $d(AB) = \sqrt{(2-2.571)^2 + (3+3.064)^2 + (-1-2)^2} = 6.790$

1.2.2 Unit vectors :

In cylindrical coordinate system, the three unit vectors are $\pm\ \bar{a}_\rho$, $\pm\ \bar{a}_\phi$,

$\pm\ \bar{a}_z$. All the three unit vectors are mutually perpendicular to each other at a point.

The unit vectors $\pm\ \bar{a}_\rho$ are perpendicular where as the unit vectors $\pm\ \bar{a}_\phi$ and $\pm\ \bar{a}_z$ are tangent to the cylinder represented by $\rho =$ constant .

The unit vectors $\pm\ \bar{a}_\phi$ are perpendicular where as the unit vectors $\pm\ \bar{a}_\rho$ and $\pm\ \bar{a}_z$ are tangent to the plane represented by $\phi =$ constant.

The unit vectors $\pm\ \bar{a}_z$ are perpendicular where as the unit vectors $\pm\ \bar{a}_\rho$ and $\pm\ \bar{a}_\phi$ are tangent to the plane represented by $z =$ constant.

1.2.3 Position vector :

In cylindrical coordinate system, a vector directed from (0, 0, 0) towards (ρ, ϕ, z) is given by

$$\bar{r} = \rho\,\bar{a}_\rho + z\,\bar{a}_z$$

1.2.4 Expression for the dot product :

At any point we have

$$\bar{a}_\rho \cdot \bar{a}_\rho = \bar{a}_\phi \cdot \bar{a}_\phi = \bar{a}_z \cdot \bar{a}_z = 1$$

as the angle between the vectors is zero and

$$\bar{a}_\rho \cdot \bar{a}_\phi = \bar{a}_\phi \cdot \bar{a}_\rho = \bar{a}_\phi \cdot \bar{a}_z = \bar{a}_z \cdot \bar{a}_\phi = \bar{a}_z \cdot \bar{a}_\rho = \bar{a}_\rho \cdot \bar{a}_z = 0$$

as the angle between the vectors is 90°.

If $\bar{A} = A_\rho\,\bar{a}_\rho + A_\phi\,\bar{a}_\phi + A_z\,\bar{a}_z$ and $\bar{B} = B_\rho\,\bar{a}_\rho + B_\phi\,\bar{a}_\phi + B_z\,\bar{a}_z$

Then,

$$\bar{A} \cdot \bar{B} = A_\rho B_\rho + A_\phi B_\phi + A_z B_z$$

1.2.5 General representation of a vector :

In cylindrical coordinate system, any vector may be written as

$$\bar{A} = A_\rho\,\bar{a}_\rho + A_\phi\,\bar{a}_\phi + A_z\,\bar{a}_z$$

where $A_\rho =$ Component of $\bar{A}$ in the direction of $\bar{a}_\rho = \bar{A} \cdot \bar{a}_\rho$

$A_\phi =$ Component of $\bar{A}$ in the direction of $\bar{a}_\phi = \bar{A} \cdot \bar{a}_\phi$

$A_z =$ Component of $\bar{A}$ in the direction of $\bar{a}_z = \bar{A} \cdot \bar{a}_z$

Also $|\bar{A}| = \sqrt{A_\rho^2 + A_\phi^2 + A_z^2}$

and $$\bar{a}_A = \frac{\bar{A}}{|\bar{A}|} = \frac{A_\rho\,\bar{a}_\rho + A_\phi\,\bar{a}_\phi + A_z\,\bar{a}_z}{\sqrt{A_\rho^2 + A_\phi^2 + A_z^2}}$$

1.2.6. Expression for the cross product :

At any point, we have

$\bar{a}_\rho \times \bar{a}_\rho = \bar{a}_\phi \times \bar{a}_\phi = \bar{a}_z \times \bar{a}_z = 0$

as the angle between the vectors is zero and

$\bar{a}_\rho \times \bar{a}_\phi = \bar{a}_z$; $\bar{a}_\phi \times \bar{a}_z = \bar{a}_\rho$; $\bar{a}_z \times \bar{a}_\rho = \bar{a}_\phi$

$\bar{a}_\phi \times \bar{a}_\rho = -\bar{a}_z$; $\bar{a}_z \times \bar{a}_\phi = -\bar{a}_\rho$; $\bar{a}_\rho \times \bar{a}_z = -\bar{a}_\phi$

The cross product is given by

$$\bar{A} \times \bar{B} = \begin{vmatrix} \bar{a}_\rho & \bar{a}_\phi & \bar{a}_z \\ A_\rho & A_\phi & A_z \\ B_\rho & B_\phi & B_z \end{vmatrix}$$

1.2.7 Transformation of a vector from cartesian to cylindrical coordinate system and vice–versa :

Using the following dot product table, the transformation is carried out.

Dot Product Table 1.1

	$\bar{a}_\rho$	$\bar{a}_\phi$	$\bar{a}_z$
$\bar{a}_x$	$\cos\phi$	$-\sin\phi$	0
$\bar{a}_y$	$\sin\phi$	$\cos\phi$	0
$\bar{a}_z$	0	0	1

Problem 1.6 : Transform each of the following vectors to cylindrical coordinates at the point specified : (a) $5\,\bar{a}_x$ at $P(\rho = 4, \phi = 120^\circ, z = -1)$; (b) $5\,\bar{a}_x$ at $Q(x = 3, y = 4, z = -1)$; (c) $4\,\bar{a}_x - 2\,\bar{a}_y - 4\,\bar{a}_z$ at $A(x = 2, y = 3, z = 5)$.

Solution :

(a) Let $\bar{A} = 5\,\bar{a}_x = A_\rho\,\bar{a}_\rho + A_\phi\,\bar{a}_\phi + A_z\,\bar{a}_z$

$A_\rho = \bar{A}\cdot\bar{a}_\rho = 5\,\bar{a}_x\cdot\bar{a}_\rho = 5\cos\phi$

$A_\phi = \bar{A}\cdot\bar{a}_\phi = 5\,\bar{a}_x\cdot\bar{a}_\phi = -5\sin\phi$

$A_z = \bar{A}\cdot\bar{a}_z = 5\,\bar{a}_x\cdot\bar{a}_z = 0$

$\therefore\ \bar{A} = 5\cos\phi\,\bar{a}_\rho - 5\sin\phi\,\bar{a}_\phi$

Substituting ϕ coordinate of point P in the above vector , we have
$\overline{A}$ at point $P = -2.5\,\overline{a}_\rho - 4.33\,\overline{a}_\phi$

(b) For point Q , $\phi = \tan^{-1}(y/x) = \tan^{-1}(4/3) = 53.13^\circ$
$\therefore$ $\overline{A}$ at point $Q = 5\cos 53.13^\circ\,\overline{a}_\rho - 5\sin 53.13^\circ\,\overline{a}_\phi = 3\,\overline{a}_\rho - 4\,\overline{a}_\phi$

(c) Let $\overline{B} = 4\,\overline{a}_x - 2\,\overline{a}_y - 4\,\overline{a}_z = B_\rho\,\overline{a}_\rho + B_\phi\,\overline{a}_\phi + B_z\,\overline{a}_z$

$$B_\rho = \overline{B}\cdot\overline{a}_\rho = (4\,\overline{a}_x - 2\,\overline{a}_y - 4\,\overline{a}_z)\cdot\overline{a}_\rho$$
$$= 4\,\overline{a}_x\cdot\overline{a}_\rho - 2\,\overline{a}_y\cdot\overline{a}_\rho - 4\,\overline{a}_z\cdot\overline{a}_\rho$$
$$= 4\cos\phi - 2\sin\phi$$
$$B_\phi = \overline{B}\cdot\overline{a}_\phi = (4\,\overline{a}_x - 2\,\overline{a}_y - 4\,\overline{a}_z)\cdot\overline{a}_\phi$$
$$= 4\,\overline{a}_x\cdot\overline{a}_\phi - 2\,\overline{a}_y\cdot\overline{a}_\phi - 4\,\overline{a}_z\cdot\overline{a}_\phi$$
$$= -4\sin\phi - 2\cos\phi$$
$$B_z = \overline{B}\cdot\overline{a}_z = (4\,\overline{a}_x - 2\,\overline{a}_y - 4\,\overline{a}_z)\cdot\overline{a}_z = -4$$
$$\therefore\ \overline{B} = (4\cos\phi - 2\sin\phi)\,\overline{a}_\rho + (-4\sin\phi - 2\cos\phi)\,\overline{a}_\phi - 4\,\overline{a}_z$$

For point A, $\phi = \tan^{-1}(y/x) = \tan^{-1}(3/2) = 56.31^\circ$
Substituting ϕ coordinate of point A in the above $\overline{B}$, we have
$\overline{B}$ at point A $= 0.555\,\overline{a}_\rho - 4.438\,\overline{a}_\phi - 4\,\overline{a}_z$

1.2.8. Expression for $\overline{R}_{12}$:

Let point 1 is (ρ_1, ϕ_1, z_1) and point 2 is (ρ_2, ϕ_2, z_2)

(1) **If out of these two points, at least one lies on the z axis, then use the formula**

$$\overline{R}_{12} = \overline{r}_2 - \overline{r}_1$$

where $\overline{r}_2$ = p. v. of point 2 and $\overline{r}_1$ = p. v. of point 1

(2) Transform both the points into cartesian coordinate system, calculate $\overline{R}_{12}$ in cartesian coordinate system and then transform this vector into cylindrical coordinate system using the dot product table 1.1. Then substitute ϕ coordinate of either point 1 or point 2 so as to calculate $\overline{R}_{12}$ at point 1 or point 2 respectively.

1.3 SPHERICAL COORDINATE SYSTEM :

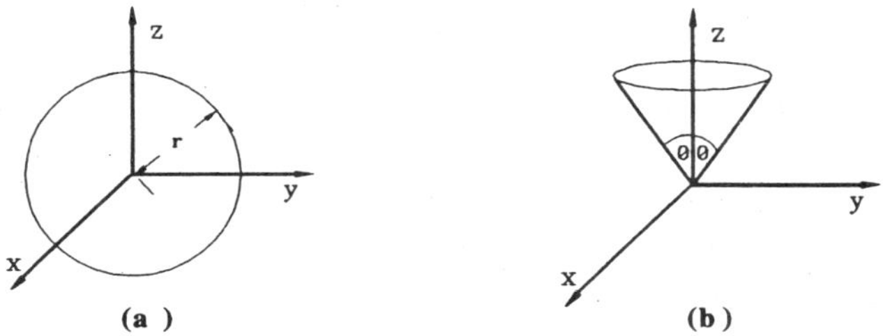

FIG. 1.9

1.3.1. Coordinates :

In spherical coordinate system a point is represented by the coordinates (r, θ, ϕ). The r coordinate represents a sphere centered at origin as shown in Fig. 1.9 (a) The r coordinate is always positive as it is the radius of the sphere. Further it varies from 0 to ∞ The θ coordinate represents the angle made by a cone with the z axis as shown in Fig. 1.9 (b). The θ coordinate is always positive and it varies from 0 to π. The ϕ coordinate is same as that of cylindrical coordinate system. These three surfaces intersect each other orthogonally to form a point as shown in Fig. 1.10.

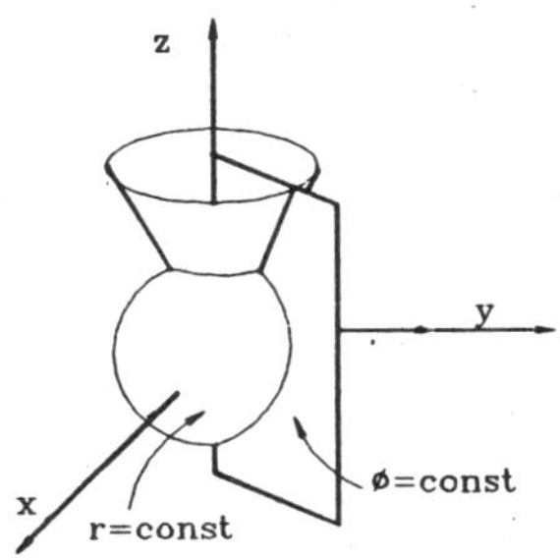

FIG.1.10

Transformation of spherical coordinates into cartesian coordinates :

The formule are

$$x = r \sin\theta \cos\phi$$

$$y = r \sin\theta \sin\phi$$

$$z = r \cos\theta$$

Transformation of cartesian coordinates into spherical coordinates :

The formulae are

$$r = \sqrt{x^2 + y^2 + z^2}$$

$$\theta = \cos^{-1}\left(\frac{z}{r}\right)$$

$$\phi = \tan^{-1}\left(\frac{y}{x}\right)$$

Problem 1.7 : Given the points A (x= 2, y= 3, z= − 1) and B (r= 4, θ= 25°, ϕ = 120°) , find : (a) the spherical coordinates of A ; (b) the e cartesian coordinates of B ; (c) the distance from A to B

Solution :

(a) $r = \sqrt{(2)^2 + (3)^2 + (-1)^2} = 3.742$

$\theta = \cos^{-1}\left(\frac{-1}{3.742}\right) = 105.5^\circ \quad \phi = \tan^{-1}\left(\frac{3}{2}\right) = 56.3^\circ$

(b) $x = 4 \sin 25^\circ \cos 120^\circ = -0.845$

$y = 4 \sin 25^\circ \sin 120^\circ = 1.464 \; ; \; z = 4 \cos 25^\circ = 3.625$

(c) $d(AB) = \sqrt{(2 + 0.845)^2 + (3 - 1.464)^2 + (-1 - 3.625)^2}$

$= 5.643$

1.3.2. Unit vectors :

In spherical coordinates system, the three unit vectors are $\pm \bar{a}_r$, $\pm \bar{a}_\theta$, $\pm \bar{a}_\phi$. All the three unit vectors are mutually perpendicular to each other at a point.

The unit vectors $\pm \bar{a}_r$ are perpendicular where as the unit vector $\pm \bar{a}_\theta$ and $\pm \bar{a}_\phi$ are tangent to the sphere represented by r = constant .

The unit vectors $\pm \bar{a}_\theta$ are perpendicular where as the unit vectors $\pm \bar{a}_r$ and $\pm \bar{a}_\phi$ are tangent to the cone represented by θ = constant.

The unit vectors $\pm \bar{a}_\phi$ are perpendicular where as the unit vectors $\pm \bar{a}_r$ and $\pm \bar{a}_\theta$ are tangent to the plane represented by ϕ = constant .

1.3.3. Position vector :

In spherical coordinate system, a vector directed from (0, 0, 0) towards (r, θ, ϕ) is given by

$$\bar{r} = r\,\bar{a}_r$$

1.3.4. Expression for the dot product :

At any point we have

$\bar{a}_r \cdot \bar{a}_r = \bar{a}_\theta \cdot \bar{a}_\theta = \bar{a}_\phi \cdot \bar{a}_\phi = 1$

as the angle between the vectors is zero and

$\bar{a}_r \cdot \bar{a}_\theta = \bar{a}_\theta \cdot \bar{a}_r = \bar{a}_\theta \cdot \bar{a}_\phi = \bar{a}_\phi \cdot \bar{a}_\theta = \bar{a}_\phi \cdot \bar{a}_r = \bar{a}_r \cdot \bar{a}_\phi = 0$

as the angle between the vectors is90^o

Let $\overline{A} = A_r \bar{a}_r + A_\theta \bar{a}_\theta + A_\phi \bar{a}_\phi$ and $\overline{B} = B_r \bar{a}_r + B_\theta \bar{a}_\theta + B_\phi \bar{a}_\phi$

Then, $\overline{\mathbf{A}} \cdot \overline{\mathbf{B}} = A_r B_r + A_\theta B_\theta + A_\phi B_\phi$

1.3.5. General representation of a vector :

In spherical coordinate system, any vector may be written as

$\overline{A} = A_r \bar{a}_r + A_\theta \bar{a}_\theta + A_\phi \bar{a}_\phi$

where A_r = Component of $\overline{A}$ in the direction of $\bar{a}_r = \overline{A} \cdot \bar{a}_r$

A_θ = Component of $\overline{A}$ in the direction of $\bar{a}_\theta = \overline{A} \cdot \bar{a}_\theta$

A_ϕ = Component of $\overline{A}$ in the direction of $\bar{a}_\phi = \overline{A} \cdot \bar{a}_\phi$

Also $|\overline{A}| = \sqrt{A_r^2 + A_\theta^2 + A_\phi^2}$

and $\bar{a}_A = \dfrac{\overline{A}}{|\overline{A}|} = \dfrac{A_r \bar{a}_r + A_\theta \bar{a}_\theta + A_\phi \bar{a}_\phi}{\sqrt{A_r^2 + A_\theta^2 + A_\phi^2}}$

1.3.6. Expression for the cross product :

At any point, we have

$\bar{a}_r \times \bar{a}_r = \bar{a}_\theta \times \bar{a}_\theta = \bar{a}_\phi \times \bar{a}_\phi = 0$

as the angle is zero and

$\bar{a}_r \times \bar{a}_\theta = \bar{a}_\phi$; $\bar{a}_\theta \times \bar{a}_\phi = \bar{a}_r$; $\bar{a}_\phi \times \bar{a}_r = \bar{a}_\theta$

$\bar{a}_\theta \times \bar{a}_r = -\bar{a}_\phi$; $\bar{a}_\phi \times \bar{a}_\theta = -\bar{a}_r$; $\bar{a}_r \times \bar{a}_\phi = -\bar{a}_\theta$. Then

$$\overline{\mathbf{A}} \times \overline{\mathbf{B}} = \begin{vmatrix} \bar{\mathbf{a}}_r & \bar{\mathbf{a}}_\theta & \bar{\mathbf{a}}_\phi \\ A_r & A_\theta & A_\phi \\ B_r & B_\theta & B_\phi \end{vmatrix}$$

1.3.7 Transformation of a vector from cartesian to spherical coordinate system and vice–versa :

Using the following dot product table, the transformation is carried out.

Dot Product Table 1.2

·	$\bar{a}_r$	$\bar{a}_\theta$	$\bar{a}_\phi$
$\bar{a}_x$	$\sin\theta\cos\phi$	$\cos\theta\cos\phi$	$-\sin\phi$
$\bar{a}_y$	$\sin\theta\sin\phi$	$\cos\theta\sin\phi$	$\cos\phi$
$\bar{a}_z$	$\cos\theta$	$-\sin\theta$	0

Problem 1.8 : Transform each of the following vectors to spherical coordinates at the point specified :(a) $5\,\bar{a}_x$ at B ($r = 4, \theta = 25^\circ, \phi = 120^\circ$), (b) $5\,\bar{a}_x$ at A ($x = 2, y = 3, z = -1$) ; (c) $4\,\bar{a}_x - 2\,\bar{a}_y - 4\,\bar{a}_z$ at P ($x = -2, y = -3, z = 4$) .

Solution :

(a) Let $\bar{A} = 5\,\bar{a}_x = A_r\,\bar{a}_r + A_\theta\,\bar{a}_\theta + A_\phi\,\bar{a}_\phi$

$A_r = \bar{A}\cdot\bar{a}_r = 5\,\bar{a}_x\cdot\bar{a}_r = 5\sin\theta\cos\phi$

$A_\theta = \bar{A}\cdot\bar{a}_\theta = 5\,\bar{a}_x\cdot\bar{a}_\theta = 5\cos\theta\cos\phi$

$A_\phi = \bar{A}\cdot\bar{a}_\phi = 5\,\bar{a}_x\cdot\bar{a}_\phi = -5\sin\phi$

$\therefore\ \bar{A} = 5\sin\theta\cos\phi\,\bar{a}_r + 5\cos\theta\cos\phi\,\bar{a}_\theta - 5\sin\phi\,\bar{a}_\phi.$

Substituting θ and ϕ coordinates of point B in the above vector, we have

$\bar{A}$ at point $B = -1.057\,\bar{a}_r - 2.266\,\bar{a}_\theta - 4.33\,\bar{a}_\phi$

(b) For point A $r = \sqrt{(2)^2 + (3)^2 + (-1)^2} = 3.742$

$\theta = \cos^{-1}\left(\frac{-1}{3.742}\right) = 105.5^\circ$; $\phi = \tan^{-1}\left(\frac{3}{2}\right) = 56.3^\circ$

Substituting θ and ϕ coordinates of point A in the above $\bar{A}$, we have

$\bar{A}$ at point $A = 2.673\,\bar{a}_r - 0.741\,\bar{a}_\theta - 4.160\,\bar{a}_\phi$

(c) Let $\bar{B} = 4\,\bar{a}_x - 2\,\bar{a}_y - 4\,\bar{a}_z = B_r\,\bar{a}_r + B_\theta\,\bar{a}_\theta + B_\phi\,\bar{a}_\phi$

$B_r = \bar{B}\cdot\bar{a}_r = (4\,\bar{a}_x - 2\,\bar{a}_y - 4\,\bar{a}_z)\cdot\bar{a}_r = 4\,\bar{a}_x\cdot\bar{a}_r - 2\,\bar{a}_y\cdot\bar{a}_r - 4\,\bar{a}_z\cdot\bar{a}_r$

$= 4\sin\theta\cos\phi - 2\sin\theta\sin\phi - 4\cos\theta$

$B_\theta = \bar{B}\cdot\bar{a}_\theta = (4\,\bar{a}_x - 2\,\bar{a}_y - 4\,\bar{a}_z)\cdot\bar{a}_\theta = 4\,\bar{a}_x\cdot\bar{a}_\theta - 2\,\bar{a}_y\cdot\bar{a}_\theta - 4\,\bar{a}_z\cdot\bar{a}_\theta$

$= 4\cos\theta\cos\phi - 2\cos\theta\sin\phi + 4\sin\theta$

$B_\phi = \bar{B}\cdot\bar{a}_\phi = (4\,\bar{a}_x - 2\,\bar{a}_y - 4\,\bar{a}_z)\cdot\bar{a}_\phi = 4\,\bar{a}_x\cdot\bar{a}_\phi - 2\,\bar{a}_y\cdot\bar{a}_\phi - 4\,\bar{a}_z\cdot\bar{a}_\phi$

$= -4\sin\phi - 2\cos\phi$

$\therefore\ \bar{B} = (4\sin\theta\cos\phi - 2\sin\theta\sin\phi - 4\cos\theta)\,\bar{a}_r +$

$(4\cos\theta\cos\phi - 2\cos\theta\sin\phi + 4\sin\theta)\,\bar{a}_\theta +$

$(-4\sin\phi - 2\cos\phi)\,\bar{a}_\phi$

For point P $r = \sqrt{(-2)^2 + (-3)^2 + (4)^2} = 5.385$

$\theta = \cos^{-1}(4/5.385) = 42.03^\circ$ and $\phi = 180^\circ + \tan^{-1}[(-3)/(-2)] = 236.3^\circ$

Substituting θ and ϕ coordinates of point P in the above $\bar{B}$, we have

$\bar{B}$ at point P $= -3.343\,\bar{a}_r + 2.266\,\bar{a}_\theta + 4.438\,\bar{a}_\phi$

1.3.8 Expression for $\overline{R}_{12}$

Transform both the points into cartesian cocrdinate system,calculate $\overline{R}_{12}$ in cartesian coordinate system and then transform this vector into spherical cooordinate system using dot product table 1.2. Then substitute coordinates of points 1 or 2 so as to cacculate $\overline{R}_{12}$ at points 1 or 2 respectively.

1.4 DIFFERENTIAL LENGTH ($\overline{dL}$), AREA ($\overline{dS}$), AND VOLUME (dv) :

Cartesian coordinate system :

$$\overline{dL} = dx\,\overline{a}_x + dy\,\overline{a}_y + dz\,\overline{a}_z$$

$$\overline{dS} = \begin{cases} \pm\; dy\; dz\; \overline{a}_x \\ \pm\; dx\; dz\; \overline{a}_y \\ \pm\; dx\; dy\; \overline{a}_z \end{cases}$$

$$dv = dx\; dy\; dz$$

Cylindrical Coordinate system :

$$\overline{dL} = d\rho\,\overline{a}_\rho + \rho\, d\phi\,\overline{a}_\phi + dz\,\overline{a}_z$$

$$\overline{dS} = \begin{cases} \pm\; \rho\, d\phi\; dz\; \overline{a}_\rho \\ \pm\; d\rho\; dz\; \overline{a}_\phi \\ \pm\; \rho\, d\rho\; d\phi\; \overline{a}_z \end{cases}$$

$$dv = \rho\, d\rho\; d\phi\; dz$$

Spherical Coordinate System :

$$\overline{dL} = dr\,\overline{a}_r + r\,d\theta\,\overline{a}_\theta + r\sin\theta\, d\phi\,\overline{a}_\phi$$

$$\overline{dS} = \begin{cases} \pm\; r^2\sin\theta\, d\theta\; d\phi\; \overline{a}_r \\ \pm\; r\sin\theta\; dr\; d\phi\; \overline{a}_\theta \\ \pm\; r\; dr\; d\theta\; \overline{a}_\phi \end{cases}$$

$$dv = r^2\sin\theta\; dr\; d\theta\; d\phi$$

Problem 1.9 : Let each of the vectors $\overline{A} = 5\,\overline{a}_x - \overline{a}_y + 3\,\overline{a}_z$, $\overline{B} = -2\,\overline{a}_x + 2\,\overline{a}_y + 4\,\overline{a}_z$, and $\overline{C} = 3\,\overline{a}_y - 4\,\overline{a}_z$ extend outward from the origin of a cartesian coordinate system to points A , B , and C , respectively. Find a unit vector directed from point A toward : (a) the origin ; (b) point B ; (c) a point equidistant from B and C on the line BC . (d) Find the length of the perimeter of the triangle ABC .

Solution :

(a) Vector directed from origin toward point $A = \overline{A}$

Hence vector directed from point A toward origin $= -\overline{A}$.

Hence the unit vector directed from point A toward the origin

$$= -\frac{\overline{A}}{|\overline{A}|} = -\frac{5\,\overline{a}_x - \overline{a}_y + 3\,\overline{a}_z}{\sqrt{(5)^2 + (-1)^2 + (3)^2}}$$

$$= -0.845\,\overline{a}_x + 0.169\,\overline{a}_y - 0.507\,\overline{a}_z$$

(b) $\overline{R}_{AB} = \overline{r}_B - \overline{r}_A = \overline{B} - \overline{A}$

(Note that $\overline{r}_A$, $\overline{r}_B$, and $\overline{r}_C$ are same as $\overline{A}$, $\overline{B}$, and $\overline{C}$ as these vectors are originating from origin)

$$\therefore\ \overline{R}_{AB} = (-2\,\overline{a}_x + 2\,\overline{a}_y + 4\,\overline{a}_z) - (5\,\overline{a}_x - \overline{a}_y + 3\,\overline{a}_z)$$

$$= -7\,\overline{a}_x + 3\,\overline{a}_y + \overline{a}_z$$

$$\therefore\ \overline{a}_{AB} = \frac{\overline{R}_{AB}}{|\overline{R}_{AB}|} = \frac{-7\,\overline{a}_x + 3\,\overline{a}_y + \overline{a}_z}{\sqrt{(-7)^2 + (3)^2 + (1)^2}}$$

$$= -0.911\,\overline{a}_x + 0.391\,\overline{a}_y + 0.130\,\overline{a}_z$$

(c) Let point M is a point equidistant from B and C, i.e. M is the midpoint of line BC.

Position vector of point M is given by $\overline{r}_M = \dfrac{\overline{r}_B + \overline{r}_C}{2} = \dfrac{\overline{B} + \overline{C}}{2}$

$$\therefore\ \overline{r}_M = \frac{-2\,\overline{a}_x + 2\,\overline{a}_y + 4\,\overline{a}_z + 3\,\overline{a}_y - 4\,\overline{a}_z}{2} = -\overline{a}_x + 2.5\,\overline{a}_y$$

$$\overline{R}_{AM} = \overline{r}_M - \overline{r}_A = (-\overline{a}_x + 2.5\,\overline{a}_y) - (5\,\overline{a}_x - \overline{a}_y + 3\,\overline{a}_z)$$

$$= -6\,\overline{a}_x + 3.5\,\overline{a}_y - 3\,\overline{a}_z$$

$$\therefore\ \overline{a}_{AM} = \frac{\overline{R}_{AM}}{|\overline{R}_{AM}|} = \frac{-6\,\overline{a}_x + 3.5\,\overline{a}_y - 3\,\overline{a}_z}{\sqrt{(-6)^2 + (3.5)^2 + (-3)^2}}$$

$$= -0.793\,\overline{a}_x + 0.463\,\overline{a}_y - 0.396\,\overline{a}_z$$

(d) $\overline{R}_{AB} = -7\,\overline{a}_x + 3\,\overline{a}_y + \overline{a}_z$

$$\therefore\ |\overline{R}_{AB}| = \sqrt{(-7)^2 + (3)^2 + (1)^2} = 7.681$$

$$\overline{R}_{BC} = \overline{r}_C - \overline{r}_B = \overline{C} - \overline{B} = (3\,\overline{a}_y - 4\,\overline{a}_z) - (-2\,\overline{a}_x + 2\,\overline{a}_y + 4\,\overline{a}_z)$$

$$= 2\bar{a}_x + \bar{a}_y - 8\bar{a}_z$$

$$\therefore |\bar{R}_{BC}| = \sqrt{(2)^2 + (1)^2 + (-8)^2} = 8.307$$

$$\bar{R}_{CA} = \bar{r}_A - \bar{r}_C = \bar{A} - \bar{C} = (5\bar{a}_x - \bar{a}_y + 3\bar{a}_z) - (3\bar{a}_y - 4\bar{a}_z)$$

$$= 5\bar{a}_x - 4\bar{a}_y + 7\bar{a}_z$$

$$\therefore |\bar{R}_{CA}| = \sqrt{(5)^2 + (-4)^2 + (7)^2} = 9.487$$

As seen fromFig.P1.9, the triangle ABC has sides represented by$\bar{R}_{AB}$, $\bar{R}_{BC}$, and $\bar{R}_{CA}$, the magnitude of these three vectors gives the lengths of three sides of the triangle, the addition of which gives perimeter of the triangle ABC .Hence,perimeter of the triangleABC .

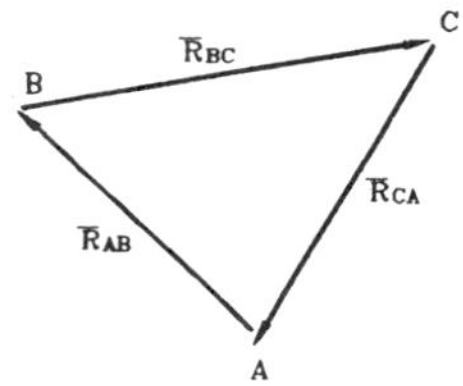

FIG.P1.9

$$= |\bar{R}_{AB}| + |\bar{R}_{BC}| + |\bar{R}_{CA}| = 7.681 + 8.307 + 9.487 = 25.475$$

Problem 1.10 : If $\bar{A} = 2\bar{a}_x + 6\bar{a}_y - 3\bar{a}_z$ & $\bar{B} = -3\bar{a}_x - 4\bar{a}_y - 5\bar{a}_z$, find : (a) a unit vector in the direction of $\bar{A} - \bar{B}$; (b) the magnitude of $\bar{A} + \bar{B}$;(c) $\left|\left[(\bar{A} + 3\bar{B}) / |\bar{A} + 3\bar{B}|\right]\right| \bar{B}$.

Solution. :

(a) $\bar{A} - \bar{B} = (2\bar{a}_x + 6\bar{a}_y - 3\bar{a}_z) - (-3\bar{a}_x - 4\bar{a}_y - 5\bar{a}_z)$

$$= 5\bar{a}_x + 10\bar{a}_y + 2\bar{a}_z$$

Unit vector in the direction of$\bar{A} - \bar{B} = (\bar{A} - \bar{B}) / |\bar{A} - \bar{B}|$

$$= \frac{5\bar{a}_x + 10\bar{a}_y + 2\bar{a}_z}{\sqrt{(5)^2 + (10)^2 + (2)^2}} = 0.440\,\bar{a}_x + 0.880\,\bar{a}_y + 0.176\,\bar{a}_z$$

(b) $\bar{A} + \bar{B} = (2\bar{a}_x + 6\bar{a}_y - 3\bar{a}_z) + (-3\bar{a}_x - 4\bar{a}_y - 5\bar{a}_z)$

$$= -\bar{a}_x + 2\bar{a}_y - 8\bar{a}_z$$

$$\therefore |\bar{A} + \bar{B}| = \sqrt{(-1)^2 + (2)^2 + (-8)^2} = 8.3066$$

(c) $(\bar{A} + 3\bar{B}) / |\bar{A} + 3\bar{B}|$

gives a unit vector in the direction of a vector given by $\bar{A} + 3\bar{B}$.

$\left|\left[(\bar{A} + 3\bar{B}) / |\bar{A} + 3\bar{B}|\right]\right|$

gives the magnitude of above unit vector which is equal to one. Hence

$$\left|\left[(\bar{A} + 3\bar{B}) / |\bar{A} + 3\bar{B}|\right]\right| \bar{B} = 1\bar{B} = -3\bar{a}_x - 4\bar{a}_y - 5\bar{a}_z$$

Problem 1.11 : The vector $\overline{R}_{AB}$ extends from *A* (1, 2, 3) to *B* . If the length of $\overline{R}_{AB}$ is 10 units and its direction is given by $\bar{a} = 0.6\,\bar{a}_x + 0.64\,\bar{a}_y + 0.48\,\bar{a}_z$, find the coordinates of *B* .

Solution :

$|\overline{R}_{AB}| = 10$; $\bar{a}_{AB} = \bar{a} = 0.6\,\bar{a}_x + 0.64\,\bar{a}_y + 0.48\,\bar{a}_z$

$$\therefore\ \overline{R}_{AB} = |\overline{R}_{AB}|\ \bar{a}_{AB} = 10(0.6\,\bar{a}_x + 0.64\,\bar{a}_y + 0.48\,\bar{a}_z) = 6\,\bar{a}_x + 6.4\,\bar{a}_y + 4.8\,\bar{a}_z$$

Let the coordinates of point *B* are (x, y, z) . Then

$$\overline{R}_{AB} = (x-1)\,\bar{a}_x + (y-2)\,\bar{a}_y + (z-3)\,\bar{a}_z$$

comparing the coefficients of unit vectors from both the vectors $\overline{R}_{AB}$, we have

$x-1 = 6$ or $x = 7$; $y-2 = 6.4$ or $y = 8.4$; $z-3 = 4.8$ or $z = 7.8$

Hence the coordinates of point *B* are (7, 8.4, 7.8)

Problem 1.12 : Given two points, *M* (2, 5, –3) and *N* (–3, 1, 4) : (a) find their separation ; (b) find the distance from the origin to the midpoint of the line *MN* ; (c) find a unit vector in the direction of $\overline{R}_{MN}$; (d) find the point of intersection of the line *MN* and the $x = 0$ plane.

Solution :

(a) $d(MN) = \sqrt{(2+3)^2 + (5-1)^2 + (-3-4)^2} = 9.487$

(b) Let *A* be the midpoint of line *MN* . Then

$$A = \left(\frac{2-3}{2}, \frac{5+1}{2}, \frac{-3+4}{2}\right) = (-0.5, 3, 0.5)$$

$$\therefore\ d(OA) = \sqrt{(0+0.5)^2 + (0-3)^2 + (0-0.5)^2} = 3.082$$

(c) $\overline{R}_{MN} = \bar{r}_N - \bar{r}_M = (-3\,\bar{a}_x + \bar{a}_y + 4\,\bar{a}_z) - (2\,\bar{a}_x + 5\,\bar{a}_y - 3\,\bar{a}_z)$

$$= -5\,\bar{a}_x - 4\,\bar{a}_y + 7\,\bar{a}_z$$

$$\therefore\ \bar{a}_{MN} = \frac{\overline{R}_{MN}}{|\overline{R}_{MN}|} = \frac{-5\,\bar{a}_x - 4\,\bar{a}_y + 7\,\bar{a}_z}{\sqrt{(-5)^2 + (-4)^2 + (7)^2}}$$

$$= -0.527\,\bar{a}_x - 0.422\,\bar{a}_y + 0.738\,\bar{a}_z$$

(d) Equations of line *MN* are

$$\frac{x-2}{2-(-3)} = \frac{y-5}{5-1} = \frac{z-(-3)}{-3-4} \quad \text{or} \quad \frac{x-2}{5} = \frac{y-5}{4} = \frac{z+3}{-7}$$

Substituting $x = 0$ in the above equations and solving for y and z we have

$y = -8/5 + 5 = 3.4$ and $z = 14/5 - 3 = -0.2$

Hence line MN intersects the $x = 0$ plane at a point given by (0, 3.4, –0.2)

Problem 1.13 : Given the vector field, $\overline{G} = 2x^2y\,\overline{a}_x - 2(z-x)\,\overline{a}_y + 3xyz\,\overline{a}_z$, find ; (a) $\overline{G}$ at P (2, –3, 4) ; (b) a unit vector in the direction of $\overline{G}$ at P ; (c) the (scalar) equation of the surface on which $|\overline{G}| = 100$; (d) the y coordinate of Q (– 3 , y , 5) if $|\overline{G}_Q| = 100$ and $y > 0$; (e) the distance between P and Q .

Solution :

(a) $\overline{G}$ at $P = 2(2)^2(-3)\,\overline{a}_x - 2(4-2)\,\overline{a}_y + 3(2)(-3)(4)\,\overline{a}_z$

$$= -24\,\overline{a}_x - 4\,\overline{a}_y - 72\,\overline{a}_z$$

$$\text{(b) } \overline{a} = \frac{\overline{G}}{|\overline{G}|} = \frac{-24\,\overline{a}_x - 4\,\overline{a}_y - 72\,\overline{a}_z}{\sqrt{(-24)^2 + (-4)^2 + (-72)^2}}$$

$$= -0.316\,\overline{a}_x - 0.053\,\overline{a}_y - 0.947\,\overline{a}_z$$

$$\text{(c) } |\overline{G}| = \sqrt{(2x^2y)^2 + [-2(z-x)]^2 + (3xyz)^2}$$

$$= \sqrt{4x^4y^2 + 4(z-x)^2 + 9x^2y^2z^2} = 100$$

Squaring both sides, we have scalar equation of the surface as

$4x^4y^2 + 4(z-x)^2 + 9x^2y^2z^2 = 10000$

(d) Substitute the coordinates of point Q in the above equation of the surface

$4(-3)^4y^2 + 4[5-(-3)]^2 + 9(-3)^2y^2(5)^2 = 10000$

$\therefore\ 2349\,y^2 = 9744$ or $y^2 = 4.1481$ or $y = \pm\ 2.037$

Since $y > 0$, $y = 2.037$ is selected.

$$\text{(e) } d(PQ) = \sqrt{(2+3)^2 + (-3-2.037)^2 + (4-5)^2} = 7.167$$

Problem 1.14 : A vector field is given as $\overline{G} = 2x^2y\,\overline{a}_x - 2(z-x)\,\overline{a}_y + 3xyz\,\overline{a}_z$. Prepare sketches of G_x , G_y , G_z , and $|\overline{G}|$, all evaluated along the line $x = 2$, $y = -1$, for $0 \le z \le 10$

Solution :

$\overline{G} = 2x^2y\,\overline{a}_x - 2(z - x)\,\overline{a}_y + 3xyz\,\overline{a}_z = G_x\,\overline{a}_x + G_y\,\overline{a}_y + G_z\,\overline{a}_z$

$G_x = 2x^2y \;;\; G_y = -2(z - x) \;;\; G_z = 3xyz$

Substitute $x = 2$ and $y = -1$ in the above expressions

$|\overline{G}| = \sqrt{G_x^2 + G_y^2 + G_z^2} = \sqrt{64 + (-2z + 4)^2 + 36z^2}$

$\therefore G_x = -8 \;;\; G_y = -2z + 4 \;;\; G_z = -6z$

Plot G_x, G_y, G_z, and $|\overline{G}|$ on the y-axis and z on the x-axis on a graph paper.

Problem 1.15 : A vector field is specified as $\overline{F} = 2(x + y)\sin\pi z\,\overline{a}_x - (x^2 + y)\,\overline{a}_y + [10/(x^2 + y^2)]\,\overline{a}_z$. Specify the locus of all points at which : (a) $F_x = 0$; (b) $F_y = 0$; (c) $|F_z| = 1$.

Solution :

(a) $\overline{F} = F_x\,\overline{a}_x + F_y\,\overline{a}_y + F_z\,\overline{a}_z$

$\therefore F_x = 2(x + y)\sin\pi z = 0$

$\therefore x + y = 0$ or $\sin\pi z = 0$

$\therefore y = -x$ or $\pi z = 0, \pm\pi, \pm 2\pi, \pm 3\pi, \text{-----}$

$y = -x$ represents a plane and $z = 0, \pm 1, \pm 2, \pm 3$ represents planes.

(b) $F_y = -(x^2 + y) = 0$

$\therefore y = -x^2$. It represents a parabolic cylinder

(c) $F_z = 10/(x^2 + y^2)$

$\therefore |F_z| = 10/(x^2 + y^2) = 1$ or $x^2 + y^2 = 10$

It represents a circular cylinder of radius $\sqrt{10}$ with z axis as its axis.

Problem 1.16 : Given fields $\overline{G}_1 = 5(x + y)\,\overline{a}_x + 10\,\overline{a}_y$ and $\overline{G}_2 = 5\,\overline{a}_x + 2xy\,\overline{a}_y$, find at $P(3, 2, 0)$: (a) $|\overline{G}_1|$; (b) $|\overline{G}_2|$; (c) $\overline{a}_1$, a unit vector in the direction of $\overline{G}_1$; (d) $\overline{a}_2$, a unit vector in the direction of $\overline{G}_2$; (e) $\overline{a}_{12}$, a unit vector in the direction of $(\overline{G}_1 + \overline{G}_2)$.

Solution :

(a) $\overline{G}_1$ at point $P = 5(3 + 2)\,\overline{a}_x + 10\,\overline{a}_y = 25\,\overline{a}_x + 10\,\overline{a}_y$

$\therefore |\overline{G}_1| = \sqrt{(25)^2 + (10)^2} = 26.926$

(b) $\overline{G}_2$ at point P $= 5\,\overline{a}_x + 2(3)(2)\,\overline{a}_y = 5\,\overline{a}_x + 12\,\overline{a}_y$

$\therefore |\overline{G}_2| = \sqrt{(5)^2 + (12)^2} = 13$

(c) $\bar{a}_1 = \dfrac{\bar{G}_1}{|\bar{G}_1|} = \dfrac{25\,\bar{a}_x + 10\,\bar{a}_y}{26.926} = 0.928\,\bar{a}_x + 0.371\,\bar{a}_y$

(d) $\bar{a}_2 = \dfrac{\bar{G}_2}{|\bar{G}_2|} = \dfrac{5\,\bar{a}_x + 12\,\bar{a}_y}{13} = 0.385\,\bar{a}_x + 0.923\,\bar{a}_y$

(e) $\bar{G}_1 + \bar{G}_2 = (25\,\bar{a}_x + 10\,\bar{a}_y) + (5\,\bar{a}_x + 12\,\bar{a}_y) = 30\,\bar{a}_x + 22\,\bar{a}_y$

$$\bar{a}_{12} = \frac{\bar{G}_1 + \bar{G}_2}{|\bar{G}_1 + \bar{G}_2|} = \frac{30\,\bar{a}_x + 22\,\bar{a}_y}{\sqrt{(30)^2 + (22)^2}} = 0.806\,\bar{a}_x + 0.591\,\bar{a}_y$$

Problem 1.17 : Given the field, $\bar{G} = [2x/(1+y^2)]\,\bar{a}_x + (y+z+1)\,\bar{a}_y + (5x - z^2)\,\bar{a}_z$, find : (a) a unit vector in the direction of $\bar{G}$ at P (1, 2, −3) ; (b) the angle between $\bar{G}$ and the $y = 0$ plane at Q (2, 0, 4) ; (c) the value of $\int_{y=0}^{2} \int_{x=1}^{3} \bar{G} \cdot dx\,dy\,\bar{a}_z$ at the plane z = 1.

Solution :

(a) $\bar{G}$ at point P

$= 2(1)/[1 + (2)^2]\,\bar{a}_x + (2 - 3 + 1)\,\bar{a}_y + [5(1) - (-3)^2]\,\bar{a}_z$

$= 0.4\,\bar{a}_x - 4\,\bar{a}_z$

$$\therefore\ \bar{a} = \frac{\bar{G}}{|\bar{G}|} = \frac{0.4\,\bar{a}_x - 4\,\bar{a}_z}{\sqrt{(0.4)^2 + (-4)^2}} = 0.1\,\bar{a}_x - 0.995\,\bar{a}_z$$

(b) $\bar{G}$ at point Q $= 2(2)\,\bar{a}_x + (0 + 4 + 1)\,\bar{a}_y + (5 \times 2 - 4^2)\,\bar{a}_z$

$= 4\,\bar{a}_x + 5\,\bar{a}_y - 6\,\bar{a}_z$

The unit vector $\bar{a}_y$ is normal to the $y = 0$ plane. As seen from Fig.P1.17, the angle α between $\bar{a}_y$ and $\bar{G}$ at point P is given by

$$\cos\alpha = \frac{\bar{G} \cdot \bar{a}_y}{|\bar{G}|\,|\bar{a}_y|}$$

$$= \frac{(4\,\bar{a}_x + 5\,\bar{a}_y - 6\,\bar{a}_z) \cdot \bar{a}_y}{\sqrt{(4)^2 + (5)^2 + (-6)^2}}$$

$= 5/\sqrt{77} = 0.5698$

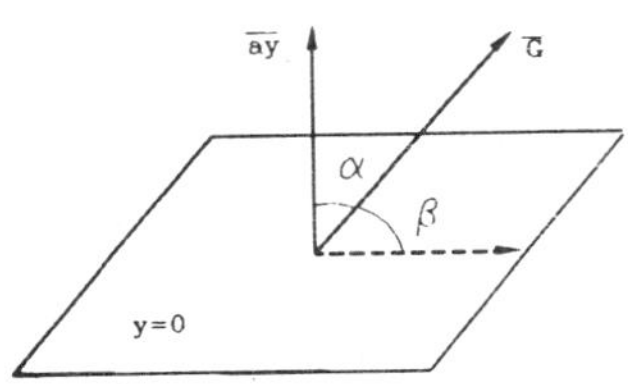

FIG.P1.17

$\therefore\ \alpha = \cos^{-1}(0.5698) = 55.264^\circ$. Now $\alpha + \beta = 90^\circ$.

$\therefore\ \beta =$ angle between $\overline{G}$ at point P and the $y = 0$ plane

$= 90^\circ - \alpha = 90^\circ - 55.264 = 34.736^\circ$

(c) $\overline{G} \cdot dx\, dy\, \bar{a}_z = (5x - z^2)\, dx\, dy$, as $\bar{a}_x \cdot \bar{a}_z = \bar{a}_y \cdot \bar{a}_z = 0$

$$\therefore \int_{y=0}^{2} \int_{x=1}^{3} \overline{G} \cdot dx\, dy\, \bar{a}_z \Big|_{z=1} = \int_{y=0}^{2} \int_{x=1}^{3} (5x - z^2)\, dx\, dy \Big|_{z=1}$$

$$= 5\left[\frac{x^2}{2}\right]_1^3 6\left[y\right]_0^2 - (1)^2\left[x\right]_1^3\left[y\right]_0^2 = 36$$

Problem 1.18 : Given points P (2, 5, –1), Q (–1, –4, 1), and T (5, 0, 2), find : (a) the vector $\overline{R}_{PQ}$; (b) a unit vector in the direction of $\overline{R}_{PQ}$; (c) the length of the perimeter of the triangle PQT ; (d) the interior angle at Q ; (e) the vector projection of $\overline{R}_{PQ}$ on $\overline{R}_{PT}$; (f) the length of the altitude of the triangle that extends from Q perpendicularly to the opposite side or its extension.

Solution :

(a) $\overline{R}_{PQ} = \bar{r}_Q - \bar{r}_P = (-\bar{a}_x - 4\bar{a}_y + \bar{a}_z) - (2\bar{a}_x + 5\bar{a}_y - \bar{a}_z)$

$= -3\bar{a}_x - 9\bar{a}_y + 2\bar{a}_z$

(b) $$\bar{a}_{PQ} = \frac{\overline{R}_{PQ}}{|\overline{R}_{PQ}|} = \frac{-3\bar{a}_x - 9\bar{a}_y + 2\bar{a}_z}{\sqrt{(-3)^2 + (-9)^2 + (2)^2}}$$

$= -0.309\,\bar{a}_x - 0.928\,\bar{a}_y + 0.206\,\bar{a}_z$

(c) The perimeter of the triangle $PQT = d(PQ) + d(QT) + d(TP)$

$= \sqrt{(2+1)^2 + (5+4)^2 + (-1-1)^2} + \sqrt{(-1-5)^2 + (-4)^2 + (1-2)^2} + \sqrt{(5-2)^2 + (0-5)^2 + (2+1)^2}$

$= 9.695 + 7.280 + 6.557 = 23.532$

(d) $\overline{R}_{QP} = \bar{r}_P - \bar{r}_Q = (2\bar{a}_x + 5\bar{a}_y - \bar{a}_z) - (-\bar{a}_x - 4\bar{a}_y + \bar{a}_z)$

$= 3\bar{a}_x + 9\bar{a}_y - 2\bar{a}_z$

$\overline{R}_{QT} = \bar{r}_T - \bar{r}_Q = (5\bar{a}_x + 2\bar{a}_z) - (-\bar{a}_x - 4\bar{a}_y + \bar{a}_z)$

$= 6\bar{a}_x + 4\bar{a}_y + \bar{a}_z$

$$\cos\alpha = \frac{\overline{R}_{QP}\,.\,\overline{R}_{QT}}{|\,\overline{R}_{QP}\,|\;|\,\overline{R}_{QT}\,|} = \frac{(3)(6)+(9)(4)+(-2)(1)}{\sqrt{(3)^2+(9)^2+(2)^2}\;\sqrt{(6)^2+(4)^2+(1)^2}}$$

$= 0.7367.$

Hence interior angle at Q , $\alpha = \cos^{-1}(0.7367) = 42.55^{\circ}$

(e) $\overline{R}_{PQ} = -3\,\overline{a}_x - 9\,\overline{a}_y + 2\,\overline{a}_z$

$\overline{R}_{PT} = \overline{r}_T - \overline{r}_P = (5\,\overline{a}_x + 2\,\overline{a}_z) - (2\,\overline{a}_x + 5\,\overline{a}_y - \overline{a}_z)$

$= 3\,\overline{a}_x - 5\,\overline{a}_y + 3\,\overline{a}_z$

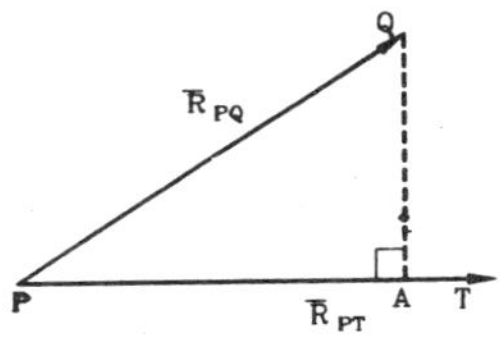

FIG.P1.18 (a)

From Fig.P1.18 (a),

$\overline{R}_{PA}$ = Vector projectio of $\overline{R}_{PQ}$ on $\overline{R}_{PT}$

$= (\overline{R}_{PQ}\cdot\overline{a}_{PT})\,\overline{a}_{PT}$

$$= \left(\overline{R}_{PQ}\cdot\frac{\overline{R}_{PT}}{|\,\overline{R}_{PT}\,|}\right)\frac{\overline{R}_{PT}}{|\,\overline{R}_{PT}\,|}$$

$$= \left[\frac{(-3)(3)+(-9)(-5)+(2)(3)}{\sqrt{(3)^2+(-5)^2+(3)^2}}\right]\left[\frac{3\,\overline{a}_x - 5\,\overline{a}_y + 3\,\overline{a}_z}{\sqrt{(3)^2+(-5)^2+(3)^2}}\right]$$

$= 2.930\,\overline{a}_x - 4.884\,\overline{a}_y + 2.930\,\overline{a}_z$

(f) $$\cos\beta = \frac{\overline{R}_{PQ}\cdot\overline{R}_{PT}}{|\,\overline{R}_{PQ}\,|\;|\,\overline{R}_{PT}\,|}$$

$$= \frac{(-3)(3)+(-9)(-5)+(2)(3)}{\sqrt{(-3)^2+(-9)^2+(2)^2}\;\sqrt{(3)^2+(-5)^2+(3)^2}}$$

$= 0.6606$

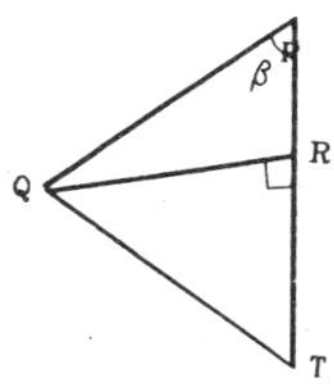

FIG.P1.18 (b)

$\therefore\ \beta = \cos^{-1}(0.6606) = 48.653^{\circ}$

.Now from Fig . P1.18(b),

$d(QR) = d(PQ)\sin\beta = |\,\overline{R}_{PQ}\,|\sin\beta = \sqrt{94}\,\sin(48.653^{\circ}) = 7.279$

Problem 1.19 : Given points E (2, 5, 1), F (–1, 4, –2) and G (3, –2, 4), find : (a) a unit vector directed from E towards F ; (b) the angle between $\overline{R}_{EF}$ and $\overline{R}_{EG}$; (c) the length of the perimeter of triangle EFG ; (d) the scalar projection of $\overline{R}_{EF}$ on $\overline{R}_{EG}$.

Solution :

(a) $\overline{R}_{EF} = \overline{r}_F - \overline{r}_E = (-\overline{a}_x + 4\overline{a}_y - 2\overline{a}_z) - (2\overline{a}_x + 5\overline{a}_y + \overline{a}_z)$

$= -3\overline{a}_x - \overline{a}_y - 3\overline{a}_z$

$$\therefore \overline{a}_{EF} = \frac{\overline{R}_{EF}}{|\overline{R}_{EF}|} = \frac{-3\overline{a}_x - \overline{a}_y - 3\overline{a}_z}{\sqrt{(-3)^2 + (-1)^2 + (-3)^2}}$$

$= -0.688\,\overline{a}_x - 0.229\,\overline{a}_y - 0.688\,\overline{a}_z$

(b) $\overline{R}_{EG} = \overline{r}_G - \overline{r}_E = (-\overline{a}_x + 4\overline{a}_y - 2\overline{a}_z) - (2\overline{a}_x + 5\overline{a}_y + \overline{a}_z)$

$= \overline{a}_x - 7\overline{a}_y + 3\overline{a}_z$

$$\cos\theta = \frac{\overline{R}_{EF} \cdot \overline{R}_{EG}}{|\overline{R}_{EF}|\,|\overline{R}_{EG}|} = \frac{(-3)(1) + (-1)(-7) + (-3)(3)}{\sqrt{(-3)^2 + (-1)^2 + (-3)^2}\,\sqrt{(1)^2 + (-7)^2 + (3)^2}}$$

$= -0.1493$. Hence $\theta = \cos^{-1}(-0.1493) = 98.59^\circ$

(c)The length of the perimeter of triangle EFG

$= d(EF) + d(FG) + d(GE)$

$= \sqrt{(2+1)^2 + (5-4)^2 + (1+2)^2} +$

$\sqrt{(-1-3)^2 + (4+2)^2 + (-2-4)^2} +$

$\sqrt{(3-2)^2 + (-2-5)^2 + (4-1)^2}$

$= 4.359 + 9.381 + 7.681 = 21.421$

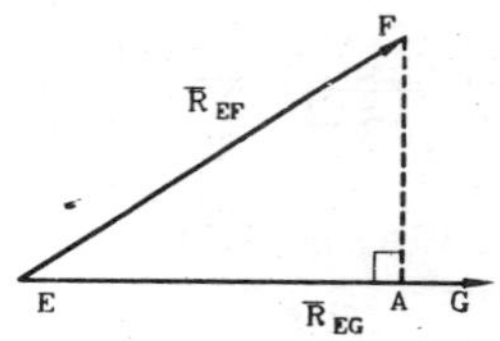

FIG.P1.19

(d) From Fig.P1.19, EA = Scalar projection of $\overline{R}_{EF}$ on $\overline{R}_{EG} = \overline{R}_{EF} \cdot \overline{a}_{EG}$

$$= \overline{R}_{EF} \cdot \frac{\overline{R}_{EG}}{|\overline{R}_{EG}|} = \frac{(-3)(1) + (-1)(-7) + (-3)(3)}{\sqrt{(1)^2 + (-7)^2 + (3)^2}} = -0.651$$

Problem 1.20 : Express in cartesian components : (a) the vector $\overline{G}$ extending from the origin to the midpoint of the line joining A (2, –3, 5) to B (6, –5, 5) ; (b) the vector $\overline{D}$ extending from C (–2, 7, 3) to the midpoint of the line joining A to B ; (c) the component of $\overline{R}_{AB}$ that is in the direction $\overline{R}_{AC}$; (d) a unit vector in the direction of $\overline{R}_{BC}$.

Solution :

(a) Let M be the midpoint of the line joining points A and B .

$$\therefore M = \left(\frac{2+6}{2}, \frac{-3-5}{2}, \frac{5+5}{2}\right) = (4, -4, 5)$$

$\therefore \overline{G} = \overline{R}_{OM} = \overline{r}_M = 4\overline{a}_x - 4\overline{a}_y + 5\overline{a}_z$

(b) $\overline{D} = \overline{R}_{CM} = \overline{r}_M - \overline{r}_C = (4\,\overline{a}_x - 4\,\overline{a}_y + 5\,\overline{a}_z) - (-2\,\overline{a}_x + 7\,\overline{a}_y + 3\,\overline{a}_z)$

$$= 6\,\overline{a}_x - 11\,\overline{a}_y + 2\,\overline{a}_z$$

(c) $\overline{R}_{AB} = \overline{r}_B - \overline{r}_A = (6\,\overline{a}_x - 5\,\overline{a}_y + 5\,\overline{a}_z) - (2\,\overline{a}_x - 3\,\overline{a}_y + 5\,\overline{a}_z)$

$$= 4\,\overline{a}_x - 2\,\overline{a}_y$$

$\overline{R}_{AC} = \overline{r}_C - \overline{r}_A = (-2\,\overline{a}_x + 7\,\overline{a}_y + 3\,\overline{a}_z) - (2\,\overline{a}_x - 3\,\overline{a}_y + 5\,\overline{a}_z)$

$$= -4\,\overline{a}_x + 10\,\overline{a}_y - 2\,\overline{a}_z$$

From Fig.P1.20 ,

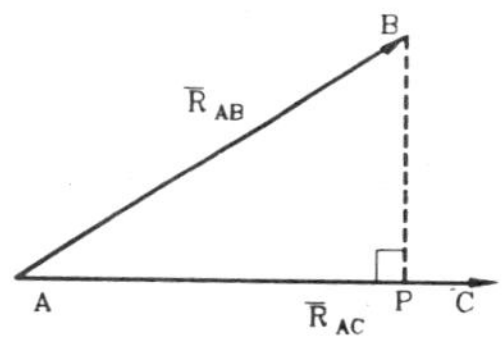

FIG.P1.20

$\overline{R}_{AP} =$ Vector projection of $\overline{R}_{AB}$ on $\overline{R}_{AC}$

$$= (\overline{R}_{AB} \cdot \overline{a}_{AC})\,\overline{a}_{AC}$$

$$= \left(\overline{R}_{AB} \cdot \frac{\overline{R}_{AC}}{|\overline{R}_{AC}|}\right)\frac{\overline{R}_{AC}}{|\overline{R}_{AC}|}$$

$$= \left[\frac{(4)(-4) + (-2)(10) + (0)(-2)}{\sqrt{(-4)^2 + (10)^2 + (-2)^2}}\right]\left[\frac{-4\,\overline{a}_x + 10\,\overline{a}_y - 2\,\overline{a}_z}{\sqrt{(-4)^2 + (10)^2 + (-2)^2}}\right]$$

$$= 1.2\,\overline{a}_x - 3\,\overline{a}_y + 0.6\,\overline{a}_z$$

(d) $\overline{R}_{BC} = \overline{r}_C - \overline{r}_B = (-2\,\overline{a}_x + 7\,\overline{a}_y + 3\,\overline{a}_z) - (6\,\overline{a}_x - 5\,\overline{a}_y + 5\,\overline{a}_z)$

$$= -8\,\overline{a}_x + 12\,\overline{a}_y - 2\,\overline{a}_z$$

$$\therefore\ \overline{a}_{BC} = \frac{\overline{R}_{BC}}{|\overline{R}_{BC}|} = \frac{-8\,\overline{a}_x + 12\,\overline{a}_y - 2\,\overline{a}_z}{\sqrt{(-8)^2 + (12)^2 + (-2)^2}}$$

$$= -0.549\,\overline{a}_x + 0.824\,\overline{a}_y - 0.137\,\overline{a}_z$$

Problem 1.21 : Given $\overline{A} = -4\,\overline{a}_x + 2\,\overline{a}_y + 3\,\overline{a}_z$ and $\overline{B} = 3\,\overline{a}_x + 4\,\overline{a}_y - \overline{a}_z$, find : (a) the magnitude of $5\overline{A} - 2\overline{B}$; (b) a unit vector in the direction of $(5\overline{A} - 2\overline{B})/|\overline{A}|$; (c) the vector component of $\overline{A}$ that is parallel to $\overline{B}$; (d) the vector component of $\overline{A}$ that is perpendicular to $\overline{B}$.

Solution :

(a) $5\overline{A} - 2\overline{B} = 5(-4\,\overline{a}_x + 2\,\overline{a}_y + 3\,\overline{a}_z) - 2(3\,\overline{a}_x + 4\,\overline{a}_y - \overline{a}_z)$

$$= -26\,\overline{a}_x + 2\,\overline{a}_y + 17\,\overline{a}_z$$

$$|5\overline{A} - 2\overline{B}| = \sqrt{(-26)^2 + (2)^2 + (17)^2} = 31.129$$

(b) $$\frac{5\overline{A} - 2\overline{B}}{|\overline{A}|} = \frac{-26\,\overline{a}_x + 2\,\overline{a}_y + 17\,\overline{a}_z}{\sqrt{(-4)^2 + (2)^2 + (3)^2}}$$

$$= -4.828\,\bar{a}_x + 0.371\,\bar{a}_y + 3.157\,\bar{a}_z$$

Hence the unit vector in the direction of vector$(5\bar{A} - 2\bar{B}) / |\bar{A}|$

$$= \frac{-4.828\,\bar{a}_x + 0.371\,\bar{a}_y + 3.157\,\bar{a}_z}{\sqrt{(-4.828)^2 + (0.371)^2 + (3.157)^2}}$$

$$= -0.835\,\bar{a}_x + 0.064\,\bar{a}_y + 0.546\,\bar{a}_z$$

(c) The vector component of $\bar{A}$ that is parallel to $\bar{B}$

$$= (\bar{A}\cdot\bar{a}_B)\,\bar{a}_B = \left(\bar{A}\cdot\frac{\bar{B}}{|\bar{B}|}\right)\frac{\bar{B}}{|\bar{B}|}$$

$$= \left[\frac{(-4)(3)+(2)(4)+(3)(-1)}{\sqrt{(3)^2+(4)^2+(-1)^2}}\right]\left[\frac{3\,\bar{a}_x + 4\,\bar{a}_y - \bar{a}_z}{\sqrt{(3)^2+(4)^2+(-1)^2}}\right]$$

$$= -0.808\,\bar{a}_x - 1.077\,\bar{a}_y + 0.269\,\bar{a}_z$$

(d) From Fig.P1.21, we have $\bar{A}$ = component of $\bar{A}$ parallel to $\bar{B}$ + component of $\bar{A}$ perpendicular to $\bar{B}$. Hence component of $\bar{A}$ perpendicular to $\bar{B}$

$= \bar{A} -$ component of $\bar{A}$ parallel to $\bar{B}$

$$= (-4\,\bar{a}_x + 2\,\bar{a}_y + 3\,\bar{a}_z) - (-0.808\,\bar{a}_x - 1.077\,\bar{a}_y + 0.269\,\bar{a}_z)$$

$$= -3.192\,\bar{a}_x + 3.077\,\bar{a}_y + 2.731\,\bar{a}_z$$

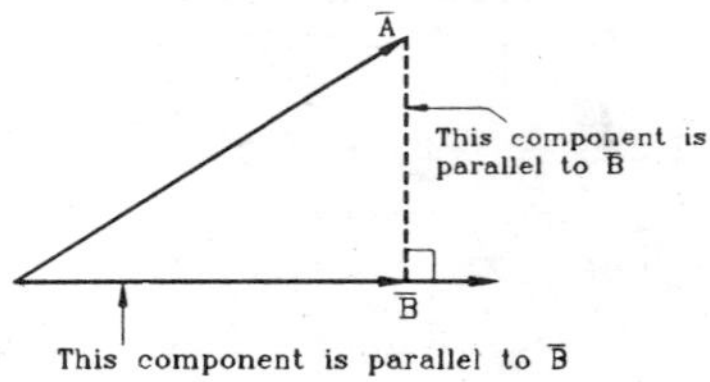

FIG.P1.21

Problem 1. 22 : Given the three points *A* (2, –1, 2), *B* (–1, 1, 4), and *C* (4, 3, –1), find : (a) the angle between $\bar{R}_{AB}$ and $\bar{R}_{AC}$; (b) the (scalar) area of triangle *ABC* ; (c) a unit vector perpendicular to *ABC* .

Solution :

(a) $\bar{R}_{AB} = \bar{r}_B - \bar{r}_A = (-\bar{a}_x + \bar{a}_y + 4\bar{a}_z) - (2\bar{a}_x - \bar{a}_y + 2\bar{a}_z)$

$$= -3\,\bar{a}_x + 2\,\bar{a}_y + 2\,\bar{a}_z$$

$\bar{R}_{AC} = \bar{r}_C - \bar{r}_A = (4\,\bar{a}_x + 3\,\bar{a}_y - \bar{a}_z) - (2\,\bar{a}_x - \bar{a}_y + 2\,\bar{a}_z)$

$$= 2\,\bar{a}_x + 4\,\bar{a}_y - 3\,\bar{a}_z$$

$$\cos\theta = \frac{\bar{R}_{AB}\cdot\bar{R}_{AC}}{|\bar{R}_{AB}|\,|\bar{R}_{AC}|} = \frac{(-3)(2)+(2)(4)+(2)(-3)}{\sqrt{(-3)^2+(2)^2+(2)^2}\,\sqrt{(2)^2+(4)^2+(-3)^2}}$$

$$= -0.18015 \;\therefore\; \theta = \cos^{-1}(-0.18015) = 100.38^\circ$$

(b) Area of triangle $ABC = 0.5 \mid \bar{R}_{AB} \times \bar{R}_{AC} \mid$

$$\bar{R}_{AB} \times \bar{R}_{AC} = \begin{vmatrix} \bar{a}_x & \bar{a}_y & \bar{a}_z \\ -3 & 2 & 2 \\ 2 & 4 & -3 \end{vmatrix} = -14\,\bar{a}_x - 5\,\bar{a}_y - 16\,\bar{a}_z$$

Hence area of triangle $ABC = 0.5\sqrt{(-14)^2 + (-5)^2 + (-16)^2} = 10.92$

$$\text{(c) } \bar{a}_n = \pm \frac{\bar{R}_{AB} \times \bar{R}_{AC}}{\mid \bar{R}_{AB} \times \bar{R}_{AC} \mid} = \pm \frac{(-14\,\bar{a}_x - 5\,\bar{a}_y - 16\,\bar{a}_z)}{\sqrt{(-14)^2 + (-5)^2 + (-16)^2}}$$

$$= \pm\,(-0.641\,\bar{a}_x - 0.229\,\bar{a}_y - 0.733\,\bar{a}_z)$$

Problem 1.23 : Find in rectangular components the unit vector which is :(a) in the direction of $\bar{E}$ at $P(2, 3, -4)$ if $\bar{E} = (x^2 + y^2 + z^2)\left[x\bar{a}_x / \sqrt{y^2 + z^2} + y\bar{a}_y / \sqrt{x^2 + z^2} + z\bar{a}_z / \sqrt{x^2 + y^2}\right]$; (b) perpendicular to the plane passing through M (1, –5, 5), N (–2, 4, 0), and Q (2, 3, 4), and having a positive x component. (c) Find the angle between the vectors $\bar{R}_{MN}$ and $\bar{R}_{MQ}$.

Solution :

$$\text{(a) } \bar{E} \text{ at } P = [(2)^2 + (3)^2 + (-4)^2]\left[\frac{2\,\bar{a}_x}{\sqrt{(3)^2 + (-4)^2}} + \frac{3\,\bar{a}_y}{\sqrt{(2)^2 + (-4)^2}} - \frac{4\,\bar{a}_z}{\sqrt{(2)^2 + (3)^2}}\right]$$

$$= 11.6\,\bar{a}_x + 19.454\,\bar{a}_y - 32.173\,\bar{a}_z$$

$$\therefore\ \bar{a} = \frac{\bar{E}}{\mid \bar{E} \mid} = \frac{11.6\,\bar{a}_x + 19.454\,\bar{a}_y - 32.173\,\bar{a}_z}{\sqrt{(11.6)^2 + (19.454)^2 + (-32.173)^2}}$$

$$= 0.295\,\bar{a}_x + 0.494\,\bar{a}_y - 0.818\,\bar{a}_z$$

$$\text{(b) } \bar{R}_{MN} = \bar{r}_N - \bar{r}_M = (-2\,\bar{a}_x + 4\,\bar{a}_y) - (\bar{a}_x - 5\,\bar{a}_y + 5\,\bar{a}_z)$$

$$= -3\,\bar{a}_x + 9\,\bar{a}_y - 5\,\bar{a}_z$$

$$\bar{R}_{MQ} = \bar{r}_Q - \bar{r}_M = (2\,\bar{a}_x + 3\,\bar{a}_y + 4\,\bar{a}_z) - (\bar{a}_x - 5\,\bar{a}_y + 5\,\bar{a}_z)$$

$$= \bar{a}_x + 8\,\bar{a}_y - \bar{a}_z$$

$$\bar{R}_{MN} \times \bar{R}_{MQ} = \begin{vmatrix} \bar{a}_x & \bar{a}_y & \bar{a}_z \\ -3 & 9 & -5 \\ 1 & 8 & -1 \end{vmatrix} = 31\,\bar{a}_x - 8\,\bar{a}_y - 33\,\bar{a}_z$$

$$\therefore\ \bar{a}_n = \pm \frac{\bar{R}_{MN} \times \bar{R}_{MQ}}{\mid \bar{R}_{MN} \times \bar{R}_{MQ} \mid} = \pm \frac{31\,\bar{a}_x - 8\,\bar{a}_y - 33\,\bar{a}_z}{\sqrt{(31)^2 + (-8)^2 + (-33)^2}}$$

$$= \pm\,(0.674\,\bar{a}_x - 0.174\,\bar{a}_y - 0.718\,\bar{a}_z)$$

The unit vector with positive x component is $0.674\,\bar{a}_x - 0.174\,\bar{a}_y - 0.718\,\bar{a}_z$

$$\text{(c) } \sin\theta = \frac{|\bar{R}_{MN} \times \bar{R}_{MQ}|}{|\bar{R}_{MN}|\;|\bar{R}_{MQ}|} = \frac{\sqrt{(31)^2 + (-8)^2 + (-33)^2}}{\sqrt{(-3)^2 + (9)^2 + (-5)^2}\;\sqrt{(1)^2 + (8)^2 + (-1)^2}}$$

$$= 0.52775 \quad \text{or} \quad \theta = \sin^{-1}(0.52775) = 31.854^{\circ}$$

Problem 1.24 : Given vectors from the origin $\bar{r}_A = 3\,\bar{a}_x + 8\,\bar{a}_y + \bar{a}_z$, $\bar{r}_B = \bar{a}_x - 6\,\bar{a}_y + 5\,\bar{a}_z$, and $\bar{r}_C = 7\,\bar{a}_x - 4\,\bar{a}_z$, find : (a) a unit vector perpendicular to both $\bar{r}_A$ and $\bar{r}_B$; (b) a unit vector perpendicular to all the vectors, $\bar{r}_A - \bar{r}_B$, $\bar{r}_B - \bar{r}_C$, and $\bar{r}_C - \bar{r}_A$; (c) the area of the traiangle defined by the heads of the vectors $\bar{r}_A$, $\bar{r}_B$, and $\bar{r}_C$.

Solution :

$$\text{(a)} \quad \bar{r}_A \times \bar{r}_B = \begin{vmatrix} \bar{a}_x & \bar{a}_y & \bar{a}_z \\ 3 & 8 & 1 \\ 1 & -6 & 5 \end{vmatrix} = 46\,\bar{a}_x - 14\,\bar{a}_y - 26\,\bar{a}_z$$

$$\therefore\; \bar{a}_n = \pm\frac{\bar{r}_A \times \bar{r}_B}{|\bar{r}_A \times \bar{r}_B|} = \pm\frac{46\,\bar{a}_x - 14\,\bar{a}_y - 26\,\bar{a}_z}{\sqrt{(46)^2 + (-14)^2 + (-26)^2}}$$

$$= \pm\,(0.842\,\bar{a}_x - 0.256\,\bar{a}_y - 0.476\,\bar{a}_z)$$

$$\text{(b) } \bar{R}_{BA} = \bar{r}_A - \bar{r}_B = (3\,\bar{a}_x + 8\,\bar{a}_y + \bar{a}_z) - (\bar{a}_x - 6\,\bar{a}_y + 5\,\bar{a}_z)$$
$$= 2\,\bar{a}_x + 14\,\bar{a}_y - 4\,\bar{a}_z$$

$$\bar{R}_{CB} = \bar{r}_B - \bar{r}_C = (\bar{a}_x - 6\,\bar{a}_y + 5\,\bar{a}_z) - (7\,\bar{a}_x - 4\,\bar{a}_z)$$
$$= -6\,\bar{a}_x - 6\,\bar{a}_y + 9\,\bar{a}_z$$

$$\bar{R}_{AC} = \bar{r}_C - \bar{r}_A = (7\,\bar{a}_x - 4\,\bar{a}_z) - (3\,\bar{a}_x + 8\,\bar{a}_y + \bar{a}_z)$$
$$= 4\,\bar{a}_x - 8\,\bar{a}_y - 5\,\bar{a}_z$$

$$\bar{R}_{BA} \times \bar{R}_{AC} = \begin{vmatrix} \bar{a}_x & \bar{a}_y & \bar{a}_z \\ 2 & 14 & -4 \\ 4 & -8 & -5 \end{vmatrix}$$

$$\bar{a}_n = \pm\frac{\bar{R}_{BA} \times \bar{R}_{AC}}{|\bar{R}_{BA} \times \bar{R}_{AC}|} = \pm\frac{-102\,\bar{a}_x - 6\,\bar{a}_y - 72\,\bar{a}_z}{\sqrt{(-102)^2 + (-6)^2 + (-72)^2}}$$

$$= \pm\,(-0.816\,\bar{a}_x - 0.048\,\bar{a}_y - 0.576\,\bar{a}_z)$$

This unit vector is normal to the vectors $\bar{R}_{BA}$ and $\bar{R}_{AC}$ or to the plane defined by these vectors. Since $\bar{R}_{CB}$ also lies in the same plane, the unit vector is perpendicular to it.

(c) Area of triangle ABC = $0.5 \mid \overline{R}_{BA} \times \overline{R}_{AC} \mid$

$$= 0.5 \sqrt{(-102)^2 + (-6)^2 + (-72)^2} = 62.498$$

Problem 1.25 : Given the three points *M* (6, 2, – 3), *N* (– 2, 3, 0), and *P* (– 4, 6, 5), find : (a) the area of the triangle they define ; (b) a unit vector perpendicular to this triangular surface ; (c) a unit vector bisecting the interior angle of the triangle at *M*.

Solution :

(a) $\overline{R}_{MN} = \overline{r}_N - \overline{R}_M = (-2\overline{a}_x + 3\overline{a}_y) - (6\overline{a}_x + 2\overline{a}_y - 3\overline{a}_z)$

$$= -8\overline{a}_x + \overline{a}_y + 3\overline{a}_z$$

$\overline{R}_{MP} = \overline{r}_P - \overline{r}_M = (-4\overline{a}_x + 6\overline{a}_y + 5\overline{a}_z) - (6\overline{a}_x + 2\overline{a}_y - 3\overline{a}_z)$

$$= -10\overline{a}_x + 4\overline{a}_y + 8\overline{a}_z$$

$$\overline{R}_{MN} \times \overline{R}_{MP} = \begin{vmatrix} \overline{a}_x & \overline{a}_y & \overline{a}_z \\ -8 & 1 & 3 \\ -10 & 4 & 8 \end{vmatrix} = -4\overline{a}_x + 34\overline{a}_y - 22\overline{a}_z$$

Area of triangle $MNP = 0.5 \mid \overline{R}_{MN} \times \overline{R}_{MP} \mid$

$$= 0.5 \sqrt{(-4)^2 + (34)^2 + (-22)^2} = 20.347$$

$$\text{(b) } \overline{a}_n = \pm \frac{\overline{R}_{MN} \times \overline{R}_{MP}}{\mid \overline{R}_{MN} \times \overline{R}_{MP} \mid} = \pm \frac{-4\overline{a}_x + 34\overline{a}_y - 22\overline{a}_z}{\sqrt{(-4)^2 + (34)^2 + (-22)^2}}$$

$$= \pm(-0.098\overline{a}_x + 0.836\overline{a}_y - 0.541\overline{a}_z)$$

(c) Let line MA bisect the angle NMP. That is $\angle NMA = \angle PMA$

Applying law of sine to triangle MAN we have

$$\frac{MN}{\sin(\angle MAN)} = \frac{NA}{\sin(\angle NMA)} = \frac{MA}{\sin(\angle MNA)} \quad \text{-------(1)}$$

and to triangle MAP we have

$$\frac{MP}{\sin(\angle MAP)} = \frac{PA}{\sin(\angle PMA)} = \frac{MA}{\sin(\angle MPA)} \quad \text{----(2)}$$

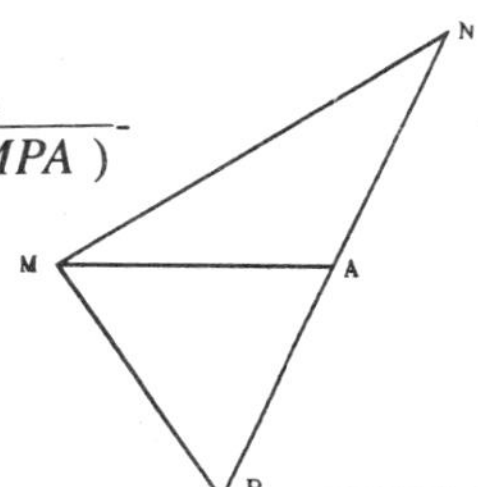

FIG.P1.25 (a)

As seen from above Fig.P1.25(a),

$\angle MAP + \angle MAN = 180^o$ or

$\angle MAP = 180^o - \angle MAN$

Taking sine of both sides we have

$\sin(\angle MAP) = \sin(180^\circ - \angle MAN) = \sin(\angle MAN)$

Considering first two terms from equations (1) and (2), we have

$$\frac{NA}{MN} = \frac{\sin(\angle NMA)}{\sin(\angle MAN)} \quad \text{and} \quad \frac{PA}{MP} = \frac{\sin(\angle PMA)}{\sin(\angle MAP)}$$

Since $\angle NMA = \angle PMA$ and $\sin(\angle MAN) = \sin(\angle MAP)$

we have $\dfrac{NA}{MN} = \dfrac{PA}{MP}$ or $\dfrac{NA}{PA} = \dfrac{MN}{MP} = \dfrac{|\bar{R}_{MN}|}{|\bar{R}_{MP}|}$

$$\therefore \frac{NA}{PA} = \frac{\sqrt{(-8)^2 + (1)^2 + (3)^2}}{\sqrt{(-10)^2 + (4)^2 + (8)^2}} = 0.6412$$

Hence point A divides NP in a ratio 0.6412 : 1 as shown in Fig.P1.25 (b) .

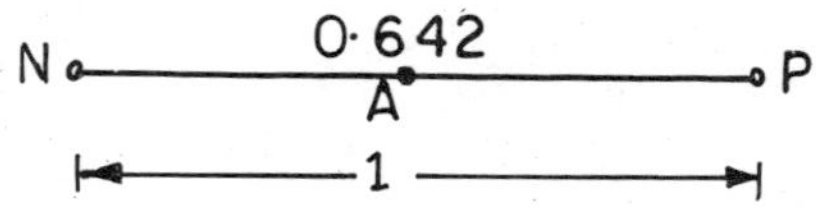

FIG.P1.25 (b)

Coordinates of A

$$= \left(\frac{-4\times 0.6412 - 2\times 1}{0.6412 + 1} = -2.781, \frac{6\times 0.6412 + 3\times 1}{0.6412 + 1} = 4.172, \frac{5\times 0.6412}{0.6412 + 1} = 1.953 \right)$$

Hence $\bar{R}_{MA} = \bar{r}_A - \bar{r}_M$

$$= (-2.781\,\bar{a}_x + 4.172\,\bar{a}_y + 1.953\,\bar{a}_z) - (6\,\bar{a}_x + 2\,\bar{a}_y - 3\,\bar{a}_z)$$

$$= -8.781\,\bar{a}_x + 2.172\,\bar{a}_y + 4.953\,\bar{a}_z$$

$$\bar{a}_{MA} = \frac{\bar{R}_{MA}}{|\bar{R}_{MA}|} = \frac{-8.781\,\bar{a}_x + 2.172\,\bar{a}_y + 4.953\,\bar{a}_z}{\sqrt{(-8.781)^2 + (2.172)^2 + (4.953)^2}}$$

$$= -0.851\,\bar{a}_x + 0.211\,\bar{a}_y + 0.480\,\bar{a}_z$$

Problem 1.26 : (a) Express the field $\bar{F} = 2xyz\,\bar{a}_x - 5(x + y + z)\,\bar{a}_z$ in cylindrical coordinates (variables and components). (b) Find $|\bar{F}|$ at $P(\rho = 2, \phi = 60^\circ, z = 3)$.

Solution :

(a) $\bar{F} = 2xyz\,\bar{a}_x - 5(x + y + z)\,\bar{a}_z = F_\rho\,\bar{a}_\rho + F_\phi\,\bar{a}_\phi + F_z\,\bar{a}_z$

$$F_\rho = \bar{F}\cdot\bar{a}_\rho = 2xyz\,\bar{a}_x\cdot\bar{a}_\rho - 5(x + y + z)\,\bar{a}_z\cdot\bar{a}_\rho$$

$$= 2xyz\cos\phi - 5(x + y + z)(0)$$

$$= 2(\rho\cos\phi)(\rho\sin\phi)(z)\cos\phi = 2\rho^2 z\cos^2\phi\sin\phi$$

$$F_\phi = \bar{F}\cdot\bar{a}_\phi = 2xyz\,\bar{a}_x\cdot\bar{a}_\phi - 5(x + y + z)\,\bar{a}_z\cdot\bar{a}_\phi$$

$$= 2xyz(-\sin\phi) - 5(x + y + z)(0)$$

$$= -2(\rho\cos\phi)(\rho\sin\phi)(z)(\sin\phi)$$

$$= -2\rho^2 z \cos\phi \sin^2\phi$$

$$F_z = \overline{F} \cdot \overline{a}_z = 2xyz\,\overline{a}_x \cdot \overline{a}_z - 5(x+y+z)\,\overline{a}_z \cdot \overline{a}_z$$

$$= 2xyz(0) - 5(x+y+z)(1) = -5(x+y+z)$$

$$= -5(\rho\cos\phi + \rho\sin\phi + z)$$

$$\therefore \overline{F} = 2\rho^2 z\cos^2\phi\sin\phi\,\overline{a}_\rho - 2\rho^2 z\cos\phi\sin^2\phi\,\overline{a}_\phi - 5(\rho\cos\phi + \rho\sin\phi + z)\,\overline{a}_z$$

(b) Substituting the coordinates of point P in the above $\overline{F}$ we have

$\overline{F} = 5.196\,\overline{a}_\rho - 9\,\overline{a}_\phi - 28.660\,\overline{a}_z$

$$\therefore |\overline{F}| \text{ at } P = \sqrt{(5.196)^2 + (-9)^2 + (-28.660)^2} = 30.486$$

Problem 1.27 : Given the points $P(\rho = 5, \phi = 60°, z = 2)$ and $Q(\rho = 2, \phi = 110^o, z = -1)$; (a) find the distance $|\overline{R}_{PQ}|$; (b) give a unit vector in cartesian coordinates at P that is directed towards Q ; (c) give a unit vector in cylindrical coordinates at P that is directed towards Q .

Solution :

(a)Cartesian coordinates of point P are $x = 5\cos 60^o = 2.5$, $y = 5\sin 60^o = 4.33$, $z = 2$, i.e. P (2.5, 4.33,2) and that of point Q are $x = 2\cos 110^o = -0.684$, $y = 2\sin 110^o = 1.879$, $z = -1$, i.e. Q (- 0.684, 1.879, - 1). Hence

$$\therefore d(PQ) = \sqrt{(2.5 + 0.684)^2 + (4.33 - 1.879)^2 + (2+1)^2} = 5.015$$

(b) $\overline{R}_{PQ} = \overline{r}_Q - \overline{r}_P$

$$= (-0.684\,\overline{a}_x + 1.879\,\overline{a}_y - \overline{a}_z) - (2.5\,\overline{a}_x + 4.33\,\overline{a}_y + 2\,\overline{a}_z)$$

$$= -3.184\,\overline{a}_x - 2.451\,\overline{a}_y - 3\,\overline{a}_z$$

$$\therefore \overline{a}_{PQ} = \frac{\overline{R}_{PQ}}{|\overline{R}_{PQ}|} = \frac{-3.184\,\overline{a}_x - 2.451\,\overline{a}_y - 3\,\overline{a}_z}{5.015}$$

$$= -0.635\,\overline{a}_x - 0.489\,\overline{a}_y - 0.598\,\overline{a}_z$$

(c)Let $\overline{A} = \overline{R}_{PQ} = -3.184\,\overline{a}_x - 2.451\,\overline{a}_y - 3\,\overline{a}_z = A_\rho\,\overline{a}_\rho + A_\phi\,\overline{a}_\phi + A_z\,\overline{a}_z$

$$A_\rho = \overline{A} \cdot \overline{a}_\rho = -3.184\,\overline{a}_x \cdot \overline{a}_\rho \cdot 2.451\,\overline{a}_y \cdot \overline{a}_\rho - 3\,\overline{a}_z \cdot \overline{a}_\rho$$

$$= -3.184\cos\phi - 2.451\sin\phi$$

$$A_\phi = \overline{A} \cdot \overline{a}_\phi = -3.184\,\overline{a}_x \cdot \overline{a}_\phi - 2.451\,\overline{a}_y \cdot \overline{a}_\phi - 3\,\overline{a}_z \cdot \overline{a}_\phi$$

$$= 3.184 \sin \phi - 2.451 \cos \phi$$

$$A_z = \overline{A} \cdot \overline{a}_z = -3.184\, \overline{a}_x \cdot \overline{a}_z - 2.451\, \overline{a}_y \cdot \overline{a}_z - 3\, \overline{a}_z \cdot \overline{a}_z = -3$$

$$\therefore \overline{A} = \overline{R}_{PQ} = (-3.184 \cos \phi - 2.451 \sin \phi)\, \overline{a}_\rho + (3.184 \sin \phi - 2.451 \cos \phi)\, \overline{a}_\phi - 3\, \overline{a}_z$$

Substituting the coorodinates of point P in the above $\overline{A}$ we have

$\overline{R}_{PQ}$ at point $P = -0.741\, \overline{a}_\rho + 0.305\, \overline{a}_\phi - 0.598\, \overline{a}_z$

Problem 1.28. : Find a cylindrical components : (a) a unit vector $P(\rho = 5, \phi = 53.13°, z = -2)$ in the direction of $\overline{F} = \rho z \cos \phi\, \overline{a}_\rho - \rho z \sin \phi\, \overline{a}_\phi + \rho\, a\, \overline{a}_z$; (b) a unit vector at P parallel to $\overline{a}_x$; (c) a unit vector at $Q(\rho = 5, \phi = -36.87°, z = -2)$ parallel to $\overline{a}_x$; (d) $\overline{G} = 2\, \overline{a}_x - 4\, \overline{a}_y + 4\, \overline{a}_z$ at P.

Solution :

(a) $\overline{F}$ at $P = (5)(-2) \cos 53.13^\circ\, \overline{a}_\rho - (5)(-2) \sin 53.13^\circ\, \overline{a}_\phi + 5\, a\, \overline{a}_z$

$$= -6\, \overline{a}_\rho + 8\, \overline{a}_\phi + 5\, a\, \overline{a}_z$$

$$\overline{a}_F = \frac{\overline{F}}{|\overline{F}|} = \frac{-6\, \overline{a}_\rho + 8\, \overline{a}_\phi + 5\, a\, \overline{a}_z}{\sqrt{100 + 25\, a^2}}$$

(b) Let $\overline{A} = \overline{a}_x = A_\rho\, \overline{a}_\rho + A_\phi\, \overline{a}_\phi + A_z\, \overline{a}_z$

$$A_\rho = \overline{A} \cdot \overline{a}_\rho = \overline{a}_x \cdot \overline{a}_\rho = \cos \phi$$

$$A_\phi = \overline{A} \cdot \overline{a}_\phi = \overline{a}_x \cdot \overline{a}_\phi = -\sin \phi$$

$$A_z = \overline{A} \cdot \overline{a}_z = \overline{a}_x \cdot \overline{a}_z = 0$$

$$\therefore \overline{A} = \cos \phi\, \overline{a}_\rho - \sin \phi\, \overline{a}_\phi$$

$$\therefore \overline{A} \text{ at } P = \cos 53.13^\circ\, \overline{a}_\rho - \sin 53.13^\circ\, \overline{a}_\phi = 0.6\, \overline{a}_\rho - 0.8\, \overline{a}_\phi$$

$$\therefore \overline{a} = \frac{\overline{A}}{|\overline{A}|} = \frac{0.6\, \overline{a}_\rho - 0.8\, \overline{a}_\phi}{\sqrt{(0.6)^2 + (-0.8)^2}} = 0.6\, \overline{a}_\rho - 0.8\, \overline{a}_\phi$$

(c) $\overline{A}$ at $Q = \cos(-36.87^\circ)\, \overline{a}_\rho - \sin(-36.87^\circ)\, \overline{a}_\phi = 0.8\, \overline{a}_\rho + 0.6\, \overline{a}_\phi$

$$\therefore \overline{a} = \frac{\overline{A}}{|\overline{A}|} = 0.8\, \overline{a}_\rho + 0.6\, \overline{a}_\phi$$

(d) $\overline{G} = 2\,\overline{a}_x - 4\,\overline{a}_y + 4\,\overline{a}_z = G_\rho\,\overline{a}_\rho + G_\phi\,\overline{a}_\phi + G_z\,\overline{a}_z$

$$G_\rho = \overline{G} \cdot \overline{a}_\rho = 2\,a_x \cdot \overline{a}_\rho - 4\,\overline{a}_y \cdot \overline{a}_\rho - 4\,\overline{a}_z \cdot \overline{a}_\rho$$
$$= 2\cos\phi - 4\sin\phi$$

$$G_\phi = \overline{G} \cdot \overline{a}_\phi = 2\,\overline{a}_x \cdot \overline{a}_\phi - 4\,\overline{a}_y \cdot \overline{a}_\phi + 4\,\overline{a}_z \cdot \overline{a}_\phi$$
$$= -2\sin\phi - 4\cos\phi$$

$$G_z = \overline{G} \cdot \overline{a}_z = 2\,\overline{a}_x \cdot \overline{a}_z - 4\,\overline{a}_y \cdot \overline{a}_z + 4\,\overline{a}_z \cdot \overline{a}_z = 4$$

$$\therefore\ \overline{G} = (2\cos\phi - 4\sin\phi)\,\overline{a}_\rho + (-2\sin\phi - 4\cos\phi)\,\overline{a}_\phi + 4\,\overline{a}_z$$

Substituting the coordinates of point P in the above $\overline{G}$ we have

$$\overline{G} \text{ at } P = -2\,\overline{a}_\rho - 4\,\overline{a}_\phi + 4\,\overline{a}_z$$

Problem 1.29 : (a) Give the vector in cartesian coordinates that extends from P ($\rho = 4, \phi = 10^\circ, z = 1$) to Q ($\rho = 7, \phi = 75^\circ, z = 4$) (b) Give the vector in cylindrical coordinates at M ($x = 5, y = 1, z = 2$) that extends to N (2, 4, 6). (c) How far is it from A (110,60° , – 20) to B (30,125° , 10) ?

Solution :

(a)Cartesian coordinates of point P are $x = 4\cos 10^\circ = 3.939$, $y = 4\sin 10^\circ = 0.695$, $z = 1$, i.e., P (3.939,0.695,1) and that of point Q are $x = 7\cos 75^\circ = 1.812$, $y = 7\sin 75^\circ = 6.761$, $z = 4$, i.e. Q (1.812, 6.761, 4)

$$\therefore\ \overline{R}_{PQ} = \overline{r}_Q - \overline{r}_P$$
$$= (1.812\,\overline{a}_x + 6.761\,\overline{a}_y + 4\,\overline{a}_z) - (3.939\,\overline{a}_x + 0.695\,\overline{a}_y + \overline{a}_z)$$
$$= -2.127\,\overline{a}_x + 6.066\,\overline{a}_y + 3\,\overline{a}_z$$

(b) $\overline{R}_{MN} = \overline{r}_N - \overline{r}_M = (2\,\overline{a}_x + 4\,\overline{a}_y + 6\,\overline{a}_z) - (5\,\overline{a}_x + \overline{a}_y + 2\,\overline{a}_z)$

$$= -3\,\overline{a}_x + 3\,\overline{a}_y + 4\,\overline{a}_z$$

Let $\overline{A} = \overline{R}_{MN} = -3\,\overline{a}_x + 3\,\overline{a}_y + 4\,\overline{a}_z = A_\rho\,\overline{a}_\rho + A_\phi\,\overline{a}_\phi + A_z\,\overline{a}_z$

$$A_\rho = \overline{A} \cdot \overline{a}_\rho = -3\,\overline{a}_x \cdot \overline{a}_\rho + 3\,\overline{a}_y \cdot \overline{a}_\rho + 4\,\overline{a}_z \cdot \overline{a}_\rho$$
$$= -3\cos\phi + 3\sin\phi$$

$$A_\phi = \overline{A} \cdot \overline{a}_\phi = -3\,\overline{a}_x \cdot \overline{a}_\phi + 3\,\overline{a}_y \cdot \overline{a}_\phi + 4\,\overline{a}_z \cdot \overline{a}_\phi$$
$$= 3\sin\phi + 3\cos\phi$$

$$A_z = \overline{A} \cdot \overline{a}_z = -3\,\overline{a}_x \cdot \overline{a}_z + 3\,\overline{a}_y \cdot \overline{a}_z + 4\,\overline{a}_z \cdot \overline{a}_z = 4$$

$$\therefore\ \overline{A} = \overline{R}_{MN} = (-3\cos\phi + 3\sin\phi)\,\overline{a}_\rho + (3\sin\phi + 3\cos\phi)\,\overline{a}_\phi + 4\,\overline{a}_z$$

For point M , $\phi = \tan^{-1}(1/5) = 11.31^{\circ}$. Substituting this value of ϕ in the above vector we have $\overline{R}_{MN}$ at point $M = -2.353\,\overline{a}_\rho + 3.530\,\overline{a}_\phi + 4\,\overline{a}_z$

(c) Cartesian coordinates of point A are $x = 110\cos 60^{\circ} = 55$, $y = 110 \sin 60^{\circ} = 95.263$, $z = -20$, i.e. A (55, 95.263, –20) and that of point B are $x = 30\cos 125^{\circ} = -17.207$, $y = 30\sin 125^{\circ} = 24.575$, $z = 10$, i.e. B (-17.207, 24.575, 10)

$$d(AB) = \sqrt{(55 + 17.207)^2 + (95.263 - 24.575)^2 + (-20 - 10)^2} = 115.407$$

Problem 1.30 : Given points A ($x = 2, y = 3, z = -1$) and B ($\rho = 4, \phi = -50^{\circ}, z = 2$) , find a unit vector in cylindrical coordinates : (a) at point B directed toward point A ; (b) at point A directed toward point B .

Solution :

(a) Cartesian coordinates of point B are $x = 4\cos(-50^{\circ}) = 2.571$, $y = 4\sin(-50^{\circ}) = -3.064$, $z = 2$, i.e. B (2.571, -3.064, 2).

$$\overline{R}_{BA} = \overline{r}_A - \overline{r}_B = (2\,\overline{a}_x + 3\,\overline{a}_y - \overline{a}_z) - (2.571\,\overline{a}_x - 3.064\,\overline{a}_y + 2\,\overline{a}_z)$$

$$\overline{R}_{BA} = -0.571\,\overline{a}_x + 6.064\,\overline{a}_y - 3\,\overline{a}_z$$

Let $\overline{A} = \overline{R}_{BA} = -0.571\,\overline{a}_x + 6.064\,\overline{a}_y - 3\,\overline{a}_z = A_\rho\,\overline{a}_\rho + A_\phi\,\overline{a}_\phi + A_z\,\overline{a}_z$

$$A_\rho = \overline{A}\cdot\overline{a}_\rho = -0.571\,\overline{a}_x\cdot\overline{a}_\rho + 6.064\,\overline{a}_y\cdot\overline{a}_\rho - 3\,\overline{a}_z\cdot\overline{a}_\rho$$
$$= -0.571\cos\phi + 6.064\sin\phi$$

$$A_\phi = \overline{A}\cdot\overline{a}_\phi = -0.571\,\overline{a}_x\cdot\overline{a}_\phi + 6.064\,\overline{a}_y\cdot\overline{a}_\phi - 3\,\overline{a}_z\cdot\overline{a}_\phi$$
$$= 0.571\sin\phi + 6.064\cos\phi$$

$$A_z = \overline{A}\cdot\overline{a}_z = -0.571\,\overline{a}_x\cdot\overline{a}_z + 6.064\,\overline{a}_y\cdot\overline{a}_z - 3\,\overline{a}_z\cdot\overline{a}_z = -3$$

$$\therefore\ \overline{A} = \overline{R}_{BA} = (-0.571\cos\phi + 6.064\sin\phi)\,\overline{a}_\rho + (0.571\sin\phi + 6.064\cos\phi)\,\overline{a}_\phi - 3\,\overline{a}_z$$

Substituting ϕ coordinate of point B in the above vector we have $\overline{R}_{BA}$ at point $B = -5.012\,\overline{a}_\rho + 3.460\,\overline{a}_\phi - 3\,\overline{a}_z$

$$\therefore\ \overline{a}_{BA} = \frac{\overline{R}_{BA}}{|\,\overline{R}_{BA}\,|} = \frac{-5.012\,\overline{a}_\rho + 3.460\,\overline{a}_\phi - 3\,\overline{a}_z}{\sqrt{(-5.012)^2 + (3.460)^2 + (-3)^2}}$$
$$= -0.738\,\overline{a}_\rho + 0.510\,\overline{a}_\phi - 0.442\,\overline{a}_z$$

(b)$\overline{R}_{AB} = \overline{r}_B - \overline{r}_A = (2.571\,\overline{a}_x - 3.064\,\overline{a}_y + 2\,\overline{a}_z) - (2\,\overline{a}_x + 3\,\overline{a}_y - \overline{a}_z)$

$= 0.571\,\overline{a}_x - 6.064\,\overline{a}_y + 3\,\overline{a}_z$

Let$\overline{B} = \overline{R}_{AB} = 0.571\,\overline{a}_x - 6.064\,\overline{a}_y + 3\,\overline{a}_z = B_\rho\,\overline{a}_\rho + B_\phi\,\overline{a}_\phi + B_z\,\overline{a}_z$

$B_\rho = \overline{B}\cdot\overline{a}_\rho = 0.571\,\overline{a}_x\cdot\overline{a}_\rho - 6.064\,\overline{a}_y\cdot\overline{a}_\rho + 3\,\overline{a}_z\cdot\overline{a}_\rho$

$= 0.571\cos\phi - 6.064\sin\phi$

$B_\phi = \overline{B}\cdot\overline{a}_\phi = 0.571\,\overline{a}_x\cdot\overline{a}_\phi - 6.064\,\overline{a}_y\cdot\overline{a}_\phi + 3\,\overline{a}_z\cdot\overline{a}_\phi$

$= -0.571\sin\phi - 6.604\cos\phi$

$B_z = \overline{B}\cdot\overline{a}_z = 0.571\,\overline{a}_x\cdot\overline{a}_z - 6.064\,\overline{a}_y\cdot\overline{a}_z + 3\,\overline{a}_z\cdot\overline{a}_z = 3$

$\overline{B} = \overline{R}_{AB} = (0.571\cos\phi - 6.064\sin\phi)\,\overline{a}_\rho +$

$(-0.571\sin\phi - 6.064\cos\phi)\,\overline{a}_\phi + 3\,\overline{a}_z$

For Point A, $\phi = \tan^{-1}(3/2) = 56.31^\circ$. Substituting this value of ϕ in the above vector we have $\overline{R}_{AB}$ at point $A = -4.729\,\overline{a}_\rho - 3.839\,\overline{a}_\phi + 3\,\overline{a}_z$

$$\therefore\ \overline{a}_{AB} = \frac{\overline{R}_{AB}}{|\overline{R}_{AB}|} = \frac{-4.729\,\overline{a}_\rho - 3.839\,\overline{a}_\phi + 3\,\overline{a}_z}{\sqrt{(-4.729)^2 + (-3.839)^2 + (3)^2}}$$

$$= -0.696\,\overline{a}_\rho - 0.565\,\overline{a}_\phi + 0.442\,\overline{a}_z$$

Problem 1.31 : (a) Find the volume defined by $4 < \rho < 6$, $30^\circ < \phi < 60^\circ$, $2 < z < 5$, . (b) What is the length of the longest straight line that lies entirely within the volume ? (c) Find the total area of the surface.

Solution :

$$\text{(a) } V = \int dv = \int_4^6 \int_{30^\circ}^{60^\circ} \int_2^5 \rho\, d\rho\, d\phi\, dz$$

$$= \left[\frac{\rho^2}{2}\right]_4^6 \left[\phi\right]_{30^\circ}^{60^\circ} \left[z\right]_2^5 = 15.708$$

(b) Suppose points P and Q are dimetrically opposite corners of the volume. Then the lower limits of ρ, ϕ and z gives the coordinates of point P where as the higher limits of ρ, ϕ, and z gives coordinates of point Q

$\therefore$ $P(4, 30^\circ, 2)$ and $Q(6, 60^\circ, 5)$

$$d(PQ) = \sqrt{(4\cos 30^\circ - 6\cos 60^\circ)^2 + (4\sin 30^\circ - 6\sin 60^\circ)^2 + (2-5)^2}$$

$$= 4.408$$

(c) Six surfaces of the volume are located by the coordinates

$\rho = 4$ and 6, $\phi = 30^\circ$ and 60°, $z = 2$ and 5

$$S = S_1 + S_2 + S_3 + S_4 + S_5 + S_6$$

$$= \int_{30^\circ}^{60^\circ}\int_2^5 \rho\, d\phi\, dz_{\rho=4} + \int_{30^\circ}^{60^\circ}\int_2^5 \rho\, d\phi\, dz_{\rho=6} + \int_4^6\int_2^5 d\rho\, dz_{\phi=30^\circ}$$

$$+ \int_4^6\int_2^5 d\rho\, dz_{\phi=60^\circ} + \int_4^6\int_{30^\circ}^{60^\circ} \rho\, d\rho\, d\phi_{z=2} + \int_4^6\int_{30^\circ}^{60^\circ} \rho\, d\rho\, d\phi_{z=5}$$

$$= 4\left[\phi\right]_{30^\circ}^{60^\circ}\left[z\right]_2^5 + 6\left[\phi\right]_{30^\circ}^{60^\circ}\left[z\right]_2^5 + \left[\rho\right]_4^6\left[z\right]_2^5$$

$$+ \left[\rho\right]_4^6\left[z\right]_2^5 + \left[\frac{\rho^2}{2}\right]_4^6\left[\phi\right]_{30^\circ}^{60^\circ} + \left[\frac{\rho^2}{2}\right]_4^6\left[\phi\right]_{30^\circ}^{60^\circ} = 38.180$$

Problem 1.32 : Using the the coordinate system named, give the vector at point A (2,–1,–3) that extends to B (1, 3, 4) (a) cartesian ; (b) cylindrical ; (c) spherical.

Solution :

(a) $\overline{R}_{AB} = \overline{r}_B - \overline{r}_A = (\overline{a}_x + 3\,\overline{a}_y + 4\,\overline{a}_z) - (2\,\overline{a}_x - \overline{a}_y - 3\,\overline{a}_z)$

$= -\overline{a}_x + 4\,\overline{a}_y + 7\,\overline{a}_z$

(b) Let $\overline{A} = \overline{R}_{AB} = -\overline{a}_x + 4\,\overline{a}_y + 7\,\overline{a}_z = A_\rho\,\overline{a}_\rho + A_\phi\,\overline{a}_\phi + A_z\,\overline{a}_z$

$A_\rho = \overline{A}\cdot\overline{a}_\rho = -\overline{a}_x\cdot\overline{a}_\rho + 4\,\overline{a}_y\cdot\overline{a}_\rho + 7\,\overline{a}_z\cdot\overline{a}_\rho = -\cos\phi + 4\sin\phi$

$A_\phi = \overline{A}\cdot\overline{a}_\phi = -\overline{a}_x\cdot\overline{a}_\phi + 4\,\overline{a}_y\cdot\overline{a}_\phi + 7\,\overline{a}_z\cdot\overline{a}_\phi = \sin\phi + 4\cos\phi$

$A_z = \overline{A}\cdot\overline{a}_z = -\overline{a}_x\cdot\overline{a}_z + 4\,a_y\cdot\overline{a}_z + 7\,\overline{a}_z\cdot\overline{a}_z = 7$

$\therefore\ \overline{A} = \overline{R}_{AB} = (-\cos\phi + 4\sin\phi)\,\overline{a}_\rho + (\sin\phi + 4\cos\phi)\,\overline{a}_\phi + 7\,\overline{a}_z$

For point A, $\phi = \tan^{-1}(-1/2) = -26.565^\circ$. Substituting this value of ϕ in the above vector we have $\overline{R}_{AB}$ at point $A = -2.683\,\overline{a}_\rho + 3.130\,\overline{a}_\phi + 7\,\overline{a}_z$

(c) Let $\overline{A} = A_r\,\overline{a}_r + A_\theta\,\overline{a}_\theta + A_\phi\,\overline{a}_\phi$

$A_r = \overline{A}\cdot\overline{a}_r = -\overline{a}_x\cdot\overline{a}_r + 4\,\overline{a}_y\cdot\overline{a}_r + 7\,\overline{a}_z\cdot\overline{a}_r$

$= -\sin\theta\cos\phi + 4\sin\theta\sin\phi + 7\cos\theta$

$A_\theta = \overline{A}\cdot\overline{a}_\theta = -\overline{a}_x\cdot\overline{a}_\theta + 4\,\overline{a}_y\cdot\overline{a}_\theta + 7\,\overline{a}_z\cdot\overline{a}_\theta$

$= -\cos\theta\cos\phi + 4\cos\theta\sin\phi - 7\sin\theta$

$A_\phi = \overline{A}\cdot\overline{a}_\phi = -\overline{a}_x\cdot\overline{a}_\phi + 4\,\overline{a}_y\cdot\overline{a}_\phi + 7\,\overline{a}_z\cdot\overline{a}_\phi = \sin\phi + 4\cos\phi$

$$\therefore \bar{A} = \bar{R}_{AB} = (-\sin\theta\cos\phi + 4\sin\theta\sin\phi + 7\cos\theta)\,\bar{a}_r + (-\cos\theta\cos\phi + 4\cos\theta\sin\phi - 7\sin\theta)\,\bar{a}_\theta + (\sin\phi + 4\cos\phi)\,\bar{a}_\phi$$

For point A, $r = \sqrt{(2)^2 + (-1)^2 + (-3)^2} = 3.742$

$\theta = \cos^{-1}(-3/3.742) = 143.29^\circ \quad \phi = \tan^{-1}(-1/2) = -26.565^\circ$

Substituting the coordinates of point A in the above vector, we have

$\bar{R}_{AB}$ at point $A = -7.216\,\bar{a}_r - 2.033\,\bar{a}_\theta + 3.130\,\bar{a}_\phi$

Problem 1.33 : An electric field intensity is given as $\bar{E} = (100\cos\theta)\,\bar{a}_r / r^3 + (50\sin\theta)\,\bar{a}_\theta / r^3$. At the point whose spherical coordinates are $r = 2$, $\theta = 60^\circ$, $\phi = 20^\circ$, find : (a) $|\bar{E}|$; (b) a unit vector (in cartesian coordinates) in the direction of $\bar{E}$.

Solution :

(a) $\bar{E} = (100\cos 60)/(2)^3\,\bar{a}_r + (50\sin 60)/(2)^3\,\bar{a}_\theta = 6.25\,\bar{a}_r + 5.413\,\bar{a}_\theta$

$$\therefore |\bar{E}| = \sqrt{(6.25)^2 + (5.413)^2} = 8.268$$

(b) $\bar{E} = E_x\,\bar{a}_x + E_y\,\bar{a}_y + E_z\,\bar{a}_z$

$$E_x = \bar{E}\cdot\bar{a}_x = 6.25\,\bar{a}_r\cdot\bar{a}_x + 5.413\,\bar{a}_\theta\cdot\bar{a}_x = 6.25\sin\theta\cos\phi + 5.413\cos\theta\cos\phi$$

$$E_y = \bar{E}\cdot\bar{a}_y = 6.25\,\bar{a}_r\cdot\bar{a}_y + 5.413\,\bar{a}_\theta\cdot\bar{a}_y = 6.25\sin\theta\sin\phi + 5.413\cos\theta\sin\phi$$

$$E_z = \bar{E}\cdot\bar{a}_z = 6.25\,\bar{a}_r\cdot\bar{a}_z + 5.413\,\bar{a}_\theta\cdot\bar{a}_z = 6.25\cos\theta - 5.413\sin\theta$$

$$\therefore \bar{E} = (6.25\sin\theta\cos\phi + 5.413\cos\theta\cos\phi)\,\bar{a}_x + (6.25\sin\theta\sin\phi + 5.413\cos\theta\sin\phi)\,\bar{a}_y + (6.25\cos\theta - 5.413\sin\theta)\,\bar{a}_z$$

Substitutng $\theta = 60^\circ$ and $\phi = 20^\circ$ in the above vector we have

$\bar{E}$ at point $(r = 2, \theta = 60^\circ, \phi = 20^\circ) = 7.630\,\bar{a}_x + 2.777\,\bar{a}_y - 1.563\,\bar{a}_z$

$$\therefore \bar{a}_E = \frac{\bar{E}}{|\bar{E}|} = \frac{7.630\,\bar{a}_x + 2.777\,\bar{a}_y - 1.563\,\bar{a}_z}{\sqrt{(7.630)^2 + (2.777)^2 + (-1.563)^2}} = 0.923\,\bar{a}_x + 0.336\,\bar{a}_y - 0.189\,\bar{a}_z$$

Problem 1.34 : (a) Find $\bar{a}_x$ in spherical components at P (3, – 4, 5). (b) Find $\bar{a}_\theta$ in cartesian components at P

Solution :

(a) Let $\bar{A} = \bar{a}_x = A_r \bar{a}_r + A_\theta \bar{a}_\theta + A_\phi \bar{a}_\phi$

$A_r = \bar{A} \cdot \bar{a}_r = \bar{a}_x \cdot \bar{a}_r = \sin\theta \cos\phi$

$A_\theta = \bar{A} \cdot \bar{a}_\theta = \bar{a}_x \cdot \bar{a}_\theta = \cos\theta \cos\phi$

$A_\phi = \bar{A} \cdot \bar{a}_\phi = \bar{a}_x \cdot \bar{a}_\phi = -\sin\phi$

$\therefore \bar{A} = \sin\theta \cos\phi \; \bar{a}_r + \cos\theta \cos\phi \; \bar{a}_\theta - \sin\phi \; \bar{a}_\phi$

At point P (3, – 4, 5), $r = \sqrt{(3)^2 + (-4)^2 + (5)^2} = 7.071$

$\theta = \cos^{-1}(5/7.071) = 45^\circ$ and $\phi = \tan^{-1}(-4/3) = -53.13^\circ$

Substituting θ and ϕ coordinates of point P in the $\bar{A}$ we have

$\bar{A}$ at point $P = 0.424\,\bar{a}_r + 0.424\,\bar{a}_\theta + 0.8\,\bar{a}_\phi$

(b) Let $\bar{B} = \bar{a}_\theta = B_x \bar{a}_x + B_y \bar{a}_y + B_z \bar{a}_z$

$B_x = \bar{B} \cdot \bar{a}_x = \bar{a}_\theta \cdot \bar{a}_x = \cos\theta \cos\phi$

$B_y = \bar{B} \cdot \bar{a}_y = \bar{a}_\theta \cdot \bar{a}_y = \cos\theta \sin\phi$

$B_z = \bar{B} \cdot \bar{a}_z = \bar{a}_\theta \cdot \bar{a}_z = -\sin\theta$

$\therefore \bar{B} = \cos\theta \cos\phi \; \bar{a}_r + \cos\theta \sin\phi \; \bar{a}_\theta - \sin\theta \; \bar{a}_\phi$

Substituting the θ and ϕ coordinates of point P in the $\bar{B}$, we have

$\bar{B}$ at point $P = 0.424\,\bar{a}_r - 0.566\,\bar{a}_\theta - 0.707\,\bar{a}_\phi$

Problem 1.35 : Given the points M ($r = 5, \theta = 2\,0, \phi = 12\,0$) and N ($r = 2, \theta = 80^\circ, \phi = 30^\circ$) : (a) find the distance from M to N ; (b) give a unit vector in cartesian coordinates at M that is directed toward N ; (c) give a unit vector in spherical coordinates at M that is directed toward N.

Solution :

(a) Cartesian coordinates of point M are $x = 5 \sin 20^\circ \cos 120^\circ = -0.855$ $y = 5 \sin 20^\circ \sin 120^\circ = 1.481$, $z = 5 \cos 20^\circ = 4.698$ i.e. M (– 0.855, 1.481, 4.698) and that of point N are $x = 2 \sin 80^\circ \cos 30^\circ = 1.706$, $y = 2 \sin 80^\circ \sin 3\,0 = 0.985$, $z = 2 \cos 8\,0 = 0.347$, i.e. N (1.706, 0.985, 0.347).

$d(MN) = \sqrt{(-0.855 - 1.706)^2 + (1.481 - 0.985)^2 + (4.698 - 0.347)^2} = 5.073$

(b) $\overline{R}_{MN} = \overline{r}_N - \overline{r}_M$

$= (1.706\,\overline{a}_x + 0.985\,\overline{a}_y + 0.347\,\overline{a}_z) - (-0.855\,\overline{a}_x + 1.481\,\overline{a}_y + 4.698\,\overline{a}_z)$

$= 2.561\,\overline{a}_x - 0.496\,\overline{a}_y - 4.351\,\overline{a}_z$

$$\overline{a}_{MN} = \frac{\overline{R}_{MN}}{|\overline{R}_{MN}|} = \frac{2.561\,\overline{a}_x - 0.496\,\overline{a}_y - 4.351\,\overline{a}_z}{\sqrt{(2.561)^2 + (-0.496)^2 + (-4.351)^2}}$$

$$= 0.505\,\overline{a}_x - 0.098\,\overline{a}_y - 0.858\,\overline{a}_z$$

(c) Let $\overline{A} = \overline{R}_{MN} = 2.561\,\overline{a}_x - 0.496\,\overline{a}_y - 4.351\,\overline{a}_z = A_r\,\overline{a}_r + A_\theta\,\overline{a}_\theta + A_\phi\,\overline{a}_\phi$

$A_r = \overline{A}\cdot\overline{a}_r = 2.561\,\overline{a}_x\cdot\overline{a}_r - 0.496\,\overline{a}_y\cdot\overline{a}_r - 4.351\,\overline{a}_z\cdot\overline{a}_r$

$= 2.561\sin\theta\cos\phi - 0.496\sin\theta\sin\phi - 4.351\cos\theta$

$A_\theta = \overline{A}\cdot\overline{a}_\theta = 2.561\,\overline{a}_x\cdot\overline{a}_\theta - 0.496\,\overline{a}_y\cdot\overline{a}_\theta - 4.351\,\overline{a}_z\cdot\overline{a}_\theta$

$= 2.561\cos\theta\cos\phi - 0.496\cos\theta\sin\phi + 4.351\sin\theta$

$A_\phi = \overline{A}\cdot\overline{a}_\phi = 2.561\,\overline{a}_x\cdot\overline{a}_\phi - 0.496\,\overline{a}_y\cdot\overline{a}_\phi - 4.351\,\overline{a}_z\cdot\overline{a}_\phi$

$= -2.561\sin\phi - 0.496\cos\phi$

$\therefore\ \overline{A} = \overline{R}_{MN} = (2.561\sin\theta\cos\phi - 0.496\sin\theta\sin\phi - 4.351\cos\theta)\,\overline{a}_r +$
$(2.561\cos\theta\cos\phi - 0.496\cos\theta\sin\phi + 4.351\sin\theta)\,\overline{a}_\theta +$
$(-2.561\sin\phi - 0.496\cos\phi)\,\overline{a}_\phi$

Substituting the θ and ϕ coordinates of point M in $\overline{R}_{MN}$ we have

$\overline{R}_{MN}$ at point $M = -4.673\,\overline{a}_r - 0.119\,\overline{a}_\theta - 1.970\,\overline{a}_\phi$

$$\therefore\ \overline{a}_{MN} = \frac{\overline{R}_{MN}}{|\overline{R}_{MN}|} = \frac{-4.673\,\overline{a}_r - 0.119\,\overline{a}_\theta - 1.970\,\overline{a}_\phi}{\sqrt{(-4.673)^2 + (-0.119)^2 + (-1.970)^2}}$$

$$= -0.921\,\overline{a}_r - 0.023\,\overline{a}_\theta - 0.388\,\overline{a}_\phi$$

Problem 1.36 : A closed surface is defined in spherical coordinates by $3 < r < 5$, $0.1\pi < \theta < 0.3\pi$, $1.2\pi < \phi < 1.6\pi$. (a) Find the volume nciosed. (b) Find the distance from P_1 ($r = 3, \theta = 0.1\pi, \phi = 1.2\pi$) to . ($r = 5, \theta = 0.3\pi, \phi = 1.6\pi$) .(c) Find the total surface area.

olution :

$$\text{(a) } V = \int dv = \int_3^5 \int_{0.1\pi}^{0.3\pi} \int_{1.2\pi}^{1.6\pi} r^2 \sin\theta\, dr\, d\theta\, d\phi$$

$$= \left[\frac{r^3}{3}\right]_3^5 \left[-\cos\theta\right]_{0.1\pi}^{0.3\pi} \left[\phi\right]_{1.2\pi}^{1.6\pi} = 14.912$$

(b) $d(P_1 P_2) = \Big[(3 \sin 0.1\pi \cos 1.2\pi - 5 \sin 0.3\pi \cos 1.6\pi)^2 + (3 \sin 0.1\pi \sin 1.2\pi - 5 \sin 0.3\pi \sin 1.6\pi)^2 + (3 \cos 0.1\pi - 5 \cos 0.3\pi)^2 \Big]^{1/2} = 3.862$

(c) $S = S_1 + S_2 + S_3 + S_4 + S_5 + S_6$

$$= \int_{0.1\pi}^{0.3\pi} \int_{1.2\pi}^{1.6\pi} r^2 \sin\theta \, d\theta \, d\phi \Big|_{r=3} + \int_{0.1\pi}^{0.3\pi} \int_{1.2\pi}^{1.6\pi} r^2 \sin\theta \, d\theta \, d\phi \Big|_{r=5}$$

$$+ \int_3^5 \int_{1.2\pi}^{1.6\pi} r \sin\theta \, dr \, d\phi \Big|_{\theta = 0.1\pi} + \int_3^5 \int_{1.2\pi}^{1.6\pi} r \sin\theta \, dr \, d\phi \Big|_{\theta = 0.3\pi}$$

$$+ \int_3^5 \int_{0.1\pi}^{0.3\pi} r \, dr \, d\theta \Big|_{\phi = 1.2\pi} + \int_3^5 \int_{0.1\pi}^{0.3\pi} r \, dr \, d\phi \Big|_{\phi = 1.6\pi}$$

$$= (3)^2 \Big[-\cos\theta \Big]_{0.1\pi}^{0.3\pi} \Big[\phi \Big]_{1.2\pi}^{1.6\pi} + (5)^2 \Big[-\cos\theta \Big]_{0.1\pi}^{0.3\pi} \Big[\phi \Big]_{1.2\pi}^{1.6\pi}$$

$$+ \sin(0.1\pi) \left[\frac{r^2}{2} \right]_3^5 \Big[\phi \Big]_{1.2\pi}^{1.6\pi} + \sin(0.3\pi) \left[\frac{r^2}{2} \right]_3^5 \Big[\phi \Big]_{1.2\pi}^{1.6\pi}$$

$$+ \left[\frac{r^2}{2} \right]_3^5 \Big[\theta \Big]_{0.1\pi}^{0.3\pi} + \left[\frac{r^2}{2} \right]_3^5 \Big[\theta \Big]_{0.1\pi}^{0.3\pi} = 36.814$$

Problem 1.37 : (a) Given the vector in cartesian coordinates that extends from $P(r = 4, \theta = 20^\circ, \phi = 10^\circ)$ to $Q(r = 7, \theta = 120^\circ, \phi = 75^\circ)$. (b) Give the vector in spherical coordinates at $M(x = 5, y = 1, z = 2)$ that extends to N (2, 4, 6). (c) How far is it from $A(r = 110, \theta = 30^\circ, \phi = 60^\circ)$ to $B(r = 30, \theta = 75^\circ, \phi = 125^\circ)$?

Solution:

(a)Cartesian coordinates of$P$ are $x = 4 \sin 20^\circ \cos 10^\circ = 1.347$, $y = 4$ i.e. P (1.347 ,0.238,3.759)and that of Q are $x = 7 \sin 120^\circ \cos 75^\circ = 1.569$

$y = 7 \sin 120^\circ \sin 75^\circ = 5.856$, $z = 7 \cos 120^\circ = -3.5$, i.e. Q (1.569, 5.856 , −3.5)

$\bar{R}_{PQ} = \bar{r}_Q - \bar{r}_P$

$= (1.569\,\bar{a}_x + 5.856\,\bar{a}_y - 3.5\,\bar{a}_z) - (1.347\,\bar{a}_x + 0.238\,\bar{a}_y + 3.759\,\bar{a}_z)$

$= 0.222\,\bar{a}_x + 5.618\,\bar{a}_y - 7.259\,\bar{a}_z$

(b) $\overline{R}_{MN} = \overline{r}_N - \overline{r}_M = (2\overline{a}_x + 4\overline{a}_y + 6\overline{a}_z) - (5\overline{a}_x + \overline{a}_y + 2\overline{a}_z)$

$= -3\overline{a}_x + 3\overline{a}_y + 4\overline{a}_z$

Let $\overline{A} = \overline{R}_{MN} = -3\overline{a}_x + 3\overline{a}_y + 4\overline{a}_z = A_r\overline{a}_r + A_\theta\overline{a}_\theta + A_\phi\overline{a}_\phi$

$A_r = \overline{A}\cdot\overline{a}_r = -3\overline{a}_x\cdot\overline{a}_r + 3\overline{a}_y\cdot\overline{a}_r + 4\overline{a}_z\cdot\overline{a}_r$

$= -3\sin\theta\cos\phi + 3\sin\theta\sin\phi + 4\cos\theta$

$A_\theta = \overline{A}\cdot\overline{a}_\theta = -3\overline{a}_x\cdot\overline{a}_\theta + 3\overline{a}_y\cdot\overline{a}_\theta + 4\overline{a}_z\cdot\overline{a}_\theta$

$= -3\cos\theta\cos\phi + 3\cos\theta\sin\phi - 4\sin\theta$

$A_\phi = \overline{A}\cdot\overline{a}_\phi = -3\overline{a}_x\cdot\overline{a}_\phi + 3\overline{a}_y\cdot\overline{a}_o + 4\overline{a}_z\cdot\overline{a}_\phi = 3\sin\phi + 3\cos\phi$

$\therefore\ \overline{A} = \overline{R}_{MN} = (-3\sin\theta\cos\phi + 3\sin\theta\sin\phi + 4\cos\theta)\,\overline{a}_r +$

$(-3\cos\theta\cos\phi + 3\cos\theta\sin\phi - 4\sin\theta)\,\overline{a}_\theta +$

$(3\sin\phi + 3\cos\phi)\,\overline{a}_\phi$

For point M (5, 1, 2) $r = \sqrt{(5)^2 + (1)^2 + (2)^2} = 5.477$

$\theta = \cos^{-1}(2/5.477) = 68.583^\circ$ and $\phi = \tan^{-1}(1/5) = 11.31^\circ$

Substituting θ and ϕ coordinates of point M in $\overline{R}_{MN}$ we have

$\overline{R}_{MN}$ at point $M = -0.730\,\overline{a}_x - 4.583\,\overline{a}_\theta + 3.530\,\overline{a}_\phi$

Problem 1.38. : Points A ($r = 100, \theta = 90^\circ, \phi = 0^\circ$) andB ($r = 100, \theta = 90^\circ, \phi = 5^\circ$) are located on the surface of 100 m radius sphere. (a) What is their separation, using a path on the spherical surface ? (b) What is their separation, using a straight–line path ?

Solution :

(a) $\overline{dL} = dr\,\overline{a}_r + r\,d\theta\,\overline{a}_\theta + r\sin\theta\,d\phi\,\overline{a}_\phi$

For points A and B, r and θ coordinates are constant. Hence dr and $d\theta$ both are equal to zero. Hence $\overline{dL} = r\sin\theta\ d\phi\ \overline{a}_\phi$ or $dL = r\sin\theta\ d\phi$

$$\therefore\ L = \int dL = \int_{0^\circ}^{5^\circ} \underset{\substack{r=100\\ \theta=90^\circ}}{r\sin\theta\ d\phi} = 100\sin 90^\circ\left[\ \phi\ \right]_{0^\circ}^{5^\circ} = 8.727$$

$$\text{(b) } d(AB) = \Big[\,(100\sin 90^0\cos 0^\circ - 100\sin 90^\circ\cos 5^0)^2 + (100\sin 90^\circ\sin 0^\circ - 100\sin 90^\circ\sin 5^\circ)^2 + (100\cos 90^\circ - 100\cos 90^\circ)^2\Big]^{1/2} = 8.724$$

Problem 1.39 : (a) Express $\overline{a}_\rho$ in spherical components and variables. (b) Express $\overline{a}_r$ in cylindrical components and variables.

Solution :

(a) Let $\overline{A} = \overline{a}_\rho = A_x\,\overline{a}_x + A_y\,\overline{a}_y + A_z\,\overline{a}_z$

$A_x = \overline{A}\cdot\overline{a}_x = \overline{a}_\rho\cdot\overline{a}_x = \cos\phi$

$A_y = \overline{A}\cdot\overline{a}_y = \overline{a}_\rho\cdot\overline{a}_y = \sin\phi$

$A_z = A\cdot\overline{a}_z = \overline{a}_\rho\cdot\overline{a}_z = 0$

$\therefore\ \overline{A} = \cos\phi\,\overline{a}_x + \sin\phi\,\overline{a}_y$

Let $\overline{A} = A_r\,\overline{a}_r + A_\theta\,\overline{a}_\theta + A_\phi\,\overline{a}_\phi$

$A_r = \overline{A}\cdot\overline{a}_r = \cos\phi\,\overline{a}_x\cdot\overline{a}_r + \sin\phi\,\overline{a}_y\cdot\overline{a}_r$

$= \cos\phi\,\sin\theta\,\cos\phi + \sin\phi\,\sin\theta\,\sin\phi = \sin\theta$

$A_\theta = \overline{A}\cdot\overline{a}_\theta = \cos\phi\,\overline{a}_x\cdot\overline{a}_\theta + \sin\phi\,\overline{a}_y\cdot\overline{a}_\theta$

$= \cos\phi\,\cos\theta\,\cos\phi + \sin\phi\,\cos\theta\,\sin\phi = \cos\theta$

$A_\phi = \overline{A}\cdot\overline{a}_\phi = \cos\phi\,\overline{a}_x\cdot\overline{a}_\phi + \sin\phi\,\overline{a}_y\cdot\overline{a}_\phi$

$= \cos\phi\,(-\sin\phi) + \sin\phi\,\cos\phi = 0$

$\therefore\ \overline{A} = \sin\theta\,\overline{a}_r + \cos\theta\,\overline{a}_\theta$

(b) Let $\overline{A} = \overline{a}_r = A_x\,\overline{a}_x + A_y\,\overline{a}_y + A_z\,\overline{a}_z$

$A_x = A\cdot\overline{a}_x = \overline{a}_r\cdot\overline{a}_x = \sin\theta\,\cos\phi$

$A_y = A\cdot\overline{a}_y = \overline{a}_r\cdot\overline{a}_y = \sin\theta\,\sin\phi$

$A_z = A\cdot\overline{a}_z = \overline{a}_r\cdot\overline{a}_z = \cos\theta$

$\overline{A} = \sin\theta\cos\phi\,\overline{a}_x + \sin\theta\sin\phi\,\overline{a}_y + \cos\theta\,\overline{a}_z$

Let $\overline{A} = A_\rho\,\overline{a}_\rho + A_\phi\,\overline{a}_\phi + A_z\,\overline{a}_z$

$A_\rho = \overline{A}\cdot\overline{a}_\rho = \sin\theta\cos\phi\,\overline{a}_x\cdot\overline{a}_\rho + \sin\theta\sin\phi\,\overline{a}_y\cdot\overline{a}_\rho + \cos\theta\,\overline{a}_z\cdot\overline{a}_\rho$

$= \sin\theta\cos\phi\,\cos\phi + \sin\theta\sin\phi\,\sin\phi = \sin\theta$

$A_\phi = \overline{A}\cdot\overline{a}_\phi = \sin\theta\cos\phi\,\overline{a}_x\cdot\overline{a}_\phi + \sin\theta\sin\phi\,\overline{a}_y\cdot\overline{a}_\phi + \cos\theta\,\overline{a}_z\cdot\overline{a}_\phi$

$= -\sin\theta\,\cos\phi\,\sin\phi + \sin\theta\sin\phi\,\cos\phi = 0$

$A_z = \overline{A}\cdot\overline{a}_z = \sin\theta\cos\phi\,\overline{a}_x\cdot\overline{a}_z + \sin\theta\sin\phi\,\overline{a}_y\cdot\overline{a}_z + \cos\theta\,\overline{a}_z\cdot\overline{a}_z = \cos\theta$

$\therefore\ \overline{A} = \sin\theta\,\overline{a}_\rho + \cos\theta\,\overline{a}_z$

Now $\cos\theta = z/r = z/\sqrt{\rho^2 + z^2}$ and $\sin\theta = \rho/\sqrt{\rho^2 + z^2}$

$$\therefore\ \overline{A} = \frac{1}{\sqrt{\rho^2 + z^2}}\,(\rho\,\overline{a}_\rho + z\,\overline{a}_z)$$

—X—

CHAPTER 2

Coulomb's Law & Electric Field Intensity

2.1 : COULOMB'S LAW :

Coulomb's law states that the force between two very small objects separated in a vaccum or free space by a distance which is large compared to their size is proportional to charge on each and inversely proportional to the square of the distance between them or

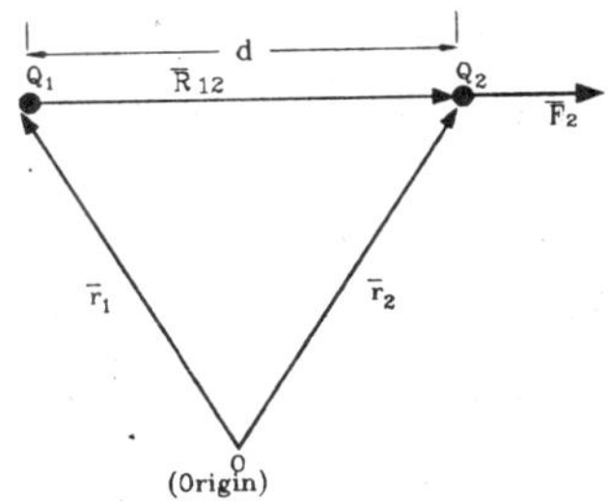

FIG. 2.1

$$F \propto Q_1 Q_2$$

$$F \propto \frac{1}{d^2}$$

$$\therefore \; F \propto \frac{Q_1 Q_2}{d^2} \quad ; \quad F = \frac{K Q_1 Q_2}{d^2}$$

where Q_1 and Q_2 are the positive or negative quantities of charge, d is the distance between them and K is a constant of proportionality. In SI system K is written as $K = 1/4\pi\varepsilon_0$, where ε_0 is the permitivity of free space and is equal to 8.854×10^{-12} F/m .

$$\therefore \; F = \frac{Q_1 Q_2}{4\pi\varepsilon_0 d^2}$$

The law also states that the force acts along the line joining the two charges and is repulsive if the charges are alike in sign and attractive if they are of opposite sign. The vector form of Coulomb's law is

$$\bar{F}_2 = \frac{Q_1 Q_2}{4 \pi \varepsilon_o d^2} \bar{a}_{12}$$

where $\bar{a}_{12}$ is a unit vector in the direction of $\bar{R}_{12}$

$$\bar{a}_{12} = \frac{\bar{R}_{12}}{|\bar{R}_{12}|} = \frac{\bar{r}_2 - \bar{r}_1}{|\bar{r}_2 - \bar{r}_1|}$$

As $|\bar{R}_{12}|$ gives the distance between the points 1 and 2, we have $d = |\bar{R}_{12}|$. Hence we can write

$$\bar{F}_2 = \frac{Q_1 Q_2}{4 \pi \varepsilon_o |\bar{R}_{12}|^2} \left(\frac{\bar{R}_{12}}{|\bar{R}_{12}|} \right) \quad \ldots\ldots (2.1)$$

F is in Newton, Q is in Coulomb and d is in meter.

Note : *Force is a vector quantity. If N is the total number of charges present, then each charge will experience (N – 1) forces and the resultant force will be the vector sum of these (N – 1) forces.*

Problem 2.1 : A 2 – mC (milli coulomb) positive charge is located in vacuum at P_1 1 (3, – 2, – 4) and a 5 – μC (micro coulomb) negative charge is at P_2 (1, – 4, 2). (a) Find the vector force on the negative charge. (b) What is the magnitude of the force on the charge at P_1 ?

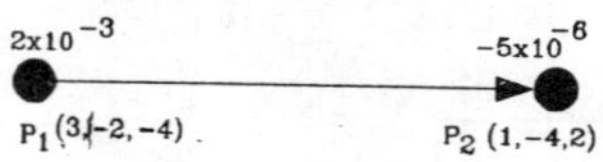

FIG. P2.1

Solution :

(a) Vector force on the -ve charge

$$= \frac{(2 \times 10^{-3})(-5 \times 10^{-6})}{4 \pi \varepsilon_o (6.633)^2} \left[\frac{(1-3)\bar{a}_x + (-4+2)\bar{a}_y + (2+4)\bar{a}_z}{\sqrt{(-2)^2 + (-2)^2 + (6)^2} = 6.633} \right]$$

$$= 0.616 \bar{a}_x + 0.616 \bar{a}_y - 1.848 \bar{a}_z \text{ N}$$

(b) Magnitude of force on the charge at P_1 = Magnitude of force on the charge at P_2 = $\sqrt{(0.616)^2 + (0.616)^2 + (-1.848)^2} = 2.043$ N

2.2 : CONCEPT OF ELECTRIC FIELD INTENSITY :

Suppose charge Q_1 is located at point 1 and charge Q_2 at point 2 as show in Fig.2.2(a). The force acting on charge Q_2 due to charge Q_1 is given by y $\quad \bar{F}_2 = \dfrac{Q_1 Q_2}{4 \pi \varepsilon_0 d^2} a_{12}$.

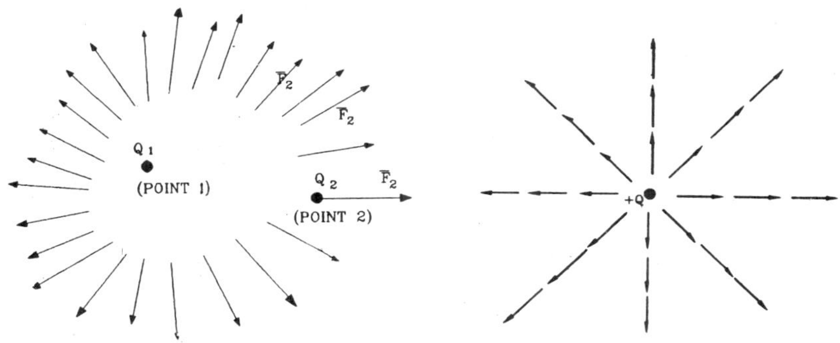

(a) (b)

FIG. 2.2

This force is shown in Fig.2.2(a) (assuming both the charges are positive). If charge Q_1 is fixed and charge Q_2 is moved around the charge Q_1, then there exists every where a force on this second charge. Thus the second charge is displaying the existance of a force field as shown in Fig.2.2(a).Writing this force as a force per unit charge,

$$\frac{\bar{F}_2}{Q_2} = \frac{Q_1}{4 \pi \varepsilon_0 d^2} \bar{a}_{12}$$

The electric field intensity is defined as the force per unit charge. It is a vector quantity and is denoted by $\bar{E}$. Hence

$$\bar{E}_2 = \frac{\bar{F}_2}{Q_2} = \frac{Q_1}{4 \pi \varepsilon_0 d^2} \bar{a}_{12}$$

Another defination of electric field intensity may be stated as : It is the force experienced by a unit positive test charge, that is $Q_2 =$ 1C. The unit of $\bar{E}$ is V/m. A sketch of electric field intensity due to a positive charge Q_1 is shown in Fig.2.2(b). The electric field intensity at point 2 due to charge Q_1 at point 1 is given by

$$\boxed{\bar{\mathbf{E}}_2 = \frac{Q}{4 \pi \varepsilon_0 |\bar{\mathbf{R}}_{12}|^2} \left(\frac{\bar{\mathbf{R}}_{12}}{|\bar{\mathbf{R}}_{12}|} \right)} \quad \ldots\ldots (2.2)$$

Note : *Electric field inteisnty is a vector quanitity. If N charges are present, then at any point, N electric field intensity vectors are acting. The vector sum of these N* *electric fields gives the resultant* $\overline{E}$.

Problem 2.2. : Calculate $\overline{E}$ at M (3, – 4, 2) in free space caused by : (a) charge Q_1 = 2 μC at P_1 (0 , 0 , 0) ; (b) a charge Q_2 = 3 μC at P_2 (–1, 2, 3) ; (c) a charge Q_1 = 2 μC at P_1 (0, 0, 0), and a charge Q_2 = 3 μC at P_2 (–1, 2, 3).

Solution :

$$\text{(a) } \overline{E} = \frac{2\times 10^{-6}}{4\pi\varepsilon_0 (5.385)^2}\left[\frac{(3-0)\,\overline{a}_x + (-4-0)\,\overline{a}_y + (2-0)\,\overline{a}_z}{\sqrt{(3)^2+(-4)^2+(2)^2} = 5.385}\right]$$

$$= 345.307\,\overline{a}_x - 460.409\,\overline{a}_y + 230.204\,\overline{a}_z \text{ V/m}$$

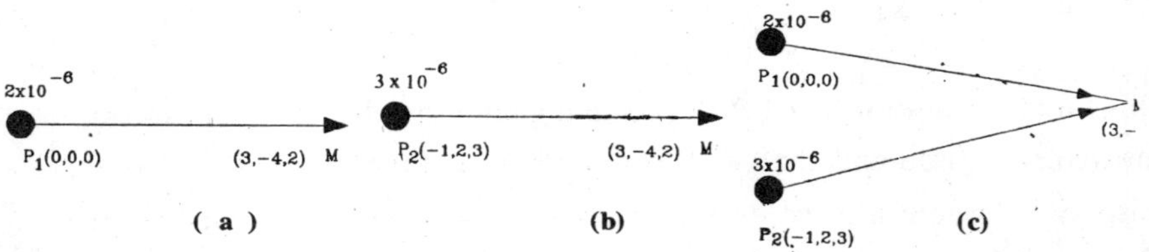

FIG. P 2.2

$$\text{(b) } \overline{E} = \frac{3\times 10^{-6}}{4\pi\varepsilon_0 (7.28)^2}\left[\frac{(3+1)\,\overline{a}_x + (-4-2)\,\overline{a}_y + (2-3)\,\overline{a}_z}{\sqrt{(4)^2+(-6)^2+(-1)^2} = 7.28}\right]$$

$$= 279.523\,\overline{a}_x - 419.285\,\overline{a}_y - 69.881\,\overline{a}_z \text{ V/m}$$

$$\text{(c) } \overline{E} = \overline{E} \text{ at } M \text{ due } Q_1 + \overline{E} \text{ at } M \text{ due } Q_2$$

$$= 345.307\,\overline{a}_x\, 460.409\,\overline{a}_y + 230 - 204\,\overline{a}_z + 279.523\,\overline{a}_x - 419.285\,\overline{a}_y - 69.881\,\overline{a}_z$$

$$= 624.83\,\overline{a}_x - 879.694\,\overline{a}_y + 160.323\,\overline{a}_z \text{ V/m}$$

2.3 : VOLUME CHARGE DENSITY :

It is denoted by ρ_v. Suppose within a very small volume dv (also known as differential volume), a total charge dQ (also known as differential charge) lies. Then volume charge density is defined by $\rho_v = \dfrac{dQ}{dv}$ and the unit is C/m^3.

Another definition of volume charge density is

$$\rho_v = \lim_{\Delta v \to 0} \frac{\Delta q}{\Delta v}$$

Note : The important formulae related with volume charge density are

$$\rho_v = \frac{dQ}{dv} \qquad \ldots\ldots (2.3.a)$$

$$dQ = \rho_v \, dv \qquad \ldots\ldots (2.3.b)$$

$$Q_{total} = \int dQ = \int \rho_v \, dv \qquad \ldots\ldots\ldots (2.3.c)$$

Problem 2.3 : Find the total charge inside each of the volumes indicated :

(a) $\rho_v = 10 z^2 e^{-0.1x} \sin \pi y, -1 \le x \le 2, 0 \le y \le 1, 3 \le z \le 3.6$;

(b) $\rho_v = 4xyz^2, 0 \le \rho \le 2, 0 \le \phi \le \pi/2, 0 \le z \le 3$;

(c) $\rho_v = 3\pi \cos^2 \theta \cos^2 \phi / [2r^2 (r^2 + 1)]$; **Universe.**

Solution :

(a) $$Q = \int dQ = \int \rho_v \, dv = \int_{-1}^{2} \int_{0}^{1} \int_{3}^{3.6} 10 z^2 e^{-0.1x} \sin(\pi y) \, dx \, dy \, dz$$

$$= 10 \left[\frac{z^3}{3} \right]_3^{3.6} \left[\frac{e^{-0.1x}}{-0.1} \right]_{-1}^{2} \left[\frac{-\cos(\pi y)}{\pi} \right]_0^1 = 119.478$$

(b) $$\rho_v = 4xyz^2 = 4(\rho \cos \phi)(\rho \sin \phi) z^2 = 2\rho^2 \sin 2\phi \, z^2$$

$$\therefore Q = \int dQ = \int_0^2 \int_0^{\pi/2} \int_0^3 2\rho^2 \sin 2\phi \, z^2 \rho \, d\rho \, d\phi \, dz$$

$$= 2 \left[\frac{\rho^4}{4} \right]_0^2 \left[\frac{-\cos 2\phi}{2} \right]_0^{\pi/2} \left[\frac{z^3}{3} \right]_0^3 = 72 \text{ C}$$

(c) For universe the limits are $0 \le r \le \infty, 0 \le \theta \le \pi$ and $0 \le \phi \le 2\pi$

$$Q = \int dQ = \int_0^{\infty} \int_0^{\pi} \int_0^{2\pi} \frac{3\pi \cos^2 \theta \cos^2 \phi}{2r^2 (r^2 + 1)} r^2 \sin \theta \, dr \, d\theta \, d\phi$$

$$= 1.5\pi \left[\tan^{-1} r \right]_0^{\infty} \left[\frac{-\cos^3 \theta}{3} \right]_0^{\pi} \left[\frac{\phi}{2} + \frac{\sin 2\phi}{4} \right]_0^{2\pi} = 15.503$$

2.4 : ELECTRIC FIELD INTENSITY DUE TO DIFFERENTIAL AMOUNT OF CHARGE dQ

Duc to differential charge dQ, differential amount of electric field intensity

($\overline{dE}$) is produced. From eqn.2.2 , we have

$$\overline{dE} = \frac{dQ}{4\pi\varepsilon_0 |\overline{R}_{12}|^2}\left(\frac{\overline{R}_{12}}{|\overline{R}_{12}|}\right) \qquad \ldots\ldots(2.4)$$

The total electric field intensity is obtained by integrating above differential electric field intensity.

2.5 : LINE CHARGE DENSITY :

It is denoted by ρ_L . Suppose a differentialcharge dQ is present on a differential length dL . Then line charge density is defined as $\rho_L = dQ / dL$ and the unit is C/m.

Note : Important formulae for line charge density are

$$\rho_L = \frac{dQ}{dL} \qquad \ldots\ldots (2.5.a)$$

$$dQ = \rho_L\, dL \qquad \ldots\ldots (2.5.b)$$

$$Q_{total} = \int dQ = \int \rho_L\, dL \qquad \ldots (2.5.c)$$

2.6 : EXPRESSION FOR ELECTRIC FIELD INTENSITY AT ANY POINT DUE TO A LINE CHARGE WITH UNIFORM CHARGE DENSITY ρ_L C/m ON THE INFINITELY LONG Z-AXIS.

Fig.2.3 shows a line charge ρ_L C/m on the z – axis. Consider at any point 1 (0, 0, z) on the z – axis, a differential amount of charge $dQ = \rho_L\, dL$ is present. Since the charge lies on the z – axis $dL = dz$. Hence $dQ = \rho_L\, dz$. Let point 2 (ρ , ϕ , z') be the point at which electric field intensity is desired. Now

$$d\overline{E} = \frac{dQ}{4\pi\varepsilon_0 |\overline{R}_{12}|^2}\left(\frac{\overline{R}_{12}}{|\overline{R}_{12}|}\right)$$

Now $\overline{R}_{12} = \overline{r}_2 - \overline{r}_1$, where

$\overline{r}_2$ = p.v. of point 2 = $\rho\,\overline{a}_\rho + z'\,\overline{a}_z$

and

$\overline{r}_1$ = p.v. of point 1 = $z\,\overline{a}_z$

Hence $\overline{R}_{12} = \rho\,\overline{a}_\rho + (z' - z)\,\overline{a}_z$

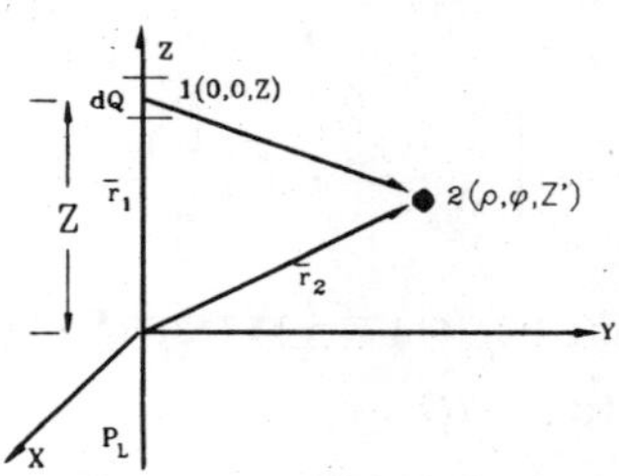

FIG. 2.3

and $|\overline{R}_{12}| = \sqrt{\rho^2 + (z' - z)^2}$

$$\therefore \quad d\overline{E} = \frac{\rho_L \, dz}{4\pi\varepsilon_o} \frac{\rho\,\overline{a}_\rho + (z' - z)\,\overline{a}_z}{[\rho^2 + (z' - z)^2]^{3/2}}$$

$$\therefore \; \overline{E} = \int d\overline{E} = \int_{-\infty}^{\infty} \frac{\rho_L \, dz}{4\pi\varepsilon_o} \frac{\rho\,\overline{a}_\rho + (z' - z)\,\overline{a}_z}{[\rho^2 + (z' - z)^2]^{3/2}}$$

$$\overline{E} = \frac{\rho_L}{4\pi\varepsilon_o}\left\{ \overline{a}_\rho \int_{-\infty}^{\infty} \frac{\rho\, dz}{[\rho^2 + (z' - z)^2]^{3/2}} + \overline{a}_z \int_{-\infty}^{\infty} \frac{(z' - z)\, dz}{[\rho^2 + (z' - z)^2]^{3/2}} \right\}$$

$$= \frac{\rho_L}{4\pi\varepsilon_o}\left[\overline{a}_\rho \frac{1}{\rho} \frac{-(z' - z)}{[\rho^2 + (z' - z)^2]^{1/2}} + \overline{a}_z \frac{1}{[\rho^2 + (z' - z)^2]^{1/2}} \right]_{-\infty}^{\infty}$$

$$= \frac{\rho_L}{2\pi\varepsilon_o\rho}\,\overline{a}_\rho$$

Note : Electric field intensity at point 2 due to uniform line charge density ρ_L C/m which is infinitely long is given by

$$\boxed{\overline{\mathbf{E}} = \frac{\rho_L}{2\pi\varepsilon_o d}\,\overline{\mathbf{a}}_{12}} \qquad \text{..........(2.6)}$$

where point 1 is the foot of perpendicular dropped from point 2 on the line charge, and d is the distance between points 1 and 2.

Problem 2.4 : An infinitely long, uniform line charge is located at $y = 3, z = 5$. If $\rho_L = 30$ nC/m , find $\overline{E}$ at : (a) the origin; (b) P_B (0, 6, 1) ; (c) P_C (5, 6, 1).

Soution :

(a) $\overline{E} = \dfrac{\rho_L}{2\pi\varepsilon_0 d}\,\overline{a}_{12}$

Point 2 is that point at which $\overline{E}$ is desired = (0,0,0)

Point 1 is the foot of perpendicular dropped from point 2 on the line charge = $(0, y = 3, z = 5)$

$\therefore \; \overline{R}_{12} = (0 - 3)\,\overline{a}_y + (0 - 5)\,\overline{a}_z = -3\,\overline{a}_y - 5\,\overline{a}_z$

$d = |\overline{R}_{12}| = \sqrt{(-3)^2 + (-5)^2} = \sqrt{34}$

$$\overline{a}_{12} = \frac{\overline{R}_{12}}{|\overline{R}_{12}|} = \frac{-3\,\overline{a}_y - 5\,\overline{a}_z}{\sqrt{34}}$$

$$\therefore\ \bar{E} = \frac{30\times 10^{-9}}{2\pi\varepsilon_0\sqrt{34}}\left[\frac{-3\bar{a}_y - 5\bar{a}_z}{\sqrt{34}}\right] = -47.582\,\bar{a}_y - 9.304\,\bar{a}_z \text{ V/m}$$

(b) Point 2 is that point at which $\bar{E}$ is desired = (0,6,1)

Point 1 is the foot of perpendicular dropped from point 2 on the line charge = $(0, y = 3, z = 5)$

$$\therefore\ \bar{R}_{12} = (6-3)\,\bar{a}_y + (1-5)\,\bar{a}_z = 3\bar{a}_y - 4\bar{a}_z$$

$$d = |\bar{R}_{12}| = \sqrt{(3)^2 + (-4)^2} = \sqrt{25}$$

$$\bar{a}_{12} = \frac{\bar{R}_{12}}{|\bar{R}_{12}|} = \frac{3\bar{a}_y - 4\bar{a}_z}{\sqrt{25}}$$

$$\therefore\ \bar{E} = \frac{30\times 10^{-9}}{2\pi\varepsilon_0\sqrt{25}}\left[\frac{3\bar{a}_y - 4\bar{a}_z}{\sqrt{25}}\right] = 64.712\,\bar{a}_y - 86.282\,\bar{a}_z \text{ V/m}$$

(c) Point 2 is that point at which $\bar{E}$ is desired = (5,6,1)

Point 1 is the foot of perpondicular dropped from point 2 on the line charge = $(5, y = 3, z = 5)$

$$\therefore\ \bar{R}_{12} = (6-3)\,\bar{a}_y + (1-5)\,\bar{a}_z = 3\bar{a}_y - 4\bar{a}_z$$

$$d = |\bar{R}_{12}| = \sqrt{(3)^2 + (-4)^2} = \sqrt{25}$$

$$\bar{a}_{12} = \frac{\bar{R}_{12}}{|\bar{R}_{12}|} = \frac{3\bar{a}_y - 4\bar{a}_z}{\sqrt{25}}$$

$$\therefore\ \bar{E} = \frac{30\times 10^{-9}}{2\pi\varepsilon_0\sqrt{25}}\left[\frac{3\bar{a}_y - 4\bar{a}_z}{\sqrt{25}}\right] = 64.712\,\bar{a}_y - 86.282\,\bar{a}_z \text{ V/m}$$

2.7 : SURFACE CHARGE DENSITY :

It is denoted by ρ_s. Suppose a differential charge dQ is present on a differential area dS. Then surface charge density is defined as $\rho_s = \frac{dQ}{dS}$ and the unit is C/m^2.

Note : Important formulae related with surface charge density are

$$\rho_s = \frac{dQ}{dS} \quad \text{..... (2.7.a)}$$

$$dQ = \rho_s\, dS \quad \text{...... (2.7.b)}$$

$$Q_{total} = \int dQ = \int \rho_s\, dS \quad \text{..... (2.7.c)}$$

2.8 : EXPRESSION FOR ELECTRIC FIELD INTENSITY DUE TO A SHEET OF CHARGE WITH UNIFORM SURFACE CHARGE DENSITY ρ_s C/m^2 ON AN INFINITE PLANE :

The cylindrical co–ordinate system is used. Fig.2.4 shows surface charge density ρ_s C/m^2 in $z = 0$ plane. Let us consider a differential amount of charge $dQ = \rho_s \, dS$ is located at point 1 (ρ , ϕ , 0). The point at which electric field intensity is desired is 2 (0 , 0 , z). Now

$$\overline{dE} = \frac{\rho_s \, dS}{4\pi\varepsilon_o |\overline{R}_{12}|^2}\left(\frac{\overline{R}_{12}}{|\overline{R}_{12}|}\right)$$

The surface charge density lies in $z = 0$ plane to which $\overline{a}_z$ is pependicular. Hence we have

$\overline{dS} = \rho \, d\rho \, d\phi \; \overline{a}_z$ or $dS = \rho \, d\rho \, d\phi$

$\overline{R}_{12} = \overline{r}_2 - \overline{r}_1$

where

$\overline{r}_2$ = p.v. of point 2 = $z\,\overline{a}_z$

$\overline{r}_1$ = p.v. of point 1 = $\rho\,\overline{a}_\rho$

$\therefore \; \overline{R}_{12} = -\rho\,\overline{a}_\rho + z\,\overline{a}_z$

and $|\overline{R}_{12}| = \sqrt{\rho^2 + z^2}$

$$\overline{a}_{12} = \frac{\overline{R}_{12}}{|\overline{R}_{12}|} = \frac{-\rho\,\overline{a}_\rho + z\,\overline{a}_z}{\sqrt{\rho^2 + z^2}}$$

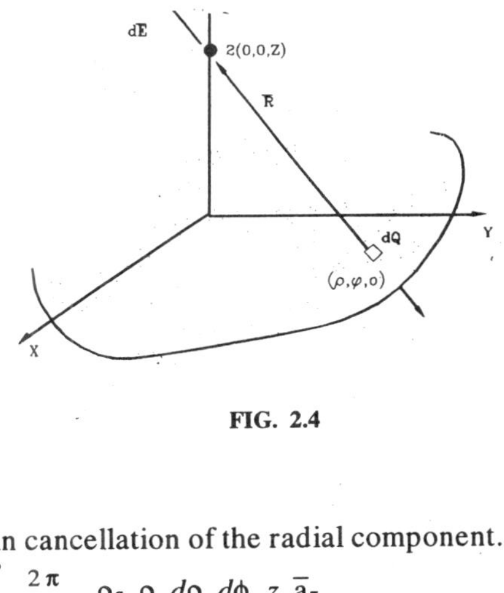

FIG. 2.4

$$\overline{dE} = \frac{\rho_s \, \rho \, d\rho \, d\phi \, [-\rho\,\overline{a}_\rho + z\,\overline{a}_z]}{4\pi\varepsilon_o \, [\rho^2 + z^2]^{3/2}}$$

Symmetry about the z – axis results in cancellation of the radial component.

$$\therefore \; \overline{E} = \int \overline{dE} = \int_0^{\infty}\int_0^{2\pi} \frac{\rho_s \, \rho \, d\rho \, d\phi \; z\,\overline{a}_z}{4\pi\varepsilon_o \, [\rho^2 + z^2]^{3/2}}$$

$$= \frac{\rho_s \, z}{2\varepsilon_o}\left[\frac{-1}{\sqrt{\rho^2 + z^2}}\right]_0^{\infty} \overline{a}_z = \frac{\rho_s}{2\varepsilon_o}\,\overline{a}_z$$

This result is for points above the xy plane. Below the xy plane the unit vector changes to $-\overline{a}_z$. The generalized form may be written using the unit normal vector.

$$\overline{E} = \frac{\rho_s}{2\varepsilon_o}\,\overline{a}_n$$

Note : Electric field intensity at point 2 due to uniform surface charge density lying on infinite plane is given by

$$\overline{E} = \frac{\rho_S}{2\varepsilon_0}\overline{a}_N \qquad \text{.........(2.8)}$$

where $\overline{a}_N$ is a unit vector perpendicular to the plane on which ρ_S lies and directed towards point 2.

Problem 2.6 : Four infinite uniform sheets of charge are located as follows : 20 pC/m^2 at $y = 7$, – 8 pC pC/m^2 at $y = 3$, 6 pC/m^2 at $y = -1$, and. – 18 pC/m^2 at $y = -4$. Find $\overline{E}$ at the point : (a) P_A (2, 6, – 4) ; (b) P_B (0, 0, 0) ; (c) P_C (–1, –1, 1.5) ; (d) P_D (10^6, 10^6, 10^6).

Solution :

Let ρ_{S1} = 20 pC/m^2 at $y = 7$ $\quad$ ρ_{S2} = – 8 pC/m^2 at $y = 3$

ρ_{S3} = 6 pC/m^2 at $y = -1$ $\quad$ ρ_{S4} = – 18 pC/m^2 at $y = -4$

(a) $\overline{E}$ at P_A = $\overline{E}$ at point P_A due to ρ_{S1} + $\overline{E}$ at point P_A due to ρ_{S2} + $\overline{E}$ at point P_A due to ρ_{S3} + $\overline{E}$ at point P_A due to ρ_{S4}

We will have to concentrate only on the y coordinate of the point at which $\overline{E}$ is desired. This is due to the fact that all the surface charge densities lies on the planes whose equations are given by y coordinate.

$$\overline{E} \text{ at } P_A = \frac{20\times 10^{-12}}{2\varepsilon_0}(-\overline{a}_y) \quad [\because y \text{ coordinate of point } P_A < y = 7]$$

$$- \frac{8\times 10^{-12}}{2\varepsilon_0}(+\overline{a}_y) \quad [\because y \text{ coordinate of point } P_A > y = 3]$$

$$+ \frac{6\times 10^{-12}}{2\varepsilon_0}(+\overline{a}_y) \quad [\because y \text{ coordinate of point } P_A > y = -1]$$

$$- \frac{18\times 10^{-12}}{2\varepsilon_0}(+\overline{a}_y) \quad [\because y \text{ coordinate of point } P_A > y = -4]$$

$$= \frac{10^{-12}}{2\varepsilon_0}(-20 - 8 + 6 - 18)\,\overline{a}_y = -\ 2.259\,\overline{a}_y \text{ V/m}$$

(b) $\bar{E}$ at $P_B = \frac{20 \times 10^{-12}}{2\varepsilon_0}(-\bar{a}_y)$ [$\because$ y coordinate of point $P_B < y = 7$]

$- \frac{8 \times 10^{-12}}{2\varepsilon_0}(-\bar{a}_y)$ [$\because$ y coordinate of point $P_B < y = 3$]

$+ \frac{6 \times 10^{-12}}{2\varepsilon_0}(+\bar{a}_y)$ [$\because$ y coordinate of point $P_B > y = -1$]

$- \frac{18 \times 10^{-12}}{2\varepsilon_0}(+\bar{a}_y)$ [$\because$ y coordinate of point $P_B > y = -4$]

$$= \frac{10^{-12}}{2\varepsilon_0}(-20 + 8 + 6 - 18)\,\bar{a}_y = -1.355\,\bar{a}_y \text{ V/m}$$

(c) $\bar{E}$ at $P_C = \frac{20 \times 10^{-12}}{2\varepsilon_0}(-\bar{a}_y)$ [$\because$ y coordinate of point $P_C < y = 7$]

$- \frac{8 \times 10^{-12}}{2\varepsilon_0}(-\bar{a}_y)$ [$\because$ y coordinate of point $P_C < y = 3$]

$+ \frac{6 \times 10^{-12}}{2\varepsilon_0}(-\bar{a}_y)$ [$\because$ y coordinate of point $P_C < y = -1$]

$- \frac{18 \times 10^{-12}}{2\varepsilon_0}(+\bar{a}_y)$ [$\because$ y coordinate of point $P_B > y = -4$]

$$= \frac{10^{-12}}{2\varepsilon_0}(-20 + 8 - 6 - 18)\,\bar{a}_y = -2.033\,\bar{a}_y \text{ V/m}$$

(d) $\bar{E}$ at $P_D = \frac{20 \times 10^{-12}}{2\varepsilon_o}(+\bar{a}_y)$ [$\because$ y coordinate of point $P_D > y = 7$]

$- \frac{8 \times 10^{-12}}{2\varepsilon_o}(+\bar{a}_y)$ [$\because$ y coordinate of point $P_D > y = 3$]

$+ \frac{6 \times 10^{-12}}{2\varepsilon_o}(+\bar{a}_y)$ [$\because$ y coordinate of point $P_D > y = -1$]

$- \frac{18 \times 10^{-12}}{2\varepsilon_o}(+\bar{a}_y)$ [$\because$ y coordinate of point $P_D > y = -4$]

$$= \frac{10^{-12}}{2\varepsilon_o}[20 - 8 + 6 - 18]\,\bar{a}_y = 0.$$

2.9 : STREAMLINES :

The lines which shows only the direction of a vector field (such as $\bar{E}$, $\bar{D}$,

$\overline{H}$ or $\overline{B}$) are known as streamlines. These lines are also known as flux lines or direction lines. These lines are drawn from the source of a vector field and are everywhere tangent to vector field. To determine the equations of streamlines for electric field intensity, the formulae are

$$\frac{E_x}{dx} = \frac{E_y}{dy} = \frac{E_z}{dz} \qquad \text{.......(2.9.a)}$$

$$\frac{E_\rho}{d\rho} = \frac{E_\phi}{\rho\, d\phi} = \frac{E_z}{dz} \qquad \text{.......(2.9.b)}$$

$$\frac{E_r}{dr} = \frac{E_\theta}{r\, d\theta} = \frac{E_\phi}{r \sin\theta\, d\phi} \qquad \text{.......(2.9. c)}$$

Problem 2.6 : Obtain the equation of the streamline that passes through the point P (–2, 7, 10) in the field $\overline{E}$ = : (a) $2(y-1)\,\overline{a}_x + 2x\,\overline{a}_y$; (b) $e^y\,\overline{a}_x + (x+1)\,e^y\,\overline{a}_y$.

Solution :

(a) $\overline{E} = 2(y-1)\,\overline{a}_x + 2x\,\overline{a}_y = E_x\,\overline{a}_x + E_y\,\overline{a}_y$

where $E_x = 2(y-1)$ and $E_y = 2x$

The equation of the stream line is given by

$$\frac{dy}{dx} = \frac{E_y}{E_x} = \frac{2x}{2(y-1)} = \frac{x}{y-1}$$

$\therefore\ (y-1)\,dy = x\,dx$ or $0.5\,y^2 - y = 0.5\,x^2 + C$

Substituting the coordinates of point P in the above expression, we have italic C = 15.5. Hence equation of the stream line passing through the point P (–2,7,10) is $0.5\,y^2 - y = 0.5\,x^2 + 15.5$ or $(y-1)^2 - x^2 = 32$

(b) $\overline{E} = e^y\,\overline{a}_x + (x+1)\,e^y\,\overline{a}_y = E_x\,\overline{a}_x + E_y\,\overline{a}_y$

The equation of the stream line is given by

$$\frac{dy}{dx} = \frac{E_y}{E_x} = \frac{(x+1)\,e^y}{e^y} = (x+1)$$

$\therefore\ dy = (x+1)\,dx$ or $y = 0.5\,x^2 + x + C$

Substituting the coordinates of point P in the above expression, we have C =

7. Hence equation of the stream line passing through the point P (-2,7,10) is $y = 0.5x^2 + x + 7$ or $2y - (x+1)^2 = 13$

Problem 2.7 : The two small plastic spheres shown in Fig. P2.7 (a) are arranged so that they can slide freely along the insulating fiber. Each has a mass of 0.1 g. The spheres are given identical positive charges. (a) If the lower sphere is resting against the lower knot while the upper sphere moves to a location 6 cm distant from that knot, find the charge on each. (b) If the charges remain the same, what angle should the fiber make with the horizontal to double the separation ?

Solution :

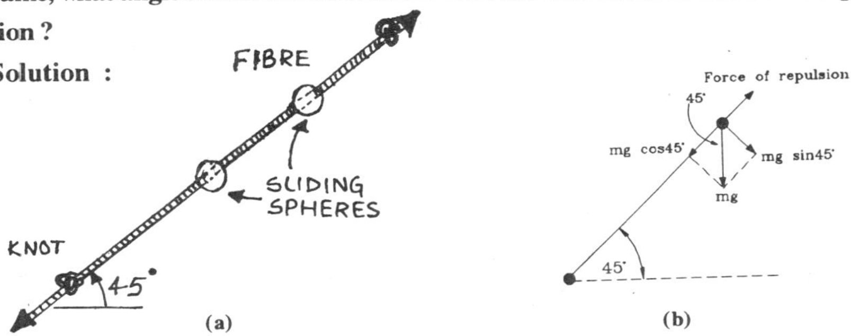

FIG. P2.7

(a) Let Q be the charge given to each sphere. Then the force of repulsion acting on each sphere is $Q^2 / 4\pi\varepsilon_0 (0.06)^2$. Since each sphere has a mass of 0.1 g, the weight of the sphere will be $0.1 \times 10^{-3} \times 9.81 = 0.981 \times 10^{-3}$ N,acting vertically downward. This weight has two components as shown in the Fig. P2.7(b). One component is along the string where as second component is perpendicular to the string. In equilibrium condition, the force of repulsion is balanced by that component of weight which is acting through the string.

$$\therefore \frac{Q^2}{4\pi\varepsilon_0 (0.06)^2} = 0.981 \times 10^{-3} \cos 45^\circ \text{ or } Q = 16.669 \text{ nC}$$

(b) Let θ be the angle required to make the separation 12 cm.

$$\therefore \frac{(16.669 \times 10^{-9})^2}{4\pi\varepsilon_0 (0.12)^2} = 0.981 \times 10^{-3} \times \cos\theta \text{ or } \theta = 79.82^\circ$$

Problem 2.8 : A charge Q_o = 1 nC is located in free space at $P(a, 0, 0)$. Prepare a sketch of the magnitude of the force on Q_o, as a function of

a, $0 \leq a \leq 5$ m, produced by two other charges, $Q_1 = 1$ C at (0, 1, 0) and Q_2 = : (a) 1 C at (0, –1, 0) ; (b) –1 C at (0 , –1, 0).

Solution :

(a)Force on Q_0 = Force on Q_0 due to Q_1 + Force on Q_0 due to Q_2

$$= \frac{(1)(1 \times 10^{-9})}{4\pi\varepsilon_0}\left\{\frac{(a-0)\,\bar{a}_x + (0-1)\,\bar{a}_y}{[(a)^2 + (-1)^2]^{3/2}}\right\}$$

$$+ \frac{(1)(1 \times 10^{-9})}{4\pi\varepsilon_0}\left\{\frac{(a-0)\,\bar{a}_x + (0+1)\,\bar{a}_y}{[(a)^2 + (1)^2]^{3/2}}\right\}$$

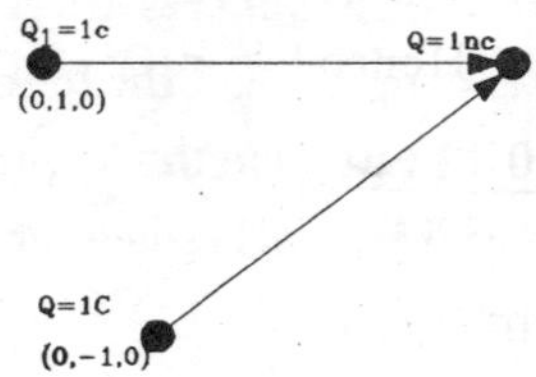

FIG. P2.8 (a)

$$= \frac{10^{-9}[a\,\bar{a}_x - \bar{a}_y + a\,\bar{a}_x + \bar{a}_y]}{4\pi\varepsilon_0(a^2 + 1)^{1.5}}$$

$$= \frac{10^{-9} \times 2\,\bar{a}_x}{4\pi\varepsilon_0(a^2 + 1)^{1.5}}$$

Magnitude of the force on $Q_0 = \dfrac{17.975}{(a^2 + 1)^{1.5}}$ N. Plot this as a function of a, for $0 \leq a \leq 5$ m. Plot is not shown .

(b)Force on Q_0 = Force on Q_0 due to Q_1 + Force on Q_0 due to Q_2

$$= \frac{(1)(1 \times 10^{-9})}{4\pi\varepsilon_0}\left\{\frac{(a-0)\,\bar{a}_x + (0-1)\,\bar{a}_y}{[(a)^2 + (-1)^2]^{3/2}}\right\}$$

$$+ \frac{(1)(-1 \times 10^{-9})}{4\pi\varepsilon_0}\left\{\frac{(a-0)\,\bar{a}_x + (0+1)\,\bar{a}_y}{[(a)^2 + (1)^2]^{3/2}}\right\}$$

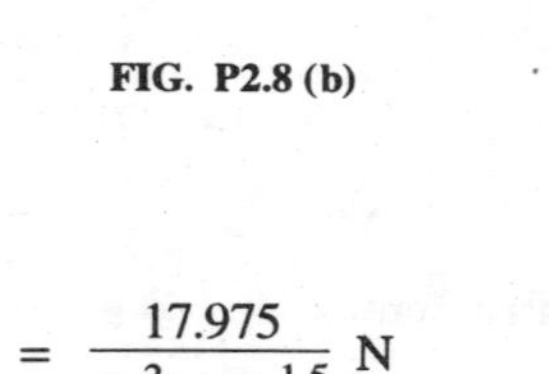

FIG. P2.8 (b)

$$= \frac{10^{-9}[a\,\bar{a}_x - \bar{a}_y - a\,\bar{a}_x - \bar{a}_y]}{4\pi\varepsilon_0(a^2 + 1)^{1.5}}$$

$$= \frac{10^{-9}(-2\,\bar{a}_y)}{4\pi\varepsilon_0(a^2 + 1)^{1.5}}$$

Magnitude of the force on $Q_0 = \dfrac{10^{-9}(2)}{4\pi\varepsilon_0(a^2 + 1)^{1.5}} = \dfrac{17.975}{(a^2 + 1)^{1.5}}$ N

Plot this as a function of a, for $0 \leq a \leq 5$ m. Plot is not shown.

Problem 2.9 : A regular tetrahedron has vertices at P_1 1 (2, 0, 0), P_2 (–1, $\sqrt{3}$, 0), P_3 (– 1, – $\sqrt{3}$, 0), and P_4 (0, 0, 2 $\sqrt{2}$). Charges of 1 mC are located at each of the four vertices. If the configuration is located in free space : (a)

find the magnitude of the force on each charge, and (b) the angle between any two edges of the tetrahedron.

Soluiton :

(a) Force on charge at P_1 = Force on charge at P_1 due to charge at P_2 + Force on charge at P_1 due to charge at P_3 + Force on charge at P_1 due to charge at P_4. Hence $\bar{F}$ =

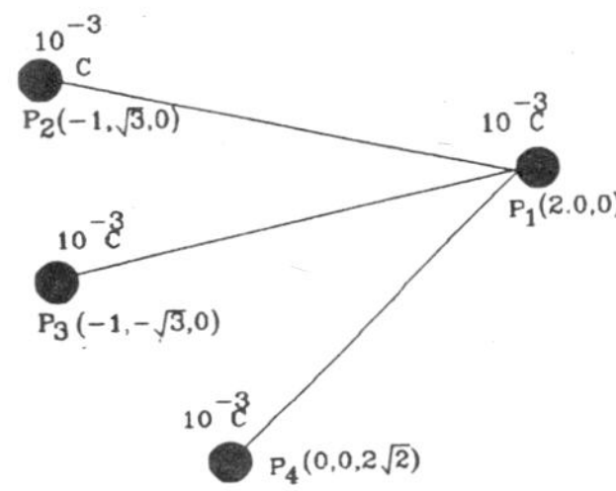

FIG. P2.9

$$\frac{(10^{-3})(10^{-3})}{4\pi\varepsilon_0}\left\{\frac{(2+1)\bar{a}_x+(0-\sqrt{3})\bar{a}_y}{[(3)^2+(-\sqrt{3})^2]^{3/2}}\right\}+$$

$$\frac{(10^{-3})(10^{-3})}{4\pi\varepsilon_0}\left\{\frac{(2+1)\bar{a}_x+(0+\sqrt{3})\bar{a}_y}{[(3)^2+(\sqrt{3})^2]^{3/2}}\right\}+$$

$$\frac{(10^{-3})(10^{-3})}{4\pi\varepsilon_0}\left\{\frac{(2+0)\bar{a}_x+(0-2\sqrt{2})\bar{a}_z}{[(2)^2+(-2\sqrt{2})^2]^{3/2}}\right\}$$

$$= \frac{10^{-6}}{4\pi\varepsilon_0(12)^{1.5}}\left[8\bar{a}_x - 2\sqrt{2}\bar{a}_z\right]$$

$$|\bar{F}| = \frac{10^{-6}}{4\pi\varepsilon_0(12)^{1.5}} \times \sqrt{(8)^2+(-2\sqrt{2})^2} = 1834.615 \text{ N}$$

The magnitude of the force on each charge remains the same as all the charges are equal (1 mC) and distance between any two charges is also equal ($\sqrt{12}$).

(b) $\bar{R}_{P_1P_2} = \bar{r}_{P_2} - \bar{r}_{P_1} = (-\bar{a}_x + \sqrt{3}\bar{a}_y) - (2\bar{a}_x) = -3\bar{a}_x + \sqrt{3}\bar{a}_y$

$\bar{R}_{P_1P_3} = \bar{r}_{P_3} - \bar{r}_{P_1} = (-\bar{a}_x - \sqrt{3}\bar{a}_y) - (2\bar{a}_x) = -3\bar{a}_x - \sqrt{3}\bar{a}_y$

$$\therefore \cos\theta = \frac{\bar{R}_{P_1P_2}\cdot\bar{R}_{P_1P_3}}{|\bar{R}_{P_1P_2}|\,|\bar{R}_{P_1P_3}|} = \frac{(-3)(-3)+(\sqrt{3})(-\sqrt{3})}{\sqrt{(-3)^2+(\sqrt{3})^2}\sqrt{(-3)^2+(-\sqrt{3})^2}}$$

$$= 0.5 \quad \therefore \theta = 60^\circ$$

Problem 2.10 : A point charge, Q_1 = 10 μC, is located at P_1 (1, 2, 3) in free space, while Q_2 = − 5 μC is at P_2 (1, 2, 10). (a) Find the vector force exerted on Q_2 by Q_1. (b) Find the coordinates of P_3 at which a point charge Q_3 experiences no force.

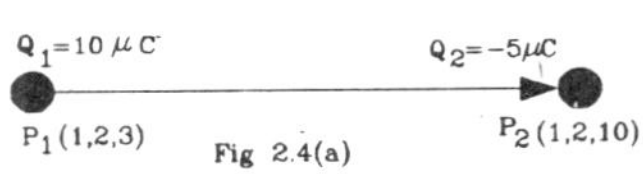

FIG. P2.10 (a)

Solution :

(a)Force on Q_2 due to Q_1 =

$$= \frac{(10\times 10^{-6})(-5\times 10^{-6})}{4\pi \varepsilon_0 (7)^2}\left[\frac{(1-1)\bar{a}_x+(2-2)\bar{a}_y+(10-3)\bar{a}_z}{\sqrt{(7)^2}=7}\right]$$

$$= -1.834\ \bar{a}_z \text{ mN}$$

(b) Let the coordinates of point P_3 are (x, y, z).

Force on Q_3 = Force on Q_3 due to charge Q_1 + Force on Q_3 due to charge Q_2

$$\bar{F} = \frac{(10\times 10^{-6})\,Q_3[(x-1)\bar{a}_x+(y-2)\bar{a}_y+(z-3)\bar{a}_z]}{4\pi\varepsilon_0[(x-1)^2+(y-2)^2+(z-3)^2]^{3/2}}$$

$$+ \frac{(-5\times 10^{-6})\,Q_3[(x-1)\bar{a}_x+(y-2)\bar{a}_y+(z-3)\bar{a}_z]}{4\pi\varepsilon_0[(x-1)^2+(y-2)^2+(z-3)^2]^{3/2}}$$

Since two forces are acting on charge Q_3, these forces must be equal and of opposite so that total force on Q_3 is zero. According to coulomb's law, the force always acts along the line joining the two charges. Hence we conclude that charge Q_3 must lie on the line joining the charge Q_1 and Q_2. Equations of the line joining Q_1 and Q_2 is given by

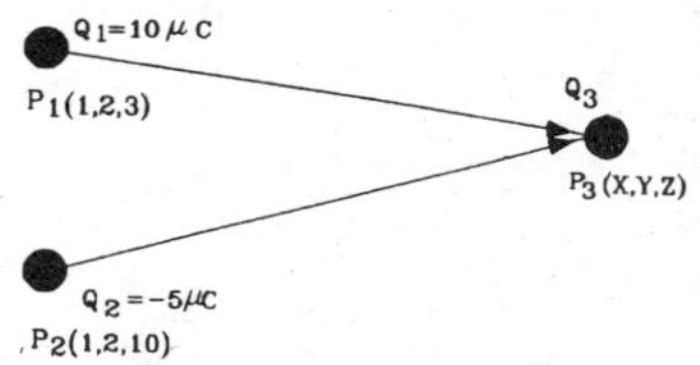

FIG.P 2.10 (b)

$$\frac{x-1}{1-1}=\frac{y-2}{2-2}=\frac{z-3}{10-3} \text{ or } \frac{x-1}{0}=\frac{y-2}{0}=\frac{z-3}{7}=K \text{ (constant)}$$

Hence we have

$$\frac{x-1}{0}=K \text{ or } x=0\ ;\ \frac{y-2}{0}=K \text{ or } y=0\ ;\ \frac{z-3}{7}=K \text{ or } z=7K+3$$

Substitute $x=1$, $y=2$, and $z=7K+3$ in the expression for $\bar{F}$, we find

$$\bar{F} = \frac{5\times 10^{-6}\,Q_3}{4\pi\varepsilon_0}\left\{\frac{14K}{[(7K)^2]^{1.5}} - \frac{7K-7}{[(7K-7)^2]^{1.5}}\right\}\bar{a}_z$$

Since $\bar{F}=0$, we have the term inside the bracket must be equal to zero.

$$\therefore \frac{14K}{[(7K)^2]^{1.5}} = \frac{7K-7}{[(7K-7)^2]^{1.5}} \qquad \text{--------(1)}$$

$$\therefore \frac{14K}{(7K)^3}=\frac{7K-7}{(7K-7)^3} \text{ or } \frac{2}{49K^2}=\frac{1}{(7K-7)^2}=\frac{1}{49(K-1)^2}$$

or $2(K-1)^2=K^2$; $K^2-4K+2=0$; $K = 3.414$ or 0.586

We must select that value of K which satisfy equation (1) .It is found that K = 3.414 satisfy equation (1). Hence $z = 7K + 3 = 26.898$.HenceQ_3 must be located at P_3 (1, 2, 26.898).

Problem 2.11 : In free space, let Q_1 = 10 nC be at P_1 (0, – 4, 0), andQ_2 = 20 nC be at P_2 (0, 0, 4). (a) Find $\bar{E}$ at the origin. (b) Where should a 30 – nC point charge be located so that $\bar{E}$ = 0 at the origin ?

Solution :

(a) $\bar{E}$ at Origin = $\bar{E}$ at origin due to charge Q_1 + $\bar{E}$ at origindue to charge Q_2

$$= \frac{10\times 10^{-9}}{4\pi\varepsilon_0 (4)^2}\left[\frac{(0+4)\,\bar{a}_y}{\sqrt{(4)^2}=4}\right]$$

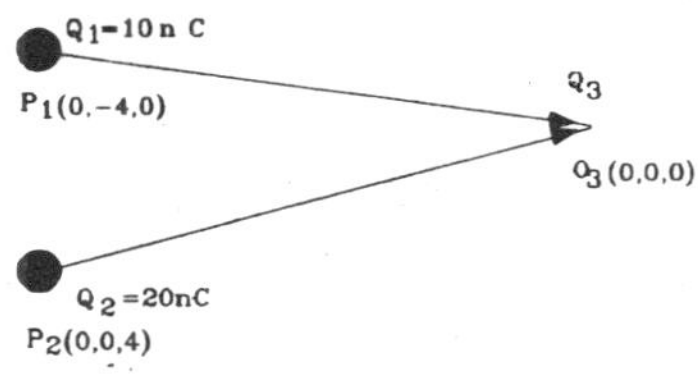

$$+ \frac{20\times 10^{-9}}{4\pi\varepsilon_0 (4)^2}\left[\frac{(0-4)\,\bar{a}_z}{\sqrt{(-4)^2}=4}\right]$$

$$= 5.617\ \bar{a}_y - 11.235\ \bar{a}_z\ \ \text{V/m}$$

FIG.P2.11 (a)

(b) Let the charge of 30 nC is located at (x, y, z) .

Then $\bar{E}$ at origin = $\bar{E}$ at origin due to charge Q_1 + $\bar{E}$ at origin due to charge Q_2 + $\bar{E}$ at origin due to charge Q_3

$$= 5.617\,\bar{a}_y - 11.235\,\bar{a}_z + \frac{30\times 10^{-9}\,[\,(0-x)\,\bar{a}_x + (0-y)\,\bar{a}_y + (0-z)\,\bar{a}_z\,]}{4\pi\varepsilon_0\,[\,x^2+y^2+z^2\,]^{3/2}}$$

$$= \frac{-269.632\,x\,\bar{a}_x}{(x^2+y^2+z^2)^{1.5}} + \left[5.617 - \frac{269.632\,y}{(x^2+y^2+z^2)^{1.5}}\right]\bar{a}_y - \left[11.235 + \frac{269.632\,z}{(x^2+y^2+z^2)^{1.5}}\right]\bar{a}_z$$

Since $\bar{E}$ is equal to zero at origin, the coefficients of unit vectors must be equal to zero. Equating coefficient of $\bar{a}_x$ equal to zero we have

$$\frac{-269.632\,x}{(x^2+y^2+z^2)^{1.5}} = 0 \text{ or } x = 0$$

Equating coefficient of $\bar{a}_y$ equal to zero

$$5.617 - \frac{269.632\,y}{(x^2+y^2+z^2)^{1.5}} = 0$$

$$\text{or } \frac{y}{(x^2+y^2+z^2)^{1.5}} = 0.0208\ldots\ldots(1)$$

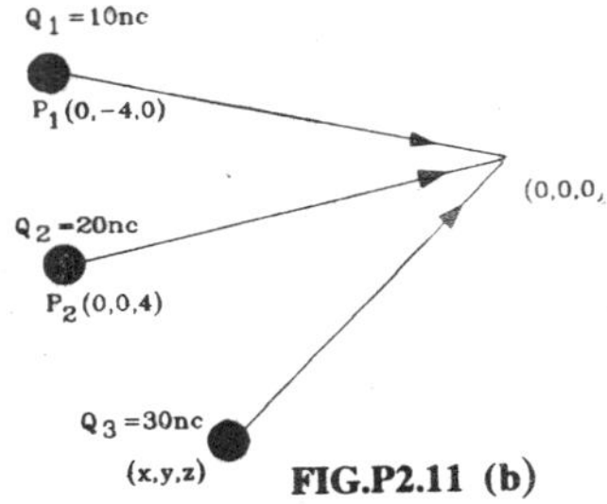

FIG.P2.11 (b)

Equating coefficient of $\bar{a}_z$ equal to zero, we have

$$-11.235 - \frac{269.632\,z}{(x^2+y^2+z^2)^{1.5}} = 0 \text{ or } \frac{z}{(x^2+y^2+z^2)^{1.5}} = -0.04166 \quad ....(2)$$

Since $x = 0$, equations (1) and (2) reduces to

$$\frac{y}{(y^2+z^2)^{1.5}} = 0.0208 \quad ....(3) \quad \text{and} \quad \frac{z}{(y^2+z^2)^{1.5}} = -0.04166 \quad ......(4)$$

Dividing equation (4) by equation (3), we have

$$\frac{z}{y} = -\frac{0.04166}{0.0208} = -2.00288 \text{ or } z = -2.00288\,y$$

Substitute this value of z in equation (3) we have

$$\frac{y}{(5.011528\,y^2)^{1.5}} = 0.0208 \text{ or } y^2 = 4.2853 \text{ or } y = \pm 2.07$$

Select that value of y which satisfy equation (1). Naturally y must be positive. Hence $y = 2.07$. Then $z = -2.00288\,(2.07) = -4.146$

Hence 30 nC point charge should be located at (0,2.07,– 4.146)

Problem 1.12 : A 5 - nC point charge is located at A (2, –1, –3) in free space. (a) Find $\bar{E}$ at the origin. (b) Plot $|\bar{E}(x, 0, 0)|$ vs. x, $-10 < x < 10$ m. (c) What is $|\bar{E}(x, 0, 0)|_{max}$?

Solution :

(a) $\bar{E}$ at origin $= \dfrac{5\times 10^{-9}}{4\pi\varepsilon_0 (3.742)^2}\left[\dfrac{(0-2)\,\bar{a}_x + (0+1)\,\bar{a}_y + (0+3)\,\bar{a}_z}{\sqrt{(-2)^2+(1)^2+(3)^2} = 3.742}\right]$

$$= -1.716\,\bar{a}_x + 0.858\,\bar{a}_y + 2.574\,\bar{a}_z \text{ V/m.}$$

FIG.P 2 .12

(b) $\bar{E}$ at $(x, 0, 0) = \dfrac{5\times 10^{-9}}{4\pi\varepsilon_0}\left[\dfrac{(x-2)\,\bar{a}_x + (0+1)\,\bar{a}_y + (0+3)\,\bar{a}_z}{[(x-2)^2+(1)^2+(3)^2]^{3/2}}\right]$

$$\therefore |\bar{E}| = \frac{5\times 10^{-9}}{4\pi\varepsilon_0 (x^2-4x+14)} = \frac{44.9387}{(x^2-4x+14)}$$

Plot this equation for $-10 \le x \le 10$. Plot is not given.

(c) $\dfrac{d\,|\,\overline{E}\,|}{dx} = \dfrac{-\,44.9387\,(2x-4)}{(x^2-4x+14)^2} = 0$

or $2x-4=0$ or $x=2$. Hence at $x=2$, $|\,\overline{E}\,|$ has a maximum value and

$$|\,\overline{E}\,|_{max} = \frac{44.9387}{(4-8+14)} = 4.494 \text{ V/m}$$

Problem 2.13 : A point charge, $Q_A = 1\ \mu C$, is at A (0, 0, 1), and $Q_B = -\ 1\ \mu C$ is at B (0, 0, –1). Find E_r, E_θ, and E_ϕ at P (1, 2, 3).

Solution :

(a) $\overline{E}$ at P = $\overline{E}$ at P due to charge Q_A + $\overline{E}$ at P due to charge Q_B

$$\overline{E} = \frac{10^{-6}}{4\pi\varepsilon_0}\left\{\frac{(1-0)\,\overline{a}_x + (2-0)\,\overline{a}_y + (3-1)\,\overline{a}_z}{[(1)^2+(2)^2+(2)^2]^{3/2}}\right\}$$

$$-\ \frac{10^{-6}}{4\pi\varepsilon_0}\left\{\frac{(1-0)\,\overline{a}_x + (2-0)\,\overline{a}_y + (3+1)\,\overline{a}_z}{[(1)^2+(2)^2+(4)^2]^{3/2}}\right\}$$

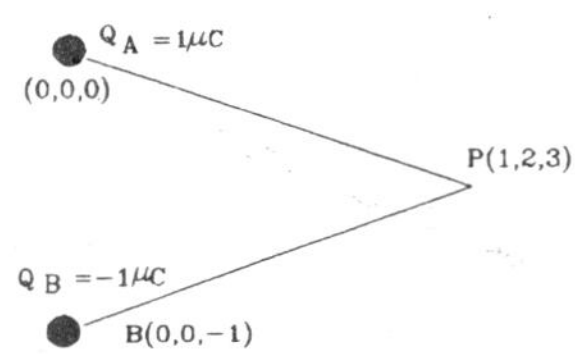

FIG. P2 . 13

$\overline{E} = 239.484\,\overline{a}_x + 478.97\,\overline{a}_y + 292.181\,\overline{a}_z$

Let $\overline{E} = E_r\,\overline{a}_r + E_\theta\,\overline{a}_\theta + E_\phi\,\overline{a}_\phi$

$$E_r = \overline{E}\cdot\overline{a}_r = 239.484\,\overline{a}_x\cdot\overline{a}_r + 478.97\,\overline{a}_y\cdot\overline{a}_r + 292.181\,\overline{a}_z\cdot\overline{a}_r$$

$$= 239.484\sin\theta\cos\phi + 478.97\sin\theta\sin\phi + 292.181\cos\theta$$

$$E_\theta = \overline{E}\cdot\overline{a}_\theta = 239.484\,\overline{a}_x\cdot\overline{a}_\theta + 478.97\,\overline{a}_y\cdot\overline{a}_\theta + 292.181\,\overline{a}_z\cdot\overline{a}_\theta$$

$$= 239.484\cos\theta\cos\phi + 478.97\cos\theta\sin\phi - 292.181\sin\theta$$

$$E_\phi = \overline{E}\cdot\overline{a}_\phi = 239.484\,\overline{a}_x\cdot\overline{a}_\phi + 478.97\,\overline{a}_y\cdot\overline{a}_\phi + 292.181\,\overline{a}_z\cdot\overline{a}_\phi$$

$$= -\,239.484\sin\phi + 478.97\cos\phi$$

$$\therefore\ \overline{E} = (239.484\sin\theta\cos\phi + 478.97\sin\theta\sin\phi + 292.181\cos\theta)\,\overline{a}_r + (239.484\cos\theta\cos\phi + 478.97\cos\theta\sin\phi - 292.181\sin\theta)\,\overline{a}_\theta + (-\,239.484\sin\phi + 478.97\cos\phi)\,\overline{a}_\phi$$

For point P (1, 2, 3) $r = \sqrt{1^2+2^2+3^2}$

$\theta = \cos^{-1}(3/\sqrt{14}) = 36.7^\circ$, $\phi = \tan^{-1}(2/1) = 63.43^\circ$

Substituting the coordinates of point P in the expression for $\overline{E}$, we have

$\overline{E}$ at point $P = 554.294\,\overline{a}_r + 254.740\,\overline{a}_\theta + 0.217\,\overline{a}_\phi$

Problem 2.14 : A 20 – nC point charge is located at P (2, 4, –3) in free space. (a) Find $\overline{E}(\,\overline{r}\,)$. (b) Find $\overline{E}$ at A (–3, 2, 0). (c) Find the locus of all points at

which $E_x = 1$ V/m .

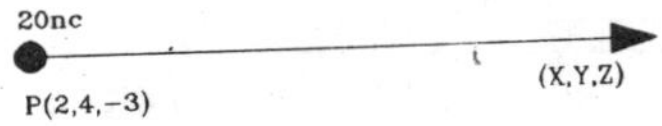

FIG.P2.14

Solution :

(a) $\overline{E}(\overline{r}) = \overline{E}(x\,\overline{a}_x + y\,\overline{a}_y + z\,\overline{a}_z)$

$$= \frac{20\times 10^{-9}[(x-2)\,\overline{a}_x + (y-4)\,\overline{a}_y + (z+3)\,\overline{a}_z]}{4\pi\varepsilon_0[(x-2)^2 + (y-4)^2 + (z+3)^2]^{3/2}}$$

$$= \frac{179.755\,[(x-2)\,\overline{a}_x + (y-4)\,\overline{a}_y + (z+3)\,\overline{a}_z]}{[(x-2)^2 + (y-4)^2 + (z+3)^2]^{1.5}}$$

(b) $\overline{E}$ at $P(-3,2,0) = \dfrac{179.755\,[(-3-2)\,\overline{a}_x + (2-4)\,\overline{a}_y + (0+3)\,\overline{a}_z]}{[(-3-2)^2 + (2-4)^2 + (0+3)^2]^{1.5}}$

$$= -3.837\,\overline{a}_x - 1.535\,\overline{a}_y + 2.302\,\overline{a}_z \text{ V/m.}$$

(c) $E_x = \dfrac{179.755\,(x-2)}{[(x-2)^2 + (y-4)^2 + (z+3)^2]^{1.5}} = 1$

or $[(x-2)^2 + (y-4)^2 + (z+3)^2]^{1.5} = 179.755\,(x-2)$

This is the equaiton of locus of all points at which $E_x = 1$ V/m.

Problem 2.15 : Eight 25 - nC point charges in free space are located symmetrically on a circle of radius 0.2 m centered at the origin in the $z = 0$. plane. (a) At what point on the z axis is $|\overline{E}|$ a maximum ? (b) What is $|\overline{E}|_{max}$?

Solution :

(a)Since eight charges are located symmetrically on a circle of radius 0.2 m in $z = 0$ plane, each charge must be located at an interval of $360^o/8 = 45^o$. The points at which these charges are located are shown in Fig.P2.15. We have to determine $\overline{E}$ at $(0,0,z)$. Let us consider two charge located at points A and E only. Cartesian coordinates of points A and E are (0.2,0,0) and

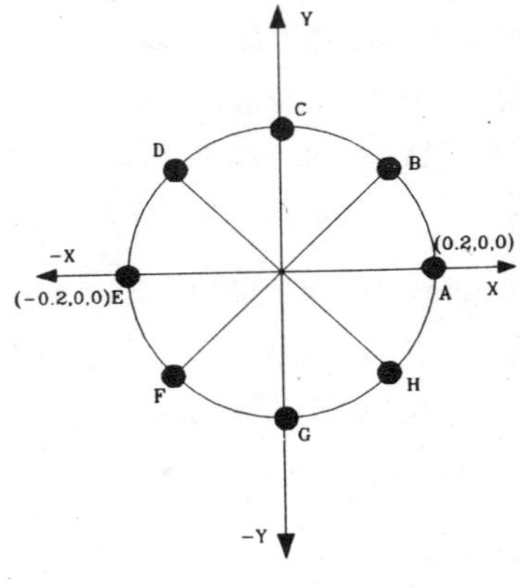

FIG. P 2.15

(-0.2, 0, 0) respectively. Due to these two charges, $\overline{E}$ at (0,0,z)

$$= \frac{25\times 10^{-9}}{4\pi\varepsilon_0}\left\{\frac{(0-0.2)\,\bar{a}_x+(z-0)\,\bar{a}_z}{[(-0.2)^2+z^2]^{3/2}}+\frac{(0+0.2)\,\bar{a}_x+(z-0)\,\bar{a}_z}{[(0.2)^2+z^2]^{3/2}}\right\}$$

$$= \frac{25\times 10^{-9}}{4\pi\varepsilon_0(0.04+z^2)^{1.5}}\times(2z)\,\bar{a}_z \qquad \text{-------(1)}$$

Hence only z component is present in $\overline{E}$ due to a pair of symmetrically located charges at points A and E. In all there are such 4 pairs of symmetrically located charges ($A-E, B-F, C-G, D-H$). Hence total electric field will be 4 times of that electric field given by equation (1).

$$\therefore\ \overline{E} = 4\times\frac{25\times 10^{-9}}{4\pi\varepsilon_0(0.04+z^2)^{1.5}}\times(2z)\,\bar{a}_z = \frac{1797.548\,z}{(0.04+z^2)^{1.5}}\,\bar{a}_z$$

$$|\overline{E}| = \frac{1797.548\,z}{(0.04+z^2)^{1.5}} \qquad \text{----------(2)}$$

$$\therefore\ \frac{d|\overline{E}|}{dz} = 1797.548\left\{\frac{(0.04+z^2)^{1.5}-z\times 1.5\,(0.04+z^2)^{0.5}\,(2z)}{(0.04+z^2)^3}\right\} = 0$$

$$\therefore\ (0.04+z^2)-3z^2 = 0 \text{ or } z = \pm\sqrt{0.02} = \pm\ 0.1414$$

Hence at points (0,0,$\pm$ 0.1414), $|\overline{E}|$ is maximum.

(b)From equation (2) we have

$$|\overline{E}|_{max} = \left|\frac{1797.548\ (\pm\ 0.1414)}{(0.04+0.02)^{1.5}}\right| = 17.2943\ \text{KV/m}$$

Problem 2.16 : Identical 1 – μC point charges are located in free space at (0, 0, 1) and (0, 0, -1). Prepare a plot showing $|\overline{E}|$ vs. z along the line $x = 0, y = 2$, for $|z| \le 4$.

Solution :

Any point on the line $x = 0, y = 2$ is given by (0, 2,z). Hence

$\overline{E}$ = $\overline{E}$ at (0,2,z) due to 1 μC at (0,0,1) + $\overline{E}$ at (0,2,z) due to 1 μC at (0,0,–1)

$$= \frac{10^{-6}}{4\pi\varepsilon_0}\left\{\frac{(2-0)\,\bar{a}_y+(z-1)\,\bar{a}_z}{[(2)^2+(z-1)^2]^{3/2}}\right\}$$

$$+\ \frac{10^{-6}}{4\pi\varepsilon_0}\left\{\frac{(2-0)\,\bar{a}_y+(z+1)\,\bar{a}_z}{[(2)^2+(z+1)^2]^{3/2}}\right\}$$

$$= E_x\,\bar{a}_x + E_y\,\bar{a}_y\ .$$

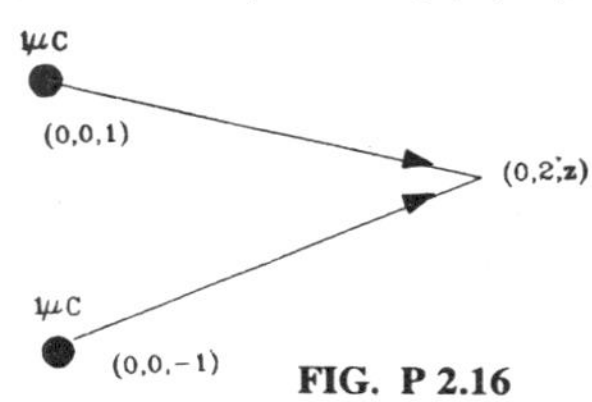

FIG. P 2.16

Hence $|\bar{E}| = \sqrt{E_y^2 + E_z^2}$

where $E_y = \dfrac{10^{-6}}{4\pi\varepsilon_0}\left\{\dfrac{2}{[4+(z-1)^2]^{1.5}} + \dfrac{2}{[4+(z+1)^2]^{1.5}}\right\}$

$E_z = \dfrac{10^{-6}}{4\pi\varepsilon_0}\left\{\dfrac{z-1}{[4+(z-1)^2]^{1.5}} + \dfrac{z+1}{[4+(z+1)^2]^{1.5}}\right\}$

Pl ot $|\bar{E}|$ vs. z for $|z| \leq 4$ i.e. for $-4 \leq z \leq 4$.

Problem 2.17 : Eight point charges of 1 nC each are located at the corners of a cube in free space that is 1 m on a side. Find $|\bar{E}|$ at the center of : (a) the cube ; (b) a face ; (c) an edge.

Solution :

(a) Fig.P2.17(a) shows a cube 1m on a side. The coordinates of corners of the cube are also shown in figure. The center of the cube is that point through which all the bodydiagonals passess. (Four body diagonals are AG, DF, CE, BH). It is also the midpoint of any body diagonal. Let P be the center of the cube. Then using the coordinates of points A and G, we have,

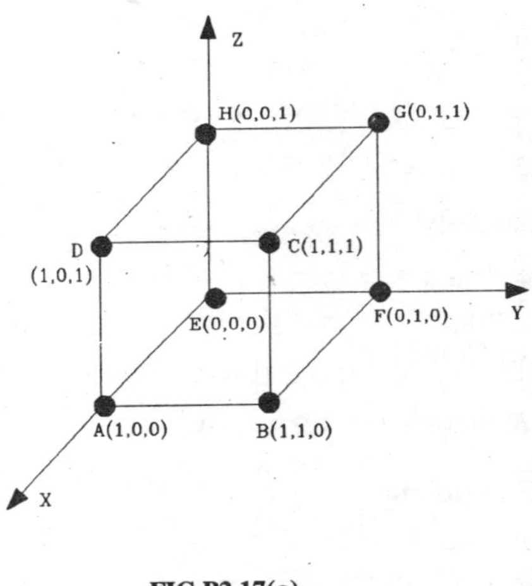

FIG.P2.17(a)

$$P \equiv \left(\frac{1+0}{2}, \frac{0+1}{2}, \frac{0+1}{2}\right) = (0.5, 0.5, 0.5)$$

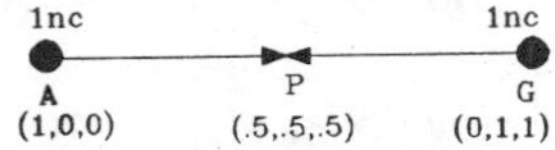

FIG. P2.17 (b)

Fig.P2.17(b) shows the directions of electric field intensity acting at point P due to charges located at points A and G. Both the electric fields are acting in opposite directions.Hence $\bar{E}$ at point P due to charges at points A and G is equal to zero.Similarly $\bar{E}$ at point P due to pair charges at points D and F, C and E, B and H are equal to zero. Hence the total electric field intensity at point P, i.e., at the center of the cube is equal to zero.

(b) Consider front face of the cube with corners $ABCD$. The center of the face is the point at which face diagonals AC and BD intersect. It is also the midpoint of

AC or BD. Let point Q is the center of front face. Then using the coordinates of points A and C we have

$$Q = \left(\frac{1+1}{2}, \frac{0+1}{2}, \frac{0+1}{2}\right) = (1, 0.5, 0.5)$$

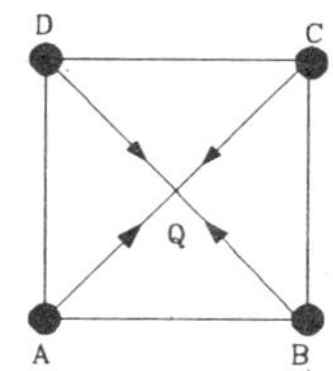

FIG.P2.17(C)

From the Fig.P2.17(c), it is seen that electric field intensity at point Q due to charges located at points $A, B, C,$ and D gets cancelled out. Hence $\bar{E} = 0$ due to charges at $A, B, C,$ and D. Now consider charges located at H and F.

$\bar{E} = \bar{E}$ at point Q due to charge at H + $\bar{E}$ at point Q due to charge at F

$$= \frac{10^{-9}}{4\pi\varepsilon_0}\left\{\frac{(1-0)\,\bar{a}_x + (0.5-0)\,\bar{a}_y + (0.5-1)\,\bar{a}_z}{[(1)^2 + (0.5)^2 + (-0.5)^2]^{3/2}}\right\} +$$

$$\frac{10^{-9}}{4\pi\varepsilon_0}\left\{\frac{(1-0)\,\bar{a}_x + (0.5-1)\,\bar{a}_y + (0.5-0)\,\bar{a}_z}{[(1)^2 + (-0.5)^2 + (0.5)^2]^{3/2}}\right\}$$

$$= 9.785\ \bar{a}_x\ \text{V/m}$$

Similarly $\bar{E}$ at point Q due to charges located at E and G = $9.785\ \bar{a}_x$ V/m

$\therefore\ \bar{E}_{total} = 9.785\ \bar{a}_x + 9.785\ \bar{a}_x = 19.569\ \bar{a}_x \quad \therefore\ |\bar{E}| = 19.569$ V/m

(c) Consider an edge such as AB. Let R be the center of this edge. Then

$$R = \left(\frac{1+1}{2}, \frac{0+1}{2}, 0\right) = (1, 0.5, 0)$$

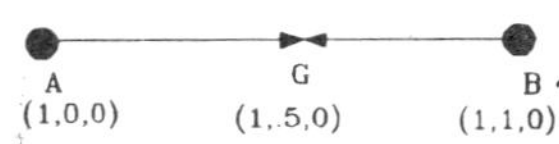

FIG.P2.17(D)

$\bar{E}$ at point R due to charges located at points A and B is zero as shown in Fig.P2.17(d).

$\bar{E} = \bar{E}$ at point R due to charges located at points $D, C, E, F, H,$ and G.

$$= \frac{1\times 10^{-9}}{4\pi\varepsilon_0}\left\{\frac{(1-1)\,\bar{a}_x + (0.5-0)\,\bar{a}_y + (0-1)\,\bar{a}_z}{[(0.5)^2 + (-1)^2]^{3/2}} + \right.$$

$$\frac{(1-1)\,\bar{a}_x + (0.5-1)\,\bar{a}_y + (0-1)\,\bar{a}_z}{[(-0.5)^2 + (-1)^2]^{3/2}} + \frac{(1-0)\,\bar{a}_x + (0.5-0)\,\bar{a}_y}{[(1)^2 + (0.5)^2]^{3/2}} +$$

$$\frac{(1-0)\,\bar{a}_x + (0.5-1)\,\bar{a}_y}{[(1)^2 + (-0.5)^2]^{3/2}} + \frac{(1-0)\,\bar{a}_x + (0.5-0)\,\bar{a}_y + (0-1)\,\bar{a}_z}{[(1)^2 + (0.5)^2 + (-1)^2]^{3/2}} +$$

$$\left.\frac{(1-0)\,\bar{a}_x + (0.5-1)\,\bar{a}_y + (0-1)\,\bar{a}_z}{[(1)^2 + (-0.5)^2 + (-1)^2]^{3/2}}\right\}$$

$$\bar{E} = 18.188\ \bar{a}_x - 18.188\ \bar{a}_z$$

$\therefore \ |\overline{E}| = \sqrt{(18.188)^2 + (-18.188)^2} = 25.722$ V/m

Problem 2.18 : A point charge Q_1, located at the origin in free space, produces a field $\overline{E} = 10\,\overline{a}_r$ V/m for $r = 1$ m. (a) Find Q_1. (b) Using this value of Q_1, find $\overline{E}$ at P (2, -1, 5) in cartesian components. (c) Using the same value of Q_1, find $\overline{E}$ at C (2, 2, 2) in cylindrical components.

Solution :

(a) $\overline{E}$ at (r, θ, ϕ) due to charge Q_1 at origin $= \dfrac{Q_1}{4\pi\varepsilon_0 r^2}\,\overline{a}_r$ V/m

At $r = 1$m, $\overline{E} = 10\,\overline{a}_r = \dfrac{Q_1}{4\pi\varepsilon_0 (1)^2}\,\overline{a}_r \qquad \therefore Q_1 = 40\,\pi\,\varepsilon_0$ C

$$(b)\ \overline{E} \text{ at } P = \frac{40\pi\varepsilon_0}{4\pi\varepsilon_0}\left\{\frac{(2-0)\,\overline{a}_x + (-1-0)\,\overline{a}_y + (5-0)\,\overline{a}_z}{[(2)^2 + (-1)^2 + (5)^2]^{3/2}}\right\}$$

$$= 0.122\,\overline{a}_x - 0.061\,\overline{a}_y + 0.304\,\overline{a}_z \text{ V/m}$$

$$(c)\ \overline{E} \text{ at } C = \frac{40\pi\varepsilon_0}{4\pi\varepsilon_0}\left\{\frac{(2-0)\,\overline{a}_x + (2-0)\,\overline{a}_y + (2-0)\,\overline{a}_z}{[(2)^2 + (2)^2 + (2)^2]^{3/2}}\right\}$$

$$= 0.481\,\overline{a}_x + 0.481\,\overline{a}_y + 0.481\,\overline{a}_z \text{ V/m}$$

Let $\overline{E} = E_\rho\,\overline{a}_\rho + E_\phi\,\overline{a}_\phi + E_z\,\overline{a}_z$

$E_\rho = \overline{E}\cdot\overline{a}_\rho = 0.481\,\overline{a}_x\cdot\overline{a}_\rho + 0.481\,\overline{a}_y\cdot\overline{a}_\rho + 0.481\,\overline{a}_z\cdot\overline{a}_\rho$

$= 0.481\cos\phi + 0.481\sin\phi$

$E_\phi = \overline{E}\cdot\overline{a}_\phi = 0.481\,\overline{a}_x\cdot\overline{a}_\phi + 0.481\,\overline{a}_y\cdot\overline{a}_\phi + 0.481\,\overline{a}_z\cdot\overline{a}_\phi$

$= -0.481\sin\phi + 0.481\cos\phi$

$E_z = \overline{E}\cdot\overline{a}_z = 0.481\,\overline{a}_x\cdot\overline{a}_z + 0.481\,\overline{a}_y\cdot\overline{a}_z + 0.481\,\overline{a}_z\cdot\overline{a}_z = 0.481$

$\therefore \overline{E} = (0.481\cos\phi + 0.481\sin\phi)\,\overline{a}_\rho + (-0.481\sin\phi + 0.481\cos\phi)\,\overline{a}_\phi$

$+ 0.481\,\overline{a}_z$

For point C (2,2,2), $\phi = \tan^{-1}(2/2) = 45^\circ$

Substituting this value of ϕ in the above expression for $\overline{E}$, we have

$\overline{E}$ at point $C = 0.680\,\overline{a}_\rho + 0.481\,\overline{a}_z$ V/m

Problem 2.19 : For $x, y,$ and z positive, let $\rho_v = 40\,xyz$ C/m^3. Find the total charge within the region defined by : (a) $0 \le x, y, z \le 2$; (b) $x = 0$, $y = 0$, $0 \le 2x + 3y \le 10$, $0 \le z \le 2$.

Solution :

$$(a)\ Q = \int_0^2 \int_0^2 \int_0^2 40\,x\,y\,z\ dx\ dy\ dz = 40\left[\frac{x^2}{2}\right]_0^2 \left[\frac{y^2}{2}\right]_0^2 \left[\frac{z^2}{2}\right]_0^2$$

$$= 40\times 2\times 2\times 4 = 320\ \text{C}$$

$$(b) Q = \int_0^5 \int_{y=0}^{\frac{10-2x}{3}} \int_0^2 40\,x\,y\,z\,dx\,dy\,dz = \int_0^5 40\,x\,dx \left[\frac{y^2}{2}\right]_0^{\frac{10-2x}{3}} \left[\frac{z^2}{2}\right]_0^2$$

$$= \int_0^5 10\,x\,dx\left(\frac{10-2x}{3}\right)^2 (4) = \frac{40}{9}\int_0^5 x\,(100-40x+4x^2)\,dx$$

$$= \frac{40}{9}\int_0^5 (100x-40x^2+4x^3)\,dx = \frac{40}{9}\left[\frac{100x^2}{2} - \frac{40x^3}{3} + \frac{4x^4}{4}\right]_0^5$$

$$= \frac{40}{9}\left[50\times 25 - \frac{40}{3}\times 125 + 625\right] = 925.926\ \text{C}$$

Problem 2.20 : Volume charge density is given by $\rho_V = 10\,e^{-1000\rho}\,e^{-100z}\ \text{C/m}^3$. (a) Find the maximum value of ρ_V in the region $0\le \rho\le 0.01$ m, $0\le \phi \le 2\pi$, $0\le z\le 0.01$ m. (b) Find the total charge contained in the first octant, where $x, y,$ and z are positive. (c) Find b if the total charge found in the volume $0\le \rho\le b$, $0\le \phi\le \pi/2$, $z\ge 0$, is half the value found in part (b) above.

Solution :

(a) ρ_V will be maximum if all the terms within ρ_V is maximum . Now $e^{-100\rho}$ will be maximum at $\rho = 0$ and e^{-100z} will be maximum at $z = 0$.

$$\therefore\ \rho_{V_{max.}} = 10\times 1\times 1 = 10\ \text{C/m}^3$$

$$(b)\ Q = \int_0^\infty \int_0^{\pi/2} \int_0^\infty 10\ e^{-1000\rho}\ e^{-100z}\ \rho\ d\rho\ d\phi\ dz$$

[∵ for the first octant ϕ varies from 0 to $\pi/2$.]

$$= 10\left[\frac{-\rho\,e^{-1000\rho}}{10^3} - \frac{e^{-1000\rho}}{10^6}\right]_0^\infty \left[\frac{-e^{-100z}}{100}\right]_0^\infty \left[\,\phi\,\right]_0^{\pi/2}$$

$$= 157.08\ \text{nC}$$

(c) $$\frac{157.08 \times 10^{-9}}{2} = \int_0^b \int_0^{\pi/2} \int_0^{\infty} 10\, e^{-1000\rho}\, e^{-100z}\, \rho\, d\rho\, d\phi\, dz$$

$$= 10 \left[\frac{-\rho\, e^{-1000\rho}}{10^3} - \frac{e^{-1000\rho}}{10^6} \right]_0^b \left[\frac{-e^{-100z}}{100} \right]_0^{\infty} \left[\phi \right]_0^{\pi/2}$$

$$\frac{157.08 \times 10^{-9}}{2} = 10 \left[\frac{-b\, e^{-1000b}}{10^3} - \frac{e^{-1000b}}{10^6} + \frac{1}{10^6} \right] (10^{-2})\, (\pi/2)$$

$\therefore\ e^{-1000b}\,(1000b + 1) = 0.5$ or $1000b + 1 = 0.5\, e^{1000b}$

By trial and error procedure, the value $b = 1.67835$ mm

Problem 2.21 : Let $\rho_v = (x + 2y + 3z)\ C/m^3$ in the cubical region $0 \le x, y, z \le 1$ mm , and $\rho_v = 0$ outside the cube. (a) What is the total charge contained within the cube ? (b) Set up the volume integral that will give $\overline{E}\,(x, 0, 0)$ for $x > 1$ mm. Do not integrate.

Solution :

(a) $$Q = \int_0^{10^{-3}} \int_0^{10^{-3}} \int_0^{10^{-3}} (x + 2y + 3z)\, dx\, dy\, dz$$

$$= \left[\frac{x^2}{2} \right]_0^{10^{-3}} \left[y \right]_0^{10^{-3}} \left[z \right]_0^{10^{-3}} + 2 \left[\frac{y^2}{2} \right]_0^{10^{-3}} \left[x \right]_0^{10^{-3}} \left[z \right]_0^{10^{-3}} +$$

$$3 \left[\frac{z^2}{2} \right]_0^{10^{-3}} \left[x \right]_0^{10^{-3}} \left[y \right]_0^{10^{-3}} = 3 \text{ pC}$$

(b) $$d\overline{E} = \frac{dQ}{4\pi \varepsilon_0 d^2}\, \overline{a}_{12}$$

$dQ = \rho_v\, dv = (x' + 2y' + 3z')\, dx'\, dy'\, dz'$

Coordinates of points 1 & 2 are (x', y', z') and $(x, 0, 0)$ respectively.

$\therefore\ \overline{R}_{12} = \overline{r}_2 - \overline{r}_1 = (x - x')\, \overline{a}_x - y'\, \overline{a}_y + -\, z'\, \overline{a}_z$

$$d\overline{E} = \frac{(x' + 2y' + 3z')\, dx'\, dy'\, dz'\, [(x - x')\, \overline{a}_x - y'\, \overline{a}_y - z'\, \overline{a}_z]}{4\pi \varepsilon_0\, [(x - x')^2 + (-y')^2 + (-z')^2]^{1.5}}$$

$$\overline{E} = \int_0^{10^{-3}} \int_0^{10^{-3}} \int_0^{10^{-3}} \frac{(x' + 2y' + 3z')\, dx'\, dy'\, dz'\, [(x - x')\, \overline{a}_x - y'\, \overline{a}_y - z'\, \overline{a}_z]}{4\pi \varepsilon_0 \left[(x - x')^2 + (y')^2 + (z')^2 \right]^{1.5}}$$

Note that x, y, and z from ρ_v are replaced by x', y', and z' respectively as these are the variables within the given volume where as x coordinate from point 2 is a constant.

Problem 2.22 : Volume charge density is given as $\rho_v = 10^{-5} e^{-100r} \sin\theta$ C/m^3 for $0 \le r \le 1$ cm , and $\rho_v = 0$ for $r > 1$ cm . Estimate $\bar{E}$ at $r = 1$ m, $\theta = 90^o$, $\phi = 0$, by thinking in terms of a point charge.

Solution : Total charge within the region $0 \le r \le 1$ cm is given by

$$Q = \int_0^{0.01}\int_0^{\pi}\int_0^{2\pi} 10^{-5} e^{-100r} \sin\theta \, r^2 \sin\theta \, dr \, d\theta \, d\phi$$

$$= 10^{-5}\left[\frac{-r^2 e^{-100r}}{100} - \frac{re^{-100r}}{5000} - \frac{e^{-100r}}{500000}\right]_0^{0.01}\left[\frac{\theta}{2} - \frac{\sin 2\theta}{4}\right]_0^{\pi}\left[\phi\right]_0^{2\pi}$$

$$= 10^{-5} \times 1.606 \times 10^{-7} \times 0.5 \times \pi \times 2\pi = 1.585 \times 10^{-11} \text{ C}$$

Now the region $0 \le r \le 1$ cm. is a spherical region centered at origin. So let us locate Q at (0,0,0) and determine $\bar{E}$ at ($r = 1$m, $\theta = 90^o$, $\phi = 0$) or $\bar{E}$ at ($x = 1, y = 0, z = 0$)

$$\therefore \; \bar{E} = \frac{1.585 \times 10^{-11}}{4\pi\varepsilon_0(\sqrt{1})^2}\left[\frac{(1-0)\,\bar{a}_x}{\sqrt{(1)^2}}\right] = 0.142 \; \bar{a}_x \text{ V/m}$$

Problem 23. : A uniform volume charge density of 10 $\mu C/m^3$ is present in the spherical shell $0.9 < r < 1$ m, and $\rho_v = 0$ elsewhere. (a) Find Q_{tot}, the total charge present. (b) This symmetrical charge distribution in free space produces an electric field for $r > 1$ m that is identical to the field that would be produced by a point charge Q_{tot} at the origin. Find $\bar{E}$ in spherical coordinates for $r > 1$ m.

Solution :

$$\text{(a)}Q_{\text{total}} = \int_{0.9}^{1.0}\int_0^{\pi}\int_0^{2\pi} 10 \times 10^{-6} \, r^2 \sin\theta \, dr d\theta \, d\phi$$

$$= 10^{-5}\left[\frac{r^3}{3}\right]_{0.9}^{1.0}\left[-\cos\theta\right]_0^{\pi}\left[\phi\right]_0^{2\pi} = 11.353 \; \mu\text{C}$$

(b) $\overline{E} = \dfrac{Q_{total}}{4\pi\varepsilon_0 r^2}\overline{a}_r = \dfrac{11.352\times 10^{-6}}{4\pi\varepsilon_0 r^2}\overline{a}_r = \dfrac{102025.450}{r^2}\overline{a}_r$ V/m for r > 1m..

Problem 2.24 : A line charge density of 24 nC / m is located in free space on the line $y = 1, z = 2$. (a) Find $\overline{E}$ at P (6, -1, 3). (b) What point charge Q_A should be located at Q (- 3, 4, 1) to cause E_y to be equal to zero at P ?

Solution : (a) $\overline{E} = \dfrac{\rho_L}{2\pi\varepsilon_0 d}\overline{a}_{12}$

Point 2 is that point at which $\overline{E}$ is desired = (6, –1, 3).

Point 1 is the foot of perpendicular dropped from point 2 on the line charge = (6, 1, 2)

$$\overline{R}_{12} = (-1-1)\,\overline{a}_y + (3-2)\,\overline{a}_z = -2\,\overline{a}_y + \overline{a}_z$$

$$d = |\overline{R}_{12}| = \sqrt{(-2)^2 + (1)^2} = \sqrt{5}$$

$$\overline{a}_{12} = \frac{\overline{R}_{12}}{|\overline{R}_{12}|} = \frac{-2\,\overline{a}_y + \overline{a}_z}{\sqrt{5}}$$

$$\overline{E} = \frac{24\times 10^{-9}}{2\pi\varepsilon_0\sqrt{5}}\left[\frac{-2\,\overline{a}_y + \overline{a}_z}{\sqrt{5}}\right] = -172.565\,\overline{a}_y + 86.282\,\overline{a}_z \text{ V/m}.$$

(b) $\overline{E}$ due to Q_A $= \dfrac{Q_A}{4\pi\varepsilon_0}\left\{\dfrac{(6+3)\,\overline{a}_x + (-1-4)\,\overline{a}_y + (3-1)\,\overline{a}_z}{[(6+3)^2 + (-1-4)^2 + (3-1)^2]^{3/2}}\right\}$

Hence $E_y\,(\text{total}) = -172.565 + \dfrac{Q_A(-5)}{4\pi\varepsilon_0(110)^{1.5}} = 0$

$\therefore\; Q_A = -4.430\;\mu C$

Problem 2.25 : A uniform line charge, $\rho_{L1} = 15$ nC/m, is at $y = -1, z = 0$,, while $\rho_{L2} = -15$ nC/m is symmetrically located at $y = 1$, $z = 0$. Find $\overline{E}$ as a function of z at $y = 0$..

Solution :

In $y = 0$ plane, any point is given by $(x, 0, z)$.

$$\overline{E}_1 = \frac{\rho_L}{2\pi\varepsilon_0 d}\overline{a}_{12}$$

Point 2 is that point at which $\overline{E}$ is desired = $(x, 0, z)$.

Point 1 is the foot of perpendicular dropped from point 2 on the line charge

$= (x, -1, 0)$.

$\therefore\ \overline{R}_{12} = (0+1)\,\bar{a}_y + (z-0)\,\bar{a}_z = \bar{a}_y + z\,\bar{a}_z$

$d = |\overline{R}_{12}| = \sqrt{1+z^2}$ and $\bar{a}_{12} = \dfrac{\bar{a}_y + z\,\bar{a}_z}{\sqrt{1+z^2}}$

$$\therefore\ \overline{E} = \frac{15\times 10^{-9}}{2\pi\varepsilon_o}\left[\frac{\bar{a}_y + z\,\bar{a}_z}{1+z^2}\right]$$

Similarly $\overline{E}_2 = \dfrac{-15\times 10^{-9}}{2\pi\varepsilon_o}\left[\dfrac{-\bar{a}_y + z\,\bar{a}_z}{1+z^2}\right]$

$$\overline{E} = \overline{E}_1 + \overline{E}_2 = \frac{15\times 10^{-9}\,[\bar{a}_y + z\,\bar{a}_z + \bar{a}_y - z\,\bar{a}_z]}{2\pi\varepsilon_o\,(1+z^2)} = \frac{539.2645}{1+z^2}\,\bar{a}_y \text{ V/m}$$

Problem 2.26 : A uniform line charge density of 20 nC / m lies on the z axis between $z = 1$ and $z = 1$ m. No other charge is present. Find $\overline{E}$ at : (a) the origin ; (b) P (4, 0, 0).

Solution : (a) $\overline{dE} = \dfrac{dQ}{4\pi\varepsilon_o d^2}\,\bar{a}_{12}$

As the line charge lies on the $z-$ axis, $\overline{dL} = dz\,\bar{a}_z$ or $dL = dz$.

$\therefore\ dQ = \rho_L\,dL = 20\times 10^{-9}\,dz$.

Point 2 is that point at which $\overline{E}$ is desired = (0, 0, 0)

Point 1 is the general point on the line charge i.e. $z-$ axis. = (0, 0, z)

$\therefore\ \overline{R}_{12} = (0-z)\,\bar{a}_z = -z\,\bar{a}_z$

$d = |\overline{R}_{12}| = \sqrt{(-z)^2} = z$; $\bar{a}_{12} = \dfrac{\overline{R}_{12}}{|\overline{R}_{12}|} = \dfrac{-z\,\bar{a}_z}{z} = -\bar{a}_z$

$$\therefore\ \overline{dE} = \frac{20\times 10^{-9}\,dz}{4\pi\varepsilon_o z^2}\,(-\bar{a}_z)$$

$$\overline{E} = \int \overline{dE} = \int_1^3 \frac{-20\times 10^{-9}\,dz}{4\pi\varepsilon_o z^2}\,\bar{a}_z = \frac{-20\times 10^{-9}}{4\pi\varepsilon_o}\,\bar{a}_z\left[-\frac{1}{z}\right]_1^3$$

$= -119.837\,\bar{a}_z$ V/m .

(b) Similarly $\overline{dE} = \dfrac{20\times 10^{-9}\,dz}{4\pi\varepsilon_o}\left\{\dfrac{(4-0)\,\bar{a}_x + (0-z)\,\bar{a}_z}{[(4-0)^2 + (0-z)^2]^{3/2}}\right\}$

As the line charge is not symmetrically located w.r.t. point 2, both the components may be present. Hence

$$\overline{E} = \int \overline{dE} = \frac{20\times 10^{-9}}{4\pi\varepsilon_o}\left\{\int_1^3 \frac{4\,dz\,\overline{a}_x}{[16+z^2]^{3/2}} - \int_1^3 \frac{z\,dz\,\overline{a}_z}{[16+z^2]^{3/2}}\right\}$$

$$= \frac{20\times 10^{-9}}{4\pi\varepsilon_o}\left\{\frac{4}{16}\left[\frac{z}{\sqrt{z^2+16}}\right]_1^3 \overline{a}_x + \left[\frac{1}{\sqrt{z^2+16}}\right]_1^3 \overline{a}_z\right\}$$

$$= 16.064\,\overline{a}_x - 8.988\,\overline{a}_z \text{ V/m}$$

Problem 2.27 : That portion of the z axis for which $-2 < z < 2$ m carries a nonuniform charge density of $z^2 + 1$ nC/m in free space. There are no other charges anywhere. Find $\overline{E}$ at : (a) P_A (1, 0, 0) ; (b) P_B (2, 0, 0).

Solution :

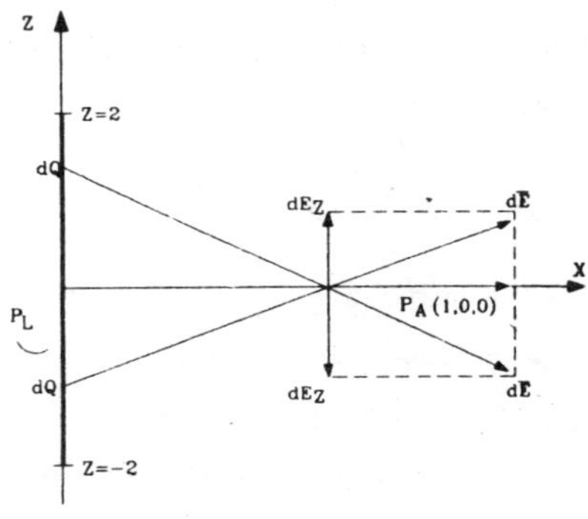

FIG.P2.27

(a) $\overline{dE} = \dfrac{dQ}{4\pi\varepsilon_o d^2}\,\overline{a}_{12}$

As the line charge lies on the z axis, $\overline{dL} = dz\,\overline{a}_z$ or $dL = dz$. Hence $dQ = \rho_L\,dL = (z^2+1)\,dz$ nC . Point 2 is that point at which $\overline{E}$ is desired = (1, 0, 0). Point 1 is the general point on the line charge i.e. z axis. (0, 0, z)

$$\therefore\ \overline{R}_{12} = (1-0)\,\overline{a}_x + (0-z)\,\overline{a}_z$$

$$= \overline{a}_x - z\,\overline{a}_z$$

$$d = |\overline{R}_{12}| = \sqrt{(1)^2+(-z)^2} = \sqrt{1+z^2}$$

$$\overline{dE} = \frac{(z^2+1)\,dz\,10^{-9}}{4\pi\varepsilon_o\left[\sqrt{1+z^2}\right]^2}\left[\frac{\overline{a}_x - z\,\overline{a}_z}{\sqrt{1+z^2}}\right]$$

As the line charge is symmetrically located w.r.t. point P_A , E_z component get cancelled and only E_x is present.(Refer to Fig.P2.27)

$$\overline{dE} = \frac{(1+z^2)\,dz\,10^{-9}}{4\pi\varepsilon_o(1+z^2)^{3/2}}\,\overline{a}_x = \frac{dz\,10^{-9}}{4\pi\varepsilon_o\sqrt{1+z^2}}\,\overline{a}_x$$

$$\overline{E} = \int \overline{dE} = \int_{-2}^{2} \frac{10^{-9}\, dz\, \overline{a}_x}{4\pi\varepsilon_o \sqrt{1+z^2}} = \frac{10^{-9}}{4\pi\varepsilon_o}\overline{a}_x\left[\ln\left(z+\sqrt{1+z^2}\right)\right]_{-2}^{2}$$

$= 25.95\,\overline{a}_x$ V/m .

(b) Similarly $\overline{E}$ at P_B (2, 0, 0) $= \displaystyle\int_{-2}^{2} \frac{2\times 10^{-9}(z^2+1)\, dz\, \overline{a}_x}{4\pi\varepsilon_o (4+z^2)^{3/2}}$

$$= \frac{10^{-9}\overline{a}_x}{2\pi\varepsilon_o}\left[z - \frac{3}{2}\tan^{-1}\left(\frac{z}{2}\right)\right]_{-2}^{2} = 29.548\,\overline{a}_x \text{ V/m}$$

Problem 2.28 : A sheet of charge, $\rho_S = 2$ nC/m^2 , is present at the plane $x = 3$ in free space, and a line charge, $\rho_L = 20$ nC/m , is located at $x = 1, z = .4$. (a) Find the magnitude of the electric field intensity at the origin. (b) Find the direction of $\overline{E}$ at P (4, 5, 6) . (c) What is the force per meter length on the line charge ?

Solution :

(a) $\overline{E}$ at (0, 0, 0) = $\overline{E}$ due ρ_S + $\overline{E}$ due ρ_L

$$= \frac{2\times 10^{-9}}{2\varepsilon_o}(-\overline{a}_x) + \frac{20\times 10^{-9}}{2\pi\varepsilon_o\sqrt{17}}\left[\frac{(0-1)\,\overline{a}_x + (0-4)\,\overline{a}_z}{\sqrt{(0-1)^2+(0-4)^2} = \sqrt{17}}\right]$$

$= -\ 112.943\,\overline{a}_x - 21.148\,\overline{a}_x - 84.591\,\overline{a}_z = -\ 134.091\,\overline{a}_x - 84.591\,\overline{a}_z$

$\therefore\ |\overline{E}| = \sqrt{(-134.091)^2 + (-84.591)^2} = 158.543$ V/m .

(b) $\overline{E}$ at P (4, 5, 6) = $\overline{E}$ due ρ_S + $\overline{E}$ due ρ_L

$$= \frac{2\times 10^{-9}}{2\varepsilon_o}(\overline{a}_x) + \frac{20\times 10^{-9}}{2\pi\varepsilon_o\sqrt{13}}\left[\frac{(4-1)\,\overline{a}_x + (6-4)\,\overline{a}_z}{\sqrt{(4-1)^2+(6-4)^2} = \sqrt{13}}\right]$$

$= 112.943\,\overline{a}_x + 82.964\,\overline{a}_x + 55.309\,\overline{a}_z = 195.907\,\overline{a}_x + 55.309\,\overline{a}_z$

$\therefore\ |\overline{E}| = \sqrt{(195.907)^2 + (55.309)^2} = 203.565$ V/m

$$\therefore\ \overline{a}_E = \frac{\overline{E}}{|\overline{E}|} = \frac{195.907\,\overline{a}_x + 55.309\,\overline{a}_z}{203.565} = 0.962\,\overline{a}_x + 0.272\,\overline{a}_z$$

(c) Any point on the line charge is (1, y, 4). At this point, $\overline{E}$ due to surface charge density is $\overline{E} = \dfrac{2\times 10^{-9}}{2\varepsilon_o}(-\overline{a}_x) = -\ 112.943\,\overline{a}_x$

As x and z coordinates are constant for the line charge, $\overline{dL} = dy\ \overline{a}_y$ or

$dL = dy$. Hence $dQ = \rho_L\, dL = 20 \times 10^{-9}\ dy$

$$\overline{dF} = dQ\,\overline{E} = 20 \times 10^{-9}\ dy\ (-\ 112.943\ \overline{a}_x)$$

$$= -\ 2.259\ dy\ \overline{a}_x\ \ \mu N$$

$$\therefore\ \frac{\overline{dF}}{dy} = -\ 2.259\,\overline{a}_x\ \ \mu N/m\,.$$

Problem 2.29 : Find $\overline{E}$ at P (1, 5, 2) in free space if a point charge of 6 μC is located at Q (0, 0, 1), a uniform line charge of 180 nC / m lies along the x axis, and a uniform sheet of charge equal to 25 nC / m^2 lies in the plane $z = -1$.

Solution :

$\overline{E} = \overline{E}$ due to point charge + $\overline{E}$ due to line charge + $\overline{E}$ du to surface charge

$$= \frac{6 \times 10^{-6}}{4\pi\varepsilon_0}\left[\frac{(1-0)\,\overline{a}_x + (5-0)\,\overline{a}_y + (2-1)\,\overline{a}_z}{[\,(1-0)^2 + (5-0)^2 + (2-1)^2\,]^{3/2}}\right] +$$

$$\frac{180 \times 10^{-9}}{2\pi\varepsilon_0}\left[\frac{(5-0)\,\overline{a}_y + (2-0)\,\overline{a}_z}{[\,(5-0)^2 + (2-0)^2\,]^{3/2}}\right] + \frac{25 \times 10^{-9}}{2\,\varepsilon_0}\,\overline{a}_z\,.$$

$$= 384.376\,\overline{a}_x + 1921.880\,\overline{a}_y + 384.376\,\overline{a}_z +$$

$$557.860\,\overline{a}_y + 223.144\,\overline{a}_z + 1411.791\,\overline{a}_z$$

$$= 384.376\,\overline{a}_x + 2479.74\,\overline{a}_y + 2019.311\,\overline{a}_z\ \ V/m\,.$$

Problem 2.30 : A line charge, $\rho_L = 50$ nC/m , is located a long the line $x = 2, y = 5$, in free space . (a) Find $\overline{E}$ at P (1, 3 ,- 4). (b) If the surface $x = 4$ contains a uniform surface charge density, $\rho_S = 18$ nC/m^2 , at what point in the $z = 0$ plane is $\overline{E}_{total} = 0$?

Solution :

(a) $\overline{E} = \dfrac{\rho_L}{2\pi\varepsilon_0 d}\,\overline{a}_{12}$ as the infinite line charge is present.

Point 2 is that point at which $\overline{E}$ is desired = (1, 3, - 4).

Point 1 is the foot of perpendicular dropped from point 2 on the line charge = (2, 5, –4).

$$\overline{R}_{12} = (1-2)\,\overline{a}_x + (3-5)\,\overline{a}_y = -\ \overline{a}_x - 2\,\overline{a}_y$$

$$d = |\bar{R}_{12}| = \sqrt{(-1)^2 + (-2)^2} = \sqrt{5}$$

$$\bar{a}_{12} = \frac{-\bar{a}_x - 2\bar{a}_y}{\sqrt{5}}$$

$$\therefore \bar{E} = \frac{50 \times 10^{-9}}{2\pi\varepsilon_0\sqrt{5}}\left[\frac{-\bar{a}_x - 2\bar{a}_y}{\sqrt{5}}\right] = -179.755\,\bar{a}_x - 359.510\,\bar{a}_y \text{ V/m.}$$

(b) Any point in the $z = 0$ plane is given by $(x, y, 0)$. Applying the procedure of part (a), the electric field intensity at point $(x, y, 0)$ due to line charge is

$$\bar{E} = \frac{50 \times 10^{-9}\,[(x-2)\,\bar{a}_x + (y-5)\,\bar{a}_y)]}{2\pi\varepsilon_0\,[(x-2)^2 + (y-5)^2]}$$

and $\bar{E}$ due to surface charge is $\bar{E} = \dfrac{18 \times 10^{-9}}{2\varepsilon_0}(\pm\,\bar{a}_x) = 1016.490\,(\pm\,\bar{a}_x)$

[$+\bar{a}_x$, if $x > 4$ and $-\bar{a}_x$, if $x < 4$]

$$\therefore \bar{E}_{total} = \left\{\frac{898.774\,(x-2)}{[(x-2)^2 + (y-5)^2]} \pm 1016.490\right\}\bar{a}_x + \frac{898.774\,(y-5)}{[(x-2)^2 + (y-5)^2]}\,\bar{a}_y$$

If $\bar{E}_{total}$ is equal to zero, then coefficient of $\bar{a}_x$ and $\bar{a}_y$ must be equal to zero. At $y = 5$, the coefficient of $\bar{a}_y$ is equal to zero. Now the coefficient of $\bar{a}_x$ is

$$\frac{898.774\,(x-2)}{(x-2)^2 + (5-5)^2} \pm 1016.490 = 0 \text{ or } x - 2 = \pm\frac{898.774}{1016.490} = \pm\,0.88419$$

Hence $x = 2.884$ and 1.1158 i.e., the points are (2.884, 5, 0) and (1.1158, 5, 0). In both the points, x coordinate is less than 4 and when $x < 4$, the x component is given by $\dfrac{898.774\,(x-2)}{(x-2)^2 + (y-5)^2} - 1016.490$ and this component is zero only when $x = 2.884$. When $x = 1.1158$, the x component is not equal to zero. Hence the correct point is (2.884, 5, 0)

Problem 2.31 : The circular region, $\rho < a$, $z = 0$, carries a uniform surface charge density ρ_s. Find $\bar{E}$ at $P(0, 0, h)$, $h > 0$.

Solution :

$dQ = \rho_s\,dS$ as surface charge density is given. Further ρ_s lies in $z = 0$ plane to which $\bar{a}_z$ is perpendicular. Hence

$\overline{dS} = \rho\, d\rho\, d\phi\, \bar{a}_z$ or

$dS = \rho\, d\rho\, d\phi$

$\therefore\ dQ = \rho_s\, \rho\, d\rho\, d\phi$.

Point 2 is that point at which $\overline{E}$ is desired = $(0, 0, h)$

Point 1 is the general point on the surface charge density ($z =$ 0plane) = $(\rho, \phi, 0)$

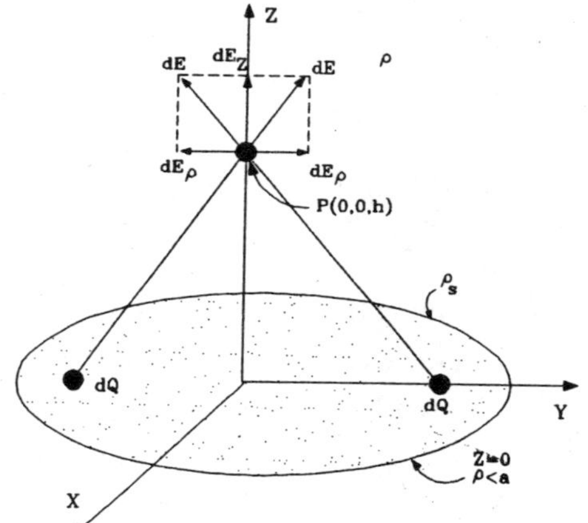

FIG.P2.31

$\overline{R}_{12} = \bar{r}_2 - \bar{r}_1$

$= h\,\bar{a}_z - \rho\,\bar{a}_\rho$

$d = |\overline{R}_{12}| = \sqrt{h^2 + \rho^2}$

$$\therefore\ \overline{dE} = \frac{\rho_s\, \rho\, d\rho\, d\phi}{4\pi\varepsilon_0}\left\{\frac{h\,\bar{a}_z - \rho\,\bar{a}_\rho}{[h^2 + \rho^2]^{3/2}}\right\}$$

As the surface charge is symmetrically located w.r.t. point 2, only E_z component of $\overline{E}$ is present and E_ρ component get cancelled(Refer to Fig.P2.31) .

$$\therefore\ \overline{E} = \int \overline{dE} = \int_0^a \int_0^{2\pi} \frac{\rho_s\, \rho\, d\rho\, d\phi\, h\,\bar{a}_z}{4\pi\varepsilon_0 (h^2 + \rho^2)^{3/2}}$$

$$= \frac{\rho_s h}{4\pi\varepsilon_0}\left[\frac{1}{\sqrt{h^2+\rho^2}}\right]_0^a [\phi]_0^{2\pi}\,\bar{a}_z = \frac{h\,\rho_s}{2\varepsilon_0}\left[\frac{1}{\sqrt{h^2+a^2}} - \frac{1}{h}\right]\bar{a}_z$$

Problem 2.32 : In the rectangular region, $-2 < x < 2, -3 < y < 3, z = 0$, the surface charge density is given by $\rho_s = (x^2 + y^2 + 1)^{3/2}$. If no other charge is present, find $\overline{E}$ at $P\ (0, 0, 1)$.

Solution :

As surface charge density is given, we have

$dQ = \rho_s\, dS$

ρ_s lies in $z = 0$ plane to which $\bar{a}_z$ is perpendicular.

Hence$\overline{dS} = dx\, dy\, \bar{a}_z$ or

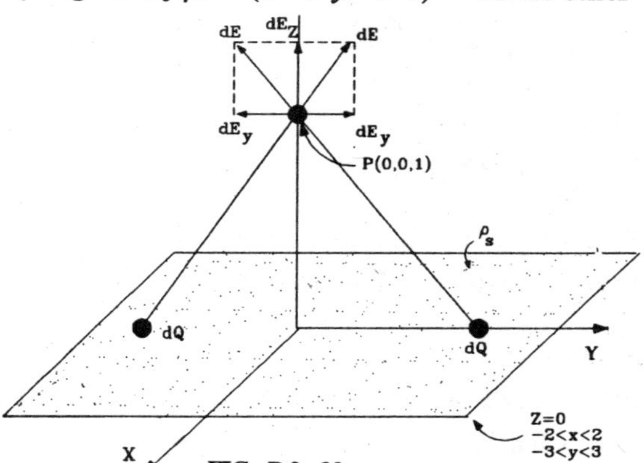

FIG . P 2 . 32

$dS = dx\ dy.$

Hence $dQ = \rho_S\, dx\ dy$

Point 2 is that point at which $\overline{E}$ is desired $= (0, 0, 1)$ and

Point 1 is the general point on the surface charge density ($z = 0$ plane) $= (x, y, 0)$.

$\therefore\ \overline{R}_{12} = (0 - x)\,\overline{a}_x + (0 - y)\,\overline{a}_y + (1 - 0)\,\overline{a}_z = -x\,\overline{a}_x - y\,\overline{a}_y + \overline{a}_z$

$d = |\overline{R}_{12}| = \sqrt{x^2 + y^2 + 1}$

$$\therefore\ \overline{dE} = \frac{(x^2 + y^2 + 1)^{3/2}\, dx\ dy}{4\pi\varepsilon_0}\left[\frac{-x\,\overline{a}_x - y\,\overline{a}_y + \overline{a}_z}{(x^2 + y^2 + 1)^{3/2}}\right]$$

As w.r.t. point 2 (0, 0, 1), the surface charge is symmetrically located, E_x and E_y components of $\overline{E}$ get cancelled and only E_z component is present.

$$\therefore\ \overline{E} = \int \overline{dE} = \int_{-2}^{2}\int_{-3}^{3} \frac{(x^2 + y^2 + 1)^{3/2}\, dx\ dy\,\overline{a}_z}{4\pi\varepsilon_0 (x^2 + y^2 + 1)^{3/2}}$$

$$= \int_{-2}^{2}\int_{-3}^{3} \frac{dx\ dy}{4\pi\varepsilon_0}\,\overline{a}_z = \frac{\overline{a}_z}{4\pi\varepsilon_0}\Big[\,x\,\Big]_{-2}^{2}\Big[\,y\,\Big]_{-3}^{3} = 0.2157\,\overline{a}_z\ \text{GV/m}$$

Problem 2.33 : Given that field $\overline{E} = 15x^2y\,\overline{a}_x + 5x^3\,\overline{a}_y$, find (a) the equation of the streamline that passes through P (2, 3, - 4) ; (b) a unit vector $\overline{a}_E$ specifying the direction of $\overline{E}$ at P ; (c) a unit vector $\overline{a}_\perp = (l, m, 0)$ that is perpendicular to $\overline{a}_E$ at P and has $l > 0$.

Solution :

(a) $\overline{E} = 15x^2y\,\overline{a}_x + 5x^3\,\overline{a}_y = E_x\,\overline{a}_x + E_y\,\overline{a}_y$

$$\frac{dy}{dx} = \frac{E_y}{E_x} = \frac{5x^3}{15x^2y} = \frac{x}{3y}$$

$\therefore\ 3y\,dy = x\,dx$ or $1.5y^2 = 0.5x^2 + C$

Substituting the coordinates of point P in the above equation, we have $C = 11.5$. Hence the equation of the stream line is

$1.5y^2 = 0.5x^2 + 11.5$ or $3y^2 = x^2 + 23$, $z = -4$

(b) $\bar{E}$ at P = $15\,(2)^2\,(3)\,\bar{a}_x + 5\,(2)^3\,\bar{a}_y = 180\,\bar{a}_x + 40\,\bar{a}_y$

$|\bar{E}|$ at P = $\sqrt{(180)^2 + (40)^2} = 184.391$ V/m

$$\bar{a}_E \text{ at } P = \frac{\bar{E} \text{ at } P}{|\bar{E} \text{ at } P|} = \frac{180\,\bar{a}_x + 40\,\bar{a}_y}{184.391} = 0.976\,\bar{a}_x + 0.217\,\bar{a}_y$$

(c) Let $f = 3y^2 - x^2 - 23$ then $\bar{a}_\perp = \pm \dfrac{\bar{\nabla} f}{|\bar{\nabla} f|}$

$$\bar{\nabla} f = \frac{\partial}{\partial x}(3y^2 - x^2 - 23)\,\bar{a}_x + \frac{\partial}{\partial y}(3y^2 - x^2 - 23)\,\bar{a}_y = -\,2x\,\bar{a}_x + 6y\,\bar{a}_y$$

$\bar{\nabla} f$ at $P(2, 3, -4) = -\,2\,(2)\,\bar{a}_x + 6\,(3)\,\bar{a}_y = -\,4\,\bar{a}_x + 18\,\bar{a}_y$

$|\bar{\nabla} f| = \sqrt{(-4)^2 + (18)^2} = 18.439$

$$\therefore\ \bar{a}_\perp = \pm \left[\frac{-\,4\,\bar{a}_x + 18\,\bar{a}_y}{18.439}\right] = \pm\,(-\,0.2169\,\bar{a}_x + 0.976\,\bar{a}_y)$$

as $l > 0$, we have $\bar{a}_\perp = 0.2169\,\bar{a}_x - 0.976\,\bar{a}_y$

Problem 2.34 : If $\bar{E} = 200 \cosh 2x \sin 2y + 200 \sinh 2x \cos 2y$ V/m, find the equation of the direction line passing through the point P (1, 0, 0) and sketch it in the $z = 0$ plane for $0 < x < 5$ and $0 < y < \pi/4$.

Solution :

$\bar{E} = 200 \cosh 2x \sin 2y\,\bar{a}_x + 200 \sinh 2x \cos 2y\,\bar{a}_y = E_x\,\bar{a}_x + E_y\,\bar{a}_y$

$$\frac{dy}{dx} = \frac{E_y}{E_x} = \frac{200 \sinh 2x \cos 2y}{200 \cosh 2x \sin 2y} = \tanh 2x \cot 2y$$

$\therefore\ \tan 2y\,dy = \tanh 2x\,dx$ or $0.5 \ln(\sec 2y) = 0.5 \ln(\cosh 2x) + C$

where C is a constant of integration. Let $C = 0.5 \ln K$. Then

$0.5 \ln(\sec 2y) = 0.5 \ln(\cosh 2x) + 0.5 \ln K$

or $\ln(\sec 2y) = \ln(\cosh 2x) + \ln K = \ln(K \cosh 2x)$

$\therefore\ \sec 2y = K \cosh 2x$

At point P (1, 0, 0), $\sec(0) = K \cosh(2)$, or $K = 0.266$

Hence the required equation is $\sec 2y = 0.266 \cosh 2x$, $z = 0$

Sketch is not given.

Problem 2.35 : The electric field intensity is given as $\bar{E} = 5\,e^{-2x}(\sin 2y\,\bar{a}_x - \cos 2y\,\bar{a}_y)$ V/m. (a) Find the equation of the streamline passing through the

point P (0.5, $\pi/10$, 0). (b) Find a unit vector tangent to the streamline at P .

Solution :

(a) $\bar{E} = 5e^{-2x}(\sin 2y\,\bar{a}_x - \cos 2y\,\bar{a}_y) = E_x\,\bar{a}_x + E_y\,\bar{a}_y$.

Hence $E_x = 5e^{-2x}\sin 2y$ and $E_y = -5e^{-2x}\cos 2y$

$$\frac{E_x}{E_y} = \frac{dx}{dy} = \frac{5e^{-2x}\sin 2y}{-5e^{-2x}\cos 2y} = -\tan 2y. \text{ Hence}$$

$x = 0.5\ln(\cos 2y) + C$ where C is a constant of integration. Hence

$2x = \ln(\cos 2y) + \ln\left(\frac{1}{K}\right)$ where $C = \ln\left(\frac{1}{K}\right)$ where K is a constant.

$\therefore\ 2x = \ln\left(\frac{\cos 2y}{K}\right)$ or $\cos 2y = K e^{2x}$

At the point $P\left(0.5, \frac{\pi}{10}, 0\right)$, $\cos\left(\frac{2\pi}{10}\right) = K e^{2(0.5)}$ or $K = 0.298$.

Hence the equation is $\cos 2y = 0.298\,e^{2x}$ and $z = 0$

(b) Direction of $\bar{E}$ is given by a unit vector which is tangent to the streamline.

$$\bar{E} \text{ at point } P = 5e^{-2(0.5)}\left[\sin\left(\frac{2\pi}{10}\right)\bar{a}_x - \cos\left(\frac{2\pi}{10}\right)\bar{a}_y\right]$$

$$= 1.081\,\bar{a}_x - 1.488\,\bar{a}_y\ ;$$

$$|\bar{E}| = \sqrt{(1.081)^2 + (-1.488)^2} = 1.839$$

$$\bar{a}_E = \frac{\bar{E}}{|\bar{E}|} = \frac{1.081\,\bar{a}_x - 1.488\,\bar{a}_y}{1.839} = 0.588\,\bar{a}_x - 0.809\,\bar{a}_y$$

Also $-(0.588\,\bar{a}_x - 0.809\,\bar{a}_y)$ is tangential .

Problem 2.36 : Given a field $\bar{E}(\rho,\phi) = E_\rho(\rho,\phi)\,\bar{a}_\rho + E_\varphi(\rho,\phi)\,\bar{a}_\phi$ in cylindrical coordinates, the equations of the streamlines in a plane $z =$ constant are obtained by solving the differential equation $E_\rho/E_\phi = d\rho/(\rho\, d\phi)$. Find the equation of the line passing through point P (2, 30°, 0) for the field $\bar{E} = \rho\sin 2\phi\,\bar{a}_\rho - \rho\cos 2\phi\,\bar{a}_\phi$.

Solution :

$\overline{E} = \rho \sin 2\phi \, \overline{a}_\rho - - \rho \cos 2\phi \, \overline{a}_\phi = E_\rho \overline{a}_\rho + E_\phi \overline{a}_\phi$

$\therefore E_\rho = \rho \sin 2\phi$ and $E_\phi = - \rho \cos 2\phi$

$$\frac{E_\rho}{E_\phi} = \frac{d\rho}{\rho \, d\phi} = \frac{\rho \sin 2\phi}{- \rho \cos 2\phi} = - \tan 2\phi$$

$$\frac{d\rho}{\rho \, d\phi} = - \tan 2\phi \text{ or } \ln \rho = \frac{\ln (\cos 2\phi)}{2} + C$$

or $2 \ln \rho = \ln (\cos 2\phi) + 2C$.

Let $\ln K = 2C$.

Then $\ln \rho^2 = \ln (\cos 2\phi) + \ln K$ or $\ln \rho^2 = \ln K \cos 2\phi$

Hence $\rho^2 = K \cos 2\phi$. Equation of the line passing through the point $P(2, 30^o, 0)$ is $(2)^2 = K \cos 60^o$, $K = 8$.

Hence the required equation is $\rho^2 = 8 \cos 2\phi$ and $z = 0$

—X—

CHAPTER 3

Electric Flux Density, Gauss's Law, Divergence & The Divergence Theorem

3.1 : CONCEPT OF ELECTRIC FLUX :

Fig.3.1 shows a concentric metallic sphere of radius a which forms the inner sphere. Suppose a positive charge Q is given to this sphere. Then this sphere is enclosed by another concentric metallic sphere of radius b ($b > a$) as shown in fig. This outer sphere is in the form of two hemispheres. Between these two spheres, a dielectric material is present. The outer sphere was discharged by connecting it momentarily to ground. Then the charge on the outer sphere is measured carefully.

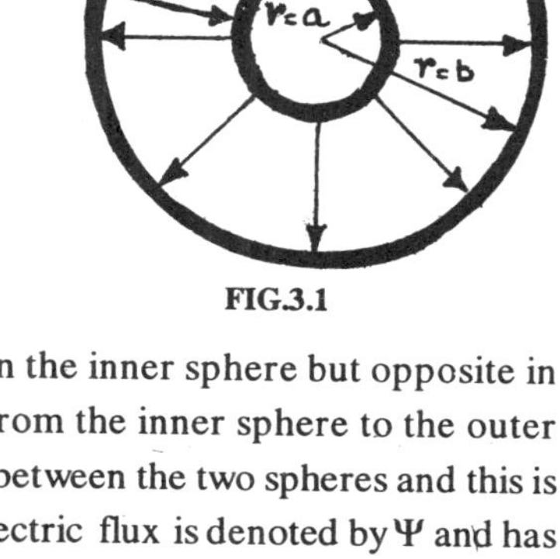

FIG.3.1

It is found that the charge on the outer sphere is equal in magnitude to the charge on the inner sphere but opposite in sign. Hence there is a sort of displacement from the inner sphere to the outer sphere which is independent of the medium between the two spheres and this is known as *displacement flux* or *electric flux*.. Electric flux is denoted by Ψ and has the units Coulomb (C).

$$\Psi = Q$$

Note :

(1) *The electric flux originates from positive charge and terminates on negative charge. In the absence of negative charge, the flux Ψ terminates at infinity.*

(2) *In the above experiment the charge enclosed by the outer sphere (radius b) is equal to Q which is same as the flux Ψ. Hence we conclude that*

$$\boxed{\begin{aligned}\Psi &= \textbf{flux leaving through a closed surface}\\ &= \textbf{charge enclosed by the closed surface} = Q\end{aligned}} \qquad \ldots\ldots (3.1)$$

Problem 3.1 : A 25 – μC point charge is located at the origin. Calculate the electric flux passing through: (a) that portion of the sphere $r = 20$ cm bounded by $\theta = 0$ and π, $\phi = 0$ and $\pi/2$; (b) the closed surface $\rho = 0.8$ m , $z = \pm\ 0.5$ m ; (c) the plane $z = 4$ m.

Solution :

(a) Area of the sphere $= \int dS = \int_0^{\pi} \int_0^{\pi/2} r^2 \sin\theta \, d\theta \, d\phi \Big|_{r = 0.2}$

$$= (0.2)^2 \Big[-\cos\theta \Big]_0^{\pi} \Big[\phi \Big]_0^{\pi/2} = 0.04\,\pi \text{ m}^2$$

If the charge 25 μC is completely enclosed by the sphere $r = 0.2\,m$ [area = $4\pi r^2 = 4\pi (0.2)^2 = 0.16\,\pi \text{ m}^2$], then the flux leaving through the complete sphere is 25 μC. But the area given to us is $0.04\pi \text{ m}^2$ which is just $\dfrac{0.04\,\pi}{0.16\,\pi} = \left(\dfrac{1}{4}\right)^{\text{th}}$ of the total area. Hence the flux leaving through the given area is $\dfrac{1}{4} \times 25\,\mu\text{C} = 6.25\ \mu\text{C}$.

(b) As seen from the Fig. P3.1(a), the charge of 25 μC is completely enclosed by the cylinder $\rho = 0.8$ m, $z = \pm\ 0.5$ m. Hence the flux leaving through the given surface = charge enclosed by that surface = 25 μC.

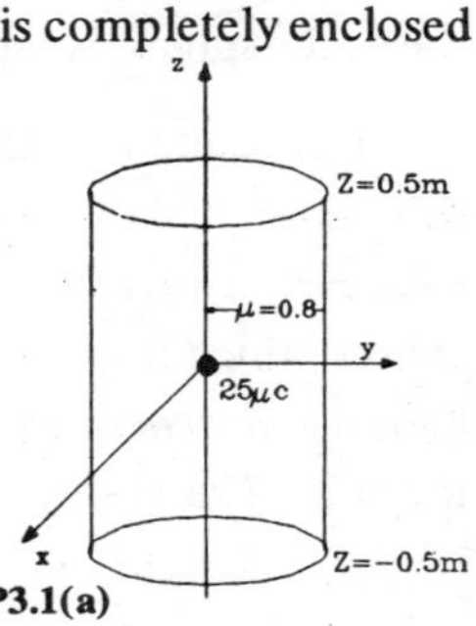

FIG. P3.1(a)

(c) The charge of 25 μC is located at origin. In all the directions, the flux is coming out from this point charge as shown in Fig.P3.1(b). Half of the total flux is above the $z = 0$ plane and remaining half of the total flux is below the $z = 0$ plane. Hence the flux crossing the infinite plane $z = 4$ m is half of the charge located at origin, i.e. 12.5μC.

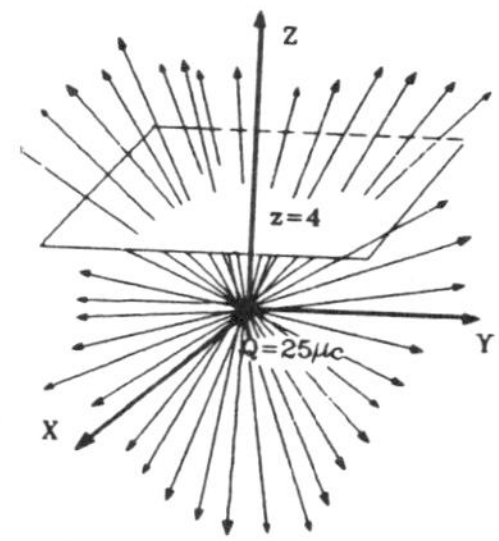

FIG. P3.1(b)

3.2 : ELECTRIC FLUX DENSITY AND ITS RELATIONSHIP WITH ELECTRIC FIELD INTENSITY

Fig.3.2(a) shows the flux lines due to some charge distribution. At point P, consider a differential area dS to which $\bar{a}$ is perpendicular. Also in the neighbourhood of point P the lines of flux are in the same direction as that of $\bar{a}$. Then electric flux density is defined as the number of flux lines crossing a surface per unit area. It is a vector quantity denoted by $\overline{D}$. Hence

$$\overline{D} = \frac{d\Psi}{dS}\,\bar{a} \;\; C/m^2$$

where $d\Psi$ is the amount of flux leaving through the area dS.

With reference to Fig.3.2 (b) consider a $+\,Q$ C charge at the origin of the coordinate system. Let point $P(r, \theta, \phi)$ lies on the spherical surface of radius r. Then at point P

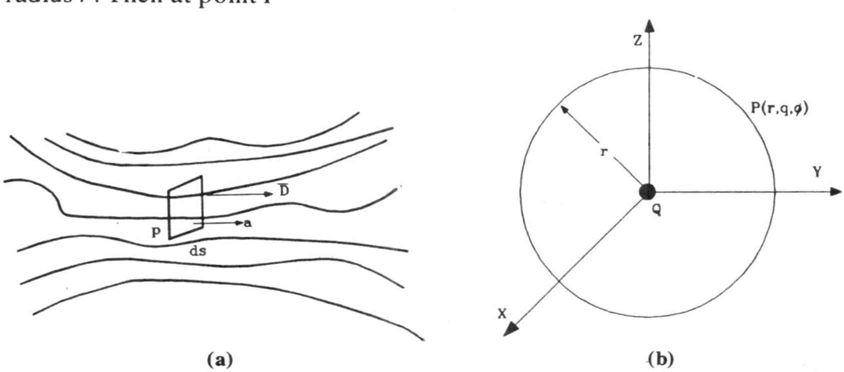

FIG. 3.2

$$\overline{E} = \frac{Q}{4\pi\varepsilon_o r^2}\,\overline{a}_r \quad \text{and}$$

$$\overline{D} = \frac{\text{charge enclosed by sphere}}{\text{area of sphere}}\,\overline{a}_r = \frac{Q}{4\pi r^2}\,\overline{a}_r = \varepsilon_o \overline{E}$$

Hence $\overline{D} = \varepsilon_o \overline{E}$ for free space. In general $\overline{D} = \varepsilon \overline{E} = \varepsilon_o \varepsilon_r \overline{E}$, where ε_r is the relative permitivity of the medium.

Note : *The relationship between electric flux density and electric field intensity is*

$$\boxed{\overline{\mathbf{D}} = \varepsilon_o \varepsilon_r \overline{\mathbf{E}}} \qquad \text{.......(3.2)}$$

Problem 3.2 : Find $\overline{D}$ (in cartesian coordinates) at P (6, 8,-10) caused by: (a) a point charge of 30 mC at the origin; (b) a uniform line charge $\rho_L = 40\ \mu C/m$ on the z axis; (c) a uniform surface charge denisty $\rho_S = 57.2\ \mu C/m^2$ on the plane $x = 9$.

Solution :

$$\text{(a)}\overline{D} = \varepsilon_0 \overline{E} = \frac{\varepsilon_0 \times 30 \times 10^{-3}}{4\pi\varepsilon_0}\left\{\frac{(6-0)\,\overline{a}_x + (8-0)\,\overline{a}_y + (-10-0)\,\overline{a}_z}{[(6)^2 + (8)^2 + (-10)^2]^{3/2}}\right\}$$

$$= 5.064\,\overline{a}_x + 6.752\,\overline{a}_y - 8.440\,\overline{a}_z\ \mu C/m^2$$

(b) Equations of the z axis are $x = 0$ and $y = 0$.

$$\overline{D} = \varepsilon_0 \overline{E} = \varepsilon_0\left[\frac{\rho_L}{2\pi\varepsilon_0 d}\,\overline{a}_{12}\right] = \frac{\rho_L}{2\pi d}\,\overline{a}_{12}$$

Point 2 is that point at which electric flux density is desired = (6,8,–10)

Point 1 is the foot of the perpendicular dropped from point 2 on the line charge = (0,0,–10)

$$\therefore \overline{R}_{12} = (2-0)\,\overline{a}_x + (8-0)\,\overline{a}_y = 2\,\overline{a}_x + 8\,\overline{a}_y$$

$$d = |\overline{R}_{12}| = \sqrt{(2)^2 + (8)^2} = \sqrt{68}$$

$$\therefore \overline{a}_{12} = \frac{\overline{R}_{12}}{|\overline{R}_{12}|} = \frac{2\,\overline{a}_x + 8\,\overline{a}_y}{\sqrt{68}}$$

$$\overline{D} = \frac{40 \times 10^{-6}}{2\pi\sqrt{68}}\left[\frac{2\,\overline{a}_x + 8\,\overline{a}_y}{\sqrt{68}}\right] = 0.187\,\overline{a}_x + 0.749\,\overline{a}_y\ \mu C/m^2$$

(c) $\overline{D} = \varepsilon_0 \overline{E} = \varepsilon_0 \left[\frac{\rho_s}{2\varepsilon_0} \overline{a}_N \right] = \frac{\rho_s}{2} \overline{a}_N$

The surface charge density lies on the plane $x = 9$. The x coordinate of point P is 6 which is less than 9. Hence $\overline{a}_N = -\overline{a}_x$

$$\therefore \overline{D} = \frac{57.2 \times 10^{-6}}{2} (-\overline{a}_x) = -28.6\, \overline{a}_x \ \mu C/m^2$$

3.3 : GAUSS'S LAW AS APPLIED TO ELECTRIC FLUX DENSITY :

The generalizations of Faraday's experiment lead to the following statement, which is known as Gauss's law. *"The electric flux passing through any closed surface is equal to the total charge enclosed by that surface".* Let us consider a distribution of charge shown as a cloud of do not charges in the Fig. 3.3 surrounded by a closed surface of any shape.

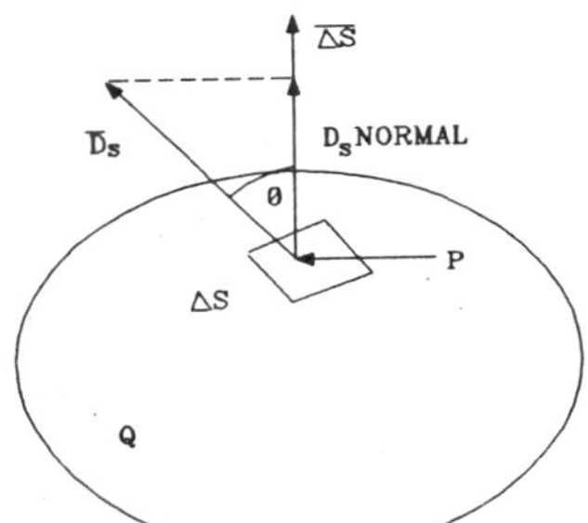

FIG. 3.3

If the total charge is Q, then Q coulombs of electric flux will pass through the enclosing surface. At every point on the surface the electric flux density vector $\overline{D}$ will have some value $\overline{D}_S$. Let at point P, $\overline{a}_N$ is a unit vector normal to differential area ΔS located at point P. Then $\overline{\Delta S} = \Delta S\, \overline{a}_N$ and $\overline{D}_S$ makes an angle θ with $\overline{a}_N$ at point P. Then

$$\Delta \Psi = \text{flux crossing } \overline{\Delta S} = D_{S,\,\text{norm}}\, \Delta S = D_S \cos\theta\, \Delta S$$
$$= \overline{D}_S \cdot \overline{a}_N\, \Delta S = \overline{D}_S \cdot \overline{\Delta S}$$

and the total flux is given by

$$\Psi = \int d\Psi = \oint_S \overline{D}_S \cdot \overline{dS}$$

Thus the mathematical formulation of Gauss's law is

$$\boxed{\Psi = Q = \oint_S \overline{D}_S \cdot \overline{dS}} \qquad \text{..........(3.3)}$$

Problem 3.3 : Let $\overline{D} = r\,\overline{a}_r/3$ nC/m² in free space. (a) Find E at $r = 0.2$ m. (b) Find the total charge within the sphere $r = 0.2$ m. (c) Find the

total electric flux leaving the sphere r = 0.3 m.

Solution :

$$\text{(a)}\overline{E} = \frac{\overline{D}}{\varepsilon_0} = \frac{r\bar{a}_r \times 10^{-9}}{3\,\varepsilon_0} = \frac{0.2 \times 10^{-9}}{3\,\varepsilon_0}\bar{a}_r = 7.53\,\bar{a}_r \text{ V/m.}$$

or $|\overline{E}| = 7.53$ V/m

(b)To the sphere ,$\bar{a}_r$ is perpendicular. Hence $\overline{dS} = r^2 \sin\theta\, d\theta\, d\phi\, \bar{a}_r$

$$Q = \oint_S \overline{D}\cdot\overline{dS} = \int_0^{\pi}\int_0^{2\pi} (r/3)\,\bar{a}_r \times 10^{-9} \cdot r^2 \sin\theta\, d\theta\, d\phi\, \bar{a}_r \Big|_{r=0.2}$$

$$= \int_0^{\pi}\int_0^{2\pi} (r^3/3) \times 10^{-9} \times \sin\theta\, d\theta\, d\phi \Big|_{r=0.2}$$

$$= \frac{(0.2)^3}{3} \times 10^{-9} \times \Big[-\cos\theta\Big]_0^{\pi}\,[\phi]_0^{2\pi} = 33.510 \text{ pC}$$

$$\text{(c) } \Psi = Q = \oint_S \overline{D}\cdot\overline{dS} = \int_0^{\pi}\int_0^{2\pi} (r/3) \times 10^{-9} \sin\theta\, d\theta\, d\phi \Big|_{r=0.3}$$

$$= \frac{(0.3)^3}{3} \times 10^{-9}\Big[-\cos\theta\Big]_0^{\pi}\,[\phi]_0^{2\pi} = 113.097 \text{ pC.}$$

Problem 3.4 : Find the total electric flux leaving the spherical surface r = 2.5 m given the charge configuration : (a) $Q = 2^{-x^2}$ nC on the x axis at x = 0, ± 1, ± 2,.... m ; (b) a line charge $\rho_L = 1/(z^2 + 1)$ nC/m on the z axis; (c) a surface charge $\rho_S = 1/(x^2 + y^2 + 4)$ nC/m^2 on the z = 0 plane.

Solution :

(a) As shown in Fig. P3.4(a) ,the sphere of radius 2.5 m intersect the x axis at x = –2.5 and x = 2.5. Hence the charges located at x = 0, ± 1, ± 2 must be considered so as to determine the charge within the sphere of radius 2.5 m.

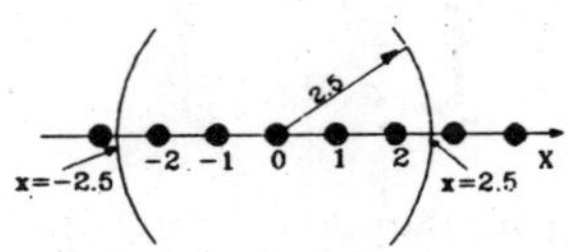

FIG.P3.4 (a)

$$Q = 2^{-(0)^2} + 2^{-(1)^2} + 2^{-(-1)^2} + 2^{-(2)^2} + 2^{-(-2)^2} = 2.125 \text{ nC}$$

(b) As shown in Fig.P3.4(b),the sphere of radius 2.5 m intersect the z axis at $z = -2.5$ and $z = 2.5$. Hence to determine the charge within the sphere of radius 2.5m, we must find out the charge on the z axis between

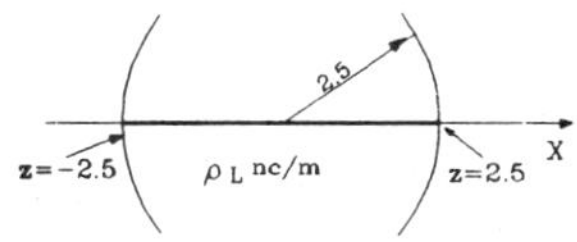

FIG.P3.4 (b)

$z = -2.5$ m to $z = 2.5$ m due to line charge $\rho_L = 1/(z^2 + 1)$ nC/m. Now $dQ = \rho_L \, dL$ and $dL = dz$ as the line charge lies on the z axis.

$$\therefore Q = \int dQ = \int_{-2.5}^{2.5} \frac{dz}{z^2 + 1} = \left[\tan^{-1} z\right]_{-2.5}^{2.5} = 2.381 \text{ nC}$$

$$\therefore \Psi = Q = 2.381 \text{ nC}$$

(c) The spherical surface $r = 2.5$ m intersect the $z = 0$ plane in a circle of radius 2.5 m as seen from the adjoining Fig.P3.4 (c). Hence to determine the flux leaving through $r = 2.5$ m, we must determine the charge present within the circular surface of radius 2.5 m centered at origin. To the surface $\theta = \pi/2$, $\bar{a}_\theta$ is perpendicular .Hence

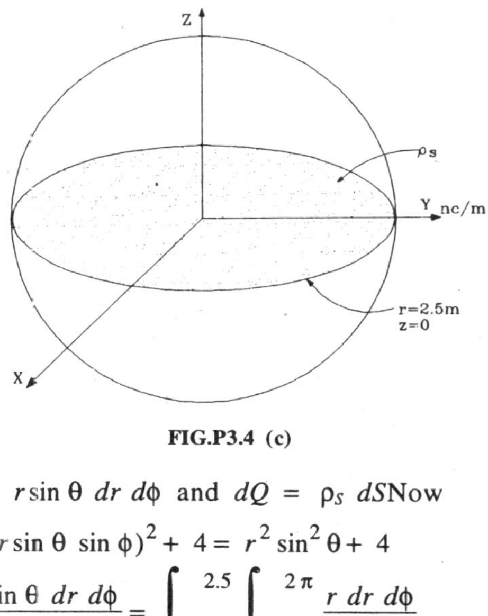

FIG.P3.4 (c)

$\overline{dS} = r \sin\theta \, dr \, d\phi \, \bar{a}_\theta$ or $dS = r \sin\theta \, dr \, d\phi$ and $dQ = \rho_S \, dS$Now $x^2 + y^2 + 4 = (r\sin\theta \cos\phi)^2 + (r\sin\theta \sin\phi)^2 + 4 = r^2 \sin^2\theta + 4$

$$\therefore Q = \int dQ = \int_0^{2.5}\int_0^{2\pi} \frac{r\sin\theta \, dr \, d\phi}{r^2 \sin^2\theta + 4} = \int_0^{2.5}\int_0^{2\pi} \frac{r \, dr \, d\phi}{r^2 + 4}$$

$$= 0.5 \left[\ln (r^2 + 4)\right]_0^{2.5} [\phi]_0^{2\pi} = 2.956 \text{ nC}$$

$$\therefore \Psi = Q = 2.956 \text{ nC}$$

Problem 3.5 : Surface charge densities of 200, - 50, and ρ_{Sx} μC/m^2 are located at $r = 3, 5$, and 7 cm, respectively. Find $\overline{D}$ at $r =$: (a) 2 cm ; (b) 4

cm; (c) 6 cm. (d) Find ρ_{Sx} if $\overline{D} = 0$ at $r = 7.32$ cm.

Solution :

(a) Fig.P3.5 (a) shows a sphere $r = 2$ cm along with spheres $r = 3$, 5, and 7 cm. As seen from this figure, inside the sphere $r = 2$ cm, no charge is present. That is, the charge enclosed by the sphere $r = 2$ cm is zero. Therefore flux leaving the sphere $r = 2$ cm is also zero.

$$\therefore \overline{D} = \frac{\text{flux}}{\text{area}} \overline{a}_r = \frac{0}{4\pi(0.02)^2} \overline{a}_r = 0$$

(b) As seen from Fig.P3.5(b), the sphere of radius 4 cm encloses the sphere of

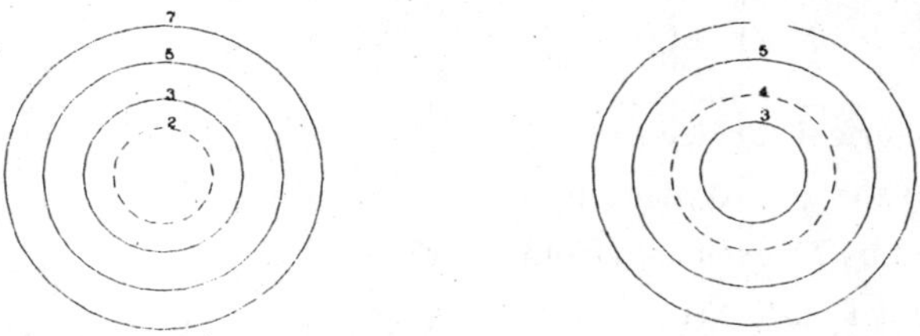

(a) (b)

FIG. P3.5

radius 3 cm. Hence charge enclosed by $r = 4$ cm is equal to charge present on $r = 3$ cm. Hence $Q = \Psi =$ Flux density $\times$ area

$$= 200 \times 10^{-6} \times \left[4\pi(0.03)^2\right] = 0.72\pi\, 10^{-6} \text{ C}$$

$$\therefore \overline{D}\,(\text{at } r = 4 \text{ cm}) = \frac{\Psi}{\text{area}} \overline{a}_r = \frac{0.72\pi\, 10^{-6}}{4\pi(0.04)^2} \overline{a}_r = 112.5\, \overline{a}_r \ \mu\text{C/m}^2$$

(c) As seen from Fig.P3.5(c),Charge enclosed by $r = 6$ cm = Charge present on $r = 3$ cm + Charge present on $r = 5$ cm.

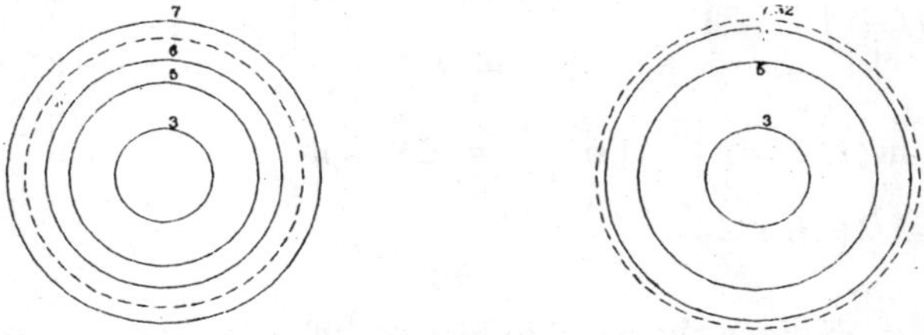

(c) (d)

FIG. P3.5

$$\therefore Q = \Psi = 0.72\,\pi\, 10^{-6} - 50\times 10^{-6}\times [4\,\pi\,(0.05)^2] = 0.22\,\pi\,10^{-6}\ \text{C}$$

$$\therefore \overline{D}\,(\text{at } r = 6 \text{ cm}) = \frac{0.22\,\pi\,10^{-6}}{4\,\pi\,(0.06)^2}\,\overline{a}_r = 15.278\,\overline{a}_r\ \mu\text{C/m}^2$$

$$\text{(d)}\overline{D}\,(\text{at } r = 7.32 \text{ cm}) = \frac{\Psi}{\text{area}}\,\overline{a}_r = \frac{Q}{4\,\pi\,(0.0732)^2}\,\overline{a}_r = 0. \ \therefore Q = 0.$$

As seen from Fig.P3.5(d), Charge enclosed by $r = 7.32$ cm = Charge present on $r = 3$ cm + Charge present on $r = 5$ cm + Charge present on $r = 7$ cm.

$$\therefore Q = 0.72\,\pi\,10^{-6} - 0.5\,\pi\times 10^{-6} + \rho_{sx}\left[4\,\pi\,(0.07)^2\right] = 0$$

$$\therefore 0.0196\,\pi\,\rho_{sx} = -0.22\,\pi\,10^{-6} \text{ or } \rho_{sx} = -11.224\ \mu\text{C/m}^2$$

3.4 : EXPRESSION FOR DIVERGENCE OF $\overline{D}$:

Let us consider a small differential volume element in the form of a rectangular box centered at the point $(P\,x_o, y_o, z_o)$ in cartesian coordinate system as shown in Fig3.4. The edges are parallel to axes with lengths $\Delta x, \Delta y$, and Δz. Apply Gauss's law to this closed surface

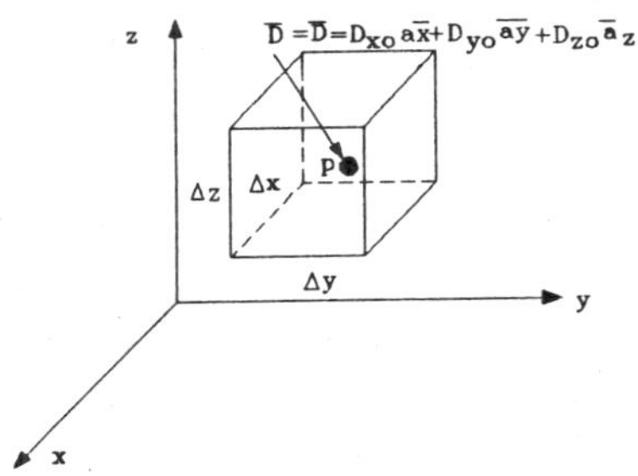

FIG. 3.4

$$Q = \oint_S \overline{D}\cdot\overline{dS}$$

The left hand side is broken up into six integrals.

$$\oint_S \overline{D}\cdot\overline{dS} = \int_{\text{front}} + \int_{\text{back}} + \int_{\text{left}} + \int_{\text{right}} + \int_{\text{top}} + \int_{\text{bottom}}$$

Since the surface element is very small, $\overline{D}$ is essentially constant and

$$\int_{\text{front}} \doteq \overline{D}_{\text{front}}\cdot\overline{\Delta S}_{\text{front}}$$

$$\doteq \overline{D}_{\text{front}}\cdot(\Delta y\,\Delta z\,\overline{a}_x) \doteq D_{x,\text{front}}\,\Delta y\,\Delta z$$

To determine $D_{x,\text{front}}$, expand D_x around the point $\left(x_o + \frac{\Delta x}{2}\right)$ by using Taylor's series and neglect higher order terms.

$$D_{x,\,\text{front}} \doteq D_x\left(x_o + \frac{\Delta x}{2}\right) = D_{xo} + \frac{\Delta x}{2}\frac{\partial D_x}{\partial x}$$

[$\overline{D}$ at point P is equal $\overline{D}_o = D_{xo}\,\overline{a}_x + D_{yo}\,\overline{a}_y + D_{zo}\,\overline{a}_z$ where D_{xo} is the value of D_x at point $P\,(x_o, y_o, z_o)$

$$\therefore \int_{\text{front}} \doteq \left(D_{xo} + \frac{\Delta x}{2}\frac{\partial D_x}{\partial x}\right)\Delta y\,\Delta z$$

Similarly

$$\int_{\text{back}} \doteq \overline{D}_{\text{back}} \cdot \overline{\Delta S}_{\text{back}} \doteq \overline{D}_{\text{back}} \cdot (-\Delta y\,\Delta z\,\overline{a}_x)$$

$$\doteq -D_{x,\,\text{back}}\,\Delta y\,\Delta z$$

$$D_{x,\,\text{back}} \doteq D_x\left(x_o - \frac{\Delta x}{2}\right) \doteq D_{xo} - \frac{\Delta x}{2}\frac{\partial D_x}{\partial x}$$

$$\therefore \int_{\text{back}} \doteq \left(-D_{xo} + \frac{\Delta x}{2}\frac{\partial D_x}{\partial x}\right)\Delta y\,\Delta z$$

Adding these integrals we get

$$\int_{\text{front}} + \int_{\text{back}} \doteq \frac{\partial D_x}{\partial x}\Delta x\,\Delta y\,\Delta z.$$

Similarly

$$\int_{\text{right}} + \int_{\text{back}} \doteq \frac{\partial D_y}{\partial y}\Delta x\,\Delta y\,\Delta z$$

and

$$\int_{\text{right}} + \int_{\text{left}} \doteq \frac{\partial D_z}{\partial z}\Delta x\,\Delta y\,\Delta z$$

and these results may be collected to yield

$$\oint_S \overline{D} \cdot \overline{dS} = Q \doteq \left(\frac{\partial D_x}{\partial x} + \frac{\partial D_y}{\partial y} + \frac{\partial D_z}{\partial z}\right)\Delta x\,\Delta y\,\Delta z$$

Volume enclosed by rectangular box = $\Delta v = \Delta x\,\Delta y\,\Delta z$

Hence

$$\oint_S \overline{D} \cdot \overline{dS} = Q \doteq \left(\frac{\partial D_x}{\partial x} + \frac{\partial D_y}{\partial y} + \frac{\partial D_z}{\partial z}\right)\Delta v$$

Dividng the above expression by Δv and taking the limit as Δv shink to zero

$$\frac{\partial D_x}{\partial x} + \frac{\partial D_y}{\partial y} + \frac{\partial D_z}{\partial z} = \lim_{\Delta v \to 0} \frac{\oint_S \overline{D} \cdot \overline{dS}}{\Delta v} = \lim_{\Delta v \to 0} \frac{Q}{\Delta v} = \rho_v$$

Now the divergence of the vector flux density $\overline{D}$ is defined as the outflow of flux from a small closed surface per unit volume as the volume shrinks to zero.

$$\text{Divergence of } \overline{D} = \text{div } \overline{D} = \lim_{\Delta v \to 0} \frac{\oint_S \overline{D} \ \overline{dS}}{\Delta v}$$

Hence we have

$$\text{div } \overline{D} = \frac{\partial D_x}{\partial x} + \frac{\partial D_y}{\partial y} + \frac{\partial D_z}{\partial z}$$

Note : *Following are the imortant formulae related with divergence*

Charge enclosed in volume Δv is given by

$$\doteq \left(\frac{\partial D_x}{\partial x} + \frac{\partial D_y}{\partial y} + \frac{\partial D_z}{\partial z} \right) \times \textbf{volume } \Delta v$$

. . . .(3.4 a)

$$\rho_v = \textbf{div } \overline{\mathbf{D}}$$

. . . .(3.4 b)

$$\textbf{div } \overline{\mathbf{D}} = \frac{\partial D_x}{\partial x} + \frac{\partial D_y}{\partial y} + \frac{\partial D_z}{\partial z}$$

. . . .(3.4 c)

$$\textbf{div } \overline{\mathbf{D}} = \frac{1}{\rho} \frac{\partial}{\partial \rho} (\rho D_\rho) + \frac{1}{\rho} \frac{\partial D_\phi}{\partial \phi} + \frac{\partial D_z}{\partial z}$$

. . . .(3.4 d)

$$\textbf{div } \overline{\mathbf{D}} = \frac{1}{r^2} \frac{\partial}{\partial r} (r^2 D_r) + \frac{1}{r \sin \theta} \frac{\partial}{\partial \theta} (D_\theta \sin \theta) + \frac{1}{r \sin \theta} \frac{\partial D_\phi}{\partial \phi}$$

. . . .(3.4 e)

Problem 3.6 : Let $\overline{\mathbf{D}} = y^2 z^3 \overline{a}_x + 2xyz^3 \overline{a}_y + 3xy^2 z^2 \overline{a}_z$ pC/m^2 in

n free space. (a) Find the total electric flux passing the surface $x = 3$, $0 \le y \le 2$, $0 \le z \le 1$ in a direction away from the origin. (b) Find $|\bar{E}|$ at P (3.2.1). (c) Find the total charge contained in an incremental sphere having a radius of 2 μm centered at P (3,2,1).

Solution :

(a) To the plane $x = 3$, $\bar{a}_x$ is perpendicular. $\therefore$ $\overline{dS} = dy\ dz\ \bar{a}_x$

$$\Psi = Q = \int \bar{D}\ \overline{dS} = \int_0^2 \int_0^1 y^2 z^3\ dy\ dz \Big|_{x=3} = \left[\frac{y^3}{3}\right]_0^2 \left[\frac{z^4}{4}\right]_0^1$$

$$= \frac{8}{3} \times \frac{1}{4} = 0.667 \text{ pC}$$

(b) $\bar{D}$ at $P = (2)^2 (1)^3 \bar{a}_x + 2(3)(2)(1)^3 \bar{a}_y + 3(3)(2)^2 (1)^1 \bar{a}_z$

$$= 4\bar{a}_x + 12\bar{a}_y + 36\bar{a}_z \text{ pC/m}^2$$

$\therefore |\bar{D}| = \sqrt{(4)^2 + (12)^2 + (36)^2} = \sqrt{1456}$ pC/m^2

Now $|\bar{D}| = \varepsilon_0 |\bar{E}|$

$$\therefore |\bar{E}| \text{ at } P(3,2,1) = \frac{\sqrt{1456} \times 10^{-12}}{8.854 \times 10^{-12}} = 4.31 \text{ V/m}$$

(c) Using eqn.(3.4c), we have

$$\rho_v = \frac{\partial}{\partial x}(y^2 z^3) + \frac{\partial}{\partial y}(2xyz^3) + \frac{\partial}{\partial y}(3xy^2z^2)$$

$$= 0 + 2xz^3 + 6xzy^2 = 2xz^3 + 6xzy^2 \text{ pC/m}^3$$

$\therefore \rho_v$ at $P(3,2,1) = 2(3)(1)^3 + 6(3)(1)(2)^2 = 78$ pC/m^3

Charge enclosed in a sphere having a radius of 2 μm

$$= \rho_v \times \text{volume of sphere}$$

$$\doteq 78 \times 10^{-12} \times \frac{4}{3} \times \pi \times (2 \times 10^{-6})^3 \doteq 2.614 \times 10^{-27} \text{ C}$$

Problem 3.7 : Find a numerical value for the divergence of $\bar{D}$ at the point indicated if (a) $\bar{D} = 20xy^2(z+1)\bar{a}_x + 20x^2y(z+1)\bar{a}_y + 10x^2y^2\bar{a}_z$ C/m^2 at P_A (0.3, 0.4, 0.5) ; (b) $\bar{D} = 4\rho z \sin\phi\ \bar{a}_\rho + 2\rho z \cos\phi\ \bar{a}_\phi + 2\rho^2 \sin\phi\ \bar{a}_z$ C/m^2 at P_B (1, π/2, 2) ; (c) $\bar{D} = \sin\theta\cos\phi\ \bar{a}_r + \cos\theta\ \cos\phi$

$\bar{\mathbf{a}}_\theta - \sin\phi\ \bar{\mathbf{a}}_\phi$ **C/m²** **at** P_C **(2,** $\theta = \pi/3, \phi = \pi/6$**).**

Solution :

(a) Using eqn.(3.4c), we have

$$\text{div}\,\overline{\text{D}} = \frac{\partial}{\partial x}\left[20\,xy^2(z+1)\right] + \frac{\partial}{\partial y}\left[20\,x^2y(z+1)\right] + \frac{\partial}{\partial z}\left[10\,x^2y^2\right]$$

$$= 20\,y^2(z+1) + 20\,x^2(z+1) + 0\ \text{C/m}^3$$

$$\therefore\ \text{div}\,\overline{\text{D}}\ \text{at}\ P_A = 20\,(0.40)^2(0.5+1) + 20\,(0.3)^2(0.5+1) = 7.5\ \text{C/m}^3$$

(b) Using eqn.(3.4d), we have

$$\text{div}\,\overline{\text{D}} = \frac{1}{\rho}\frac{\partial}{\partial\rho}(\rho\times 4\rho z\sin\phi) + \frac{1}{\rho}\frac{\partial}{\partial\rho}(2\rho z\cos\phi) + \frac{\partial}{\partial z}(2\rho^2\sin\phi)$$

$$= \frac{1}{\rho}(8\rho z\sin\phi) + \frac{1}{\rho}(-2\rho z\sin\phi) + 0$$

$$\therefore\ \text{div}\,\overline{\text{D}}\ \text{at}\ \text{P}_\text{B} = 8\times 2\times\sin\frac{\pi}{2} - 2\times 2\times\sin\frac{\pi}{2} = 12\ \text{C/m}^3$$

(c) Using eqn.(3.4e),we have

$$\text{div}\,\overline{\text{D}} = \frac{1}{r^2}\frac{\partial}{\partial r}(r^2\times\sin\theta\cos\phi) + \frac{1}{r\sin\theta}\frac{\partial}{\partial\theta}(\sin\theta\times\cos\theta\cos\phi) + \frac{1}{r\sin\theta}\frac{\partial}{\partial\theta}(-\sin\phi)$$

$$= \frac{2\sin\theta\cos\phi}{r} + \frac{\cos 2\theta\cos\phi}{r\sin\theta} - \frac{\cos\phi}{r\sin\theta}$$

$$\therefore\ \text{div}\,\overline{\text{D}}\ \text{at}\ P_C = \frac{2\sin(\pi/3)\cos(\pi/6)}{2} + \frac{\cos(2\pi/3)\cos(\pi/6)}{2\sin(\pi/3)} - \frac{\cos(\pi/6)}{2\sin(\pi/3)}$$

$$= 0.75 - 0.1443 - 0.5 = 0.1057\ \text{C/m}^3$$

3.5 : MAXWELL'S FIRST EQUATION IN ELECTROSTATICS :

According to Gauss's law

$$Q = \oint_S \overline{\text{D}}\cdot\overline{\text{dS}}$$

Dividing both sides of above equation by Δv, and take the limit as the volume shrinks to zero, we have

$$\lim_{\Delta v \to 0} \frac{Q}{\Delta v} = \lim_{\Delta v \to 0} \frac{\oint_S \overline{D} \cdot \overline{dS}}{\Delta v}$$

Now the volume charge density ρ_v is defined as

$$\rho_v = \lim_{\Delta v \to 0} \frac{Q}{\Delta v}$$

and divergence of electric flux density by

$$\text{div } \overline{D} = \lim_{\Delta v \to 0} = \frac{\oint_S \overline{D} \cdot \overline{dS}}{\Delta v}$$

Hence we have

$$\rho_v = \text{div} \overline{D} .$$

This is Maxwell's first equation in electrostatics. It states that

The electric flux per unit volume leaving a vanishingly small volume unit is exactly equal to the volume charge density, there. This equation is the point form of Gauss's law.

The del operator $\overline{\nabla}$ is defined only in cartesian coordinate system by the equation.

$$\overline{\nabla} = \frac{\partial}{\partial x} \overline{a}_x + \frac{\partial}{\partial y} \overline{a}_y + \frac{\partial}{\partial z} \overline{a}_z$$

Let $\overline{D} = D_x \overline{a}_x + D_y \overline{a}_y + D_z \overline{a}_z$.

Then. $\overline{\nabla} \cdot \overline{D} = \left(\frac{\partial}{\partial x} \overline{a}_x + \frac{\partial}{\partial y} \overline{a}_y + \frac{\partial}{\partial z} \overline{a}_z \right) \cdot (D_x \overline{a}_x + D_y \overline{a}_y + D_z \overline{a}_z)$

$$= \frac{\partial D_x}{\partial x} + \frac{\partial D_y}{\partial y} + \frac{\partial D_y}{\partial z}$$

which is same as div $\overline{D}$. Hence

$$\rho_v = \text{div } \overline{D} = \overline{\nabla} \cdot \overline{D}$$

Problem 3.8 : Find the volume charge density that is associated with each of the following fields: (a) $\overline{D} = xy^2 \overline{a}_x + yx^2 \overline{a}_y + z \overline{a}_z$ C/m^2 ; (b) $\overline{D} = \rho z^2 \sin^2 \phi \, \overline{a}_\rho + \rho z^2 \sin \phi \cos \phi \, \overline{a}_\phi + \rho^2 z \sin^2 \phi \, \overline{a}_z$ C/m^2 ; (c) $\overline{a}_r$ C/m^2.

Solution :

(a) Using eqn.(3.4 c), we have

$$\rho_v = \bar{\nabla} \cdot \bar{D} = \frac{\partial}{\partial x}(x y^2) + \frac{\partial}{\partial y}(y x^2) + \frac{\partial}{\partial z}(z) = y^2 + x^2 + 1 \ \text{C/m}^3$$

(b) Using eqn.(3.4d),we have

$$\rho_v = \bar{\nabla} \cdot \bar{D} = \frac{1}{\rho}\frac{\partial}{\partial \rho}(\rho x \rho z^2 \sin^2 \phi) + \frac{1}{\rho}\frac{\partial}{\partial \phi}(\rho z^2 \sin \phi \cos \phi) + \frac{\partial}{\partial z}(\rho^2 z \sin^2 \phi)$$

$$= 2 z^2 \sin^2 \phi + z^2 \cos 2\phi + \rho^2 \sin^2 \phi = z^2 + \rho^2 \sin^2 \phi \ \text{C/m}^3$$

(c) Using eqn.(3.4 e),we have

$$\rho_v = \bar{\nabla} \cdot \bar{D} = \frac{1}{r^2}\frac{\partial}{\partial r}(r^2 \times 1) = \frac{2}{r} \ \text{C/m}^3$$

3.6 : STATE AND PROVE DIVERGENCE THEOREM :

According to Gauss's law

$$Q = \oint_S \bar{D} \cdot \overline{dS}$$

and from the defination of volume charge density, we have

$$Q = \int_{vol} \rho_v \, dv$$

and from Maxwell's first equation in electrostatics, we have

$$\rho_v = \bar{\nabla} \ \bar{D}$$

Hence

$$\oint_S \bar{D} \ \overline{dS} = \int_{vol} \rho_v \, dv = \int_{vol} \bar{\nabla} \cdot \bar{D} \, dv$$

From the first and last expressions we have

$$\boxed{\oint_S \bar{\mathbf{D}} \cdot \overline{\mathbf{dS}} = \int_{vol} \bar{\nabla} \ \bar{\mathbf{D}} \, dv} \qquad \text{.........(3.5)}$$

which is the divergence theorem. It states that

The integral of the normal component of any vector field over a closed surface is equal to the divergence of this vector field throughout the volume enclosed by the closed surface.

Problem 3.9 : Given the flux density $\overline{D} = (2\cos\theta/r^3)\,\overline{a}_r + (\sin\theta/r^3)\,\overline{a}_\theta$ C/m², evaluate both sides of the divergence theorem for the region defined by $1 < r < 2, 0 < \theta < \pi/2, 0 < \phi < \pi/2$.

Solution :

The divergence theorem is $\oint_S \overline{D}\,.\,\overline{dS} = \int_{vol} \overline{\nabla}\cdot\overline{D}\,dv$

$$\text{L.H.S.} = \oint_S \overline{D}\cdot\overline{dS} = \int_{r=1} \overline{D}\cdot\overline{dS} + \int_{r=2} \overline{D}\cdot\overline{dS} + \int_{\theta=0} \overline{D}\cdot\overline{dS} + \int_{\theta=\pi/2} \overline{D}\cdot\overline{dS}$$

$$= \int_0^{\frac{\pi}{2}}\int_0^{\frac{\pi}{2}} \overline{D}\cdot\underset{(r=1)}{(-\,r^2\sin\theta\,d\theta\,d\phi\,\overline{a}_r)} + \int_0^{\frac{\pi}{2}}\int_0^{\frac{\pi}{2}} \overline{D}\cdot\underset{(r=2)}{(r^2\sin\theta\,d\theta\,d\phi\,\overline{a}_r)}$$

$$+ \int_1^{2}\int_0^{\frac{\pi}{2}} \overline{D}\cdot\underset{(\theta=0)}{(-\,r\sin\theta\,dr\,d\phi\,\overline{a}_\theta)} + \int_1^{2}\int_0^{\frac{\pi}{2}} \overline{D}\cdot\underset{(\theta=\pi/2)}{(r\sin\theta\,dr\,d\phi\,\overline{a}_\theta)}$$

$$= \int_0^{\frac{\pi}{2}}\int_0^{\frac{\pi}{2}} \underset{(r=1)}{\frac{-\,2\cos\theta\sin\theta}{r}}\,d\theta\,d\phi + \int_0^{\frac{\pi}{2}}\int_0^{\frac{\pi}{2}} \underset{(r=2)}{\frac{2\cos\theta\sin\theta}{r}}\,d\theta\,d\phi\,+$$

$$\int_0^{2}\int_0^{\frac{\pi}{2}} \underset{(\theta=0)}{-\frac{\sin^2\theta}{r^2}}\,dr\,d\phi + \int_0^{2}\int_0^{\frac{\pi}{2}} \underset{(\theta=\pi/2)}{\frac{\sin^2\theta}{r^2}}\,dr\,d\phi$$

$$= \left[\frac{\cos 2\theta}{2}\right]_0^{\pi/2}[\phi]_0^{\pi/2} - \frac{1}{2}\left[\frac{\cos 2\theta}{2}\right]_0^{\pi/2}[\phi]_0^{\pi/2} + 0 + \left[\frac{-1}{r}\right]_1^2[\phi]_0^{\pi/2}$$

$$= -\frac{\pi}{2} + \frac{\pi}{4} + \frac{\pi}{4} = 0$$

$$\text{RHS} = \int_{vol} \overline{\nabla}\cdot\overline{D}\,dv$$

$$\overline{\nabla}\cdot\overline{D} = \frac{1}{r^2}\frac{\partial}{\partial r}\left(r^2\times\frac{2\cos\theta}{r^3}\right) + \frac{1}{r\sin\theta}\frac{\partial}{\partial\theta}\left(\sin\theta\times\frac{\sin\theta}{r^3}\right)$$

$$= \frac{-\,2\cos\theta}{r^4} + \frac{2\cos\theta}{r^4} = 0$$

$\therefore$ RHS = 0

Hence LHS = RHS = 0

Problem 3.10 : An empty metal paint can is placed on a marble table, the lid is removed, and both parts are discharged (honorably) by touching them to ground. An insulating nylon thread is glued to the center of the lid, and a penny, a nickel, and a dime are glued to the thread so that they are not touching each other. The penny is given a charge of + 5 nC, and the nickel and dime are discharged. The assembly is lowered into the can so that the coins hang clear of all walls, and the lid is secured. The outside of the can is again touched momentarily to ground. The device is carefully disassembled with insulating gloves and tools. (a) What charges are found on each of the five metallic pieces ? (b) If the penny had been given a charge of + 5 nC, the dime a charge of –2 nC, and the nickel a charge of –1 nC, what would the final charge arrangement have been ?

Solution :

(a) The metal paint can along with the lid forms a closed surface. Inside this closed surface a penny, a nickel and a dime are present. The charge on penny is + 5 nC. The charge on nickel anddime is zero as it do not encloses penny. Further the charge on penny remains as it is i.e. 5 nC. Since can and lid encloses the penny, it carries a charge of –5 nC.

(b) The charge on penny, nickel, and dime will be 5 nC, –1 nC and –2 nC respectively. Total charge enclosed by can and lid = 5 – 1 – 2 = 2 nC. Hence charge on can and lid is –2 nC.

Problem 3.11 : Point charges of 15 nC each are symmetrically located at (4,4,0), (4,–4, 0), (–4,4,0), and (–4,–4,0), and a uniform line charge of 50 nC/m lies at $x = 0, y = 8$, all z. (a) Find $\overline{\mathbf{D}}$ at the origin. (b) How much electric flux crosses the $y = 0$ plane ? (c) How much electric flux leaves the surface of a sphere of radius 5 m centered at (0,6,0) ?

Solution : Fig.P3.11 shows the location of four point charges. These charges are located on the corners of a square, origin being the center of the square. Due to the charges located at A and C, the electric flux density vectors get cancelled out. Also due

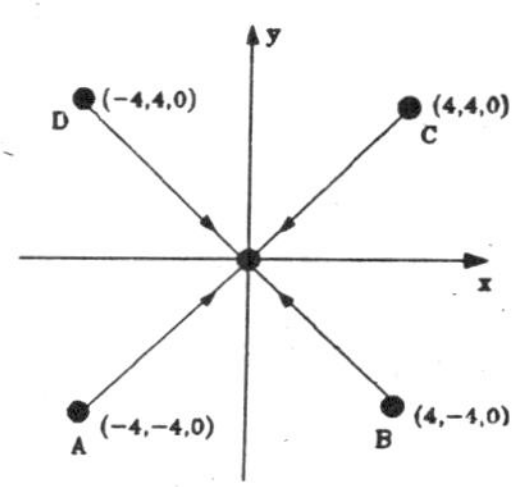

FIG.P3.11

to the charges located at B and D, the electric flux density is zero. Hence due to four point charges, $\overline{D}$ at origin is zero. Now consider the line charge.

$$\overline{D} = \varepsilon_0 \overline{E} = \varepsilon_0 \left[\frac{\rho_L}{2\pi\varepsilon_0 d} \overline{a}_{12} \right] = \frac{\rho_L}{2\pi d} \overline{a}_{12}$$

Point 2 is that point at which $\overline{D}$ is desired = (0,0,0)

Point 1 is the foot of perpendicular dropped from point 2 on the line charge = (0,8,0)

$$\therefore \overline{R}_{12} = (0-8)\,\overline{a}_y = -8\,\overline{a}_y$$

$$d = |\overline{R}_{12}| = \sqrt{(-8)^2} = 8$$

$$\overline{a}_{12} = \frac{\overline{R}_{12}}{|\overline{R}_{12}|} = -\frac{8\,\overline{a}_y}{8} = -\overline{a}_y$$

$$\therefore \overline{D} = \frac{50\times 10^{-9}}{2\pi(8)}(-\overline{a}_y) = -994.718\,\overline{a}_y \text{ pC/m}^2$$

(b) Any point in y = 0 plane is given by (x,0,z). At this point, $\overline{D}$ due to four point charges is given by

$$\overline{D} = \frac{15\times 10^{-9}}{4\pi}\left\{ \frac{(x-4)\,\overline{a}_x + (0-4)\,\overline{a}_y + (z-0)\,\overline{a}_z}{[(x-4)^2+16+z^2]^{3/2}} + \frac{(x-4)\,\overline{a}_x + (0+4)\,\overline{a}_y + (z-0)\,\overline{a}_z}{[(x-4)^2+16+z^2]^{3/2}} + \frac{(x+4)\,\overline{a}_x + (0-4)\,\overline{a}_y + (z-0)\,\overline{a}_z}{[(x-4)^2+16+z^2]^{3/2}} + \frac{(x+4)\,\overline{a}_x + (0+4)\,\overline{a}_y + (z-0)\,\overline{a}_z}{[(x+4)^2+16+z^2]^{3/2}} \right\}$$

$$= \frac{15\times 10^{-9}}{4\pi}\left\{ \frac{2[(x-4)\,\overline{a}_x + z\,\overline{a}_z]}{[(x-4)^2+16+z^2]^{3/2}} + \frac{2[(x+4)\,\overline{a}_x + z\,\overline{a}_z]}{[(x+4)^2+16+z^2]^{3/2}} \right\}$$

$$\overline{dS} = \pm\,(dx\,dy)\,\overline{a}_y$$

$\therefore \overline{D}\cdot\overline{dS} = 0$ as $\overline{a}_x \cdot \overline{a}_y$ and $\overline{a}_z \cdot \overline{a}_y$ is equal to zero. Hence the flux crossing the $y = 0$ plane due to four point charges is zero. $\overline{D}$ at $(x, 0, z)$ due to line charge 50 nC/m at $x = 0, y = 8$ is given by

$$\overline{D} = \frac{\rho_L}{2\pi d}\overline{a}_{12}$$

Point 2 is that point at which $\overline{D}$ is desired = $(x, 0, z)$

Point 1 is the foot of perpendicular dropped from point 2 on the line charge.
$= (0, 8, z)$

$$\overline{R}_{12} = (x - 0)\,\overline{a}_x + (0 - 8)\,\overline{a}_y = x\,\overline{a}_x - 8\,\overline{a}_y$$

$$d = |\overline{R}_{12}| = \sqrt{x^2 + 64}$$

$$\overline{a}_{12} = \frac{x\,\overline{a}_x - 8\,\overline{a}_y}{\sqrt{x^2 + 64}}$$

$$\therefore\ \overline{D} = \frac{50 \times 10^{-9}}{2\pi\sqrt{x^2 + 64}}\left[\frac{x\,\overline{a}_x - 8\,\overline{a}_y}{\sqrt{x^2 + 64}}\right]$$

$$\overline{dS} = \pm\,(dx\,dz\,\overline{a}_y)\ ;\ \overline{D}\cdot\overline{dS} = \frac{\pm\,50 \times 10^{-9}\,(-8\,dx\,dz)}{2\pi\,(x^2 + 64)}$$

$$\therefore\ \Psi = \int \overline{D}\cdot\overline{dS} = \int_{-\infty}^{\infty}\int_{-\infty}^{\infty} \pm\left[\frac{-400 \times 10^{-9}\,dx\,dz}{2\pi\,(x^2 + 64)}\right]$$

$$= \pm\left(\frac{-50 \times 10^{-9}}{2\pi}\right)\left[\tan^{-1}\left(\frac{x}{8}\right)\right]_{-\infty}^{\infty}\left[z\right]_{-\infty}^{\infty} = \infty$$

Since the length of line charge is infinite, infinite amount of flux is crossing through the $y = 0$ plane.

(c) Equation of the sphere of radius 5 m centered at (0,6,0) is given by

$(x - 0)^2 + (y - 6)^2 + (z - 0)^2 = (5)^2$ or

$$x^2 + (y - 6)^2 + z^2 = 25 \qquad \text{---------(1)}$$

Substitute (4,– 4,0) in equation (1)

LHS $= (4)^2 + (4 - 6)^2 + (0)^2 = 16 + 4 = 20 <$ RHS of eqn.(1)

Hence the point charge located at (4,4,0) lies inside the sphere.

$\therefore\ \Psi_1 = 15$ nC

Substitute (4,- 4,0) in equation (1)

LHS $= (4)^2 + (-4 - 6)^2 + (0)^2 = 16 + 100 = 116 >$ RHS of eqn.(1)

Hence the point charge located at (4, - 4, 0) lies outside the sphere.

$\therefore\ \Psi_2 = 0$

Substitute (- 4,4,0) in equation (1)

LHS $= (-4)^2 + (4 - 6)^2 + (0)^2 = 20 <$ RHS of eqn.(1)

Hence the charge located at (- 4, 4, 0) lies inside the sphere.

$\therefore\ \Psi_3 = 15$ nC

Substitute (- 4, - 4, 0) in equation (1)

LHS = $(-4)^2 + (-4-6)^2 + (0)^2 = 116 >$ RHS of eqn.(1)

Hence the point charge located at (- 4, - 4,0) lies outside the sphere.

$\therefore\ \Psi_4 = 0$

Substitute $x = 0, y = 8$ (equations of line charge) in equation (1).

$(0)^2 + (8-6)^2 + z^2 = 25$ or $z^2 = 21$ or $z = \pm\ 4.583$. Hence the portion of the line charge from $z = -4.583$ to $z = 4.583$ lies inside the sphere.

Total length = 4.583 - (- 4.583) = 9.166

Charge on above length = $50 \times 9.166 = 458.3$ nC.

[or use $Q = \int dQ = \int \rho_L\ dL = \int_{-4.583}^{4.583} 50\,dz$ nC]

Hence due to line charge flux leaving the sphere is $\Psi_5 = 458.3$ nC.

$\therefore$ Total flux leaving the sphere $\Psi = \Psi_1 + \Psi_2 + \Psi_3 + \Psi_4 + \Psi_5$

$= 15 + 0 + 15 + 0 + 458.3 = 488.3$ nC

Problem 3.12 : A nonuniform surface charge density of 5 $\rho/(\rho^2 + 1)$ nC/m^2 lies in the plane $z = 2$ wherever $\rho < 5$; $\rho_S = 0$ for $\rho < 5$. (a) How much electric flux leaves the circular region $\rho < 5, z = 2$? (b) How much electric flux corsses the $z = 0$ plane in the $-\ \bar{a}_z$ direction ? (c) How much electric flux leaves the cylinder $\rho = 3$ in the $\bar{a}_\rho$ direction ?

Solution :

(a) To the plane z = 2, $\pm\ \bar{a}_z$ is perpendicular.

Since circular surface is given we will use cylinderical coordinate system.

$\overline{dS} = \pm\ (\rho\ d\rho\ d\phi)\ \bar{a}_z$ or $dS = \rho\ d\rho\ d\phi$

$$\therefore\ \Psi = Q = \int dQ = \int \rho_S\ dS = \int_0^5 \int_0^{2\pi} \frac{5\rho}{\rho^2 + 1}\ \rho\ d\rho\ d\phi$$

$$= 5\,[\rho - \tan^{-1}\rho]_0^5\ [\phi]_0^{2\pi}$$

$$\therefore\ \Psi = 5 \times 3.6266 \times 2\pi = 113.933 \text{ nC}$$

(b) Half of the above flux leaves the surface z = 2 in + $\bar{a}_z$ direction where as remaining half of the flux in $-\bar{a}_z$ direction. Hence the flux crossing the z= 0 plane in $-\bar{a}_z$ direction will be 113.933/2 = 56.966 nc.

$$\text{(c) } \Psi = Q = \int dQ = \int \rho_s \, dS = \int_0^3 \int_0^{2\pi} \frac{5\rho}{\rho^2 + 1} \rho \, d\rho \, d\phi$$

$$= 5\left[\rho - \tan^{-1}\rho\right]_0^3 \left[\phi\right]_0^{2\pi} = 55.008 \text{ nC}$$

Problem 3.13 The spherical surfaces r = 1, 2, and 3 carry surface charge densities of 20, - 9, and 2 nC/m^2, respectively. (a) How much electric flux leaves the surface r = 5 ? (b) Find $\overline{D}$ at P (1, –1, 2).

Solution :

(a) Sphere r = 5 encloses the spherical surfaces r = 1, 2 and 3. Hence

Q = Charge enclosed by the spherical surface (r = 5)

= Charge present on (r= 1) + Charge present on (r= 2) + Charge present on (r= 3)

$$Q = 20 \times 4\pi (1)^2 - 9 \times 4\pi (2)^2 + 2 \times 4\pi (3)^2 = 25.1327 \text{ nC}$$

Hence flux leaving through the surface (r= 5) = $\Psi = Q$ = 25.1327 nC

(b) r coordinate of point $P = \sqrt{(1)^2 + (-1)^2 + (2)^2} = \sqrt{6} = 2.449$

Point P (1, - 1,2) lies on the spherical surface r = 2.449. This surface encloses the spherical surfaces r = 1 and r = 2. Hence

Q = Charge enclosed by (r= 2.449) surface

= Charge present at (r= 1) + Charge present on (r= 2)

$$Q = 20 \times 4\pi (1)^2 - 9 \times 4\pi (2)^2 = -64\pi \text{ nC.}$$

$$\therefore \overline{D} \text{ at } P = \frac{\Psi}{\text{area}} \bar{a}_r = \frac{Q}{4\pi r^2} \bar{a}_r = \frac{-64\pi}{4\pi(\sqrt{6})^2} \bar{a}_r = -2.667 \, \bar{a}_r \text{ nC/m}^2$$

Problem 3.14 : A point charge of 6 μC is located at the origin, a uniform line charge density of 180 nC/m lies along the x axis, and a uniform sheet of charge equal to 25 nC/m^2 lies in the z = 0 plane. (a) Find $\overline{D}$ at A (0,0,4). (b) Find $\overline{D}$ at B (1,2,4). (c) Calculate the total electric flux leaving the surface of a sphere of 4 - m radius centered at the origin.

Solution :

(a) $\overline{D}_1$, due to point charge $= \dfrac{6\times 10^{-6}}{4\pi(4)^2}\left[\dfrac{(4-0)\,\overline{a}_z}{\sqrt{(4)^2}=4}\right] = 29.842\ \overline{a}_z\ \text{nC/m}^2$

$\overline{D}$ due to line charge $= \dfrac{\rho_L}{2\pi d}\ \overline{a}_{12}$

Equations of the line charge (x axis) are $y = 0, z = 0$.

Point 2 is that point at which electric flux density is desired = (0, 0, 4)

Point 1 is the foot of perpendicular dropped from point 2 on the line charge = (0,0,0)

$\overline{R}_{12} = (4-0)\ a_z = 4\,\overline{a}_z$ and $d = |\overline{R}_{12}| = \sqrt{(4)^2} = 4$

$\overline{a}_{12} = \overline{R}_{12} / |\overline{R}_{12}| = \overline{a}_z$

$\therefore\ \overline{D}_2$ due to line charge $= \dfrac{180\times 10^{-9}}{2\pi(4)}\ \overline{a}_z = 7.162\ \overline{a}_z\ \text{nC/m}^2$

$\overline{D}$ due to surface charge $= \dfrac{\rho_S}{2}\ \overline{a}_N$

Surface charge lies in $z = 0$ plane.

Since z coordinate of point A is 4 which is greater, than $z = 0$, $\overline{a}_N = \overline{a}_z$

$\therefore\ \overline{D}_3$ due to surface charge $= \dfrac{25\times 10^{-9}}{2}\ \overline{a}_z = 12.5\ \overline{a}_z\ \text{nC/m}^2$

$\therefore\ \overline{D}$ at point $A = \overline{D}_1 + \overline{D}_2 + \overline{D}_3 = 29.842\,\overline{a}_z + 7.162\,\overline{a}_z + 12.5\,\overline{a}_z$

$= 49.504\ \overline{a}_z\ \text{nC/m}^2$

(b) Similarly

$\overline{D}_1$ at point B due point charge $= 4.962\,\overline{a}_x + 9.923\,\overline{a}_y + 19.846\,\overline{a}_z\ \text{nC/m}^2$

$\overline{D}_2$ at point B due to line charge $= 2.865\,\overline{a}_y + 5.730\,\overline{a}_z\ \text{nC/m}^2$

$\overline{D}_3$ at point B due to surface charge $= 12.5\,\overline{a}_z\ \text{nC/m}^2$

$\therefore\ \overline{D} = \overline{D}_1 + \overline{D}_2 + \overline{D}_3 = 4.962\,\overline{a}_x + 12.788\,\overline{a}_y + 38.076\,\overline{a}_z\ \text{nC/m}^2$

(c) Point charge 6 μC is located at origin. This charge is completely enclosed by the sphere of radius 4m centered at origin.

$\therefore\ \Psi_1 = 6\ \mu\text{C}$

Sphere of radius 4m intersect the x axis at $x = -4$ and $x = 4$. Hence the charge on the line charge which lies inside the sphere $= 180\times(8) = 1440$ nC [or use $Q = \int dQ = \int \rho_L \, dL = \int \rho_L \, dx = \int_{-4}^{4} 180 \, dx$ nC]

$\therefore \; \Psi_2 = 1440$ nC

Sphere of radius 4 m intersect the $z = 0$ plane in a circle of radius 4 m. Area of the circle = $\pi (4)^2 = 16\pi$ m^2. Hence charge on the surface charge which lies inside the sphere = $25 \times (16\pi) = 1256.637$ nC.

[or use $Q = \int dQ = \int \rho_s \, dS = \int_0^4 \int_0^{2\pi} 25 \rho \, d\rho \, d\phi$ nC]

$\therefore \; \Psi_3 = 1256.637$ nC

$\therefore \; \Psi = \Psi_1 + \Psi_2 + \Psi_3 = 6000 + 1440 + 1256.637 = 8696.637$ nC

Problem 3.15 : If $\overline{D} = 3y^2 \overline{a}_x + 3x^2 y \overline{a}_y + 5\overline{a}_z$ C/m^2, find the total charge enclosed within the region $0 < x, y, z < 2$ by evaluating one or more surface integrals.

Solution :

The six surfaces are located by $x = 0$, $x = 2$, $y = 0$, $y = 2$, $z = 0$, and $z = 2$

$$Q = \int_{x=0} \overline{D} \cdot \overline{dS} + \int_{x=2} \overline{D} \cdot \overline{dS} + \int_{y=0} \overline{D} \cdot \overline{dS} + \int_{y=2} \overline{D} \cdot \overline{dS} + \int_{z=0} \overline{D} \cdot \overline{dS} + \int_{z=2} \overline{D} \cdot \overline{dS}$$

$$= \int_0^2 \int_0^2 \underset{(x=0)}{\overline{D} \cdot (-\, dy \, dz \, \overline{a}_x)} + \int_0^2 \int_0^2 \underset{(x=2)}{\overline{D} \cdot (dy \, dz \, \overline{a}_x)} +$$

$$\int_0^2 \int_0^2 \underset{(y=0)}{\overline{D} \cdot (-\, dx \, dz \, \overline{a}_y)} + \int_0^2 \int_0^2 \underset{(y=2)}{\overline{D} \cdot (dx \, dz \, \overline{a}_y)} +$$

$$\int_0^2 \int_0^2 \underset{(z=0)}{\overline{D} \cdot (-\, dx \, dy \, \overline{a}_z)} + \int_0^2 \int_0^2 \underset{(z=2)}{\overline{D} \cdot (dx \, dy \, \overline{a}_z)}$$

$$= \int_0^2 \int_0^2 \underset{(x=0)}{(-\, 3y^2 dy \, dz)} + \int_0^2 \int_0^2 \underset{(x=2)}{(3y^2 dy \, dz)} +$$

$$\int_0^2 \int_0^2 \underset{(y=0)}{(-\, 3x^2 y \, dx \, dz)} + \int_0^2 \int_0^2 \underset{(y=2)}{(3x^2 y \, dx \, dz)} +$$

$$\int_0^2 \int_0^2 \underset{(z=0)}{(-\, 5 \, dx \, dy)} + \int_0^2 \int_0^2 \underset{(z=2)}{(5 \, dx \, dy)}$$

$$\therefore\ Q = -3\left[\frac{y^3}{3}\right]_0^2\left[z\right]_0^2 + 3\left[\frac{y^3}{3}\right]_0^2\left[z\right]_0^2 + 0 + 3(2)\left[\frac{x^3}{3}\right]_0^2\left[z\right]_0^2$$

$$-5\left[x\right]_0^2\left[y\right]_0^2 + 5\left[x\right]_0^2\left[z\right]_0^2$$

$$= -16 + 16 + 0 + 32 - 20 + 20 = 32\ \text{C}$$

Problem 3.16 : Uniform line charges of 20 nC/ m each lie in the $z = 0$ plane at $x = 0, \pm 1, \pm 2, \ldots, \pm 5$. (a) Find the total ˙ctric flux leaving the spherical surface $r = 2.5$. (b) Find $\overline{D}$ at P (0,2.5,4)

Solution :

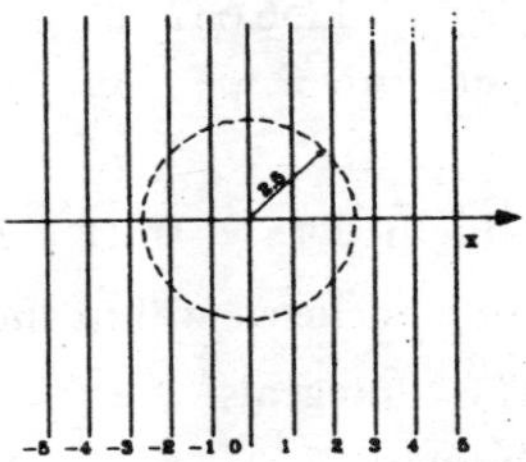

FIG.P3.16

Fig.P3.16 shows $z = 0\,(xy)$ plane along with line charges at $x = 0, \pm 1, \pm 2, \ldots, \pm 5$. The sphere $r = 2.5$ intersect $z = 0$ plane in a circle of radius 2.5 m. Equation of the circle centred at origin and of radius 2.5m is given by $x^2 + y^2 = (2.5)^2 = 6.25$ ....(1)

Substitute $x = 0$ in eqn. (1), we have $y^2 = 6.25$ or $y = \pm\sqrt{6.25} = \pm 2.5$

Hence 2.5 - (- 2.5) = 5 m of line charge located at $x = 0$ lies inside the circle.

Charge on 5 m length of line charge = 5 × 20 = 100 nC

Substitute $x = 1$ in eqn.(1), we have $(1)^2 + y^2 = 6.25$ or $y = \pm 2.29128$

Hence 2.29128 - (- 2.29128) = 4.58256 m of line charge located at $x = 1$ lies inside the circle.

Charge on 4.58256 m length of line charge = 4.58256 × 20 = 91.651 nC.

Similarly for the line charge located at $x = -1$, we have $y = \pm 2.29128$, so that charge = 91.651 nC.

Substitute $x = 2$ in eqn.(1), we have $(2)^2 + y^2 = 6.25$ or $y = \pm 1.5$

Hence 1.5 - (-1.5) = 3 m of line charge located at $x = 2$ lies inside the circle.

Charge on 3 m length of line charge = 3 × 20 = 60 nC.

Similarly for the line charge located at $x = -2$, we have $y = \pm 1.5$, so that charge = 60 nC.

Do not consider line charges located at $x = \pm 3, \pm 4, \pm 5$ as these line charges lies outside the sphere. Hence total flux leaving the sphere is equal to

$\Psi = 100 + 91.651 + 91.651 + 60 + 60 = 403.302$ nC

(b) The line charge is located by the equations $x = a, z = 0$ (where $a = 0, \pm 1, \pm 2, \ldots\ldots \pm 5$).

$\overline{\text{D}}$ due to line charge $= \dfrac{\rho_L}{2\pi d}\,\bar{a}_{12}$

Point 2 is that point at which $\overline{\text{D}}$ is desired = (0,2.5,4)

Point 1 is the foot of perpendicular dropped from point 2 on the line charge = (a ,2.5,0)

Hence $\overline{\text{R}}_{12} = (0 - a)\,\bar{a}_x + (4 - 0)\,\bar{a}_z = -a\,\bar{a}_x + 4\,\bar{a}_z$

$$d = |\,\overline{\text{R}}_{12}\,| = \sqrt{a^2 + 16} \quad \text{and} \quad \bar{a}_{12} = \frac{\overline{\text{R}}_{12}}{|\,\overline{\text{R}}_{12}\,|} = \frac{-a\,\bar{a}_x + 4\,\bar{a}_z}{\sqrt{a^2 + 16}}$$

$$\therefore\ \overline{\text{D}} = \frac{20}{2\pi\sqrt{a^2 + 16}}\left[\frac{-a\,\bar{a}_x + 4\,\bar{a}_z}{\sqrt{a^2 + 16}}\right] = \frac{10\,(-a\,\bar{a}_x + 4\,\bar{a}_z)}{\pi\,(a^2 + 16)}\ \text{nC/m}^2\text{S}$$

ubstituting $a = 0, \pm 1, \pm 2, \ldots\ldots \pm 5$ in the above expression, we have

$$\overline{\text{D}} \text{ at } P\,(0,2.5,4) = \frac{10}{\pi}\left[\frac{4\,\bar{a}_z}{16} + \frac{(\bar{a}_x + 4\,\bar{a}_z)}{17} + \frac{(-\bar{a}_x + 4\,\bar{a}_z)}{17}\right.$$
$$+ \frac{(2\,\bar{a}_x + 4\,\bar{a}_z)}{20} + \frac{(-2\,\bar{a}_x + 4\,\bar{a}_z)}{20} + \frac{(3\,\bar{a}_x + 4\,\bar{a}_z)}{25} + \frac{(-3\,\bar{a}_x + 4\,\bar{a}_z)}{25}$$
$$\left. + \frac{(4\,\bar{a}_x + 4\,\bar{a}_z)}{32} + \frac{(-4\,\bar{a}_x + 4\,\bar{a}_z)}{32} + \frac{(5\,\bar{a}_x + 4\,\bar{a}_z)}{41} + \frac{(-5\,\bar{a}_x + 4\,\bar{a}_z)}{41}\right]$$

$$\therefore\ \overline{\text{D}} = \frac{10}{\pi}\left(\frac{4}{16} + \frac{8}{17} + \frac{8}{20} + \frac{8}{25} + \frac{8}{32} - \frac{8}{41}\right)\bar{a}_z = 6.002\,\bar{a}_z\ \text{nC/m}^2$$

Problem 3.17 : A point charge of 90 nC is located at the origin, and there are also three uniform surface charge distributions : – 8 nC/m² at $r = 2$, and 1.2 nC/m² at $r = 3$. (a) Find $\overline{\text{D}}$ everywhere. (b) Plot | $\overline{\text{D}}$ | vs. r, $0 < r < 4$.

Solution :

(a) Consider a sphere of radius $r < 1$ m as shown in Fig.P3.17(a). This sphere encloses only the point charge. Hence $\Psi = Q = 90$ nC and

$$\overline{\text{D}} = \frac{90 \times 10^{-9}}{4\pi r^2}\,\bar{a}_r = \frac{7.162}{r^2}\,\bar{a}_r\ \text{nC/m}^2$$

for $0 < r < 1$.

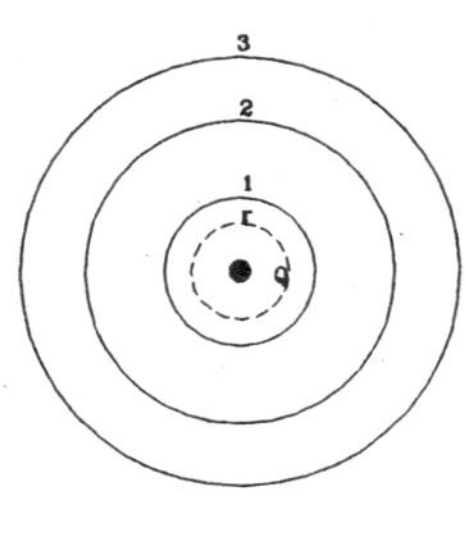

FIG. P 3.17 (a)

Consider a sphere of radius r such that $1 < r < 2$ as shown in Fig.P3.17 (b). This sphere encloses the point charge and surface charge density of - 8 nC / m^2 which is located at $r =$ 1 m. Hence $\Psi =$ charge enclosed by sphere of radius $r =$ 90 + (- 8) 4 π (1)^2 = - 10.531 nC

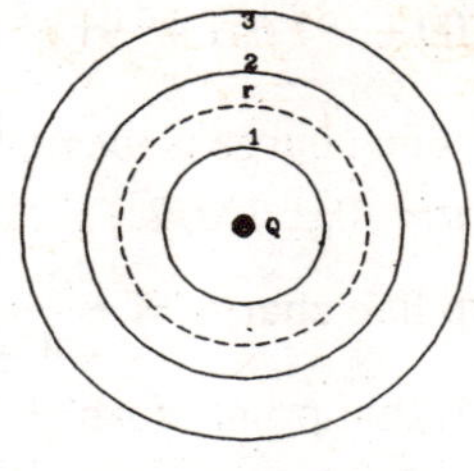

FIG.P 3.17 (b)

$$\therefore\ \overline{\text{D}} = \frac{-10.531}{4\pi r^2}\,\bar{a}_r = \frac{-0.838}{r^2}\,\bar{a}_r \text{ nC/m}^2 \text{ fo}\cdot\ < r < 2$$

Consider a sphere of radius r such that $2 < r < 3$ as shown in Fig.P3.17(c). This sphere encloses the point charge and the surface charge densities of - 8 nC /m^2 (at $r =$ 1) and 4.5 nC/m^2 (at $r =$ 2). Hence $\Psi =$ charge enclosed by the sphere of radius $r =$ 90 + (- 8) 4 π (1)^2 + (4.5) 4π (2)^2 = 215.664 nC

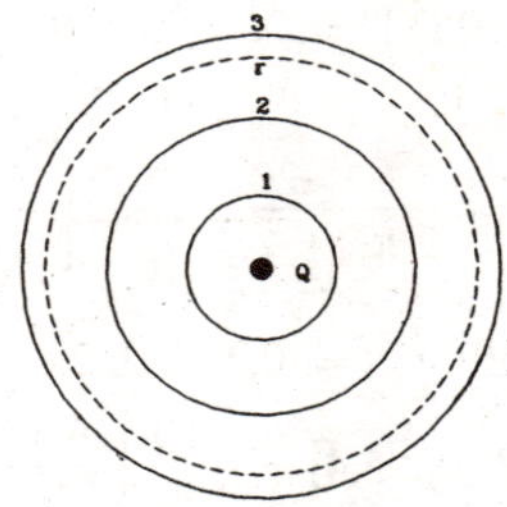

FIG.P3.17 (c)

$$\therefore\ \overline{\text{D}} = \frac{215.664}{4\pi r^2}\,\bar{a}_r = \frac{17.162}{r^2}\,\bar{a}_r \text{ nC/m}^2 \text{ for } 2 < r < 3$$

Consider a sphere of radius r such that $r > 3$ as shown in Fig.P3.17(d). This sphere encloses the point charge and the surface charge densities of - 8 nC/m^2 (at $r =$ 1), 4.5 nC/m^2 (at $r =$ 2) and 1.2 nC/m^2 (at $r =$ 3). Hence $\Psi =$ charge enclosed by the sphere of radius $r =$ 90 + (-8)4 π (1)^2 + (4.5) 4 π (2)^2 + (1.2) 4 π (3)^2 = 351.381 nC.

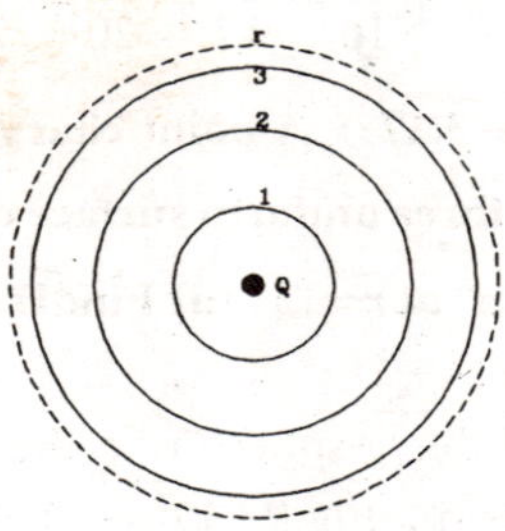

FIG.P3.17 (d)

$$\therefore\ \overline{\text{D}} = \frac{351.381}{4\pi r^2}\,\bar{a}_r = \frac{27.962}{r^2}\,\bar{a}_r \text{ nC/m}^2 \text{ for } r > 3$$

Hence we have

$$\overline{D}(r) = \begin{cases} 7.162/r^2 \ \bar{a}_r & r < 1 \\ -0.838/r^2 \ \bar{a}_r & 1 < r < 2 \\ 17.162/r^2 \ \bar{a}_r & 2 < r < 3 \\ 27.962/r^2 \ \bar{a}_r & r > 3 \end{cases}$$

where r is in meters and $\overline{D}$ in nC/m^2.

(b) Plot $|\overline{D}|$ on the y axis and r on the x axis. Plot is not given.

Probnlem 3.18 : The spherical region, $0 < r < 10$ cm, contains a uniform volume charge density $\rho_v = 4\ \mu C/m^3$. (a) Find Q_{tot}, $0 < r < 10$ cm. (b) Find D_r, $0 < r < 10$ cm. (c) The nonuniform volume charge density, $\rho_v = -3/(r^3 + 0.001)$ nC/m^3, exists for 10 cm $< r < r_o$. Find r_o so that the total charge, $0 < r < r_o$, is zero.

Solution :

$$\text{(a)}\ Q = \int dQ = \int \rho_v\, dv = \int_0^{0.1}\int_0^{\pi}\int_0^{2\pi} 4\, r^2 \sin\theta\, dr\, d\theta\, d\phi$$

$$= 4\left[\frac{r^3}{3}\right]_0^{0.1}\left[-\cos\theta\right]_0^{\pi}\left[\phi\right]_0^{2\pi} = 0.016755\ \mu C$$

(b) Figure shows a sphere of radius r such that $0 < r < 10$ cm. It encloses the volume charge density between 0 to r. Hence

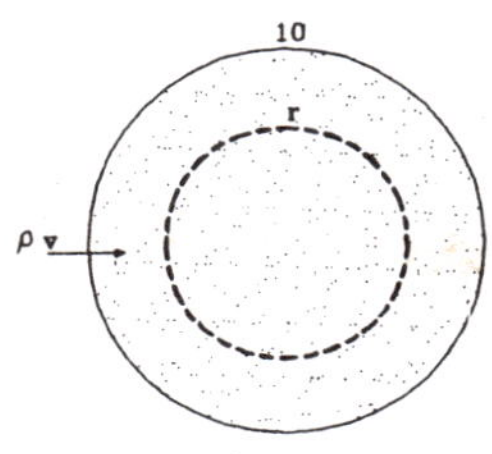

FIG.P3.18

$$\Psi = \int_0^{r}\int_0^{\pi}\int_0^{2\pi} 4 \times r^2 \sin\theta\, dr\, d\theta\, d\phi$$

$$= 4\left[\frac{r^3}{3}\right]_0^{r}\left[-\cos\theta\right]_0^{\pi}\left[\phi\right]_0^{2\pi}$$

$$= 16.75516\ r^3\ \mu C$$

$$\therefore\ D_r = \frac{16.75516\, r^3}{4\pi r^2} = 1.333\, r\ \mu C/m^2$$

(c) Q = Charge present between 10 cm $< r < r_o$

$$Q = \int_{0.1}^{r_0}\int_0^{\pi}\int_0^{2\pi} \frac{-3\, r^2}{r^3 + 0.001} \sin\theta\, dr\, d\theta\, d\phi$$

$$Q = -\left[\ln(r^3 + 0.001)\right]_{0.1}^{r_o}\left[-\cos\theta\right]_0^{\pi}\left[\phi\right]_0^{2\pi}$$

$$= -\ln\left[\frac{r_o^{\ 3} + 0.001}{0.002}\right]\left[2\right]\left[2\pi\right] \text{ nC}$$

Hence total charge present between $0 < r < r_o$ is

$$Q_{tot} = 16.755 \text{ nC} - 4\pi \ln\left[\frac{r_o^{\ 3} + 0.001}{0.002}\right] \text{nC} = 0.0$$

$$\therefore \ln\left[\frac{r_o^{\ 3} + 0.001}{0.002}\right] = 1.33332 \text{ or } \frac{r_o^{\ 3} + 0.001}{0.002} = 1.33332$$

or $r_o = 0.187456 = 18.746$ cm

Problem 3.19 : A uniform sheet of charge ρ_{so} lies in that portion of the $z = 0$ plane for which $|x| < 5$. Find $\overline{D}(x, 0, z)$ for $z > 0$

Solution : Fig.P3.19 shows the location of surface charge density. It lies in $z = 0$ plane. Consider a strip of width dx as shown in figure which is infinite in y direction. The width dx is so small, so that it can be considered as a line charge with $\rho_L = \rho_{so}\, dx$ C/m. The point at which $\overline{D}$ is desired is $(x, 0, z)$ for $z > 0$. Let us call this point as $(a, 0, b)$ so that $x = a$ and $z = b$. Equations of the line charge is given by $z = 0$, x [width dx can be located at any value of x provided $|x| < 5$].

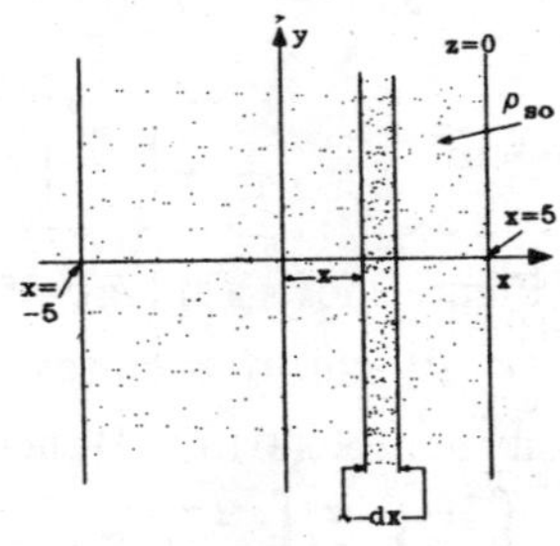

FIG.P3.18

$$\overline{D} \text{ due to line charge} = \frac{\rho_L}{2\pi d}\,\overline{a}_{12}$$

Point 2 is that point at which $\overline{D}$ is desired $= (a, 0, b)$

Point 1 is the foot of perpendicular dropped from point 2 on the line charge $= (x, 0, 0)$

$$\overline{R}_{12} = (a - x)\,\overline{a}_x + b\,\overline{a}_z \quad \text{and} \quad d = |\overline{R}_{12}| = \sqrt{(a-x)^2 + b^2}$$

$$\therefore \overline{dD} = \frac{\rho_{so}\, dx}{2\pi\sqrt{(a-x)^2 + b^2}}\left[\frac{(a-x)\,\overline{a}_x + b\,\overline{a}_z}{\sqrt{(a-x)^2 + b^2}}\right]$$

$$= \frac{\rho_{so}\, dx\,[(a-x)\,\overline{a}_x + b\,\overline{a}_z]}{2\pi[(a-x)^2 + b^2]}$$

$$\therefore \ \overline{D} = \int \overline{dD} = \int_{-5}^{5} \frac{\rho_{so}\, dx\, [(a-x)\,\bar{a}_x + b\,\bar{a}_z]}{2\pi\,[(a-x)^2 + b^2]}$$

$$= \frac{\rho_{so}}{2\pi}\left\{\int_{-5}^{5} \frac{(a-x)\, dx\, \bar{a}_x}{(a-x)^2 + b^2} + \int_{-5}^{5} \frac{b\, dx\, \bar{a}_z}{(a-x)^2 + b^2}\right\}$$

$$\overline{D} = \frac{\rho_{so}}{2\pi}\left\{\frac{-1}{2}\ln\left[(a-x)^2 + b^2\right]\bar{a}_x - \tan^{-1}\left(\frac{a-x}{b}\right)\bar{a}_z\right\}_{-5}^{5}$$

$$= \frac{\rho_{so}}{2\pi}\left\{\frac{-1}{2}\ln\left[\frac{(a-5)^2 + b^2}{(a+5)^2 + b^2}\right]\bar{a}_x - \left[\tan^{-1}\left(\frac{a-5}{b}\right) - \tan^{-1}\left(\frac{a+5}{b}\right)\right]\bar{a}_z\right\}$$

Using $\tan^{-1} m - \tan^{-1} n = \tan^{-1}\left(\frac{m-n}{1+mn}\right)$ we have

$$\overline{D} = \frac{\rho_{so}}{2\pi}\left\{\ln\left[\frac{(a+5)^2 + b^2}{(a-5)^2 + b^2}\right]\bar{a}_x + \tan^{-1}\left[\frac{10b}{a^2 + b^2 - 25}\right]\bar{a}_z\right\}$$

Resubstitute $a = x$ and $b = z$ we have

$$\overline{D} = \frac{\rho_{so}}{2\pi}\left\{\ln\left[\frac{(x+5)^2 + z^2}{(x-5)^2 + z^2}\right]\bar{a}_x + \tan^{-1}\left[\frac{10z}{x^2 + z^2 - 25}\right]\bar{a}_z\right\} \ C/m^2$$

Problem 3.20 : Let $\rho_v = 0$ for $\rho < 1$ cm and also for $\rho > 3$ cm. In the region $1 < \rho < 3$ cm, $\rho_v = 10^{-8} \cos 50\pi\rho$ C/m^3. Find $\overline{D}$ every where.

Solution :

Consider a cylinder of radius $\rho < 1$ cm. as shown in Fig.P3.20 (a). This cylinder do not encloses any charge. Hence flux leaving through this cylinder is zero. Hence $\overline{D} = 0$ for $\rho < 1$ cm

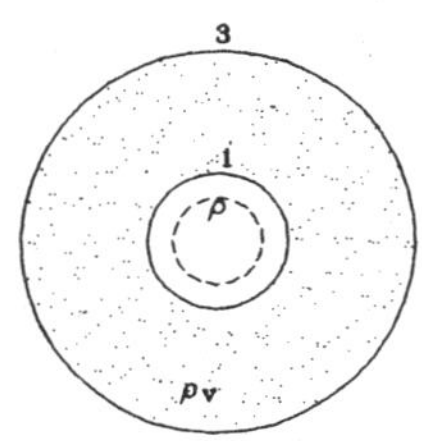

FIG.P3.20 (a)

Consider a cylinder of radius ρ such that 1 cm $< \rho < 3$ cm .as shown in Fig.P3.20 (b).This cylinder encloses some part of the volume charge density. Let us assume that height of the cylinder and that of volume charge density is equal to 2L i.e. z varies from –L to L. Charge enclosed by this cylinder is equal to

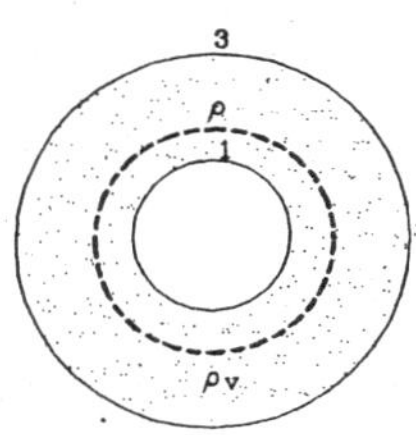

FIG.P3.20 (b)

$$Q = \int dQ = \int \rho_v \, dv = \int_{0.01}^{\rho} \int_{0}^{2\pi} \int_{-L}^{L} 10^{-8} \cos(50\pi\rho)\, \rho \, d\rho \, d\phi \, dz$$

$$= 10^{-8} [\phi]_0^{2\pi} \left[z\right]_{-L}^{L} \left[\frac{\rho \sin(50\pi\rho)}{50\pi} + \frac{\cos(50\pi\rho)}{2500\pi^2}\right]_{0.01}^{\rho}$$

$$= 10^{-8} (2\pi)(2L) \left[\frac{\rho \sin(50\pi\rho)}{50\pi} + \frac{\cos(50\pi\rho)}{2500\pi^2} - 6.366 \times 10^{-5}\right]$$

Area of cylinder = $2\pi\rho \times 2L$

$$\therefore \bar{D} = \frac{Q}{\text{Area}} \bar{a}_\rho = \frac{10^{-8}}{\rho}\left[\frac{\rho \sin(50\pi\rho)}{50\pi} + \frac{\cos(50\pi\rho)}{2500\pi^2} - 6.366 \times 10^{-5}\right] \bar{a}_\rho$$

$$= \left[\frac{4\cos(50\pi\rho) + 200\pi\rho\sin(50\pi\rho) - 2\pi}{\pi^2\rho}\right] \bar{a}_\rho \ \text{C/m}^2$$

for $0.01 < \rho < 0.03\, m$.

Consider a cylinder of radius ρ such that ρ > 3 cm. as shown in Fig. P 3.20 (c). This cylinder encloses volume charge density completely. Hence

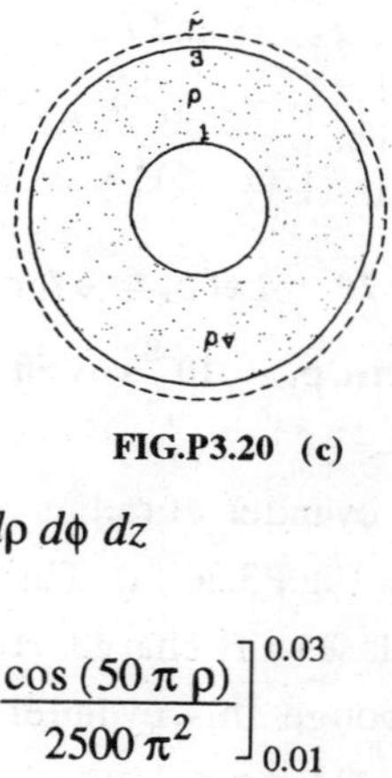

FIG.P3.20 (c)

$$Q = \int dQ = \int \rho_v \, dv$$

$$= \int_{0.01}^{0.03} \int_{0}^{2\pi} \int_{-L}^{L} 10^{-8} \cos(50\pi\rho)\, \rho \, d\rho \, d\phi \, dz$$

$$= 10^{-8} [\phi]_0^{2\pi} \left[z\right]_{-L}^{L} \left[\frac{\rho \sin(50\pi\rho)}{50\pi} + \frac{\cos(50\pi\rho)}{2500\pi^2}\right]_{0.01}^{0.03}$$

$$= 10^{-8} (2\pi)(2L)(-2.546479 \times 10^{-4})$$

Area of cylinder = $(2\pi\rho) \times (2L)$

$$\therefore \bar{D} = \frac{Q}{\text{Area}} \bar{a}_\rho = \frac{-2.546}{\rho} \bar{a}_\rho = \frac{-8}{\pi\rho} \bar{a}_\rho \ \text{C/m}^2 \ \text{for} \ \rho > 0.03 \text{ m}.$$

Problem 3.21 : A uniform volume charge density, $\rho_v = 60\rho$ C/m^3, lies wthin the cylinder ρ = 1 m, and $\rho_v = 0$ elsewhere. (a) Find D_ρ everywhere. (b) What surface charge density ρ_{s2} should be on the cylinder ρ = 2 so that

$D_\rho |_{\rho = 2^-} = 2 D_\rho |_{\rho = 2^+}$? **(c) Plot D_ρ vs. ρ for $0 < \rho < 5$ with both charge distributions present.**

Solution :

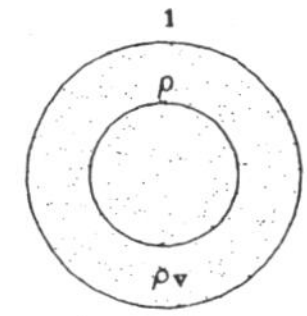

FIG. P3.21 (a)

(a) Fig.P3.21(a) shows a cylinder of radius ρ such that ρ < 1 m. This cylinder encloses some part of the volume charge density. Let us assume that the length of cylinder and that of volume charge density is 2L i.e. limits of z are from - L to L. Charge enclosed by this cylinder

$$Q = \int dQ = \int_0^\rho \int_0^{2\pi} \int_{-L}^{L} 60\,\rho\; d\rho\; d\phi\; dz$$

$$= 60 \left[\frac{\rho^3}{3} \right]_0^\rho [\phi]_0^{2\pi} \left[z \right]_{-L}^{L} = 20\,\rho^3 (2\,\pi)\,(2\,L)\ \ C$$

Area of cylinder = $(2\,\pi\,\rho)\ (2L)$

$$\therefore\ \overline{D} = \frac{Q}{\text{Area}}\ \bar{a}_\rho \ \text{ or } D_\rho = \frac{20\,\rho^3 (2\,\pi)\,(2L)}{(2\,\pi\,\rho)\,(2L)} = 20\,\rho^2\ \ C/m^2$$

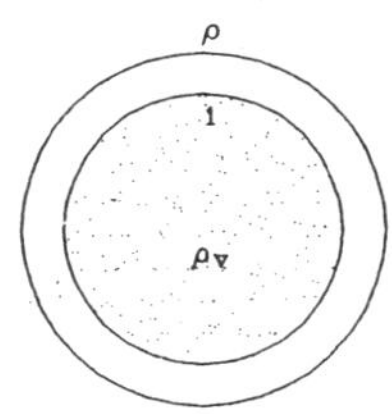

FIG. P3.21 (b)

Fig.P3.21(b) shows a cylinder of radius ρ such that ρ > 1 m. This cylinder encloses the volume charge density completely. Hence

$$Q = \int_0^1 \int_0^{2\pi} \int_{-L}^{L} 60\,\rho\; d\rho\; d\phi\; dz$$

$$= 60 \left[\frac{\rho^3}{3} \right]_0^1 [\phi]_0^{2\pi} \left[z \right]_{-L}^{L} = 20\ (2\,\pi)\ (2L)\ \ C$$

Area of cylinder = $(2\,\pi\,\rho)\ (2\,L)$

$$\therefore\ D_\rho = \frac{20\,(2\,\pi)\ (2\,L)}{(2\,\pi\,\rho)\ (2L)} = \frac{20}{\rho}\ C/m^2$$

Hence

$$D_\rho = \begin{cases} 20\,\rho^2 & 0 < \rho < 1\text{m} \\ \dfrac{20}{\rho} & \rho > 1\text{m} \end{cases}$$

where ρ is in meters and D_ρ is in C/m^2

(b) $\rho = 2^-$ indicates a cylinder just before $\rho = 2$ and $\rho = 2^+$ indicates a cylinder just after $\rho = 2$. Fig.P3.21(c) shows both the charge densities. Consider a cylinder of radius ρ such that $\rho > 2$.

Area of cylinder = $(2\pi\rho)(2L)$

$$\therefore D_\rho = \frac{[20 + \rho_{s2} \times 2](2\pi)(2L)}{(2\pi\rho)(2L)} = \frac{20 + 2\rho_{s2}}{\rho} \text{ C/m}^2$$

Hence with surface charge density, the expressions for D_ρ will b

$$D_\rho = \begin{cases} 20\rho^2 & 0 < \rho < 1 \quad .....(1) \\ \dfrac{20}{\rho} & 1 < \rho < 2 \quad ......(2) \\ \dfrac{20 + 2\rho_{s2}}{\rho} & \rho > 2 \quad .....(3) \end{cases}$$

where ρ is in meters and D_ρ is in C/m^2

At $\rho = 2^-$, using equation (2), we have

$D_\rho |_{\rho = 2^-} = 20/2 = 10$. At $\rho = 2^+$, using equation (3), we have

$$D_\rho |_{\rho = 2^+} = \frac{20 + 2\rho_{s2}}{2} = 10 + \rho_{s2}$$

As $D_\rho |_{\rho = 2^-} = 2 D_\rho |_{\rho = 2^+}$ we have $10 = 2(10 + \rho_{s2})$

$\therefore \rho_{s2} = -5$ C/m^2

(c) Substituting $\rho_{s2} = -5$ C/m^2 in equation (3) we have

$$D_\rho = \begin{cases} 20\rho^2 & 0 < \rho < 1 \\ \dfrac{20}{\rho} & 1 < \rho < 2 \\ \dfrac{10}{\rho} & \rho > 2 \end{cases}$$

where ρ is in meters and D_ρ is in C/m^2. Plot of D_ρ vs ρ is not given.

Problem 3.22 : Cylindrical surfaces at $\rho = 2$, 4 and 6 m carry uniform charge densities of 20 nC/m^2, – 4 nC/m^2 and ρ_{so}, respectively. (a) Find $\overline{D}$ at $\rho = 1$, 3, and 5 m. (b) Determine ρ_{so} such that $\overline{D} = 0$ at $\rho = 7$ m.

Solution : (a) Through out the problem, assume 2L as the length of all the cylinders i.e. $\int_{-L}^{L} dz = 2L$.

Consider a cylinder of radius ρ such that $\rho < 2$ as shown in Fig.P3.22(a). This cylinder do not enclose any charge as seen from figure. Hence flux leaving through this cylinder is zero.

$\therefore\ \overline{D} = 0$ when $\rho < 2$ m.

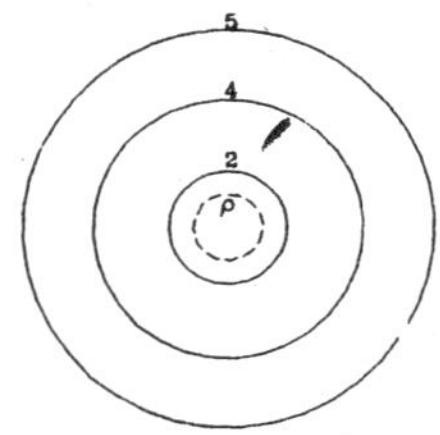

FIG.P3.22(a)

Consider a cylinder of radius ρ such that $2 < \rho < 4$ m as shown in Fig.P3.22 (b). This cylinder encloses the surface charge density present on $\rho = 2$ m. Hence

Q = charge enclosed by the cylinder = charge present on $\rho = 2$ m

$Q = (\rho_s \text{ on } \rho = 2)$ (area of $\rho = 2$ cylinder)

$= (20)\ (2\pi \times 2 \times 2L)$ nC

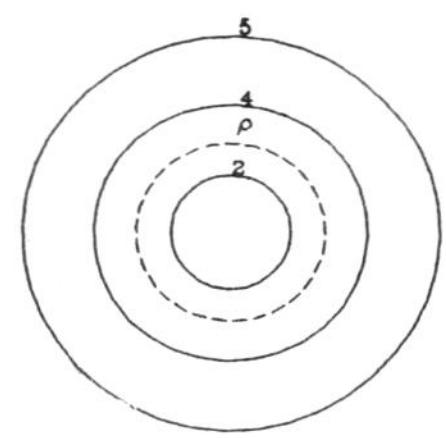

FIG.P3.22(b)

Area of the cylinder = $(2\pi\rho)\ (2L)$

$$\therefore\ \overline{D} = \frac{Q}{\text{Area}}\,\bar{a}_\rho = \frac{(20)\ (4\pi)\ (2L)}{(2\pi\rho)\ (2L)}\,\bar{a}_\rho = \frac{40}{\rho}\,\bar{a}_\rho\ \text{nC/m}^2$$

Consider a cylinder of radius ρ such that $4 < \rho < 6$ m as shown in Fig.P3.22 (c). This cylinder encloses the surface charge densities of 20 nC/m^2 and - 4 nC/m^2 which are present on $\rho = 2$ m and $\rho = 4$ m respectively. Hence

Q = Charge present on $\rho = 2$ + charge present on $\rho = 4$

$= 20 \times (2\pi \times 2 \times 2L) - 4 \times (2\pi \times 4 \times 2L)$

$= 96\pi L$ nC

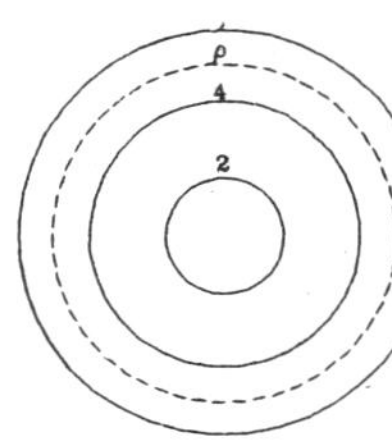

FIG.P3.22 (c)

Area of cylinder = $(2\pi\rho)\ (2L) = 4\pi\rho L$

$$\therefore\ \overline{D} = \frac{Q}{\text{Area}}\,\bar{a}_\rho = \frac{96\pi L}{4\pi\rho L}\,\bar{a}_\rho = \frac{24}{\rho}\,\bar{a}_\rho\ \text{nC/m}^2$$

Consider a cylinder of radius ρ such that $\rho > 6$ m as shown in Fig.P3.22 (d). This cylinder encloses all the three surface charge densities. Hence

Q = charge present on $\rho = 2$ + charge present on $\rho = 4$ + charge present on $\rho = 6$

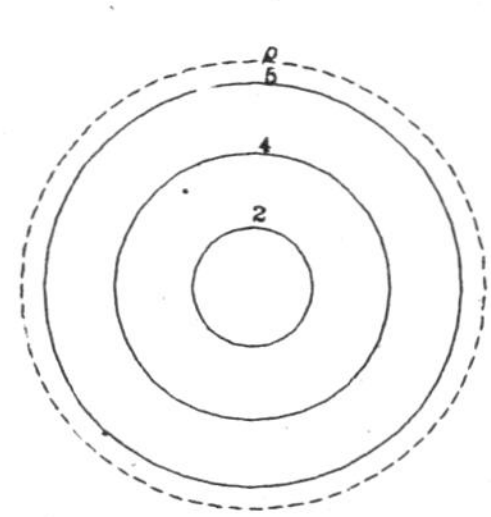

FIG.P3.22 (d)

$Q = 96\pi L + \rho_{so}(2\pi \times 6)(2L) = 96\pi L + 24\rho_{so}\pi L$ nC

Area of cylinder $= (2\pi\rho)(2L) = 4\pi\rho L$

$$\therefore \ \overline{D} = \left(\frac{96\pi L + 24\rho_{so}\pi L}{4\pi\rho L}\right)\overline{a}_\rho = \frac{24 + 6\rho_{so}}{\rho}\overline{a}_\rho \ \text{nC/m}^2$$

$$\overline{D} = \begin{cases} 0 & \rho < 2 & \ldots\ldots(1) \\ (40/\rho)\,\overline{a}_\rho & 2 < \rho < 4 & \ldots\ldots(2) \\ (24/\rho)\,\overline{a}_\rho & 4 < \rho < 6 & \ldots\ldots(3) \\ \dfrac{(24 + 6\rho_{so})}{\rho}\,\overline{a}_\rho & \rho > 6 & \ldots\ldots(4) \end{cases}$$

where ρ is in meters and $\overline{D}$ is in nC/m^2

From equation (1), for $\rho = 1$ m $\overline{D} = 0$

From equation (2), for $\rho = 3$ m $\overline{D} = \dfrac{40}{3}\overline{a}_\rho = 13.333\,\overline{a}_\rho$ nC/m^2

From equation (3), for $\rho = 5$ m $\overline{D} = \dfrac{24}{5}\overline{a}_\rho = 4.8\,\overline{a}_\rho$ nC/m^2

(b) From equation (4), for $\rho = 7$ m $\overline{D} = \dfrac{(24 + 6\rho_{so})}{7}\overline{a}_\rho$ nC/m$^2 = 0$

$$\therefore \ \frac{24 + 6\rho_{so}}{7} = 0 \ \text{or} \ \rho_{so} = -4 \ \text{nC/m}^2$$

Problem 3.23 : Let $\rho_v = 8z(1-z)$ C/m^3 for $0 < z < 1$ m, $8z(1+z)$ for $-1 < z < 0$, and 0 for $|z| > 1$. (a) Find $\overline{D}$ everywhere. (b) Sketch D_z vs. z, $-2 < z < 2$.

Solution :

(a) A plot of ρ_v vs z is shown in Fig.P3.23(a). It has odd symmetry. ρ_v is a function of z. It means it is infinite in x and y direction. Therefore $\overline{D}$ has only z component. [OR $\rightarrow$ If we consider an infinite plane of thickness dz

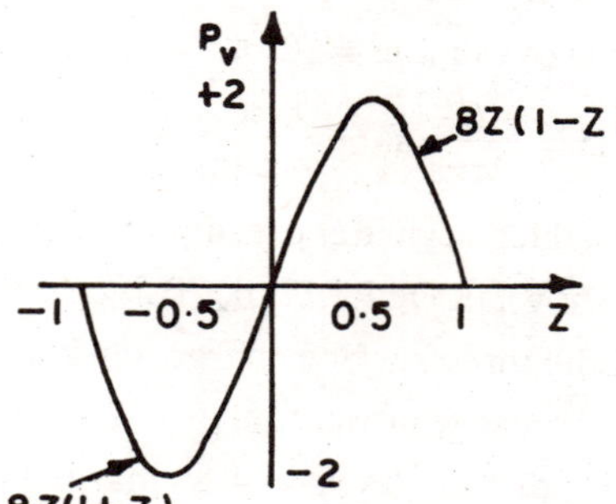

FIG.P3.23 (a)

which is located at any value of z between -1 and 1, then the surface charge density on this plane will be $\rho_s = \rho_v\, dz$ C/m^2. Now due to uniform surface charge density on an infinite plane, $\overline{D} = 0.5\, \rho_s\, \overline{a}_N$, and since ρ_s is located on a plane whose equation is given by z coordinate, $\overline{a}_N = \pm\, \overline{a}_Z$. Hence at any point, $\overline{D}$ has only z component.]

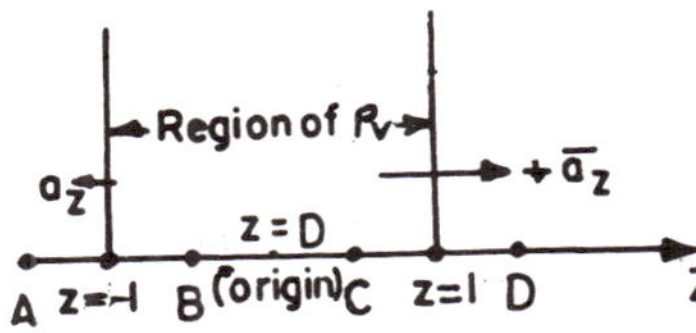

FIG.P3.23 (b)

Four different regions must be considered as per the z coordinate of a point.

Region 1 : Suppose a point is located at A as shown in Fig.P 3.23 (b) at which z coordinate of a point is less than or equal to -1. Now ρ_v has two different regions which are $-1 < z < 0$ and $0 < z < 1$

$\overline{dD}$ at point A $= 0.5\, \rho_s\, \overline{a}_N = 0.5\, \rho_v\, dz\, \overline{a}_N$

$$\therefore\ \overline{D} = \int \overline{dD} = \int_{-1}^{1} 0.5\, \rho_v\, dz\, \overline{a}_N$$

As point A is on LHS of both ρ_v's, $\overline{a}_N = -\,\overline{a}_Z$

$$\therefore\ \overline{D} = \int_{-1}^{0} 0.5 \times 8z(1+z)\, dz\,(-\overline{a}_Z) \int_{0}^{1} 0.5 \times 8z(1+z)\, dz\,(-\overline{a}_Z)$$

$$= -4\,\overline{a}_Z \left\{ \left[\frac{z^2}{2} + \frac{z^3}{3}\right]_{-1}^{0} + \left[\frac{z^2}{2} - \frac{z^3}{3}\right]_{0}^{1} \right\}$$

$$= 0$$

Region 2 : Suppose a point is located at B as shown in Fig.P3.23 (b) at which z coordinate of a point is in between -1 and 0. Now ρ_v itself has three different regions.

$$\therefore\ \overline{D} = \int \overline{dD} = \int_{-1}^{1} 0.5\, \rho_v\, dz\, \overline{a}_N$$

$$= \int_{-1}^{z} 0.5\, \rho_v\, dz\, \overline{a}_N + \int_{z}^{0} 0.5\, \rho_v\, dz\, \overline{a}_N + \int_{0}^{1} 0.5\, \rho_v\, dz\, \overline{a}_N$$

where z is the z coordinate of point B .

For ρ_v in between - 1 and z (point B), $\bar{a}_N = \bar{a}_z$, as the point B is on RHS of this ρ_v.

For ρ_v in between z (point B) and 0 (origin) , $\bar{a}_N = -\bar{a}_z$, as the point B is on LHS of this ρ_v.

Same statement is true for ρ_v in between 0 (origin) and 1

$$\therefore \bar{D} = \int_{-1}^{z} 0.5 \times 8z(1+z)\,dz\,(\bar{a}_z) + \int_{z}^{0} 0.5 \times 8z(1+z)\,dz\,(-\bar{a}_z) + \int_{0}^{1} 0.5 \times 8z(1-z)\,dz\,(-\bar{a}_z)$$

$$= 4\,\bar{a}_z \left\{ \left[\frac{z^2}{2} + \frac{z^3}{3}\right]_{-1}^{z} - \left[\frac{z^2}{2} + \frac{z^3}{3}\right]_{z}^{0} - \left[\frac{z^2}{2} - \frac{z^3}{3}\right]_{0}^{1} \right\}$$

$$= (-1.333 + 4z^2 + 8z^3/3)\,\bar{a}_z$$

Region 3 : Suppose a point is located at C as shown in Fig.P3.23 (b) at which z coordinate of a point is in between 0 and 1. Again ρ_v is divided into three different parts.

$$\therefore \bar{D} = \int \overline{dD} = \int_{-1}^{1} 0.5\,\rho_v\,dz\,\bar{a}_N$$

$$= \int_{-1}^{0} 0.5\,\rho_v\,dz\,\bar{a}_N + \int_{0}^{z} 0.5\,\rho_v\,dz\,\bar{a}_N + \int_{z}^{1} 0.5\,\rho_v\,dz\,\bar{a}_N$$

where z is the z coordinate of point C .

For ρ_v in between - 1 and 0, $\bar{a}_N = \bar{a}_z$, as the point C is on RHS of this ρ_v.

Same statement is true for ρ_v in between 0 (origin) and z (point C)

For ρ_v in between z (point C) and 1 , $\bar{a}_N = -\bar{a}_z$, as the point C is on LHS of this ρ_v.

$$\therefore \bar{D} = \int_{-1}^{0} 0.5 \times 8z(1+z)\,dz\,(\bar{a}_z) + \int_{0}^{z} 0.5 \times 8z(1-z)\,dz\,(\bar{a}_z)$$

$$+ \int_{z}^{1} 0.5 \times 8z\,(1 - z)\,dz\,(-\bar{a}_z)$$

$$= 4\,\bar{a}_z \left\{ \left[\frac{z^2}{2} + \frac{z^3}{3}\right]_{-1}^{0} + \left[\frac{z^2}{2} - \frac{z^3}{3}\right]_{0}^{z} - \left[\frac{z^2}{2} - \frac{z^3}{3}\right]_{z}^{1} \right\},$$

$$= (-1.333 + 4z^2 - 8z^3/3)\,\bar{a}_z$$

Region 4 : Suppose a point is located at D as shown in Fig.P3.23 (b) at which z coordinate of a point is greater than 1. Here ρ_v is divided into two parts.

$$\therefore\ \bar{D} = \int \overline{dD} = \int_{-1}^{1} 0.5\,\rho_v\,dz\,\bar{a}_N$$

$$= \int_{-1}^{0} 0.5\,\rho_v\,dz\,\bar{a}_N + \int_{0}^{1} 0.5\,\rho_v\,dz\,\bar{a}_N$$

As point D is on RHS of both ρ_v's , $\bar{a}_N = \bar{a}_z$

$$\therefore\ \bar{D} = \int_{-1}^{0} 0.5 \times 8z\,(1 + z)\,dz\,\bar{a}_z + \int_{0}^{1} 0.5 \times 8z\,(1 - z)\,dz\,\bar{a}_z$$

$$= 4\,\bar{a}_z \left\{ \left[\frac{z^2}{2} + \frac{z^3}{3}\right]_{-1}^{0} + \left[\frac{z^2}{2} - \frac{z^3}{3}\right]_{0}^{1} \right\} = 0$$

$$= 0$$

Hence we have

$$\bar{D} = \begin{cases} 0 & z \le -1 \\ (-1.333 + 4z^2 + 8z^3/3)\,\bar{a}_z & -1 \le z \le 0 \\ (-1.333 + 4z^2 - 8z^3/3)\,\bar{a}_z & 0 \le z \le 1 \\ 0 & z \ge 1 \end{cases}$$

where z is in meters and $\bar{D}$ is in C/m^2

(b) D_z can be obtained $\bar{D}$ as $\bar{D} = D_z\,\bar{a}_z$. Sketch of D_z vs z for $-2 < z < 2$

is not given .

Problem 3.24 : Uniform surface charge densities of 100, - 60, and 50 $\mu C/m^2$ lie in the $x = 0$, $y = 0$, and $z = 0$ planes, respectively. Find $|\bar{D}|$ at points 2.5 m from every coordinate plane in each of the eight octants.

Solution :

Eight points in eight octants are given below

1(2.5, 2.5, 2.5), 2(-2.5, 2.5, 2.5), 3(-2.5, -2.5, 2.5), 4(2.5, -2.5, 2.5),
5(2.5, 2.5, -2.5), 6(-2.5, 2.5, -2.5), 7(-2.5, -2.5, -2.5), 8(2.5, -2.5, -2.5)

At point 1

$$\overline{D} = \frac{100}{2}\,\overline{a}_x \text{ (} x \text{ coordinate of point 1 is greater than } x = 0\text{)} +$$
$$\frac{-60}{2}\,\overline{a}_y \text{ (} y \text{ coordinate of point 1 is greater than } y = 0\text{)} +$$
$$\frac{50}{2}\,\overline{a}_z \text{ (} z \text{ coordinate of point 1 is greater than } z = 0\text{)}$$

$\therefore\ \overline{D} = 50\,\overline{a}_x - 30\,\overline{a}_y + 25\,\overline{a}_z\ \ \mu C/m^2$

or $|\overline{D}| = \sqrt{(50)^2 + (-30)^2 + (25)^2} = 63.443\ \ \mu C/m^2$

The magnitude of $\overline{D}$ remains the same at all the above points.

Problem 3.25 : Find $\overline{D}$ and plot D_r vs. r for each of the following charge distributions: (a) 1 C at the origin; (b) $1/(4\pi)$ C/m^2 at $r = 1$ m ; (c) $3/(4\pi)$ C/m^3 for $0 < r < 1$, 0 elsewhere.

Solution : (a) $\overline{D} = \dfrac{Q}{4\pi d^2}\,\overline{a}_{12}$

where $\overline{a}_{12} = \overline{a}_r$ { a unit vector directed from point 1 i.e., location of charge (origin) towards point 2 i.e., a point at which $\overline{D}$ is desired.} and $d = r$ { distance of point 2 from point 1}. Hence

$$\overline{D} = \frac{1}{4\pi r^2}\,\overline{a}_r \text{ or } D_r = \frac{1}{4\pi r^2}\ \text{C/m}^2$$

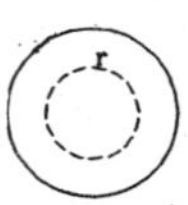

FIG.P3.25 (a)

(b) Consider a sphere of radius r such that $r < 1$ m as shown in Fig.P3.25 (a). This sphere do not enclose any charge. Hence $\Psi = 0$ and $\overline{D} = 0$

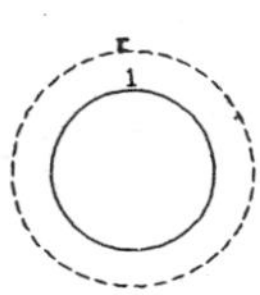

FIG.P3.25 (b)

Consider a sphere of radius r such that $r > 1$ m as shown in Fig.P3.25 (b). This sphere encloses the surfaces charge density of $(1/4\pi)$ C/m^2 located at $r = 1$ m. Hence

$$Q = \frac{1}{4\pi} \times [4\pi \times (1)^2] = 1 \text{ C}$$

$$\therefore \ \overline{D} = \frac{Q}{\text{area}} \bar{a}_r = \frac{1}{4\pi r^2} \bar{a}_r \ \text{C/m}^2 \ \text{ or } \ D_r = \frac{1}{4\pi r^2}$$

Hence we have

$$D_r = \begin{cases} 0 & r < 1 \\ \dfrac{1}{4\pi r^2} & r > 1 \end{cases}$$

where r is in meter and D_r is in C/m^2.

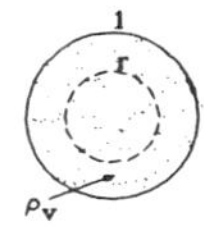

FIG.P3.25 (c)

(c) Consider a sphere of radius r such that $r < 1$ m as shown in Fig.P3.25 (c). This sphere encloses the volume charge density partially. Hence

$$\Psi = Q = \text{charge enclosed by sphere of radius } r = \int dQ = \int \rho_v \, dv$$

$$= \int_0^r \int_0^\pi \int_0^{2\pi} \frac{3}{4\pi} r^2 \sin\theta \, dr \, d\theta \, d\phi = r^3$$

Area of sphere $= 4\pi r^2$

$$\therefore \ \overline{D} = \frac{\Psi}{\text{Area}} \bar{a}_r = \frac{r^3}{4\pi r^2} \bar{a}_r = \frac{r}{4\pi} \bar{a}_r \ \text{ or } \ D_r = \frac{r}{4\pi}$$

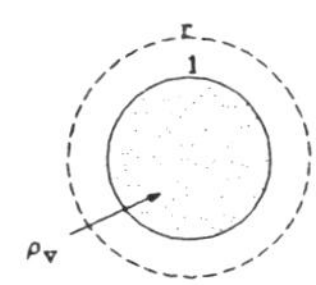

FIG.P3.25 (d)

Consider a sphere of radius r such that $r > 1$ m as shown in Fig. P3.25(d). Sphere encloses the volume charge density completely.Hence

$Q =$ charge enclosed by sphere of radius r

$$= \int dQ = \int \rho_v \, dv = \int_0^1 \int_0^\pi \int_0^{2\pi} \frac{3}{4\pi} r^2 \sin\theta \, dr \, d\theta \, d\phi = 1$$

$$\therefore \ \overline{D} = \frac{Q}{\text{area}} \bar{a}_r = \frac{1}{4\pi r^2} \bar{a}_r \ \text{ or } \ D_r = \frac{1}{4\pi r^2}$$

Hence we have

$$D_r = \begin{cases} \dfrac{r}{4\pi} & 0 < r < 1 \text{ m} \\ \dfrac{1}{4\pi r^2} & r > 1 \text{ m} \end{cases}$$

where r is in meter and D_r is in C/m^2.

Problem 3.26 : Let $\overline{D} = (8x + 4x^3)\,\overline{a}_x - 2y\,\overline{a}_y + 2z\,\overline{a}_z$ C/m². (a) Use Gauss's law to determine the charge enclosed in the cubical region, $-a < x, y, z < a$. (b) Obtain an approximate value for the above charge. Evaluate the derivatives at the center of the volume. (c) Show that your answers agree in the limit as a becomes very small.

Solution :

The six surfaces are located by $x = -a$, $x = a$, $y = -a$, $y = a$, $z = -a$, and $z = a$

$$Q = \int_{x=-a} \overline{D}\cdot\overline{dS} + \int_{x=a} \overline{D}\cdot\overline{dS} + \int_{y=-a} \overline{D}\cdot\overline{dS} + \int_{y=a} \overline{D}\cdot\overline{dS} + \int_{z=-a} \overline{D}\cdot\overline{dS} + \int_{z=a} \overline{D}\cdot\overline{dS}$$

$$= \int_{-a}^{a}\int_{-a}^{a} \underset{(x=-a)}{\overline{D}\cdot(-\,dy\,dz\,\overline{a}_x)} + \int_{-a}^{a}\int_{-a}^{a} \underset{(x=a)}{\overline{D}\cdot(dy\,dz\,\overline{a}_x)} +$$

$$\int_{-a}^{a}\int_{-a}^{a} \underset{(y=-a)}{\overline{D}\cdot(-\,dx\,dz\,\overline{a}_y)} + \int_{-a}^{a}\int_{-a}^{a} \underset{(y=a)}{\overline{D}\cdot(dx\,dz\,\overline{a}_y)} +$$

$$\int_{-a}^{a}\int_{-a}^{a} \underset{(z=-a)}{\overline{D}\cdot(-\,dx\,dy\,\overline{a}_z)} + \int_{-a}^{a}\int_{-a}^{a} \underset{(z=a)}{\overline{D}\cdot(dx\,dy\,\overline{a}_z)}$$

$$= \int_{-a}^{a}\int_{-a}^{a} \underset{(x=-a)}{-\,(8x + 4x^3)\,dy\,dz} + \int_{-a}^{a}\int_{-a}^{a} \underset{(x=a)}{(8x + 4x^3)\,dy\,dz} +$$

$$\int_{-a}^{a}\int_{-a}^{a} \underset{(y=-a)}{-\,(-\,2y\,dx\,dz)} + \int_{-a}^{a}\int_{-a}^{a} \underset{(y=a)}{(-\,2y\,dx\,dz)} +$$

$$\int_{-a}^{a}\int_{-a}^{a} \underset{(y=-a)}{-\,(2z\,dx\,dz)} + \int_{-a}^{a}\int_{-a}^{a} \underset{(y=a)}{(2z\,dx\,dz)}$$

$$\therefore Q = 16a^3(2 + a^2) + 16a^3(2 + a^2) - 8a^3 - 8a^3 + 8a^3 + 8a^3$$

$$= 32a^3(2 + a^2)\ \text{C}$$

(b) Δv = Volume of cube = $2a \times 2a \times 2a = 8a^3$

Centre of the volume is = $\left(\dfrac{-a+a}{2}, \dfrac{-a+a}{2}, \dfrac{-a+a}{2}\right) = (0,0,0)$

$$\left[\frac{\partial D_x}{\partial x} + \frac{\partial D_y}{\partial y} + \frac{\partial D_z}{\partial z}\right]_{\text{at the centre of volume}} = \left[8 + 12x^2 - 2 + 2\right]_{(0,0,0)} = 8$$

Using eqn.(3.4a), we have

$\therefore\ \Delta Q \doteq 8\times 8a^3 = 64a^3$ C.

(c) From Gauss's law we have $Q = 32a^3(2+a^2) = 64a^3 + 32a^5$

As a becomes very small, we can neglect a^5 in comparison with a^3 so that $Q \doteq 64a^3$. Hence both the answers are same.

Problem 3.27 : Let $\bar{D} = (10xyz^2 + 4x)\,\bar{a}_x + 5x^2z^2\,\bar{a}_y + 10x^2yz\,\bar{a}_z$ nC/m^2 . (a) Find the total charge enclosed in a cube having a volume of 10^{-10} m^3 located at (2, 3, 4) . (b) How much flux leaves this volume ? (c) What is the average volume charge density within this volume ?

Solution :

(a) $\Delta v = 10^{-10}$ m^3

$$\left[\frac{\partial D_x}{\partial x} + \frac{\partial D_y}{\partial y} + \frac{\partial D_z}{\partial z}\right]_{(2,3,4)} = \left[10yz^2 + 4 + 10x^2y\right]_{(2,3,4)}$$

$$= (10\times 3\times 16 + 4 + 10\times 4\times 3)$$

$$= 604 \text{ nC/m}^3$$

Using eqn.(3.4a), we have

$\therefore\ \Delta Q \doteq 604\times 10^{-9}\times 10^{-10} = 604\times 10^{-18}$ C

(b) charge enclosed = flux leaving = 604×10^{-18} C

(c) $\left(\dfrac{\partial D_x}{\partial x} + \dfrac{\partial D_y}{\partial y} + \dfrac{\partial D_z}{\partial z}\right)$ at the location of volume gives average value of volume charge density. Hence ρ_v (avg.) = 604 nC/m^3

Problem 3.28 : Let $\bar{D} = 12xyz\,\bar{a}_x + (6x^2z + 6yz)\,\bar{a}_y + (6x^2y + 3y^2)\,\bar{a}_z$ C/m^2 and find the incremental amount of charge in a volume of 10^{-10} m^3 located at : (a) (0,0,0); (b) (4, 2, -3); (c) (4, y ,- 3). (d) At what location in the cubical region $0 \le x, y, z \le 2$ should the small volume be located to contain a maximum charge ?

Solution : (a)

$$\Delta Q \doteq \left[\frac{\partial D_x}{\partial x} + \frac{\partial D_y}{\partial y} + \frac{\partial D_z}{\partial z}\right]_{(0,0,0)} \Delta v \doteq \left[12yz + 6z\right]_{(0,0,0)} \times 10^{-10} = 0$$

(b) $\Delta Q \doteq \left[12yz + 6z \right]_{(4,2,-3)} \times 10^{-10}$

$= (-12 \times 2 \times 3 - 6 \times 3)\, 10^{-10} = -9$ nC

(c) $\Delta Q \doteq \left[12yz + 6z \right]_{(4,y,-3)} \times 10^{-10}$

$= (-12 \times y \times 3 - 6 \times 3)\, 10^{-10} = -(3.6y + 1.8)$ nC

(d) ΔQ will be maximum if $12yz + 6z$ is a maximum. Now $12yz + 6z$ will be maximum if y and z is maximum. With the region $0 \le x, y, z \le 2$, the maximum value of y and z is 2. Hence volume must be located at $(x, 2, 2)$.

Problem 3.29 : A cubical region of space is defined by the surfaces $x = 1.8$, $y = 1.8$, $z = 1.8$, $x = 2$, $y = 2$, and $z = 2$. If $\overline{D} = 3y^2\,\overline{a}_x + 3x^2y\,\overline{a}_y$ C/m^2: (a) find the exact value of the total charge enclosed within the cube by surface intergration; (b) find an approximate value for the enclosed charge by evaluation the derivatives of at the center of the cube.

Solution :

(a)The six surfaces are located by $x = 1.8$, $x = 2$, $y = 1.8$, $y = 2$, $z = 1.8$, and $z = 2$

$$Q = \int_{x=1.8} \overline{D}\cdot\overline{dS} + \int_{x=2} \overline{D}\cdot\overline{dS} + \int_{y=1.8} \overline{D}\cdot\overline{dS} + \int_{y=2} \overline{D}\cdot\overline{dS} + \int_{z=1.8} \overline{D}\cdot\overline{dS} + \int_{z=2} \overline{D}\cdot\overline{dS}$$

$$= \int_{1.8}^{2}\int_{1.8}^{2} \overline{D}\cdot\underset{(x=1.8)}{(-dy\,dz\,\overline{a}_x)} + \int_{1.8}^{2}\int_{1.8}^{2} \overline{D}\cdot\underset{(x=2)}{(dy\,dz\,\overline{a}_x)} +$$

$$\int_{1.8}^{2}\int_{1.8}^{2} \overline{D}\cdot\underset{(y=1.8)}{(-dx\,dz\,\overline{a}_y)} + \int_{1.8}^{2}\int_{1.8}^{2} \overline{D}\cdot\underset{(y=2)}{(dx\,dz\,\overline{a}_y)} +$$

$$\int_{1.8}^{2}\int_{1.8}^{2} \overline{D}\cdot\underset{(z=1.8)}{(-dx\,dy\,\overline{a}_z)} + \int_{1.8}^{2}\int_{1.8}^{2} \overline{D}\cdot\underset{(z=2)}{(dx\,dy\,\overline{a}_z)}$$

$$= \int_{1.8}^{2}\int_{1.8}^{2} -\underset{(x=1.8)}{(y^2\,dy\,dz)} + \int_{1.8}^{2}\int_{1.8}^{2} \underset{(x=2)}{(y^2\,dy\,dz)} +$$

$$\int_{1.8}^{2}\int_{1.8}^{2} -\underset{(y=1.8)}{(3x^2y\,dx\,dz)} + \int_{1.8}^{2}\int_{1.8}^{2} \underset{(y=2)}{(3x^2y\,dx\,dz)} + 0 + 0$$

$Q = -0.4336 + 0.4336 - 0.78048 + 1.7344 = 0.95392$ C

(b) $\left(\dfrac{\partial D_x}{\partial x} + \dfrac{\partial D_y}{\partial y} + \dfrac{\partial D_z}{\partial z}\right) = \dfrac{\partial}{\partial x}(3y^2) - \dfrac{\partial}{\partial y}(3x^2 y) = 0 + 3x^2 = 3x^2$

The center of the cube is $\left(\dfrac{1.8+2}{2}, \dfrac{1.8+2}{2}, \dfrac{1.8+2}{2}\right)$ i.e.(1.9, 1.9, 1.9)

$$\Delta Q = \left[\frac{\partial D_x}{\partial x} + \frac{\partial D_y}{\partial y} + \frac{\partial D_z}{\partial z}\right]_{\text{at the center of the volume}} \Delta v = \left[3x^2\right]_{(1.9,1.9,1.9)} \Delta v$$

$$= \left[3(1.9)^2\right](0.2 \times 0.2 \times 0.2) = 0.08664 \text{ C}$$

Problem 3.30 : Calculate the divergence of $\overline{G}$ at $P(2, -3, 4)$ if $\overline{G}$ = : (a) $x\,\bar{a}_x + y\,\bar{a}_y + z\,\bar{a}_z$; (b) $\rho\,\bar{a}_\rho$; (c) $r\,\bar{a}_r$ (d) $6r^2 \sin\theta\,\bar{a}_r + 2r^2\cos\theta\,\bar{a}_\theta$.

Solution :

(a)Using eqn.(3.4c),we have

Divergence of $\overline{G} = \dfrac{\partial}{\partial x}(x) + \dfrac{\partial}{\partial y}(y) + \dfrac{\partial}{\partial z}(z) = 1 + 1 + 1 = 3$

(b)Using eqn.(3.4d),we have

Divergence of $\overline{G} = \dfrac{1}{\rho}\dfrac{\partial}{\partial \rho}(\rho \times \rho) = 2$

(c)Using eqn.(3.4d),we have

Divergence of $\overline{G} = \dfrac{1}{r^2}\dfrac{\partial}{\partial r}(r^2 \times r) = 3$

(d)Using eqn.(3.4d),we have

$$\text{div. of } \overline{G} = \frac{1}{r^2}\frac{\partial}{\partial r}\left[r^2(6r^2\sin\theta)\right] + \frac{1}{r\sin\theta}\frac{\partial}{\partial\theta}\left[\sin\theta(2r^2\cos\theta)\right]$$

$$= 24\,r\sin\theta + \frac{2\,r\cos 2\theta}{\sin\theta}$$

For point P, $r = 5.385$ and $\theta = 42.03^\circ$. Hence at point P,

$$\text{div. of } \overline{G} = 24(5.385)\sin(42.03^\circ) + \frac{2(5.385)\cos(84.062^\circ)}{\sin(42.03^\circ)} = 8.193$$

Problem 3.31 : If $\overline{D} = 12x^2\,\bar{a}_x - 3z^3\,\bar{a}_y - 9yz^2\,\bar{a}_z$ C/m^2 in free space, specify the point within the cube, $1 \leq x, y, z \leq 2$ at which the following quantity is a maximum and give that maximum value : (a) $|\overline{D}|$; (b) $|\rho_v|$; (c) ρ_v

Solution :

(a) $|\overline{D}| = \sqrt{(12x^2)^2 + (-3z^3)^2 + (-9yz^2)^2}$

$= \sqrt{144x^4 + 9z^9 + 81y^2z^4}$

$|\overline{D}|$ will be maximum if x, y, and z are maximum. Hence at point (2, 2, 2), $|\overline{D}|$ has a maximum value which is equal to

$\sqrt{144(2)^4 + 9(2)^9 + 81(2)^2(2)^4} = \sqrt{12096} = 109.982 \text{ C/m}^2$

(b) $\rho_v = \text{div.}\overline{D} = \frac{\partial}{\partial x}(12x^2) + \frac{\partial}{\partial y}(-3z^3) + \frac{\partial}{\partial z}(-9yz^2)$

$= 24x + 0 - 18yz = 24x - 18yz \text{ C/m}^3$

Hence at point (1, 2, 3), $|\rho_v|$ has a maximum value and the maximum value is $|24 \times 1 - 18 \times 2 \times 2| = 48 \text{ C/m}^3$

(c) $\rho_v = 24x - 18yz$

At point (2, 1, 1), ρ_v has a maximum value which is equal to

$\rho_v = 24 \times 2 - 18 \times 1 \times 1 = 30 \text{ C/m}^3$

Problem 3.32 : The electric flux density $\overline{D} = 10r^3\,\overline{a}_r$ C/m² in the region $0 \leq r < 1$ m. (a) Find ρ_v at $r = 0.4$ m. (b) Find D_r at $r = 0.4$ m. (c) What total charge is contained in the region $0 \leq r \leq 0.4$ m? (d) If $\rho_v = 0$ for $r > 1$ m, find D_r for $r > 1$ m.

Solution :

(a) $\rho_v = \overline{\nabla} \cdot \overline{D} = \frac{1}{r^2}\frac{\partial}{\partial r}(r^2 \times 10r^3) = \frac{1}{r^2}(10 \times 5r^4) = 50r^2 \text{ C/m}^3$

$\therefore$ ρ_v at $r = 0.4$ m is equal to $50(0.4)^2 = 8 \text{ C/m}^3$

(b) $D_r = 10r^3$. Hence at $r = 0.4$ m, $D_r = 10(0.4)^3 = 0.64 \text{ C/m}^2$

(c) $Q = \int \rho_v \, dv = \int_0^{0.4}\int_0^{\pi}\int_0^{2\pi} 50r^2 \times r^2 \sin\theta \, dr \, d\theta \, d\phi$

$= 50\left[\frac{r^5}{5}\right]_0^{0.4}\left[-\cos\theta\right]_0^{\pi}\left[\phi\right]_0^{2\pi} = 1.28679 \text{ C}$

(d) Fig. P3.32 shows sphere of radius r such that $r > 1$ m. Charge enclosed sphere is

$$Q = \int_0^1 \int_0^{\pi} \int_0^{2\pi} 50\, r^2 \times r^2 \sin\theta \; dr \; d\theta \; d\phi \{$$

$$= 50\left[\frac{r^5}{5}\right]_0^1 \left[-\cos\theta\right]_0^{\pi} \left[\phi\right]_0^{2\pi} = 125.6637 \text{ C}$$

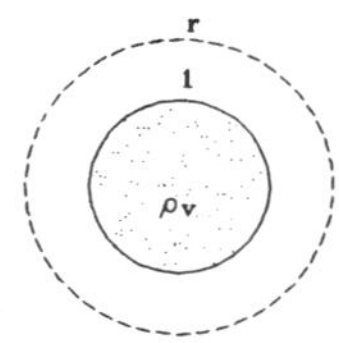

FIG. P3.32

Area of the sphere = $4\pi r^2$ m^2

$$\therefore\; D_r = \frac{\text{flux}}{\text{Area}} = \frac{125.6637}{4\pi r^2} = \frac{10}{r^2} \text{ C/m}^2 \text{ for } r > 1\text{m}$$

Problem 3.33 : Let $\overline{D} = (10r^2 + 5e^{-r})\,\overline{a}_r$ C/m^2. (a) Find ρ_v as a function of r. (b) Find the total charge lying within a sphere of radius a centered at the origin.

Solution :

$$\text{(a) } \rho_v = \overline{\nabla}\cdot\overline{D} = \frac{1}{r^2}\frac{\partial}{\partial r}(r^2 D_r) = \frac{1}{r^2}\frac{\partial}{\partial r}\left[r^2(10r^2 + 5e^{-r})\right]$$

$$= \frac{1}{r^2}\left[40r^3 + 10re^{-r} - 5r^2e^{-r}\right] \text{ C/m}^3$$

$$\text{(b)} Q = \int_0^a \int_0^{\pi} \int_0^{2\pi} \frac{1}{r^2}\left[40r^3 + 10re^{-r} - 5r^2e^{-r}\right] r^2 \sin\theta \, dr\, d\theta\, d\phi$$

$$= \left[\frac{40r^4}{4} + 5r^2e^{-r}\right]_0^a \left[-\cos\theta\right]_0^{\pi}\left[\phi\right]_0^{2\pi}$$

$$= (10a^4 + 5a^2e^{-a})(2)(2\pi) = 20\pi a^2(2a^2 + e^{-a}) \text{ C}$$

Problem 3.34 : If $\overline{D} = 4x^3\,\overline{a}_x - 2z\,\overline{a}_y - 2y\,\overline{a}_z$ C/m^2, find: (a) $\overline{\nabla}\cdot\overline{D}$; (b) ρ_v at $P(x, y, z)$; (c) the total charge lying within the region $-1 < x, y, z < 1$; (d) the total charge lying within that region without finding ρ_v first .

Solution :

$$\text{(a)} \overline{\nabla}\cdot\overline{D} = \frac{\partial}{\partial x}(4x^3) + \frac{\partial}{\partial y}(-2z) + \frac{\partial}{\partial z}(-2y) = 12x^2 \text{ C/m}^3$$

(b) ρ_v at $P(x, y, z) = 12x^2$ C/m^3

(c) $Q = \int \rho_v \, dv = \int_{-1}^{1}\int_{-1}^{1}\int_{-1}^{1} 12x^2 \, dx \, dy \, dz$

$$= 12\left[\frac{x^3}{3}\right]_{-1}^{1}\left[y\right]_{-1}^{1}\left[z\right]_{-1}^{1} = 32 \text{ C}$$

(d) The six surfaces are located by $x = -1$, $x = 1$, $y = -1$, $y = 1$, $z = -1$, and $z = 1$

$$Q = \int_{x=-1} \overline{D}\cdot\overline{dS} + \int_{x=1} \overline{D}\cdot\overline{dS} + \int_{y=-1} \overline{D}\cdot\overline{dS} + \int_{y=1} \overline{D}\cdot\overline{dS} + \int_{z=-1} \overline{D}\cdot\overline{dS} + \int_{z=1} \overline{D}\cdot\overline{dS}$$

$$= \int_{-1}^{1}\int_{-1}^{1} \underset{(x=-1)}{\overline{D}\cdot(-dy\,dz\,\overline{a}_x)} + \int_{-1}^{1}\int_{-1}^{1} \underset{(x=1)}{\overline{D}\cdot(dy\,dz\,\overline{a}_x)} +$$

$$\int_{-1}^{1}\int_{-1}^{1} \underset{(y=-1)}{\overline{D}\cdot(-dx\,dz\,\overline{a}_y)} + \int_{-1}^{1}\int_{-1}^{1} \underset{(y=1)}{\overline{D}\cdot(dx\,dz\,\overline{a}_y)} +$$

$$\int_{-1}^{1}\int_{-1}^{1} \underset{(z=-1)}{\overline{D}\cdot(-dx\,dy\,\overline{a}_z)} + \int_{-1}^{1}\int_{-1}^{1} \underset{(z=1)}{\overline{D}\cdot(dx\,dy\,\overline{a}_z)}$$

$$= \int_{-1}^{1}\int_{-1}^{1} \underset{(x=-1)}{-4x^3\,dy\,dz} + \int_{-1}^{1}\int_{-1}^{1} \underset{(x=1)}{4x^3\,dy\,dz} +$$

$$\int_{-1}^{1}\int_{-1}^{1} \underset{(y=-1)}{2z\,dx\,dz} + \int_{-1}^{1}\int_{-1}^{1} \underset{(y=1)}{-2z\,dx\,dz} +$$

$$\int_{-1}^{1}\int_{-1}^{1} \underset{(z=-1)}{2y\,dx\,dy} + \int_{-1}^{1}\int_{-1}^{1} \underset{(z=1)}{-2y\,dx\,dy}$$

$$= 16 + 16 + 0 + 0 + 0 + 0 = 32 \text{ C}$$

Problem 3.35 : (a) A point charge Q lies at the origin. Show that $\overline{\nabla}\cdot\overline{D} = 0$ everywhere except at the origin. (b) Replace the point charge with a uniform volume charge density ρ_{vo} for $0 \le r \le a$. Relate ρ_{vo} Q and a so that the total charge is the same. Find $\overline{\nabla}\cdot\overline{D}$ everywhere.

Solution :

(a) $\overline{D} = \dfrac{Q}{4\pi r^2}\,\overline{a}_r$ if Q lies at origin.

$$\therefore\ \bar{\nabla}\cdot\bar{D} = \frac{1}{r^2}\frac{\partial}{\partial r}(r^2 D_r) = \frac{1}{r^2}\frac{\partial}{\partial r}\left(r^2 \times \frac{Q}{4\pi r^2}\right) = \frac{1}{r^2}\frac{\partial}{\partial r}\left(\frac{Q}{4\pi}\right) = \frac{0}{r^2}$$

Hence $\bar{\nabla}\cdot\bar{D} = 0$ for any value of r except at origin where it is undefined because $r = 0$. .

(b) Volume of sphere of radius a is equal to $(4/3)\,\pi\, a^3$. Inside this sphere, total charge is Q . Therefore

$$\rho_{vo} = \frac{\text{Charge}}{\text{Volume}} = \frac{Q}{(4/3)\,\pi\, a^3} = \frac{3\,Q}{4\,\pi\, a^3}\ \text{C/m}^3$$

Consider a sphere of radius r such that $r \le a$ as shown in Fig.P3.35 (a). Charge enclosed by this sphere

$$= \int_0^r \int_0^\pi \int_0^{2\pi} \frac{3Q}{4\pi a^3} r^2 \sin\theta\, dr\, d\theta\, d\phi$$

$$= \frac{3Q}{4\pi a^3}\left[\frac{r^3}{3}\right]_0^r \left[-\cos\theta\right]_0^\pi \left[\phi\right]_0^{2\pi} = \frac{Q\,r^3}{a^3}$$

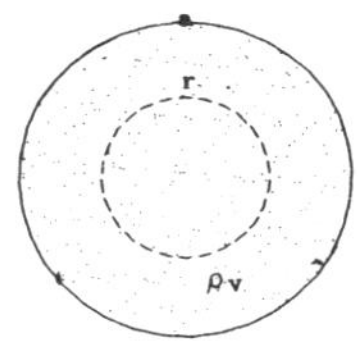

FIG.P3.35 (a)

Volume of this sphere $= (4/3)\,\pi\, r^3\ \text{m}^3$. Hence

$$\bar{\nabla}\cdot\bar{D}\ (\text{for } r \le a) = \frac{Q\, r^3/a^3}{(4/3)\,\pi\, r^3} = \frac{3Q}{4\,\pi\, a^3}\ \text{C/m}^3$$

Consider a sphere of radius r such that $r \ge a$ as shown in Fig.P3.35 (b).

Charge enclosed by this sphere is equal to Q .

Volume of this sphere $= (4/3)\,\pi\, r^3\ \text{m}^3$. Hence

$$\bar{\nabla}\cdot\bar{D}\ (\text{for } r \ge a) = \frac{Q}{(4/3)\,\pi\, r^3} = \frac{3Q}{4\,\pi\, r^3}\ \text{C/m}^3$$

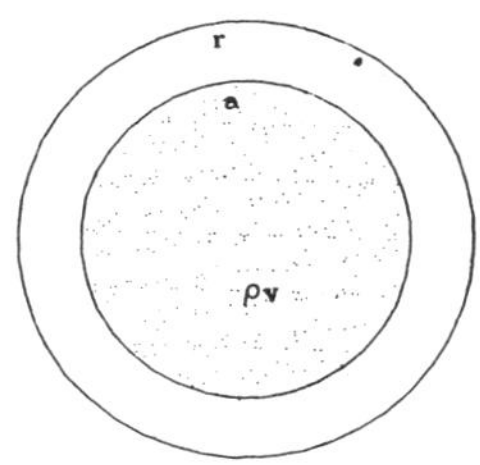

FIG. P3.35 (b)

Problem 3.36 : Within the cylindrical region $\rho \le 4$ m, the electric flux density is given as $5\,\rho^3\,\bar{a}_\rho$ C/m^2. (a) What is the volume charge density at $\rho = 3$ m ? (b) What is the electric flux density at $\rho = 3$? (c) How much electric flux leaves the cylinder, $\rho = 3$, $|\,z\,| \le 2.5$? (d) How much charge is contained within the cylinder, $\rho = 3$, $|\,z\,| \le 2.5$?

Solution : (a) $\rho_v = \frac{1}{\rho}\frac{\partial}{\partial \rho}(\rho\, D_\rho) = \frac{1}{\rho}\frac{\partial}{\partial \rho}(\rho \times 5\,\rho^3) = \frac{1}{\rho}(5 \times 4\,\rho^3) = 20\,\rho^2$

Hence ρ_v at $\rho = 3$ m is $20(3)^2 = 180$ C/m^3

(b) $\overline{D}$ at $\rho = 3$ m is $5(3)^3 \bar{a}_\rho = 135 \bar{a}_\rho$ C/m^2

(c) $\Psi = Q = \int \rho_v \, dv = \int_0^3 \int_0^{2\pi} \int_{-2.5}^{2.5} 20\rho^2 \rho \, d\rho \, d\phi \, dz$

$$= 20\left[\frac{\rho^4}{4}\right]_0^3 \left[\phi\right]_0^{2\pi} \left[z\right]_{-2.5}^{2.5} = 12723.450 \text{ C}$$

(d) $Q = \Psi = 12723.450$ C

Problem 3.37 : Inside the sphere $r \leq 4$ m, the electric flux density is given as $5r^3 \bar{a}_r$ C/m^2. (a) What is the volume charge density at $r = 3$ m? (b) What is the electric flux density at $r = 3$? (c) How much electric flux leaves the sphere $r = 3$? (d) How much charge is contained within the sphere $r = 3$?

Solution :

(a) $\rho_v = \bar{\nabla} \cdot \overline{D} = \frac{1}{r^2}\frac{\partial}{\partial r}(r^2 D_r) = \frac{1}{r^2}\frac{\partial}{\partial r}(r^2 \times 5r^3) = \frac{1}{r^2}(5 \times 5r^4) = 25r^2$

At $r = 3$ m, $\rho_v = 25(3)^2 = 225$ C/m^3

(b) $\overline{D}$ at $r = 3$ m is $5(3)^3 \bar{a}_r = 135 \bar{a}_r$ C/m^2

(c) $\Psi = Q = \int \rho_v \, dv = \int_0^3 \int_0^{2\pi} \int_{-2.5}^{2.5} 25r^2 r^2 \sin\theta \, dr \, d\theta \, d\phi$

$$= 25\left[\frac{r^5}{5}\right]_0^3 \left[-\cos\theta\right]_0^{\pi} \left[\phi\right]_0^{2\pi} = 15268.1403 \text{ C}$$

(d) $Q = \Psi = 15268.1403$ C

Problem 3.38 : If $\overline{D} = 0.1 r \bar{a}_r$ C/m^2 for $r \leq 0.4$ m, and $\overline{D} = (0.0064/r^2)$ $\bar{a}_r$ for $r \geq 0.4$: (a) find ρ_v at $r = 0.2$ and 0.5 m. (b) What point charge could be placed at the origin to cause $\overline{D}$ to be zero for $r \geq 0.4$?

Solution :

(a) $\rho_v = \bar{\nabla} \cdot \overline{D} = \frac{1}{r^2}\frac{\partial}{\partial r}(r^2 D_r) = \frac{1}{r^2}(0.1 \times 3r^2) = 0.3$ for $r \leq 0.4$ m)

$$\rho_v = \frac{1}{r^2}\frac{\partial}{\partial r}\left(r^2 \times \frac{0.0064}{r^2} \right) = 0.0 \text{ for } r \geq 0.4 \text{ m}$$

$\therefore$ ρ_v at $r = 0.2$ m is 0.3 C/m^3 and at $r = 0.5$ m is 0.0 C/m^3

(b) $\overline{D}$ due to a point charge Q located origin is given by $\overline{D} = \dfrac{Q}{4\pi r^2}\,\overline{a}_r$.

Hence for $r \geq 0.4$, we have

$$\overline{D}_{total} = \frac{0.0064}{r^2}\,\overline{a}_r + \frac{Q}{4\pi r^2}\,\overline{a}_r = 0$$

$$\therefore \frac{Q}{4\pi r^2} + \frac{0.0064}{r^2} = 0 \text{ or } Q = -0.08042 \text{ C}.$$

Problem 3.39 : Let $\overline{D} = 6xyz^2\,\overline{a}_x + 3x^2z^2\,\overline{a}_y + 6x^2yz\,\overline{a}_z$ C/m^2. Find the total charge lying within the region bounded by $x = 1$ and 3, $y = 0$ and 1, and $z = -1$ and 1 by separately evaluating each side of the divergence theorem.

Solution : LHS $= \oint \overline{D}\cdot\overline{dS}$

$$= \int_{x=1} \overline{D}\cdot\overline{dS} + \int_{x=3} \overline{D}\cdot\overline{dS} + \int_{y=0} \overline{D}\cdot\overline{dS} + \int_{y=1} \overline{D}\cdot\overline{dS} + \int_{z=-1} \overline{D}\cdot\overline{dS} + \int_{z=1} \overline{D}\cdot\overline{dS}$$

$$= \int_0^1\int_{-1}^1 \underset{(x=1)}{\overline{D}\cdot(-\,dy\,dz\,\overline{a}_x)} + \int_0^1\int_{-1}^1 \underset{(x=3)}{\overline{D}\cdot(dy\,dz\,\overline{a}_x)} +$$

$$\int_1^3\int_{-1}^1 \underset{(y=0)}{\overline{D}\cdot(-\,dx\,dz\,\overline{a}_y)} + \int_1^3\int_{-1}^1 \underset{(y=1)}{\overline{D}\cdot(dx\,dz\,\overline{a}_y)} +$$

$$\int_1^3\int_0^1 \underset{(z=-1)}{\overline{D}\cdot(-\,dx\,dy\,\overline{a}_z)} + \int_1^3\int_0^1 \underset{(z=1)}{\overline{D}\cdot(dx\,dy\,\overline{a}_z)}$$

$$= \int_0^1\int_{-1}^1 \underset{(x=1)}{-\,6xyz^2\,dy\,dz} + \int_0^1\int_{-1}^1 \underset{(x=3)}{6xyz^2\,dy\,dz} +$$

$$\int_1^3\int_{-1}^1 \underset{(y=0)}{-\,3x^2z^2\,dx\,dz} + \int_1^3\int_{-1}^1 \underset{(y=1)}{3x^2z^2\,dx\,dz} +$$

$$\int_1^3\int_0^1 \underset{(z=-1)}{-\,6x^2yz\,dx\,dy} + \int_1^3\int_0^1 \underset{(z=1)}{6x^2yz\,dx\,dy}$$

$$\text{LHS} = -6\left[\frac{y^2}{2}\right]_0^1\left[\frac{z^3}{3}\right]_{-1}^1 + 18\left[\frac{y^2}{2}\right]_0^1\left[\frac{z^3}{3}\right]_{-1}^1 - 3\left[\frac{x^3}{3}\right]_1^3\left[\frac{z^3}{3}\right]_{-1}^1$$

$$+ 3\left[\frac{x^3}{3}\right]_1^3\left[\frac{z^3}{3}\right]_{-1}^1 + 6\left[\frac{x^3}{3}\right]_1^3\left[\frac{y^2}{2}\right]_0^1 + 6\left[\frac{x^3}{3}\right]_1^3\left[\frac{y^2}{2}\right]_0^1$$

$$= -2 + 6 - 17.333 + 17.333 + 26 + 26 = 56 \text{ C}$$

$$\text{RHS} = \int_V \bar{\nabla}\cdot\bar{D}\, dv$$

$$\bar{\nabla}\cdot\bar{D} = \frac{\partial}{\partial x}(6xyz^2) + \frac{\partial}{\partial y}(3x^2z^2) + \frac{\partial}{\partial z}(6x^2yz) = 6yz^2 + 6x^2y$$

$$\therefore \text{RHS} = \int_1^3\int_0^1\int_{-1}^1 (6yz^2 + 6x^2y)\, dx\, dy\, dz$$

$$= 6[x]_1^3\left[\frac{y^2}{2}\right]_0^1\left[\frac{z^3}{3}\right]_{-1}^1 + 6\left[\frac{x^3}{3}\right]_1^3\left[\frac{y^2}{2}\right]_0^1[z]_{-1}^1 = 56 \text{ C}$$

Problem 3.40 : Given the field $\bar{D} = (20/\rho^2)(-\sin^2\phi\, \bar{a}_\rho + \sin 2\phi\, \bar{a}_\phi)$ C/m^2, find the total charge lying within the volume, $1 < \rho < 2$, $0 < \phi < \pi/2$, $0 < z < 1$ by two different methods.

Solution :

Method 1 : $Q = \oint_S \bar{D}\cdot\overline{dS}$

$$= \int_{\rho=1}\bar{D}\cdot\overline{dS} + \int_{\rho=2}\bar{D}\cdot\overline{dS} + \int_{\phi=0}\bar{D}\cdot\overline{dS} + \int_{\phi=\pi/2}\bar{D}\cdot\overline{dS} + \int_{z=0}\bar{D}\cdot\overline{dS} + \int_{z=1}\bar{D}\cdot\overline{dS}$$

$$= \int_0^{\pi/2}\int_0^1 \bar{D}\cdot(-\rho\, d\phi\, dz\, \bar{a}_\rho)_{(\rho=1)} + \int_0^{\pi/2}\int_0^1 \bar{D}\cdot(\rho\, d\phi\, dz\, \bar{a}_\rho)_{(\rho=2)} +$$

$$\int_1^2\int_0^1 \bar{D}\cdot(-d\rho\, dz\, \bar{a}_\phi)_{(\phi=0)} + \int_1^2\int_0^1 \bar{D}\cdot(d\rho\, dz\, \bar{a}_\phi)_{(\phi=\pi/2)} +$$

$$\int_1^3\int_0^1 \bar{D}\cdot(-\rho\, d\rho\, d\phi\, \bar{a}_z)_{(z=0)} + \int_1^3\int_0^1 \bar{D}\cdot(\rho\, d\rho\, d\phi\, \bar{a}_z)_{(z=1)}$$

$$= \int_0^{\pi/2}\int_0^1 \frac{20}{\rho}\sin^2\phi\, d\phi\, dz_{(\rho=1)} + \int_0^{\pi/2}\int_0^1 -\frac{20}{\rho}\sin^2\phi\, d\phi\, dz_{(\rho=2)} +$$

$$\int_1^2 \int_0^1 -\frac{20}{\rho^2} \sin 2\phi \, d\rho \, dz \Big|_{(\phi = 0)} + \int_1^2 \int_0^1 \frac{20}{\rho^2} \sin 2\phi \, d\rho \, dz \Big|_{(\phi = \pi/2)} + 0 + 0$$

$$= \frac{20}{1} \times \frac{1}{2}\left[\phi - \frac{\sin 2\phi}{2}\right]_0^{\pi/2} \left[z\right]_0^1 - \frac{20}{2} \times \frac{1}{2} \times \left[\phi - \frac{\sin 2\phi}{2}\right]_0^{\pi/2} \left[z\right]_0^1$$

$= 7.85398 \ \text{C}$

Method 2 : $$Q = \int_V \bar{\nabla} \cdot \bar{D} \, dv$$

$$\bar{\nabla} \cdot \bar{D} = \frac{1}{\rho}\frac{\partial}{\partial \rho}\left[\rho\left(-\frac{20 \sin^2 \phi}{\rho^2}\right)\right] + \frac{1}{\rho}\frac{\partial}{\partial \phi}\left(\frac{20 \sin 2\phi}{\rho^2}\right)$$

$$= \frac{20 \sin^2 \phi}{\rho^3} + \frac{40 \cos 2\phi}{\rho^3} = \frac{1}{\rho^3}\left[10 + 30 \cos 2\phi\right]$$

$$\therefore \ Q = \int_1^2 \int_0^{\pi/2} \int_0^2 \frac{1}{\rho^3}\left[10 + 30 \cos 2\phi\right] \rho \, d\rho \, d\phi \, dz$$

$$= \left[\frac{-1}{\rho}\right]_1^2 \left[10\phi + \frac{30 \sin 2\phi}{2}\right]_0^{\pi/2} \left[z\right]_0^1 = 7.85398 \ \text{C}$$

Problem 3.41 : If $\bar{G} = 5 r \sin^2 \theta \cos^2 \phi \ \bar{a}_r$, evaluate both sides of the divergence theorem for the region $r \leq 2$.

Solution : According to divergence theorem

$$\oint_S \bar{G} \cdot \overline{dS} = \int_V \bar{\nabla} \cdot \bar{G} \, dv$$

$$\text{LHS} = \oint_S \bar{G} \cdot \overline{dS}.$$

To the surface $r = 2$, $\pm \ \bar{a}_r$ is perpendicular. Hence

$\overline{dS} = \pm \ (r^2 \sin\theta \ d\theta \ d\phi) \ \bar{a}_r$ and

$$\bar{G} \cdot \overline{dS} = 5 r \sin^2 \theta \cos^2 \phi \ \bar{a}_r \cdot (r^2 \sin\theta \ d\theta \ d\phi \ \bar{a}_r)$$

$\left[\because + \bar{a}_r \text{ is assumed}\right]$

$$= 5r^3 \sin^3\theta \cos^2\phi \; d\theta \; d\phi$$

$$\therefore \text{ LHS} = \int_0^{\pi} \int_0^{2\pi} \underset{(r=2)}{5r^3 \sin^3\theta \cos^2\phi} \; d\theta \; d\phi$$

$$= 5(2)^3 \left[\frac{\cos^3\theta}{3} - \cos\theta \right]_0^{\pi} \frac{1}{2} \left[\phi + \frac{\sin 2\phi}{2} \right]_0^{2\pi} = 167.548$$

$$\text{RHS} = \int_V \bar{\nabla} \cdot \bar{G} \; dv$$

$$\bar{\nabla} \cdot \bar{G} = \frac{1}{r^2} \frac{\partial}{\partial r} (r^2 G_r) = \frac{1}{r^2} \frac{\partial}{\partial r} (r^2 \times 5r \sin^2\theta \cos^2\phi) = 15 \sin^2\theta \cos^2\phi$$

$$\therefore \text{ RHS} = \int_0^2 \int_0^{\pi} \int_0^{2\pi} 15 \sin^2\theta \cos^2\phi \times \cdot r^2 \sin\theta \; dr \; d\theta \; d\phi$$

$$= 15 \left[\frac{r^3}{3} \right]_0^2 \left[\frac{\cos^3\theta}{3} - \cos\theta \right]_0^{\pi} \frac{1}{2} \left[\phi + \frac{\sin 2\phi}{2} \right]_0^{2\pi} = 167.548$$

Hence LHS = RHS and divergence theorem is verified.

Problem 3.42 : Let $\bar{D} = x\bar{a}_x$ and find the value of $\oint_S \bar{D} \cdot \overline{dS}$ over the surface of the sphere $r = 1$.

Solution : To the sphere $r = 1$ m, $\pm\ \bar{a}_r$ is perpendicular .

Hence $\overline{dS} = \pm\ (r^2 \sin\theta \; d\theta \; d\phi)\ \bar{a}_r$

$\bar{D} \cdot \overline{dS} = x\bar{a}_x \cdot (r^2 \sin\theta \; d\theta \; d\phi)\ \bar{a}_r$ [$\because$ + ve $\bar{a}_r$ is assumed]

$$= x r^2 \sin\theta \; d\theta \; d\phi \; (\sin\theta \cos\phi)$$

$$= (r \sin\theta \cos\phi) \; r^2 \sin\theta \; d\theta \; d\phi \; (\sin\theta \cos\phi)$$

$$= r^3 \sin^3\theta \cos^2\phi \; d\theta \; d\phi$$

$$\therefore \Psi = \oint_S \bar{D} \cdot \overline{dS} = \int_0^{\pi} \int_0^{2\pi} \underset{(r=1)}{r^3 \sin^3\theta \cos^2\phi} \; d\theta \; d\phi$$

$$= (1)^3 \left[\frac{\cos^3\theta}{3} - \cos\theta \right]_0^{\pi} \frac{1}{2} \left[\phi + \frac{\sin 2\phi}{2} \right]_0^{2\pi}$$

$$= 4.1887$$

—X—

CHAPTER 4

Work, Energy And Potential

4.1 ENERGY EXPANDED IN MOVING A POINT CHARGE IN AN ELECTRIC FIELD :

If a charge Q is placed in an electric field $\overline{E}$, then the force due to electric field is given by $\overline{F}_E = Q\,\overline{E}$. Suppose this charge has to be moved through a distance dL in the direction of unit vector $\overline{a}_L$ in the presence of electric field. The force which we must overcome is the component of the force $\overline{F}_E$ in $\overline{a}_L$ direction. It is given by $F_{EL} = \overline{F}_E \cdot \overline{a}_L = Q\,\overline{E} \cdot \overline{a}_L$. The external force required to move the charge is equal and opposite to F_{EL} .Hence

$$F_{appl.} = -\,Q\,\overline{E} \cdot \overline{a}_L$$

and therefore the differential amount of work done which is the product of force applied and the distance transversed is given by

$$dW = -\,Q\,\overline{E} \cdot \overline{dL}$$

where $dL\,\overline{a}_L$ is replaced by $\overline{dL}$. The work required to move the charge through a finite distance is

$$W = -\,Q \int_{\text{initial}}^{\text{final}} \overline{E} \cdot \overline{dL}$$

Note : (1) In the presence of electric field intensity $\overline{E}$, differential amount of work (dW) required to move a charge Q through a distance dL in the direction of $\overline{a}_L$, is given by

$$dW = -Q\,\overline{\mathrm{E}} \cdot \overline{\mathrm{dL}} \qquad \ldots\ldots(4.1)$$

where $\overline{\mathrm{dL}} = dL\ \overline{\mathrm{a}}_{\mathrm{L}}$

(2) In the presence of electric field intensity $\overline{\mathrm{E}}$, the total work (W) required to move a charge Q from an initial point to a final point is given by

$$W = -Q \int_{\text{initial}}^{\text{final}} \overline{\mathrm{E}} \cdot \overline{\mathrm{dL}} \qquad \ldots\ldots(4.2)$$

where

$$\overline{\mathrm{dL}} = dx\ \overline{\mathrm{a}}_x + dy\ \overline{\mathrm{a}}_y + dz\ \overline{\mathrm{a}}_z \quad \text{(cartesian)} \qquad \ldots(4.3a)$$

$$\overline{\mathrm{dL}} = d\rho\ \overline{\mathrm{a}}_\rho + \rho\, d\phi\ \overline{\mathrm{a}}_\phi + dz\ \overline{\mathrm{a}}_z \quad \text{(cylindrical)} \qquad \ldots(4.3b)$$

$$\overline{\mathrm{dL}} = dr\ \overline{\mathrm{a}}_r + r\,d\theta\ \overline{\mathrm{a}}_\theta + r\sin\theta\, d\phi\ \overline{\mathrm{a}}_\phi \quad \text{(spherical)} \qquad \ldots(4.3c)$$

Problem 4.1 : An electric field is given as $\overline{\mathrm{E}} = 6y^2z\ \overline{\mathrm{a}}_x + 12xyz\ \overline{\mathrm{a}}_y + 6xy^2\ \overline{\mathrm{a}}_z$ V/m. An incremental path is represented by $\overline{\Delta \mathrm{L}} = -3\,\overline{\mathrm{a}}_x + 5\,\overline{\mathrm{a}}_y - 2\,\overline{\mathrm{a}}_z$ µm. Find the work done in moving a 2 µC charge along this path if the location of the path is at (a) P_A (0, 2, 5) (b) P_B (1, 1, 1) (c) P_C (− 0.7, − 2, − 0.3) .

Solution :

(a) $\overline{\mathrm{E}}$ at $P_A = 6\,(2)^2\,(5)\,\overline{\mathrm{a}}_x + 12\,(0)\,(2)^2\,(5)\,\overline{\mathrm{a}}_y + 6\,(0)\,(2)^2\,\overline{\mathrm{a}}_z = 120\,\overline{\mathrm{a}}_x$ V/m

Using eqn.(4.1), we have

$$dW = -2\times 10^{-6}(120\,\overline{\mathrm{a}}_x)\cdot(-3\,\overline{\mathrm{a}}_x + 5\,\overline{\mathrm{a}}_y - 2\,\overline{\mathrm{a}}_z)\,10^{-6}$$

$$= -2\times 10^{-12}(-360 + 0 + 0) = 720 \text{ pJ}$$

(b) $\overline{\mathrm{E}}$ at $P_B = 6\,(1)^2\,(1)\,\overline{\mathrm{a}}_x + 12\,(1)\,(1)\,(1)\,\overline{\mathrm{a}}_y + 6\,(1)\,(1)^2\,\overline{\mathrm{a}}_z$

$$= 6\,\overline{\mathrm{a}}_x + 12\,\overline{\mathrm{a}}_y + 6\,\overline{\mathrm{a}}_z \text{ V/m}$$

Using eqn.(4.1), we have

$$dW = -2\times 10^{-6}(6\,\overline{\mathrm{a}}_x + 12\,\overline{\mathrm{a}}_y + 6\,\overline{\mathrm{a}}_z)\cdot(-3\,\overline{\mathrm{a}}_x + 5\,\overline{\mathrm{a}}_y - 2\,\overline{\mathrm{a}}_z)\,10^{-6}$$

$$= -2\times 10^{-12}(-18 + 60 - 12) = -60 \text{ pJ}$$

(c) Substituting the coordinates of point P_C in the expression for $\overline{E}$ we have

$\overline{E}$ at $P_C = -7.2\,\overline{a}_x - 5.04\,\overline{a}_y - 16.8\,\overline{a}_z$ V/m

$$\overline{E}\cdot\overline{dL} = (-7.2\,\overline{a}_x - 5.04\,\overline{a}_y - 16.8\,\overline{a}_z)\cdot(-3\,\overline{a}_x + 5\,\overline{a}_y - 2\,\overline{a}_z)\,10^{-6}$$

$$= 10^{-6}(21.6 - 25.2 + 33.6) = 30\times 10^{-6}$$

$$\therefore\ dW = -Q\,\overline{E}\cdot\overline{dL} = -2\times 10^{-6}\times 30\times 10^{-6} = -60\ \text{pJ}$$

Problem 4.2 : Find the work done in moving a 5 μC charge from the origin to P (2 , – 1, 4) through the field $\overline{E} = 2xyz\,\overline{a}_x + x^2z\,\overline{a}_y + x^2y\,\overline{a}_z$ V/m via the path : (a) straight line segments : (0, 0, 0) to (2, 0, 0) to (2, –1, 0) to (2, –1, 4) ; (b) straight line : $x = -2y, z = 2x$; (c)curve : $x = -2y^3, z = 4y^2$.

Solution : (a) $W = -Q\int_O^P \overline{E}\cdot\overline{dL}$ **where O is the origin.**

$$\overline{E}\cdot\overline{dL} = (2xyz\,\overline{a}_x + x^2z\,\overline{a}_y + x^2y\,\overline{a}_z)\cdot(dx\,\overline{a}_x + dy\,\overline{a}_y + dz\,\overline{a}_z)$$

$$= 2xyz\,dx + x^2z\,dy + x^2y\,dz$$

$$\therefore\ W = -Q\left\{\int_0^2 2xyz\,dx + \int_0^{-1} x^2z\,dy + \int_0^4 x^2y\,dz\right\} \qquad \textbf{.........(1)}$$

Path of integration is (0, 0, 0) to (2, 0, 0 to (2, –1, 0) to (2, –1, 4) . Between first and second point, y and z coordinates are constant and x coordinate varies from $x = 0$ to $x = 1$. Hence in the first integral substitute $y = 0$ and $z = 0$. Between second and third point,x and z coordinates are constant and y coordinate varies from $y = 0$ to $y = -1$. Hence in the second integral substitute $x = 2$ and $z = 0$. Between third and fourth point, x and y coordinates are constant and z coordinate varies from $z = 0$ to $z = 4$. Hence in the third integral substitute $x = 2$ and $y = -1$. Hence

$$W = -Q\left\{\int_0^2 2(0)(0)\,x\,dx + \int_0^{-1}(2)^2(0)\,dy + \int_0^4 (2)^2(-1)\,dz\right\}$$

$$= -5\times 10^{-6}\left\{0 + 0 - 4\Big[z\Big]_0^4\right\} = -5\times 10^{-6}(-4\times 4) = 80\ \mu\text{J}$$

$$\text{(b)}\ \therefore\ W = -Q\left\{\int_0^2 2xyz\,dx + \int_0^{-1} x^2z\,dy + \int_0^4 x^2y\,dz\right\} \qquad .....(1)$$

Equations of the straight line are $x = -2y$, $z = 2x$.. In the first integral, integration is carried out w.r.t. x. Hence we substitute y and z interms of x. Hence in the first integral we substitute $y = -0.5x$ and $z = 2x$.

In the second integral, integration is carried out w.r.t. y. Hence we substitute x and z in terms of y. Hence in the second integral we substitute $x = -2y$ and $z = 2x = 2(-2y) = -4y$.

In the third integral, integration is carried out w.r.t. z. Hence we substitute x and y in terms of z. Hence in the third integral we substitute $x = 0.5z$ and $y = -0.5x = -0.5(0.5z) = -0.25z$.

$$W = -Q\left[\int_0^2 2x(-0.5x)(2x)\,dx + \int_0^{-1}(-2y)^2(-4y)\,dy + \int_0^4 (0.5z)^2(-0.25)\,dz\right]$$

$$= -5\times 10^{-6}\left\{-2\left[\frac{x^4}{4}\right]_0^2 - 16\left[\frac{y^4}{4}\right]_0^{-1} - 0.0625\left[\frac{z^4}{4}\right]_0^4\right\}$$

$$= -5\times 10^{-6}(-8-4-4) = 80\,\mu\text{J}$$

$$\text{(c)} \therefore W = -Q\left\{\int_0^2 2xyz\,dx + \int_0^{-1} x^2 z\,dy + \int_0^4 x^2 y\,dz\right\} \quad .....(1)$$

Equations of the curve are $x = -2y^3$, $z = 4y^2$. For the reasons mentioned in part (b), we have to make proper substitutions.

In the first integral, $2xyz\,dx = \dfrac{2xy^3 z\,dx}{y^2} = \dfrac{2x(-0.5x)z\,dx}{0.25z} = -4x^2\,dx$

In the second integral, $x^2 z\,dy = (-2y^3)^2(4y^2)\,dy = 16y^8\,dy$

In the third integral, $x^2 y\,dz = (-2y^3)^2 y\,dz = 4y^7\,dz$

Now $z = 4y^2$, $y^2 = 0.25z$, $y = \pm\sqrt{0.25z}$

(Selection of proper sign is very important)

At origin, when $z = 0$, we must have $y = 0$ which is possible with any sign. At point P, when $z = 4$, we must have $y = -1$ which is possible only if we consider – ve sign . Hence $y = -\sqrt{0.25z} = -0.5\sqrt{z}$

Hence in the third integral, $x^2 y\,dz = 4(-0.5\sqrt{z})^7\,dz = -0.03125\,z^{3.5}\,dz$

$$\therefore W = -Q\left\{\int_0^2 -4x^2\,dx + \int_0^{-1} 16y^8\,dy + \int_0^4 -0.03125\,z^{3.5}\,dz\right\}$$

$$= -Q\left\{-4\left[\frac{x^3}{3}\right]_0^2 + 16\left[\frac{y^9}{9}\right]_0^{-1} - 0.03125\left[\frac{z^{4.5}}{4.5}\right]_0^4\right\}$$

$$= -5\times 10^{-6}\left[-32/3 - 16/9 - 16/4.5\right] = 80\ \mu\text{J}$$

Problem 4.3 : If $\overline{\mathrm{E}} = x\,\overline{\mathrm{a}}_y$ V/m at a certain instant of time, then calculate the work required to move a 3 - C point charge from (1, 3, 5) to (2, 0, 3) along the straight - line segments joining : (a) (1, 3, 5) to (2, 3, 5) to (2, 0, 5) to (2, 0, 3); (b) (1, 3, 5) to (1, 3, 3) to (1, 0, 3) to (2, 0, 3).

Solution :(a) $W = -Q\int_{(1,3,5)}^{(2,0,3)} \overline{\mathrm{E}}\cdot\overline{\mathrm{dL}}$

$$\overline{\mathrm{E}}\cdot\overline{\mathrm{dL}} = (x\,\overline{\mathrm{a}}_y)\cdot(dx\,\overline{\mathrm{a}}_x + dy\,\overline{\mathrm{a}}_y + dz\,\overline{\mathrm{a}}_z) = x\,dy$$

$$\therefore W = -3\int_3^0 x\,dy \qquad \ldots\ldots(1)$$

Path of integration is (1, 3, 5) to (2, 3, 5) to (2, 0, 5) to (2, 0, 3). Between second and third point, x and z coordinates are constant and y coordinate varies from $y = 3$ to $y = 0$. Hence substitute $x = 2$ in equation 1.

$$\therefore W = -3\int_3^0 2\,dy = -6\left[y\right]_3^0 = -6(0-3) = 18\ \text{J}$$

(b) Path of integration is (1, 3, 5) to (1, 3, 3) to (1, 0, 3) to (2, 0, 3). Between second and third point, x and z coordinates are constant whereas y coordinate varies from $y = 3$ to $y = 0$. Hence substitute $x = 1$ in equation 1.

$$\therefore W = -3\int_3^0 1\cdot dy = -3(0-3) = 9\ \text{J}$$

4.2 DEFINATION OF POTENTIAL DIFFERENCE AND POTENTIAL

4.2.1 Defination :

The potential difference is defined as the work done by an external source in moving a unit positive charge from one point to another in an electric field. Substituting $Q = 1$ C in eqn.4.2 we have

$$V = -\int_{\text{initial}}^{\text{final}} \overline{\text{E}} \cdot \overline{\text{dL}}$$

The unit of potential difference is V (volt). If point P is a final point and point Q is the initial point, then the potential difference between these two points is written as

$$\boxed{V_{PQ} = -\int_{Q}^{P} \overline{\mathbf{E}} \cdot \overline{\mathbf{dL}}} \qquad \text{.....(4.4)}$$

The potential or absolute potential of a point is the potential difference between that point and a reference point which is considered to have zero potential. Generally this reference point is located at infinity. If the potential of point P is V_P and that of point Q is V_Q, then

$$\boxed{V_{PQ} = V_P - V_Q} \qquad \text{.....(4.5)}$$

Problem 4.4 : Let $\overline{\mathbf{E}} = (-6y/x^2)\,\overline{\mathbf{a}}_x + (6/x)\,\overline{\mathbf{a}}_y + 5\,\overline{\mathbf{a}}_z$ V/m and calculate : (a) V_{PQ} given P (–7, 2, 1) and Q (4, 1, 2) (b) V_P if $V = 0$ at Q ; (c) V_P if $V = 0$ at (2, 0, –1)

Solution :(a) $V_{PQ} = -\int_{Q}^{P} \overline{\text{E}} \cdot \overline{\text{dL}}$

$$\overline{\text{E}} \cdot \overline{\text{dL}} = \left[(-6y/x^2)\,\overline{\text{a}}_x + (6/x)\,\overline{\text{a}}_y + 5\,\overline{\text{a}}_z\right] \cdot \left[dx\,\overline{\text{a}}_x + dy\,\overline{\text{a}}_y + dz\,\overline{\text{a}}_z\right]$$

$$= (-6y/x^2)\,dx + (6/x)\,dy + 5\,dz$$

$$\therefore V_{PQ} = -\left[\int_{4}^{-7} (-6y/x^2)\,dx + \int_{1}^{2} (6/x)\,dy + \int_{2}^{1} 5\,dz\right]$$

The path of integration is Q (4, 1, 2) to (–7, 1, 2) to (–7, 2, 2) to P (–7, 2, 1). Between first and second point, y and z coordinates are constant whereas x coordinate varies from $x = 4$ to $x = -7$. Hence in the first integral, substitute $y = 1$ and $z = 2$. Between second and third point, x and z coordinates are constant whereas y coordinate varies from $y = 1$ to $y = 2$. Hence in the second integral, substitute $x = -7$ and $z = 2$.

$$\therefore V_{PQ} = -\left\{\int_{4}^{-7} -\frac{6(1)}{x^2}\,dx + \int_{1}^{2} -\frac{6}{7}\,dy + \int_{2}^{1} 5\,dz\right\}$$

$$= -\left\{6\left[\frac{1}{x}\right]_4^{-7} - \frac{6}{y}\left[y\right]_1^2 + 5\left[z\right]_2^1\right\} = 8.214 \text{ V}$$

(b) $V_{PQ} = V_P - V_Q$, $V_{PQ} = 8.214$, $V_Q = 0$

$\therefore V_P = V_{PQ} + V_Q = 8.214 + 0 = 8.214$ V

(c) Let A be the point with coordinates (2, 0, –1).Then

$$V_{PA} = -\int_A^P \overline{\mathrm{E}} \cdot \overline{\mathrm{dL}} = -\left\{\int_{2}^{-7} -\frac{6y}{x^2}\,dx + \int_{0}^{2} \frac{6}{x}\,dy + \int_{-1}^{1} 5\,dz\right\}$$

The path of integration is broken as given below.

A (2. 0, –1) to (–7, 0, –1) to (–7, 2, –1) to P (–7, 2, 1)

Between first and second point, y and z coordinates are constant whereas x coordinate varies from $x = 2$ to $x = -7$. Hence in the first integral, substitute $y = 0$ and $z = -1$. Between second and third point, x and z coordinates are constant whereas y coordinate varies from $y = 0$ to $y = 2$. Hence in the second integral, substitute $x = -7$ and $z = -1$.

$$\therefore V_{PA} = -\left\{\int_{2}^{-7} -\frac{6(0)}{x^2}\,dx + \int_{0}^{2} -\frac{6}{7}\,dy + \int_{-1}^{1} 5\,dz\right\}$$

$$= -\left\{0 - \frac{6}{7}\left[y\right]_0^2 + 5\left[z\right]_{-1}^1\right\} = -8.286 \text{ V}$$

Now $V_{PA} = V_P - V_A$ and $V_A = 0$

$\therefore V_P = V_{PA} + V_A = -8.286 + 0 = -8.286$ V

4.2.2 Expression for the potential difference V_{AB} in the field of a point charge

Suppose the charge Q is located at the origin as shown Fig. 4.1. Electric field intensity $\overline{\mathrm{E}}$ at any point (r, θ, ϕ) due to charge Q at origin is given by

$$\overline{\mathrm{E}} = \frac{Q}{4\pi\varepsilon_0 r^2}\,\overline{a}_r \quad \text{V/m}$$

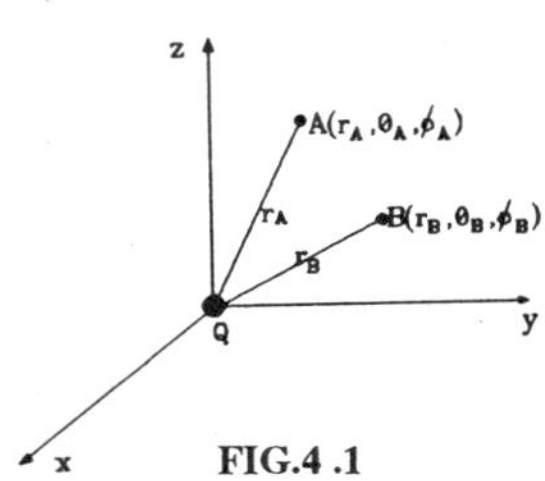

FIG.4 .1

In spherical coordinate system,

$\overline{dL} = dr\,\bar{a}_r + r\,d\theta\,\bar{a}_\theta + r\sin\theta\,d\phi\,\bar{a}_\phi$. Hence

$$\bar{E}\cdot\overline{dL} = \frac{Q\,dr}{4\pi\varepsilon_o r^2} \quad \text{and} \quad V_{AB} = -\int_B^A \bar{E}\cdot\overline{dL} = -\int_{r_B}^{r_A}\frac{Q\,dr}{4\pi\varepsilon_o r^2}$$

$$\boxed{V_{AB} = \frac{Q}{4\pi\varepsilon_o}\left[\frac{1}{r_A} - \frac{1}{r_B}\right]} \quad \ldots\ldots(4.6)$$

where r_A and r_B are distances of points A and B respectively from the charge Q. Further while applying the above formula, it is not necessary that the charge Q should be always located at origin. The interpretation of r_A and r_B remains the same.

If point B is located at ∞, then $r_B = \infty$ and if zero reference is also at ∞, then $V_B = 0$, so that $V_{AB} = V_A - V_B$ or

$$V_A = V_{AB} + V_B = V_{AB} = \frac{Q}{4\pi\varepsilon_o}\left(\frac{1}{r_A} - \frac{1}{\infty}\right) = \frac{Q}{4\pi\varepsilon_o r_A}$$

Hence potential at point A due to charge Q is given by

$$\boxed{V_A = \frac{Q}{4\pi\varepsilon_o r_A}} \quad \ldots\ldots(4.7)$$

where r_A is the distance of point A from the charge Q. Again the charge Q may be located at any point.

Problem 5.5 : A point charge of 6 nC is located at the origin in free space. Find V_P if point P is located at P (0.2, – 0.4, 0.4) and : (a) $V = 0$ at infinity ; (b) $V = 0$ at (1, 0, 0) ; (c) $V = 20$ V at (– 0.5, 1, –1).

Solution :

(a) $V_{PQ} = \frac{Q}{4\pi\varepsilon_o}\left(\frac{1}{r_P} - \frac{1}{r_Q}\right)$ where r_P and r_Q are the distances of points P and Q respectively from the charge Q. Charge is located at origin.

Hence $r_P = \sqrt{(0.2)^2 + (-0.4)^2 + (0.4)^2} = 0.6$

Let Q is located at infinity. Then $r_Q = \infty$

$$\therefore\ V_{PQ} = \frac{6\times 10^{-9}}{4\pi\varepsilon_o}\left(\frac{1}{0.6} - \frac{1}{\infty}\right) = 89.877 \text{ V}$$

Now $V_{PQ} = V_P - V_Q$ and $V_Q = 0$ (given)

$\therefore\ V_P = V_{PQ} + V_Q = 89.877$ V

(b) Let point Q is (1, 0 , 0).Then $r_Q = \sqrt{(1)^2 + (0)^2 + (0)^2} = 1$

$$\therefore\ V_{PQ} = \frac{6 \times 10^{-9}}{4\pi\varepsilon_o}\left(\frac{1}{0.6} - \frac{1}{1}\right) = 35.951 \text{ V}$$

Now $V_{PQ} = V_P - V_Q$ and $V_Q = 0$ (given)

$\therefore\ V_P = V_{PQ} + V_Q = 35.951$ V

(c) Let point Q is (– 0.5,1 ,–1).Then $r_Q = \sqrt{(-0.5)^2 + (1)^2 + (-1)^2} = 1.5$

$$\therefore\ V_{PQ} = \frac{6 \times 10^{-9}}{4\pi\varepsilon_o}\left(\frac{1}{0.6} - \frac{1}{1.5}\right) = 53.926 \text{ V}$$

Now $V_{PQ} = V_P - V_Q$ and $V_Q = 20$ V (given)

$\therefore\ V_P = V_{PQ} + V_Q = 53.926 + 20 = 73.926$ V

4.2.3 Expression for the potential difference V_{AB} in the field of a line charge:

Suppose the line charge ρ_L C/m is present on the z axis as shown in Fig. 4.2. Electric field intensity $\overline{E}$ at any point (ρ, ϕ, z) due to this line charge is given by

$$\overline{E} = \frac{\rho_L}{2\pi\varepsilon_o\rho}\,\overline{a}_\rho$$

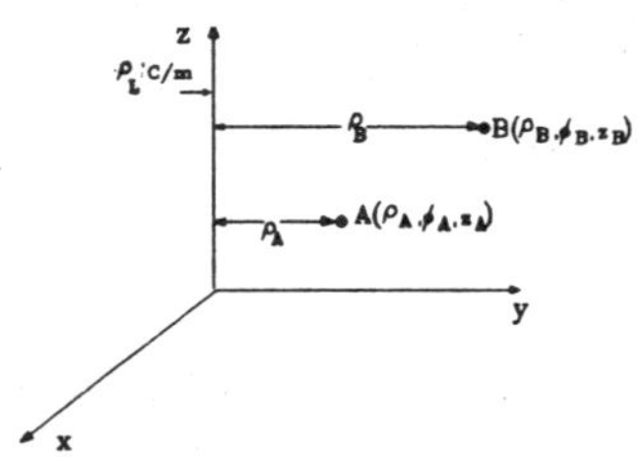

FIG.4.2

In cylindrical coordinate system ,

$$\overline{dL} = d\rho\,\overline{a}_\rho + \rho\,d\phi\,\overline{a}_\phi + dz\,\overline{a}_z$$

Hence $\overline{E} \cdot \overline{dL} = \dfrac{\rho_L\,d\rho}{2\pi\varepsilon_o\rho}$

$$\therefore\ V_{AB} = -\int_B^A \overline{E} \cdot \overline{dL} = -\int_{\rho_B}^{\rho_A} \frac{\rho_L\,d\rho}{2\pi\varepsilon_o\rho} = -\frac{\rho_L}{2\pi\varepsilon_o}\ln\left(\frac{\rho_A}{\rho_B}\right)$$

$$\boxed{V_{AB} = -\frac{\rho_L}{2\pi\varepsilon_o}\ln\left(\frac{\rho_A}{\rho_B}\right)} \qquad \ldots\ldots(4.8)$$

This formula is applicable to a uniform line charge ρ_L C/m whose length is infinite in both the directions. Further ρ_A and ρ_B represents shortest distance of points A and B respectively from the line charge. Also it is not necessary that the line charge should always lie along the z axis.

4.2.4 Expression for the potential difference V_{PQ} in the field of a surface charge

Electric field intensity due to surface charge density ρ_S C/m^2 is given by

$$\overline{E} = \frac{\rho_S}{2\,\varepsilon_o}\,\overline{a}_N$$

where $\overline{a}_N$ is a unit vector perpendicular to the plane on which surface charge density lies. In any problem both the points P and Q will lie on one side of the plane on which ρ_S lies. So that $\overline{a}_N$ will be same for both the points P and Q. After $\overline{E}$ is determined, calculate $\overline{E} \cdot \overline{dL}$ and then use the equation

$$V_{PQ} = -\int_{Q}^{P} \overline{E} \cdot \overline{dL}$$

so as to determine potential difference V_{PQ}. No general formula can be derived as $\overline{a}_N$ is unknown.

4.2.5 Potential due to line charge, surface charge and volume charge density with respect to zero reference at infinity

Suppose charge Q is located at point 1. If zero reference of potential is located at infinity, then due to a point charge Q, the potential at point 2 can be obtained by using the equation.

$$V = \frac{Q}{4\,\pi\,\varepsilon_o\ (\text{ distance between point 2 and charge } Q\ )}$$

$$= \frac{Q}{4\,\pi\,\varepsilon_o\ (\text{ distance between point 2 and point 1 })}$$

Now distance between point 2 and point 1 $= |\,\overline{R}_{12}\,| = |\,\overline{r}_2 - \overline{r}_1\,|$, where $\overline{r}_2$ is the position vector of point 2 and $\overline{r}_1$ is the position vector of point 1. Hence,

$$V = \frac{Q}{4\,\pi\,\varepsilon_o\,|\,\overline{r}_2 - \overline{r}_1\,|}$$

If charge Q is replaced by differential charge dQ, then differential amount of potential dV is given by

$$dV = \frac{dQ}{4\,\pi\,\varepsilon_o\,|\,\overline{r}_2 - \overline{r}_1\,|}$$

For a line charge, $dQ = \rho_L\,dL$. Hence,

$$dV = \frac{\rho_L\,dL}{4\,\pi\,\varepsilon_o\,|\,\overline{r}_2 - \overline{r}_1\,|}$$

and

$$V = \int \frac{\rho_L \, dL}{4 \pi \varepsilon_0 \, | \, \bar{r}_2 - \bar{r}_1 \, |} \quad \ldots\ldots (4.9)$$

For a surface charge, $dQ = \rho_S \, dS$. Hence

$$dV = \frac{\rho_S \, dS}{4 \pi \varepsilon_0 \, | \, \bar{r}_2 - \bar{r}_1 \, |}$$

and

$$V = \int \frac{\rho_S \, dS}{4 \pi \varepsilon_0 \, | \, \bar{r}_2 - \bar{r}_1 \, |} \quad \ldots\ldots (4.10)$$

For a volume charge density, $dQ = \rho_v \, dv$. Hence

$$dV = \frac{\rho_v \, dv}{4 \pi \varepsilon_0 \, | \, \bar{r}_2 - \bar{r}_1 \, |}$$

and

$$V = \int \frac{\rho_v \, dv}{4 \pi \varepsilon_0 \, | \, \bar{r}_2 - \bar{r}_1 \, |} \quad \ldots\ldots (4.11)$$

In all these three formulae, $\bar{r}_1$ is the position vector of point 1 at which differential charge dQ is located. $\bar{r}_2$ is the position vector of point 2 at which potential is desired.

Problem 4.6 : Assume a zero reference at infinity, and find the potential at P (0, 0,10) that is caused by this charge configuration in free space : (a) 20 nC at the origin ; (b) 10 nC / m along the line $x = 0, z = 0, -1 < y < 1$; (c) 10 nC / m along the line $x = 0, y = 0, -1 < z < 1$.

Solution :

(a) $V_P = \dfrac{Q}{4 \pi \varepsilon_o r_P}$ where r_P is the distance of point P from charge Q.

Charge is located at origin, Hence $r_P = \sqrt{(0)^2 + (0)^2 + (10)^2} = 10$

$$V_P = \frac{20 \times 10^{-9}}{4 \pi \varepsilon_o \, (10)} = 17.975 \text{ V}$$

(b) $V_2 = \displaystyle\int \frac{dQ}{4 \pi \varepsilon_o \, | \, \bar{r}_2 - \bar{r}_1 \, |}$

where $\bar{r}_2$ is the position vector of point 2 and $\bar{r}_1$ is the position vector of point

1. Now point 2 is that point at which potential is desired = (0, 0, 10) and point 1 is the general point on the line charge = ($x = 0, y, z = 0$)

$\therefore\ \bar{r}_2 = 10\,\bar{a}_z$ and $\bar{r}_1 = y\,\bar{a}_y$

$\therefore\ \bar{r}_2 - \bar{r}_1 = 10\,\bar{a}_z - y\,\bar{a}_y$ or $|\,\bar{r}_2 - \bar{r}_1\,| = \sqrt{100 + y^2}$

As the equations of the line charge are $x = 0, z = 0$ which are constants, we have $dx = 0$ and $dz = 0$. Hence $\overline{dL} = dx\,\bar{a}_x + dy\,\bar{a}_y + dz\,\bar{a}_z = dy\,\bar{a}_y$.

Hence for the line charge $dL = dy$ and $dQ = \rho_L\,dL = 10 \times 10^{-9}\,dy$.

$$\therefore\ V_2 = \int_{-1}^{1} \frac{10 \times 10^{-9}\,dy}{4\pi\varepsilon_o\sqrt{100 + y^2}}$$

$$= \frac{10^{-8}}{4\pi\varepsilon_o}\Big[\ln\left(y + \sqrt{y^2 + 100}\right)\Big]_{-1}^{1} = 17.9456\ \text{V}$$

(c) Point 2 is (0, 0, 10) and point 1 is ($x = 0, y = 0, z$)..

Hence $\bar{r}_2 = 10\,\bar{a}_z$ and $\bar{r}_1 = z\,\bar{a}_z$

$\bar{r}_2 - \bar{r}_1 = 10\,\bar{a}_z - z\,\bar{a}_z = (10 - z)\,\bar{a}_z$ and $|\,\bar{r}_2 - \bar{r}_1\,| = (10 - z)$

As the equations of the line charge are $x = 0, y = 0$ which are constants, we have $dx = 0$ and $dy = 0$. Hence $\overline{dL} = dx\,\bar{a}_x + dy\,\bar{a}_y + dz\,\bar{a}_z = dz\,\bar{a}_z$.

Hence for the line charge $dL = dz$ and $dQ = \rho_L\,dL = 10 \times 10^{-9}\,dz$.

$$V_2 = \int_{-1}^{1} \frac{10 \times 10^{-9}\,dz}{4\pi\varepsilon_o(10 - z)} = \frac{10^{-8}}{4\pi\varepsilon_o}\Big[-\ln(10 - z)\Big]_{-1}^{1} = 18.035\ \text{V}$$

4.3 THE CONCEPT OF POTENTIAL GRADIENT AND THE RELATIONSHIP BETWEEN $\bar{E}$ AND V.

We know that

$$V = -\int \bar{E} \cdot \overline{dL}$$

When this equation is applied to a very short element of length $\overline{\Delta L}$ along which $\bar{E}$ is constant, then an incremental potential difference ΔV is given by

$$\Delta V = -\bar{E} \cdot \overline{\Delta L} = -E\,\Delta L\cos\theta$$

where θ is the angle between $\overline{\Delta L}$ and $\bar{E}$. Since V is a function of coordinates of any point and it is a single valued function, we have

$$\frac{dV}{dL} = -E\cos\theta$$

If $\bar{a}_L$ ($\overline{\Delta L} = \Delta L\ \bar{a}_L$) is placed in a opposite direction to $\bar{E}$ ($\theta = 180^o$) then

$$\frac{dV}{dL} = -E\cos 180^o = E$$

Hence

$$\left.\frac{dV}{dL}\right|_{max} = E$$

This shows that

1. The magnitude of the electric field intensity is given by the maximum value of the rate of change of potential with distance.
2. This maximum value is obtained when the direction of the distance increment is opposite to $\bar{E}$ or in other words, the direction of $\bar{E}$ is opposite to the direction in which the potential is increasing the most rapidly.

Now $\bar{E}$ is perpendicular to the equipotential surface and is directed towards the decreasing value of the potential. Let $\bar{a}_N$ is the unit vector towards the increasing value of the potential. Then

$$\bar{E} = -\left.\frac{dV}{dL}\right|_{max}\bar{a}_N$$

Since $\left.\frac{dV}{dL}\right|_{max}$ occurs when $\overline{\Delta L}$ is in the direction of $\bar{a}_N$,

$$\left.\frac{dV}{dL}\right|_{max} = \frac{dV}{dN}$$

and

$$\bar{E} = -\frac{dV}{dN}\bar{a}_N$$

The gradient of a scalar field T is defined as

$$\text{Gradient of } T = \text{grad}.T = \frac{dT}{dN}\bar{a}_N$$

where $\bar{a}_N$ is a unit vector normal to the equipotential surfaces and the normal is chosen which points in the direction of increasing values of T . So

$$\bar{E} = -\text{grad } V$$

Since V is a function of (x, y, z) , we have

$$dV = \frac{\partial V}{\partial x}dx + \frac{\partial V}{\partial y}dy + \frac{\partial V}{\partial z}dz$$

Let $\bar{E} = E_x\,\bar{a}_x + E_y\,\bar{a}_y + E_z\,\bar{a}_z$ and $\overline{dL} = dx\,\bar{a}_x + dy\,\bar{a}_y + dz\,\bar{a}_z$

Then $dV = -\bar{E} \cdot \overline{dL} = -E_x\, dx - E_y\, dy - E_z\, dz$

Comparing the coefficients of *dx*, *dy*, and *dz* from both the expressions for *dV*, we get.

$$E_x = -\frac{\partial V}{\partial x} \; ; \; E_y = -\frac{\partial V}{\partial y} \; ; \; E_z = -\frac{\partial V}{\partial z}$$

Hence

$$\bar{E} = -\left(\frac{\partial V}{\partial x}\bar{a}_x + \frac{\partial V}{\partial y}\bar{a}_y + \frac{\partial V}{\partial z}\bar{a}_z \right)$$

As

$$\bar{E} = -\,\text{grad}\, V$$

we have

$$\text{grad}\, V = \frac{\partial V}{\partial x}\bar{a}_x + \frac{\partial V}{\partial y}\bar{a}_y + \frac{\partial V}{\partial z}\bar{a}_z$$

Now del ($\bar{\nabla}$) operator is given by

$$\bar{\nabla} = \frac{\partial}{\partial x}\bar{a}_x + \frac{\partial}{\partial y}\bar{a}_y + \frac{\partial}{\partial x}\bar{a}_z$$

Hence

$$\bar{\nabla} V = \frac{\partial V}{\partial x}\bar{a}_x + \frac{\partial V}{\partial y}\bar{a}_y + \frac{\partial V}{\partial z}\bar{a}_z$$

or

$$\bar{E} = -\bar{\nabla} V$$

Note :

(1) Electric field intensity is perpendicular to the equipotential surface (a surface on which potential is constant) and is directed from higher to lower potential i.e. towards the decreasing value of potential.

(2) Maximum rate of change of potential with respect to distance gives the magnitude of electric field intensity.

(3) Note the following important formulae

$$\boxed{|\bar{\mathbf{E}}| = \left.\frac{dV}{dL}\right|_{\text{max.}}} \qquad \text{......(4.12a)}$$

$$\boxed{\bar{\mathbf{E}} = -\,\textbf{grad V}} \qquad \text{......(4.12b)}$$

$$\bar{\mathbf{E}} = -\bar{\nabla} V \qquad \text{.....(4.12c)}$$

$$\textbf{grad } V = \bar{\nabla} V \qquad \text{.....(4.12d)}$$

$$\bar{\mathbf{E}} = -\left.\frac{dV}{dL}\right|_{\max} \bar{\mathbf{a}}_N \qquad \text{.....(4.12e)}$$

$$\bar{\mathbf{E}} = -\frac{dV}{dN}\,\bar{\mathbf{a}}_N \qquad \text{.....(4.12f)}$$

where $\bar{a}_N$ is a unit vector perpendicular to the equipotential surface and directed from lower to higher potential. That is, towards the increasing value of the potential. In other words $\bar{a}_N$ is in opposite direction to $\bar{E}$.

$$\bar{\nabla} V = \frac{\partial V}{\partial x}\bar{\mathbf{a}}_x + \frac{\partial V}{\partial y}\bar{\mathbf{a}}_y + \frac{\partial V}{\partial z}\bar{\mathbf{a}}_z \quad \textbf{(Cartesian)} \qquad \text{..(4.13 a)}$$

$$\bar{\nabla} V = \frac{\partial V}{\partial \rho}\bar{\mathbf{a}}_\rho + \frac{1}{\rho}\frac{\partial V}{\partial \phi}\bar{\mathbf{a}}_\phi + \frac{\partial V}{\partial z}\bar{\mathbf{a}}_z \quad \textbf{(Cylindrical)} \qquad \text{..(4.13 b)}$$

$$\bar{\nabla} V = \frac{\partial V}{\partial r}\bar{\mathbf{a}}_r + \frac{1}{r}\frac{\partial V}{\partial \theta}\bar{\mathbf{a}}_\theta + \frac{1}{r\sin\theta}\frac{\partial V}{\partial \phi}\bar{\mathbf{a}}_\phi \quad \textbf{(Spherical)} \qquad \text{(4.13 c)}$$

Problem 4.7 : If $V = 60 \sin\theta / r^2$ V in free space and point P is located at $r = 3$ m, $\theta = 60°$, $\phi = 25°$, find : (a) V_P ; (b) E_P ; (c) dV/dN at P ; (d) $\bar{\mathbf{a}}_N$ at at P ; (e) ρ_v at P .

Solution :

(a) $V_P = 60 \sin 60° / (3)^2 = 5.774$ V

(b) $\bar{E} = -\bar{\nabla} V = -\left[\frac{\partial V}{\partial r}\bar{a}_r + \frac{1}{r}\frac{\partial V}{\partial \theta}\bar{a}_\theta\right]$

$$= -\left[\frac{-120\sin\theta}{r^3}\bar{a}_r + \frac{60\cos\theta}{r^3}\bar{a}_\theta\right]$$

$$\therefore\ \bar{E}_P = \frac{120\sin 60^\circ}{(3)^3}\bar{a}_r - \frac{60\cos 60^\circ}{(3)^3}\bar{a}_\theta = 3.849\,\bar{a}_r - 1.111\,\bar{a}_\theta \text{ V/m.}$$

$$\text{(c) } \left.\frac{dV}{dN}\right|_P = |\bar{E}_P| = \sqrt{(3.849)^2 + (-1.111)^2} = 4.006 \text{ V/m}$$

$$\text{(d) } \bar{a}_N\Big|_P = -\frac{\bar{E}_P}{|\bar{E}_P|} = -\frac{(3.849\,\bar{a}_r - 1.111\,\bar{a}_\theta)}{4.006} = -0.961\,\bar{a}_r + 0.277\,\bar{a}_\theta$$

$$\text{(e) } \bar{D} = \varepsilon_o\bar{E} = \varepsilon_o\left[\frac{120\sin\theta}{r^3}\bar{a}_r - \frac{60\cos\theta}{r^3}\bar{a}_\theta\right]$$

$$\rho_v = \bar{\nabla}\cdot\bar{D} = \frac{1}{r^2}\frac{\partial}{\partial r}\left(\frac{120\,\varepsilon_o\sin\theta\, r^2}{r^3}\right) + \frac{1}{r\sin\theta}\frac{\partial}{\partial\theta}\left(\frac{-60\,\varepsilon_o\cos\theta\sin\theta}{r^3}\right)$$

$$= -\frac{120\,\varepsilon_o\sin\theta}{r^4} - \frac{60\,\varepsilon_o\cos 2\theta}{r^4\sin\theta}$$

$$\therefore\ \rho_v \text{ at } P = -\frac{120\,\varepsilon_o\sin 60^\circ}{(3)^4} - \frac{60\,\varepsilon_o\cos 120^\circ}{(3)^4\sin 60^\circ} = -7.573 \text{ pC/m}^3$$

4.4. THE POTENTIAL FIELD OF A DIPOLE :

An electric dipole is the name given to two point charges of equal magnitude and opposite in sign separated by a distance which is small com -

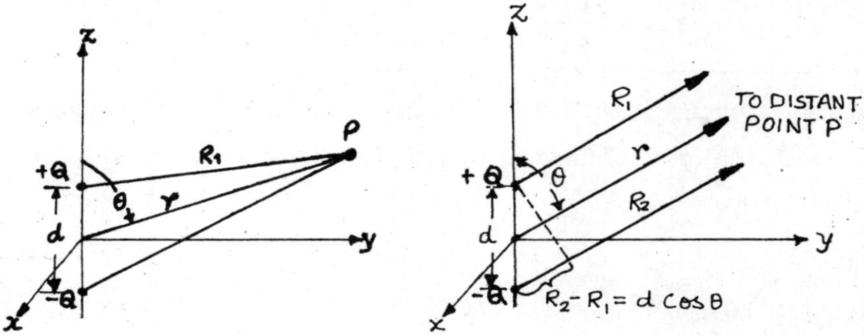

(a) (b)

FIG.4.3

pared to the distance of the point P at which electric and potential fields are required.

Exact expression for potential at point P :

Let $+Q$ charge is located a $(0, 0, d/2)$ and $-Q$ charge is located at $(0, 0, -d/2)$ as shown in Fig.4.3 (a). Then at point P

$$V = \frac{Q}{4\pi\varepsilon_o}\left[\frac{1}{R_2} - \frac{1}{R_1}\right] \qquad \ldots\ldots(4.14)$$

where R_1 and R_2 are the distances of $+Q$ and $-Q$ charges from the point P respectively.

Exact Expression for $\overline{\mathrm{E}}$ at point P :

$$\overline{\mathrm{E}} = \overline{\mathrm{E}}_1 + \overline{\mathrm{E}}_2$$

where $\overline{\mathrm{E}}_1$ is the electric field at point P due to $+Q$ charge and $\overline{\mathrm{E}}_2$ is the electric field due to $-Q$ charge.

Approximate expression for potential at point P :

If point P is at a large distance then $R_1 = R_2$ and $R_1 R_2$ is replaced by r^2. From Fig.4.3 b, $R_2 - R_1 = d\cos\theta$ .Substituting thes values in eqn.4.14,we have

$$V = \frac{Q}{4\pi\varepsilon_o}\frac{d\cos\theta}{r^2} \qquad \ldots\ldots(4.15)$$

Approximate Expression for $\overline{\mathrm{E}}$ at point P :

Using eqn 4.13 c and eqn. 4.15 ,we have

$$\overline{\mathrm{E}} = -\overline{\nabla} V = -\left(-\frac{Q\,d\cos\theta}{2\pi\varepsilon_o r^3}\,\overline{\mathrm{a}}_r - \frac{Q\,d\sin\theta}{4\pi\varepsilon_o r^3}\,\overline{\mathrm{a}}_\theta\right)$$

$$\overline{\mathbf{E}} = \frac{Q\,d}{4\pi\varepsilon_o r^3}\left(2\cos\theta\,\overline{\mathbf{a}}_r + \sin\theta\,\overline{\mathbf{a}}_\theta\right) \qquad \ldots\ldots(4.16)$$

Dipole moment is defined as $\overline{\mathrm{p}} = Q\,\overline{\mathrm{d}}$, where $\overline{\mathrm{d}}$ is a vector from $-Q$ to $+Q$.

Now $\overline{\mathrm{d}} = (0-0)\,\overline{\mathrm{a}}_x + (0-0)\,\overline{\mathrm{a}}_y + [d/2 - (-d/2)]\,\overline{\mathrm{a}}_z = d\,\overline{\mathrm{a}}_z$

Hence $\overline{\mathrm{p}} = Q\,d\,\overline{\mathrm{a}}_z$. As the angle between $\overline{\mathrm{p}}$ and $\overline{\mathrm{a}}_r$ is θ, we have

$\overline{\mathrm{p}}\cdot\overline{\mathrm{a}}_r = \mathrm{Q\,d}\;\overline{\mathrm{a}}_z\cdot\overline{\mathrm{a}}_r = Q\,d\cos\theta$

From eqn. 4.15, we have

$$V = \frac{\bar{p} \cdot \bar{a}_r}{4 \pi \varepsilon_0 r^3}.$$

This result may be generalized as

$$\boxed{V = \frac{\bar{p} \cdot (\bar{r}_2 - \bar{r}_1)}{4 \pi \varepsilon_0 |\bar{r}_2 - \bar{r}_1|^3}} \quad \ldots\ldots (4.17)$$

where $\bar{r}_2$ locates the field point P and $\bar{r}_1$ locates dipole center.

Problem 4.8 : A dipole of moment $\bar{p} = -4\bar{a}_x + 5\bar{a}_y + 3\bar{a}_z$ nC– m is located at D (1, 2, –1) in free space. Find V at : (a) P_A (0, 0, 0) ; (b) P_B (1, 2, 0) ; (c) P_C (1, 2, –2) ; (d) P_D (2, 6, 1).

Solution :

(a) $V = \dfrac{\bar{p} \cdot (\bar{r}_2 - \bar{r}_1)}{4 \pi \varepsilon_0 |\bar{r}_2 - \bar{r}_1|^3}$

where $\bar{r}_2$ is the position vector of point 2 at which potential is desired and $\bar{r}_1$ is the position vector of point 1 which is the dipole center.

Point 2 = (0, 0, 0) $\therefore \bar{r}_2 = 0\bar{a}_x + 0\bar{a}_y + 0\bar{a}_z$

Point 1 = (1, 2, –1) $\therefore \bar{r}_1 = \bar{a}_x + 2\bar{a}_y - \bar{a}_z$

$\therefore \bar{r}_2 - \bar{r}_1 = -\bar{a}_x - 2\bar{a}_y + \bar{a}_z$ and

$|\bar{r}_2 - \bar{r}_1| = \sqrt{(-1)^2 + (-2)^2 + (1)^2} = 2.449$

$\bar{p} \cdot (\bar{r}_2 - \bar{r}_1) = (-4\bar{a}_x + 5\bar{a}_y + 3\bar{a}_z)\,10^{-9} \cdot (-\bar{a}_x - 2\bar{a}_y + \bar{a}_z)$

$= (4 - 10 + 3)\,10^{-9} = -3 \times 10^{-9}$

$$\therefore V = \frac{-3 \times 10^{-9}}{4 \pi \varepsilon_0 (2.449)^3} = -1.835 \text{ V}$$

(b) Point 2 = (1, 2, 0) $\bar{r}_2 = \bar{a}_x + 2\bar{a}_y$

Point 1 = (1, 2, –1) $\bar{r}_1 = \bar{a}_x + 2\bar{a}_y - \bar{a}_z$

$\bar{r}_2 - \bar{r}_1 = \bar{a}_z$, $|\bar{r}_2 - \bar{r}_1| = 1$, $\bar{p} \cdot (\bar{r}_2 - \bar{r}_1) = 3 \times 10^{-9}$

$$\therefore V = \frac{3 \times 10^{-9}}{4 \pi \varepsilon_0 (1)^3} = 26.963 \text{ V}$$

(c) Point 2 = (1, 2, –2) $\bar{r}_2 = \bar{a}_x + 2\bar{a}_y - 2\bar{a}_z$

Point 1 = (1, 2, –1) ; $\bar{r}_1 = \bar{a}_x + 2\,\bar{a}_y - \bar{a}_z$

$\bar{r}_2 - \bar{r}_1 = -\bar{a}_z$; $|\bar{r}_2 - \bar{r}_1| = 1$,

$\bar{p} \cdot (\bar{r}_2 - \bar{r}_1) = -3 \times 10^{-9}$

$$\therefore V = \frac{-3 \times 10^{-9}}{4\pi\varepsilon_o (1)^3} = -26.963 \text{ V}$$

(d) Point 2 = (2, 6, 1) ; $\bar{r}_2 = 2\,\bar{a}_x + 6\,\bar{a}_y + \bar{a}_z$

Point 1 = (1, 2, –1) ; $\bar{r}_1 = \bar{a}_x + 2\,\bar{a}_y - \bar{a}_z$

$\therefore \bar{r}_2 - \bar{r}_1 = \bar{a}_x + 4\,\bar{a}_y + 2\,\bar{a}_z$ and

$|\bar{r}_2 - \bar{r}_1| = \sqrt{(1)^2 + (4)^2 + (2)^2} = 4.583$

$$\bar{p} \cdot (\bar{r}_2 - \bar{r}_1) = (-4\,\bar{a}_x + 5\,\bar{a}_y + 3\,\bar{a}_z)\, 10^{-9} \cdot (\bar{a}_x + 4\,\bar{a}_y + 2\,\bar{a}_z)$$

$$= (-4 + 20 + 6)\, 10^{-9} = 22 \times 10^{-9}$$

$$\therefore V = \frac{22 \times 10^{-9}}{4\pi\varepsilon_o (4.583)^3} = 2.055 \text{ V}$$

Problem 4.9 : Point charges of + 3 μC and – 3 μC are located at (0, 0, 1 mm) and (0, 0, –1 mm), respectively, in free space. (a) Find $\bar{p}$. (b) Find $\bar{E}$ in spherical components at $P\,(r = 2, \theta = 40^o, \phi = 50^o)$. (c) Find $\bar{E}$ in spherical components at (1, 2, 1.5).

Solution :

(a) $\bar{p} = Q\,\bar{d}$, where $\bar{d}$ is a vector directed from negative charge towards positive charge. Hence

$\bar{d} = (0.001\,\bar{a}_z) - (-0.001\,\bar{a}_z) = 0.002\,\bar{a}_z$

$\therefore \bar{p} = 3 \times 10^{-6}\,(0.002\,\bar{a}_z) = 6 \times 10^{-9}\,\bar{a}_z$ C– m = $6\,\bar{a}_z$ nC– m

(b) Since the dipole center is located at origin, we may use

$$\bar{E} = \frac{Q\,d}{4\pi\varepsilon_o r^3}\,(2\cos\theta\,\bar{a}_r + \sin\theta\,\bar{a}_\theta)$$

$$\therefore \bar{E} = \frac{3 \times 10^{-6} \times 0.002}{4\pi\varepsilon_o (2)^3} \times (2\cos 40^o\,\bar{a}_r + \sin 40^o\,\bar{a}_\theta)$$

$$= 10.328\,\bar{a}_r + 4.333\,\bar{a}_\theta \text{ V/m}$$

(c) At point (1, 2, 1.5) we have $r = \sqrt{(1)^2 + (2)^2 + (1.5)^2} = 2.6926$

$\theta = \cos^{-1}(1.5/2.69258) = 56.145^{\circ}$, $\phi = \tan^{-1}(2/1) = 63.435^{\circ}$

$$\therefore \bar{E} = \frac{3\times 10^{-6}\times 0.002}{4\pi\varepsilon_o(2.69258)^3}(2\cos 56.145^{\circ}\,\bar{a}_r + \sin 56.145^{\circ}\,\bar{a}_\theta)$$

$$= 3.078\,\bar{a}_r + 2.294\,\bar{a}_\theta \ \text{V/m}$$

4.5 THE ENERGY DENSITY IN THE ELECTROSTATIC FIELD :

Suppose three charges Q_1, Q_2 and Q_3 are located at ∞ as shown in Fig.4.4(a) and consider the region which is charge free so that initially $\bar{E} = 0$.

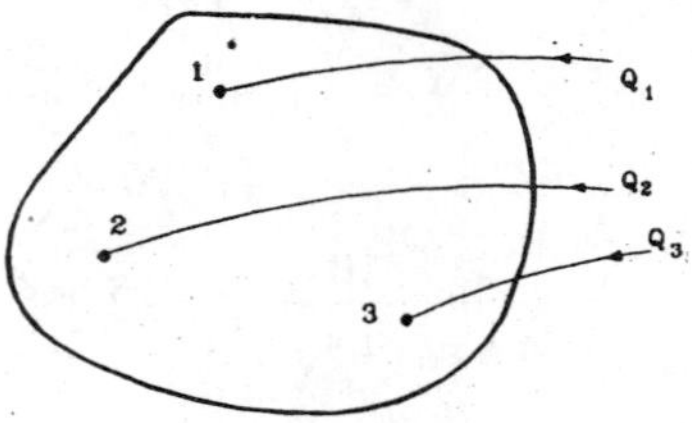

FIG. 4.4 (a)

The work required to place the charge Q_1 into position 1 is zero. Then when Q_2 is moved toward the region, work equal to the product of this charge and the potential due to Q_1 is required. The total work to position the three charges is

$$W_E = W_1 + W_2 + W_3 = 0 + Q_2V_{2,1} + Q_3(V_{3,1} + V_{3,2})$$

where $V_{2,1}$ means the potential at point 2 due to charge Q_1 at point 1

Now if the three charges were brought into place in reverse order, the total work would be

$$W_E = W_3 + W_2 + W_1 = 0 + Q_2V_{2,3} + Q_1(V_{1,3} + V_{1,2})$$

$$\therefore 2W_E = Q_1(V_{1,2} + V_{1,3}) + Q_2(V_{2,1} + V_{2,3}) + Q_3(V_{3,1} + V_{3,2})$$

Now $V_{1,2} + V_{1,3} = V_1$; $V_{2,1} + V_{2,3} = V_2$; $V_{3,1} + V_{3,2} = V_3$

Hence

$$2W_E = Q_1V_1 + Q_2V_2 + Q_3V_3$$

If n is the total number of charges, then

$$W_E = \frac{1}{2}\sum_{m=1}^{m=n} Q_m V_m \qquad \text{......(4.18)}$$

For a region with a charge density ρ_v C/m^3, the summation becomes an integration

$$W_E = \frac{1}{2}\int \rho_v V\, dv \qquad \ldots\ldots(4.19)$$

Fig.4.4(b) shows charge containing volume v enclosed within a large sphere of radius R .As $\rho_v = \overline{\nabla} \cdot \overline{D}$, eqn.4.19 may br written as

$$W_E = \frac{1}{2}\int (\overline{\nabla} \cdot \overline{D})\, V\, dv$$

Now using the vector identity

$\overline{\nabla} \cdot V\overline{A} = \overline{A} \cdot \overline{\nabla} V + V(\overline{\nabla} \cdot \overline{A})$, weget

$$W_E = \frac{1}{2}\int (\overline{\nabla} \cdot V\overline{D})\, dv - \frac{1}{2}\int (\overline{D} \cdot \overline{\nabla} V)\, dv$$

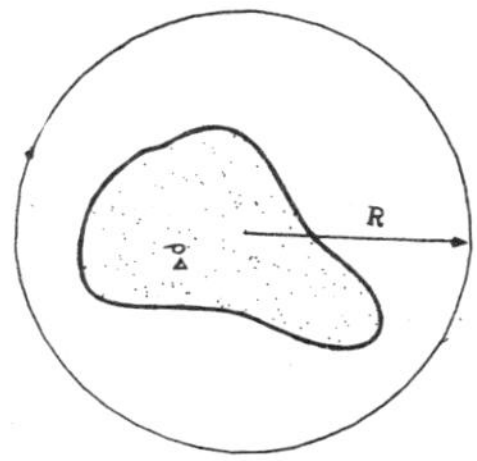

FIG. 4.4 (b)

This expression holds for an arbitrarily large radius R . The first integral on the right equals, by the divergence theorem , $\frac{1}{2}\oint V\,\overline{D} \cdot \overline{dS}$. Now as the enclosing sphere becomes very large, the enclosed volume charge looks like a point charge. Thus, at the surface, D appears as K_2/R^2 and V appears as K_2/R . So the integrand is decreasing as $1/R^3$. Since the surface area increases only as R^2 , it follows that

$$\lim_{R\to\infty} \oint V\,\overline{D} \cdot \overline{dS} = 0$$

The remaining integral gives, in the limit,

$$W_E = -\frac{1}{2}\int (\overline{D} \cdot \overline{\nabla} V)\, dV = \frac{1}{2}\int (\overline{D} \cdot \overline{E})\, dV$$

And since $\overline{D} = \varepsilon\overline{E}$, the energy stored is also given by

$$W_E = \frac{1}{2}\int \varepsilon\,|\overline{E}|^2\, dV \qquad \ldots\ldots(4.20)$$

or

$$W_E = \frac{1}{2}\int \frac{|\overline{D}|^2}{\varepsilon}\, dV \qquad \ldots\ldots(4.21)$$

The unit of W_E is J (joule). Energy density is given by

$$\boxed{\frac{dW_E}{dV} = \frac{1}{2}\varepsilon\,|\,\bar{E}\,|^2} \qquad \ldots\ldots.(4.22)$$

or

$$\boxed{\frac{dW_E}{dV} = \frac{1}{2}\frac{|\,\bar{D}\,|^2}{\varepsilon}} \qquad \ldots\ldots.(4.23)$$

The unit of energy density is J/m^3.

Problem 4.10 : Find the energy stored in free space for the region $0 < \rho < a$, $0 < \phi < \pi$, $0 < z < 2$, given the potential field $V =$: (a) $V_o(\rho/a)$; (b) $V_o(\rho/a)\cos^2\phi$.

Solution :

(a) $$\bar{E} = -\bar{\nabla}V = -\left[\frac{\partial V}{\partial \rho}\bar{a}_\rho\right] = -\frac{\partial}{\partial\rho}\left(\frac{V_o\rho}{a}\right)\bar{a}_\rho = -\frac{V_o}{a}\bar{a}_\rho$$

$$\therefore |\,\bar{E}\,| = \frac{V_o}{a}$$

Using eqn.4.20 , we have

$$W_E = \frac{\varepsilon_o}{2}\int_0^a\int_0^\pi\int_0^2 \frac{V_o^2}{a^2}\,\rho\,d\rho\,d\phi\,dz = \frac{\varepsilon_o V_o^2}{2a^2}\left[\frac{\rho^2}{2}\right]_0^2\left[\phi\right]_0^\pi\left[z\right]_0^2$$

$$= 1.571\,\varepsilon_o V_o^2\ \text{J}$$

(b) $$\bar{E} = -\bar{\nabla}V = -\left[\frac{\partial}{\partial\rho}\left(\frac{V_o\rho\cos^2\phi}{a}\right)\bar{a}_\rho + \frac{1}{\rho}\frac{\partial}{\partial\phi}\left(\frac{V_o\rho\cos^2\phi}{a}\right)\bar{a}_\phi\right]$$

$$= -\frac{V_o}{a}\left(\cos^2\phi\,\bar{a}_\rho - 2\cos\phi\sin\phi\,\bar{a}_\phi\right)$$

$$\therefore |\,\bar{E}\,| = \frac{V_o}{a}\sqrt{(\cos^2\phi)^2 + (-2\cos\phi\sin\phi)^2} = \frac{V_o}{a}\cos\phi\sqrt{1+3\sin^2\phi}$$

$$\therefore W_E = \frac{\varepsilon_o}{2}\int_0^a\int_0^\pi\int_0^{2\pi}\frac{V_o^2}{a^2}\cos^2\phi\,(1+3\sin^2\phi)\times\rho\,d\rho\,d\phi\,dz$$

$$= \frac{\varepsilon_o V_o^2}{2a^2}\left[\frac{\rho^2}{2}\right]_0^2\left[\frac{1}{2}\left(\phi+\frac{\sin 2\phi}{2}\right)+\frac{3}{8}\left(\phi-\frac{\sin 4\phi}{4}\right)\right]_0^\pi\left[z\right]_0^2$$

$$= \frac{\varepsilon_o V_o^2}{2a^2} \times \frac{a^2}{2} \times 2.74889 \times 2 = 1.374\, \varepsilon_o V_o^2 \text{ J}$$

Problem 4.11 : A field is given in spherical coordinates at P ($r = 5, \theta = 30^o, \phi = 60^o$) as $\overline{E} = 20\,\bar{a}_r - 30\,\bar{a}_\theta + 60\,\bar{a}_\phi$ V/m . Find the incremental work done in moving a 10 – μC charge a distance of 0.8μC in the direction : (a) $\bar{a}_r$; (b) $\bar{a}_\theta$; (c) $\bar{a}_\phi$; (d) of $\overline{E}$; (e) of $\overline{G} = 2\,\bar{a}_x + 4\,\bar{a}_y - 3\,\bar{a}_z$.

Solution :

$dW = -Q\,\overline{E} \cdot \overline{dL}$ where $\overline{dL} = dL\,\bar{a}_L$

$dL = 0.8\ \mu\text{m} = 0.8 \times 10^{-6}\text{m}$ and $\bar{a}_L = \bar{a}_r$

$\therefore\ \overline{dL} = 0.8 \times 10^{-6}\,\bar{a}_r$

$\therefore\ dW = -10 \times 10^{-6} (20\,\bar{a}_r - 30\,\bar{a}_\theta - 60\,\bar{a}_\phi) \cdot (0.8 \times 10^{-6}\,\bar{a}_r)$

$= -10^{-5} (20 \times 0.8 \times 10^{-6}) = -160$ pJ

(b) $\therefore\ \overline{dL} = 0.8 \times 10^{-6}\,\bar{a}_\theta$

$\therefore\ dW = -10 \times 10^{-6} (20\,\bar{a}_r - 30\,\bar{a}_\theta + 60\,\bar{a}_\phi) \cdot (0.8 \times 10^{-6}\,\bar{a}_\theta)$

$= -10^{-5} (-30 \times 0.8 \times 10^{-6}) = 240$ pJ

(c) $\therefore\ \overline{dL} = 0.8 \times 10^{-6}\,\bar{a}_\phi$

$\therefore\ dW = -10 \times 10^{-6} (20\,\bar{a}_r - 30\,\bar{a}_\theta + 60\,\bar{a}_\phi) \cdot (0.8 \times 10^{-6}\,\bar{a}_\phi)$

$= -10^{-5} (60 \times 0.8 \times 10^{-6}) = -480$ pJ .

$$\text{(d) } \bar{a}_E = \frac{\overline{E}}{|\overline{E}|} = \frac{20\,\bar{a}_r - 20\,\bar{a}_\theta + 60\,\bar{a}_\phi}{\sqrt{(20)^2 + (-30)^2 + (60)^2}} = \frac{20\,\bar{a}_r - 30\,\bar{a}_\theta + 60\,\bar{a}_\phi}{70}$$

$$\therefore\ \overline{dL} = dL\,\bar{a}_L = dL\,\bar{a}_E = 0.8 \times 10^{-6} \left(\frac{20\,\bar{a}_r - 30\,\bar{a}_\theta + 60\,\bar{a}_\phi}{70} \right)$$

$$\overline{E} \cdot \overline{dL} = (20\,\bar{a}_r - 30\,\bar{a}_\theta + 60\,\bar{a}_\phi) \cdot 0.8 \times 10^{-6} \left(\frac{20\,\bar{a}_r - 30\,\bar{a}_\theta + 60\,\bar{a}_\phi}{70} \right)$$

$$= 0.011428 \times 10^{-6} (400 + 900 + 3600) = 56 \times 10^{-6}$$

$\therefore\ dW = -10 \times 10^{-6} \times 56 \times 10^{-6} = -560$ pJ

$$\text{(e) } \bar{a}_L = \frac{\overline{G}}{|\overline{G}|} = \frac{2\,\bar{a}_x + 4\,\bar{a}_y - 3\,\bar{a}_z}{\sqrt{(2)^2 + (4)^2 + (-3)^2}} = \frac{2\,\bar{a}_x + 4\,\bar{a}_y - 3\,\bar{a}_z}{5.385}$$

$$\overline{E}\cdot\overline{dL} = (20\,\overline{a}_r - 30\,\overline{a}_\theta + 60\,\overline{a}_\phi)\cdot 0.8\times 10^{-6}\left(\frac{2\,\overline{a}_x - 4\,\overline{a}_y + 3\,\overline{a}_z}{5.385}\right)$$

$$= \Big(40\sin\theta\cos\phi + 80\sin\theta\sin\phi - 60\cos\theta - 60\cos\theta\cos\phi - 120\cos\theta\times \sin\phi - 90\sin\theta - 120\sin\phi + 240\cos\phi\Big)\,0.14856\times 10^{-6}$$

Substituting coordinates of point $P\,(r = 5, \theta = 30^\circ, \phi = 60^\circ)$ in the above equation, we get $\overline{E}\cdot\overline{dL} = -22.612\times 10^{-6}$. Hence

$$dW = -10\times 10^{-6}(-22.612\times 10^{-6}) = 226.12 \text{ pJ}$$

Problem 4.12 : The vector electric field intensity in the neighborhood of points P and Q is $\overline{E} = -5\,\overline{a}_x + 3\,\overline{a}_y - 2\,\overline{a}_z$ kV/m. Find an approximate value for the incremental work done in moving a 5 - mC charge from P (2, –1, 3) to : (a) Q (2.01, –1, 3) ; (b) Q (2.01, – 0.98, 2.97).

Solution :

(a) $\overline{dL} = \overline{R}_{PQ} = (2.01 - 2)\,\overline{a}_x + (-1 + 1)\,\overline{a}_y + (3 - 3)\,\overline{a}_z = 0.01\,\overline{a}_x$

$$\therefore\ dW = -5\times 10^{-3}(-5\,\overline{a}_x + 3\,\overline{a}_y - 2\,\overline{a}_z)\,10^3\cdot(0.01\,\overline{a}_x)$$

$$= -5\times 10^{-3}(-5\times 0.01)\,10^3 = 0.25 \text{ J}$$

(b) $\overline{dL} = \overline{R}_{PQ} = (2.01 - 2)\,\overline{a}_x + (-0.98 + 1)\,\overline{a}_y + (2.97 - 3)\,\overline{a}_z$

$$= 0.01\,\overline{a}_x + 0.02\,\overline{a}_y - 0.03\,\overline{a}_z$$

$$dW = -5\times 10^{-3}(-5\,\overline{a}_x + 3\,\overline{a}_y - 2\,\overline{a}_z)\,10^3\cdot(0.01\,\overline{a}_x + 0.02\,\overline{a}_y - 0.03\,\overline{a}_z)$$

$$= -5(-0.05 + 0.06 + 0.06) = -0.35 \text{ J}$$

Problem 4.13 : If $\overline{E} = -50\,y\,\overline{a}_x - 50\,x\,\overline{a}_y + 30\,\overline{a}_z$ V/m, find the differential amount of work done in moving a 2 – μC charge a distane of 5 μm from : (a) P (1, 2, 3) toward Q (2, 4, 1) ; (b) Q (2, 4, 1) toward P (1, 2, 3) .

Solution :

(a) $\overline{E}_P = -50(2)\,\overline{a}_x - 50(1)\,\overline{a}_y + 30\,\overline{a}_z = -100\,\overline{a}_x - 50\,\overline{a}_y + 30\,\overline{a}_z$

$$\overline{a}_L = \frac{\overline{R}_{PQ}}{|\overline{R}_{PQ}|} = \frac{(2-1)\,\overline{a}_x + (4-2)\,\overline{a}_y + (1-3)\,\overline{a}_z}{\sqrt{(2-1)^2 + (4-2)^2 + (1-3)^2}} = \frac{\overline{a}_x + 2\,\overline{a}_y - 2\,\overline{a}_z}{3}$$

$$dL = 5\times 10^{-6} \text{ m}$$

$$dW = -2\times 10^{-6}(-100\,\overline{a}_x - 50\,\overline{a}_y + 30\,\overline{a}_z)\cdot 5\times 10^{-6}\left(\frac{\overline{a}_x + 2\,\overline{a}_y - 2\,\overline{a}_z}{3}\right)$$

$$= -\frac{10^{-11}}{3}(-100-100-60) = 866.667 \text{ pJ}$$

(b) $\overline{E}_Q = -50(4)\,\overline{a}_x - 50(2)\,\overline{a}_y + 30\,\overline{a}_z = -200\,\overline{a}_x - 100\,\overline{a}_y + 20\,\overline{a}_z$

$$\overline{a}_L = \frac{\overline{R}_{QP}}{|\overline{R}_{QP}|} = \left(\frac{\overline{a}_x + 2\,\overline{a}_y - 2\,\overline{a}_z}{3}\right); \; dL = 5\times 10^{-6} \text{ m}$$

$$dW = -2\times 10^{-6}(-200\,\overline{a}_x - 100\,\overline{a}_y + 30\,\overline{a}_z)\cdot 5\times 10^{-6}\left(\frac{\overline{a}_x + 2\,\overline{a}_y - 2\,\overline{a}_z}{3}\right)$$

$$= -\frac{10^{-11}}{3}(200+200+60) = -1533.333 \text{ pJ}$$

Problem 4.14 : Find the value of $\int_{P_1}^{P_2} \overline{G}\cdot\overline{dL}$ if $\overline{G} = x^2\,\overline{a}_x - xyz\,\overline{a}_z$, given points P_1 (1,2,3) and P_2 (5,0,4) and a path consisting of : (a) the straight line segments connecting P_1 to (5, 2, 3) to (5, 0, 3) to (5, 0, 4) ; (b) the straight line which is the intersection of the planes $x = 5 - 2y$ and $y = 8 - 2z$.

Solution :

(a) $\overline{G}\cdot\overline{dL} = (x^2\,\overline{a}_x - xyz\,\overline{a}_z)\cdot(dx\,\overline{a}_x + dy\,\overline{a}_y + dz\,\overline{a}_z)$

$$= x^2\,dx - xyz\,dz \therefore \int_{P_1}^{P_2}\overline{G}\cdot\overline{dL} = \int_1^5 x^2\,dx - \int_3^4 xyz\,dz$$

Path of integration is P (1, 2, 3) to (5, 2, 3) to (5, 0. 3) to (5, 0, 4). Between first and second point, y and z coordinates are constant where as x coordinate varies from $x = 1$ to $x = 5$. Hence in the first integral, if required substitute $y = 2$ and $z = 3$.

Between third and fourth point, x and y coordinates are constant where as z coordinate varies from $z = 3$ to $z = 4$. Hence in the second integral, substitute $x = 5$ and $y = 0$

$$\therefore \int_{P_1}^{P_2}\overline{G}\cdot\overline{dL} = \int_1^5 x^2\,dx = \left[\frac{x^3}{3}\right]_1^5 = \frac{125-1}{3} = 41.333$$

$$\text{(b)} \int_{P_1}^{P_2}\overline{G}\cdot\overline{dL} = \int_1^5 x^2\,dx - \int_3^4 xyz\,dz$$

In the second integral substitute $y = 8 - 2z$ and $x = 5 - 2y = 5 - 2(8-2z) = -11 + 4z$

$$\therefore \int_{P_1}^{P_2} \overline{G} \cdot \overline{dL} = \int_1^5 x^2\, dx - \int_3^4 (-11 + 4z)(8 - 2z)\, z\, dz$$

$$= \left[\frac{x^3}{3} \right]_1^5 - \left[-\frac{88z^2}{2} + \frac{54z^3}{3} - \frac{8z^4}{4} \right]_3^4 = 33.333$$

Problem 4.15 : Compute the line integral of $\overline{E}$ from B (0, 0, 0) to A (2, 1, 3) if $\overline{E} = -4xy\,\bar{a}_x - (4y + 2x^2)\,\bar{a}_y + 8z\,\bar{a}_z$. Use the path : (a) straight line segments : (0, 0, 0) to (2, 0, 0) to (2, 1, 0) to (2, 1, 3) ; (b) direct straight line.

Solution :

(a) $\overline{E} \cdot \overline{dL} = [-4xy\,\bar{a}_x - (4y + 2x^2)\,\bar{a}_y + 8z\,\bar{a}_z] \cdot [dx\,\bar{a}_x + dy\,\bar{a}_y + dz\,\bar{a}_z]$

$$= -4xy\,dx - (4y + 2x^2)\,dy + 8z\,dz$$

$$\int_B^A \overline{E} \cdot \overline{dL} = -4\int_0^2 xy\,dx - \int_0^1 (4y + 2x^2)\,dy + \int_0^3 8z\,dz$$

Path of integration is (0, 0, 0) to (2, 0, 0) to (2, 1, 0) to (2, 1, 3). Between first and second point, y and z coordinates are constant and x coordinates varies from $x = 0$ to $x = 2$. Hence in the first integral we substitute $y = 0$ and $z = 0$.

Between second and third point, x and z coordinates are constant and y coordinate varies from $y = 0$ to $y = 1$. Hence in the second integral we substitute $x = 2$ and $z = 0$. Three is no problem with the last integral as the integration is w.r.t. z and only z is present.

$$\therefore \int_B^A \overline{E} \cdot \overline{dL} = -\int_0^1 [4y + 2(2)^2)]\,dy + \int_0^3 8z\,dz$$

$$= -\left[\frac{4y^2}{2} + 8y \right]_0^1 + \left[\frac{8z^2}{2} \right]_0^3 = -10 + 36 = 26$$

(b) Equations of the straight line passing through the points B (0, 0, 0) and A (2, 1, 3) are

$$\frac{x - 0}{0 - 2} = \frac{y - 0}{0 - 1} = \frac{z - 0}{0 - 3} \quad \text{i.e. } \frac{x}{2} = \frac{y}{1} = \frac{z}{3}$$

In the first integral, replace y in terms of x with the substitution $y = 0.5\,x$.
In the second integral, replace x in terms of y with the substitution $x = 2\,y$.

$$\int_B^A \overline{E}\cdot\overline{dL} = -4\int_0^2 x\,(0.5x)\,dx - \int_0^1 [4y + 2\,(2y)^2]\,dy + \int_0^3 8z\,dz$$

$$= -2\left[\frac{x^3}{3}\right]_0^2 - \left[\frac{4y^2}{2} + \frac{8y^3}{3}\right]_0^1 + 8\left[\frac{z^2}{2}\right]_0^3 = 26$$

Problem 4.16 : If $\overline{G} = 3x^2 y\,\overline{a}_x + (2x - z)\,\overline{a}_y$, evaluate $\int \overline{G}\cdot\overline{dL}$ from an initial point P (2, 1, 1) to a final point Q (4, 3, 1) using the path : (a) a straight line : $y = x - 1,\ z = 1$; (b) a parabola : $6y = x^2 + 2,\ z = 1$.

Solution :

(a) $\overline{G}\cdot\overline{dL} = \left[\,3x^2 y\,\overline{a}_x - (2x - z)\,\overline{a}_y\,\right]\cdot\left[\,dx\ \overline{a}_x + dy\ \overline{a}_y + dz\ \overline{a}_z\,\right]$

$$= 3x^2 y\ dx - (2x - z)\ dy$$

$$\int_P^Q \overline{G}\cdot\overline{dL} = \int_2^4 3x^2 y\,dx - \int_1^3 (2x - z)\,dy$$

Path of integration is $y = x - 1,\ z = 1$. In the first integral substitute $y = x - 1$. In the second integral substitute $x = y + 1$ and $z = 1$

$$\therefore \int_P^Q \overline{G}\cdot\overline{dL} = \int_2^4 3x^2\,(x - 1)\,dx - \int_1^3 (2y + 2 - 1)\,dy$$

$$= 3\left[\frac{x^4}{4} - \frac{x^3}{3}\right]_2^4 - \left[\frac{2y^2}{2} + y\right]_1^3 = 114$$

(b) The path of integration is $6y = x^2 + 2,\ z = 1$. In the first integral substitute $y = (x^2 + 2)/6$. Also from the path of integration ,we have $x^2 = 6y - 2$ or $x = \pm\sqrt{6y - 2}$. At point P, $y = 1, x = 2$ and these values can be obtained with positive sign in the expression for x. At point Q, $y = 3, x = 4$ and again positive sign is in use. Hence we will substitute $x = \sqrt{6y - 2}$ in the second integral along with $z = 1$.

$$\therefore \int_P^Q \overline{G}\cdot\overline{dL} = \int_2^4 3x^2\left[\frac{x^2 + 2}{6}\right]dx - \int_1^3 \left(2\sqrt{6y - 2} - 1\right)dy$$

$$= \frac{1}{2}\left[\frac{x^5}{5}+\frac{2x^3}{3}\right]_2^4 - \left[\frac{2\times 3}{2}\times\frac{(6y-2)^{3/2}}{6} - y\right]_1^3 = 91.867$$

Problem 4.17 : If $\overline{E} = -8xy\,\overline{a}_x - 4x^2\,\overline{a}_y + \overline{a}_z$ V/m, find the work done in carrying a 6 – C charge $M(1, 8, 5)$ to $N(2, 18, 6)$ along the path : $y = 3x^2 + z, z = x + 4$.

Solution :

$$\overline{E}\cdot\overline{dL} = (-8xy\,\overline{a}_x - 4x^2\,\overline{a}_y + \overline{a}_z)\cdot(dx\,\overline{a}_x + dy\,\overline{a}_y + dz\,\overline{a}_z)$$

$$= -8xy\,dx - 4x^2\,dy + dz$$

$$W = -Q\int_M^N \overline{E}\cdot\overline{dL} = -6\left[-\int_1^2 8xy\,dx - \int_8^{18} 4x^2\,dy + \int_5^6 dz\right]$$

The path of integration is $y = 3x^2 + z, z = x + 4$. In the first integral, substitute $y = 3x^2 + z = 3x^2 + x + 4$. In the second integral, substitute the value of x in terns of y as follows.

$3x^2 + x + 4 = y$ or $3x^2 + x + (4 - y) = 0$ or

$$x = \frac{-1 \pm \sqrt{-47 + 12y}}{6} \quad \ldots\ldots (1)$$

At point M, $x = 1, y = 8$ which are obtained if we consider positive sign in the equation (1). Similarly at point N, $x = 2, y = 18$ which are obtained if we consider positive sign in equation (1). Hence

$$W = -6\left[-\int_1^2 8x(3x^2 + x + 4)\,dx - \int_8^{18} 4\left(\frac{-1+\sqrt{-47+12y}}{6}\right)^2 dy + \int_5^6 dz\right]$$

$$= -6\left\{-8\left[\frac{3x^4}{4}+\frac{x^3}{3}+\frac{4x^2}{2}\right]_1^2 - \frac{1}{9}\left[-46y + \frac{12y^2}{2} - \frac{2}{12}\times\frac{2}{3}(12y-47)^{1.5}\right]_8^{18} + [z]_5^6\right\}$$

$$= -6\left[-156.667 - 99.333 + 1\right] = 1530 \text{ J}$$

Problem 4.18 : A point charge Q_1 is located at origin in free space. Find the work done in carrying a charge Q_2 : (a) $B(r_B, \theta_B, \phi_B)$ to $C(r_A, \theta_B, \phi_B)$ with θ and ϕ held constant ; (b) $C(r_A, \theta_B, \phi_B)$ to $D(r_A, \theta_A, \phi_B)$ with r and ϕ held constant ; (c) $D(r_A, \theta_A, \phi_B)$ to $A(r_A, \theta_A, \phi_A)$ with r and θ held constant

Solution :

(a) $\overline{E}$ at any point due to charge Q_1 located at origin= $\dfrac{Q_1}{4\pi\varepsilon_o r^2}\,\overline{a}_r$ V/m

$$\overline{E}\cdot\overline{dL} = \frac{Q_1}{4\pi\varepsilon_o r^2}\,\overline{a}_r\cdot(dr\,\overline{a}_r + r\,d\theta\,\overline{a}_\theta + r\sin\theta\,d\phi\,\overline{a}_\phi) = \frac{Q_1}{4\pi\varepsilon_o}\frac{dr}{r^2}$$

$$W = -Q_2\int_B^C \overline{E}\cdot\overline{dL} = -Q_2\int_{r_B}^{r_A}\frac{Q_1}{4\pi\varepsilon_o}\frac{dr}{r^2}$$

$$= \frac{Q_2 Q_1}{4\pi\varepsilon_o}\left[\frac{1}{r}\right]_{r_B}^{r_A} = \frac{Q_2 Q_1}{4\pi\varepsilon_o}\left[\frac{1}{r_A} - \frac{1}{r_B}\right]$$

$$\text{(b) } W = \frac{-Q_2 Q_1}{4\pi\varepsilon_o}\int_C^D \frac{dr}{r^2} = \frac{-Q_2 Q_1}{4\pi\varepsilon_o}\int_{r_A}^{r_A}\frac{dr}{r^2} = 0$$

$$\text{(c) } W = -\frac{Q_2 Q_1}{4\pi\varepsilon_o}\int_D^A \frac{dr}{r^2} = -\frac{Q_2 Q_1}{4\pi\varepsilon_o}\int_{r_A}^{r_A}\frac{dr}{r^2} = 0$$

Problem 4.19 : Each coordinate axis carries a uniform charge density of 20 nC / m in free space. Given the points A (1, 2, 3) and B (6, 8, 10), find V_{AB} due : (a) solely to ρ_L on the x axis ; (b) solely to ρ_L on the y – axis ; (c) solely to ρ_L on z axis ; (d) to all three line charges.

Solution :

$$\text{(a) } V_{AB} = -\frac{\rho_L}{2\pi\varepsilon_o}\left[\ln d_A - \ln d_B\right]$$

If the line charge ρ_L lies on the x axis, then the equations of the line charge are $y = 0$ and $z = 0$. Coordinates of point A are (1, 2, 3). Coordinates of the foot of perpendicular dropped from point A on the line charge are (1, 0, 0)

$$\therefore\ d_A = \sqrt{(1-1)^2+(2-0)^2+(3-0)^2} = \sqrt{13}$$

Coordinates of point B are (6, 8, 10). Coordinates of the foot of perpendicular dropped from point B on the line charge are (6, 0, 0).

$$\therefore\ d_B = \sqrt{(6-6)^2+(8-0)^2+(10-0)^2} = \sqrt{164}$$

$$\therefore\ V_{AB} = -\frac{20\times 10^{-9}}{2\pi\varepsilon_o}[\ln\sqrt{13} - \ln\sqrt{164}] = 455.664 \text{ V}$$

(b) If the line charge lies on the y – axis, then the equations of the line charge are $x = 0$ and $z = 0$. Coordinates of point A are (1, 2, 3). Coordinates of the foot of perpendicular dropped from point A on the line charge are (0, 2, 0).

$$\therefore\ d_A = \sqrt{(1-0)^2+(2-2)^2+(3-0)^2} = \sqrt{10}$$

Coordinates of point B are (6, 8, 10). Coordinates of the foot of perpendicular dropped from point B on the line charge are (0, 8, 0).

$$d_B = \sqrt{(6-0)^2+(8-8)^2+(10-0)^2} = \sqrt{136}$$

$$\therefore\ V_{AB} = -\frac{20\times 10^{-9}}{2\pi\varepsilon_o}\left[\ln\sqrt{10} - \ln\sqrt{136}\right] = 469.173 \text{ V}$$

(c) If the line charge lies on the z axis, then the equations of the line charge are $x = 0$ and $y = 0$. Coordinates of point A are (1, 2, 3). Coordinates of the foot of perpendicular dropped from point A on the line charge are (0, 0, 3).

$$\therefore\ d_A = \sqrt{(1-0)^2+(2-0)^2+(3-3)^2} = \sqrt{5}$$

Coordinates of point B are (6, 8, 10). Coordinates of the foot of perpendicular dropped from point B on the line charge are (0, 0, 10)

$$\therefore\ d_B = \sqrt{(6-0)^2+(8-0)^2+(10-10)^2} = \sqrt{100}$$

$$V_{AB} = -\frac{20\times 10^{-9}}{2\pi\varepsilon_o}[\ln\sqrt{5} - \ln\sqrt{100}] = 538.497 \text{ V}$$

(d) $V_{AB} = 455.664 + 469.173 + 538.497 = 1463.334$ V

Problem 4.20 : Given a point charge of $200\pi\varepsilon_o$ C at (3, –1, 2), a line charge of $40\pi\varepsilon_o$ C/m on the x axis, and a surface charge of $8\varepsilon_o$ C/m^2 on the plane $x = -3$, all in the free space, find the potential at P (5, 6, 7) if $V = 0$ at Q (0, 0, 1).

Solution :

Due to point charge, $V_{PQ} = \dfrac{Q}{4\pi\varepsilon_o}\left[\dfrac{1}{d(PC)} - \dfrac{1}{d(QC)}\right]$

$d(PC)$ = distance of charge from the point P

$$= \sqrt{(5-3)^2 + (6+1)^2 + (7-2)^2} = 8.832$$

$d(QC)$ = distance of charge from the point Q

$$= \sqrt{(0-3)^2 + (0+1)^2 + (1-2)^2} = 3.317$$

$$\therefore\ V_{PQ} = \frac{200\pi\varepsilon_o}{4\pi\varepsilon_o}\left[\frac{1}{8.832} - \frac{1}{3.317}\right] = -9.414 \text{ V}$$

Due to line charge, $V_{PQ} = -\dfrac{\rho_L}{2\pi\varepsilon_o}\left[\ln d_P - \ln d_Q\right]$

Equations of the line charge are $y = 0$ and $z = 0$. Coordinates of point P are (5, 6, 7). Coordinates of the foot of perpendicular dropped from point P on the line charge are (5, 0, 0)

$$d_P = \sqrt{(5-5)^2 + (6-0)^2 + (7-0)^2} = \sqrt{85}$$

Coordinates of point Q are (0, 0, 1). Coordinates of the foot of perpendicular dropped from point Q on the line charge are (0, 0, 0).

$$\therefore\ d_Q = 1$$

$$\therefore\ V_{PQ} = -\frac{40\pi\varepsilon_o}{2\pi\varepsilon_o}(\ln\sqrt{85} - \ln 1) = -44.427 \text{ V} = -44.427 \text{ V}$$

Surface charge density lies on the plane $x = -3$. At point P, $\overline{E} = \dfrac{\rho_S}{2\varepsilon_o}\,\overline{a}_N$ & $\overline{a}_N = \overline{a}_x$ as x coordinate of point P is greater than $x = -3$. Similarly at point Q, $\overline{E} = \dfrac{\rho_S}{2\varepsilon_o}\,\overline{a}_x$. Now

$$\overline{E}\cdot\overline{dL} = \frac{\rho_S}{2\varepsilon_o}\,\overline{a}_x\cdot(dx\,\overline{a}_x + dy\,\overline{a}_y + dz\,\overline{a}_z) = \frac{\rho_S}{2\varepsilon_o}\,dx$$

$$\therefore\ V_{PQ} = -\int_Q^P \overline{E}\cdot\overline{dL} = -\frac{\rho_S}{2\varepsilon_o}\int_0^5 dx = -\frac{8\varepsilon_o}{2\varepsilon_o}(5-0) = -20 \text{ V}$$

$$\therefore\ V_{PQ} = -9.414 - 44.427 - 20 = -73.841 \text{ V}$$

Also $V_{PQ} = V_P - V_Q$ and $V_Q = 0$ (given)

$$\therefore\ V_P = V_{PQ} = -73.841 \text{ V}$$

Problem 4.21 : In free space, a line charge $\rho_L = 80$ nC/m lies along the entire z axis, while point charges of 100 nC each are located at (1, 0, 0) and (0, 1, 0). Find the potential difference V_{PQ} given P (2, 1, 0) and Q (3, 2, 5).

Solution :

Equations the line charge are $x = 0$ and $y = 0$. Coordinates of point P are (2, 1, 0). Coordinates of the foot of perpendicular dropped from point P on the line charge are (0,0,0).

$$\therefore\ d_P = \sqrt{(2-0)^2 + (1-0)^2} = \sqrt{5}$$

Coordinates of point Q are (3, 2, 5). Coordinates of the foot of perpendicular dropped from point Q on the line charge are (0, 0, 5).

$$\therefore\ d_Q = \sqrt{(3-0)^2 + (2-0)^2} = \sqrt{14}$$

$$\therefore\ V_{PQ} = -\frac{\rho_L}{2\pi\varepsilon_o}\left[\ln d_P - \ln d_Q\right]$$

$$= -\frac{80 \times 10^{-9}}{2\pi\varepsilon_o}\left[\ln\sqrt{5} - \ln\sqrt{14}\right] = 740.316\ \text{V}$$

Due to 100 nC located at A (1, 0, 0),

$$V_{PQ} = \frac{100 \times 10^{-9}}{4\pi\varepsilon_o}\left[\frac{1}{d(PA)} - \frac{1}{d(QA)}\right]$$

$d(PA)$ = distance of point P from the charge

$$= \sqrt{(2-1)^2 + (1-0)^2} = \sqrt{2}$$

$d(QA)$ = distance of point Q from the charge

$$= \sqrt{(3-1)^2 + (2-0)^2 + (5-0)^2} = \sqrt{33}$$

$$\therefore\ V_{PQ} = \frac{10^{-7}}{4\pi\varepsilon_o}\left[\frac{1}{\sqrt{2}} - \frac{1}{\sqrt{33}}\right] = 479.073\ \text{V}$$

Due to 100 nC located at B (0, 1, 0),

$$V_{PQ} = \frac{100 \times 10^{-9}}{4\pi\varepsilon_o}\left[\frac{1}{d(PB)} - \frac{1}{d(QB)}\right]$$

$d(PB)$ = distance of point P from the charge

$$= \sqrt{(2-0)^2 + (1-1)^2} = \sqrt{4} = 2$$

$d(QB)$ = distance of point Q from the charge

$$= \sqrt{(3-0)^2 + (2-1)^2 + (5-0)^2} = \sqrt{35}$$

$$\therefore\ V_{PQ} = \frac{10^{-7}}{4\pi\varepsilon_o}\left[\frac{1}{2} - \frac{1}{\sqrt{35}}\right] = 297.467 \text{ V}$$

$$\therefore\ V_{PQ} = 740.316 + 479.073 + 297.467 = 1516.856 \text{ V}$$

Problem 4.22 : A uniform sheet of charge, $\rho_{s1} = 50\,\varepsilon_o$ C/m^2, is located at $z = 1.5$ m, while $\rho_{s2} = -50\,\varepsilon_o$ C/m^2 is at $z = -0.5$ m. (a) Find $\overline{E}$ everywhere. (b) Find and sketch $V(z)$ as a function of z for $-0.5 < z < 1.5$ if $V = 0$ at $z = -0.5$ m. (c) Repeat if $V = 0$ at $z = 0$.

Solution :

Consider a point located at a for which $z > 1.5$ and $z > -0.5$ as shown in Fig.P4.22. At this point $\overline{a}_N = \overline{a}_z$ for both the surface charge densities. Hence

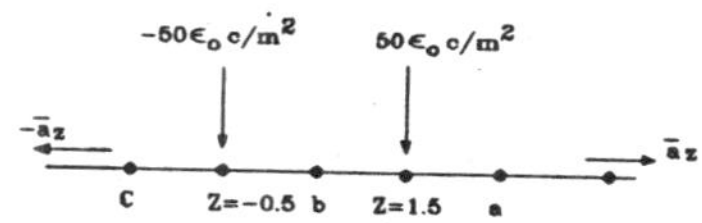

FIG.P4.22

$$\overline{E}(z) = \frac{50\,\varepsilon_o}{2\,\varepsilon_o}\,\overline{a}_z + \left(\frac{-50\,\varepsilon_o}{2\,\varepsilon_o}\right)\overline{a}_z = 0 \text{ V/m}$$

Consider a point located at b for which $1.5 > z > -0.5$. For surface charge density of $50\,\varepsilon_o$ C/m^2, $\overline{a}_N = -\overline{a}_z$ where as for $-50\,\varepsilon_o$ C/m^2, $\overline{a}_N = \overline{a}_z$

$$\therefore\ \overline{E}(z) = \frac{50\,\varepsilon_o}{2\,\varepsilon_o}(-\overline{a}_z) + \left(\frac{-50\,\varepsilon_o}{2\,\varepsilon_o}\right)(\overline{a}_z) = -50\,\overline{a}_z \text{ V/m}$$

Consider a point located at c for which $z < 1.5$ and $z < -0.5$. At this point $\overline{a}_N = -\overline{a}_z$.

$$\therefore\ \overline{E}(z) = \frac{50\,\varepsilon_o}{2\,\varepsilon_o}(-\overline{a}_z) + \left(\frac{-50\,\varepsilon_o}{2\,\varepsilon_o}\right)(-\overline{a}_z) = 0 \text{ V/m}$$

$$\text{Hence we have } \overline{E} = \begin{cases} 0 & z > 1.5 \\ -50\,\overline{a}_z & 1.5 > z > -0.5 \\ 0 & z < -0.5 \end{cases}$$

(b) In the region $-0.5 < z < 1.5$ $(1.5 > z > -0.5)$ $\overline{E} = -50\,\overline{a}_z$ and let $A = (x, y, z)$ and $B = (x, y, -0.5)$.

$$V_{AB} = -\int_B^A \overline{E}\cdot\overline{dL} = -\int_B^A (-50\,\overline{a}_z)\cdot(dx\,\overline{a}_x + dy\,\overline{a}_y + dz\,\overline{a}_z)$$

$$= 50\int_{-0.5}^{z} dz = 50\Big[\,z\,\Big]_{-0.5}^{z} = 50\,(z + 0.5)$$

But $V_{AB} = V_A - V_B$ and $V_B = 0$ (given)

$\therefore\ V_A = V_{AB} = V(z) = 50\,(z + 0.5) = 50\,z + 25$

Plot $V(z)$ on the y axis and z on the x axis.

(c) Let $A = (x, y, z)$ and $B = (x, y, 0)$

$$\therefore\ V_{AB} = 50\int_{0}^{z} dz = 50\Big[\,z\,\Big]_{0}^{z} = 50\,z$$

Now $V_{AB} = V_A - V_B$ and $V_B = 0$ (given)

$\therefore\ V_{AB} = V_A = V(z) = 50\,z$

Plot $V(z)$ on the y axis and z on the x axis.

Problem 4.23 : Four identical point charges of 4 pC each are located at the corners of a square 0.5 mm on a side in free space. How much work must be done to move one charge to a point equidistant from the other three and in the same plane ?

Solution :

Suppose the square is located in $z = 0$ plane i.e. xy plane. The coordinates of the vertices are O (0, 0, 0), A (0.0005, 0, 0), B (0.0005, 0.0005, 0) and C (0, 0.0005, 0). Suppose the charge located at O is moved to a point which is equidistant from the other three vertices. The centre of the square i.e. point D is equidistant from points A, B, and C. The coordinates of point D are (0.00025, 0.00025, 0).

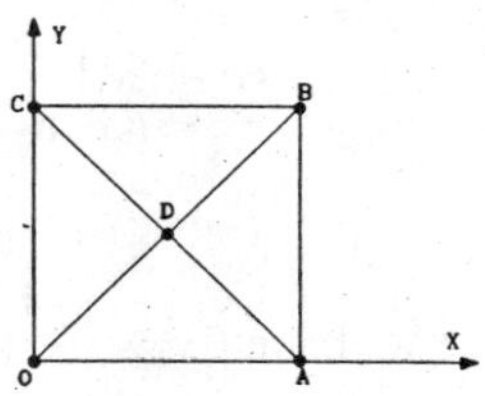

FIG.P4.23

V_{DO} due to charge of 4 pC located at point A

$$= \frac{4\times 10^{-12}}{4\,\pi\,\varepsilon_o}\left[\frac{1}{d\,(DA)} - \frac{1}{d\,(OA)}\right]$$

$$= \frac{4\times 10^{-12}}{4\,\pi\,\varepsilon_o}\left[\frac{1}{\sqrt{2}\,(0.00025)} - \frac{1}{0.0005}\right] = 29.783\ \text{V}$$

V_{DO} due to charge of 4 pC located at point B

$$= \frac{4\times 10^{-12}}{4\pi\varepsilon_o}\left[\frac{1}{d(DB)} - \frac{1}{d(OB)}\right]$$

$$= \frac{4\times 10^{-12}}{4\pi\varepsilon_o}\left[\frac{1}{\sqrt{2}\,(0.00025)} - \frac{1}{\sqrt{2}\,(0.0005)}\right] = 50.842 \text{ V}$$

V_{DO} due to charge of 4 pC located at point C

$$= \frac{4\times 10^{-12}}{4\pi\varepsilon_o}\left[\frac{1}{d(DC)} - \frac{1}{d(OC)}\right]$$

$$= \frac{4\times 10^{-12}}{4\pi\varepsilon_o}\left[\frac{1}{\sqrt{2}\,(0.00025)} - \frac{1}{0.0005}\right] = 29.783 \text{ V}$$

$\therefore\ V_{DO}\,(\text{total}) = 29.783 + 50.842 + 29.783 = 110.408 \text{ V}$

Now $W = QV$

$\therefore\ W = 4\times 10^{-12}\times 110.408 = 441.632 \text{ pJ}$.

[Note :– We should not determine V_{DO} due to charge at O as the charge at O has to be moved to point D]

Problem 4.24 : Point charges of 6 nC each are located at A (1, 0 , 0),B (0, 1, 0), C (–1, 0, 0) , and D (0, –1, 0) in free space. Assuming a zero reference at infinity : (a) find V at P (0, 0, z), and (b) give its maximum value for any z . (c) Use the result of part (a) to calculate $\left| dV/dz \right|$ on the z axis, and (d) give its maximum value.

Solution :

$$\text{(a) } V = \frac{6\times 10^{-9}}{4\pi\varepsilon_o d(PA)} + \frac{6\times 10^{-9}}{4\pi\varepsilon_o d(PB)} + \frac{6\times 10^{-9}}{4\pi\varepsilon_o d(PC)} + \frac{6\times 10^{-9}}{4\pi\varepsilon_o d(PD)}$$

$$= \frac{6\times 10^{-9}}{4\pi\varepsilon_o}\left[\frac{1}{\sqrt{1+z^2}} + \frac{1}{\sqrt{1+z^2}} + \frac{1}{\sqrt{1+z^2}} + \frac{1}{\sqrt{1+z^2}}\right]$$

$$= 215.7058\ (1+z^2)^{-0.5}$$

$$\text{(b) } \frac{dV}{dz} = 215.7058\,(-0.5)\,(1+z^2)^{-1.5}\,(2z) = 0 \text{ or } z\,(1+z^2)^{-1.5} = 0$$

Hence $z = 0$ or $(1+z^2)^{-1.5} = 0$ or $1+z^2 = 0$ or $z = \pm j$

Since z is a real number, $z = 0$ is selected. At this value of z , V is maximum and the maximum value is $V_{max} = 215.7058\ (1+0)^{-0.5} = 215.7058 \text{ V}$.

(c) $\left|\frac{dV}{dz}\right| = \left|215.7058(-0.5)(1+z^2)^{-1.5}(2z)\right| = \left|\frac{215.7058\,z}{(1+z^2)^{1.5}}\right|$

(d) Let $M = \frac{dV}{dz} = \frac{215.7058\,z}{(1+z^2)^{1.5}}$. Then

$$\frac{dM}{dz} = 215.7058\left\{\frac{(1+z^2)^{1.5} - z\times 1.5(1+z^2)^{0.5}(2z)}{\left[(1+z^2)^{1.5}\right]^2}\right\} = 0 \text{ or}$$

$$(1+z^2)^{0.5}\left(1+z^2-3z^2\right) = 0 \text{ or } 1-2z^2 = 0 \text{ or } z = \pm\, 0.7071$$

Hence $\left.\frac{dV}{dz}\right|_{max} = \frac{215.7058\,(0.7071)}{(1+0.5)^{1.5}} = 83.024$ V/m

Problem 4.25 : A point charge of 16 nC is located at Q (2, 3, 5) in free space, and a uniform line charge of 5 nC / m is at the intersection of the planes $x = 2$ and $y = 4$. If the potential at the origin is 100 V, find V at P (4, 1, 3).

Solution :

$$V_{PO} = \frac{16\times 10^{-9}}{4\pi\varepsilon_o}\left[\frac{1}{d(PQ)} - \frac{1}{d(OQ)}\right] = \frac{16\times 10^{-9}}{4\pi\varepsilon_o}\left[\frac{1}{\sqrt{12}} - \frac{1}{\sqrt{38}}\right]$$

$$= 18.185 \text{ V}$$

$$V_{PO} = \frac{-\rho_L}{2\pi\varepsilon_o}\left[\ln d_P - \ln d_O\right]$$

Coordinates of point P are (4, 1, 3). Coordinates of the foot of perpendicular dropped from point P on the line charge (2, 4, 3).

$\therefore\ d_P = \sqrt{(4-2)^2 + (1-4)^2} = \sqrt{13}$

Coordinates of point O are (0, 0, 0). Coordinates of the foot of perpendicular dropped from point O on the line charge (2, 4, 0).

$\therefore\ d_O = \sqrt{((0-20)^2 + (0-4)^2} = \sqrt{20}$

$$\therefore\ V_{PO} = \frac{5\times 10^{-9}}{2\pi\varepsilon_o}\left[\ln\sqrt{13} - \ln\sqrt{20}\right] = 19.3588 \text{ V}$$

$\therefore\ V_{PO}$ (total) $= 18.185 + 19.3588 = 37.544$ V

Now $V_{PO} = V_P - V_O$ and $V_O = 100$

Hence $V_P = V_{PO} + V_O = 37.544 + 100 = 137.544$ V

Problem 4.26 : Given these uniform surface– charge distributions : 100 pC / m^2 at $r = 2$ m, 40 pC / m^2 at $r = 3$ m, and – 47.5 pC / m^2 at $r = 4$ m.

(a) find V at $r = 5$ m if $V = 0$ at infinity. (b) Plot V vs. r, $1 < r < 6$ m, if $V = 0$ at infinity.

Solution :

Consider a sphere of radius r ($r < 2$ m) which passes through point A as shown in above Fig.P4.26. This sphere do not encloses any charge. Hence $Q = 0, \overline{D} = 0, \overline{E} = 0$

$\therefore\ V = 0$ for $r < 2$ m.

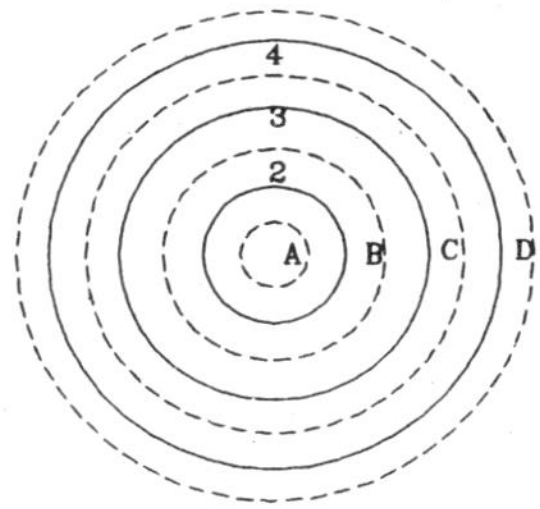

FIG.P4.26

Consider a sphere of radius r ($2 < r < 3$) which passes through point B as shown in Fig.P4.26. This sphere encloses the surface charge density of 100 pC / m^2. Hence charge enclosed by the sphere

$$Q = 100 \times 4\pi (2)^2 = 1600\ \pi\ \text{pC}$$

$$\therefore\ \overline{D} = \frac{Q}{4\pi r^2}\,\bar{a}_r = \frac{1600\ \pi}{4\pi r^2}\,\bar{a}_r\ \text{pC/m}^2 \quad \text{or}$$

$$\overline{E} = \overline{D}/\varepsilon_o = \frac{1600\ \pi}{4\pi\varepsilon_o r^2}\,\bar{a}_r\ \text{pV/m} = \frac{400 \times 10^{-12}}{\varepsilon_o r^2}\,\bar{a}_r\ \ \text{V/m}$$

$$\overline{E}\cdot\overline{dL} = \frac{400 \times 10^{-12}}{\varepsilon_o r^2}\,dr \ \text{as } \overline{E} \text{ has only } \bar{a}_r$$

Let point B lies on the sphere of radius r and point P at ∞. Then

$$V_{BP} = -\int_P^B \overline{E}\cdot\overline{dL} = -\int_\infty^r \frac{400 \times 10^{-12}}{\varepsilon_o r^2}\,dr = \frac{400 \times 10^{-12}}{\varepsilon_o}\left[\frac{1}{r}\right]_\infty^r$$

$$= \frac{400 \times 10^{-12}}{\varepsilon_o r}$$

Now $V_{BP} = V_B - V_P$ and $V_P = 0\ \therefore\ V_B = V_{BP} = V$

$$V = \frac{400 \times 10^{-12}}{\varepsilon_o r} = \frac{45.177}{r} \quad \text{for } 2 < r < 3$$

Consider a sphere of radius r ($3 < r < 4$) which passes through point C as shown in the Fig.P4.26. It encloses surface charge densities of 100 pC / m^2 and 40 pC / m^2

Hence $Q = 100 \times 4\pi (2)^2 + 40 \times 4\pi (3)^2 = 3040\ \pi\ $ pC

$$\therefore\ \overline{D} = \frac{Q}{4\pi r^2}\,\bar{a}_r = \frac{3040\,\pi}{4\pi r^2}\,\bar{a}_r \ \text{pC/m}^2 \quad \text{or}$$

$$\overline{E} = \overline{D}/\varepsilon_o = \frac{3040\,\pi}{4\pi\varepsilon_o r^2}\,\bar{a}_r \ \text{pV/m} = \frac{760\times 10^{-12}}{\varepsilon_o r^2}\,\bar{a}_r \ \text{V/m}$$

$$\overline{E}\cdot\overline{dL} = \frac{760\times 10^{-12}}{\varepsilon_o r^2}\,dr \quad \text{as } \overline{E} \text{ has only } \bar{a}_r$$

Let point C lies on the sphere of radius r and point P at ∞ then

$$V_{CP} = -\int_P^C \overline{E}\cdot\overline{dL} = \frac{-760\times 10^{-12}}{\varepsilon_o}\int_\infty^r \frac{dr}{r^2} = \frac{760\times 10^{-12}}{\varepsilon_o}\left[\frac{1}{r}\right]_\infty^r$$

$$= \frac{760\times 10^{-12}}{\varepsilon_o r}$$

Now $V_{CP} = V_C - V_P$ and $V_P = 0 \ \therefore\ V_{CP} = V_C = V$

$$\therefore\ V = \frac{760\times 10^{-12}}{\varepsilon_o r} = \frac{85.837}{r} \quad \text{for } 3 < r < 4$$

Consider a sphere of radius r ($r > 4$) which passes through point D as shown in Fig.P4.26. This sphere enclose all the three surface charge densities. Hence charge enclosed by this sphere

$$Q = 100\times 4\pi(2)^2 + 40\times 4\pi(3)^2 - 47.5\times 4\pi(4)^2 = 0$$

$$\therefore\ \overline{D} = \frac{Q}{4\pi r^2}\,\bar{a}_r = \frac{Q}{4\pi r^2}\,\bar{a}_r = 0 \ \therefore\ \overline{E} = 0 \ \therefore\ V = 0 \text{ for } r > 4 \text{ m.}$$

Hence we have

$$V(r) = \begin{cases} 0 & \text{for } r < 2m \\ 45.177/r & \text{for } 2 < r < 3m \\ 85.837/r & \text{for } 3 < r < 4m \\ 0 & \text{for } r > 4m \end{cases}$$

Plot $V(r)$ on the y – axis and r on the x – axis

Problem 4.27 : A non uniform line charge density, $\rho_L = 10/z^2$ nC/m , lies along the z axis for $|z| > 1$; $\rho_L = 0$ for $|z| < 1$. Find the potential at P (2, 3, 0) in free space if $V(\infty) = 0$.

Solution : $V = \displaystyle\int \frac{\rho_L\, dL}{4\pi\varepsilon_o\,|\bar{r}_2 - \bar{r}_1|}$

$\rho_L = 10/z^2$ nC/m and $dL = dz$ (as the line charge lies on the z axis).Point 2 is that point at which V has to be determined. Hence 2 (2, 3, 0). Point 1 is the general point on the line charge. Hence 1 (0, 0, z).

$\bar{R}_{12} = \bar{r}_2 - \bar{r}_1 = (2\,\bar{a}_x + 3\,\bar{a}_y) - (z\,\bar{a}_z) = 2\,\bar{a}_x + 3\,\bar{a}_y - z\,\bar{a}_z$

$|\,\bar{r}_2 - \bar{r}_1\,| = \sqrt{(2)^2 + (3)^2 + (-z)^2} = \sqrt{13 + z^2}$

$$\therefore V = \frac{1}{4\pi\varepsilon_o}\left[\int_{-\infty}^{-1} \frac{10\,dz}{z^2\sqrt{13+z^2}} + \int_{1}^{\infty} \frac{10\,dz}{z^2\sqrt{13+z^2}}\right] 10^{-9}$$

$$= \frac{10^{-9}}{4\pi\varepsilon_o}\left\{10\left[-\frac{1}{13}\operatorname{cosec}\theta\right]_{-90^\circ}^{-15.5^\circ} + 10\left[-\frac{1}{13}\operatorname{cosec}\theta\right]_{15.5^\circ}^{90^\circ}\right\}$$

$$= 37.91 \text{ V}$$

(Note :– In both the integrals , use the substitution $z = \sqrt{13}\,\tan\theta$)

Problem 4.28 : The charge density $\rho_L = 10/\,|\,z\,|$ nC/m exists along the z axis for $|\,z\,| > 1$. Find V at P if $V = 0$ at $\rho = \infty$.

Solution :

$$V = \int \frac{\rho_L\,dL}{4\pi\varepsilon_o\,|\,\bar{r}_2 - \bar{r}_1\,|}$$

$\rho_L = 10/|\,z\,|$ nC/m and it lies on the z axis for $|\,z\,| > 1$,i.e., for $-\infty < z < -1$ and for $1 < z < \infty$. Also and $dL = dz$ (as the line charge lies on the z axis). Point 2 (ρ , 0 , 0) and point 1 (0, 0, z). Hence

$\bar{R}_{12} = \bar{r}_2 - \bar{r}_1 = \rho\,\bar{a}_\rho - z\,\bar{a}_z$ and $|\,\bar{r}_2 - \bar{r}_1\,| = \sqrt{\rho^2 + z^2}$

$$\therefore V = \frac{1}{4\pi\varepsilon_o}\left\{\int_{-\infty}^{-1} \frac{10\,dz}{|\,z\,|\sqrt{\rho^2+z^2}} + \int_{1}^{\infty} \frac{10\,dz}{|\,z\,|\sqrt{\rho^2+z^2}}\right\} 10^{-9}$$

The expression for ρ_L contains $|\,z\,|$ so that when ρ_L lies between $-\infty$ to -1 , its value will be positive . To remove the modulus sign from first integral, replace $|\,z\,|$ by $-z$ so that when z is negative, $-z$ will be positive. In the second integral, modulus sign can be removed as z is already positive.

$$\therefore V = \frac{1}{4\pi\varepsilon_o}\left\{\int_{-\infty}^{-1} \frac{-10\,dz}{z\sqrt{\rho^2+z^2}} + \int_{1}^{\infty} \frac{10\,dz}{z\sqrt{\rho^2+z^2}}\right\} 10^{-9}$$

$$\therefore\ V = \frac{1}{4\pi\varepsilon_o}\left\{\int_1^{\infty}\frac{10\,dz}{z\sqrt{\rho^2+z^2}}+\int_1^{\infty}\frac{10\,dz}{z\sqrt{\rho^2+z^2}}\right\}10^{-9}$$

$$= \frac{10\times10^{-9}}{4\pi\varepsilon_o}\left\{2\int_1^{\infty}\frac{10\,dz}{z\sqrt{\rho^2+z^2}}\right\}$$

$$= \frac{10\times10^{-9}}{4\pi\varepsilon_o}\times 2\left\{-\frac{1}{\rho}\ln\left[\frac{\rho+\sqrt{z^2+\rho^2}}{z}\right]_1^{\infty}\right\}$$

$$= -\frac{10^{-8}}{2\pi\varepsilon_0\rho}\left\{\ln(1)-\ln\left[\frac{\rho+\sqrt{1+\rho^2}}{1}\right]\right\}$$

$$= \frac{179.755}{\rho}\ln\left[\rho+\sqrt{1+\rho^2}\right]$$

Problem 4.29 : A circular disc, $\rho \le 1$ cm, $z = 0$, carries a surface charge density $\rho_s = 1/\rho$ nC/m^2. Find V at $P(0, 0, z)$ if $V = 0$ at infinity.

Solution :

$$V = \int \frac{\rho_s\,dS}{4\pi\varepsilon_o\,|\,\bar r_2 - \bar r_1\,|}$$

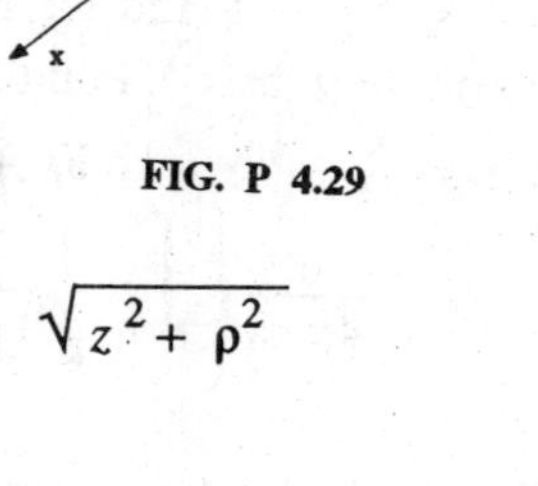

FIG. P 4.29

$\rho_s = 1/\rho$ nC/m^2. The surface charge density lies in $z = 0$ plane. To this plane $\pm\,\bar a_z$ are perpendicular. Hence

$\overline{dS} = \pm\,(\rho\,d\rho\,d\phi)\,\bar a_z$ or $dS = \rho\,d\rho\,d\phi$ Point 2 is that point at which potential is desired. Hence 2 (0, 0, z). Point 1 is the general point in $z = 0$ plane. Hence 1 (ρ, ϕ, 0)

$\bar R_{12} = \bar r_2 - \bar r_1 = z\,\bar a_z - \rho\,\bar a_\rho$ and $|\,\bar r_2 - \bar r_1\,| = \sqrt{z^2+\rho^2}$

$$\therefore\ V = \frac{1}{4\pi\varepsilon_o}\int_0^{0.01}\int_0^{2\pi}\frac{10^{-9}\rho\,d\rho\,d\phi}{\rho\sqrt{z^2+\rho^2}}$$

$$= \frac{10^{-9}}{4\pi\varepsilon_o}\left[\ln\left(\frac{\rho+\sqrt{\rho^2+z^2}}{z}\right)\right]_0^{0.01}\left[\phi\right]_0^{2\pi}$$

$$= 56.472\ \ln\left(\frac{0.01+\sqrt{0.0001+z^2}}{z}\right)$$

Problem 4.30 : Fig.P4.30 shows three separste charge distributions in the $z = 0$ plane in free space. (a) Find the total charge for each distribution. (b) Find the potential at P (0, 0, 5) caused by each of the three charge distributions acting alone. (c) Find V_P.

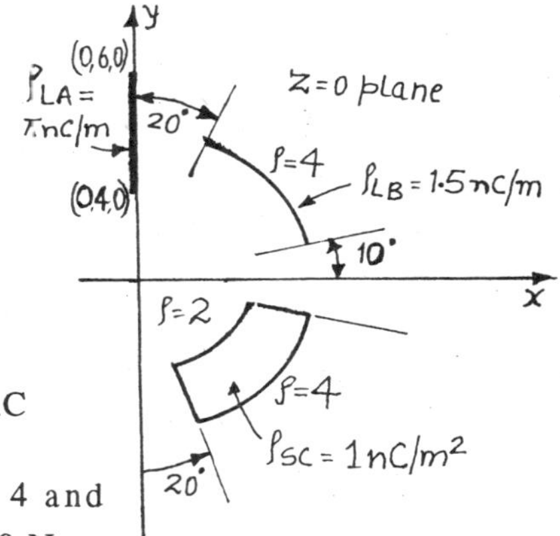

FIG.P4.30

Solution :

(a)For the straight line charge

$$Q = \int \rho_L \, dL = \int_4^6 \rho_L \, dy$$

($\because$ $dL = dy$ as ρ_L lies on y axis)

$$Q = \int_4^6 \pi \, dy = \pi \left[y \right]_4^6 = 2\pi \text{ nC}$$

For the circular line charge, $\rho = 4$ and $z = 0$ are constants. Hence $d\rho = dz = 0$ Now $\overline{dL} = d\rho \, \bar{a}_\rho + \rho \, d\phi \, \bar{a}_\phi + dz \, \bar{a}_z$.

Hence $\overline{dL} = \rho \, d\phi \, \bar{a}_\phi$ **or** $dL = \rho \, d\phi = 4 \, d\phi$

$$Q = \int \rho_L \, dL = \int_{10^\circ}^{70^\circ} 1.5 \times 4 \, d\phi$$

$$= 6\left[\phi \right]_{10}^{70^\circ} = 6\left[(70^\circ - 10^\circ) \frac{\pi}{180} \right]$$

$$= 2\pi \text{ nC}$$

The surface charge lies in $z = 0$ plane. Hence $\overline{dS} = \rho \, d\rho \, d\phi \, (\pm \, \bar{a}_z)$ or $dS = \rho \, d\rho \, d\phi$

$$Q = \int \rho_S \, dS = \int_2^4 \int_{290^\circ}^{350^\circ} \rho \, d\rho \, d\phi = \left[\frac{\rho^2}{2} \right]_2^4 \left[\phi \right]_{290^\circ}^{380^\circ}$$

$$= 0.5 \, [16 - 4] \left[(350^\circ - 290^\circ) \frac{\pi}{180} \right] = 2\pi \text{ nC}$$

(b) Potential due to straight line charge $V = \int \dfrac{\rho_L \, dL}{4\pi\varepsilon_0 \, | \bar{r}_2 - \bar{r}_1 |}$

Point 2 is that point at which potential is desired = (0, 0, 5)

Point 1 is the general point on the line charge = (0, y , 0)

$\bar{r}_2 - \bar{r}_1 = 5\,\bar{a}_z - y\,\bar{a}_y$ and $|\,\bar{r}_2 - \bar{r}_1\,| = \sqrt{25 + y^2}$

$$V = \int_4^6 \frac{\pi\, dy \times 10^{-9}}{4\pi\varepsilon_o \sqrt{25 + y^2}} = \frac{10^{-9}}{4\,\varepsilon_o}\left[\ln\left(\frac{y + \sqrt{y^2 + 25}}{5}\right)\right]_4^6 = 8\text{ V}$$

Potential due to circular line charge

Point 2 = $(0, 0, 5)$ and point 1 is $(\rho = 4, \phi, z = 0)$

$\bar{r}_2 - \bar{r}_1 = 5\,\bar{a}_z - 4\,\bar{a}_\rho$ and $|\,\bar{r}_2 - \bar{r}_1\,| = \sqrt{25 + 16} = \sqrt{41}$

$$V = \int_{10^o}^{70^o} \frac{1.5 \times 4\, d\phi \times 10^{-9}}{4\pi\varepsilon_o \sqrt{41}} = \frac{1.5 \times 10^{-9}}{\pi\,\varepsilon_o \sqrt{41}}\left[\,\phi\,\right]_{10^o}^{70^o} = 8.819\text{ V}$$

Potential due to surface charge density

Point 2 = $(0, 0, 5)$ and Point 1 = $(\rho, \phi, 0)$

$\bar{r}_2 - \bar{r}_1 = 5\,\bar{a}_z - \rho\,\bar{a}_\rho$ and $|\,\bar{r}_2 - \bar{r}_1\,| = \sqrt{25 + \rho^2}$

$$V = \int_2^4 \int_{290^o}^{350^o} \frac{1 \times \rho\, d\rho\, d\phi \times 10^{-9}}{4\pi\varepsilon_o \sqrt{25 + \rho^2}} = \frac{10^{-9}}{4\pi\varepsilon_o}\left[\sqrt{25 + \rho^2}\right]_2^4 \left[\,\phi\,\right]_{290^o}^{350^o}$$

$$= 9.581\text{ V}$$

(c) $V_{total} = 8 + 8.819 + 9.581 = 26.4\text{ V}$

Problem 4.31 : Let $V = 2\,(x + 1)^2\,(y + 2)^2\,(z + 3)^2$ V in free space. At P (2, –1, 4) find : (a) V ; (b) $\bar{E}$; (c) $|\,\bar{E}\,|$; (d) $\bar{D}$; (e) $|\,\bar{D}\,|$; (f) ρ_v ; (g) $\bar{a}_N$.

Solution :

(a) $V = 2\,(2 + 1)^2\,(-1 + 2)^2\,(4 + 3)^2 = 882\text{ V}$

$$\text{(b) } \bar{E} = -\,\bar{\nabla} V = -\left[\frac{\partial V}{\partial x}\,\bar{a}_x + \frac{\partial V}{\partial y}\,\bar{a}_y + \frac{\partial V}{\partial z}\,\bar{a}_z\right]$$

$$= -\Big[4\,(x + 1)\,(y + 2)^2\,(z + 3)^2\,\bar{a}_x + 4\,(x + 1)^2\,(y + 2)\,(z + 3)^2\,\bar{a}_y + 4\,(x + 1)^2\,(y + 2)^2\,(z + 3)\,\bar{a}_z\Big]$$

Substituting the coordinates of point P in the above expression for $\bar{E}$, we have

$\bar{E} = -\,588\,\bar{a}_x - 1764\,\bar{a}_y - 252\,\bar{a}_z$ V/m

(c) $|\,\bar{E}\,| = \sqrt{(-588)^2 + (-1764)^2 + (-252)^2} = 1876.418\text{ V/m}$

(d) $\overline{D} = \varepsilon_o \overline{E} = \varepsilon_o(-588\,\overline{a}_x - 1764\,\overline{a}_y - 252\,\overline{a}_z)$

$$= -5.206\,\overline{a}_x - 15.618\,\overline{a}_y - 2.231\,\overline{a}_z \ \ nC/m^2$$

(e) $|\overline{D}| = \varepsilon_o|\overline{E}| = \varepsilon_o \times 1876.418 = 16.614 \ \ nC/m^2$

(f) $\overline{D} = \varepsilon_o \overline{E} = -4\,\varepsilon_o(x+1)(y+2)^2(z-3)^2\,\overline{a}_x - 4\,\varepsilon_o(x+1)^2(y+2)$

$$(z+3)^2\,\overline{a}_y - 4\,\varepsilon_o(x+1)^2(y+2)^2(z+3)\,\overline{a}_z \ \ C/m^2$$

$$= \frac{\partial}{\partial x}[-4\,\varepsilon_o(y+2)^2(z+3)^2] + \frac{\partial}{\partial y}[-4\,\varepsilon_o(x+1)^2(z+3)^2] + \frac{\partial}{\partial z}[-4\,\varepsilon_o(x+1)^2(y+2)^2]$$

$$= -4\,\varepsilon_o\left[(y+2)^2(z+3)^2 + (x+1)^2(z+3)^2 + (x+1)^2(y+2)^2\right]$$

Substituting the coordinates of point P in the above expression of ρ_v, we have

$\rho_v = -17.673 \ \ nC/m^3$

(g) $\overline{a}_N = -\dfrac{\overline{E}}{|\overline{E}|} = \dfrac{-(-588\,\overline{a}_x - 1764\,\overline{a}_y - 252\,\overline{a}_z)}{1876.418}$

$$= 0.313\,\overline{a}_x + 0.94\,\overline{a}_y + 0.134\,\overline{a}_y$$

Problem 4.32 : Given the potential field $V = 100\sqrt{r}$ V in free space, find : (a) $\overline{E}$; (b) $\overline{D}$. (c) How much charge lies within the surface $r = 0.5$?

Solution :

(a) $\overline{E} = -\overline{\nabla} V = -\dfrac{\partial V}{\partial r}\,\overline{a}_r$ as r is present in the expression of potential.

$$= -\frac{\partial}{\partial r}(100\sqrt{r})\,\overline{a}_r = \frac{-50}{\sqrt{r}}\,\overline{a}_r \ \ V/m$$

(b) $\overline{D} = \varepsilon_o \overline{E} = \dfrac{-50\,\varepsilon_o}{\sqrt{r}}\,\overline{a}_r \ \ C/m^2$

(c) To the sphere $r = 0.5$, $\pm\,\overline{a}_r$ are perpendicular. Hence

$\overline{dS} = r^2 \sin\theta\, d\theta\, d\phi\, \overline{a}_r$ and $\overline{D}\cdot\overline{dS} = -50\,r^{1.5}\sin\theta\, d\theta\, d\phi$

$$Q = \int \overline{D}\cdot\overline{dS} = \int_0^{\pi}\int_0^{2\pi} -50\,r^{1.5}\sin\theta\, d\theta\, d\phi \ \ {}_{(r=0.5)}$$

$$= - 50\,(0.5)^{1.5}\Big[-\cos\theta \Big]_0^{\pi} \Big[\phi \Big]_0^{2\pi} = - 222.144 \text{ C}$$

Problem 4.33 : For the point $P\,(3, 60^o, 2)$ in cylindrical coordinates and the potential field $V = 10\,(\rho + 1)\,z^2 \cos\phi$ V in free space, find at P : (a) V ; (b) $\overline{E}$; (c) $\overline{D}$; (d) ρ_v ; (e) dV/dN ; (f) $\bar{a}_N$.

Solution :

(a) $V = 10\,(3 + 1)\,(2)^2 \cos 60^o = 80$ V

(b)we have

$$\overline{E} = -\left\{ \frac{\partial}{\partial\rho}[\,10\,(\rho + 1)\,z^2 \cos\phi\,]\,\bar{a}_\rho + \frac{1}{\rho}\frac{\partial}{\partial\phi}[\,10\,(\rho + 1)\,z^2 \cos\phi\,]\,\bar{a}_\phi + \frac{\partial}{\partial z}[\,10\,(\rho + 1)\,z^2 \cos\phi\,]\,\bar{a}_z \right\}$$

$$\overline{E} = -10\left[z^2 \cos\phi\,\bar{a}_\rho - \frac{(\rho + 1)\,z^2 \sin\phi}{\rho}\,\bar{a}_\phi + 2\,(\rho + 1)\,z \cos\phi\,\bar{a}_z \right]$$

Substituting the coordinates of point P in the above expression,we have

$\overline{E}$ at $P = -20\,\bar{a}_\rho + 46.188\,\bar{a}_\phi - 80\,\bar{a}_z$ V/m

(c) $\overline{D}$ at $P = \varepsilon_o \overline{E} = -177.08\,\bar{a}_\rho + 408.949\,\bar{a}_\phi - 708.32\,\bar{a}_z$ pC/m^2

(d)$\overline{D} = -10\,\varepsilon_o\left[z^2 \cos\phi\,\bar{a}_\rho - \frac{(\rho + 1)}{\rho} z^2 \sin\phi\,\bar{a}_\phi + 2\,(\rho + 1)\,z \cos\phi\,\bar{a}_z \right]$

we have

$$\rho_v = -10\,\varepsilon_o\left\{ \frac{\partial}{\partial\rho}(\,z^2 \cos\phi\,) - \frac{1}{\rho}\frac{\partial}{\partial\phi}\left[\left(\frac{\rho + 1}{\rho}\right) z^2 \sin\phi \right] + \frac{\partial}{\partial z}\Big[\,2\,(\rho + 1)\,z \cos\phi\,\Big] \right\}$$

$$= -10\,\varepsilon_o\left[\frac{z^2 \cos\phi}{\rho} - \frac{(\rho + 1)\,z^2 \cos\phi}{\rho^2} + 2\,(\rho + 1)\cos\phi \right]$$

Substituting the coordinates of point P in the above expression,we have

$\rho_v = -334.484$ pC/m^3

(e) $\frac{dV}{dN} = |\,\overline{E}\,| = \sqrt{(-20)^2 + (46.188)^2 + (-80)^2} = 94.516$ V/m

(f) $\bar{a}_N = -\frac{\overline{E}}{|\,\overline{E}\,|} = \frac{-(-20\,\bar{a}_\rho + 46.188\,\bar{a}_\phi - 80\,\bar{a}_z)}{94.516}$

$$\doteq 0.212\,\bar{a}_\rho - 0.489\,\bar{a}_\phi + 0.846\,\bar{a}_z$$

Problem 4.34 : A potential field is expressed as $V = (50\,\rho^2 \cos\phi) / (z + 1)$ V . Given a point P (4, 30°, 2) in free space, find at P : (a) V ; (b) $\overline{E}$; (c) ρ_v ; (d) dV/dN ; (e) $\bar{a}_N$.

Solution :

Same as Problem 4.33 . Only answers are given below.

(a) 230.940 V

(b) $-\ 115.47\,\bar{a}_\rho + 33.333\,\bar{a}_\phi + 76.98\,\bar{a}_z$ V/m

(c) $-\ 837.777$ pC/m^3

(d) 142.725 V/m

(e) $0.809\,\bar{a}_\rho - 0.234\,\bar{a}_\phi - 0.539\,\bar{a}_z$

Problem 4.35 : Within the sphere $r = 1$, the potential is given as $V = 100 + 50\,r + 150\,r \sin\theta \sin\phi$ V. (a) Find V, $\overline{E}$, $\overline{D}$, and ρ_v at $r = 1, \theta = \pi/2$, $\phi = 0$. (b) How much charge lies within the sphere $r = 1$?

Solution : Same as Problem 4.7. Only answers are given below.

(a) 150 V, $-\ 50\,\bar{a}_r - 150\,\bar{a}_\phi$ V/m, $-\ 0.443\,\bar{a}_r - 1.328\,\bar{a}_\phi$ nC/m^2; $-\ 0.885$ nC/m^3.

(b) $-\ 5.56$ nC

Problem 4.36 : Point charge of 1 μC and $-$ 1 μC are located at (0, 0, 0.5) and (0, 0, $-$ 0.5), respectively. Treating these two charges as a dipole at the origin, calculate : (a) V at P (3, 0, 4) ; (b) $|\,\overline{E}\,|$ at P . Now find the exact values for : (c) V at P ; (d) $|\,\overline{E}\,|$ at P .

Solution :

(a) $V = \dfrac{Q\,d\cos\theta}{4\,\pi\,\varepsilon_o\,r^2}$ where $Q = 1\,\mu\text{C} = 10^{-6}\,\text{C}$, $d = 1.0$ m

For point P (3, 0, 4) , $r = \sqrt{9 + 16} = 5$, $\theta = \cos^{-1}\left(\dfrac{4}{5}\right) = 36.87^o$

$$\therefore\ V = \frac{10^{-6} \times 1 \times \cos 36.87^o}{4\,\pi\,\varepsilon_o\,(5)^2} = 287.607$$

$$\text{(b) } \overline{E} = \frac{Q\,d}{4\pi\varepsilon_o r^3}\,(2\cos\theta\,\overline{a}_r + \sin\theta\,\overline{a}_\theta)$$

$$= \frac{10^{-6}\times 1}{4\pi\varepsilon_o (5)^3}\,(2\cos 36.87^\circ\,\overline{a}_r + \sin 36.87^\circ\,\overline{a}_\theta)$$

$$= 115.043\,\overline{a}_r + 43.141\,\overline{a}_\theta \text{ V/m}$$

$$|\overline{E}| = \sqrt{(115.043)^2 + (43.141)^2} = 122.866 \text{ V/m}$$

$$\text{(c) } V = \frac{Q}{4\pi\varepsilon_o}\left(\frac{1}{R_1} - \frac{1}{R_2}\right) \text{ where}$$

R_1 = distance of point P from the positive charge

$$= \sqrt{(3-0)^2 + (4-0.5)^2} = \sqrt{21.25}$$

R_2 = distance of point P from the negative charge

$$= \sqrt{(3-0)^2 + (4+0.5)^2} = \sqrt{29.25}$$

$$\therefore V = \frac{10^{-6}}{4\pi\varepsilon_o}\left(\frac{1}{\sqrt{21.25}} - \frac{1}{\sqrt{29.25}}\right) = 287.881 \text{ V}$$

(d) $\overline{E}$ at point P = $\overline{E}$ due to charge at point A + $\overline{E}$ due to charge at point B

$$= \frac{10^{-6}}{4\pi\varepsilon_o}\left\{\frac{(3-0)\,\overline{a}_x + (4-0.5)\,\overline{a}_z}{[(3)^2 + (3.5)^2]^{3/2}}\right\} - \frac{10^{-6}}{4\pi\varepsilon_o}\left\{\frac{(3-0)\,\overline{a}_x + (4+0.5)\,\overline{a}_z}{[(3)^2 + (4.5)^2]^{3/2}}\right\}$$

$$= 104.809\,\overline{a}_x + 65.463\,\overline{a}_z$$

$$|\overline{E}| = \sqrt{(104.809)^2 + (65.463)^2} = 123.573 \text{ V/m}$$

Problem 4.37 : In Fig.4.3 let Q = 50 nC , d = 0.5 m ,and the field point be P (2, 0, 3) in cartesian coordinates.Find V_P by : (a) the approximate dipole relationship ,Eq.(4.15) ; (b) exact methods.

Solution : $V_P = \dfrac{Q\,d\cos\theta}{4\pi\varepsilon_o r^2}$

At point P , $r = \sqrt{(2)^2 + (3)^2} = \sqrt{13}$, $\theta = \cos^{-1}\left(3/\sqrt{13}\right) = 33.69^\circ$

$$V_P = \frac{50\times 10^{-9}\times 0.5\times\cos(33.69^\circ)}{4\pi\varepsilon_o(\sqrt{13})^2} = 14.381 \text{ V}$$

(b) With reference to Fig.4.3, the positive charge is located at (0, 0, 0.25) and the negative charge is located at (0, 0, – 0.25)

$$V_P = \frac{Q}{4\pi\varepsilon_o}\left[\frac{1}{R_1} - \frac{1}{R_2}\right]$$

R_1 = distance of point P from the positive charge

$$= \sqrt{(2-0)^2 + (3-0.25)} = \sqrt{11.5625}$$

R_2 = distance of point P from the negative charge

$$= \sqrt{(2-0)^2 + (3+0.25)^2} = \sqrt{14.5625}$$

$$\therefore\ V_P = \frac{50\times 10^{-9}}{4\pi\varepsilon_o}\left[\frac{1}{\sqrt{11.5625}} - \frac{1}{\sqrt{14.5625}}\right] = 14.397 \text{ V}$$

Problem 4.38 : A dipole having a moment $\bar{p} = 10\,\bar{a}_x - 3\,\bar{a}_y + 5\,\bar{a}_z$ nC-m is located at Q (2, 4, 1) in free space. (a) Find V at $P(x, y, z)$. (b) Find V at (5, 1, 0).

Solution :

(a) $V = \dfrac{\bar{p}\cdot(\bar{r}_2 - \bar{r}_1)}{4\pi\varepsilon_o\,|\bar{r}_2 - \bar{r}_1|^3}$ where $\bar{p} = 10\,\bar{a}_x - 3\,\bar{a}_y + 5\,\bar{a}_z$ nC-m

Point 1 is that point at which dipole is located = (2, 4, 1)

Point 2 is that point at which potential is desired = (x, y, z)

$$\bar{r}_2 - \bar{r}_1 = (x\,\bar{a}_x + y\,\bar{a}_y + z\,\bar{a}_z) - (2\,\bar{a}_x + 4\,\bar{a}_y + \bar{a}_z)$$

$$= (x-2)\,\bar{a}_x + (y-4)\,\bar{a}_y + (z-1)\,\bar{a}_z$$

$$|\bar{r}_2 - \bar{r}_1| = \sqrt{(x-2)^2 + (y-4)^2 + (z-1)^2}$$

$$\bar{p}\cdot(\bar{r}_2 - \bar{r}_1) = [10(x-2) - 3(y-4) + 5(z-1)]^{-9}$$

$$\therefore\ V = \frac{[10(x-2) - 3(y-4) + 5(z-1)]\,10^{-9}}{4\pi\varepsilon_o\sqrt{[(x-2)^2 + (y-4)^2 + (z-1)^2]^3}}$$

$$= \frac{8.98774\,(10x - 3y + 5z - 13)}{[(x-2)^2 + (y-4)^2 + (z-1)^2]^{3/2}}$$

(b) Point 1 = (2, 4, 1) and point 2 = (5, 1, 0)

$$\bar{r}_2 - \bar{r}_1 = (5\,\bar{a}_x + \bar{a}_y) - (2\,\bar{a}_x + 4\,\bar{a}_y + \bar{a}_z) = 3\,\bar{a}_x - 3\,\bar{a}_y - \bar{a}_z$$

$$\bar{p}\cdot\bar{r}_2 - \bar{r}_1 = [(10)(3) + (-3)(-3) + (5)(-1)]\,10^{-9} = 37\times 10^{-9}$$

$$|\bar{r}_2 - \bar{r}_1| = \sqrt{(3)^2 + (-3)^2 + (-1)^2} = \sqrt{19}$$

$$V = \frac{37\times 10^{-9}}{4\pi\varepsilon_o(\sqrt{19})^3} = 4.015 \text{ V}$$

Problem 4.39 : A dipole at the origin in free space has $\bar{p} = 80\pi\varepsilon_o \bar{a}_z$ C · m. (a) Give V at $P(x, y, z)$ in cartesian coordinates (b) Give $\bar{E}$ at $P(x, y, z)$ in cartesian coordinates.

Solution :

(a) $V = \dfrac{\bar{p}\cdot(\bar{r}_2 - \bar{r}_1)}{4\pi\varepsilon_o|\bar{r}_2 - \bar{r}_1|^3}$ where $\bar{p} = 80\pi\varepsilon_o\bar{a}_z$ C· m

Point 1 is that point at which dipole is located = (0, 0, 0)

Point 2 is that point at which potential is desired = (x, y, z)

$$\bar{r}_2 - \bar{r}_1 = (x\bar{a}_x + y\bar{a}_y + z\bar{a}_z) - (0\bar{a}_x + 0\bar{a}_y + 0\bar{a}_z)$$

$$= x\bar{a}_x + y\bar{a}_y + z\bar{a}_z$$

$$\bar{p}\cdot(\bar{r}_2 - \bar{r}_1) = 80\pi\varepsilon_o z$$

$$|\bar{r}_2 - \bar{r}_1| = \sqrt{x^2 + y^2 + z^2}$$

$$\therefore\ V = \frac{80\pi\varepsilon_o z}{4\pi\varepsilon_o(x^2 + y^2 + z^2)^{3/2}} = \frac{20z}{(x^2 + y^2 + z^2)^{1.5}}$$

(b) we have

$$\bar{E} = -\left\{\frac{\partial}{\partial x}\left[\frac{20z}{(x^2 + y^2 + z^2)^{1.5}}\right]\bar{a}_x + \frac{\partial}{\partial y}\left[\frac{20z}{(x^2 + y^2 + z^2)^{1.5}}\right]\bar{a}_y + \frac{\partial}{\partial z}\left[\frac{20z}{(x^2 + y^2 + z^2)^{1.5}}\right]\bar{a}_z\right\}$$

$$= \frac{60xz\,\bar{a}_x + 60zy\,\bar{a}_y + 20(x^2 + y^2 - 2z^2)\,\bar{a}_z}{(x^2 + y^2 + z^2)^{2.5}}$$

Problem 4.40 : (a) Given the potential field $V = 50xyz$ V in free space, find the total energy stored within the cube $0 < x, y, z < 2$. (b) What value would be obtained by assuming a uniform energy density equal to the value at the center of the cube ?

Solution :

(a)we have

$$\bar{E} = -\left[\frac{\partial}{\partial x}(50xyz)\bar{a}_x + \frac{\partial}{\partial y}(50xyz)\bar{a}_y + \frac{\partial}{\partial z}(50xyz)\bar{a}_z\right]$$

$$= -50(yz\,\bar{a}_x + xz\,\bar{a}_y + xy\,\bar{a}_z)$$

$$\therefore\ |\bar{E}| = 50\sqrt{y^2z^2 + x^2z^2 + x^2y^2}$$

$$W_E = \frac{\varepsilon_o}{2}\int |\bar{E}|^2\, dV$$

$$= \frac{\varepsilon_o}{2}\int_0^2\int_0^2\int_0^2 2500\,(y^2z^2 + x^2z^2 + x^2y^2)\,dx\,dy\,dz$$

$$= 1250\,\varepsilon_o\left\{\left[x\right]_0^2\left[\frac{y^3}{3}\right]_0^2\left[\frac{z^3}{3}\right]_o^2 + \left[\frac{x^3}{3}\right]_0^2\left[y\right]_0^2\left[\frac{z^3}{3}\right]_0^2 + \left[\frac{x^3}{3}\right]_0^2\left[\frac{y^3}{3}\right]_0^2\left[z\right]_0^2\right\}$$

$$= 1250\,\varepsilon_o\,(14.222 + 14.222 + 14.222) = 0.472\ \mu J$$

(b) Energy density $= \frac{\varepsilon_o}{2}|\bar{E}|^2 = \frac{\varepsilon_o}{2}\times 2500\,(y^2z^2 + x^2z^2 + x^2y^2)$

The center of the cube is $\left(\frac{0+2}{2}, \frac{0+2}{2}, \frac{0+2}{2}\right) = (1 ,. 1 , 1)$

Hence energy density at the center of the cube is equal to

$$1250\,\varepsilon_o\,(1 + 1 + 1) = 3750\,\varepsilon_o\ J/m^3$$

$$W_E = \int \text{Energy density} \times dV = \int_0^2\int_0^2\int_0^2 3750\,\varepsilon_o\,dx\,dy\,dz$$

$$= 3750\,\varepsilon_o\left[x\right]_0^2\left[y\right]_0^2\left[z\right]_0^2 = 0.2656\ \mu J$$

Problem 4.41 : In the region of free space where $2 < \rho < 3$, $0 < \phi < \pi/2$ and $5 < z < 7$, let $\bar{E} = (k/\rho)\,\bar{a}_\phi$. (a) Find a positive value for k so that the total energy stored is 1 µJ. (b) Show that the surfaces $\phi = 0$ and $\phi = \pi/2$ are equipotential surfaces. (c) What is the potential difference between the surfaces at $\phi = 0$ and $\phi = \pi/2$?

Solution :

(a) $|\bar{E}| = k/\rho$

$$W_E = \frac{\varepsilon_o}{2}\int |E|^2\, dV = \frac{\varepsilon_o}{2}\int_2^3\int_0^{\pi/2}\int_5^7 \frac{k^2}{\rho^2}\times \rho^2\, d\rho\, d\phi\, dz$$

$$= \frac{\varepsilon_o k^2}{2} \left[\ln \rho \right]_2^3 \left[\phi \right]_0^{\pi/2} \left[z \right]_5^7$$

$\therefore\ W_E = k^2 (5.63914 \times 10^{-12}) = 10^{-6}$

or $k^2 = 177332.0024$ or $k = 421.108$

(b) $\bar{E} = -\bar{\nabla} V = -\left[\frac{\partial V}{\partial \rho} \bar{a}_\rho + \frac{1}{\rho} \frac{\partial V}{\partial \phi} \bar{a}_\phi + \frac{\partial V}{\partial z} \bar{a}_z \right] = \left(\frac{k}{\rho} \right) \bar{a}_\phi$

$\therefore\ -\frac{1}{\rho} \frac{\partial V}{\partial \phi} \bar{a}_\phi = \frac{k}{\rho} \bar{a}_\phi$ or $\frac{1}{\rho} \frac{\partial V}{\partial \phi} = -k$ or $V = -k\phi + C$

where C is the constant of integration. Since V is a function of ϕ only, any value of ϕ represents an equipotential surface.

(c) $V_1 = -k(0) + C = C$

and $V_2 = -k\left(\frac{\pi}{2} \right) + C = -0.5 k\pi + C$

$\therefore$ Potential difference between the surfaces $\phi = 0$ and $\phi = \frac{\pi}{2}$ is equal to

$$V_1 - V_2 = C - (-0.5 k\pi + C)$$
$$= 0.5 k\pi = 0.5 \times 421.108 \times \pi = 661.475 \text{ V}$$

Problem 4.42 : A copper sphere of radius 4 cm carries a uniformly distributed total charge of 5 μC on its surface in free space. (a) Use Gauss's law to find $\bar{D}$ external to the sphere. (b) Calculate the total energy stored in the electrostatic field. (c) Use $W_E = Q^2/2C$ to calculate the capacitance of the isolated sphere.

Solution :

Consider a sphere of radius $r > 4$ cm. as shown in Fig.P4.42 . It encloses the sphere of radius 4 cm on which total charge present is 5 μC. Hence

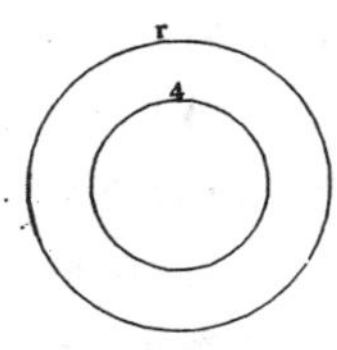

FIG.P4.42

$\Psi = Q = 5\ \mu\text{C}$

or $\bar{D} = \frac{\Psi}{4\pi r^2} \bar{a}_r = \frac{5 \times 10^{-6}}{4\pi r^2} \bar{a}_r\ \text{C/m}^2$ (for $r > 4$ cm)

$$W_E = \frac{\varepsilon_o}{2} \int |\bar{E}|^2 dv = \frac{\varepsilon_o}{2} \int \left| \frac{\bar{D}}{\varepsilon_o} \right|^2 dv = \frac{1}{2\varepsilon_o} \int |\bar{D}|^2 dv$$

$$= \frac{1}{2\,\varepsilon_o} \int_{0.04}^{\infty} \int_{0}^{\pi} \int_{0}^{2\pi} \left(\frac{5 \times 10^{-6}}{4\pi r^2} \right)^2 r^2 \sin\theta \, dr \, d\theta \, d\phi$$

$$= \frac{25 \times 10^{-12}}{2\,\varepsilon_o\,(16\pi^2)} \left[-\frac{1}{r} \right]_{0.04}^{\infty} \left[-\cos\theta \right]_0^{\pi} \left[\phi \right]_0^{2\pi} = 2.809 \text{ J/m}^3$$

[**Note** : Inside the sphere of radius 4 cm, charge is zero. Hence $\overline{E} = 0$, $W_E = 0$. Therefore the limits of r is from 0.04 to ∞ .]

(c) $W_E = \dfrac{Q^2}{2C}$ or $C = \dfrac{Q^2}{2W_E} = \dfrac{(5 \times 10^{-6})^2}{2 \times 2.809} = 4.449$ pF

Problem 4.43 : A spherical surface in free space, $r = 4$ cm, contains a uniform surface charge density of 20 μC/m^2. Determine r_A if the region $0.06 < r < r_A$ contains precisely 1 mJ of energy.

Solution :

The region $0.06 < r < r_A$ lies outside to the sphere of radius 0.04 m. Consider a sphere of radius r such that $r > 0.04$ m. The charge enclosed by this sphere

$$= 20 \times 10^{-6} \times 4\pi\,(0.04)^2 = 4.021 \times 10^{-7}\text{ C}$$

$$\therefore\ \Psi = 4.021 \times 10^{-7}\text{ C}$$

or $$\overline{D} = \frac{4.021 \times 10^{-7}}{4\pi r^2}\,\bar{a}_r = \frac{3.2 \times 10^{-8}}{r^2}\,\bar{a}_r \text{ C/m}^2$$

$$W_E = \frac{1}{2\,\varepsilon_o} \int |\overline{D}|^2 \, dv$$

$$= \frac{1}{2\,\varepsilon_o} \int_{0.06}^{r_A} \int_{0}^{\pi} \int_{0}^{2\pi} \left(\frac{3.2 \times 10^{-8}}{r^2} \right)^2 r^2 \sin\theta \, dr \, d\theta \, d\phi$$

$$= \frac{1.024 \times 10^{-15}}{2\,\varepsilon_o} \left[-\frac{1}{r} \right]_{0.06}^{r_A} \left[-\cos\theta \right]_0^{\pi} \left[\phi \right]_0^{2\pi} = 10^{-3}$$

Hence $$\left(-\frac{1}{r_A} + \frac{1}{0.06} \right) = 1.37613 \quad \text{or} \quad r_A = 0.0654 \text{ m}$$

Problem 4.44 : Two concentric conducting spheres of radii $r = 6$ cm and $b = 16$ cm have equal and opposite charges, 10^{-8} C on the inner and -10^{-8} C on the outer. Assuming $\varepsilon = \varepsilon_o$ for the region between the spheres, find : (a)

the maximum value of $|\bar{E}|$ between the spheres ; (b) the potential difference between them ; (c) the total energy stored.

Solution : (a)Consider a sphere of radius r such that $0.06 < r < 0.16$. Charge enclosed by this sphere is equal to 10^{-8} C. Hence $\Psi = 10^{-8}$ C

$$\therefore \bar{D} = \frac{10^{-8}}{4\pi r^2}\bar{a}_r \text{ C/m}^2 \text{ or } \bar{E} = \frac{10^{-8}}{4\pi\varepsilon_o r^2}\bar{a}_r \text{ or } |\bar{E}| = \frac{10^{-8}}{4\pi\varepsilon_o r^2}$$

When r is minimum, $|E|$ will be maximum.

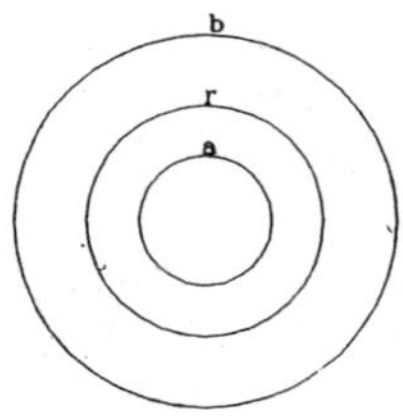

Now minimum value of r is 0.06

$$|\bar{E}| = \frac{10^{-8}}{4\pi\varepsilon_o (0.06)^2} = 24.966 \text{ kV/m}$$

(b)Using eqn. , we have

$$\bar{E} = -\bar{\nabla} V = -\frac{\partial V}{\partial r}\bar{a}_r = \frac{10^{-8}}{4\pi\varepsilon_o r^2}\bar{a}_r \text{ or } \frac{dV}{dr} = -\frac{10^{-8}}{4\pi\varepsilon_o r^2}$$

$$\therefore V = \frac{10^{-8}}{4\pi\varepsilon_o r} + C$$

$$\text{At } r = 0.06,\ V = V_1 = \frac{10^{-8}}{4\pi\varepsilon_o (0.06)} + C = 1497.957 + C$$

$$\text{At } r = 0.16,\ V = V_2 = \frac{10^{-8}}{4\pi\varepsilon_o (0.16)} + C = 561.7339 + C$$

Potential difference between the spheres $= V_1 - V_2$

$$= (1497.957 + C) - (561.7339 + C) = 936.223 \text{ V}$$

$$\text{(c) } W_E = \frac{\varepsilon_o}{2}\int |\bar{E}|^2\, dv$$

$$= \frac{\varepsilon_o}{2}\int_{0.06}^{0.16}\int_0^{\pi}\int_0^{2\pi}\left(\frac{10^{-8}}{4\pi\varepsilon_o r^2}\right)^2 r^2 \sin\theta\, dr\, d\theta\, d\phi$$

$$W_E = \frac{\varepsilon_o}{2}\times\frac{10^{-16}}{16\pi^2\varepsilon_o^2}\left[-\frac{1}{r}\right]_{0.06}^{0.16}\left[-\cos\theta\right]_0^{\pi}\left[\phi\right]_0^{2\pi} = 4.681\ \mu\text{J}$$

—X—

CHAPTER 5

Conductors, Dielectrics & Capacitance

5.1 CURRENT AND CURRENT DENSITY

Current is defined as the rate of change of charge with respect to time. Hence

$$I = \frac{dQ}{dt}$$

The unit of current is ampere (A). Current density is denoted by the letter $\bar{J}$ and is defined as the current (I) crossing an area (S) perpendicularly. It is a vector quantity with the units of A/m^2 . Hence

$$\bar{J} = \frac{I}{S}\,\bar{a}$$

where a is a unit vector perpendicular to the area S . From the current density, current can be obtained by using the equation

$$\boxed{I = \int_S \bar{J} \cdot \overline{dS}} \qquad \ldots\ldots (5.1)$$

Problem 5.1 : The vector current density is given as $\bar{J} = (4/r^2)\cos\theta\ \bar{a}_r + 20\,e^{-2r}\sin\theta\ \bar{a}_\theta - r\sin\theta\cos\phi\ \bar{a}_\phi\ A/m^2$. (a) Find $\bar{J}$ at $r = 3, \theta = 0,$. $\phi = \pi$. (b) Find the total current passing through the spherical cap $r = 3,$ $0 < \theta < 20^o, 0 < \phi < 2\pi$ in the $\bar{a}_r$ direction.

Solution :

(a) Substituting $r = 3, \theta = 0, \phi = \pi$ in the expression of $\bar{J}$, we have

$$\bar{J} = (4/9)\cos 0\, \bar{a}_r + 20\, e^{-6} \sin 0\, \bar{a}_\theta - 3 \sin 0 \cos \pi\, \bar{a}_\phi = 0.444\, \bar{a}_r \text{ A/m}^2$$

(b) To the surface $r = 3$, $\bar{a}_r$ is perpendicular. Hence $\overline{dS} = r^2 \sin\theta\, d\theta\, d\phi\, \bar{a}_r$

$$\therefore I = \int \bar{J} \cdot \overline{dS} = \int_0^{20^\circ} \int_0^{2\pi} (4/r^2) \cos\theta \times \underset{(r=3)}{r^2} \sin\theta\, d\theta\, d\phi$$

$$= 2 \left[\frac{-\cos 2\theta}{2} \right]_0^{20^\circ} \left[\phi \right]_0^{2\pi} = 1.47 \text{ A}$$

5.2 RELATIONSHIP BETWEEN CURRENT DENSITY ($\bar{J}$), VOLUME CHARGE DENSITY (ρ_v) AND ITS VELOCITY ($\bar{v}$)

Fig. 5.1 shows a rectangular box with edges parallel to coordinate axes. Inside the box, volume charge density ρ_v is present and is moving in the x direction, i.e. , only x component of velocity is present. In a time interval Δt , the element of charge has moved a distance Δx as shown in Fig.5.1 . The charge ΔQ which is moved through a reference plane perpendicular to the direction of motion is given by

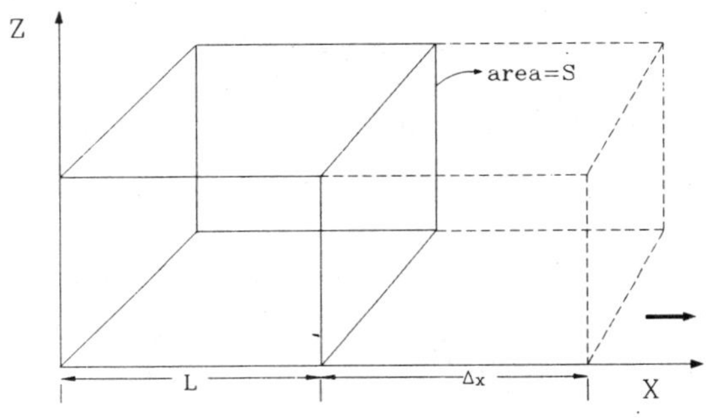

FIG.5.1

$$\Delta Q = \rho_v \times \text{volume} = \rho_v \times \Delta x \times S$$

Divide both sides of this equation by Δt as the movement of charge takes place within time Δt .

$$\frac{\Delta Q}{\Delta t} = \rho_v \times S \times \frac{\Delta x}{\Delta t}$$

Now ,

$$I = \frac{\Delta Q}{\Delta t} \quad \text{and} \quad v_x = \frac{\Delta x}{\Delta t}$$

Hence,

$$I = \rho_v \times S \times v_x$$

Hence the current density is

$$J_x = \frac{I}{S} = \rho_v \, v_x$$

or in general

$$\boxed{\bar{\mathbf{J}} = \rho_v \, \bar{\mathbf{v}}} \qquad \text{........(5.2)}$$

5.3 THE CONTINUITY EQUATION FOR CURRENT :

The current through the closed surface is given by

$$I = \oint_S \bar{\mathrm{J}} \cdot \overline{\mathrm{dS}}$$

According to law of conservation of charge, the outward flow of positive charge must be balanced by a decrease of positive charge within the closed surface. If Q_i is the charge inside the closed surface, then the rate of decrease is $- dQ_i / dt$. Hence

$$\boxed{I = \oint_S \bar{\mathbf{J}} \cdot \overline{\mathbf{dS}} = -\frac{dQ_i}{dt}} \qquad \text{........(5.3)}$$

Above equation is the integral form of continuity equation. According to divergence theorem

$$\oint_S \bar{\mathrm{J}} \cdot \overline{\mathrm{dS}} = \int_{vol} (\bar{\nabla} \cdot \bar{\mathrm{J}}) \, dv$$

Also

$$-\frac{d}{dt}(Q_i) = -\frac{d}{dt}\int dQ_i = -\frac{d}{dt}\int_{vol} \rho_v \, dv = -\frac{d}{dt}\int_{vol} \rho_v \, dv$$

Hence $\displaystyle\int_{vol} (\bar{\nabla} \cdot \bar{\mathrm{J}}) \, dv = -\frac{d}{dt}\int_{vol} \rho_v \, dv = -\int_{vol} \frac{\partial \rho_v}{\partial t} \, dv$

Hence we have

$$\boxed{\bar{\nabla} \cdot \bar{\mathbf{J}} = -\frac{\partial \rho_v}{\partial t}} \qquad \text{......(5.4)}$$

which is the differential or point form of continuity equation. This equation indicates that the current, or charge per second, diverging from a small volume per unit volume is equal to the time rate of decrease of charge per unit volume at every point.

Problem 5.2 : Assume that an electron beam carries a total current of $-500\ \mu A$ in the $\bar{a}_z$ direction, and has a current density J_z that is not a function of ρ or ϕ in the region $0 \leq \rho < 10^{-4}$ m and is zero for $\rho > 10^{-4}$. m. If the electron velocities are given by $v_z = 8 \times 10^7 z$ m/s, calculate ρ_v at $\rho = 0$ and $z =$: (a) 1 mm ; (b) 2 cm ; (c) 1 m.

Solution :

Using the equation $\bar{J} = \rho_v \bar{v}$, we have $J_z = \rho_v v_z$ or $\rho_v = J_z / v_z$.

$$\text{Now } J_z = \frac{-500 \times 10^{-6}}{\pi (10^{-4})^2} = -15915.494 \text{ A/m}^2$$

$$\therefore \rho_v = \frac{-15915.493}{8 \times 10^7 z}$$

$$\text{(a) At } \rho = 0, z = 1\text{mm}, \rho_v = \frac{-15915.4931}{8 \times 10^7 \times 10^{-3}} = -0.1989 \text{ C/m}^3$$

$$\text{(b) At } \rho = 0, z = 2 \text{ cm}, \rho_v = \frac{-15915.4931}{8 \times 10^7 \times 2 \times 10^2} = -9.947 \text{ mC/m}^3$$

$$\text{(c) At } \rho = 0, z = 1 \text{ m}, \rho_v = \frac{-15915.4931}{8 \times 10^7 \times 1} = -198.944\ \mu\text{C/m}^3$$

5.4 CONDUCTORS

In conductors valence electrons (conduction or free electrons) are present. When an electric field is applied, then these electrons start moving which constitute an electric current.

When a charge Q, is placed in an electric field $\bar{E}$, it experiences a force given by $\bar{F} = Q\bar{E}$.If this charge is an electron, then $Q = -e$ so that $\bar{F} = -e\bar{E}$. Due to negative sign of electronic charge, it is clear that the force acting on the electron is in opposite direction to that of electric field intensity $\bar{E}$. If there is no opposition or resistance to the motion of electron, then the velocity of electron will increase indefinitely with the time t provided the electric field $\overline{E}$ is constant.

However in conductors, the electron collides repeatedly with other electrons and thermally excited crystalline lattice structure. As a result of this, electron looses some of its energy and causes random changes in its direction of motion. If $\overline{E}$ is constant and the medium is homogeneous, the net effect of the collisions is to restrain the charged particle to a constant average velocity which is linearly related to the electric field intensity by the mobility of the electron in the given material. Mobility is denoted by μ (mu). Hence

$$\boxed{\overline{V}_d = -\mu_e \overline{\mathbf{E}}} \qquad \ldots\ldots (5.5)$$

where μ_e = mobility of electron. The unit of mobility is m^2/v- sec. The negative sign in the above equation indicates that the velocity (motion) of the electron is in opposite direction to the electric field intensity $\overline{E}$.

If ρ_e indicates free electron charge density (as the charge on the electron is a negative number, ρ_e is a negative number), then using equation (5.2) we have

$$\overline{J} = \rho_v \overline{v} = \rho_e \overline{v}_d = \rho_e(-\mu_e \overline{E})$$

Hence

$$\boxed{\overline{\mathbf{J}} = -\rho_e \mu_e \overline{\mathbf{E}}} \qquad \ldots\ldots (5.6)$$

From the above relationship, for a metallic conductor, the current density $\overline{J}$ is proportional to electric field intensity $\overline{E}$ and the constant of proportionality is known as conductivity (σ) . Hence

$$\boxed{\overline{\mathbf{J}} = \sigma \overline{\mathbf{E}}} \qquad \ldots\ldots (5.7)$$

and

$$\boxed{\sigma = -\rho_e \mu_e} \qquad \ldots\ldots (5.8)$$

The unit of conductivity is mho/m.

Few relations from basic electricity :

Ohm's law

$$\boxed{V = IR} \qquad \text{........(5.9.a)}$$

Resistance

$$\boxed{R = \frac{L}{\sigma S}} \qquad \text{......(5.9.b)}$$

where L = length of conductor, σ = conductivity , and S = cross sectional area.

Power consumed

$$\boxed{P_{av} = I^2 R} \qquad \text{......(5.9.c)}$$

where I is the rms value of the current.

5.5 PROPERTIES OF PERFECT CONDUCTORS

In electrostatic fields, following principles are applicable to conductors :

(1) Within a conductor, charge or charge density (ρ_v C/m^3) is zero and a surface charge density (ρ_S C/m^3) is present on the outer surface of the conductor.

(2) The electric field intensity within the conductor is zero.

(3) The tangential component of external electric field intensity as well as electric flux density is zero.

(4) The conductor surface is an equipotential surface.

(5) The normal component of electric flux density is equal to surface charge density on the conductor surface.

Problem 5.3 : Find the magnitude of the electric field intensity in a sample of silver having $\sigma = 6.17 \times 10^7$ mho/m and $\mu_e = 0.0056$ m^2/V · s if : (a) the drift velocity is 1 mm / s ; (b) the current density is 10^7 A/m^2 ; (c) the sample is a cube, 3mm on a side, carrying a total current of 80A ; (d) the sample is a cube, 3 mm on a side, having a potential difference of 0.5 mV between opposite faces.

Solution :

(a) $\overline{V}_d = -\mu_o \overline{E}. \quad \therefore \overline{E} = -\dfrac{\overline{V}_d}{\mu_e}$

$$|\overline{E}| = \frac{10^{-3}}{0.0056} = 0.1786 \text{ V/m}$$

(b) $\overline{J} = \sigma \overline{E}. \quad \therefore \overline{E} = \dfrac{\overline{J}}{\sigma}$

$$|\overline{E}| = \frac{10^{7}}{6.17\times 10^{7}} = 0.1621 \text{ V/m}$$

(c) $J = \dfrac{80}{\text{Area of one face of a cube}} = \dfrac{80}{(0.003)^2} = \dfrac{80}{9\times 10^{-6}} \text{ A/m}^2$

$$\therefore |E| = \frac{80}{9\times 10^{-16}\times 6.17\times 10^{7}} = 0.1441 \text{ V/m}.$$

(d) $|E| = \dfrac{0.5\times 10^{-3}}{3\times 10^{-3}} = 0.1667 \text{ V/m}$

Problem 5.4 : An aluminium conductor is 1000 ft long and has a circular cross section with a diameter of 0.8 in. If there is a dc voltage of 1.2 V between the ends, find : (a) the current density ; (b) the current ; (c) the power dissipated, using your vast knowledge of circuit theory.

Solution :

For Aluminium, $\sigma = 3.82 \times 10^7$ mho/m

(a) Length of conductor = 1000 ft = 1000 × 0.3048 m = 304.8 m

$$E = \frac{1.2}{304.8} = 3.937\times 10^{-3} \text{ V/m}$$

$$J = \sigma E = 3.82\times 10^{7}\times 3.937\times 10^{-3} = 1.504\times 10^{5} \text{ A/m}^2$$

(b) $R = \dfrac{L}{\sigma S}$

where S = Cross sectional area of the conductor = πr^2

r = radius of conductor = 0.4 × 0.0254 m = 0.01016 m

$$\therefore\ R = \frac{304.8}{3.82\times 10^7 \times \pi \times (0.01016)^2} = 0.024604\ \Omega$$

$$I = \frac{V}{R} = \frac{1.2}{0.024604} = 48.772\ \text{A}$$

Second Method $I = J\times S = 1.504\times 10^5 \times \pi\,(0.01016)^2 = 48.774$ A

(c) Power dissipated $= I^2 R = (48.772)^2 \times 0.024604 = 58.526$ W

5.6 THE BOUNDARY CONDITIONS AT A CONDUCTOR FREE SPACE BOUNDARY

5.6.1 Tangential boundary conditions

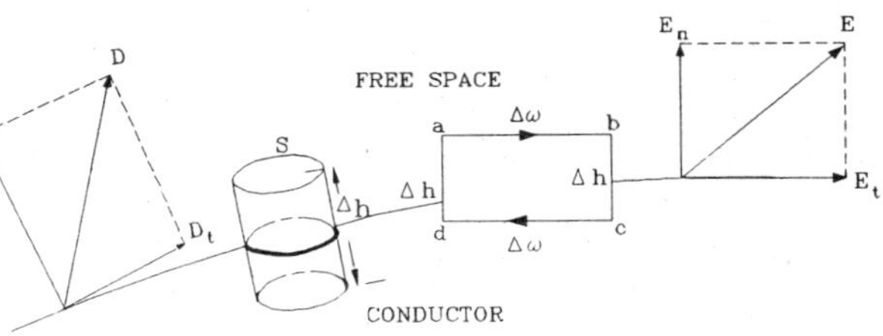

FIG.5.2

These conditions maybe determined by applying $\oint \overline{E}\cdot\overline{dL} = 0$ around the small closed path $a - b - c - d - a$ as shown in Fig. 5.2.

$$\therefore \int_a^b \overline{E}\cdot\overline{dL} + \int_b^c \overline{E}\cdot\overline{dL} + \int_c^d \overline{E}\cdot\overline{dL} + \int_d^a \overline{E}\cdot\overline{dL} = 0$$

As the electric field intensity within the conductor is zero,we have

$$\int_c^d \overline{E}\cdot\overline{dL} = 0$$

as the path $c - d$ lies inside the conductor.

$$\int_a^b \overline{E}\cdot\overline{dL} = \Delta\,\omega\,E_t$$

$$\int_b^c \overline{E} \cdot \overline{dL} = -\frac{\Delta h}{2} E_N \text{, at } b$$

$$\int_d^a \overline{E} \cdot \overline{dL} = \frac{\Delta h}{2} E_N \text{, at } a$$

Hence,

$$\Delta \omega E_t - \frac{\Delta h}{2} E_N \text{, at } a + \frac{\Delta h}{2} E_N \text{, at } a = 0$$

As Δh approaches to zero, path $a - b$ lies on the conductor surface. Hence

$$\Delta w E_t = 0$$

or

$$\boxed{\boldsymbol{E_t = 0}} \quad \text{.........(5.10.a)}$$

and

$$\boxed{\boldsymbol{D_t = 0}} \quad \text{...........(5.10.b)}$$

5.6.2 Normal boundary conditions

The determine normal boundary conditions, apply $\oint \overline{D} \cdot \overline{dS} = Q$ to the cylinder of Fig. 5.2. Hence

$$\int_{top} \overline{D} \cdot \overline{dS} + \int_{bottom} \overline{D} \cdot \overline{dS} + \int_{sides} \overline{D} \cdot \overline{dS} = 0$$

As the electric flux density is zero within the conductor,

$$\int_{bottom} \overline{D} \cdot \overline{dS} = 0$$

As the tangential component of electric flux density is zero,

$$\int_{sides} \overline{D} \cdot \overline{dS} = 0$$

$$\int_{top} \overline{D} \cdot \overline{dS} = D_N \Delta S \quad \text{and} \quad Q = \rho_s \Delta S$$

Hence

$$D_N \Delta S = \rho_s \Delta S$$

or

$$\boxed{D_N = \rho_s} \qquad \text{.........(5.10.c)}$$

As $D_N = \varepsilon_o E_N$, we have $\varepsilon_o E_N = \rho_s$

or

$$\boxed{E_N = \rho_s / \varepsilon_o} \qquad \text{.........(5.10.d)}$$

Problem 5.5 : A potential field is given as $V = 100\, e^{-5x} \sin 3y \cos 4z$ V. Let point P (0.1 , π/12 , π/24) be located at a conductor – free space boundary. At point P, find the magnitude of : (a) V ; (b) $\bar{E}$; (c) E_N ; (d) E_t ; (e) ρ_s.

Solution :

(a) $V(\text{ at } P) = 100\, e^{-0.5} \sin\left(\frac{3\pi}{12}\right) \cos\left(\frac{4\pi}{24}\right) = 37.142$ V

(b)Using equation (4.*****),we have

$$\bar{E} = -\bar{\nabla} V = 500\, e^{-5x} \sin 3y \cos 4z\, \bar{a}_x - 300\, e^{-5x} \cos 3y \cos 4z\, \bar{a}_y + 400\, e^{-5x} \sin 3y \sin 4z\, \bar{a}_z$$

Substituting the coordinates of point P in the above $\bar{E}$,we have

$$\bar{E}(\text{ at } P) = 185.711\, \bar{a}_x - 111.427\, \bar{a}_y + 85.776\, \bar{a}_z \text{ V/m}$$

$$\therefore |\bar{E}| = \sqrt{(185.711)^2 + (-111.427)^2 + (85.776)^2} = 232.942 \text{ V/m.}$$

(c) Electric field intensity is always normal to the conductor surface.

Hence $|E_N| = 232.942$ V/m

(d) Tangential component of electricfield intensity is always zero.

Hence $|E_t| = 0$

(e) $\rho_s = |D_N| = |\varepsilon_o \varepsilon_r E_N| = \varepsilon_o \times 232.942 = 2.062 \text{ nC/m}^2$

5.7 THE METHOD OF IMAGES

Fig.5.3(a) shows an electric dipole with charges Q and $-Q$ located on the z axis.In Fig. 5.3 (b), is shown the electric field map for the charge configuration of Fig.5.3 (a) . At any point P, the potential due to the dipole is given by

$$V = \frac{Q}{4\pi\varepsilon_o}\left(\frac{1}{R_1} - \frac{1}{R_2}\right)$$

If point P lies in xy plane i.e. $z = 0$ plane, then R_1 will be equal to R_2, so that $V = 0$. Hence xy plane or $z = 0$ plane is an equipotential surface as the potential at any point on this plane is constant (equal to 0). The xy plane may be replaced by a thin conducting plane which is infinite in extent. The conductor is an equipotential surface at a potential equal to zero. Thus the charge configuration of Fig. 5.3(a) if replaced by a single charge Q and conducting plane as shown in Fig. 5.3 (c), the fields above the plane remains unchanged and below the plane, all the fields are zero as shown in Fig. 5.3(d). However the fields below the plane can be obtained by placing a single negative charge ($-Q$) below the conducting plane.

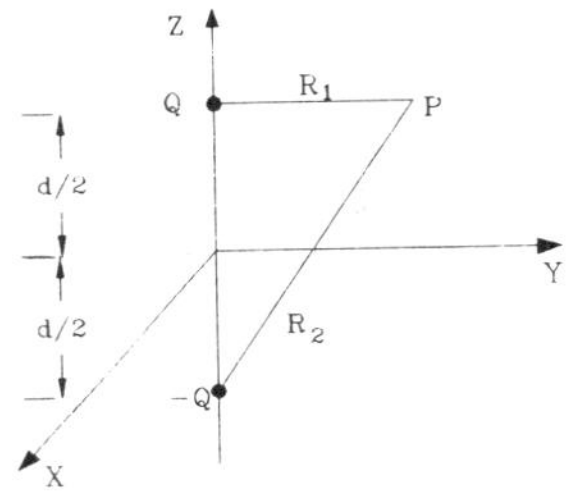

FIG. 5.3 (a)

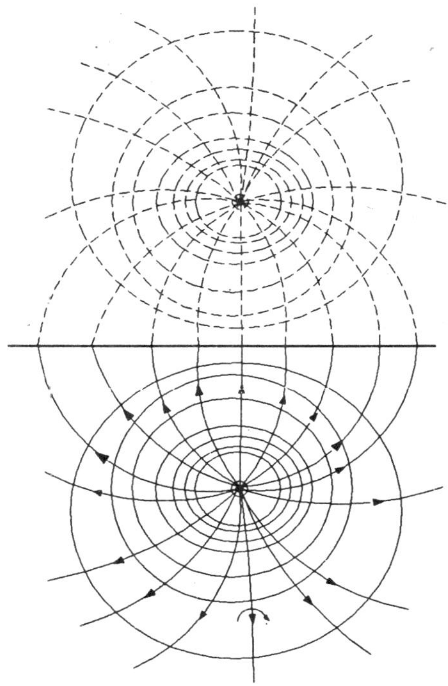

FIG. 5.3 (b)

Now let us consider reverse process. Fig. 5.3 (c) shows a single charge Q above a perfectly conducting plane. Suppose we wish to determine the fields due to this charge at any point. The field map

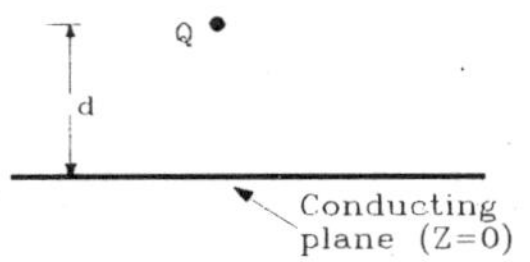

FIG. 5.3 (c)

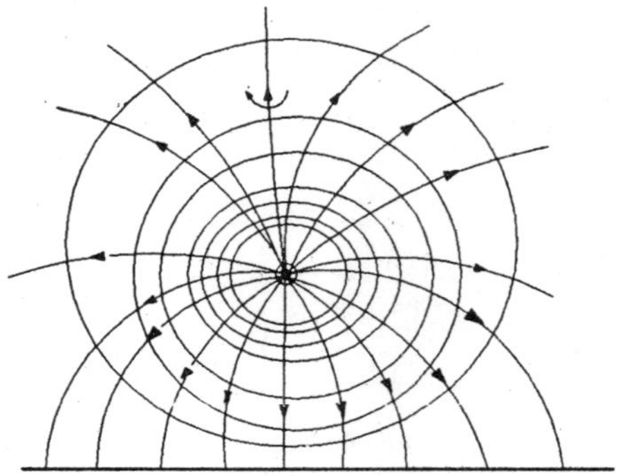

FIG. 5.3 (d)

of the charge configuration of Fig.5.3(c) will be as shown in Fig.5.3(d). The perfectly conducting plane of Fig. 5.3(c) can be removed by placing the charge

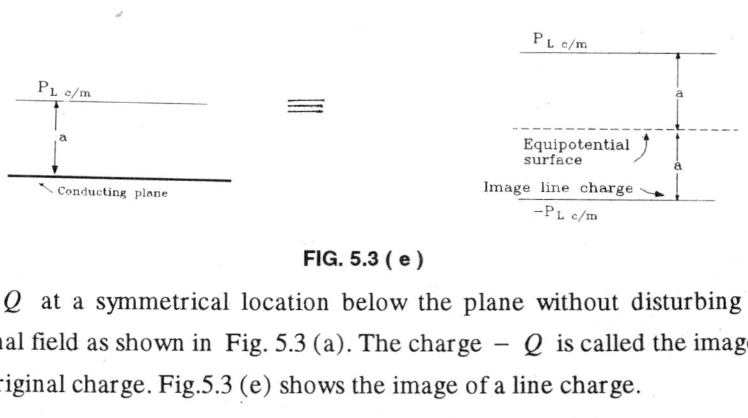

FIG. 5.3 (e)

of $-Q$ at a symmetrical location below the plane without disturbing the original field as shown in Fig. 5.3 (a). The charge $-Q$ is called the image of the original charge. Fig.5.3 (e) shows the image of a line charge.

Problem 5.6 : A point charge of 25 nC is located in free space at P (2, –3, 5), and a perfectly conducting plane is at $z = 2$. Find : (a) V at (3, 2, 4) ; (b) $\overline{\mathbf{E}}$ at (3, 2, 4) ; (c) ρ_S at (3, 2, 2).

Solution :

Point P is above the conducting plane because z coordinate of point P is greater than $z = 2$. If we drop a perpendicular from point P on the plane $z = 2$, the coordinates of the foot of the perpendicular will be (2, –3, 2). Hence the distance of point P from the $z = 2$ plane is

$$\sqrt{(2-2)^2 + (-3+3)^2 + (5-2)^2} = 3$$

Consider a point P' which is the mirror image of point P. The distance of point P' from the plane $z = 2$ will be 3. Hence the coordinates of point P' be (2, –3, –1). If a perpendicular is dropped from point P' on the plane $z = 2$, the coordinates of the foot of perpendicular will be (2, –3, 2). At the point P', the charge of – 25 nC (which is the image of 25 nC) is located.

(a) V at (3, 2, 4) = V due to 25 nC charge + V due to – 25 nC charge

$$= \frac{25 \times 10^{-9}}{4 \pi \varepsilon_o \sqrt{(3-2)^2 + (2+3)^2 + (4-5)^2}} + \frac{-25 \times 10^{-9}}{4 \pi \varepsilon_o \sqrt{(3-2)^2 + (2+3)^2 + (4+1)^2}}$$

$$= 11.7789 \text{ V}$$

(b) $\bar{E}$ at (3, 2, 4) = $\bar{E}$ due to 25 nC charge + $\bar{E}$ due to – 25 nC charge

$$= \frac{25 \times 10^{-9}}{4 \pi \varepsilon_o} \left\{ \frac{(3-2)\,\bar{a}_x + (2+3)\,\bar{a}_y + (4-5)\,\bar{a}_z}{[(3-2)^2 + (2+3)^2 + (4-5)^2]^{3/2}} \right\} + \frac{-25 \times 10^{-9}}{4 \pi \varepsilon_o} \left\{ \frac{(3-2)\,\bar{a}_x + (2+3)\,\bar{a}_y + (4+1)\,\bar{a}_z)}{[(3-2)^2 + (2+3)^2 + (4+1)^2]^{3/2}} \right\}$$

$$= 1.602\,\bar{a}_x + 8.008\,\bar{a}_y - 1.602\,\bar{a}_z - 0.617\,\bar{a}_x - 3.085\,\bar{a}_y - 3.085\,\bar{a}_z$$

$$= 0.985\,\bar{a}_x + 4.923\,\bar{a}_y - 4.687\,\bar{a}_z$$

(c) $\bar{E}$ at (3 , 2 , 2) =

$$\frac{25 \times 10^{-9}}{4 \pi \varepsilon_o} \left\{ \frac{(3-2)\,\bar{a}_x + (2+3)\,\bar{a}_y + (2-5)\,\bar{a}_z)}{[(3-2)^2 + (2+3)^2 + (2-5)^2]^{3/2}} \right\} + \frac{-25 \times 10^{-9}}{4 \pi \varepsilon_o} \left\{ \frac{(3-2)\,\bar{a}_x + (2+3)\,\bar{a}_y + (2+1)\,\bar{a}_z}{[(3-2)^2 + (2+3)^2 + (2+1)^2]^{3/2}} \right\}$$

$$= 1.085\,\bar{a}_x + 5.426\,\bar{a}_y - 3.255\,\bar{a}_z - 1.085\,\bar{a}_x - 5.426\,\bar{a}_y - 3.255\,\bar{a}_z$$

$$= -6.51\,\bar{a}_z \text{ V/m}$$

$$\therefore \bar{D} = \varepsilon_o \bar{E} = -6.51\,\varepsilon_o\,\bar{a}_z \text{ C/m}^2$$

$\therefore \ \rho_s = |\overline{D}_N| = |\overline{D}| = 6.51\,\varepsilon_o = 57.639 \text{ pC/m}^2$

ρ_s will be negative as $\overline{D}$ is directed towards the conducting plane.

Hence $\rho_s = -57.639 \text{ pC/m}^2$

5.8 SEMICONDUCTORS

In semiconductors, two types of current carries are present. These are electrons and holes. Both carriers move in an electric field, and they move in opposite directions. Hence the total current is the sum of currents due to electrons and holes. The conductivity is therefore a function of both hole and electron concentrations and mobilities.

$$\sigma = \sigma_e + \sigma_h$$

where

$$\sigma_e = -\rho_e \mu_e \quad \text{and} \quad \sigma_h = \rho_h \mu_h$$

where ρ_e and μ_e are electron charge density (a negative number) and electron mobility respectively whereas ρ_h and μ_h are the corresponding values for holes. For pure, or intrinsic, silicon the electron and hole mobilities are 0.12 and 0.025, respectively, while for germanium, the mobilities are, respectively, 0.36 and 0.17. The mobilities listed above are given for a temperature of 300 K.

Problem 5.7 : Using the values given in this section for the electron and hole mobilities in germanium at 300 K, and assuming hole and electron concentrations of 2.7×10^{19} m^{-3}, find : (a) the component of the conductivity due to holes ; (b) the component of the conductivity due to electrons ; (c) the conductivity.

Solution :

(a) Component of the conductivity due to holes = $\rho_h \mu_h$

$$= (2.7 \times 10^{19} \times 1.6 \times 10^{-19})(0.17)$$

$$= 0.7344 \text{ mho/m}$$

(b) Component of the conductivity due to electrons = $-\rho_e \mu_e$

$$= -2.7 \times 10^{19} (-1.6 \times 10^{-19})(0.36)$$

$\sigma = 1.5552$ mho/m

(c) $\sigma = \sigma_h + \sigma_e = 0.7344 + 1.5552 = 2.2896$ mho/m .

5.9 DIELECTRIC MATERIALS

Two charges equal in magnitude but opposite in sign and separated by a distance constitute an electric dipole. Fig. 5.4 shows an electric dipole with charges $+Q$ and $-Q$ and separated by a distance d. The dipole moment is denoted by the letter $\bar{p}$ and is defined as $\bar{p} = Q\,\bar{d}$

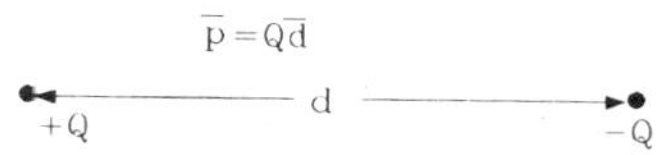

FIG.5.4

where $\bar{d}$ is a vector directed from negative charge towards the positive charge. The unit of dipole moment is C – m..

Suppose N is the no. of dipoles per m^3 . Then in a volume Δv , the total number of dipoles will be $N \Delta v$. The polarization is defined as the dipole moment per unit volume. It is denoted by $\bar{P}$ and the unit is $\dfrac{C - m}{m^3} = C/m^2$

Hence

$$\bar{P} = \lim_{\Delta v \to 0} \frac{\sum_{i=1}^{N \Delta v} \bar{p}_i}{\Delta v}$$

Relationship between polarization($\bar{P}$) and ($\bar{E}$) is

$$\boxed{\bar{\mathbf{P}} = \chi_e \varepsilon_o \bar{\mathbf{E}}} \qquad \ldots\ldots (5.11\,.\,a)$$

where χ_e is electric suspectibility. Also relationship between electric flux density ($\bar{D}$) and electric field intensity ($\bar{E}$) is

$$\boxed{\bar{\mathbf{D}} = \varepsilon_o \varepsilon_R \bar{\mathbf{E}}} \qquad \ldots\ldots (5.11\,.\,b)$$

Relationship between electric field intensity($\bar{E}$) , electric field density($\bar{D}$) , and polarization($\bar{P}$) is

$$\boxed{\bar{\mathbf{D}} = \varepsilon_o \bar{\mathbf{E}} + \bar{\mathbf{P}}} \qquad \ldots\ldots (5.11\,.\,c)$$

Relationship between relative permittivity (ε_R) and electric suspectibility (χ_e) is

$$\boxed{\varepsilon_R = \chi_e + 1} \qquad \ldots\ldots (5.11.d)$$

Bound charge density(ρ_b) is given by

$$\boxed{\rho_b = -\overline{\nabla} \cdot \overline{\mathbf{P}}} \qquad \ldots\ldots (5.11.e)$$

Total charge density (ρ_T) is given by

$$\boxed{\rho_T = \overline{\nabla} \cdot \varepsilon_o \overline{\mathbf{E}}} \qquad \ldots\ldots (5.11.f)$$

Free charge density(ρ_v) is given by

$$\boxed{\rho_v = \overline{\nabla} \cdot \overline{\mathbf{D}}} \qquad \ldots\ldots (5.11.g)$$

Problem 5.8 : A certain homogeneous slab of lossless dielectric material is characterized by an electric susceptibility of 0.12 and carries a uniform electric flux density within it of 1.6 nC / m². Find : (a) *E* ; (b) *P* ; (c) the average dipole moment if there are 2 × 10¹⁹ dipoles per cubic meter ; (d) the voltage between two equipotentials 1 in. apart.

Solution :

(a) $\chi_e = \varepsilon_r - 1 = 0.12$. Hence $\varepsilon_r = 1.12$

$$D = \varepsilon_o \varepsilon_r E \text{ or } E = \frac{D}{\varepsilon_o \varepsilon_r} = \frac{1.6 \times 10^{-9}}{(1.12\, \varepsilon_o)} = 161.348 \text{ V/m}$$

(b) $P = \chi_e \varepsilon_o E = 0.12 \times 161.348\, \varepsilon_o = 171.429 \text{ pC/m}^2$

$$\text{(c) Average dipole moment} = \frac{171.429 \times 10^{-12}}{2 \times 10^{19}}$$

$$= 8.57145 \times 10^{-30} \text{ C} - \text{m}$$

(d) $E \times d = 161.3 \times (0.0254 \text{ m}) = 4.097 \text{ V}$

5.10 BOUNDARY CONDITIONS FOR PERFECT DIELECTRIC MATERIALS

5.10.1 Boundary Conditions for tangential components

These conditions are obtained by applying $\oint \overline{E} \cdot \overline{dL} = 0$ to a small rectangular closed path as shown in Fig. 5.5.

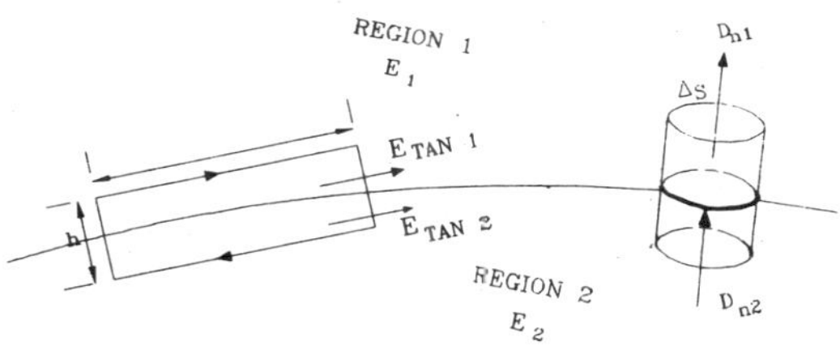

FIG. 5.5

$$\therefore \int_a^b \overline{E} \cdot \overline{dL} + \int_b^c \overline{E} \cdot \overline{dL} + \int_c^d \overline{E} \cdot \overline{dL} + \int_d^a \overline{E} \cdot \overline{dL} = 0$$

$$\int_a^b \overline{E} \cdot \overline{dL} = \Delta\omega\, E_{\tan 1} \quad \text{and} \quad \int_c^d \overline{E} \cdot \overline{dL} = -\Delta\omega\, E_{\tan 2}$$

As $\Delta h \to 0$, $\int_b^c \overline{E} \cdot \overline{dL}$ and $\int_d^a \overline{E} \cdot \overline{dL}$ are equal to zero.

Hence

$$\Delta\omega\, E_{\tan 1} - \Delta\omega\, E_{\tan 2} = 0$$

or

$$\boxed{E_{\tan 1} = E_{\tan 2}} \qquad \text{.....(5.12.a)}$$

Now $D_{\tan 1} = \varepsilon_o \varepsilon_{r1} E_{\tan 1}$ and $D_{\tan 2} = \varepsilon_o \varepsilon_{r2} E_{\tan 2}$. Hence

$$\frac{D_{\tan 2}}{\varepsilon_o \varepsilon_{r1}} = \frac{D_{\tan 2}}{\varepsilon_o \varepsilon_{r2}}$$

or

$$\boxed{\frac{D_{\tan 1}}{D_{\tan 2}} = \frac{\varepsilon_{r1}}{\varepsilon_{r2}}} \qquad \text{.....(5.12. b)}$$

5.10.2 Boundary conditions for Normal components

These condition are obtained by applying $\oint \overline{D} \cdot \overline{dS} = Q$ to the cylinder of Fig. 5.5.

$$\text{L.H.S.} = \oint \overline{D} \cdot \overline{dS} = \int_{top} \overline{D} \cdot \overline{dS} + \int_{bottom} \overline{D} \cdot \overline{dS} + \int_{sides} \overline{D} \cdot \overline{dS}$$

Now

$$\int_{top} \overline{D} \cdot \overline{dS} = D_{N1} \, \Delta S$$

and

$$\int_{bottom} \overline{D} \cdot \overline{dS} = - D_{N2} \, \Delta S$$

As the sides are assumed to be very short,we have

$$\int_{sides} \overline{D} \cdot \overline{dS} = 0$$

Hence

$$\text{L.H.S.} = D_{N1} \, \Delta S - D_{N2} \, \Delta S$$

and

$$\text{RHS} = Q = \rho_s \, \Delta S$$

where ρ_s is the surface charge density on the interface. Hence

$$D_{N1} \, \Delta S - D_{N2} \, \Delta S = \rho_s \, \Delta S$$

or

$$\boxed{D_{N1} - D_{N2} = \rho_s} \qquad \text{......(5.12. c)}$$

If $\rho_s = 0$, then

$$\boxed{D_{N1} = D_{N2}} \qquad \text{......(5.12.d)}$$

Now $D_{N1} = \varepsilon_o \varepsilon_{r1} E_{N1}$ and $D_{N2} = \varepsilon_o \varepsilon_{r2} E_{N2}$. Hence

$$\varepsilon_o \varepsilon_{r1} E_{N1} = \varepsilon_o \varepsilon_{r2} E_{N2}$$

or

$$\boxed{\frac{E_{N1}}{E_{N2}} = \frac{\varepsilon_{r2}}{\varepsilon_{r1}}} \qquad \ldots\ldots (5.12.e)$$

5.10.3 At the boundary of two perfect dielectric materials, ε_1 and ε_2, D_1 is incident at an angle θ_1 with respect to normal to the boundary surface. It can be proved that

$$D_2 = D_1 \sqrt{\cos^2\theta_1 + \left(\frac{\varepsilon_2}{\varepsilon_1}\right)^2 \sin^2\theta_1}$$

and

$$E_2 = E_1 \sqrt{\sin^2\theta_1 + \left(\frac{\varepsilon_1}{\varepsilon_2}\right)^2 \cos^2\theta_1}$$

In Fig.5.6, $\bar{a}_N$ is a unit vector perpendicular to the dielectric interface and directed from region 1 to region 2. Let us assume $\rho_s = 0$. Then $D_{N1} = D_{N2}$

and $\frac{D_{\tan 1}}{D_{\tan 2}} = \frac{\varepsilon_{r1}}{\varepsilon_{r2}} = \frac{\varepsilon_1}{\varepsilon_2}$

or $D_{\tan 2} = \frac{\varepsilon_2}{\varepsilon_1} D_{\tan 1}$

From Fig. 5.6.

$D_{N1} = D_1 \cos\theta_1$ and $D_{\tan 1} = D_1 \sin\theta_1$

where $D_1 = |\bar{D}_1|$

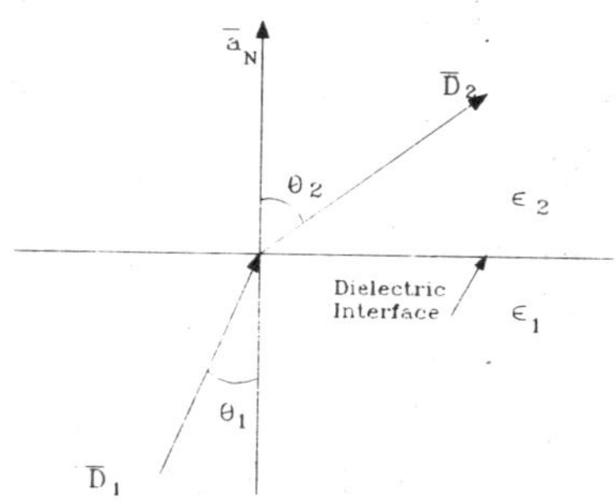

FIG. 5.6

Hence $D_{N2} = D_1 \cos\theta_1$ and $D_{\tan 2} = \frac{\varepsilon_2}{\varepsilon_1} D_1 \sin\theta_1$

Now,

$$D_2 = \sqrt{(D_{N2})^2 + (D_{\tan 2})^2} = \sqrt{(D_1 \cos\theta_1)^2 + \left(\frac{\varepsilon_2}{\varepsilon_1} D_1 \sin\theta_1\right)^2}$$

$$= D_1 \sqrt{\cos^2\theta_1 + \left(\frac{\varepsilon_2}{\varepsilon_1}\right)^2 \sin^2\theta_2}$$

Hence

$$\varepsilon_2 E_2 = \varepsilon_1 E_1 \sqrt{\cos^2\theta_1 + \left(\frac{\varepsilon_2}{\varepsilon_1}\right)^2 \sin^2\theta_1}$$

or

$$E_2 = E_1 \sqrt{\sin^2\theta_1 + \left(\frac{\varepsilon_1}{\varepsilon_2}\right)^2 \cos^2\theta_1}$$

Problem 5 .9 : The region $y < 0$ contains a dielectric material for which $\varepsilon_{R1} = 2.5$, while the region $y > 0$ is charaterized by $\varepsilon_{R2} = 4$. Let $\bar{E}_1 = -30\,\bar{a}_x + 50\,\bar{a}_y + 70\,\bar{a}_z$ V/m , and find : (a) E_{N1} ; (b) $\bar{E}_{t1}$; (c) E_{t1} ; (d) E_1 ; (e) θ_1 .

Solution :

(a) The plane $y = 0$ separates two dielectric materials. To this plane $\pm\,\bar{a}_y$ are perpendicular. Hence $\bar{E}_{N1}$ is that part of $\bar{E}_1$ which contains $\pm\,\bar{a}_y$ .Hence $\bar{E}_{N1} = 50\,\bar{a}_y$ or $E_{N1} = 50$ V/m .

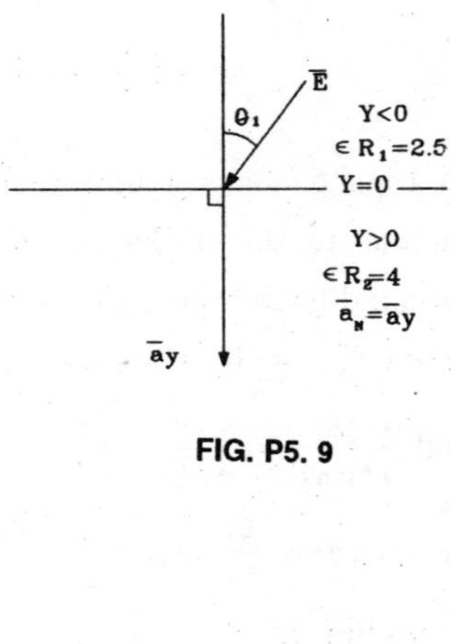

FIG. P5. 9

(b) $\bar{E}_1 = \bar{E}_{t1} + \bar{E}_{N1}$.

Hence $\bar{E}_{t1} = \bar{E}_1 - \bar{E}_{N1}$

$$= -30\,\bar{a}_x + 70\,\bar{a}_z \text{ V/m}$$

(c) $E_{t1} = \sqrt{(-30)^2 + (70)^2} = 76.158$ V/m

(d) $E_1 = \sqrt{(-30)^2 + (50)^2 + (70)^2} = 91.104$ V/m

(e) θ_1 is the angle made by $\bar{E}_1$ with the normal

$$\cos\theta_1 = \frac{\bar{E}_1 \cdot \bar{a}_y}{|\,\bar{E}_1\,|\ |\,\bar{a}_y\,|} = \frac{50}{91.104} = 0.5488\,,\ \theta_1 = 56.7^\circ$$

Problem 5 . 10 : The region $y < 0$ contains a dielectric material for which $\varepsilon_{R1} = 2.5$, while the region $y > 0$ is charaterized by $\varepsilon_{R2} = 4$ Let $\bar{E}_1 = -30\,\bar{a}_x + 50\,\bar{a}_y + 70\,\bar{a}_z$ V/m , and find : (a) D_{N2} ; (b) D_{t2} ; (c) $\bar{D}_2$; (d) $\bar{P}_2$; (e) θ_2 .

Solution :

(a) $D_{N1} = D_{N2} = \varepsilon_o \varepsilon_{r1} E_{N1}$

$$= \varepsilon_o \times 2.5 \times 50 = 1.107 \text{ nC/m}^2$$

(b) $\overline{E}_{t2} = \overline{E}_{t1} = -30\,\overline{a}_x + 70\,\overline{a}_z$

$\therefore \overline{D}_{t2} = \varepsilon_o \varepsilon_{r2} \overline{E}_{t2} = 4\varepsilon_o(-30\,\overline{a}_x + 70\,\overline{a}_z)$

$D_{t2} = 4\varepsilon_o \sqrt{(-30)^2 + (70)^2} = 2.697 \text{ nC/m}^2$

(c) $\overline{D}_2 = \overline{D}_{t2} + \overline{D}_{N2} = 4\varepsilon_o(-30\,\overline{a}_x + 70\,\overline{a}_z) + 2.5\,\varepsilon_o(50\,\overline{a}_y)$

$$= -1.062\,\overline{a}_x + 1.107\,\overline{a}_y + 2.479\,\overline{a}_z \text{ nC/m}^2$$

(d) $\overline{D}_2 = \varepsilon_o \overline{E}_2 + \overline{P}_2$

$\therefore \overline{P}_2 = -\varepsilon_o \overline{E}_2 + \overline{D}_2$

$\overline{E}_2 = \overline{E}_{t2} + \overline{E}_{N2} = \overline{E}_{t2} + \left(\frac{\varepsilon_{r1}}{\varepsilon_{r2}}\right) \overline{E}_{N1}$

$$= -30\,\overline{a}_x + 70\,\overline{a}_z + \left(\frac{2.5}{4}\right) 50\,\overline{a}_y$$

$$= -30\,\overline{a}_x + 31.25\,\overline{a}_y + 70\,\overline{a}_z$$

$\therefore \overline{P}_2 = -\varepsilon_0(-30\,\overline{a}_x + 31.25\,\overline{a}_y + 70\,\overline{a}_z) + (-120\,\varepsilon_o\,\overline{a}_x + 125\,\varepsilon_o\,\overline{a}_y + 280\,\varepsilon_o\,\overline{a}_z)$

$$= -0.797\,\overline{a}_x + 0.830\,\overline{a}_y + 1.859\,\overline{a}_z \text{ nC/m}^2$$

(e) $\frac{\tan\theta_1}{\tan\theta_2} = \frac{\varepsilon_1}{\varepsilon_2}$

$\therefore \tan\theta_2 = \frac{\varepsilon_2 \tan\theta_1}{\varepsilon_1} = \frac{4}{2.5}\tan(56.7^\circ) = 2.43576$

or $\theta_2 = 67.68^\circ$

5.11 CAPACITANCE

5.11.1 Definition

Capacitance is defined as the ratio of the magnitudes of the total charge (Q) on either conductor to the potential difference (V) between the conductors.

$$C = \frac{Q}{V}$$

The unit of capacitance is Farad.

5.11.2 Expression for capacitance of a parallel plate capacitor of area S m^2 and a separation of d m :

Fig. 5.7 shows a parallel plate capacitor. Let V be the potential difference between the two plates. Then the electric field intensity between the plates will be

$$E = \frac{V}{d}$$. Hence

$$D = \varepsilon E = \frac{\varepsilon V}{d}$$

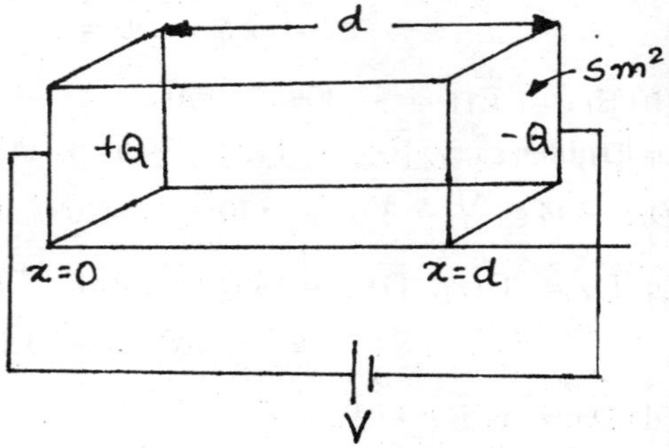

FIG. 5.7

This electric flux density it normal to the conductor i.e. plates. Hence.

$$D_N = D = \frac{\varepsilon V}{d}$$

$$\rho_s = D_N = \frac{\varepsilon V}{d} \ \text{C/m}^2$$

$$\text{Charge} = Q = \rho_s \times \text{Area} = \frac{\varepsilon V S}{d} .$$

$$C = \frac{Q}{V} = \frac{\varepsilon V S}{V d} = \frac{\varepsilon_o \varepsilon_r S}{d}$$

Further total energy stored in the capacitor is

$$W_E = \frac{\varepsilon}{2} \int E^2 \, dv = \frac{\varepsilon}{2} \int_0^d \int_0^S \frac{V^2}{d^2} \, dx \, ds$$

(It is assumed that the plates of the capacitor are located at $x = 0$ and $x = d$)

$$W_E = \frac{\varepsilon}{2} \times \frac{V^2}{d^2} \left[x \right]_0^d \left[s \right]_0^S = \frac{\varepsilon V^2 S}{2d}$$

Now $V = Q/C$. Hence ,

$$W_E = \frac{\varepsilon S Q^2}{2 d C^2}$$

As $C = \varepsilon \frac{S}{d}$, we have

$$W_E = \frac{1}{2} C V^2 = \frac{1}{2} Q V = \frac{1}{2} \frac{Q^2}{C}$$

Problem 5.11 : Find the relative permittivity of the dielectric material used in a parallelplate capacitor if : (a) $C = 40$ nF , $d = 0.1$ mm, and $S = 0.15\ \text{m}^2$; (b) $d = 0.2$ mm, $E = 500$ kV / m, and $\rho_s = 10\ \mu\text{C/m}^2$; (c) $D = 50\ \mu\text{C/m}^2$ and the energy density is $20\ \text{J/m}^3$.

Solution :

(a) $C = \dfrac{\varepsilon_o \varepsilon_r S}{d}$

$$\therefore \varepsilon_r = \frac{Cd}{\varepsilon_o S} = \frac{40 \times 10^{-9} \times 0.1 \times 10^{-3}}{\varepsilon_o \times 0.15} = 3.012$$

(b) $|\overline{D}_N| = \rho_s = 10\ \mu\text{C/m}^2$

or $\varepsilon_o \varepsilon_r E_N = 10\ \mu\text{C/m}^2$

$$\therefore \varepsilon_r = \frac{10 \times 10^{-6}}{\varepsilon_o E_N} = \frac{10^{-5}}{\varepsilon_o \times 500 \times 10^3} = 2.259$$

(c) Energy density $= \dfrac{\varepsilon_o \varepsilon_r}{2} |\overline{E}|^2 = \dfrac{D^2}{2 \varepsilon_o \varepsilon_r}$

$$\therefore \varepsilon_r = \frac{(50 \times 10^{-6})^2}{2 \varepsilon_o (20)} = 7.059$$

Problem 5.13. : A parallel - plate capacitor shown in Fig. P5.13 contains three dielectric layers.Let $\varepsilon_{R1} = 1$, $d_1 = 0.2$ mm, $\varepsilon_{R2} = 2$, $d_2 = 0.3$ mm, $\varepsilon_{R3} = 3$, $d_3 = 0.4$ mm , and $S = 20\ \text{cm}^2$

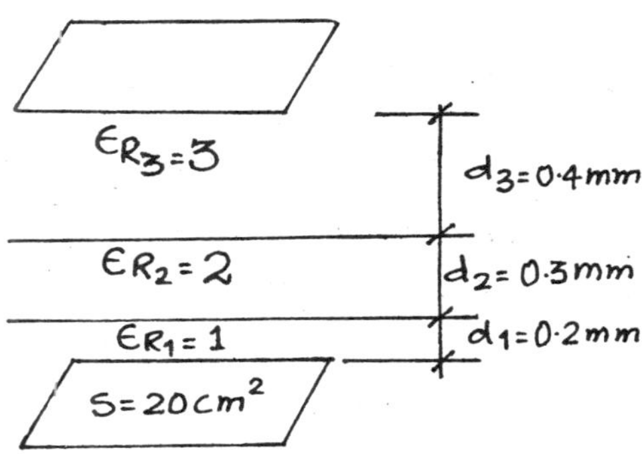

FIG . P 5.13

(a) Find C. (b) Find the percentage of the total stored energy located in each of the three regions.

Solution :

$$\text{(a) } C_1 = \varepsilon_o \varepsilon_{R1} \frac{S}{d_1} = \varepsilon_o (1) \frac{20 \times 10^{-4}}{0.2 \times 10^{-3}} = 10\, \varepsilon_o \text{ F}$$

$$C_2 = \varepsilon_o \varepsilon_{R2} \frac{S}{d_2} = \varepsilon_o (2) \frac{20 \times 10^{-4}}{0.3 \times 10^{-3}} = 13.333\, \varepsilon_o \text{ F}$$

$$C_3 = \varepsilon_o \varepsilon_{R3} \frac{S}{d_3} = \varepsilon_o (3) \frac{20 \times 10^{-4}}{0.4 \times 10^{-3}} = 15\, \varepsilon_o \text{ F}$$

These capacitors are connected in series

$$\text{Hence } \frac{1}{C} = \frac{1}{C_1} + \frac{1}{C_2} + \frac{1}{C_3} = \frac{C_1 C_2 + C_2 C_3 + C_3 C_1}{C_1 C_2 C_3}$$

$$\therefore \quad C = \frac{C_1 C_2 C_3}{C_1 C_2 + C_2 C_3 + C_3 C_1} = 36.636 \text{ pF}.$$

$$\text{(b) } W_E \text{ (total)} = \frac{1}{2} \frac{Q^2}{C} \qquad W_{E1} = \frac{1}{2} \frac{Q^2}{C_1}$$

$$\therefore \quad \frac{W_{E1}}{W_E} = \frac{C}{C_1} = 0.41377$$

$\therefore$ % of the total stored energy located in region 1 = 41.377 %.

$$\frac{W_{E2}}{W_E} = \frac{C}{C_2} = 0.3103$$

$\therefore$ % of the total stored energy located in region 2 = 31.03%

$$\frac{W_{E3}}{W_E} = \frac{C}{C_3} = 0.27585$$

$\therefore$ % of the total stored energy located in region 2 = 27.58%

5.11.3 Important Formulae for Capacitance

The capacitance of a coaxial capacitor is given by

$$\boxed{C = \frac{2 \pi \varepsilon L}{\ln (b/a)}} \qquad \text{..... (5.13)}$$

where L is the length, b is the outer radius and a is the inner radius.

The capacitance of a spherical capacitor is given by

$$C = \frac{4\pi\varepsilon}{\frac{1}{a} - \frac{1}{b}} \qquad \ldots\ldots (5.14)$$

where a and b is the radius of inner and outer concentric spherical conducting shells respectively. The capacitance of an isolated spherical conductor is

$$C = 4\pi\varepsilon a \qquad \ldots\ldots (5.15)$$

where a as the radius of spherical conductor. If the above spherical conductor is coated with a dielectric for which $\varepsilon = \varepsilon_1$, extending from $r = a$ to $r = r_1$, then the capacitance is given by

$$C = \frac{4\pi}{\frac{1}{\varepsilon_1}\left(\frac{1}{a} - \frac{1}{r_1}\right) + \frac{1}{\varepsilon_o r_1}} \qquad \ldots\ldots (5.16)$$

In Fig.5.8 a zero potential conducting plane is located at $x = 0$, and a conducting cylinder of radius b and potential V_o with its axis located a distance h from the plane. The location of the equivalent line charge as shown in Fig. 5.8 is given by

$$a = \sqrt{h^2 - b^2}$$

and the strength of the equivalent line charge by

$$\rho_L = \frac{4\pi\varepsilon V_o}{\ln K_1}$$

where K_1 is given by

$$\sqrt{K_1} = \frac{h + \sqrt{h^2 + b^2}}{b}$$

The location of a cylinder which represents an equipotential surface at a potential of V_1 is given by $K_1 = e^{4\pi\varepsilon V_1/\rho_L}$

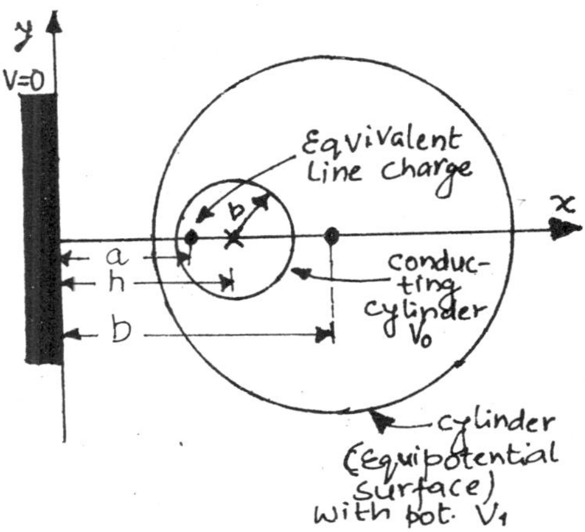

FIG. 5 . 8

$$b\,(\text{radius}) = \frac{2a\sqrt{K_1}}{K_1 - 1} \quad \text{and} \quad h = a\,\frac{K_1 + 1}{K_1 - 1}$$

The capacitance between cylinder and plane is given by

$$\boxed{C = \frac{2\pi\varepsilon}{\cosh^{-1}(h/b)}} \quad \ldots\ldots (5.17)$$

In case of a conductor for which $b << h$, then the capacitance is given by

$$\boxed{C = \frac{2\pi\varepsilon L}{\ln(2h/b)}} \quad \ldots\ldots (5.18)$$

Problem 5.12 : Find the capacitance of : (a) 20 cm of 58 C / U coaxial cable having an inner conductor 0.0295 in. in diameter, an outer conductor having an inside diameter of 0.116 in, and a polyethylene dielectric ; (b) a conducting sphere 1 cm in diameter, covered with a layer of polyethylene 1 cm thick, in free space ; (c) a conducting sphere 1 cm in diameter, covered with a layer of polyethylene 1 cm thick, and surrounded by a concentric conducting sphere 1.5 cm in radius.

Solution :

(a) $C = \dfrac{2\pi\varepsilon L}{\ln(b/a)}$

Here $L = 20 \text{ cm} = 0.2 \text{ m}$;

$$a = \frac{0.0295}{2} \text{ in} = \frac{0.0295}{2 \times 39.37} \text{ m} = 3.7465 \times 10^{-4} \text{ m}$$

$$b = \frac{0.116}{2} \text{ in} = \frac{0.116}{2 \times 39.37} \text{ m} = 1.4732 \times 10^{-3} \text{ m}$$

$$\therefore C = \frac{2\pi\varepsilon_o \times 2.26 \times 0.2}{\ln\left(\dfrac{1.4732 \times 10^{-3}}{3.7465 \times 10^{-4}}\right)} = 18.365 \text{ pF}$$

(b) $C = \dfrac{4\pi}{\dfrac{1}{\varepsilon_1}\left(\dfrac{1}{a} - \dfrac{1}{r_1}\right) + \dfrac{1}{\varepsilon_o r_1}}$

where $\varepsilon_1 = \varepsilon_o \varepsilon_{r1} = 2.26\,\varepsilon_o$ F/m

$$a = \frac{1 \text{ cm}}{2} = 0.005 \text{ m}$$

and $r_1 = \dfrac{1 \text{ cm}}{2} + 1 \text{ cm} = 1.5 \text{ cm} = 0.015 \text{ m}$

$$\therefore\ C = \frac{4\pi}{\dfrac{1}{2.26\,\varepsilon_o}\left(\dfrac{1}{0.005} - \dfrac{1}{0.015}\right) + \dfrac{1}{0.015\,\varepsilon_o}} = 0.885\text{ pF}.$$

$$\text{(c) } C = \frac{4\pi\,\varepsilon_o\,\varepsilon_r}{\dfrac{1}{a} - \dfrac{1}{b}} = \frac{4\pi\,\varepsilon_o\,(2.25)}{\dfrac{1}{0.005} - \dfrac{1}{0.015}} = 1.8859\text{ pF}$$

Problem 5.14 : Find the capacitance between a circular conducting cylinder in air, 2 mm in radius, and : (a) a conducting plane, 1 cm distant from the cylinder axis, by Eq.(5.17) ; (b) a conducting plane, 1 cm distant from the cylinder axis, by Eq.(5.18) ; (c) a similar cylinder, axes separated by 1 cm.

Solution :

(a) $h = 1\text{ cm} = 10^{-2}\text{ m}$ and $b = 2\text{ mm} = 2\times 10^{-3}\text{ m}$

$$C = \frac{2\pi\,\varepsilon\,L}{\cosh^{-1}(h/b)}$$

$$\therefore\ \frac{C}{L} = \frac{2\pi\,\varepsilon_o}{\cosh^{-1}(10^{-2}/2\times 10^{-3})} = 24.267\text{ pF/m}$$

$$\text{(b) } \frac{C}{L} = \frac{2\pi\,\varepsilon}{\ln(2h/b)} = \frac{2\pi\,\varepsilon_o}{\ln\left(\dfrac{1\times 10^{-2}}{2\times 10^{-3}}\right)} = 24.160\text{ pF/m}$$

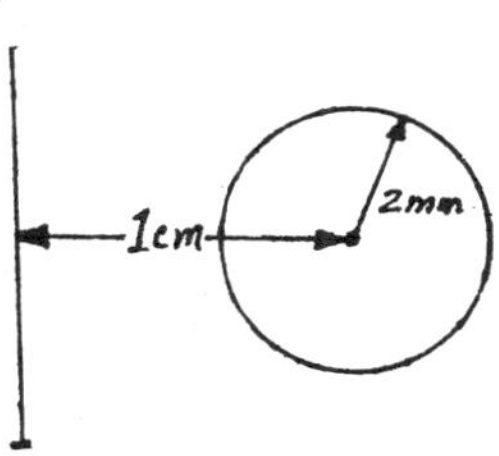

FIG.P5.14 (a)

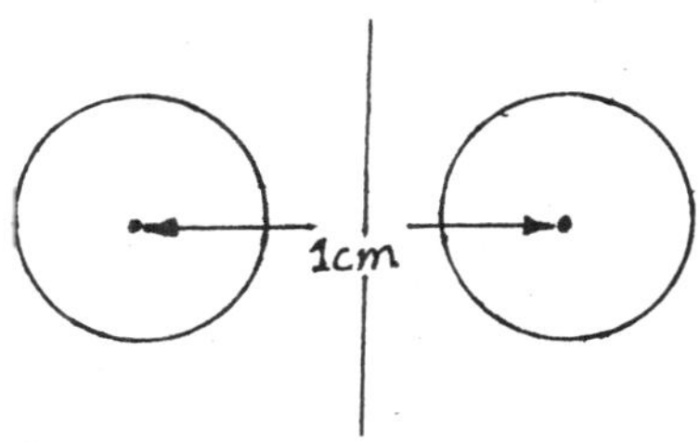

FIG.P5.14 (b)

(c) $h = 0.5\text{ cm} = 0.5\times 10^{-2}\text{ m}$ and $b = 2\times 10^{-3}\text{ m}$

The capacitance between two circular conductors separated by a distance is one half of the capacitance given by

$$\frac{C_1}{L} = \frac{2\pi\,\varepsilon_o}{\cosh^{-1}\left(\dfrac{0.5\times 10^{-2}}{2\times 10^{-3}}\right)} = 35.506\text{ pF/m}.$$

Hence required capacitance = $\frac{C}{L} = \frac{1}{2}\left(\frac{C_1}{L}\right)$

$$= \frac{35.506}{2} = 17.753 \text{ pF/m}.$$

Problem 5 .15 : The current density in a certain region is expressed as $\bar{J} = -10^5 \bar{\nabla} V$ A/m^2, where $V = 10 e^{-x} \sin y$ V. (a) Find the current in the $\bar{a}_x$ direction crossing the surface $x = 1$ and bounded by $y = 0, y = 1$; $z = 0$, and $z = 1$. Find the total current leaving the cube $0 < x, y, z < 1$ by : (b) integrating $\bar{J} \cdot \overline{dS}$ over the surface of the cube ; (c) using the divergence theorem on the closed surface integral.

Solution :

(a)Using Eq.

$$\bar{\nabla} V = -10 e^{-x} \sin y\, \bar{a}_x + 10 e^{-x} \cos y\, \bar{a}_y$$

$$\therefore \bar{J} = -10^5 (-10 e^{-x} \sin y\, \bar{a}_x + 10 e^{-x} \cos y\, \bar{a}_y) \text{ A/m}^2$$

To the plane $x = 1$, $\bar{a}_x$ is perpendicular. Hence $\overline{dS} = dy\, dz\, \bar{a}_x$

$$\therefore I = \int \bar{J} \cdot \overline{dS} = \int_0^1 \int_0^1 \underset{(x=1)}{10^6 e^{-x} \sin y\, dy\, dz}$$

$$= 10^6 e^{-1} \Big[-\cos y \Big]_0^1 \Big[z \Big]_0^1 = 169113.331 \text{ A}$$

$$\text{(b) } I = \oint \bar{J} \cdot \overline{dS} = \int_{(x=0)} \bar{J} \cdot \overline{dS} + \int_{(x=1)} \bar{J} \cdot \overline{dS} + \int_{(y=0)} \bar{J} \cdot \overline{dS}$$

$$\int_{(y=1)} \bar{J} \cdot \overline{dS} + \int_{(z=0)} \bar{J} \cdot \overline{dS} + \int_{(z=1)} \bar{J} \cdot \overline{dS}$$

$$I = \int_0^1 \int_0^1 \underset{(x=0)}{-10^6 e^{-x} \sin y\, dy\, dz} + \int_0^1 \int_0^1 \underset{(x=1)}{10^6 e^{-x} \sin y\, dy\, dz} +$$

$$\int_0^1 \int_0^1 \underset{(y=0)}{10^6 e^{-x} \cos y\, dx\, dz} - \int_0^1 \int_0^1 \underset{(y=1)}{10^6 e^{-x} \cos y\, dx\, dz}$$

$$I = -10^6 \Big[-\cos y \Big]_0^1 \Big[z \Big]_0^1 + 10^6 e^{-1} \Big[-\cos y \Big]_0^1 \Big[z \Big]_0^1 +$$

$$10^6 \cos 0\left[- e^{-x} \right]_0^1 \left[z \right]_0^1 - 10^6 \cos 1\left[- e^{-x} \right]_0^1 \left[z \right]_0^1$$

$$I = -459697.6941 + 169113.3308 + 632120.5588 - 341536.1955 = 0$$

(c) $I = \int (\overline{\nabla} \cdot \overline{J})\, dv$

Using Eq.,we have

$$\overline{\nabla} \cdot \overline{J} = \frac{\partial}{\partial x}(10^6 e^{-x} \sin y) + \frac{\partial}{\partial y}(-10^6 e^{-x} \cos y)$$

$$= -10^6 e^{-x} \sin y + 10^6 e^{-x} \sin y = 0$$

$\therefore\ I = 0$

Problem 5.16 : Within the region $1 < \rho < 5$ cm, $0 < \phi < 0.3\pi$, $0 < z < 2$ cm, current density is given as $\overline{J} = (200 \cos \phi)\, \overline{a}_\phi / (\rho + 0.01)$ A / m^2. What total current in the $\overline{a}_\phi$ direction crosses the surface : (a) $\phi = 0, 1 < \rho < 5$ cm, ? (b) $\phi = 0.3\pi, 1 < \rho < 5$ cm, and $0 < z < 2$ cm ? (c) Calculate $\overline{\nabla} \cdot \overline{J}$ at $\rho = 2$ cm, $\phi = 0.15\pi, z = 1$ cm.

Solution :

(a) To the plane $\phi = 0$, $\overline{a}_\phi$ is perpendicular. Hence $\overline{dS} = d\rho\ dz\ \overline{a}_\phi$

$$I = \int \overline{J} \cdot \overline{dS} = \int_{0.01}^{0.05} \int_0^{0.02} \frac{200 \cos \phi\ d\rho\ dz}{\rho + 0.01} \quad (\phi = 0)$$

$$= 200 \cos 0 \left[\ln(\rho + 0.01) \right]_{0.01}^{0.05} \left[z \right]_0^{0.02}$$

$$= 4.394 \text{ A}$$

(b) $$I = \int \overline{J} \cdot \overline{dS} = \int_{0.01}^{0.05} \int_0^{0.02} \frac{200 \cos \phi\ d\rho\ dz}{\rho + 0.01} \quad (\phi = 0.3\pi)$$

$$= 200 \cos(0.3\pi) \left[\ln(\rho + 0.01) \right]_{0.01}^{0.05} \left[z \right]_0^{0.02}$$

$$= 2.583 \text{ A}$$

(c)Using Eq.(3.), we have

$$\overline{\nabla} \cdot \overline{J} = \frac{1}{\rho} \frac{\partial}{\partial \phi} \left[\frac{200 \cos \phi}{\rho + 0.01} \right] = \frac{-200 \sin \phi}{\rho(\rho + 0.01)}$$

$\overline{\nabla} \cdot \overline{J}$ (at $\rho = 2$ cm , $\phi = 0.15\pi$, $z = 1$ cm)

$$= \frac{-200 \sin(0.15\pi)}{0.02(0.02 + 0.01)} = -151330.167 \text{ A/m}^3$$

Problem 5.17 : A certain current density is given as $\overline{J} = [100 \cos\theta / (r^2 + 1)]\ \overline{a}_r$ A/m^2. (a) How much current flows through the spherical cap, $r = 3, 0 < \theta < \pi/6, 0 < \phi < 2\pi$. (b) What value would $\overline{J}_{av}$ have if the same total current were produced by a uniform current density $\overline{J}_{av}$ over the cap ? (c) The same total current flows through the cap, $r = 10, 0 < \theta < \alpha, 0 < \phi < 2\pi$. Find α.

Solution :

(a) To the surface $r = 3$ m , $\overline{a}_r$ is perpendicular.

Hence $\overline{dS} = (r^2 \sin\theta$ ital;icd$\theta\ d\phi)\ \overline{a}_r$.

$$I = \int \overline{J} \cdot \overline{dS} = \int_0^{\pi/6} \int_0^{2\pi} \frac{100 \cos\theta}{r^2 + 1} r^2 \sin\theta\, d\theta\, d\phi \quad (r = 3)$$

$$= 50 \left[\frac{-\cos 2\theta}{2} \right]_0^{\pi/6} [\phi]_0^{2\pi} \left[\frac{9}{9 + 1} \right] = 70.686 \text{ A}$$

(b) Area of the cap $= \int_0^{\pi/6} \int_0^{2\pi} r^2 \sin\theta\, d\theta\, d\phi \quad (r = 3)$

$$= 9 \left[-\cos\theta \right]_0^{\pi/6} [\phi]_0^{2\pi} = 7.576 \text{ m}^2$$

Hence $\overline{J}_{av} = \frac{I}{Area} \overline{a}_r = \frac{70.686}{7.576} \overline{a}_r = 9.33\, \overline{a}_r$ A/m^2

(c) $70.686 = \int_0^{\alpha} \int_0^{2\pi} \frac{100 \cos\theta}{r^2 + 1} r^2 \sin\theta\, d\theta\, d\phi \quad (r = 10)$

$$= 50 \left[\frac{-\cos 2\theta}{2} \right]_0^{\alpha} [\phi]_0^{2\pi} \left[\frac{100}{100 + 1} \right]$$

$\therefore -\cos 2\alpha + 1 = 0.4545$ or $\cos 2\alpha = 0.5455$

$\therefore\ \alpha = 0.4969 = 28.471^\circ$

Problem 5.18 : Electrons are emitted from a planar cathode at $z = 0$ with zero initial velocity. The region $z > 0$ is a vacuum in which $\overline{E} = -2.5 \times 10^6\,\overline{a}_z$ V/m. Letting $e = 1.602 \times 10^{-19}$ C and $m = 9.11 \times 10^{-31}$ kg. : (a) find $v(t)$ for an electron leaving the cathode at $t = 0$; (b) find $z(t)$; (c) find $v(z)$. (d) Assuming that electrons are emitted continuously as a beam of radius 0.2 mm and total current 100 μA, find $\overline{J}(z)$ and $\rho_v(z)$.

Solution :

(a) Force acting on the electron

$$= -e\overline{E} = -1.602 \times 10^{-19}(-2.5 \times 10^6\,\overline{a}_z)$$

$$= 4.005 \times 10^{-13}\,\overline{a}_z\ \text{N}$$

Now $F = m\,a = m\dfrac{dv}{dt}$ or $dv = \dfrac{F}{m}\,dt$ or

$v = \displaystyle\int \frac{F}{m}\,dt + C$ where C is a constant of integration.

$$= \int \frac{4.005 \times 10^{-13}}{9.11 \times 10^{-31}}\,dt + C = 4.39626 \times 10^{17}\,t + C$$

Since electrons are leaving the cathode with zero initial velocity, we have at $t = 0$, $v = 0$ so that $C = 0$

$$\therefore\ v(t) = 4.39626 \times 10^{17}\,t$$

(b) Now $v(t) = \dfrac{dz(t)}{dt} = 4.39626 \times 10^{17}\,t$

$\therefore\ z(t) = \displaystyle\int 4.39626 \times 10^{17}\,t + C_1$, where C_1 is the integration constant.

$$\therefore\ z(t) = \frac{4.39626 \times 10^{17}}{2}\,t^2 + C_1$$

Now at $t = 0$, $z(0) = 0$, $\therefore\ C_1 = 0$

$$z(t) = 2.1981 \times 10^{17}\,t^2$$

(c) Now $v(t) = 4.39626 \times 10^{17} t$ and $z(t) = 2.1981 \times 10^{17} t^2$

$$\therefore\ t = \sqrt{\frac{z(t)}{2.1981 \times 10^{17}}}$$

$$\therefore\ v(z) = 4.39626 \times 10^{17} \sqrt{\frac{z(t)}{2.1981 \times 10^{17}}}$$

$$= 937.6924 \times 10^6 \sqrt{z(t)}$$

(d) $$\bar{J}(z) = \frac{I}{Area}\,\bar{a}_z = \frac{100 \times 10^{-6}}{\pi(0.2 \times 10^{-3})^2}\,\bar{a}_z = 795.7747\ \bar{a}_z \cdot \text{A/m}^2$$

$\bar{J} = \rho_v\,\bar{v}$ or $J = \rho_v\, v$

$$795.7747 = \rho_v\,[937.6924 \times 10^6 \sqrt{z(t)}\,]$$

$$\therefore\ \rho_v(z) = \frac{795.7747}{937.6924 \times 10^6 \sqrt{z(t)}} = \frac{0.84865}{\sqrt{z(t)}}\ \mu\text{C/m}^3$$

Problem 5.19 : Let $V = 150\,x^{4/3}$ V for $x > 0$ with $\varepsilon = \varepsilon_o$. (a) Find $\bar{E}$, $\bar{D}$, and ρ_v as functions of x. (b) If the velocity of the charge density is $v_x = 6 \times 10^6 x^{2/3}$ m/s, find J_x at $x = 0$. (b) Repeat for $x = 1$ m.

Solution :

(a) $$\bar{E} = -\bar{\nabla} V = -150 \times (4/3)\,x^{1/3}\,\bar{a}_x = -200\,x^{1/3}\,\bar{a}_x \ \text{V/m}$$

$$\bar{D} = \varepsilon_o \bar{E} = -200\,\varepsilon_o\,x^{1/3}\,\bar{a}_x = -1.7708\,x^{1/3}\,\bar{a}_x\ \text{nC/m}^2$$

$$\rho_v = \bar{\nabla} \cdot \bar{D} = -1.7708 \times (1/3) \times x^{-2/3} \times 10^{-9}$$

$$= -0.5903\,x^{-2/3}\ \text{nC/m}^3$$

(b) $$J_x = \rho_{vx}\,v_x = (-0.5903\,x^{-2/3} \times 10^{-9})\,(6 \times 10^6\,x^{2/3})$$

$$= -3.5418\ \text{mA/m}^2$$

(c) Same answer

Problem 5.20 : Current density in a certain region is given as $\bar{J} = (5/\rho)\,\bar{a}_\rho + \left[\,10/(\rho^2 + 1)\,\right]\bar{a}_z$ A/m². (a) Find the total current crossing the surface $z = 3$, $\rho < 6$, in the $\bar{a}_z$ direction. (b) Find $\partial \rho_v / \partial t$.

(c) Calculate the total current crossing the closed surface bounded by $z = 0$, $z = 3$, $\rho = 1$, and $\rho = 6$. (d) Is $\int_{vol} \bar{\nabla} \cdot \bar{J}\, dv = \oint_s \bar{J} \cdot \overline{dS}$?

Solution :

(a) To the surface $z = 3$, $\bar{a}_z$ is perpendicular. Hence $\overline{dS} = \rho\, d\rho\, d\phi\, \bar{a}_z$.

$$\therefore I = \int \bar{J} \cdot \overline{dS} = \int_0^6 \int_0^{2\pi} \frac{10\rho\, d\rho\, d\phi}{\rho^2 + 1}$$

$$= \frac{10}{2}\Big[\ln(\rho^2 + 1)\Big]_0^6 \Big[\phi\Big]_0^{2\pi} = 113.440 \text{ A}$$

(b) $\bar{\nabla} \cdot \bar{J} = -\dfrac{\partial \rho_v}{\partial t}$

$$\bar{\nabla} \cdot \bar{J} = \frac{1}{\rho}\frac{\partial}{\partial \rho}\left(\rho \times \frac{5}{\rho}\right) + \frac{\partial}{\partial z}\left(\frac{10}{\rho^2 + 1}\right) = 0$$

Hence $\dfrac{\partial \rho_v}{\partial t} = 0$

(c) $$\oint \bar{J} \cdot \overline{dS} = \int_{(z=0)} \bar{J} \cdot \overline{dS} + \int_{(z=3)} \bar{J} \cdot \overline{dS} + \int_{(\rho=1)} \bar{J} \cdot \overline{dS} + \int_{(\rho=6)} \bar{J} \cdot \overline{dS}$$

$$= \int_{(z=0)} \bar{J} \cdot (-\rho\, d\rho\, d\phi\, \bar{a}_z) + \int_{(z=3)} \bar{J} \cdot (\rho\, d\rho\, d\phi\, \bar{a}_z) +$$

$$\int_{(\rho=1)} \bar{J} \cdot (-\rho\, d\phi\, dz\, \bar{a}_\rho) + \int_{(\rho=6)} \bar{J} \cdot (\rho\, d\phi\, dz\, \bar{a}_z)$$

$$= -\int_0^6 \int_0^{2\pi} \left.\frac{10\rho\, d\rho\, d\phi}{\rho^2 + 1}\right|_{(z=0)} + \int_0^6 \int_0^{2\pi} \left.\frac{10\rho\, d\rho\, d\phi}{\rho^2 + 1}\right|_{(z=3)} -$$

$$\int_0^{2\pi} \int_0^3 \left.\frac{5}{\rho} \times \rho\, d\phi\, dz\right|_{(\rho=1)} + \int_0^{2\pi} \int_0^3 \left.\frac{5}{\rho} \times \rho\, d\phi\, dz\right|_{(\rho=6)}$$

$$= -113.440 + 113.440 - 188.495 + 188.495 = 0$$

(d) $$\int (\bar{\nabla} \cdot \bar{J})\, dv = \int 0\, dv = 0$$

$$\therefore \int_{vol} (\overline{\nabla} \cdot \overline{J})\, dv = \oint_S \overline{J} \cdot \overline{dS}$$

Problem : 5.21 : Assuming that there is no transformation of mass to energy or vice versa, it is possible to write a continuity equation for mass. (a) If we use the continuity equation for charge as our model, what quantities correspond to $\overline{J}$ and ρ_v ? (b) Given a cube 1 cm on a side, experimental data show that the rates at which mass is leaving each of the six faces are 10.25, – 9.85, 1.75, 2.00, – 4.05, and 4.45 mg / s. If we assume that the cube is an incremental volume element, determine an approximate value for the time rate of change of density at its center.

Solution :

The continuity equation for charge is $\overline{\nabla} \cdot \overline{J} = -\dfrac{\partial \rho_v}{\partial t}$

Where ρ_v is charge density in C / m^3 and $\overline{J}$ is current density in A / m^2 . Since A (Ampere)= C / sec, the unit of current density may be expressed as C / m^2 sec. If the continuity equation has to be written for mass, then ρ_v will represents density in kg/m^3 and $\overline{J}$ will represent mass per second per square meter (kg / m^2 sec) or mass rate of flow.

(b) Now $\overline{\nabla} \cdot \overline{J} = -\dfrac{\partial \rho_v}{\partial t}$

$$\therefore \frac{\partial \rho_v}{\partial t} = -\left[\frac{10.25 - 9.85 + 1.75 - 2.00 - 4.05 + 4.45}{10^{-2} \times 10^{-2} \times 10^{-2}}\right] 10^{-6}$$

(Note :– The unit of $\overline{\nabla} \cdot \overline{J}$ will be A / m^2 or C / m^3 sec. Hence in the continuity equation for mass it will represent kg / m^3 sec)

$$\therefore \frac{\partial \rho_v}{\partial t} = -\ 0.55 \text{ kg/m}^3 \text{ sec}$$

Problem 5.22 : After setting up a cartesian coordinate system inside a star, Captain Kirk and his intrepid crew make measurements over the faces of a cube centered at the origin with edges 40 km long and parallel to the coordinate axes. They find the mass rate of flow of material outward across the

six faces to be – 1112, 1183, 201, –196, 1989, and – 1920 kg / km^2.s. (a) Estimate the divergence of the mass rate of flow at the origin. (b) Estimate the rate of change of the density at the origin.

Solution :

(a) $\overline{\nabla} \cdot \overline{J}$ = divergence of the mass rate of flow at the origin.

= mass leaving through all the surfaces of cube per second per unit volume

$$= \frac{-112(40 \times 40) + 1183(40 \times 40) + 201(40 \times 40) - 196(40 \times 40) + 1989(40 \times 40) - 1920(40 \times 40)}{40 \times 40 \times 40}$$

$= 3.625$ kg/sec km^3

(b) $\dfrac{\partial \rho_v}{\partial t} = -\overline{\nabla} \cdot \overline{J} = -3.625$ kg/km^3 sec

Problem 5.23 : (a) Determine the length of 0.3 - mm - diameter Nichrome wire required to dissipate an average power of 500 W when 120 - V rms at 60 Hz is applied to it. (b) Find the maximum amplitude of the electric field intensity in the wire. (c) What is the rms current density inside the wire ?

Solution :

For Nichrome $\sigma = 0.1 \times 10^7$ mho/m

$P = VI$. Hence $I = P/V = 500/120 = 4.16667$ A

$$R = \frac{V}{I} = \frac{120}{4.16667} = 28.8\ \Omega$$

$$R = \frac{L}{\sigma A}$$ where A is cross sectional area

$A = \pi d^2/4$ where d is the diameter of the wire.

$= \pi(0.3 \times 10^{-3})^2/4 = 7.06858 \times 10^{-8}$ m^2

$\therefore L = R\sigma A = 28.8 \times 0.1 \times 10^7 \times 7.06858 \times 10^{-8} = 2.0357$ m

(b) $E = V/d = 120/2.0357 = 58.946$ rms

Maximum amplitude of $E = 58.946\sqrt{2} = 83.3626$ V/m

(c) $J = \sigma E = 0.1 \times 10^7 \times 58.946 = 5.8946 \times 10^7$ A/m^2 rms

Problem 5.24 : A conductor of circular cross section is constructed of steel ($\sigma = 6 \times 10^6$ mho/m) in the region $0 < \rho < 1$ mm, and copper ($\sigma = 5.8 \times 10^7$ mho/m), $1 < \rho < 1.5$ mm. The total current carried by the conductor is 100 A. Calculate : (a) J_{Cu} ; (b) J_{st} ; (c) E_{Cu} ; (d) E_{st} ; (e) the voltage drop in 1 km of conductor.

Solution :

Consider L as the length of the conductor.

$$R_{st} = \frac{L}{\sigma s} = \frac{L}{6 \times 10^6 \times \pi \times (10^{-3})^2} = 0.05305\ L\ \Omega$$

$$R_{cu} = \frac{L}{\sigma s} = \frac{L}{5.8 \times 10^7 \times \pi [(1.5 \times 10^{-3})^2 - (10^{-3})^2]}$$

$$= 4.39048 \times 10^{-3} L\Omega$$

$$I_{st} = I \times \frac{R_{cu}}{R_{st} + R_{cu}} = 100 \times \frac{4.39048 \times 10^{-3} L}{0.05305 L + 4.39048 \times 10^{-3} L}$$

$$= 7.84529\ \text{A}$$

$$I_{cu} = I \times \frac{R_{st}}{R_{st} + R_{cu}} = 100 \times \frac{0.05305\ L}{0.05305 L + 4.39048 \times 10^{-3} L}$$

$$= 92.35647\ \text{A}$$

$$\text{(a) } J_{cu} = \frac{I_{cu}}{Area} = \frac{92.35647}{\pi [(1.5 \times 10^{-3})^2 - (10^{-3})^2]}$$

$$= 23.51838 \times 10^6\ \text{A/m}^2$$

$$\text{(b) } J_{st} = \frac{I_{st}}{Area} = \frac{7.643529}{\pi \times (10^{-3})^2} = 2.433 \times 10^6\ \text{A/m}^2$$

$$\text{(c) } J_{cu} = \sigma E_{cu}\ .\ \therefore\ E_{cu} = \frac{J_{cu}}{\sigma} = \frac{23.51838 \times 10^6}{5.8 \times 10^7} = 0.405489\ \text{V/m}$$

$$\text{(d) } E_{st} = \frac{J_{st}}{\sigma} = \frac{2.433 \times 10^6}{6 \times 10^6} = 0.4055\ \text{V/m}$$

(e) $V = E \times L = 0.4055 \times 1000 = 405.5$ V

(Note :– $E_{cu} = E_{st}$. Difference is due to approximation is calculations.)

Problem 5.25 : Two perfectly conducting spherical surfaces are located at $r = 2$ and $r = 10$ cm. The total current passing radially outward through the medium between the spheres is 2.5 A. (a) Find the voltage and resistance between the spheres, and $\overline{E}$ in the region between the spheres, if a conducting material having $\sigma = 0.02$ mho/m is present for $2 < r < 10$ cm. (b) Repeat if $\sigma = 0.001/r$ for $2 < r < 10$ cm. (c) Show that integrating the power dissipated per unit volume in part (b) over the volume gives the total dissipated power.

Solution :

As the current is flowing radially outward its direction is given by $\bar{a}_r$. Hence

$$\text{(a)}\ \overline{J} = \frac{I}{Area}\,\bar{a}_r = \frac{2.5}{4\pi r^2}\,\bar{a}_r\ \text{A/m}^2 = \frac{198.944\,\bar{a}_r}{r^2}\ \text{m A/m}^2$$

$$\therefore\ \overline{E} = \frac{\overline{J}}{\sigma} = \frac{198.944\,\bar{a}_r \times 10^{-3}}{r^2 \times 0.02} = \frac{9.94718}{r^2}\,\bar{a}_r\ \text{V/m}$$

$$V = -\int_{0.1}^{0.02} \overline{E}\cdot\overline{dL} = -\int_{0.1}^{0.02}\left(\frac{9.94718}{r^2}\,\bar{a}_r\right)\cdot(dr\,\bar{a}_r)$$

$$= -\int_{1}^{0.02}\frac{9.94718\,dr}{r^2} = 9.94718\left[\frac{1}{r}\right]_{0.1}^{0.02} = 397.8872\ \text{V}$$

$$R = V/I = 397.8872/2.5 = 159.155\ \Omega$$

$$\text{(b)}\ \overline{E} = \frac{\overline{J}}{\sigma} = \frac{198.944\times 10^{-3}\,\bar{a}_r}{r^2}\times\frac{r}{0.001} = \frac{198.944}{r}\,\bar{a}_r\ \text{V/m}$$

$$V = -\int_{0.1}^{0.02}\overline{E}\cdot\overline{dL} = -\int_{0.1}^{0.02}\frac{198.944\,dr}{r} = -198.944\left[\ln r\right]_{0.1}^{0.02}$$

$$= 320.188\ \text{V}$$

$$R = \frac{V}{I} = \frac{320.188}{2.5} = 128.075\ \Omega$$

$$\text{(c)}\ P = I^2 R = (2.5)^2 \times 128.075 = 800.47\ \text{Watts}$$

$$\frac{dP}{dV} = \text{Power per unit volume} = JE$$

$$\therefore\ dP = JE\,dV \quad \text{or} \quad P = \int JE\,dv$$

$$P = \int_{0.02}^{0.1}\int_{0}^{\pi}\int_{0}^{2\pi} \frac{198.944\times 10^{-3}}{r^2} \times \frac{198.944}{r}\, r^2 \sin\theta\, dr\, d\theta\, d\phi$$

$$= (198.944)^2 \cdot \Big[\ln r\Big]_{0.02}^{0.1} \Big[-\cos\theta\Big]_0^{\pi} \Big[\phi\Big]_0^{2\pi} \times 10^{-3}$$

$$= 800.47 \text{ Watts}$$. Hence proved.

Problem 5.26 : The cylindrical surfaces, $\rho = 2$ and 6 cm, are perfectly conducting, and the region between them is filled with a conducting material for which $\sigma = 80$ mho/m . If the current density is $\bar{J} = (10/\pi\rho)\,\bar{a}_\rho$ A/m^2 for $2 < \rho < 6$ cm, find : (a) the current flowing from one perfect conductor to the other per meter length ; (b) $\bar{E}$; (c) the potential difference between perfect conductors ; (d) the total power being dissipated in the conducting material per meter length.

Solution :

(a) Consider a cylinder of radius ρ and length L. To this cylinder $\bar{a}_\rho$ is perpendicular. Hence $\overline{dS} = \rho\, d\phi\, dz\, \bar{a}_\rho$.

$$I = \int \bar{J}\cdot\overline{dS} = \int_0^{2\pi}\int_0^{L} \frac{10}{\pi\rho} \times \rho\, d\phi\, dz = \frac{10}{\pi}\Big[\phi\Big]_0^{2\pi}\Big[z\Big]_0^{L}$$

$= 20L$ Amp.

Hence current per unit length (I/L) is 20 Amp / m.

$$\text{(b) } \bar{E} = \bar{J}/\sigma = \frac{(10/\pi\rho)\,\bar{a}_\rho}{80} = \frac{39.7887}{\rho}\,\bar{a}_\rho \text{ mV/m}$$

$$\text{(c) } V = -\int_{0.06}^{0.02} \bar{E}\cdot\overline{dL} = -\int_{0.06}^{0.02} \bar{E}\cdot(d\rho\,\bar{a}_\rho)$$

$$= -\int_{0.06}^{0.02} \frac{39.7887}{\rho} \times 10^{-3}\, d\rho$$

$$= -39.7887 \times 10^{-3}\Big[\ln\rho\Big]_{0.06}^{0.02} = 43.712 \text{ mV}$$

(d) $P = \int dP = \int JE\,dv$

$$= \int_{0.02}^{0.06} \int_0^{2\pi} \int_0^L \frac{10}{\pi\rho} \times \frac{39.7887}{\rho} \times 10^{-3} \times \rho\, d\rho\, d\phi\, dz$$

$$= 126.651 \left[\ln \rho \right]_{0.02}^{0.06} \left[\phi \right]_0^{2\pi} \left[z \right]_0^L 10^{-3} = 0.874247\ L \text{ Watts}$$

Hence power dissipated per meter length (P/L) = 0.874274 watts / m.

Problem 5.27 : The spherical surfaces, $r = 2$ and 6 cm, are perfectly conducting, and the region between them is filled with a conducting material for which $\sigma = 80$ mho/m. If the current density is $\bar{J} = (10/\pi r^2)\,\bar{a}_r$ A/m^2 for $2 < r < 6$ cm, find : (a) the current flowing from one perfect conductor to the other ; (b) $\bar{E}$; (c) the potential difference between perfect conductors ; (d) the total power being dissipated in the conducting material.

Solution :

(a) Consider a sphere of radius r. To this sphere $\bar{a}_r$ is perpendicular. Hence $\overline{dS} = r^2 \sin\theta\, d\theta\, d\phi\, \bar{a}_r$

$$I = \int \bar{J} \cdot \overline{dS} = \int_0^{\pi} \int_0^{2\pi} \frac{10}{\pi r^2} \times r^2 \sin\theta\, d\theta\, d\phi$$

$$= \frac{10}{\pi} \left[-\cos\theta \right]_0^{\pi} \left[\phi \right]_0^{2\pi} = 40 \text{ Amp}$$

(b) $\bar{E} = \dfrac{\bar{J}}{\sigma} = \dfrac{10}{\pi r^2 \times 80}\,\bar{a}_r = \dfrac{39.7887}{r^2}\,\bar{a}_r$ mV/m

(c) $V = -\displaystyle\int_{0.06}^{0.02} \bar{E} \cdot \overline{dL} = -\int_{0.06}^{0.02} \bar{E} \cdot (dr\,\bar{a}_r)$

$$= -\int_{0.06}^{0.02} \frac{39.7887 \times 10^{-3}\, dr}{r^2}$$

$$= -39.7887 \times 10^{-3} \left[-\frac{1}{r} \right]_{0.06}^{0.02} = 1.32629 \text{ V}$$

(d) $P = VI = 1.32629 \times 40 = 53.0516$ Watts

Problem 5.28 : Find the magnitude of the current density in a conductor if : (a) the electric field intensity is 0.09 V / m, the volume charge density is -2×10^{10} C/m^3, and the electron mobility is 2.5×10^{-3} m^2/V · s ; (b) there are 5×10^{28} conduction electrons / m^3 and the electron drift velocity is 0.05 mm / s ; (c) E = 0.05 V / m and the resistivity is 2×10^{-8} Ω · m.

Solution :

(a) $J = -\rho_e \mu_e = -(-2\times 10^{10})(2.5\times 10^{-3})(0.09)$

$= 4.5\times 10^{6}$ A/m^2

(b) $J = -\rho_e \mu_e E$ and $V_d = -\mu_e E$. Hence $J = \rho_e V_d$

ρ_e = (Total No. of electrons/m^3) × Charge on one electron

$= 5\times 10^{28}\times(-1.6\times 10^{-19}) = -8\times 10^{9}$ C/m^3

$\therefore J = -8\times 10^{9}\times 0.05\times 10^{-3} = -400$ KA/m^2

$|J| = 400$ KA/m^2

(c) $J = \sigma E = \dfrac{1}{2\times 10^{-8}}\times 0.05 = 2.5\times 10^{6}$ A/m^2

Problem 5.29 : Given the potential field $V = (200\sin\theta\cos\phi)/r^2$ V : (a) find the equation of the conductor surface on which $V = 100$ V ; (b) find $\overline{E}$ at point $P(r, 30^\circ, 30^\circ)$ on the conductor surface ; (c) find ρ_s at point P. Assume $\varepsilon = \varepsilon_0$ adjacent to the surface.

Solution :

(a) $100 = \dfrac{200\sin\theta\cos\phi}{r^2}$

Hence the equation of the conducting surface on which $V = 100$ V is $r^2 = 2\sin\theta\cos\phi$.

(b)

$$\overline{E} = -\overline{\nabla}V = -200\left[\frac{-2\sin\theta\cos\phi}{r^3}\,\overline{a}_r + \frac{\cos\theta\cos\phi}{r^3}\,\overline{a}_\theta - \frac{\sin\phi}{r^3}\,\overline{a}_\phi\right]$$

Since $r^2 = 2 \sin \theta \cos \phi$, at point P, we have

$r^2 = 2 \sin 30^o \cos 30^o = 0.866025$ or $r^3 = 0.805927$

Substituting the coordinates of point P in the above expression of $\overline{E}$, we have

$$\overline{E} = \frac{1}{0.805929} (173.205\ \overline{a}_r - 150\ \overline{a}_\theta + 100\ \overline{a}_\phi)$$

$$= 214.913\ \overline{a}_r - 186.121\ \overline{a}_\theta + 124.080\ \overline{a}_\phi \quad V/m$$

$$(c) \rho_s = |\overline{D}| = \varepsilon_o |\overline{E}|$$

$$= \varepsilon_o \sqrt{(214.913)^2 + (-186.121)^2 + (124.08)^2}$$

$$= 2.7465 \quad nC/m^2$$

Problem 5.30 : The potential field, $V = 2x^2 + 4y - 2z^2$ V, exists in the free space surrounding a perfectly conducting surface. Point P (4, 3, 2) lies on the surface. (a) Give the equation of the surface. (b) Find a unit outward normal to the surface at P, assuming the origin is inside the surface. (c) Find ρ_s on the surface at P. (d) Find $|\overline{E}_P|$.

Solution :

(a) V at $P(4, 3, 2) = 2(4)^2 + 4(3) - 2(2)^2 = 36$

$\therefore$ Equation of the surface is $2x^2 + 4y - 2z^2 = 36$

or $x^2 + 2y - z^2 = 18$

(b) we have

$$\overline{\nabla} V = 4x\ \overline{a}_x + 4\ \overline{a}_y - 4z\ \overline{a}_z$$

$$\overline{\nabla} V \text{ at point } V = 4(4)\ \overline{a}_x + 4\ \overline{a}_y - 4(2)\ \overline{a}_z = 16\ \overline{a}_x + 4\ \overline{a}_y - 8\ \overline{a}_z$$

$$|\overline{\nabla} V \text{ at point } P| = \sqrt{(160)^2 + (40)^2 + (-8)^2} = 18.3303$$

$$\therefore\ \overline{a}_N = \frac{16\ \overline{a}_x + 4\ \overline{a}_y - 8\ \overline{a}_z}{18.3303} = 0.873\ \overline{a}_x + 0.218\ \overline{a}_y - 0.436\ \overline{a}_z$$

$$(c)\ \overline{E} = -\overline{\nabla} V = -(16\ \overline{a}_x + 4\ \overline{a}_y - 8\ \overline{a}_z) = -16\ \overline{a}_x - 4\ \overline{a}_y + 8\ \overline{a}_z$$

$$\therefore\ \rho_s = |\overline{D}| = |\varepsilon_o \overline{E}| = \varepsilon_o \sqrt{(-16)^2 + (-4)^2 + (8)^2}$$

$$= 162.297 \quad pC/m^2$$

(d) $|\overline{E}_P| = \sqrt{(-160)^2 + (-4)^2 + (8)^2} = 18.330$ V/m

Problem 5.31 : Given the potential field $V = 10y(x^3 + 5)$ V : (a) Find $\overline{E}$ at the surface $y = 0$. (b) Show that the surface $y = 0$ is an equipotential surface. (c) If it is known that the surface $y = 0$ is a conductor, find the total charge in the region, $0 < x < 2, y = 0, 0 < z < 1$. Assume that $\varepsilon = \varepsilon_o$ and that $V > 0$ in the region outside the conductor.

Solution :

(a) We have

$$\overline{E} = -\overline{\nabla} V = -[30yx^2\overline{a}_x + 10(x^3 + 5)\overline{a}_y]$$

$$\therefore \overline{E} \text{ (at the surface } y = 0) = -10(x^3 + 5)\overline{a}_y \text{ V/m.}$$

(b) On the plane $y = 0$ the coordinates which can vary are x and z. Now at $y = 0$, $V = 10(0)(x^3 + 5) = 0$ V, which is constant and independent of x and z coordinate. Hence $y = 0$ is a equipotential surface.

(c) $\overline{D} = \varepsilon_o \overline{E} = -10\varepsilon_o(x^3 + 5)\overline{a}_y$ C/m^2

$$\rho_s = |\overline{D}| = 10\varepsilon_o(x^3 + 5)$$

$$\therefore Q = \int dQ = \int \rho_s \, dS = \int_0^2 \int_0^1 10\varepsilon_o(x^3 + 5)\, dx\, dy$$

$$= 10\varepsilon_o \left[\frac{x^4}{4} + 5x\right]_0^2 \left[y\right]_0^1 = 1.23956 \text{ nC.}$$

This charge should be negative because $V > 0$ in the region outside the conductor.

Problem 5.32 : Given the field, $\overline{E} = 8xy(z + 1)\overline{a}_x + 4x^2(z + 1)\overline{a}_y + 4x^2y\,\overline{a}_z$ V/m in free space, and the information that point P (2, 3, 5) lies on a conducting surface, find : (a) the magnitude of ρ_s at P ; (b) ρ_v adjacent to the conductor surface at P ; (c) V_{PQ} if Q is at the origin and $\overline{E}$ is a conservative field.

Solution :

(a) $\overline{E}$ at $P = 8(2)(3)(5+1)\,\bar{a}_x + 4(2)^2(5+1)\,\bar{a}_y + 4(2)^2(3)\,\bar{a}_z$

$= 288\,\bar{a}_x + 96\,\bar{a}_y + 48\,\bar{a}_z$ V/m

$|\overline{E}| = \sqrt{(288)^2 + (96)^2 + (48)^2} = 307.35$ V/m

$\rho_s = |\overline{D}| = \varepsilon_o |\overline{E}| = \varepsilon_o \times 307.35 = 2.7213$ nC/m^2

(b) $\rho_v = \overline{\nabla} \cdot \overline{D} = \varepsilon_o(\overline{\nabla} \cdot \overline{E})$

$$= \varepsilon_o \left[\frac{\partial Ex}{\partial x} + \frac{\partial Ey}{\partial y} + \frac{\partial Ez}{\partial z} \right] = \varepsilon_o \left[8y(z+1) \right]$$

ρ_v at $P = \varepsilon_o [8(3)(5+1)] = 1.275$ nC/m^3

$$\text{(c) } V_{PQ} = -\int_Q^P \overline{E} \cdot \overline{dL}$$

$$= -\left[\int_0^2 8xy(z+1)\,dx + \int_0^3 4x^2(z+1)\,dy + \int_0^5 4x^2y\,dz \right] \text{T}$$

he path from Q to P is splitted as given below

$Q(0,0,0) \rightarrow A(2,0,0) \rightarrow B(2,3,0) \rightarrow P(2,3,5)$

Between the points Q and A, x coordinates varies from $x = 0$ to $x = 2$ where as y and z both are constants equal to 0. Hence in the first integral substitute $y = z = 0$.

Between the points A and B, y coordinate varies from $y = 0$ to $y = 3$ where as x and z both are constants. Hence in the second integral substitute $x = 2$ and $z = 0$.

Between the points B and P, z coordinate varies from $z = 0$ to $z = 5$ where as x and y both are constants. Hence in the third integral substitute $x = 2$ and $y = 3$.

$$V_{PQ} = -\int_0^2 8x(0)(0+1)\,dx - \int_0^3 4(2)^2(0+1)\,dy - \int_0^5 4(2)^2(3)\,dz$$

$$= -\left\{ 16\Big[y \Big]_0^3 + 48\Big[z \Big]_0^5 \right\} = -288$$

Problem 5.33 : Given the potential field $V = 100\,e^{-50x} \sin 50y$ V, in free space : (a) show that $\overline{\nabla} \cdot \overline{D} = 0$; (b) show that $y = 0$ is an equipotential surface ; (c) show that $\overline{E}$ is perpendicular to the $y = 0$ plane ; (d) find the total charge on the $y = 0$ plane, $0 < x < \infty$, $0 < z < 1$. Assume that $y < 0$ is the conductor interior. (e) How much energy is stored in the cube, $0 < x, y, z < 1$?

Solution :

(a) we have

$$\overline{E} = -\overline{\nabla} V = -\left[-5000\,e^{-50x} \sin 50y\,\overline{a}_x + 5000\,e^{-50x} \cos 50y\,\overline{a}_y\right]$$

$$= 5000\,(e^{-50x} \sin 50y\,\overline{a}_x - e^{-50x} \cos 50y\,\overline{a}_y)$$

$$\overline{\nabla} \cdot \overline{D} = \varepsilon_o\,(\overline{\nabla} \cdot \overline{E})$$

$$= 5000\,\varepsilon_o\left[-50\,e^{-50x} \sin 50y + 50\,e^{-50x} \sin 50y\right] = 0$$

(b) $V = 100\,e^{-50x} \sin(50 \times 0) = 0$

Since V is a constant and independent of x and z, $y = 0$ is an equipotential surface.

(c) $\overline{E}$ (at $y = 0$) $= 5000\left[e^{-50x} \sin(50 \times 0)\,\overline{a}_x - e^{-50x} \cos(50 \times 0)\,\overline{a}_y\right]$

$$= -5000\,e^{-50x}\,\overline{a}_y$$

Hence $\overline{E}$ (at $y = 0$) is perpendicular to the plane $y = 0$.

(d) $\overline{D} = \varepsilon_o\,\overline{E} = -5000\,\varepsilon_o\,e^{-50x}\,\overline{a}_y$ C/m^2

$\rho_S = |\overline{D}| = 5000\,\varepsilon_o\,e^{-50x}$ C/m^2

$$Q = \int dQ = \int \rho_S\,dS = \int_0^{\infty}\int_0^{1} 5000\,\varepsilon_o\,e^{-50x}\,dx\,dz$$

$$= 5000\,\varepsilon_o\left[\frac{e^{-50x}}{-50}\right]_0^{\infty}\left[z\right]_0^1 = 0.8854 \text{ nC}$$

Since $y < 0$ is the conductor interior, this charge should be negative.

(e) $W_E = \frac{\varepsilon_o}{2} \int | \overline{E} |^2 \, dv$

$$= \frac{\varepsilon_o}{2} \int_0^1 \int_0^1 \int_0^1 (5000 \, e^{-50x})^2 (\sin^2 50y + \cos^2 50y) \, dx \, dy \, dz$$

$$= \frac{\varepsilon_o}{2} \times 25 \times 10^6 \int_0^1 \int_0^1 \int_0^1 e^{-100x} \, dx \, dy \, dz$$

$$= 12.5 \times 10^6 \, \varepsilon_o \left[\frac{e^{-100x}}{-100} \right]_0^1 \left[y \right]_0^1 \left[z \right]_0^1 = 1.10675 \ \mu J$$

Problem 5.34 : A potential field in free space is given as $V = 10xyz - z + 5x$ V. (a) Find the equations of the equipotential surfaces on which $V = 0$ and 100 V. (b) These surfaces are conducting surfaces. Find the surface charge density on the $V = 100$ V surface at the point where $x = 2, x = 3$. Assume that $0 \leq V \leq 100$ V in the field containing region. (c) Give a unit vector at the above point, normal to the conducting surface, and directed toward the $V = 0$ surface.

Solution :

(a) Equation of the equipotential surface on which $V = 0$ is given by $10xyz - z + 5x = 0$ and for $V = 100$ is given by $10xyz - z + 5x = 100$.

(b) Substitute $x = 2$ and $y = 3$ in the equation of $V = 100$ V equipotential surface.

$10\,(2)\,(3)\,z - z + 5\,(2) = 100$ or $z = 1.5254$

Hence the point is (2, 3, 1.5254).

$$\overline{E} = - \overline{\nabla} V = - \left[(10yz + 5)\,\overline{a}_x + 10xz\,\overline{a}_y + (10xy - 1)\,\overline{a}_z \right]$$

$$\overline{E} \text{ at } (2, 3, 1.5254) = - 50.762\,\overline{a}_x - 30.508\,\overline{a}_y - 59\,\overline{a}_z \ \text{V/m}$$

$$\therefore \rho_s = | \overline{D} | = \varepsilon_o | \overline{E} | = \varepsilon_o \sqrt{(-50.762)^2 + (-30.508)^2 + (-59)^2}$$

$$= 0.74017 \ \text{nC/m}^2$$

(c) $\overline{E} = - \overline{\nabla} V = - | \overline{E} | \, \overline{a}_N$ where $\overline{a}_N$ is a unit vector perpendicular to the equipotential surface and directed towards increasing value of potential. Here we must determine $-\bar{a}_N$ so that it will be directed towards decreasing value of potential. Hence

$$\text{Unit vector required} = \frac{\bar{E}}{|\bar{E}|} = \frac{-50.762\,\bar{a}_x - 30.508\,\bar{a}_y - 59\,\bar{a}_z}{\sqrt{(-50.762)^2 + (-30.508)^2 + (-59)^2}}$$

$$= -0.607\,\bar{a}_x - 0.365\,\bar{a}_y - 0.706\,\bar{a}_z$$

Problem 5.35 : An infinite uniform line charge, $\rho_L = 25$ nC/m, lies along the line $x = 0, z = 1$ m in free space. The surface $z = 0$ is a perfect conductor. (a) Find $\bar{E}$ at P (1, 2, 3). (b) Find V at P if $V = 0$ at the origin. (c) What is the maximum magnitude of ρ_s on the conducting plane ?

Solution :

The image of $\rho_L = 25$ nC/m will be a line charge of -25 nC / m located at $x = 0, y = 1$ as shown in Fig.P. 5.35.

(a) $\bar{E}$ at $P(1, 2, 3) = \bar{E}_1$ at $P(1, 2, 3)$ due to $\rho_L = 25$ nC/m +
$\bar{E}_2$ at $P(1, 2, 3)$ due to $\rho_L = -25$ nC/m

$$\bar{E}_1 = \frac{\rho_L}{2\pi\varepsilon_o d}\,\bar{a}_{12} = \frac{25 \times 10^{-9}}{2\pi\varepsilon_o d}\,\bar{a}_{12}$$

Point 2 is that point at which $\bar{E}$ has to determine = (1, 2, 3)

Point 1 is the foot of the perpendicular dropped from point 2 on the line charge = (0, 2, 1)

$$\therefore \bar{R}_{12} = (1-0)\,\bar{a}_x + (3-1)\,\bar{a}_z$$

$$= \bar{a}_x + 2\,\bar{a}_z$$

$$d = |\bar{R}_{12}| = \sqrt{(1)^2 + (2)^2} = \sqrt{5}$$

$$\bar{a}_{12} = \frac{\bar{R}_{12}}{|\bar{R}_{12}|} = \frac{\bar{a}_x + 2\,\bar{a}_z}{\sqrt{5}}$$

$$\bar{E}_1 = \frac{28 \times 10^{-3}}{2\pi\varepsilon_o}\left[\frac{\bar{a}_x + 2\,\bar{a}_z}{\sqrt{5}}\right]$$

$$= 89.877\,\bar{a}_x + 179.755\,\bar{a}_z \text{ V/m}$$

$$\bar{E}_2 = \frac{\rho_L}{2\pi\varepsilon_o d}\,\bar{a}_{12} = \frac{-25 \times 10^{-9}}{2\pi\varepsilon_o d}\,\bar{a}_{12}$$

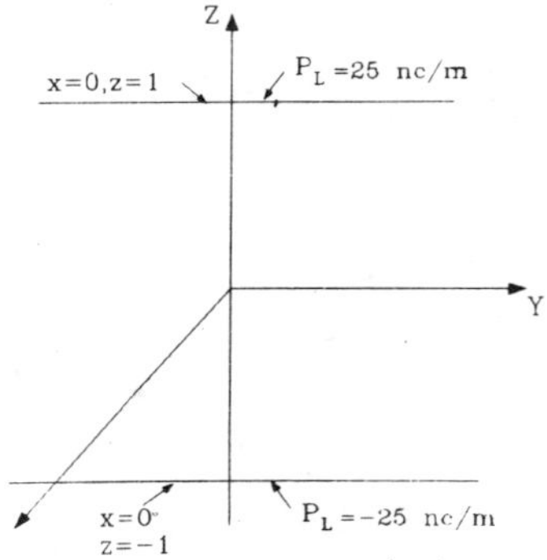

FIG. P 5.35

Point 2 is that point at which $\overline{E}$ has to be determine = (1, 2, 3)

Point 1 is the foot of the perpendicular dropped from point 2 on the line charge = (0, 2, –1)

$$\therefore \overline{R}_{12} = (1 - 0)\,\overline{a}_x + (3 + 1)\,\overline{a}_z = \overline{a}_x + 4\,\overline{a}_z$$

$$\therefore d = |\overline{R}_{12}| = \sqrt{(1)^2 + (4)^2} = \sqrt{17}$$

$$\overline{a}_{12} = \frac{\overline{R}_{12}}{|\overline{R}_{12}|} = \frac{\overline{a}_x + 4\,\overline{a}_z}{\sqrt{17}}$$

$$\therefore \overline{E}_2 = \frac{-25 \times 10^{-9}}{2\pi\varepsilon_o\sqrt{17}}\left[\frac{\overline{a}_x + 4\,\overline{a}_z}{\sqrt{17}}\right] = -26.435\,\overline{a}_x - 105.738\,\overline{a}_z \text{ V/m}$$

$$\therefore \overline{E} = \overline{E}_1 + \overline{E}_2 = 63.442\,\overline{a}_x + 74.017\,\overline{a}_z$$

(b) V_{PO} = V_{PO} due to ρ_L = 25 nC/m + V_{PO} due to ρ_L = – 25 nC/m

$$V_{PO}\,(\text{due } \rho_L = 25 \text{ nC/m}) = -\frac{\rho_L}{2\pi\varepsilon_o}\ln\frac{\rho_P}{\rho_O}$$

Coordinate of point P = (1, 2, 3)

Coordinate of the foot of perpendicular dropped from point P on the line charge = (0, 2, 1)

$$\rho_P = \sqrt{(1-0)^2 + (3-1)} = \sqrt{5}$$

Coordinates of point O = (0, 0, 0)

Coordinates of the foot of perpendicular dropped from point O on the line charge = (0, 0, 1)

$$\therefore \rho_O = \sqrt{(0-1)^2} = 1$$

$$V_{PO}\,(\text{due } \rho_L = 25 \text{ nC/m}) = \frac{-25 \times 10^{-9}}{2\pi\varepsilon_o}\ln\left(\frac{\sqrt{5}}{1}\right) = -361.630$$

$$V_{PO}\,(\text{due to } \rho_L = -25 \text{ nC/m}) = \frac{-(-25 \times 10^{-9})}{2\pi\varepsilon_o}\ln\left(\frac{\rho_P}{\rho_O}\right)$$

Coordinates of point P (1, 2, 3)

Coordinates of the foot of perpendicular dropped from point P on the line charge = (0, 2, –1)

$$\therefore \rho_P = \sqrt{(1-0)^2 + (3+1)^2} = \sqrt{17}$$

Coordinates of point O = (0, 0, 0)

Coordinates of the foot of perpendicular dropped from point O on the line charge = (0, 0, –1)

$$\therefore \rho_o = \sqrt{(0 + 1)^2} = 1$$

$$\therefore V_{PO} = -\frac{(-25 \times 10^{-9})}{2\pi\varepsilon_o} \ln\left(\frac{\sqrt{17}}{1}\right) = 636.605 \text{ V}$$

$$\therefore V_{PO}(\text{total}) = -361.63 + 636.605 = 274.975 \text{ V}$$

(c) The general point on the conducting plane is $(x, y, 0)$

At this point,
$$\overline{E} = \frac{25 \times 10^{-9}}{2\pi\varepsilon_o\sqrt{x^2 + 1}}\left[\frac{(x-0)\,\overline{a}_x + (0-1)\,\overline{a}_z}{\sqrt{x^2 + 1}}\right] - \frac{25 \times 10^{-9}}{2\pi\varepsilon_o\sqrt{x^2 + 1}}\left[\frac{(x-0)\,\overline{a}_x + (0+1)\,\overline{a}_z}{\sqrt{x^2 + 1}}\right]$$

Note : Same steps of part (a) are applicable to determine $\overline{E}_1$ and $\overline{E}_2$]

$$\therefore \overline{E} = \frac{25 \times 10^{-9}}{2\pi\varepsilon_o(x^2 + 1)}[-2\,\overline{a}_z]$$

$$\rho_s = |\overline{D}| = \varepsilon_o|\overline{E}| = \frac{\varepsilon_o \times 25 \times 10^{-9} \times 2}{2\pi\varepsilon_o(x^2 + 1)} = \frac{25 \times 10^{-9}}{\pi(x^2 + 1)} \text{ C/m}$$

At $x = 0$, ρ_s will be maximum and the maximum value is

$$\rho_{s_{max}} = \frac{25 \times 10^{-9}}{\pi} = 7.958 \text{ nC/m}^2.$$

This is the maximum magnitude of the surface charge density. Actually this charge density must be negative because in the problem ρ_L is a positive line charge.

Problem 5.36 : The $z = 0$ plane is a perfectly conducting surface. A point charge of 5 nC is located at A (2, –5, 4), and a point charge of – 8nC is at B (4, 3, 1), all in free space. (a) Find V at a point midway between the two point charges. (b) Find y if $V = 0$ at C (5, y, 1).

Solution :

Fig P. 5.36, shows the location of charges 5 nC and – 8 nC along with their images. Note that point A and A′ lies on a line which is perpendicular to the plane $z = 0$ and the distances of these points from the $z = 0$ plane is same. This is also true for the points B and B'.

(a) Let C be the midpoint of A and B, then

$$C = \left(\frac{2+4}{2}, \frac{-5+3}{2}, \frac{4+1}{2} \right)$$
$$= (3, -1, 2.5)$$

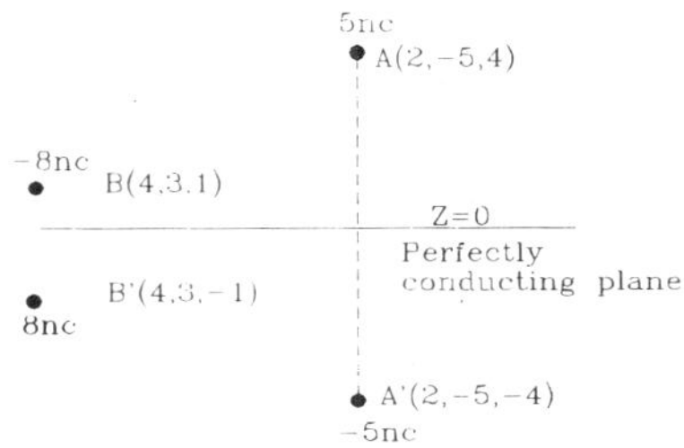

FIG.P.5.36

$$V \text{ at } C(3, -1, 2.5) = \frac{5 \times 10^{-9}}{4\pi\varepsilon_o d(AC)} - \frac{8 \times 10^{-9}}{4\pi\varepsilon_o d(BC)} - \frac{5 \times 10^{-9}}{4\pi\varepsilon_o d(A'C)} + \frac{8 \times 10^{-9}}{4\pi\varepsilon_o d(B'C)}$$

$$= \frac{5 \times 10^{-9}}{4\pi\varepsilon_o \sqrt{(3-2)^2 + (-1+5)^2 + (2.5-4)^2}}$$
$$- \frac{8 \times 10^{-9}}{4\pi\varepsilon_o \sqrt{(3-4)^2 + (-1-3)^2 + (2.5-1)^2}}$$
$$- \frac{5 \times 10^{-9}}{4\pi\varepsilon_o \sqrt{(3-2)^2 + (-1+5)^2 + (2.5+4)^2}}$$
$$+ \frac{8 \times 10^{-9}}{4\pi\varepsilon_o \sqrt{(3-4)^2 + (-1-3)^2 + (2.5+1)^2}}$$
$$= 1.311 \text{ V}$$

$$\text{(b) } V \text{ at } C(5, y, 1) = \frac{5 \times 10^{-9}}{4\pi\varepsilon_o \sqrt{(5-2)^2 + (y+5)^2 + (1-4)^2}}$$
$$- \frac{8 \times 10^{-9}}{4\pi\varepsilon_o \sqrt{(5-4)^2 + (y-3)^2 + (1-1)^2}}$$
$$- \frac{5 \times 10^{-9}}{4\pi\varepsilon_o \sqrt{(5-2)^2 + (y+5)^2 + (1+4)^2}}$$
$$+ \frac{8 \times 10^{-9}}{4\pi\varepsilon_o \sqrt{(5-4)^2 + (y-3)^2 + (1+1)^2}}$$

$$\therefore\ 0 = 5\left[\frac{1}{\sqrt{18 + (y+5)^2}} - \frac{1}{\sqrt{34 + (y+5)^2}}\right] +$$
$$8\left[\frac{1}{\sqrt{5 + (y-3)^2}} - \frac{1}{\sqrt{1 + (y-3)^2}}\right]$$

By trial and error procedure the value of y is equal to

Problem 5.37 : A dipole with $\bar{p} = 0.1\,\bar{a}_z\ \mu C\,.\,m$ is located at point A (0, 0, 1) in free space, and the $z = 0$ plane is perfectly conducting. (a) Find V at P (0, 1, 2). (b) Find the equation of the 200 – V equipotential surface in cartesian coordinates.

Solution :

(a)Fig.P. 5.37 shows the point dipole located at point A with its image in $z = 0$ conducting plane. In case of a dipole, the direction of image dipole remains the same as that of original dipole.

V at $P(0, 1, 2) = V_1$ at P (0, 1, 2) due to dipole at A + V_2 at P (0, 1, 2) due to dipole at A'

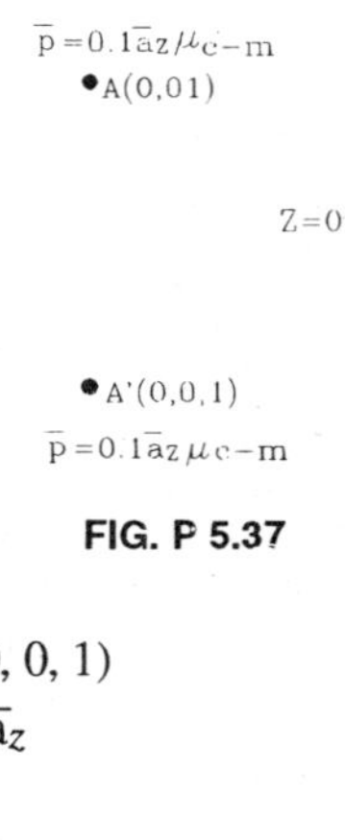

FIG. P 5.37

Now
$$V = \frac{\bar{p}\cdot(\bar{r}_2 - \bar{r}_1)}{4\pi\varepsilon_0\,|\bar{r}_2 - \bar{r}_1|^3}$$

Point 2 is that point at which potential has to be determined = (0, 1, 2)

Point 1 is that point at which dipole is located = (0, 0, 1)

$$\therefore\ \bar{r}_2 - \bar{r}_1 = (1\,\bar{a}_y + 2\,\bar{a}_z) - (\bar{a}_z) = 1\,\bar{a}_y + \bar{a}_z$$

$$\therefore\ |\bar{r}_2 - \bar{r}_1| = \sqrt{(1)^2 + (1)^2} = \sqrt{2}$$

$$\therefore\ V_1 = \frac{(0.1\,\bar{a}_z\,10^{-6})\cdot(1\,\bar{a}_y + \bar{a}_z)}{4\pi\varepsilon_0(\sqrt{2})^3} = \frac{10^{-7}}{4\pi\varepsilon_0(\sqrt{2})^3} = 317.765$$

For second dipole located at A', we have

$$\bar{r}_2 - \bar{r}_1 = (1\,\bar{a}_y + 2\,\bar{a}_z) - (-\bar{a}_z) = 1\,\bar{a}_y + 3\,\bar{a}_z$$

$$\therefore\ |\bar{r}_2 - \bar{r}_1| = \sqrt{(1)^2 + (3)^2} = \sqrt{10}$$

$$\therefore\ V_2 = \frac{(0.1\,\bar{a}_z\,10^{-6})\cdot(1\,\bar{a}_y + 3\,\bar{a}_z)}{4\pi\varepsilon_0(\sqrt{10})^3} = 0.3\times\frac{10^{-6}}{4\pi\varepsilon_0(\sqrt{10})^3}$$

$V_2 = 85.265$ V

$\therefore V$ at $P(0, 1, 2) = 317.765 + 85.265 = 403.03$ V

Let point (x, y, z) lies on the 200 V equipotential surface.

For first dipole located at A, we have

$$\bar{r}_2 - \bar{r}_1 = (x\,\bar{a}_x + y\,\bar{a}_y + z\,\bar{a}_z) - (1\,\bar{a}_z) = x\,\bar{a}_x + y\,\bar{a}_y + (z-1)\,\bar{a}_z$$

$$|\bar{r}_2 - \bar{r}_1| = \sqrt{x^2 + y^2 + (z-1)^2}$$

$$\therefore V_1 = \frac{(0.1\,\bar{a}_z\,10^{-6}) \cdot [x\,\bar{a}_x + y\,\bar{a}_y + (z-1)\,\bar{a}_z]}{4\pi\varepsilon_o[x^2 + y^2 + (z-1)^2]^{3/2}}$$

$$= \frac{(z-1)\,10^{-7}}{4\pi\varepsilon_o[x^2 + y^2 + (z-1)^2]^{3/2}}$$

For the second dipole located at A', we have

$$\bar{r}_2 - \bar{r}_1 = (x\,\bar{a}_x + y\,\bar{a}_y + z\,\bar{a}_z) - (-\bar{a}_z) = x\,\bar{a}_x + y\,\bar{a}_y + (z+1)\,\bar{a}_z$$

$$|\bar{r}_2 - \bar{r}_1| = \sqrt{x^2 + y^2 + (z+1)^2}$$

$$\therefore V_2 = \frac{(0.1\,\bar{a}_z\,10^{-6}) \cdot [x\,\bar{a}_x + y\,\bar{a}_y + (z+1)\,\bar{a}_z]}{4\pi\varepsilon_o[x^2 + y^2 + (z+1)^2]^{3/2}}$$

$$= \frac{(z+1)\,10^{-7}}{4\pi\varepsilon_o[x^2 + y^2 + (z+1)^2]^{3/2}}$$

$$V = V_1 + V_2 = 200$$

$$= \frac{10^{-7}}{4\pi\varepsilon_o}\left[\frac{z-1}{[x^2+y^2+(z-1)^2]^{1.5}} + \frac{z+1}{[x^2+y^2+(z+1)^2]^{1.5}}\right]$$

$$0.22 = \frac{z-1}{[x^2+y^2+(z-1)^2]^{1.5}} + \frac{z+1}{[x^2+y^2+(z+1)^2]^{1.5}}$$

Problem 5.38 : A circular ring, $\rho = 2$ m, $z = 1$ m, carries a uniform charge density of 8 nC/m, and the plane $z = 0$ is perfectly conducting. Assume the region $z > 0$ is free space. Find V at $P(0, 0, z)$, $z > 0$.

Solution :

Fig.P. 5.38 shows the given charge distribution along with its image

V at $P(0, 0, z) = V_1$ at $P(0, 0, z)$ due $\rho_L = 8$ nC/m +

V_2 at $P(0, 0, z)$ due $\rho_L = -8$ nC/m

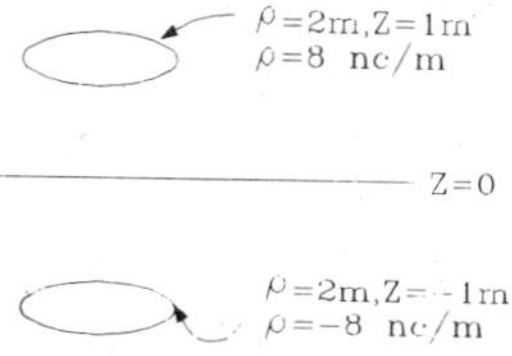

FIG. P 5.38

Now $V = \int \dfrac{\rho_L \, dL}{4 \pi \varepsilon_o \, | \bar{r}_2 - \bar{r}_1 |}$

For $\rho_L = 8$ nC/m we have

$dL = \rho \, d\phi = 2 \, d\zeta$

Point 2 is that point at which potential has to be determined = $(0, 0, z)$

Point 1 is the general point on the line charge = $(2, \phi, 1)$

$\therefore \bar{r}_2 - \bar{r}_1 = z \bar{a}_z - (2 \bar{a}_\rho + \bar{a}_z) = -2 \bar{a}_\rho + (z - 1) \bar{a}_z$

$$| \bar{r}_2 - \bar{r}_1 | = \sqrt{(-2)^2 + (z-1)^2} = \sqrt{z^2 - 2z + 5}$$

$$\therefore V_1 = \int_0^{2\pi} \frac{8 \times 10^{-9} \times 2 \, d\phi}{4 \pi \varepsilon_o \sqrt{z^2 - 2z + 5}} = \frac{903.546}{\sqrt{z^2 - 2z + 5}}$$

For $\rho_L = -8$ nC/m we have $dL = \rho \, d\phi = 2 \, d\phi$

Point 2 = $(0, 0, z)$.

Point 1 is the general point on the line charge = $(2, \phi, -1)$

$\therefore \bar{r}_2 - \bar{r}_1 = z \bar{a}_z - (2 \bar{a}_\rho - \bar{a}_z) = -2 \bar{a}_\rho + (z + 1) \bar{a}_z$

$$| \bar{r}_2 - \bar{r}_1 | = \sqrt{(-2)^2 + (z+1)^2} = \sqrt{z^2 + 2z + 5}$$

$$\therefore V_2 = \int_0^{2\pi} \frac{-8 \times 10^{-9} \times 2 \, d\phi}{4 \pi \varepsilon_o \sqrt{z^2 + 2z + 5}} = \frac{-903.546}{\sqrt{z^2 + 2z + 5}}$$

$$\therefore V = 903.546 \left[\frac{1}{\sqrt{z^2 - 2z + 5}} - \frac{1}{\sqrt{z^2 + 2z + 5}} \right]$$

Problem 5.39 : Let $\mu_e = 0.45 \, m^2/V \cdot s$ and $\mu_h = 0.25 \, m^2/V \cdot s$ in intrinsic germanium at a certain temperature. Both the electron and hole concentrations are $2.5 \times 10^{19} \, m^{-3}$. (a) Determine σ and the resistivity for the germanium sample. (b) Calculate the resistance of a germanium block that is 1 cm long with a 4 mm by 1mm cross section.

Solution :

(a) $\sigma = -\rho_e \mu_e + \rho_h \mu_h$

ρ_e = No. of electrons per m^3 × charge on one electron

$= (2.5 \times 10^{19}) (-1.6 \times 10^{-19}) = -4 \, C/m^3$

$\therefore \rho_h = (2.5 \times 10^{19})(1.6 \times 10^{-19}) = 4\ C/m^3$

$\therefore \sigma = -(-4)(0.45) + (4)(0.25) = 2.8$ mho/m

$\rho = \frac{1}{\sigma} = \frac{1}{2.8} = 0.3571\ \Omega -$ m

(b) $R = \frac{L}{\sigma S} = \frac{0.01}{2.8(0.004 \times 0.001)} = 892.857\ \Omega$

Problem 5.40 : Electron and hole concentrations increase with temperature. For pure silicon suitable expressions are $\rho_h = -\rho_e = 6200\ T^{1.5}\ e^{-7000/T}$ C/m³ . The functional dependence of the mobilities on temperature is given by $\mu_h = 2.3 \times 10^5\ T^{-2.7}\ m^2/V \cdot s$ and $\mu_e = 2.1 \times 10^5\ T^{-2.5}\ m^2/V \cdot s$. Find σ at : (a) 0° C; (b) 25 °C ; (c) 50° C.

Solution :

T	ρ_h	ρ_e	μ_h	μ_e
0° C = 273° K	2.04587×10^{-4}	-2.04587×10^{-4}	0.06082592	0.170534666
25° C = 298° K	$2.005246897 \times 10^{-3}$	$-2.005246897 \times 10^{-3}$	0.04801135	0.136986772
50° C = 323° K	0.013939293	-- 0.013939293	0.038626114	0.111998701

(a) σ at 0° C $= -(-2.04587 \times 10^{-4})(0.170534666) +$

$(2.04587 \times 10^{-4})(0.06082592)$

$= 4.733 \times 10^{-5}$ mho/m

(b) σ at 25° C $= -(-2.005246897 \times 10^{-3})(0.136986772) +$

$(2.00524689 \times 10^{-3})(0.04801135)$

$= 3.71 \times 10^{-4}$ mho/m

(c) σ at 50° C $= -(-0.01339293)(0.111998701) +$

$(0.013939293)(0.038626114)$

$= 2.1 \times 10^{-3}$ mho/m

Problem 5.41 : A small amount of phosphorus is added to silicon so that there are 2.5×10^{17} conduction electrons per cubic meter and 9.0×10^{14} holes per cubic meter. Let $\mu_e = 0.13\ m^2/V \cdot s$ for this sample, while $\mu_h = 0.05\ m^2/V \cdot s$. Determine the conductivity and resistivity.

Solution :

$$\sigma = -\left[(2.5\times 10^{17})(-1.6\times 10^{-19})\right](0.13) + \left[(9.0\times 10^{14})(1.6\times 10^{-19})\right]\times(0.05)$$

$$= 5.2072 \text{ m mho/m}$$

$$\rho = \frac{1}{\sigma} = 192.042\ \Omega - \text{m}$$

Problem 5.42 : At a certain temperature and pressure, atomic hydrogen contains 5.67×10^{25} atoms/m^3. If the electric susceptibility is 4.1×10^{-4} with $E = 3$ kV/m, find : (a) P ; (b) p ; (c) the effective dipole length d ; (d) ε_R.

Solution :

(a) $P = \chi_e \varepsilon_o E = 4.1 \times 10^{-4} \times 8.854 \times 10^{-12} \times 3000$

$$= 10.89042 \text{ pC/m}^2$$

(b) $P = Np$ or $p = \dfrac{P}{N} = \dfrac{10.89042\times 10^{-12}}{5.67\times 10^{25}} = 1.920709\times 10^{-37}$ C/m

(c) In one atom of the hydrogen, there is only one proton and one electron.

Hence $p = Q\,d$ or $d = \dfrac{p}{Q} = \dfrac{1.920709\times 10^{-37}}{1.6 \times 10^{-19}} = 1.20044\times 10^{-18}$ m

(d) $\varepsilon_R = \chi_e + 1 = 4.1 \times 10^{-4} + 1 = 1.00041$

Problem 5.43 : A certain linear, homogeneous, isotropic, dielectric material has a relative permittivity of 1.8. If $V = -4000\,y$ V in the material, find : (a) $\overline{E}$, $\overline{D}$, and $\overline{P}$; (b) ρ_v, ρ_b and ρ_T.

Solution :

(a) $\overline{E} = -\overline{\nabla} V = -(-4000\,\overline{a}_y) = 4000\,\overline{a}_y$ V/m

$\overline{P} = \chi_e\,\varepsilon_o\,\overline{E} = (\varepsilon_r - 1)\,\varepsilon_o\,\overline{E} = (1.8 - 1)\,\varepsilon_o \times 4000\,\overline{a}_y$

$= 3200\,\varepsilon_o\,\overline{a}_y\,C/m^2 = 28.3328\,\overline{a}_y\;nC/m^2$

$\overline{D} = \varepsilon_o\,\overline{E} + \overline{P} = \varepsilon_o \times 4000\,\overline{a}_y + 3200\,\varepsilon_o\,\overline{a}_y = 7200\,\varepsilon_o\,\overline{a}_y$

$= 63.7488\,\overline{a}_y\,nC/m^2$

(b) $\rho_v = \overline{\nabla} \cdot \overline{D} = 0$, $\rho_b = -(\overline{\nabla} \cdot \overline{P}) = 0$, and $\rho_T = \overline{\nabla} \cdot \varepsilon_o\,\overline{E} = 0$

Problem 5.44 : A coax with $a = 1$ mm and $b = 5$ mm has a Teflon dielectric. If $E_\rho = 80/\rho$ V/m , neglect any losses in the dielectric, assume 4×10^{10} molecules / m^3 in the Teflon, and find : (a) $\overline{E}$, $\overline{D}$, $\overline{P}$, and $\overline{p}$ as functions of ρ ; (b) V_{ab} , ε_R , and χ_e

Solution :

(a) $\overline{E} = E_\rho\,\overline{a}_\rho = \dfrac{80}{\rho}\,\overline{a}_\rho$ V/m and for Teflon $\varepsilon_r = 2.1$

$\therefore\ \chi_e = \varepsilon_R - 1 = 2.1 - 1 = 1.1$

$\therefore\ \overline{P} = \chi_e\,\varepsilon_o\,\overline{E} = 1.1\,\varepsilon_o\,\dfrac{80}{\rho}\,\overline{a}_\rho = \dfrac{779.152}{\rho}\,\overline{a}_\rho\ pC/m^2$

$\overline{D} = \varepsilon_o\,\overline{E} + \overline{P} = \dfrac{80\,\varepsilon_o}{\rho}\,\overline{a}_\rho + \dfrac{88\,\varepsilon_o}{\rho}\,\overline{a}_\rho = \dfrac{1.487472}{\rho}\,\overline{a}_\rho\ nC/m^2$

$\overline{P} = N\,\overline{p}$ or $\overline{p} = \dfrac{\overline{P}}{N} = \dfrac{779.152 \times 10^{-12}\,\overline{a}_\rho}{\rho \times 4 \times 10^{19}} = \dfrac{1.94788}{\rho}\,\overline{a}_\rho\ C - m$

(b) $V_{ab} = -\displaystyle\int_b^a \overline{E} \cdot \overline{dL} = -\int_b^a \overline{E} \cdot (d\rho\,\overline{a}_\rho) = -\int_{0.005}^{0.001} \frac{80}{\rho}\,d\rho$

$= -80\Big[\ln\rho\Big]_{0.005}^{0.001} = 128.755$ V

$\varepsilon_R = 2.1$, $\chi_e = 1.1$

Problem 5.45 : The polarization within a region having $\varepsilon_R = 2.7$ has the uniform value $\overline{P} = -0.2\,\overline{a}_x + 0.7\,\overline{a}_y + 0.3\,\overline{a}_z\ \mu C/m^2$. Find : (a) $\overline{E}$

; (b) $\overline{D}$; (c) the magnitude of the voltage gradient.

Solution :

$\chi_e = \varepsilon_R - 1 = 2.7 - 1 = 1.7$

(a) $\overline{P} = \chi_e \varepsilon_o \overline{E}$

$$\therefore \overline{E} = \frac{\overline{P}}{\chi_e \varepsilon_o} = \frac{(-0.2\,\overline{a}_x + 0.7\,\overline{a}_y + 0.3\,\overline{a}_z)\,10^{-6}}{1.7\,\varepsilon_o}$$

$$= -13287.447\,\overline{a}_x + 46506.066\,\overline{a}_y + 19931.171\,\overline{a}_z \text{ V/m}$$

(b) $\overline{D} = \varepsilon_o \overline{E} + \overline{P}$

$$= \varepsilon_o(-13287.447\,\overline{a}_x + 46506.066\,\overline{a}_y + 19931.171\,\overline{a}_z) + (-0.2\,\overline{a}_x + 0.7\,\overline{a}_y + 0.3\,\overline{a}_z)\,10^{-6}$$

$$= -0.3176\,\overline{a}_x + 1.11176\,\overline{a}_y + 0.47647\,\overline{a}_z \ \mu\text{C/m}^2$$

(c) $|\overline{E}| = \sqrt{(-13287.447)^2 + (46506.066)^2 + (19931.171)^2}$

$$= 52312.733 \text{ V/m}$$

Problem 5.46 : A unit vector directed from region 1 to region 2 at the planar boundary between two perfect dielectrics is given as $\overline{a}_{N12} = -\frac{2}{7}\overline{a}_x + \frac{3}{7}\overline{a}_y + \frac{6}{7}\overline{a}_z$. Assume $\varepsilon_{R1} = 3$, $\varepsilon_{R2} = 2$, and $\overline{E}_1 = 100\,\overline{a}_x + 80\,\overline{a}_y + 60\,\overline{a}_z$ V/m . Find $\overline{E}_2$.

Solution :

$\overline{E}_{N1} = (\overline{E}_1 \cdot \overline{a}_{N12})\,\overline{a}_{N12}$

$$= \left(-\frac{200}{7} + \frac{240}{7} + \frac{360}{7}\right)\left(-\frac{2}{7}\overline{a}_x + 3\,\overline{a}_y + \frac{6}{7}\overline{a}_z\right)$$

$$= -16.327\,\overline{a}_x + 24.490\,\overline{a}_y + 48.98\,\overline{a}_z \text{ V/m}$$

Now $\overline{E}_1 = \overline{E}_{N1} + \overline{E}_{t1}$. Hence

$\overline{E}_{t1} = \overline{E}_1 - \overline{E}_{N1} = 116.327\,\overline{a}_x + 55.51\,\overline{a}_y + 11.02\,\overline{a}_z$ V/m

Now the boundary conditions are $\overline{E}_{t2} = \overline{E}_{t1}$ and

$$\overline{E}_{N2} = \frac{\varepsilon_{r1}}{\varepsilon_{r2}}\overline{E}_{N1} = \frac{3}{2}(-16.327\,\overline{a}_x + 24.490\,\overline{a}_y + 48.98\,\overline{a}_z)$$

$$= -24.491\,\overline{a}_x + 36.735\,\overline{a}_y + 73.47\,\overline{a}_z \text{ V/m}$$

$$\therefore\ \bar{E}_2 = \bar{E}_{t2} + \bar{E}_{N2} = 116.327\,\bar{a}_x + 55.510\,\bar{a}_y + 11.02\,\bar{a}_z$$
$$- 24.491\,\bar{a}_x + 36.735\,\bar{a}_y + 73.47\,\bar{a}_z$$
$$= 91.836\,\bar{a}_x + 92.245\,\bar{a}_y + 84.49\,\bar{a}_z$$

Problem 5.47 : The region $z < 0$ contains a perfect dielectric for which $\varepsilon_{R1} = 2.5$, while the region $z > 0$ is charaterized by $\varepsilon_{R2} = 4$. Let $\bar{E}_1 = -30\,\bar{a}_x + 50\,\bar{a}_y + 70\,\bar{a}_z$ V/m and find : (a) E_{N1} ; (b) $\bar{E}_{t1}$; (c) E_{t1} ; (d) E_1 ; (e) the angle $\theta_1 \leq 90^o$ between $\bar{E}_1$ and a normal to the surface ; (f) D_{N2} ; (g) D_{t2} ; (h) $\bar{D}_2$; (i) $\bar{P}_2$; (j) the angle θ_2 between $\bar{E}_2$ and a normal to the surface.

Solution :

(a) $\pm\ \bar{a}_z$ is perpendicular to the dielectric interface $z = 0$.

Hence $\bar{E}_{N1} = 70\,\bar{a}_z$ (that part of $\bar{E}_1$ which has either $+\ \bar{a}_z$ or $-\ \bar{a}_z$)

or $E_{N1} = |\,\bar{E}_{N1}\,| = 70$ V/m .

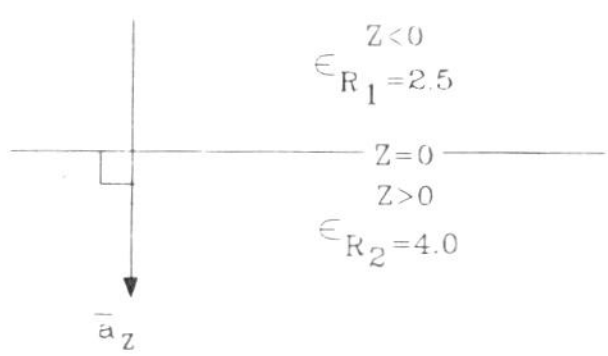

FIG. P 5.47

(b) $\bar{E}_1 = \bar{E}_{t1} + \bar{E}_{N1}$

$\therefore\ \bar{E}_{t1} = \bar{E}_1 - \bar{E}_{N1}$

$= -30\,\bar{a}_x + 50\,\bar{a}_y$ V/m

(c) $E_{t1} = |\,\bar{E}_{t1}\,| = \sqrt{(-30)^2 + (50)^2} = 58.31$ V/m

(d) $E_1 = |\,\bar{E}_1\,| = \sqrt{(-30)^2 + (50)^2 + (70)^2} = 91.105$ V/m

(e) $\cos\theta = \dfrac{\bar{E}_1 \cdot \bar{a}_z}{|\,\bar{E}_1\,|\ |\,\bar{a}_z\,|} = \dfrac{70}{91.105} = 0.768344$

or $\theta = \cos^{-1}(0.768344) = 39.7945^o$

(f) $\bar{D}_{N2} = \bar{D}_{N1} = \varepsilon_o\,\varepsilon_{r1}\,\bar{E}_{N1} = \varepsilon_o\,(2.5)\,(70\,\bar{a}_z) = 1.54945\,\bar{a}_z$ nC/m^2

$D_{N2} = |\,\bar{D}_{N2}\,| = 1.54945$ nC/m^2

(g) $\bar{E}_{t2} = \bar{E}_{t1} = -30\,\bar{a}_x + 50\,\bar{a}_y$

$\bar{D}_{t2} = \varepsilon_o\,\varepsilon_{r2}\,\bar{E}_{t2} = \varepsilon_o\,(4)\,[-30\,\bar{a}_x + 50\,\bar{a}_y]$

$$\overline{D}_{t2} = -1.06248\,\overline{a}_x + 1.7708\,\overline{a}_y \ \text{nC/m}^2$$

$$D_{t2} = |\overline{D}_{t2}| = \sqrt{(-1.06248)^2 + (1.7708)^2} = 2.0651 \ \text{nC/m}^2$$

(h) $\overline{D}_2 = \overline{D}_{t2} + \overline{D}_{N2} = -1.06248\,\overline{a}_x + 1.7708\,\overline{a}_y + 1.54945\,\overline{a}_z \ \text{nC/m}^2$

(i) $\overline{P}_2 = \chi_e \varepsilon_o \overline{E}_2$ and $\chi_{e2} = \varepsilon_{r2} - 1 = 4.0 - 1 = 3$

$$\overline{E}_2 = \overline{E}_{t2} + \overline{E}_{N2} = -30\,\overline{a}_x + 50\,\overline{a}_y + \frac{1.54945 \times 10^{-9}}{4\,\varepsilon_o}\overline{a}_z$$

$$= -30\,\overline{a}_x + 50\,\overline{a}_y + 43.75\,\overline{a}_z \ \text{V/m}$$

$$\therefore \ \overline{P}_2 = 3\,\varepsilon_o(-30\,\overline{a}_x + 50\,\overline{a}_y + 43.75\,\overline{a}_z)$$

$$= -796.86\,\overline{a}_x + 1328.1\,\overline{a}_y + 1162.0875\,\overline{a}_z \ \text{pC/m}^2$$

(j) $\cos\theta_2 = \dfrac{\overline{E}_2 \cdot \overline{a}_z}{|\overline{E}_2|\,|\overline{a}_z|} = \dfrac{43.75}{\sqrt{(-30)^2 + (50)^2 + (43.75)^2}} = 0.6$

$$\therefore \ \theta_2 = \cos^{-1}(0.6) = 53.119^o$$

Problem 5.48 : Two infinite slabs of perfect dielectric are parallel and surrounded by free space, as shown in Fig.P5.48. In region 3, $\overline{E}_3 = 3\,\overline{a}_z$ V/m and $\overline{P}_3 = 15\,\overline{a}_z$ pC/m^2. If $\overline{P}_2 = 20\,\overline{a}_z$ pC/m^2, find : (a) ε_{R3} ; (b) ε_{R2} ; (c) $\overline{E}_1$; (d) the total voltage across regions 2 and 3.

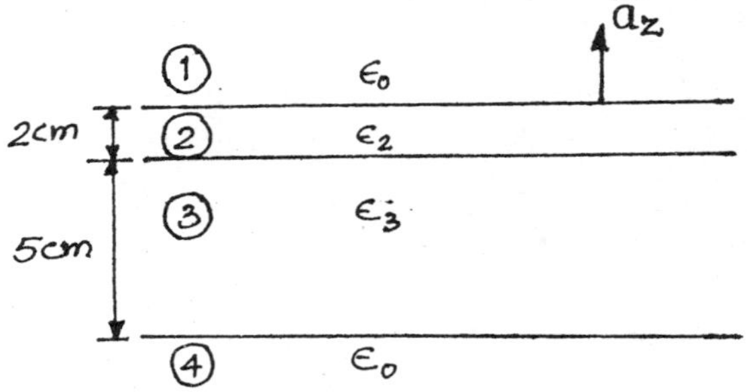

FIG.P.5.48

Solution :

(a) $\overline{P}_3 = \chi_{e3}\,\varepsilon_o\,\overline{E}_3$

$$\therefore \ 15 \times 10^{-12}\,\overline{a}_z = 3\,\varepsilon_o\,\chi_{e3}\,\overline{a}_z \quad \text{or} \quad \chi_{e3} = \frac{15 \times 10^{-12}}{3\,\varepsilon_o} = 0.5647165$$

$\varepsilon_{R3} = \chi_{e3} + 1 = 0.5647165 + 1 = 1.5647165$

(b) $\overline{D}_{N2} = \overline{D}_{N3}$ or $\varepsilon_o \varepsilon_{r2} \overline{E}_{N2} = \varepsilon_o \varepsilon_{r3} \overline{E}_{N3}$ or $\overline{E}_{N2} = \dfrac{\varepsilon_{r3}}{\varepsilon_{r2}} \overline{E}_{N3}$

No tangential component of $\overline{E}$ or $\overline{P}$ is given .Hence, $\overline{E}_{N2} = \overline{E}_2$ and

$\overline{E}_{N3} = \overline{E}_3$, so that $\overline{E}_2 = \dfrac{\varepsilon_{r3}}{\varepsilon_{r2}} \overline{E}_3$ and $\overline{P}_2 = \chi_{e2}\, \varepsilon_o \overline{E}_2$

Hence, $20 \times 10^{-12} \overline{a}_z = \chi_{e2}\, \varepsilon_o \times \dfrac{\varepsilon_{r3}}{\varepsilon_{r2}} \times 3\, \overline{a}_z$

$$\frac{\chi_{e2}}{\varepsilon_{R2}} = \frac{20 \times 10^{-12}}{3\, \varepsilon_{R3}\, \varepsilon_o} = \frac{20 \times 10^{-12}}{3 \times 1.5647165 \times 8.854 \times 10^{-12}}$$

$$\frac{\varepsilon_{R2} - 1}{\varepsilon_{R2}} = 0.4812088 \quad \text{or} \quad \varepsilon_{R2} = 1.92756$$

(c) $\overline{D}_{N1} = \overline{D}_{N2}$ or $\varepsilon_o \varepsilon_{R1} \overline{E}_{N1} = \varepsilon_o \varepsilon_{R2} \overline{E}_{N2}$

$$\therefore \overline{E}_1 = \frac{\varepsilon_{R2}}{\varepsilon_{R1}} \overline{E}_2 = \frac{\varepsilon_{R2}}{\varepsilon_{R1}} \times \frac{\varepsilon_{R3}}{\varepsilon_{R2}} \overline{E}_3 = \frac{1.5647165 \times 3\, \overline{a}_z}{1}$$

$$= 4.694\, \overline{a}_z \ \text{V/m}$$

(d) $$V = E_2 \times (0.002) + E_3 (0.05) = \frac{\varepsilon_{R3}}{\varepsilon_{R2}} E_3 \times 0.02 + 0.05\, E_3$$

$$= \frac{1.5647165 \times 0.02 \times 3}{1.92756} + 0.05 \times 3 = 0.1987 \ \text{V}$$

Problem 5.49 : In Fig.P.5.49 let $\overline{D}_1 = 2\,\overline{a}_x + 5\,\overline{a}_y - 3\,\overline{a}_z$ nC/m^2 and find : (a) $\overline{D}_2$; (b) $\overline{D}_{N2}$; (c) $\overline{D}_{t2}$; (d) the energy density in each region; (e) the angle that $\overline{D}_2$ makes with $\overline{a}_z$; (f) D_2 / D_1 ; (g) P_2 / P_1 .

Solution :

(a) $\pm\ \overline{a}_z$ is perpendicular to the dielectric interface. Hence

$\overline{D}_{N1} = -3\,\overline{a}_z$ nC/m^2

(that part of $\overline{D}_1$ which contains $+\overline{a}_z$ or $-\overline{a}_z$)

$\therefore \overline{D}_{t1} = \overline{D}_1 = \overline{D}_{N1}$

$= 2\,\overline{a}_x + 5\,\overline{a}_y$ nC/m^2

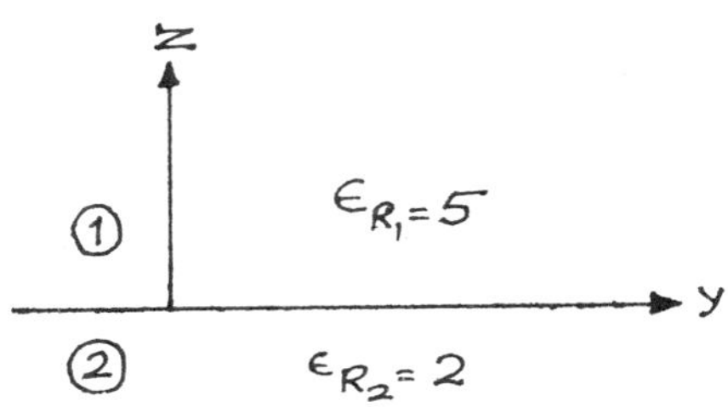

FIG. P 5. 49

Now $\overline{D}_{N1} = \overline{D}_{N2}$ and $\overline{E}_{t2} = \overline{E}_{t1}$

$$\therefore \overline{E}_{t2} = \frac{\overline{D}_{t1}}{\varepsilon_o \varepsilon_{R1}} = \frac{(2\bar{a}_x + 5\bar{a}_y)\,10^{-9}}{5\varepsilon_o}$$

$$\therefore \overline{D}_{t2} = \varepsilon_o \varepsilon_{R2} \overline{E}_{t2} = \frac{2\varepsilon_o}{5\varepsilon_o}(2\bar{a}_x + 5\bar{a}_y)\,10^{-9} = 0.8\,\bar{a}_x + 2\,\bar{a}_y \text{ nC/m}^2$$

$$\therefore \overline{D}_2 = \overline{D}_{t2} + \overline{D}_{N2} = 0.8\,\bar{a}_x + 2\,\bar{a}_y - 3\,\bar{a}_z \text{ nC/m}^2$$

(b) $\overline{D}_{N2} = -3\,\bar{a}_z \text{ nC/m}^2$

(c) $\overline{D}_{t2} = 0.8\,\bar{a}_x + 2\,\bar{a}_y \text{ nC/m}^2$

$$\text{(d)}\, W_{E1} = \frac{|\overline{D}_1|^2}{2\varepsilon_o \varepsilon_{R1}} = \frac{[(2)^2 + (5)^2 + (-3)^2] \times 10^{-18}}{2\varepsilon_o 5}$$

$$= 0.42918\ \mu\text{J/m}^3$$

$$W_{E2} = \frac{|\overline{D}_2|^2}{2\varepsilon_o \varepsilon_{R2}} = \frac{[(0.8)^2 + (2)^2 + (-3)^2] \times 10^{-18}}{2\varepsilon_o 2}$$

$$= 0.3851\ \mu\text{J/m}^3$$

$$\text{(e)}\cos\theta_2 = \frac{\overline{D}_2 \cdot \bar{a}_z}{|\overline{D}_2|\,|\bar{a}_z|} = \frac{-3 \times 10^{-9}}{\sqrt{(0.8)^2 + (2)^2 + (-3)^2}\ 10^{-9}} = -0.81229$$

$$\therefore \theta_2 = 144.3208^o$$

$$\text{(f)}\ \frac{D_2}{D_1} = \frac{\sqrt{(0.8)^2 + (2)^2 + (-3)^2}\ 10^{-9}}{\sqrt{(2)^2 + (5)^2 + (-3)^2}\ 10^{-9}} = 0.599$$

$$\text{(g)}\ \frac{P_2}{P_1} = \frac{D_2 \frac{(\varepsilon_{R2} - 1)}{\varepsilon_{R2}}}{D_1 \frac{(\varepsilon_{R1} - 1)}{\varepsilon_{R1}}} = 0.599\ \frac{\frac{(2-1)}{2}}{\frac{(5-1)}{5}} = 0.3743$$

Problem 5.50 : Two concentric conducting spheres of radii $a = 5$ cm, $b = 20$ cm, have dielectrics $\varepsilon_{R1} = 2$, $0 < \theta < \pi/2$, and $\varepsilon_{R2} = 5$, $\pi/2 < \theta < \pi$, separating them. Assume $E_{r1} = 200$ V/m at $r = 10$ cm, $\theta = 28^o$, $\phi = 106^o$, and find : (a) $\overline{E}_1$; (b) $\overline{E}_2$; (c) V_{ab} ; (d) the total charge on the inner sphere.

Solution :

(a)Let Q be the charge on the inner sphere and $\overline{E} = E_r \overline{a}_r$ be the electric field intensity. Consider a sphere of radius r such that $0.05 < r < 0.2$. Apply Gauss's law to this sphere $\oint \overline{D} \cdot \overline{dS} = Q$

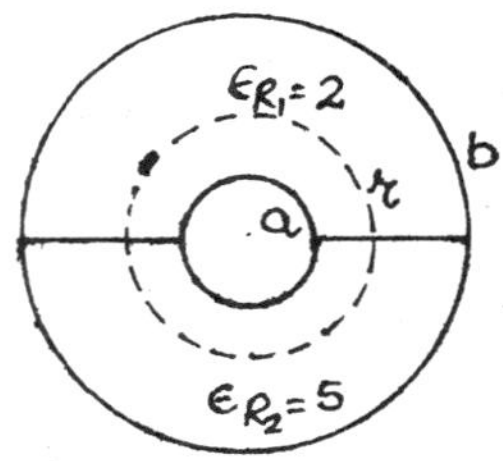

FIG. P 5.50

Let $\overline{D}_1$ and $\overline{D}_2$ be the electric flux densities in regions 1 and 2 respectively.

$$\text{LHS} = \oint \overline{D} \cdot \overline{dS} = \int \overline{D}_1 \cdot \overline{dS} + \int \overline{D}_2 \cdot \overline{dS}$$

$$= \int (\varepsilon_o \varepsilon_{R1} E_r \overline{a}_r) \cdot (r^2 \sin\theta \, d\theta \, d\phi \, \overline{a}_r) + \int (\varepsilon_o \varepsilon_{R2} E_r \overline{a}_r) \cdot (r^2 \sin\theta \, d\theta \, d\phi \, \overline{a}_r)$$

$$= \varepsilon_o \varepsilon_{R1} r^2 E_r \int_0^{\pi/2} \int_0^{2\pi} \sin\theta \, d\theta \, d\phi + \varepsilon_o \varepsilon_{R2} r^2 E_r \int_{\pi/2}^{\pi} \int_0^{2\pi} \sin\theta \, d\theta \, d\phi$$

$$= 4\pi \varepsilon_o r^2 E_r + 10\pi \varepsilon_o r^2 E_r = 14\pi \varepsilon_o r^2 E_r$$

$$\therefore \; 14\pi \varepsilon_o r^2 E_r = Q \quad \text{or} \quad E_r = \frac{Q}{14\pi \varepsilon_o r^2}$$

At $r = 10$ cm , $E_{r1} = 200$ V/m . Hence $200 = \dfrac{Q}{14\pi \varepsilon_o (0.1)^2}$

$$\therefore \; Q = 7.788 \times 10^{-10} \text{ C}$$

$$\therefore \; \overline{E}_1 = E_r \overline{a}_r = \frac{7.788 \times 10^{-10}}{14\pi \varepsilon_o r^2} \overline{a}_r = \frac{2}{r^2} \overline{a}_r \text{ V/m}$$

(b) $\overline{E}_2 = \overline{E}_1 = \dfrac{2}{r^2} \overline{a}_r$ V/m

(c) $V_{ab} = -\displaystyle\int_b^a \overline{E} \cdot \overline{dL} = -\int_b^a (E_r \overline{a}_r) \cdot (dr \, \overline{a}_r) = -\int_b^a E_r \, dr$

$$V_{ab} = -\int_{0.2}^{0.05} \frac{2}{r^2}\, dr = -2\left[-\frac{1}{r}\right]_{0.2}^{0.05} = 30 \text{ V}$$

(d) $Q = 7.788 \times 10^{-10} = 778.838$ pC .

Problem 5.51 :A dielectric interface is defined by the plane, $4x - 2y + 3z = 4$. Region 1, which includes the point P_1 (0,0,100), has $\varepsilon_{R1} = 3$, while region 2, which contains P_2 (0,0, - 100), has $\varepsilon_{R2} = 8$. (a) Find the unit vector that is perpendicular to the plane and directed into region 1. (b) Let point $A(1,3,z)$ lie on the plane. At A, $\bar{E}_1 = 10\,\bar{a}_x + 20\,\bar{a}_y + 40\,\bar{a}_z$ V/m. Find $\bar{E}_2$ at A.

Solution :

(a) Let $f = 4x - 2y + 3z - 4$, then $\bar{a}_n = \pm \dfrac{\bar{\nabla} f}{|\bar{\nabla} f|}$ gives the possible unit vectors which are perpendicular to f.. The unit vector with – ve sign gives the unit vector which is directed from higher value of f towards the lower value of f . The unit vector with + ve sign gives the unit vector which is directed from lower value of f towards the higher value of f .We have to determine $\bar{a}_{21}$.

In region 1, at point P_1 $(0, 0, 100)$, $f_1 = 296$

In region 2, at point P_2 $(0, 0, -100)$, $f_2 = -304$

Hence we have to determine the unit vector from lower value of f ($f_2 = -304$) torwards higher value of f ($f_1 = 296$) . That is with + ve sign.. Hence

$$\bar{a}_n = \frac{\bar{\nabla} f}{|\bar{\nabla} f|} = \frac{4\,\bar{a}_x - 2\,\bar{a}_y + 3\,\bar{a}_z}{\sqrt{(4)^2 + (-2)^2 + (3)^2}} = 0.743\,\bar{a}_x - 0.371\,\bar{a}_y + 0.557\,\bar{a}_z$$

(b) Let point $A(1, 3, z)$ lie on the plane. Then

$4(1) - 2(3) + 3z = 4$, i.e. $z = 2$

$$\bar{E}_{N1} = (\bar{E}_1 \cdot \bar{a}_{21})\,\bar{a}_{21}$$
$$= (7.43 - 7.42 + 22.28)\,(0.743\,\bar{a}_x - 0.371\,\bar{a}_y + 0.557\,\bar{a}_z)$$
$$= 16.56147\,\bar{a}_x - 8.26959\,\bar{a}_y + 12.41553\,\bar{a}_z \text{ V/m} .$$

$\bar{E}_1 = \bar{E}_{t1} + \bar{E}_{N1}$, $\bar{E}_{t1} = \bar{E}_1 - \bar{E}_{N1}$

$$\bar{E}_{t1} = (10\,\bar{a}_x + 20\,\bar{a}_y + 40\,\bar{a}_z) - (16.56147\,\bar{a}_x - 8.26959\,\bar{a}_y + 12.41553\,\bar{a}_z)$$
$$= -6.56147\,\bar{a}_x + 28.26959\,\bar{a}_y + 27.58447\,\bar{a}_z \text{ V/m}$$

Now $\bar{D}_{N1} = \bar{D}_{N2}$ or $\varepsilon_o\,\varepsilon_{R1}\,\bar{E}_{N1} = \varepsilon_o\,\varepsilon_{R2}\,\bar{E}_{N_2}$

$$\therefore\ \bar{E}_{N2} = \frac{\varepsilon_{R1}}{\varepsilon_{R2}}\,\bar{E}_{N1} = \frac{3}{8}\,(16.56147\,\bar{a}_x - 8.26959\,\bar{a}_y + 12.41553\,\bar{a}_z)$$
$$= 6.2015\,\bar{a}_x - 3.101\,\bar{a}_y + 4.656\,\bar{a}_z \text{ V/m}$$

$$\therefore\ \bar{E}_2 = \bar{E}_{t2} + \bar{E}_{N2} = \bar{E}_{t1} + \bar{E}_{N2}$$
$$= -6.56147\,\bar{a}_x + 28.26959\,\bar{a}_y + 27.58447\,\bar{a}_z + 6.2105\,\bar{a}_x - 3.101\,\bar{a}_y + 4.656\,\bar{a}_z$$
$$= -0.351\,\bar{a}_x + 25.169\,\bar{a}_y + 32.240\,\bar{a}_z \text{ V/m}$$

Problem 5.52 : A parallel - plate capacitor utilizes plates of area 0.06 m^2 and 0.4 mm separation. (a) If the region between the plates is filled with a dielectric for which ε_R = 200 and $\sigma = 10^{-5}$ mho/m, find C and the Q (quality factor) of the capacitor at $\omega = 10^7$ rad/s . (b) If the maximum voltage gradient allowable in the dielectric is 10^5 V / m, what is the breakdown voltage for the capacitor? (c) If 10 V dc is applied across the capacitor, find the energy stored and the power being dissipated.

Solution :

$A = 0.06\text{ m}^2$, $d = 0.4\text{ mm} = 0.4 \times 10^{-3}\text{ m}$

$$\text{(a) } C = \frac{\varepsilon_o\,\varepsilon_r\,A}{d} = \frac{\varepsilon_o \times 200 \times 0.06}{0.4 \times 10^{-3}} = 0.26562\ \mu\text{F}$$

$$Q = \frac{\omega\,\varepsilon}{\sigma} = \frac{10^7 \times \varepsilon_o \times 200}{10^{-5}} = 1770.8$$

(b) Breakdown voltage $= 10^5 \times d = 10^5 \times 0.4 \times 10^{-3} = 40$ V

$$\text{(c) Energy} = \frac{1}{2}\,C\,V^2 = \frac{1}{2} \times 0.26562 \times 10^{-6} \times (10)^2 = 13.281\ \mu\text{J}$$

$$R = \frac{L}{\sigma S} = \frac{0.4 \times 10^{-3}}{10^{-5} \times 0.06} = 666.6667\ \Omega$$

$$P = \frac{V^2}{R} = \frac{(10)^2}{666.6667} = 0.15 \text{ watts.}$$

Problem 5.53 : Two parallel conducting plates are each 10 cm by 10 cm and separated 2 mm. The region between the plates is filled with a perfect dielectric for which $\varepsilon_r = (1 + 500\,x)^2$, where x is the distance from one plate. Assume a uniform surface charge density of 10 nC / m^2 on the positive plate, and determine : (a) Q_{total} ; (b) D_x ; (c) E_x ; (d) V_o ; (e) C .

Solution :

$A = 0.1 \times 0.1 = 0.01\ \text{m}^2$ and $d = 2\ \text{mm} = 2 \times 10^{-3}\ \text{m}$

(a) $Q_{total} = \rho_S \times Area = 10 \times 10^{-9} \times 0.01 = 100\ \text{pC}$

(b) As $D_N = \rho_S = \varepsilon E$ we have $D_x = 10\ \text{nC/m}^2$

(c) $$E_x = \frac{D_x}{\varepsilon_o \varepsilon_r} = \frac{10 \times 10^{-9}}{\varepsilon_o (1 + 500\,x)^2} = \frac{1129.433}{(1 + 500\,x)^2}\ \text{V/m}$$

(d) $$V_o = -\int_0^{0.002} E_x\, dx = -\int_0^{0.002} \frac{1129.433}{(1 + 500\,x)^2}\, dx$$

$$= \left[\frac{1129.433}{500\,(1 + 500\,x)} \right]_0^{0.002} = -\ 1.129433$$

$| V_o | =$ Voltage applied to the capacitor plates $= 1.129$

(e) $$C = \frac{Q}{V} = \frac{\rho_S A}{V} = \frac{10 \times 10^{-9} \times 0.01}{1.129} \doteq 88.574\ \text{pF}$$

Problem 5 .54 : Consider a parallel - plate capacitor having $d = 5$ mm, $S = 80$ cm^2 , and $\varepsilon_r = 10$. (a) Find the capacitance. (b) If a dc voltage source of 50 V is now placed across the capacitor, calculate E , D , Q , and the total stored energy. (c) The source is now disconnected and the dielectric is carefully removed from between the plates. Find Q , D , E , and the total stored energy. (d) Find V_o .

Solution :

(a) $$C = \frac{\varepsilon_o \varepsilon_r S}{d} = \frac{\varepsilon_o \times 10 \times 80 \times 10^{-4}}{5 \times 10^{-3}} = 141.664\ \text{pF}$$

(b) $E = \frac{V}{d} = \frac{50}{5 \times 10^{-3}} = 10000$ V/m

$D = \varepsilon_o \varepsilon_R E = \varepsilon_o \times 10 \times 10000 = 0.8854\ \mu C/m^2$

$Q = D \times S = 0.8854 \times 10^{-6} \times 80 \times 10^{-4} = 7.0832$ nC .

$W_E = \frac{1}{2} C V^2 = \frac{1}{2} \times 141.664 \times 10^{-12} \times (50)^2 = 0.17708\ \mu J$

(c) $Q = 7.0832$ nC.

$D = \rho_s = \frac{Q}{A} = \frac{7.0832 \times 10^{-9}}{80 \times 10^{-4}} = 0.8854\ \mu C/m^2$

$E = \frac{D}{\varepsilon_o} = \frac{0.8854 \times 10^{-6}}{\varepsilon_o} = 100000$ V/m

$W_E = \frac{1}{2} C V^2 = \frac{1}{2} C (E d)^2 = \frac{1}{2} \times \frac{\varepsilon_o S}{d} \times E^2 \times d^2 = \frac{\varepsilon_o S E^2 d}{2}$

$= \frac{\varepsilon_o \times 80 \times 10^{-4} \times (10^5)^2 \times 5 \times 10^{-3}}{2} = 1.7708\ \mu J$

(d) $V_o = E \times d = 100000 \times 5 \times 10^{-3} = 500$ V

Problem 5. 55 : Capacitors tend to be more expensive as their capacitance C and maximum voltage $V_{max.}$ increase . The voltage V_{max} depends on the field strength at which the dielectric breaks down, E_{BD} . Which of these dielectrics will give the largest $C\ V_{max}$ product for equal plate areas :

(a) air : $\varepsilon_R = 1$, $E_{BD} = 3 \times 10^6$ V/m ;

(b) barium titanate : $\varepsilon_R = 1200$, $E_{BD} = 3 \times 10^6$ V/m ;

(c) silicon dioxide : $\varepsilon_R = 3.78$, $E_{BD} = 1.6 \times 10^7$ V/m ;

(d) polyethylene : $\varepsilon_R = 2.26$, $E_{BD} = 4.7 \times 10^6$?

Solution :

$C\,V_{max} = \frac{\varepsilon_o \varepsilon_r S}{d} \times E_{BD} \times d = \varepsilon_o \varepsilon_r S E_{BD}$

Since ε_o and S are constants, $C\,V_{max} = \varepsilon_r E_{BD}$

(a) $CV_{max} = 1 \times 3 \times 10^6$ V/m $= 3 \times 10^6$ C

(b) $CV_{max} = 1200 \times 3 \times 10^6 = 3600 \times 10^6$ C

(c) $CV_{max} = 3.78 \times 1.6 \times 10^7 = 60.48 \times 10^6$ C

(d) $CV_{max} = 2.26 \times 4.7 \times 10^6 = 10.622 \times 10^6$ C

Hence Barium titanate gives largest $C\,V_{max}$.

Problem 5.56 : (a) Find the dimensions of the two dielectric slabs shown in Fig. 5.56 if the total stored energy in region 1 is three times that stored in region 2, and $D_{1\,max} = 2D_{2\,max}$. (b) For these dimensions, find ε_{R1} and ε_{R2} if $C_{total} =$ 50 nF.

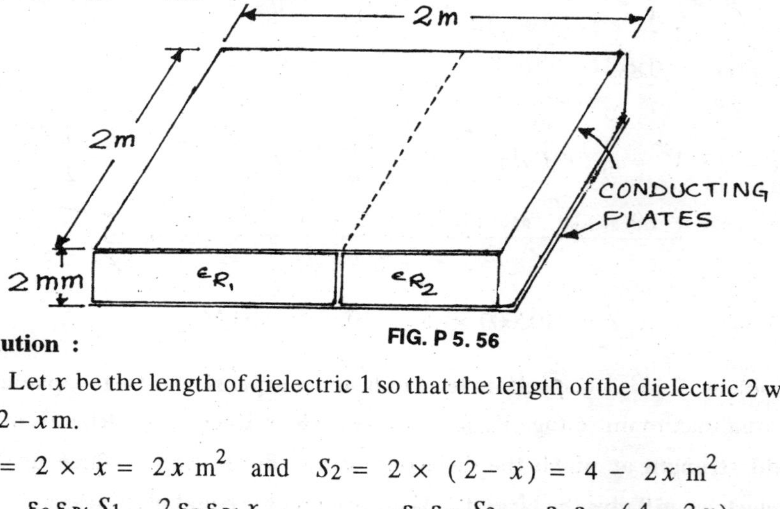

FIG. P 5. 56

Solution :

(a) Let x be the length of dielectric 1 so that the length of the dielectric 2 will be $2 - x$ m.

$$S_1 = 2 \times x = 2x \text{ m}^2 \quad \text{and} \quad S_2 = 2 \times (2 - x) = 4 - 2x \text{ m}^2$$

$$C_1 = \frac{\varepsilon_o \varepsilon_{R1} S_1}{d} = \frac{2\varepsilon_o \varepsilon_{R1} x}{d} \quad \text{and } C_2 = \frac{\varepsilon_o \varepsilon_{R2} S_2}{d} = \frac{\varepsilon_o \varepsilon_{R2}(4 - 2x)}{d}$$

$$W_{E1} = \frac{1}{2} C_1 V^2 \quad \text{and} \quad W_{E2} = \frac{1}{2} C_2 V^2 \quad \text{and} \quad W_{E1} = 3W_{E2}$$

$$\therefore \frac{1}{2} C_1 V^2 = 3 \times \frac{1}{2} C_2 V^2 \text{ or } C_1 = 3C_2 \qquad \ldots\ldots\ldots(1)$$

Now $D_{1\,max} = 2D_{2\,max}$ or

$$\varepsilon_o \varepsilon_{R1} E_{max} = 2\varepsilon_o \varepsilon_{R2} E_{max} \text{ or } \varepsilon_{R1} = 2\varepsilon_{R2} \qquad \ldots\ldots(2)$$

Using equation (1), we have $\dfrac{2\varepsilon_o \varepsilon_{R1} x}{d} = \dfrac{3\varepsilon_o \varepsilon_{R2}(4 - 2x)}{d}$

$$\therefore \varepsilon_{R1} = 3\varepsilon_{R2}(2 - x) \text{ or } 2\varepsilon_{R2} = 3\varepsilon_{R2}(2 - x)$$

Hence $x = 1.333$ m. Hence length of dielectric 1 (with ε_{R1}) is equal to 4 / 3 m where as that of dielectric 2 (with ε_{R2}) is equal to 2 / 3 m

(b) These capacitors are connected in parallel because the voltage drop across them are equal. Hence $C = C_1 + C_2 = 50$ nF

$\therefore\ 3C_2 + C_2 = 50$ nF or $C_2 = 12.5$ nF and $C_1 = 37.5$ nF .

$$\therefore\ 37.5 \times 10^{-9} = \frac{2\,\varepsilon_o\,\varepsilon_{R1}\,(4/3)}{2 \times 10^{-3}}$$

Hence $\varepsilon_{R1} = 3.17653$ and $\varepsilon_{R2} = \dfrac{\varepsilon_{R1}}{2} = 1.588265$

Problem 5. 57 : Two conducting planes are located at $z = 0$ and 6 mm. In the region $0 < z < 2$ mm, there is a perfect dielectric for which $\varepsilon_{R1} = 2$; for $2 < z < 5$ mm, let $\varepsilon_{R2} = 5$. Find the capacitance per square meter of surface area if the region for which $5 < z < 6$ mm is : (a) air ; (b) ε_{R1} ;(c) ε_{R2} ; (d) copper.

Solution :

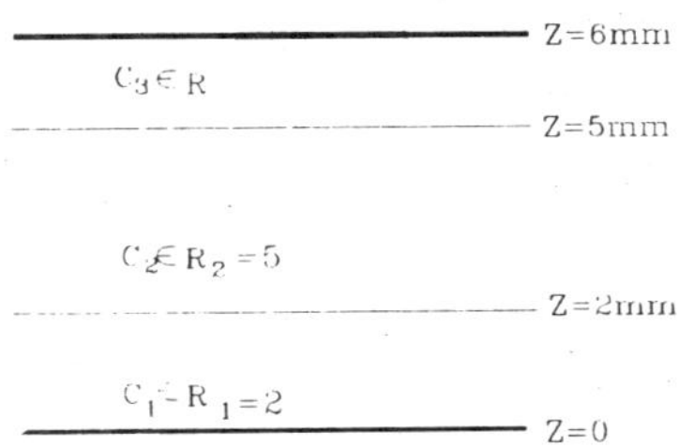

FIG. P 5. 57

(a) All the three capacitors are connected in series as the voltage drop across each section is different

$$\therefore\ \frac{1}{C} = \frac{1}{C_1} + \frac{1}{C_2} + \frac{1}{C_3}$$

$$C = \frac{C_1\,C_2\,C_3}{C_1\,C_2 + C_2\,C_3 + C_3\,C_1}$$

$$C_1 = \frac{\varepsilon_o\,\varepsilon_{r1}\,A}{d_1} = \frac{\varepsilon_o \times 2 \times 1}{2 \times 10^{-3}} = 1000\,\varepsilon_o\ \text{F/m}^2$$

$$C_2 = \frac{\varepsilon_o\,\varepsilon_{r2}\,A}{d_2} = \frac{\varepsilon_o \times 5 \times 1}{3 \times 10^{-3}} = 1666.6667\,\varepsilon_o\ \text{F/m}^2$$

$$C_3 = \frac{\varepsilon_o \times \varepsilon_{r3}\,A}{d_3} = \frac{\varepsilon_o \times 1 \times 1}{10^{-3}} = 1000\,\varepsilon_o\ \text{F/m}^2$$

$\therefore\ C = 3.405$ nF

$$\text{(b) } C_3 = \frac{\varepsilon_o \times 2 \times 1}{10^{-3}} = 2000\,\varepsilon_o\ \text{F/m}^2$$

$\therefore\ C = 4.216\ \text{nF/m}^2$

(c) $C_3 = \dfrac{\varepsilon_o \times 5 \times 1}{10^{-3}} = 5000\,\varepsilon_o\ \text{F/m}^2$

$\therefore\ C = 4.919\ \text{nF/m}^2$

(d) $\dfrac{1}{C} = \dfrac{1}{C_1} + \dfrac{1}{C_2}$

$\therefore\ C = 5.53375\ \text{nF/m}^2$

Problem 5.58 : A conducting cylinder of 1 - cm radius carries a uniformly distributed charge of 1 nC per meter length. The cylinder is coated with a material having a dielectric constant $\varepsilon_{R1} = 2$ in a layer of thickness a_1 . A second layer, ε_{R2} with thickness a_2, continues to the outer conductor of the coax, $b = 4$ cm in radius. Find ε_{R2}, a_1, and a_2 so that the voltage across each layer is the same and the capacitance of the coax is 75 pF per meter length.

Solution :

Consider a cylinder of radius ρ such that $0.01 < \rho < 0.04$. Apply Gauss's law to this cylinder $\oint \overline{D} \cdot \overline{dS} = Q$

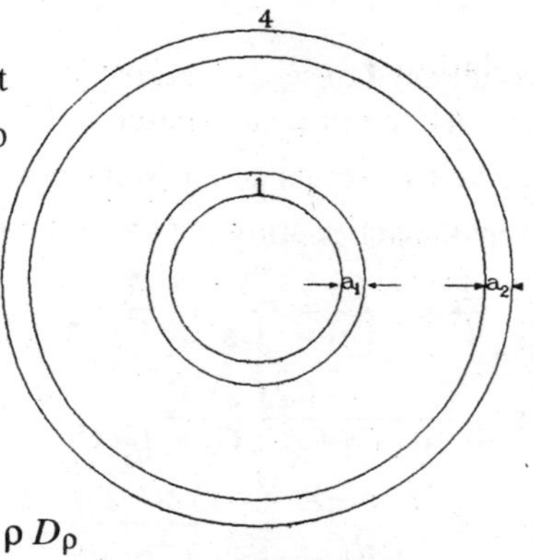

FIG. P 5 . 58

$$\text{L.H.S.} = \oint \overline{D} \cdot \overline{dS}$$

$$= \int (D_\rho\,\overline{a}_\rho) \cdot (\rho\, d\phi\, dz\,\overline{a}_\rho)$$

$$= \int_0^{2\pi}\int_0^{1} \rho\, D_\rho\, d\phi\, dz = 2\pi\rho D_\rho$$

$\text{RHS} = 10^{-9}$.

Hence $2\pi\rho D_\rho = 10^{-9}$ or $D_\rho = \dfrac{10^{-9}}{2\pi\rho}$ or $E_\rho = \dfrac{10^{-9}}{2\pi\varepsilon_o\varepsilon_R\,\rho}$

$$V = -\int \overline{E}\cdot\overline{dL} = -\int (E_\rho\,\overline{a}_\rho)\cdot(d\rho\,\overline{a}_\rho) = -\int E_\rho\, d\rho$$

Voltage across first layer $= -\displaystyle\int_{0.01}^{0.01+a_1} \frac{10^{-9}\,d\rho}{2\pi\varepsilon_o(2)\rho}$

Voltage across first layer = $-\dfrac{10^{-9}}{4\pi\varepsilon_o}\ln\left(\dfrac{0.01+a_1}{0.01}\right)$

Voltage across second layer $= -\displaystyle\int_{0.01+a_1}^{0.04-a_2}\dfrac{10^{-9}\,d\rho}{2\pi\varepsilon_o(1)\rho}$

$= -\dfrac{10^{-9}}{2\pi\varepsilon_o\varepsilon_{R2}}\ln\left(\dfrac{0.04}{0.04-a_2}\right)$

Voltage across third layer $= -\displaystyle\int_{0.04-a_2}^{0.04}\dfrac{10^{-9}\,d\rho}{2\pi\varepsilon_o\varepsilon_{R2}\rho}$

$= -\dfrac{10^{-9}}{4\pi\varepsilon_o}\ln\left(\dfrac{0.01+a_1}{0.01}\right)$

Now capacitance per meter length is given by $= \dfrac{2\pi\varepsilon_o\varepsilon_R}{\ln(b/a)}$

Capacitance of first layer $= C_1 = \dfrac{2\pi\varepsilon_o(2)}{\ln\left(\dfrac{0.01+a_1}{0.01}\right)}$

Capacitance of second layer $= C_2 = \dfrac{2\pi\varepsilon_o}{\ln\left(\dfrac{0.04-a_2}{0.0+a_1}\right)}$

Capacitance of first layer $= C_3 = \dfrac{2\pi\varepsilon_o\varepsilon_{R2}}{\ln\left(\dfrac{0.04}{0.04-a_2}\right)}$

These capacitors are connected in series. Hence total capacitance per meter length is given by $\dfrac{1}{C} = \dfrac{1}{C_1}+\dfrac{1}{C_2}+\dfrac{1}{C_3}$

$$\frac{1}{75\times10^{-12}} = \frac{1}{4\pi\varepsilon_o}\ln\left(\frac{0.01+a_1}{0.01}\right)+\frac{1}{2\pi\varepsilon_o}\ln\left(\frac{0.04-a_2}{0.01+a_1}\right)$$

$$+\frac{1}{2\pi\varepsilon_o\varepsilon_{R2}}\ln\left(\frac{0.04}{0.04-a_2}\right) \quad \ldots.(1)$$

As the voltage across each layer is same, we have

$$-\frac{10^{-9}}{4\pi\varepsilon_o}\ln\left(\frac{0.01+a_1}{0.01}\right) = -\frac{10^{-9}}{2\pi\varepsilon_o}\ln\left(\frac{0.04-a_2}{0.01+a_1}\right) = -\frac{10^{-9}}{2\pi\varepsilon_o\varepsilon_{R2}}\ln\left(\frac{0.04}{0.04-a_2}\right)$$

or

$$\frac{1}{2}\ln\left(\frac{0.01+a_1}{0.01}\right) = \ln\left(\frac{0.04-a_2}{0.01+a_1}\right) = \frac{1}{\varepsilon_{R2}}\ln\left(\frac{0.04}{0.04-a_2}\right) \quad \ldots..(2)$$

From first and second term of equation (2) we have ,

$$\ln\left(\frac{0.04 - a_2}{0.01 + a_1}\right) = \frac{1}{2}\ln\left(\frac{0.01 + a_1}{0.01}\right) \qquad .....(3)$$

and from first and third term we have ,

$$\frac{1}{\varepsilon_{R2}}\ln\left(\frac{0.04}{0.04 - a_2}\right) = \frac{1}{2}\ln\left(\frac{0.01 + a_1}{0.01}\right) \qquad .....(4)$$

From Eq.(1), (3) and (4) we get

$$\frac{1}{75 \times 10^{-12}} = \frac{1}{2\pi\varepsilon_o}\left[\frac{1}{2}\ln\left(\frac{0.01 + a_1}{0.01}\right) + \frac{1}{2}\ln\left(\frac{0.01 + a_1}{0.01}\right) + \frac{1}{2}\ln\left(\frac{0.01 + a_1}{0.01}\right)\right]$$

$\therefore \ln(0.01 + a_1) - \ln(0.01) = 0.4945$ or $a_1 = 6.396 \times 10^{-3}$ m

Substitute this value of a_1 in Eq. (3) we get $a_2 = 0.01900$ m.

Substitute the values of a_1 and a_2 in Eq.(4) we have $\varepsilon_{R2} = 2.606$

Hence the values are $a_1 = 0.639$ cm , $a_2 = 1.9$ cm , and $\varepsilon_{R2} = 2.607$

Problem 5.59 : Two conducting spherical shells have radii of $a = 2$ cm and $b = 5$ cm. The interior is a perfect dielectric for which $\varepsilon_R = 10$. (a) Find C . (b) A portion of the dielectric is now removed so that $\varepsilon_R = 1, 0 < \theta < \pi/6$, and $\varepsilon_R = 10, \pi/6 < \theta < \pi$. Find C .

Solution :

$$\text{(a) } C = \frac{4\pi\varepsilon_o\varepsilon_R}{\frac{1}{a} - \frac{1}{b}} = \frac{4\pi\varepsilon_o(10)}{\frac{1}{0.02} - \frac{1}{0.05}} = 37.087 \text{ pF}$$

(b)Let Q be the charge on the inner sphere and $\overline{E} = E_r\,\overline{a}_r$ be the electric field intensity. Consider a sphere of radius r such that $0.02 < r < 0.05$.

Apply Gauss's law to this sphere $\oint \overline{D}\cdot\overline{dS} = Q$

Let $\overline{D}_1$ be the electric flux density in the region for which $\varepsilon_R = 1$ and $\overline{D}_2$ in the region for which $\varepsilon_R = 10$.

$$\text{LHS} = \oint \overline{D}\cdot\overline{dS} = \int \overline{D}_1\cdot\overline{dS} + \int \overline{D}_2\cdot\overline{dS}$$

$$= \int (\varepsilon_o E_r \bar{a}_r) \cdot (r^2 \sin\theta\, d\theta\, d\phi\, \bar{a}_r) + \int (10\,\varepsilon_o E_r \bar{a}_r) \cdot (r^2 \sin\theta\, d\theta\, d\phi\, \bar{a}_r)$$

$$\text{L.H.S.} = \int_0^{\pi/6} \int_0^{2\pi} \varepsilon_o E_r\, r^2 \sin\theta\, d\theta\, d\phi + \int_{\pi/6}^{\pi} \int_0^{2\pi} 10\,\varepsilon_o E_r\, r^2 \sin\theta\, d\theta\, d\phi$$

$$= \varepsilon_o E_r r^2 \Big[-\cos\theta\Big]_0^{\pi/6} \Big[\phi\Big]_0^{2\pi} + 10\,\varepsilon_o E_r r^2 \Big[-\cos\theta\Big]_{\pi/6}^{\pi} \Big[\phi\Big]_0^{2\pi} = 118.087\,\varepsilon_o r^2 E_r$$

$$\therefore\ 118.087\,\varepsilon_o r^2 E_r = Q \quad \text{or} \quad E_r = \frac{Q}{118.087\,\varepsilon_o r^2}$$

$$\text{Now, } V = -\int \bar{E} \cdot \overline{dL} = -\int (E_r \bar{a}_r) \cdot (dr\,\bar{a}_r) = -\int E_r\, dr$$

$$= -\int_{0.05}^{0.02} \frac{Q\, dr}{118.087\,\varepsilon_o r^2} = -\frac{Q}{118.087\,\varepsilon_o}\left[-\frac{1}{r}\right]_{0.05}^{0.02}$$

$$= 2.869 \times 10^{-10} Q$$

$$\therefore\ C = \frac{Q}{V} = \frac{1}{2.869 \times 10^{-10}} = 34.851 \text{ pF}$$

Problem 5.60 : A coaxial capacitor has dimensions $a = 3$ mm, $b = 12$ mm, and a length of 1 m. The region between the conducting cylinders contains three different dielectrics : $\varepsilon_{R1} = 5, 3 < \rho < 6$ mm ; $\varepsilon_{R2} = 3, 6 < \rho < 9$ mm ; and $\varepsilon_{R3} = 1, 9 < \rho < 12$ mm. (a) calculate C. (b) Plot E_ρ vs. ρ if $E_{\rho(max)} = 100$ V/m.

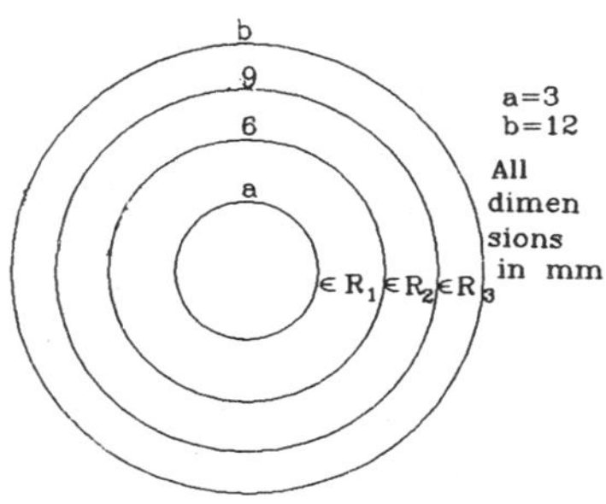

FIG.P.5.60

Solution :

$$C_1 = \frac{2\pi\varepsilon_1 L}{\ln(6/a)} = \frac{2\pi\varepsilon_o(5)\,1}{\ln(6/3)} = 401.295 \text{ pF}$$

$$C_2 = \frac{2\pi\varepsilon_2 L}{\ln(9/6)} = \frac{2\pi\varepsilon_o(3)\,1}{\ln(9/6)} = 411.611 \text{ pF}$$

$$C_3 = \frac{2\pi\varepsilon_3 L}{\ln(b/9)} = \frac{2\pi\varepsilon_o(1)\,1}{\ln(12/9)} = 193.378 \text{ pF}$$

These three capacitors are connected in series.

$$\text{Hence } C = \frac{C_1 C_2 C_3}{C_1 C_2 + C_2 C_3 + C_3 C_1} = 99.082 \text{ pF}.$$

(b) The boundary condition for electric flux density is

$D_{N1} = D_{N2} = D_{N3}$ or $\varepsilon_o \varepsilon_{R1} E_{N1} = \varepsilon_o \varepsilon_{R2} E_{N2} = \varepsilon_o \varepsilon_{R3} E_{N3}$

i.e. $\varepsilon_{R1} E_{N1} = \varepsilon_{R2} E_{N2} = \varepsilon_{R3} E_{N3}$ ..... (1)

Consider a cylinder of radius ρ such that $a < \rho < 6$ mm.

Apply Gauss's law to this cylinder. $\oint \overline{D} \cdot \overline{dS} = Q$

$$\text{L.H.S.} = \oint \overline{D} \cdot \overline{dS} = \int D_\rho \overline{a}_\rho \cdot \rho \, d\phi \, dz \, \overline{a}_\rho$$

$$= \int_0^{2\pi} \int_0^{2} \rho \, D_\rho \, d\phi \, dz = 2\pi \rho L D_\rho$$

RHS $= Q$

$$\therefore \; 2\pi \rho L D_\rho = Q \quad \text{or} \quad D_\rho = \frac{Q}{2\pi \rho L}$$

$$\text{Now } D_\rho = \varepsilon_o \varepsilon_r E_N = \frac{Q}{2\pi \rho L}$$

$$\therefore \; E_N = \frac{Q}{2\pi \rho L \varepsilon_o \varepsilon_R} \qquad \text{......(2)}$$

In this expression Q, L and $2\pi \varepsilon_o$ are constants. Hence E_N depends only on the product $\rho \varepsilon_R$. If this product is minimum, E_N will be maximum. Now with ε_{R1}, we have, $a \varepsilon_{R1} = 0.003 \times 5 = 0.015$;
with ε_{R2}, $0.006 \varepsilon_{R2} = 0.006 \times 3 = 0.018$;
and with ε_{R3}, $0.009 \varepsilon_{R3} = 0.009 \times 1 = 0.009$.

$$\text{Hence } E_{N(\max)} = E_{\rho(\max)} = \frac{Q}{2\pi L \varepsilon_o (0.009)} = 100 \text{ (given)}$$

$$\therefore \; \frac{Q}{L} = 1.8\pi \varepsilon_o \text{ C/m}.$$

From Eq.(2), we have

$$E_{N1} = \frac{1.8\pi \varepsilon_o}{2\pi \rho \varepsilon_o \varepsilon_{R1}} = \frac{1.8\pi \varepsilon_o}{2\pi \rho \varepsilon_o (5)} = \frac{0.18}{\rho}$$

$$E_{N2} = \frac{1.8\pi \varepsilon_o}{2\pi \rho \varepsilon_o \varepsilon_{R2}} = \frac{1.8\pi \varepsilon_o}{2\pi \rho \varepsilon_o (3)} = \frac{0.3}{\rho}$$

$$E_{N3} = \frac{1.8\pi \varepsilon_o}{2\pi \rho \varepsilon_o \varepsilon_{R3}} = \frac{1.8\pi \varepsilon_o}{2\pi \rho \varepsilon_o (1)} = \frac{0.9}{\rho}$$

$$\text{Hence } E_\rho = \begin{cases} 0.18 / \rho & 0.003 < \rho < 0.006 \\ 0.3 / \rho & 0.006 < \rho < 0.009 \\ 0.9 / \rho & 0.009 < \rho < 0.012 \end{cases}$$

Problem 5.61 : Change the conductor radius from 5 to 4 m in Fig. , maintaining $h = 13$ m and $V = 100$ V. Find new values for K_1 , ρ_L , a , and C

Solution :

$b = 4$ m , $h = 13$, $V_o = 100$

$$\sqrt{K_1} = \frac{h + \sqrt{h^2 - b^2}}{b} \qquad \therefore \quad K_1 = 40.225$$

$$a = \sqrt{h^2 - b^2} = \sqrt{169 - 16} = 12.369$$

$$\rho_L = \frac{4\pi\varepsilon_o\varepsilon_R V_o}{\ln K_1} = \frac{4\pi\varepsilon_o (100)}{\ln (40.225)} = 3.012 \text{ nC/m}$$

$$C = \frac{2\pi\varepsilon_o}{\cosh^{-1}(h/b)} = \frac{2\pi\varepsilon_o}{\cosh^{-1}(13/4)} = 30.116 \text{ pF/m}$$

Problem 62. : Given the potential function $V = -300 + 100 \ln\left\{[(8+x)^2 + y^2] / [(8-x)^2 + y^2]\right\}$ V, sketch the 0 - V and 200 - V equipotential surfaces in the region $x > 0$.

Solution :

$$V = 0 = -300 + 100 \ln\left[\frac{(8+x)^2 + y^2}{(8-x)^2 + y^2}\right] \text{ or } \frac{(8+x)^2 + y^2}{(8-x)^2 + y^2} = 20.0855$$

$$64 + 16x + x^2 + y^2 = 20.085 y^2 + 20.085 x^2 - 321.368 x + 1285.472$$

$$19.085 y^2 + 19.085 x^2 - 337.368 x + 1221.472 = 0$$

$$y^2 + x^2 - 17.677 x + 64 = 0 \text{ or}$$

$$y = \pm\sqrt{17.677 x - x^2 - 64} \qquad \ldots\ldots (1)$$

Similarly for $V = 200$,

$$y = \pm\sqrt{16.217 x - x^2 - 64} \qquad \ldots\ldots (2)$$

Plot these equations in $x - y$ plane for $x > 0$ i.e. in first and fourth quadrant. Plots are not given.

Problem 5.63 : A 2 - cm - diameter conductor is suspended in air with its axis 5 cm form a conducting plane. (a) Find the capacitance per unit length

between cylinder and plane. (b) Let the potential of the cylinder be 2 V and that of the plane be 0 V, and find the shortest distance from the cylinder to the 1 - V equipotential surface.

Solution :

$$C = \frac{2\pi\varepsilon_o L}{\cosh^{-1}(h/b)}$$

$$= \frac{2\pi\varepsilon_o L}{\cosh^{-1}(5\times 10^{-2}/10^{-2})}$$

$$= 24.267\,L \text{ pF}$$

$$\therefore \frac{C}{L} = 24.267 \text{ pF/m}.$$

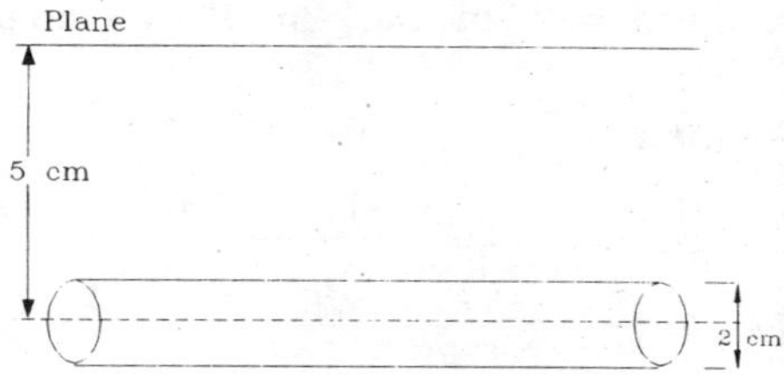

FIG .P 5. 63 (a)

(b)

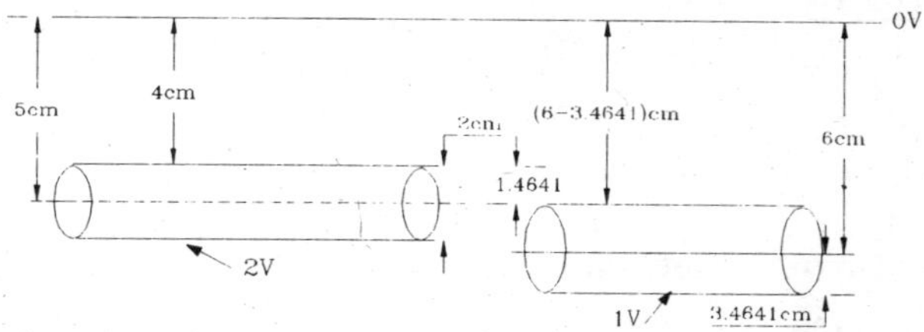

FIG. P 5 . 63 (b)

$$\sqrt{K_1} = \frac{h + \sqrt{h^2 - b^2}}{b} \qquad \therefore K_1 = 97.989$$

$$a = \sqrt{h^2 - b^2} = 0.489 \quad \text{and} \quad V_o = 2 \text{ V}$$

$$\rho_L = \frac{4\pi\varepsilon_o V_o}{\ln K_1} = 48.534 \text{ pC/m}$$

For $V = 1$ Volt , $K_1 = e^{4\pi\varepsilon_o V/\rho_L} = 9.898$

$$b_1 = \frac{2a\sqrt{K_1}}{K_1 - 1} = \frac{2(0.0489)\sqrt{9.898}}{9.898 - 1} = 0.034 \text{ m}$$

$$h_1 = a\frac{K_1 + 1}{K_1 - 1} = (0.048)\frac{9.898 + 1}{0.989 - 1} = 0.06$$

From Fig.P5.63 (b), the shortest distance between 1V equipotential and the cylinder = 1.4641 cm.

— X —

$$\text{Hence } E_\rho = \begin{cases} 0.18/\rho & 0.003 < \rho < 0.006 \\ 0.3/\rho & 0.006 < \rho < 0.009 \\ 0.9/\rho & 0.009 < \rho < 0.012 \end{cases}$$

Problem 5.61 : Change the conductor radius from 5 to 4 m in Fig. , maintaining $h = 13$ m and $V = 100$ V. Find new values for K_1, ρ_L, a, and C

Solution :

$b = 4$ m , $h = 13$, $V_o = 100$

$$\sqrt{K_1} = \frac{h + \sqrt{h^2 - b^2}}{b} \quad \therefore \quad K_1 = 40.225$$

$$a = \sqrt{h^2 - b^2} = \sqrt{169 - 16} = 12.369$$

$$\rho_L = \frac{4\pi\varepsilon_o\varepsilon_R V_o}{\ln K_1} = \frac{4\pi\varepsilon_o(100)}{\ln(40.225)} = 3.012 \text{ nC/m}$$

$$C = \frac{2\pi\varepsilon_o}{\cosh^{-1}(h/b)} = \frac{2\pi\varepsilon_o}{\cosh^{-1}(13/4)} = 30.116 \text{ pF/m}$$

Problem 62. : Given the potential function $V = -300 + 100 \ln\left\{[(8+x)^2 + y^2]/[(8-x)^2 + y^2]\right\}$ V, sketch the 0 - V and 200 - V equipotential surfaces in the region $x > 0$.

Solution :

$$V = 0 = -300 + 100 \ln\left[\frac{(8+x)^2 + y^2}{(8-x)^2 + y^2}\right] \text{ or } \frac{(8+x)^2 + y^2}{(8-x)^2 + y^2} = 20.0855$$

$$64 + 16x + x^2 + y^2 = 20.085y^2 + 20.085x^2 - 321.368x + 1285.472$$

$$19.085y^2 + 19.085x^2 - 337.368x + 1221.472 = 0$$

$$y^2 + x^2 - 17.677x + 64 = 0 \text{ or}$$

$$y = \pm\sqrt{17.677x - x^2 - 64} \quad \ldots\ldots(1)$$

Similarly for $V = 200$,

$$y = \pm\sqrt{16.217x - x^2 - 64} \quad \ldots\ldots(2)$$

Plot these equations in $x - y$ plane for $x > 0$ i.e. in first and fourth quadrant. Plots are not given.

Problem 5.63 : A 2 - cm - diameter conductor is suspended in air with its axis 5 cm form a conducting plane. (a) Find the capacitance per unit length

between cylinder and plane. (b) Let the potential of the cylinder be 2 V and that of the plane be 0 V, and find the shortest distance from the cylinder to the 1 - V equipotential surface.

Solution :

$$C = \frac{2\pi\varepsilon_o L}{\cosh^{-1}(h/b)}$$

$$= \frac{2\pi\varepsilon_o L}{\cosh^{-1}(5\times 10^{-2}/10^{-2})}$$

$$= 24.267\,L \text{ pF}$$

$$\therefore \frac{C}{L} = 24.267 \text{ pF/m}.$$

FIG .P 5. 63 (a)

(b)

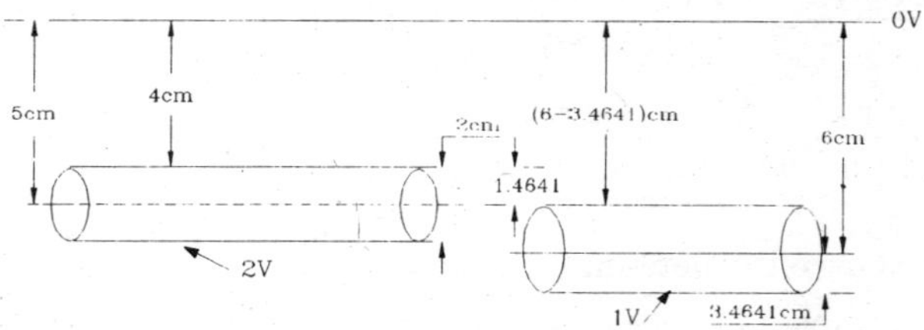

FIG. P 5 . 63(b)

$$\sqrt{K_1} = \frac{h + \sqrt{h^2 - b^2}}{b} \quad \therefore K_1 = 97.989$$

$$a = \sqrt{h^2 - b^2} = 0.489 \quad \text{and} \quad V_o = 2 \text{ V}$$

$$\rho_L = \frac{4\pi\varepsilon_o V_o}{\ln K_1} = 48.534 \text{ pC/m}$$

$$\text{For } V = 1 \text{ Volt}, \; K_1 = e^{4\pi\varepsilon_o V/\rho_L} = 9.898$$

$$b_1 = \frac{2a\sqrt{K_1}}{K_1 - 1} = \frac{2(0.0489)\sqrt{9.898}}{9.898 - 1} = 0.034 \text{ m}$$

$$h_1 = a\frac{K_1 + 1}{K_1 - 1} = (0.048)\frac{9.898 + 1}{0.989 - 1} = 0.06$$

From Fig.P5.63 (b), the shortest distance between 1V equipotential and the cylinder = 1.4641 cm.

— X —

CHAPTER 6

Poisson's & Laplace's Equations

6.1 POISSON'S AND LAPLACE'S EQUATIONS :

Point form of Gauss's law

$$\overline{\nabla} \cdot \overline{D} = \rho_v$$

Relation between $\overline{D}$ and $\overline{E}$

$$\overline{D} = \varepsilon \overline{E}$$

Relation between $\overline{E}$ and V

$$\overline{E} = -\overline{\nabla} V$$

Hence

$$\overline{\nabla} \cdot \overline{D} = \overline{\nabla} \cdot (\varepsilon \overline{E}) = -\overline{\nabla} \cdot (\varepsilon \overline{\nabla} V) = \rho_v$$

or

$$\overline{\nabla} \cdot \overline{\nabla} V = -\frac{\rho_v}{\varepsilon}$$

for a homogeneous region in which ε is a constant. The last equation is known as Poisson's equation. In cartesian coordinate system

$$\overline{\nabla} \cdot \overline{A} = \frac{\partial A_x}{\partial x} + \frac{\partial A_y}{\partial y} + \frac{\partial A_z}{\partial z}$$

$$\overline{\nabla} V = \frac{\partial V}{\partial x} \overline{a}_x + \frac{\partial V}{\partial y} \overline{a}_y + \frac{\partial V}{\partial z} \overline{a}_z$$

and therefore

$$\bar{\nabla}\cdot\bar{\nabla}V = \frac{\partial}{\partial x}\left(\frac{\partial V}{\partial x}\right) + \frac{\partial}{\partial y}\left(\frac{\partial V}{\partial y}\right) + \frac{\partial}{\partial z}\left(\frac{\partial V}{\partial z}\right)$$

$$\bar{\nabla}\cdot\bar{\nabla}V = \frac{\partial^2 V}{\partial x^2} + \frac{\partial^2 V}{\partial y^2} + \frac{\partial^2 V}{\partial z^2}$$

Usually the operation $\bar{\nabla}\cdot\bar{\nabla}$ is abbreviated $\bar{\nabla}^2 V$ (and pronounced "del squared") . Hence

$$\bar{\nabla}^2 V = \frac{\partial^2 V}{\partial x^2} + \frac{\partial^2 V}{\partial y^2} + \frac{\partial^2 V}{\partial z^2}$$

and

$$\boxed{\bar{\nabla}^2 V = -\frac{\rho_v}{\varepsilon}} \qquad \ldots\ldots.(6.1)$$

If $\rho_v = 0$, then,

$$\boxed{\bar{\nabla}^2 V = 0} \qquad \ldots\ldots(6.2)$$

which is Laplace's equation. The $\bar{\nabla}^2 V$ operation is called the Laplacian of V.

The expressions for $\bar{\nabla}^2 V$ in cartesian , cylindrical, and spherical coordinate systems are given by Eq.(6.3a), (6.3b),and (6.3c) respectively.

$$\boxed{\bar{\nabla}^2 V = \frac{\partial^2 V}{\partial x^2} + \frac{\partial^2 V}{\partial y^2} + \frac{\partial^2 V}{\partial z^2}} \qquad \ldots\ldots\ldots(6.3a)$$

$$\boxed{\bar{\nabla}^2 V = \frac{1}{\rho}\frac{\partial}{\partial \rho}\left(\rho\frac{\partial V}{\partial \rho}\right) + \frac{1}{\rho^2}\frac{\partial^2 V}{\partial \phi^2} + \frac{\partial^2 V}{\partial z^2}} \qquad \ldots.(6.3b)$$

$$\boxed{\bar{\nabla}^2 V = \frac{1}{r^2}\frac{\partial}{\partial r}\left(r^2\frac{\partial V}{\partial r}\right) + \frac{1}{r^2\sin\theta}\frac{\partial}{\partial \theta}\left(\sin\theta\frac{\partial V}{\partial \theta}\right) + \frac{1}{r^2\sin^2\theta}\frac{\partial^2 V}{\partial \phi^2}}$$

$$\ldots\ldots(6.3c)$$

Problem 6.1 : Find the potential and the volume charge density at P (0.5

,1.5, 1) in free space given the potential field : (a) $V = 2x^2 - y^2 - z^2$ V ; (b) $V = 6\rho\phi z$ V ; (c) $V = 5(2r^2 - 7)\cos\theta\cos\phi$ V ; (d) $V = 3x - y$ V

Solution :

(a) V at $P = 2(0.5)^2 - (1.5)^2 - (1)^2 = -2.75$ V

$$\frac{\partial V}{\partial x} = 4x, \frac{\partial^2 V}{\partial x^2} = 4 \; ; \; \frac{\partial V}{\partial y} = -2y, \frac{\partial^2 V}{\partial y^2} = -2 \; ; \; \frac{\partial V}{\partial z} = -2z, \frac{\partial^2 V}{\partial z^2} = -2$$

Using Eq.(6.1), we have ρ_v at $P = -\varepsilon_o \overline{\nabla}^2 V = -\varepsilon_o(4 - 2 - 2) = 0$

(b) Cylindrical coordinates of point P are (1.581, 71.565°, 1). Hence

$$V \text{ at } P = 6 \times 1.581\left(71.565^\circ \times \frac{\pi}{180}\right) \times 1 = 11848 \text{ V}$$

$$\frac{\partial V}{\partial \rho} = 6\phi z \, , \; \rho\frac{\partial V}{\partial \rho} = 6\rho\phi z \, , \; \frac{\partial}{\partial \rho}\left(\rho\frac{\partial V}{\partial \rho}\right) = 6\phi z \, , \; \frac{1}{\rho}\frac{\partial}{\partial \rho}\left(\rho\frac{\partial V}{\partial \rho}\right) = \frac{6\phi z}{\rho}$$

$$\frac{\partial V}{\partial \phi} = 6\rho z \, , \; \frac{\partial^2 V}{\partial \phi^2} = 0 \; ; \; \frac{\partial V}{\partial z} = 6\rho\phi \, , \; \frac{\partial^2 V}{\partial z^2} = 0$$

Using Eq.(6.1), we have ρ_v at $P = -\varepsilon_o \overline{\nabla}^2 V = -\varepsilon_o\left(\frac{6\phi z}{\rho} + 0 + 0\right)$

$$= -\frac{6\varepsilon_o}{1.581} \times 71.565 \times \frac{\pi}{180} = -41.970 \text{ pC/m}^3$$

(c) Spherical coordinates of point P are ($\sqrt{3.5}$, 57.688°, 71.565°)

$$V \text{ at } P = 5[2(\sqrt{3.5})^2 - 7]\cos 57.688^\circ \cos 71.565^\circ = 0$$

$$\frac{\partial V}{\partial r} = 20r\cos\theta\cos\phi \; ; \; r^2\frac{\partial V}{\partial r} = 20r^3\cos\theta\cos\phi$$

$$\frac{\partial}{\partial r}\left(r^2\frac{\partial V}{\partial r}\right) = 60r^2\cos\theta\cos\phi \; ; \; \frac{1}{r^2}\frac{\partial}{\partial r}\left(r^2\frac{\partial V}{\partial r}\right) = 60\cos\theta\cos\phi \; ;$$

$$\frac{\partial V}{\partial \theta} = -5(2r^2 - 7)\sin\theta\cos\phi \; ; \; \sin\theta\frac{\partial V}{\partial \theta} = -5(2r^2 - 7)\sin^2\theta\cos\phi$$

$$\frac{\partial}{\partial \theta}\left(\sin\theta\frac{\partial V}{\partial \theta}\right) = -10(2r^2 - 7)\sin\theta\cos\theta\cos\phi \; ;$$

$$\frac{1}{r^2\sin\theta}\frac{\partial}{\partial \theta}\left(\sin\theta\frac{\partial V}{\partial \theta}\right) = \frac{-10(2r^2 - 7)\cos\theta\cos\phi}{r^2}$$

$$\frac{\partial V}{\partial \phi} = -5(2r^2 - 7)\cos\theta\sin\phi \; ; \; \frac{\partial^2 V}{\partial \phi^2} = -5(2r^2 - 7)\cos\theta\cos\phi$$

Using Eq.(6.1), we have

$$\therefore \rho_v = -\varepsilon_o\Big[60\cos\theta\cos\phi - 10\frac{(2r^2 - 7)}{r^2}\cos\theta\cos\phi$$
$$- 5(2r^2 - 7)\cos\theta\cos\phi\Big]$$

Substitute $r = \sqrt{3.5}$, $\theta = 57.688^\circ$ and $\phi = 71.565^\circ$ in above expression, we get ρ_v at $P = -89.797 \text{ pC/m}^3$

(d) V at $P = 3(0.5) - 1.5 = 0$ V

$$\frac{\partial V}{\partial x} = 3 \, , \, \frac{\partial^2 V}{\partial x^2} = 0 \; ; \; \frac{\partial V}{\partial y} = -1 \, , \, \frac{\partial^2 V}{\partial y^2} = 0$$

Using Eq.(6.1), we have ρ_v at $P = -\varepsilon_o \bar{\nabla}^2 V = -\varepsilon_o(0) = 0$.

Problem 6.2 : Find V at P (2, 1, 3) for the field of : (a) two coaxial conducting cylinders, $V = 50$ V at $\rho = 2$m, and $V = 20$ V at $\rho = 3$m ; (b) two radial conducting planes, $V = 50$ V at $\phi = 10^\circ$, & $V = 20$ V at $\phi = 30^\circ$; (c) two concentric conducting spheres, $V = 50$ V at $r = 3$m, and $V = 20$ V at $r = 5$m ; (d) two coaxial conducting cones, $V = 50$ V at $\theta = 30^\circ$, and $V = 20$ V at $\theta = 50^\circ$.

Solution :

(a) $V = 50$ V at $\rho = 2$ m and $V = 20$ V at $\rho = 3$ m

As the potential varies w.r.t. ρ coordinate, $\frac{\partial V}{\partial \phi} = \frac{\partial V}{\partial z} = 0$

$$\text{Hence } \bar{\nabla}^2 V = \frac{1}{\rho}\frac{\partial}{\partial \rho}\left(\rho\frac{\partial V}{\partial \rho}\right) = 0$$

Assuming ρ is not equal to zero, we have $\frac{d}{d\rho}\left(\rho\frac{dV}{d\rho}\right) = 0$

Integrating this equation twice we get

$$\rho\frac{dV}{d\rho} = A \, , \, dV = A\frac{d\rho}{\rho} \, , \, V = A\ln\rho + B \quad \ldots\ldots (1)$$

Substitute the boundary conditions in Eq.(1)

$$50 = A \ln(2) + B \qquad \ldots\ldots (2)$$

$$20 = A \ln(3) + B \qquad \ldots\ldots (3)$$

Solving Eq. (2) and (3), we get $A = -73.989$ and $B = 101.285$

Hence $V = -73.989 \ln \rho + 101.285$

At point P (2, 1, 3), $\rho = \sqrt{(2)^2 + (1)^2} = 2.236$

Hence V at $P = -73.989 \ln(2.236) + 101.285 = 41.745$ V

(b) $V = 50$ V at $\phi = 10^\circ$ and $V = 20$ V at $\phi = 30^\circ$

As the potential varies w.r.t. ϕ coordinate, $\dfrac{\partial V}{\partial \rho} = \dfrac{\partial V}{\partial z} = 0$

Hence $\bar{\nabla}^2 V = \dfrac{1}{\rho} \dfrac{\partial^2 V}{\partial \phi^2} = 0$

Assuming ρ is not equal to zero, we have $\dfrac{d^2 V}{d\phi^2} = 0$

Integrating this equation twice we get,

$$\frac{dV}{d\phi} = A, \quad dV = A\, d\phi, \quad V = A\phi + B \qquad \ldots\ldots (1)$$

Substitute the boundary conditions in Eq.(1)

$$50 = A\left(10 \times \frac{\pi}{180}\right) + B \qquad \ldots\ldots (2)$$

$$20 = A\left(30 \times \frac{\pi}{180}\right) + B \qquad \ldots\ldots (3)$$

Solving Eq.(2) and (3), we get $A = -85.944$ and $B = 65$

Hence $V = -85.944\,\phi + 65$

At point P (2, 1, 3), $\phi = \tan^{-1}(1/2) = 26.565^\circ$

Hence V at point $P = -85.944\left(26.565 \times \dfrac{\pi}{180}\right) + 65 = 25.152$ V

(c) $V = 50$ V at $r = 3$ m and $V = 20$ V at $r = 5$ m

As the potential varies w.r.t. r coordinate, $\dfrac{\partial V}{\partial \theta} = \dfrac{\partial V}{\partial \phi} = 0$

Hence $\bar{\nabla}^2 V = \dfrac{1}{r^2} \dfrac{\partial}{\partial r}\left(r^2 \dfrac{\partial V}{\partial r}\right) = 0$

Assuming r is not equal to zero, we have $\dfrac{d}{dr}\left(r^2 \dfrac{dV}{dr}\right) = 0$

Integrating this equation twice, we get

$$r^2 \frac{dV}{dr} = A \ , \ dV = A \frac{dr}{r^2} \ , \ V = -\frac{A}{r} + B \qquad \ldots\ldots (1)$$

Substitute initial conditions in Eq.(1)

$$50 = -\frac{A}{3} + B \qquad \ldots\ldots (2)$$

$$20 = -\frac{A}{5} + B \qquad \ldots\ldots (3)$$

Solving Eq.(2) and (3), we get $A = -225$ and $B = -25$

Hence $V = \frac{225}{r} - 25$

At point P (2, 1, 3), $r = \sqrt{(2)^2 + (1)^2 + (3)^2} = 3.7416$

Hence V at $P = \frac{225}{3.7416} - 25 = 35.134$ V

(d) $V = 50$ V at $\theta = 30^\circ$ and $V = 20$ V at $\theta = 50^\circ$

As the potential varies w.r.t. θ coordinate, $\frac{\partial V}{\partial r} = \frac{\partial V}{\partial \phi} = 0$

Hence $\bar{\nabla}^2 V = \frac{1}{r^2 \sin\theta} \frac{\partial}{\partial \theta}\left(\sin\theta \frac{\partial V}{\partial \theta} \right) = 0$

Assuming $r^2 \sin\theta \neq 0$, we have $\frac{d}{d\theta}\left(\sin\theta \frac{dV}{d\theta} \right) = 0$

Integrating this equation twice we get

$$\sin\theta \frac{dV}{d\theta} = A \ , \ dV = A \operatorname{cosec}\theta \, d\theta \ ; \ V = A \ln\left(\tan\frac{\theta}{2} \right) + B \qquad \ldots\ldots (1)$$

Substitute initial conditions in Eq.(1)

$$50 = A \ln(\tan 15^o) + B \qquad \ldots\ldots (2)$$

$$20 = A \ln(\tan 25^o) + B \qquad \ldots\ldots (3)$$

Solving Eq.(2) and (3), we get $A = -54.147$ and $B = -21.309$

Hence $V = -54.147 \ln\left(\tan\frac{\theta}{2} \right) - 21.309$

At P (2, 1, 3), $\theta = \cos^{-1}\left(\frac{z}{r} \right) = \cos^{-1}\left(\frac{3}{3.7416} \right) = 36.698^o$

Hence V at $P = -54.147 \ln\left(\tan\frac{36.698^o}{2} \right) - 21.309 = 38.449$ V

Problem 6.3 : The volume charge density at a certain semiconductor junction is expressed as $\rho_v = 500 \operatorname{sech} 10^6 x \tanh 10^6 x$ C/m^3. If $\varepsilon_r = 12$ for the semiconductor material and the area of the junction is 10^{-7} m^2, find : (a) V_o ; (b) C ; (c) E at the junction.

Solution :

(a) $\rho_v = 500 \operatorname{sech} 10^6 x \tanh 10^6 x$ C/m^3

Let $a = 10^6$. Then $\rho_v = 500 \operatorname{sech} ax \tanh ax$ C/m^3

As ρ_v is a function of x only, $\dfrac{\partial V}{\partial y} = \dfrac{\partial V}{\partial z} = 0$

$$\therefore \bar{\nabla}^2 V = \frac{d^2V}{dx^2} = -\frac{\rho_v}{\varepsilon_o \varepsilon_r} = \frac{-500 \operatorname{sech} ax \tanh ax}{12\,\varepsilon_o}$$

Integrating this equation ,we have

$$\frac{dV}{dx} = \frac{500}{12\,\varepsilon_o\, a} \operatorname{sech} ax + C_1$$ where C_1 is a constant of integration.

$$\text{Now } E_x = -\frac{dV}{dx} = \frac{-500}{12\,\varepsilon_o\, a} \operatorname{sech} ax - C_1$$

As $x \to \pm \infty$, $E_x = 0$. $\therefore C_1 = 0$

$$\therefore E_x = -\frac{500}{12\,\varepsilon_o\, a} \operatorname{sech} ax \qquad \ldots\ldots (1)$$

Integrating again,

$$V = \frac{1000}{12\,\varepsilon_o\, a^2} \tan^{-1}(e^{ax}) + C_2$$ where C_2 is a integration constant.

Suppose center of the junction is located at $x = 0$ which is also zero reference of potential.

$$\therefore 0 = \frac{1000}{12\,\varepsilon_o\, a^2} \times \tan^{-1}(e^0) + C_2 \text{ or } C_2 = -\frac{1000}{12\,\varepsilon_o\, a^2} \times \frac{\pi}{4}$$

$$\text{Hence } V = \frac{1000}{12\,\varepsilon_o\, a^2}\left(\tan^{-1} e^{ax} - \frac{\pi}{4}\right)$$

The total potential difference V_o across the junction is given by

$$V_o = V_{x\to\infty} - V_{x\to-\infty} = \frac{1000}{12\,\varepsilon_o\, a^2}\left(\frac{\pi}{2} - \frac{\pi}{4} + \frac{\pi}{4}\right) = \frac{500\,\pi}{12\,\varepsilon_o\, a^2} \qquad \ldots\ldots (2)$$

Substitute $a = 10^6$ in Eq. (2), we get $V_o = 14.784$ V

$$\text{(b) } Q = S \int_0^{\infty} 500 \text{ sech } ax \tanh ax \, dx = \frac{500}{a} S \qquad \ldots\ldots (3)$$

where S is the junction area. Substitute the value of a from Eq. (2) in Eq. (3)

$$Q = 500\, S \sqrt{\frac{12\, \varepsilon_o V_o}{500\, \pi}}$$

$$\text{Hence } C = \frac{dQ}{dV_o} = \frac{500\, S}{2} \sqrt{\frac{12\, \varepsilon_o}{500\, \pi\, V_o}}$$

Substitute $S = 10^{-7}\ \text{m}^2$ and $V_o = 14.784$ in the expression for capacitance, we

$$\text{get } C = \frac{500}{2} \times 10^{-7} \sqrt{\frac{12 \times 8.854 \times 10^{-12}}{500\, \pi \times 14.784}} = 1.691 \text{ pF}$$

(c) Substitute $a = 10^6$, $x = 0$ (at which junction is located) in Eq. (1) we have

$$E_x = -\frac{500}{12\, \varepsilon_o\, a} \text{ sech } (0) = \frac{-500}{12 \times 8.854 \times 10^{-12} \times 10^6} = -4.706 \text{ MV/m}$$

$$|E_x| = 4.706 \text{ MV/m}$$

Problem 6.4 : If $V = 2$ V at $x = 1$ mm, and $V = 0$ at $x = 0$, find E_x at $x = 1$ mm in free space for the volume charge density $\rho_v =$: (a) $-10^6\, \varepsilon_0$ C/m^3 ; (b) $-3 \times 10^8\, \varepsilon_0\, x$ C/m^3.

Solution :

(a) Potential varies as a function of x. Hence $\frac{\partial V}{\partial y} = \frac{\partial V}{\partial z} = 0$,

$$\therefore\ \nabla^2 V = \frac{d^2 V}{dx^2} = -\frac{\rho_v}{\varepsilon_o} = -\left(\frac{-10^6\, \varepsilon_o}{\varepsilon_o}\right) = 10^6$$

Integrating this equation , we get

$$\frac{dV}{dx} = 10^6 x + A \qquad \ldots\ldots (1)$$

where A is an integration constant.

Integrating Eq.(1), we get

$$V = \frac{10^6 x^2}{2} + Ax + B \qquad \ldots\ldots (2)$$

where B is an integration constant. Substitute the boundary conditions in Eq. (2), we get

$$2 = 10^6 \frac{(10^{-3})^2}{2} + A(10^{-3}) + B \qquad \ldots\ldots (3)$$

$$0 = 10^6 \frac{(0)^2}{2} + A(0) + B \qquad \ldots\ldots (4)$$

From Eq. (4), we have $B = 0$ and from Eq. (3), $A = 1500$.

Hence from Eq. (1) we have

$$E_x = -\frac{dV}{dx} = -10^6 x - A = -10^6 x - 1500$$

At $x = 1$ mm, $E_x = -10^6(10^{-3}) - 1500 = -2500$ V/m

(b) $\bar{\nabla}^2 V = \dfrac{d^2V}{dx^2} = -\dfrac{\rho_v}{\varepsilon_o} = -\dfrac{(-3\times 10^8 \varepsilon_o x)}{\varepsilon_o} = 3\times 10^8 x$

Integrating this equation, we get

$$\frac{dV}{dx} = 3\times 10^8 \frac{x^2}{2} + A \qquad \ldots\ldots (1)$$

where A is an integration constant. Integrating Eq.(1), we get

$$V = 3\times 10^8 \frac{x^3}{6} + Ax + B \qquad \ldots\ldots (2)$$

where B is an integration constant. Substitute the boundary conditions in Eq. (2) , we get

$$2 = 3\times 10^8 \frac{(10^{-3})^3}{6} + A(10^{-3}) + B \qquad \ldots\ldots (3)$$

$$0 = 3\times 10^8 \frac{(0)^3}{6} + A(0) + B \qquad \ldots\ldots (4)$$

From Eq. (4), we get $B = 0$ and from Eq.(3), $A = 1950$ .Hence from Eq.(1),

$$E_x = -\frac{dV}{dx} = -1.5\times 10^8 x^2 - A = -1.5\times 10^8 x^2 - 1950$$

At $x = 1$ mm, $E_x = -1.5\times 10^8(10^{-3})^2 - 1950 = -2100$ V/m .

6.2 SOLUTION OF LAPLACE'S EQUATION IN TWO DIMENSIONS :

6.2.1 It can be proved that whenever the potential varies as a function of x and y coordinates, it can be expressed as

$$V = (A\cosh\alpha x + B\sinh\alpha x)(C\cos\alpha y + D\sin\alpha y)$$

The proof is given below.

As the potential is a function of x and y, $\frac{\partial V}{\partial z} = 0$. Hence

$$\bar{\nabla}^2 V = \frac{\partial^2 V}{\partial x^2} + \frac{\partial^2 V}{\partial y^2} = 0 \qquad \ldots\ldots.(1)$$

Let us assume

$$V = X\,Y \qquad \ldots\ldots(2)$$

as the solution of this partial differential equation, where X is a function of x alone and Y is a function of y alone. Substituting Eq. (2) in Eq. (1), we get

$$Y\frac{\partial^2 X}{\partial x^2} + X\frac{\partial^2 Y}{\partial y^2} = 0$$

As X is independent of y and y is independent of x, above differential equation may be written as

$$Y\frac{d^2 X}{dx^2} + X\frac{d^2 Y}{dy^2} = 0 \qquad \ldots\ldots(3)$$

Hence

$$\frac{1}{X}\frac{d^2 X}{dx^2} + \frac{1}{Y}\frac{d^2 Y}{dy^2} = 0$$

or

$$\frac{1}{X}\frac{d^2 X}{dx^2} = -\frac{1}{Y}\frac{d^2 Y}{dy^2}$$

Since $(1/X)\,d^2X/dx^2$ involves no y and $-(1/Y)\,d^2Y/dy^2$ involves no x, and since the two quantities are equal, then $(1/X)\,d^2X/dx^2$ cannot be a function of x either and similarly, $-(1/Y)\,d^2Y/dy^2$ cannot be a function of y. That is, each of these terms must be a constant. Let this constant be α^2 Then

$$\frac{1}{X}\frac{d^2 X}{dx^2} = \alpha^2 \qquad \ldots\ldots.(4)$$

$$-\frac{1}{Y}\frac{d^2 Y}{dy^2} = \alpha^2 \qquad \ldots\ldots.(5)$$

The general solution for X (for a given α) is

$$X = A_1 e^{\alpha x} + B_1 e^{-\alpha x}$$

or, equivalently

$$X = A \cosh \alpha x + B \sinh \alpha x$$

and the general solution for Y (for a given α) is

$$Y = C_1 e^{j\alpha y} + D_1 e^{-j\alpha y}$$

or , equivalently,

$$Y = C \cos \alpha y + D \sin \alpha y$$

Therefore, the potential function may be written as

$$\boxed{V = (A \cosh \alpha x + B \sinh \alpha x)\,(C \cos \alpha y + D \sin \alpha y).} \quad ..(6.4)$$

6.2.2 Fig. 6.1 shows a rectangular trough with different potentials on its sides. Derive and expression for $V(x, y)$ and $\overline{E}(x, y)$

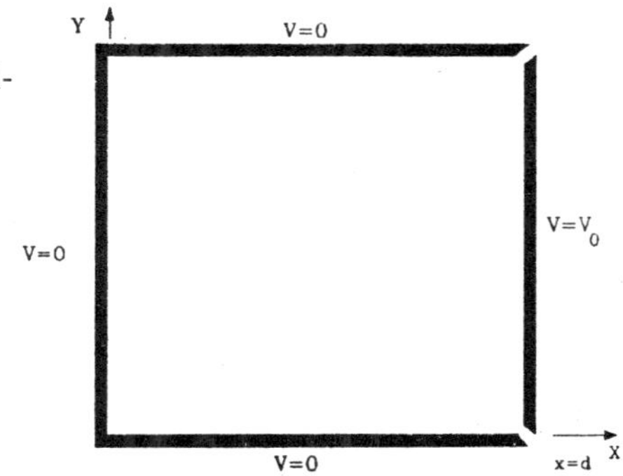

FIG. 6.1

From Fig. 6.1, the boundary conditions are

$V = 0$ for $y = 0,\ 0 < x < d$

$V = 0$ for $x = 0,\ 0 < y < b$

$V = 0$ for $y = b,\ 0 < x < d$

$V = V_o$ for $x = d,\ 0 < y < b$

The potential is given by

$$V = (A \cosh \alpha x + B \sinh \alpha x) \times (C \cos \alpha y + D \sin \alpha y)$$

As $V = 0$ for $y = 0$, we get $C = 0$. As $V = 0$ for $x = 0$, we get $A = 0$

Hence $V = BD \sinh \alpha x \sin \alpha y = V_1 \sinh \alpha x \sin \alpha y$.

As $V = 0$ for $y = b$, we have $\sin \alpha y = \sin \alpha b = 0$

$\therefore \alpha b = m\pi \qquad (m = 1, 2, 3, 4, \ldots.)$ or $\alpha = \dfrac{m\pi}{b}$. Hence

$$V = V_1 \sinh\left(\frac{m\pi x}{b}\right) \sin\left(\frac{m\pi y}{b}\right)$$

It is impossible to choose m or V_1 in such a way that $V = V_o$ at $x = d$ for each and every value of y between 0 and b. Hence potential is determined by combining an infinite number of these fields, each with a different value of

m and a corresponding value of V_1,

$$V = \sum_{m=1}^{\infty} V_{1m} \sinh\left(\frac{m\pi x}{b}\right) \sin\left(\frac{m\pi y}{b}\right) \qquad \ldots\ldots (1)$$

Applying the last boundary condition we get,

$$V_o = \sum_{m=1}^{\infty} V_{1m} \sinh\left(\frac{m\pi d}{b}\right) \sin\left(\frac{m\pi y}{b}\right)$$

$$(0 < y < b,\ m = 1, 2, 3 \ldots)$$

Let $C_m = V_{1m} \sinh\left(\frac{m\pi d}{b}\right)$ which is a function of m only.Then,

$$V_o = \sum_{m=1}^{\infty} C_m \sin\left(\frac{m\pi y}{b}\right)$$

$$(0 < y < b,\ m = 1, 2, 3, \ldots\ldots)$$

Let us determine the Fourier series for the function.

$$V = \begin{cases} V_o & (x = d, 0 < y < b) \\ -V_o & (x = d, b < y < 2b) \end{cases}$$

$$C_m = \frac{2}{2b}\left\{\int_o^b V_o \sin\frac{m\pi y}{b}\,dy + \int_b^{2b} (-V_o) \sin\frac{m\pi y}{b}\,dy\right\}$$

$$= \frac{4V_o}{m\pi} \quad (m \text{ is odd integer})$$

$$= 0 \quad (m \text{ is even integer})$$

As $C_m = V_{1m} \sinh\left(\frac{m\pi d}{b}\right)$, we have $V_{1m} = \dfrac{4V_o}{m\pi \sinh\left(\frac{m\pi d}{b}\right)}$

Substituting this value of V_{1m} in Eq.(1), we have

$$\boxed{V = \frac{4V_o}{\pi} \sum_{1,\,odd}^{\infty} \frac{1}{m} \frac{\sinh(m\pi x/b)}{\sinh(m\pi d/b)} \sin\left(\frac{m\pi y}{b}\right)} \qquad \ldots\ldots(6.5)$$

Now $\bar{E} = -\bar{\nabla} V = -\left[\frac{\partial V}{\partial x}\bar{a}_x + \frac{\partial V}{\partial y}\bar{a}_y\right]$

From Eq.(6.5), we get

$$\frac{\partial V}{\partial x} = \frac{4V_o}{b} \sum_{1,\,odd}^{\infty} \frac{\cosh(m\pi x/b)}{\sinh(m\pi d/b)} \sin\left(\frac{m\pi y}{b}\right)$$

and
$$\frac{\partial V}{\partial y} = \frac{4V_o}{b} \sum_{1, odd}^{\infty} \frac{\sinh(m\pi x/b)}{\sinh(m\pi d/b)} \cos\left(\frac{m\pi y}{b}\right)$$

Hence

$$\overline{\mathbf{E}} = \frac{-4V_o}{b}\left[\sum_{1, odd}^{\infty} \frac{\cosh(m\pi x/b)}{\sinh(m\pi d/b)} \sin\left(\frac{m\pi y}{b}\right)\overline{\mathbf{a}}_x + \sum_{1, odd}^{\infty} \frac{\sinh(m\pi x/b)}{\sinh(m\pi d/b)} \cos\left(\frac{m\pi y}{b}\right)\overline{\mathbf{a}}_y\right] \quad (6.6)$$

Note : Using Eq. (6.5), it is possible to write an expression for the potential $V(x, y)$ for the rectangular troughs of Fig. 6.2, 6.3, and 6.4.

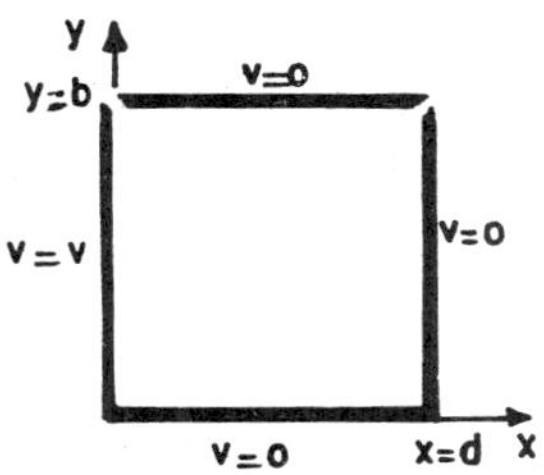

FIG.6.2

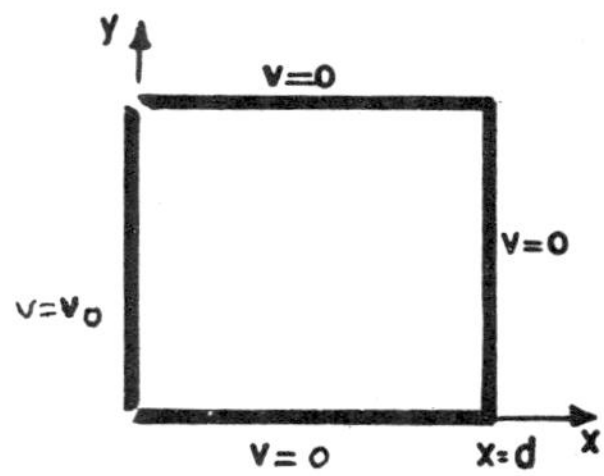

FIG.6.3

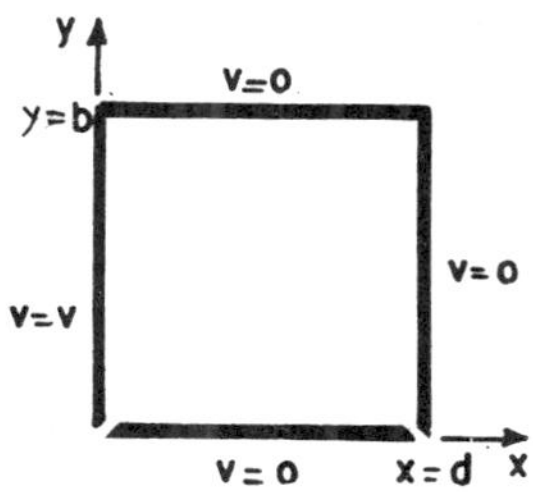

FIG. 6.4

For Fig. 6.2, replace x by $d - x$ in Eq. (6.5.). We get

$$V(x, y) = \frac{4V_o}{\pi} \sum_{1, odd}^{\infty} \frac{1}{m} \frac{\sinh[m\pi(d - x)/b]}{\sinh(m\pi d/b)} \sin\left(\frac{m\pi y}{b}\right) \quad \ldots\ldots(6.7)$$

For Fig. 6.3, replace x by y, y by x, b by d and d by b in Eq.(6.5). We get

$$V(x, y) = \frac{4V_o}{\pi} \sum_{1, odd}^{\infty} \frac{1}{m} \frac{\sinh(m\pi y/d)}{\sinh(m\pi b/d)} \sin\left(\frac{m\pi y}{b}\right) \quad \ldots(6.8)$$

For Fig. 6.4, replace y by $b - y$ in Eq.(6.8.). We get

$$V(x,y) = \frac{4V_o}{\pi} \sum_{1,\text{odd}}^{\infty} \frac{1}{m} \frac{\sinh[m\pi(b-y)/d]}{\sinh(m\pi b/d)} \sin\left(\frac{m\pi y}{b}\right) \quad \ldots\ldots(6.9)$$

Problem 6.5 : The dimensions of a square trough are $b = d = 2$ cm. If $V_o = 100$ V, find : (a) V at $x = y = 1.2$ cm ; (b) $\bar{E}$ at $x = y = 1$ cm.

Solution :

(a) Fig. P 6.5 shows a square trough with $b = d = 2$ cm. The boundary conditions are

$V = 0 \; ; \; x = 0, 0 < y < b$

$V = 0 \; ; \; y = 0, 0 < x < d$

$V = 0 \; ; \; y = b, 0 < x < d$

$V = V_o \; ; \; x = d, 0 < y < b$

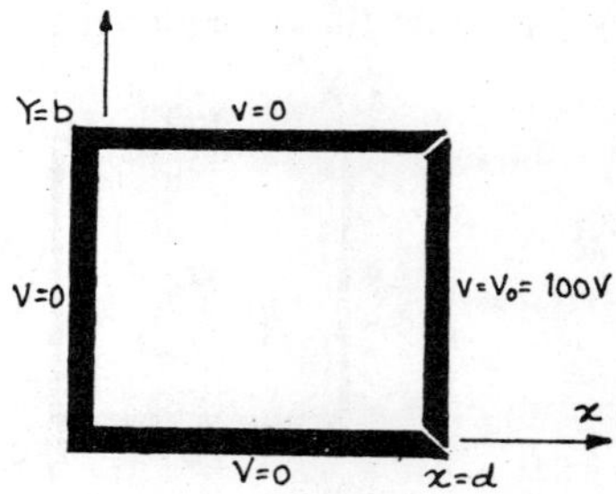

FIG.P6.5

As the potential varies with two coordinates x and y, it is given by Eq.(6.4)

$$V = XY = (A\cosh\alpha x + B\sinh\alpha x)(C\cos\alpha y + D\sin\alpha y)$$

For the given boundary conditions, the solution of above equation is given by Eq.(6.5)

$$V = \frac{4V_o}{\pi} \sum_{1,\text{odd}}^{\infty} \frac{1}{m} \frac{\sinh(m\pi x/b)}{\sinh(m\pi d/b)} \sin\left(\frac{m\pi y}{b}\right)$$

Substitute the values $V_o = 100$ V , and $b = d = 0.02$ m, in the expression for potential, we get

$$V = \frac{4(100)}{\pi} \sum_{1,\text{odd}}^{\infty} \frac{1}{m} \times \frac{\sinh(m\pi x/0.02)}{\sinh(m\pi\, 0.02/0.02)} \sin\left(\frac{m\pi y}{0.02}\right)$$

V at $(x = y = 1.2 \text{ cm})$

$$= \frac{4(100)}{\pi} \sum_{1,\text{odd}}^{\infty} \frac{1}{m} \times \frac{\sinh(0.012\pi m/0.02)}{\sinh(m\pi)} \sin\left(\frac{0.012\, m\pi}{0.02}\right)$$

$$= \frac{400}{\pi}\left[\frac{\sinh(0.6\pi)}{\sinh(\pi)}\sin(0.6\pi) + \frac{1}{3}\frac{\sinh(1.8\pi)}{\sinh(3\pi)}\sin(1.8\pi) + \frac{1}{5}\frac{\sinh(3\pi)}{\sinh(5\pi)}\sin(3\pi) + \frac{1}{7}\frac{\sinh(4.2\pi)}{\sinh(7\pi)}\sin(4.2\pi) + \ldots\right]$$

$$= 127.324\,(\,0.265 - 4.517\times 10^{-3} + 0 + 1.270\times 10^{-5} + \ldots .)$$
$$= 33.159 \text{ V}$$

(b) Substitute $x = y = 1$ cm $= 0.01$ m, $b = d = 0.02$ m and $V_o = 100$ V in the Eq.(6.6).

$$\overline{E} = -\frac{400}{0.02}\left[\sum_{1,\,odd}^{\infty}\frac{\cosh(0.5\,\pi\,m)}{\sinh(m\,\pi)}\sin(0.5\,m\,\pi)\,\overline{a}_x + \sum_{1,\,odd}^{\infty}\frac{\sinh(0.5\,\pi\,m)}{\sinh(m\,\pi)}\cos(0.5\,m\,\pi)\,\overline{a}_y\right]$$

Substitute $m = 1, 3, 5, 7, 9, 11 \ldots\ldots$ in the above expression. Then

$$\overline{E} = -4173.134\,\overline{a}_x + 0\,\overline{a}_y = -4173.134\,\overline{a}_x \text{ V/m}$$

Problem 6.6 : Given the potential function, $V = 6\cosh\alpha x\cos\alpha y$: (a) let $X = 3\cosh\alpha x$ and show that this potential function satisfies Eq. (3) of section 6 . 2 .1 ; (b) find α if $V = 0$ at $y = \pm 1$ and $V > 0$ for $-1 < y < 1$.

Solution :

(a) $V = X\,Y = 6\cosh\alpha x\cos\alpha y$

If $X = 3\cosh\alpha x$ then $Y = 2\cos\alpha y$.

We want to show that V satisfies the equation $Y\dfrac{d^2X}{dx^2} + X\dfrac{d^2Y}{dy^2} = 0$

$$\frac{dX}{dx} = 3\,\alpha\sinh\alpha x \;;\; \frac{d^2X}{dx^2} = 3\,\alpha^2\cosh\alpha x = \alpha^2 X$$

$$\frac{dY}{dy} = -2\,\alpha\sin\alpha y \;;\; \frac{d^2Y}{dy^2} = -2\,\alpha^2\cos\alpha y = -\alpha^2 Y$$

Hence $Y(\alpha^2 X) + X(-\alpha^2 Y) = 0 =$ RHS . Hence proved.

(b) As $V = 0$ at $y = \pm 1$, we have $\cos\alpha(\pm 1) = \cos(\pm\alpha) = 0$

$$\therefore \pm\alpha = \frac{n\pi}{2} \text{ where } n = 1, 3, 5, 7 \ldots \text{or } \alpha = \pm\frac{n\pi}{2}$$

As V is greater than zero for $-1 < y < 1$, we will select $n = 1$ so that the angle (αy) will always in first or fourth quadrant where cosine is positive.

$$\text{Hence } \alpha = \pm\frac{\pi}{2} = \pm 1.570796$$

6 .3 UNIQUENESS THEOREM :

Uniqueness theorem states that once a method is adopted to solve Laplace's or Poisson's equation subject to given boundary conditions, the solutions we are getting is unique even if we solve it by any other method.

To prove we assume that we have two solutions of Laplace equation V_1 and V_2 both general function of co- ordinates used. Therefore $\overline{\nabla}^2 V_1 = 0$ and $\overline{\nabla}^2 V_2 = 0$ from which $\overline{\nabla}^2 (V_1 - V_2) = 0$. Each solution must also satisfy the boundary conditions, and if we represent the given potential values on the boundaries by V_b, then the value of V_1 on the boundary V_{1b} and the value of V_2 on the boundary dV_{2b} must both be identical to V_b, $V_{1b} = V_{2b} = V_b$ or $V_{1b} - V_{2b} = 0$. We are going to use the vector identity $\overline{\nabla} \cdot (V\overline{D}) \equiv V(\overline{\nabla} \cdot \overline{D}) + \overline{D} \cdot (\overline{\nabla} V)$ which holds for any scalar V and any vector $\overline{D}$. Let us select $V_1 - V_2$ as the scalar and $\overline{\nabla}(V_1 - V_2)$ as the vector, giving

$$\overline{\nabla} \cdot \left[(V_1 - V_2)\, \overline{\nabla}(V_1 - V_2) \right] \equiv (V_1 - V_2) \left[\overline{\nabla} \cdot \overline{\nabla}(V_1 - V_2) \right] + \overline{\nabla}(V_1 - V_2) \cdot \overline{\nabla}(V_1 - V_2)$$

and integrating throughout the volume enclosed by the boundary surfaces

$$\int_{vol} \overline{\nabla} \cdot \left[(V_1 - V_2)\, \overline{\nabla}(V_1 - V_2) \right] dV \equiv \int_{vol} (V_1 - V_2)\left[\overline{\nabla} \cdot \overline{\nabla}(V_1 - V_2) \right] dV + \int_{vol} \left[\overline{\nabla}(V_1 - V_2) \right]^2 dV \qquad \ldots\ldots (1)$$

Apply the divergence theorem to the left hand side and since $V_{1b} = V_{2b}$

$$\int_{vol} \overline{\nabla} \cdot \left[(V_1 - V_2)\, \overline{\nabla}(V_1 - V_2) \right] dV \equiv \oint_S \left[(V_{1b} - V_{2b})\, \overline{\nabla}(V_{1b} - V_{2b}) \right] \cdot \overline{dS} = 0$$

i.e., L.H.S. of Eq.(1) is equal to zero. On the R.H.S. of Eq.(1), in the first integral $\overline{\nabla} \cdot \overline{\nabla}(V_1 - V_2)$ or $\overline{\nabla}^2 (V_1 - V_2)$ is present which is zero by hypothesis, and therefore that integral is zero. Hence the remaining volume integral must be zero.

$$\int_{vol} \left[\bar{\nabla} (V_1 - V_2) \right]^2 dV = 0$$

If an integral is zero, then either the integrand is zero every where or the integrand is positive in some region and negative in others. Since second can not be case because $\left[\bar{\nabla} (V_1 - V_2) \right]^2$ is always positive. Therefore $\left[\bar{\nabla} (V_1 - V_2) \right]^2 = 0$ and $\bar{\nabla} (V_1 - V_2) = 0$. If the gradient of $V_1 - V_2$ is every where zero, then $V_1 - V_2$ cannot change with any coordinates and $V_1 - V_2$ = constant. Consider any point on the boundary. Here $V_1 - V_2 = V_{1b} - V_{2b} = 0$, hence the constant is zero, i.e. $V_1 = V_2$ giving two identical solutions.

Problem 6.7 : Consider the two potential fields $V_1 = x$ and $V_2 = x +$ $\sinh y \sin x$. (a) Is $\bar{\nabla}^2 V_1 = 0$? (b) Is $\bar{\nabla}^2 V_2 = 0$? (c) Is $V_1 = 0$ at $x = 0$? (d) Is $V_2 = 0$ at $x = 0$? (e) Is $V_1 = \pi$ at $x = \pi$? (f) Is $V_2 = \pi$ at $x = \pi$? (g) Are V_1 and V_2 identical ? (h) Why does the uniqueness theorem not apply ?

Solution :

(a) $V_1 = x \; ; \; \dfrac{\partial V_1}{\partial x} = 1 \; ; \; \dfrac{\partial^2 V_1}{\partial x^2} = 0$. Hence $\bar{\nabla}^2 V_2 = 0$

(b) $V_2 = x + \sinh y \sin x$;

$$\frac{\partial V_2}{\partial x} = 1 + \sinh y \cos x \; ; \; \frac{\partial^2 V_2}{\partial x^2} = - \sinh y \sin x$$

$$\frac{\partial V_2}{\partial y} = \cosh y \sin x \; ; \; \frac{\partial^2 V_2}{\partial y^2} = \sinh y \sin x$$

$$\therefore \; \bar{\nabla}^2 V_2 = - \sinh y \sin x + \sinh y \sin x = 0$$

(c) $V_1 (\text{at } x = 0) = 0$

(d) $V_2 (\text{at } x = 0) = 0 + \sinh y \sin 0 = 0$

(e) $V_1 (\text{at } x = \pi) = \pi$

(f) $V_2 (\text{at } x = \pi) = \pi + \sinh y \sin \pi = \pi$

(g) No.

(h) Boundary conditions not given for a closed surface.

Problem 6.8 : In a certain region of a perfect dielectric, it is known that $\overline{\nabla}^2 V = 0$ and $\rho_v = 0$. If the dielectric is not homogeneous, what constraints are placed on the variation of ε with x, y, and z ?

Solution :

Point form of Gauss's law is $\overline{\nabla} \cdot \overline{D} = \rho_v$. As $\overline{D} = \varepsilon_o \varepsilon_r \overline{E}$, we have, $\overline{\nabla} \cdot (\varepsilon_o \varepsilon_r \overline{E}) = \rho_v$. As $\overline{E} = -\overline{\nabla} V$, we have $\overline{\nabla} \cdot (-\varepsilon_o \varepsilon_r \overline{\nabla} V) = \rho_v$. Hence $\overline{\nabla} \cdot (\varepsilon_r \overline{\nabla} V) = -\rho_v / \varepsilon_o$. Now using the identity

$$\overline{\nabla} \cdot (A \overline{B}) \equiv \overline{B} \cdot \overline{\nabla} A + A \overline{\nabla} \cdot \overline{B}$$

where A is a scaler and $\overline{B}$ is a vector . Hence $\overline{\nabla} \cdot (\varepsilon_r \overline{\nabla} V) = \overline{\nabla} V \cdot \overline{\nabla} \varepsilon_r + \varepsilon_r \overline{\nabla} \cdot \overline{\nabla} V = \overline{\nabla} V \cdot \overline{\nabla} \varepsilon_r + \varepsilon_r \overline{\nabla}^2 V$ or

$$\overline{\nabla} V \cdot \overline{\nabla} \varepsilon_r + \varepsilon_r \overline{\nabla}^2 V = -\frac{\rho_v}{\varepsilon_o}$$

Since $\overline{\nabla}^2 V = 0$ and $\rho_v = 0$ are given, we have $\overline{\nabla} V \cdot \overline{\nabla}^2 \varepsilon_r = 0$

If we assume that $\overline{\nabla} V$ exists, then $\overline{\nabla} \varepsilon_r = 0$, or

$$\frac{\partial \varepsilon_r}{\partial x} \overline{a}_x + \frac{\partial \varepsilon_r}{\partial y} \overline{a}_y + \frac{\partial \varepsilon_r}{\partial z} \overline{a}_z = 0 \quad \text{or} \quad \frac{\partial \varepsilon_r}{\partial x} = \frac{\partial \varepsilon_r}{\partial y} = \frac{\partial \varepsilon_r}{\partial z} = 0$$

Problem 6.9 : In a charge – free region of free space, a potential field is given as $V(x, y) = 5x^3 + f(x) - 2y^2$. Find $f(x)$ if E_x and V are both zero at the origin.

Solution :

At origin, $V(0, 0) = f(0) = 0$ ..... (1)

$$\text{Now } \overline{E} = -\overline{\nabla} V = -\left(\frac{\partial V}{\partial x} \overline{a}_x + \frac{\partial V}{\partial y} \overline{a}_y + \frac{\partial V}{\partial z} \overline{a}_z \right)$$

$$= E_x \overline{a}_x + E_y \overline{a}_y + E_z \overline{a}_z$$

$$\therefore E_x = -\frac{\partial V}{\partial x} = -\left[15x^2 + \frac{\partial f(x)}{\partial x} \right]$$

At origin, $E_x \Big|_{\text{origin}} = \dfrac{\partial f(x)}{\partial x}\Big|_{\text{origin}} = 0$ ..... (2)

Now $\overline{\nabla}^2 V = -\dfrac{\rho_v}{\varepsilon_o} = 0$ (for a charge free region $\rho_v = 0$). Hence

$\overline{\nabla}^2 V = \dfrac{\partial^2 V}{\partial x^2} + \dfrac{\partial^2 V}{\partial y^2} + \dfrac{\partial^2 V}{\partial z^2} = 0$. Now $V(x, y) = 5x^3 + f(x) - 2y^2$

$$\frac{\partial V}{\partial x} = 15x^2 + \frac{\partial f(x)}{\partial x} \; ; \; \frac{\partial^2 V}{\partial x^2} = 30x + \frac{\partial^2 f(x)}{\partial x^2}$$

$$\frac{\partial V}{\partial y} = -4y \; ; \; \frac{\partial^2 V}{\partial y^2} = -4 \; ; \; \frac{\partial^2 V}{\partial z^2} = 0$$

$$\therefore \; 30x + \frac{\partial^2 f(x)}{\partial x^2} - 4 = 0 \quad \text{or} \quad \frac{\partial^2 f(x)}{\partial x^2} = 4 - 30x$$

Integrating w.r.t. x, we have $\dfrac{\partial f(x)}{\partial x} = 4x - \dfrac{30x^2}{2} + C_1$, where C_1 is a constant of integration.From Eq.(2), at origin, $\dfrac{\partial f(x)}{\partial x} = 0$. Hence $0 = 4(0) - 15(0)^2 + C_1$ or $C_1 = 0$ or $\dfrac{\partial f(x)}{\partial x} = 4x - 15x^2$. Integrating w.r.t. x, we have $f(x) = \dfrac{4x^2}{2} - \dfrac{15x^3}{3} + C_2$, where C_2 is a constant of integration. From Eq.(1), at origin, $f(x) = 0$. Hence we have $0 = 2(0)^2 - 5(0)^3 + C_2$ or $C_2 = 0$. Hence $f(x) = 2x^2 - 5x^3$

Problem 6.10 : If $\varepsilon = \varepsilon_o$ and $V = 8x^2yz$, find : (a) V at P (2, –1, 3) ; (b) ρ_v at P ; (c) $\overline{E}$ at P ; (d) the equation of the equipotential surface through P ; (e) the equation of the streamline through P. (f) Does V satisfy Laplace's equation ?

Solution :

(a) $V = 8(2)^2(-1)(3) = -96$ V

(b) $\overline{\nabla}^2 V = \dfrac{\partial^2 V}{\partial x^2} + \dfrac{\partial^2 V}{\partial y^2} + \dfrac{\partial^2 V}{\partial z^2} = 16yz + 0 + 0 = -\dfrac{\rho_v}{\varepsilon_o}$

$\therefore \; \rho_v = -\varepsilon_o(16yz) = -16\varepsilon_o(-1)(3) = 424.992 \text{ pC/m}^3$

(c) Using Eq. ,we have

$$\overline{E} = -\left[\frac{\partial(8x^2yz)}{\partial x}\,\overline{a}_x + \frac{\partial(8x^2yz)}{\partial y}\,\overline{a}_y + \frac{\partial(8x^2yz)}{\partial z}\,\overline{a}_z\right]$$

$$= -(16xyz\,\overline{a}_x + 8x^2z\,\overline{a}_y + 8x^2y\,\overline{a}_z)$$

$$\overline{E} \text{ at } P(2,-1,3) = -16(2)(-1)(3)\,\overline{a}_x - 8(2)^2(3)\,\overline{a}_y - 8(2)^2(-1)\,\overline{a}_z$$

$$= 96\,\overline{a}_x - 96\,\overline{a}_y + 32\,\overline{a}_z \text{ V/m}$$

(d) V at P (2, –1, 3) = –96.. Hence the equation of the equipotential surface through P is $8x^2yz = -96$ or $x^2yz = -12$

(e) Equation of stream line is given by $\dfrac{E_x}{dx} = \dfrac{E_y}{dy} = \dfrac{E_z}{dz}$,

where $E_x = -16xyz$, $E_y = -8x^2z$, $E_z = -8x^2y$

Consider $\dfrac{dy}{dx} = \dfrac{E_y}{E_x} = \dfrac{-8x^2z}{-16xyz} = \dfrac{x}{2y}$ or $2y\,dy = x\,dx$

Integrating this equation, we have $\dfrac{2y^2}{2} = \dfrac{x^2}{2} + C$ where C is a constant of integration . At point P (2, –1, 3), $(-1)^2 = \dfrac{(2)^2}{2} + C$ or $C = -1$

$\therefore\ y^2 = x^2 - 1$ ....(1)

Consider $\dfrac{dz}{dy} = \dfrac{E_z}{E_y} = \dfrac{-8x^2y}{-8x^2z} = \dfrac{y}{z}$ or $z\,dz = y\,dy$. Integrating this equation, we have $\dfrac{z^2}{2} = \dfrac{y^2}{2} + C$. At point P, $\dfrac{(3)^2}{2} = \dfrac{(-1)^2}{2} + C$ or $C = 4$

$\therefore\ \dfrac{z^2}{2} = \dfrac{y^2}{2} + 4$ or $z^2 = y^2 + 8$ ......(2)

Consider $\dfrac{dz}{dx} = \dfrac{E_z}{E_x} = \dfrac{-8x^2y}{-16xyz} = \dfrac{x}{2z}$ or $2z\,dz = x\,dx$. Integrating this equation, we have $\dfrac{2z^2}{2} = \dfrac{x^2}{2} + C$. At point P (2,-1,3) , we have

$(3)^2 = \dfrac{(2)^2}{2} + C$ or $C = 7$

$\therefore\ z^2 = x^2 + 7$ ...... (3)

Any two equations out of 1, 2, and 3 gives equation of streamline.

(f) $\bar{\nabla}^2 V = 16yz \neq 0$

Hence V do not satisfy Laplace's equation.

Problem 6.11 : A potential field in free space is given as $V = 100[2\rho - \ln(\rho + 1)]$. For $0 < \rho < 5$, sketch curves of : (a) V vs. ρ ; (b) E_ρ vs. ρ ; (c) ρ_v vs. ρ.

Solution :

(a) $V = 100[2\rho - \ln(\rho + 1)]$

(b) we have E_ρ i.e., the coefficient of $\bar{a}_\rho$ as

$$E_\rho = -\frac{\partial V}{\partial \rho} = -100\left[2 - \frac{1}{\rho + 1}\right] = \frac{-100(2\rho + 1)}{\rho + 1} \text{ V/m}$$

(c) Using Eq.(6.3b), we have

$$\bar{\nabla}^2 V = \frac{1}{\rho}\frac{\partial}{\partial \rho}\left\{\rho \times 100\left[2 - \frac{1}{\rho + 1}\right]\right\} = \frac{100}{\rho}\left[2 - \frac{1}{(\rho + 1)^2}\right]$$

Using Eq.(6.1),we have

$$\rho_v = -\frac{\varepsilon_0 100}{\rho}\left[2 - \frac{1}{(\rho + 1)^2}\right] \text{ C/m}^3$$

Plot V, E_ρ, ρ_v vs ρ for $0 < \rho < 5$. Plots are not shown.

Problem 6.12 : Let $V = \ln[B(1 - \cos\theta)/(1 + \cos\theta)]$ and : (a) show that V satisfies Laplace's equation in spherical coordinates ; (b) select A and B so that $V = 100$ V and $|\bar{E}| = 500$ V/m at $r = 5$, $\theta = 90^\circ$, $\phi = 60^\circ$.

Solution : (a)

$$V = A\ln\left[\frac{B(1 - \cos\theta)}{(1 + \cos\theta)}\right] = A\ln\left[\frac{B \times 2\sin^2(\theta/2)}{2\cos^2(\theta/2)}\right]$$

$$= A\ln\left[B \times \tan^2\left(\frac{\theta}{2}\right)\right] = A\left\{\ln B + 2\ln\left[\tan\left(\frac{\theta}{2}\right)\right]\right\}$$

$$\therefore \frac{\partial V}{\partial \theta} = A\left[0 + \frac{2}{\tan(\theta/2)} \times \sec^2\left(\frac{\theta}{2}\right) \times \frac{1}{2}\right] = A\left[2 \operatorname{cosec}\theta\right]$$

Now $\frac{\partial V}{\partial r} = \frac{\partial V}{\partial \phi} = 0$ as r and ϕ are absent in V.

Using Eq.(6.3c),we have

$$\overline{\nabla}^2 V = \frac{1}{r^2 \sin\theta} \frac{\partial}{\partial \theta}\left(\sin\theta \frac{\partial V}{\partial \theta} \right) = \frac{1}{r^2 \sin\theta} \frac{\partial}{\partial \theta} (\sin\theta \times 2A \operatorname{cosec}\theta)$$

$$= \frac{1}{r^2 \sin\theta} \frac{\partial}{\partial \theta} (2A) = 0$$

Hence given V represents the solution of Laplace's equation in spherical coordinates.

(b)we have

$$\overline{\mathrm{E}} = -\overline{\nabla} V = -\frac{1}{r}\frac{\partial V}{\partial \theta}\,\bar{a}_\theta = -\frac{2A}{r} \operatorname{cosec}\theta\,\bar{a}_\theta$$

$$\overline{\mathrm{E}}\,(r = 5, \theta = 90^\circ, \phi = 60^\circ) = -\frac{2A}{5} \operatorname{cosec} 90^\circ\,\bar{a}_\theta = -\frac{2A}{5}\,\bar{a}_\theta$$

$$\therefore\ |\overline{\mathrm{E}}| = \frac{2A}{5} = 500 \quad \text{or} \quad A = 1250$$

$$V(r = 5, \theta = 90^\circ, \phi = 60^\circ) = 1250 \ln\left[\frac{B(1 - \cos 90^\circ)}{(1 + \cos 90^\circ)}\right] = 100$$

$$\therefore\ 1250 \ln(B) = 100 \quad \text{or} \quad B = 1.083$$

Problem 6.13 : Given the potential field $V = 100\,[\,\rho - (1/\rho)\,]\cos\phi$ V: (a) show that $\overline{\nabla}^2 V = 0$; (b) sketch the intersections of the 0 – V, 100 – V, and 200 – V equipotential surfaces with the $z = 0$ plane for $1 \le \rho \le 5$ and $0 \le \phi \le 90^\circ$.

Solution :

(a) $\frac{\partial V}{\partial \rho} = 100\left[1 + \frac{1}{\rho^2}\right]\cos\phi$, $\frac{\partial V}{\partial \phi} = -100\left[\rho - \frac{1}{\rho}\right]\sin\phi$, $\frac{\partial V}{\partial z} = 0$

Using Eq.(6.3b), we have

$$\overline{\nabla}^2 V = \frac{1}{\rho}\frac{\partial}{\partial \rho}\left[\rho \times 100\cos\phi\,(1 + \frac{1}{\rho^2})\right] + \frac{1}{\rho^2}\frac{\partial}{\partial \phi}\left[-100\,(\rho - \frac{1}{\rho})\sin\phi\right]$$

$$\bar{\nabla}^2 V = \frac{1}{\rho}\left[100\cos\phi\left(1-\frac{1}{\rho^2}\right)\right] + \frac{1}{\rho^2}\left[-100\left(\rho-\frac{1}{\rho}\right)\cos\phi\right]$$

$$= 100\cos\phi\left[\frac{1}{\rho}-\frac{1}{\rho^3}\right] - 100\cos\phi\left[\frac{1}{\rho}-\frac{1}{\rho^3}\right] = 0$$

Problem 6 .14 : Given the potential field $V = (50\sin\theta)/r^2$ V in free space : (a) determine whether V satisfies Laplace's equation ; (b) find the total charge stored inside the spherical shell $1 < r < 2$

Solution :

(a) Laplace's equation is $\bar{\nabla}^2 V = 0$

$$\frac{\partial V}{\partial r} = \frac{-100\sin\theta}{r^3}, \quad \frac{\partial V}{\partial\theta} = \frac{50\cos\theta}{r^2}, \quad \frac{\partial V}{\partial\phi} = 0$$

Using Eq.(6.3c),we have

$$\bar{\nabla}^2 V = \frac{1}{r^2}\frac{\partial}{\partial r}\left(\frac{-100\sin\theta\, r^2}{r^3}\right) + \frac{1}{r^2\sin\theta}\frac{\partial}{\partial\theta}\left(\frac{50\cos\theta\sin\theta}{r^2}\right) + 0$$

$$= \frac{1}{r^2}\left(\frac{100\sin\theta}{r^2}\right) + \frac{1}{r^2\sin\theta}\left(\frac{50\cos 2\theta}{r^2}\right) = \frac{100\sin\theta}{r^4} + \frac{50\cos 2\theta}{r^4\sin\theta}$$

$$= \frac{100\sin\theta}{r^4} + \frac{50(1-2\sin^2\theta)}{r^4\sin\theta} = \frac{50}{r^4\sin\theta} \neq 0$$

Hence Laplace's equation is not satisfied by the given potential field.

(b) $\rho_v = -\varepsilon_o\bar{\nabla}^2 V = -\varepsilon_o\dfrac{50}{r^4\sin\theta}$. Hence

$$Q = \int \rho_v\, dv = \int_1^2\int_0^\pi\int_0^{2\pi} -\varepsilon_o\frac{50}{r^4\sin\theta}\, r^2\sin\theta\, dr\, d\theta\, d\phi$$

$$= -50\,\varepsilon_o\left[\frac{-1}{r}\right]_1^2\left[\theta\right]_0^\pi\left[\phi\right]_0^{2\pi} = -4.369 \text{ nC}$$

Problem 6 :15 : Let $V_1(r,\theta,\phi) = 6/r$ and $V_2(r,\theta,\phi) = 3$. (a) State whether V_1 and V_2 satisfy Laplace's equation. (b) Evaluate V_1 and V_2 at $r = 2$. (c) Conciliate your results with the statement of the uniqueness

theorem.

Solution :

$$(a)\bar{\nabla}^2 V_1 = \frac{1}{r^2}\frac{\partial}{\partial r}\left(r^2\frac{\partial V_1}{\partial r}\right) = \frac{1}{r^2}\frac{\partial}{\partial r}\left[r^2\left(\frac{-6}{r^2}\right)\right] = \frac{1}{r^2}\frac{\partial}{\partial r}(-6) = 0$$

and $\bar{\nabla}^2 V_2 = 0$. Both V_1 and V_2 satisfy Laplace's equation .

(b) V_1 (at $r = 2$) $= 6/2 = 3$ and V_2 (at $r = 2$) $= 3$

(c) A single conducting surface with a voltage specified and no reference given does not form the complete boundary of a properly defined region. Hence uniqueness theorem cannot be applied.

Problem 6 .16 : The functions $V_1(\rho, z)$ and $V_2(\rho, z)$ both satisfy Laplace's equation in the region $a < \rho < b, 0 < \phi < 2\pi, -L < z < L$; each is zero on the surfaces $\rho = b$ for $-L < z < L$; $z = -L$ for $a < \rho < b$ and $z = L$ for $a < \rho < b$ and each is 100 V on the surface $\rho = a$ for $-L < z < L$. (a) In the region specified above, is Laplace'e equation satisfied by the functions $V_1 + V_2$, $V_1 - V_2$, $V_1 + 3$, and $V_1 V_2$? (b) On the boundary surfaces specified, are the potential values given above obtained from the functions $V_1 + V_2$, $V_1 - V_2$, $V_1 + 3$, and $V_1 V_2$? (c) Are the functions $V_1 + V_2$, $V_1 - V_2$, $V_1 + 3$, and $V_1 V_2$ identical with V_1 ?

Solution :

(a) $\bar{\nabla}^2 V_1 = 0$ and $\bar{\nabla}^2 V_2 = 0$

Then $\bar{\nabla}^2 (V_1 + V_2) = \bar{\nabla}^2 V_1 + \bar{\nabla}^2 V_2 = 0$ $\therefore$ Yes

$\bar{\nabla}^2 (V_1 - V_2) = \bar{\nabla}^2 V_1 - \bar{\nabla}^2 V_2 = 0$ $\therefore$ Yes

$\bar{\nabla}^2 (V_1 + 3) = \bar{\nabla}^2 V_1 + 0 = 0$ $\therefore$ Yes

$\bar{\nabla}^2 (V_1 V_2) \neq 0$ $\therefore$ No

(b) On the boundary surfaces specified, the potential values will be obtained from V_1 and V_2 and not from their combinations. Hence the answer is No , No, No, No.

(c) According to uniqueness theorem, the two solutions V_1 and V_2 must be identical. Hence the answer is Yes, No, No, No.

Problem 6 .17 : Conducting plates at $z =$ 1 cm and $z =$ 5 cm are held at potentials of – 8 V and 6 V, respectively. If the region between the plates is a homogeneous dielectric with $\varepsilon = 5\,\varepsilon_o$, find : (a) the capacitance between the plates per unit area ; (b) $V(z)$; (c) $\overline{D}(z)$.

Solution :

(a) At $z =$ 1 cm. = 0.01 m, $V = -8$ V

At $z =$ 5 cm. = 0.05 m, $V =$ 6 V. The potential changes as z changes.

Hence w.r.t. x and y, the change of potential is zero i.e. $\frac{\partial V}{\partial x} = \frac{\partial V}{\partial y} = 0$

$$\therefore\ \overline{\nabla}^2 V = \frac{\partial^2 V}{\partial x^2} + \frac{\partial^2 V}{\partial y^2} + \frac{\partial^2 V}{\partial z^2} = \frac{\partial^2 V}{\partial z^2} = \frac{d^2 V}{dz^2} = 0$$

Integrating this equation twice, we have

$$\frac{dV}{dz} = A \quad \text{or} \quad dV = A\,dz \quad \text{or} \quad V = Az + B$$

where A and B are constants of integration.. Substituting the boundary conditions in the above expression of potential, we have

$$-8 = A\,(0.01) + B \qquad \ldots\ldots(1)$$

$$6 = A\,(0.05) + B \qquad \ldots\ldots(2)$$

Solving Eq. (1) and (2), we have $A =$ 350 and $B = -11.5$ .Hence

$$V(z) = 350\,z - 11.5 \qquad \ldots\ldots(3)$$

$$\overline{E} = -\overline{\nabla} V = -\frac{\partial V}{\partial z}\,\bar{a}_z = -(A)\,\bar{a}_z = -350\,\bar{a}_z$$

$$\overline{D} = \varepsilon\overline{E} = 5\,\varepsilon_o(-350\,\bar{a}_z) = -1750\,\varepsilon_o\,\bar{a}_z \ \ C/m^2$$

Conductors are located at $z =$ 1 cm and $z =$ 5 cm to which $\pm\,\bar{a}_z$ are perpendicular. Hence $\overline{D}$ is also normal to the conductor surface.

$$\therefore\ \overline{D}_N = \overline{D} = -1750\,\varepsilon_o\,\bar{a}_z,\ C/m^2 \quad \text{or} \quad \rho_s = |\,\overline{D}_N\,| = 1750\,\varepsilon_o \ \ C/m^2$$

If a m^2 is the area of the plates, then

$Q =$ charge on positive plate $= \rho_s \times a = 1750\,\varepsilon_o\,a$

Potential difference between the plates $= 6-(-8) =$ 14 V

$$\therefore\ \text{Capacitance} = \frac{Q}{V} = \frac{1750\,\varepsilon_o\,a}{14}$$

or $\dfrac{\text{Capacitance}}{a} = \text{Capacitance per unit area} = \dfrac{1750\,\varepsilon_o}{14} = 1.10675\ \text{nF/m}^2$

(b) $V(z) = 350z - 11.5$

(c) $\overline{D}(z) = -1750\,\varepsilon_o\,\overline{a}_z = -15.4945\,\overline{a}_z\ \text{nC/m}^2$

Problem 6 .18 : The potential on the plane, $x - 2y + 5z = 2$, is 50 V. Point P (2, 3, –7) lies on a parallel conducting plane having a potential of –360 V. (a) Find V at A (–1, 4, 6). (b) Find $\overline{E}(x, y, z)$.

Solution : (a) &(b)

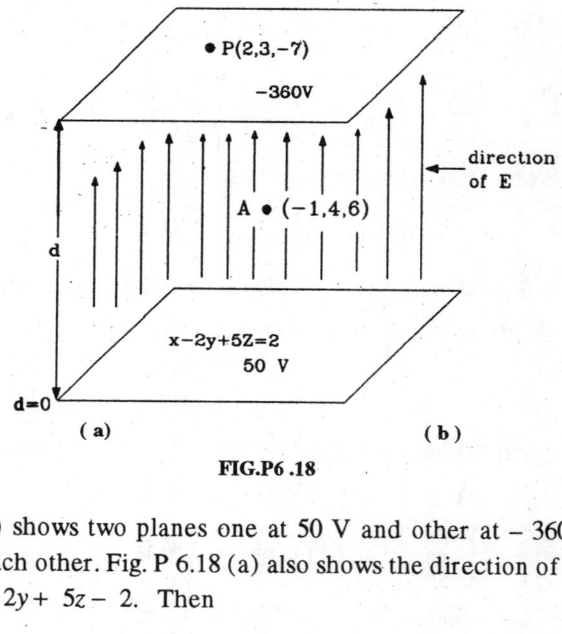

FIG.P6 .18

Fig. P6.18(a) shows two planes one at 50 V and other at – 360 V, both are parallel to each other. Fig. P 6.18 (a) also shows the direction of $\overline{E}$.

Let $f = x - 2y + 5z - 2$. Then

$$\overline{a}_n = \frac{\overline{\nabla} f}{|\,\overline{\nabla} f\,|} = \frac{\overline{a}_x - 2\,\overline{a}_y + 5\,\overline{a}_z}{\sqrt{(1)^2 + (-2)^2 + (5)^2}} = 0.183\,\overline{a}_x - 0.365\,\overline{a}_y + 0.913\,\overline{a}_z$$

is a unit vector directed towards the increasing values of f. On the plane $x - 2y + 5z = 2$, the value of f will be zero as it is the equation of plane itself, where as at point P (2, 3, – 7), $f = 2 - 2(3) + 5(-7) - 2 = -41$, i.e. the value of f is decreasing between the two parallel conducting planes as shown in Fig.

P6.18(b). Hence $\overline{a}_n = -\dfrac{\overline{\nabla} f}{|\,\overline{\nabla} f\,|} = -0.183\,\overline{a}_x + 0.365\,\overline{a}_y - 0.913\,\overline{a}_z$

is a unit vector towards decreasing value of f. That is, $\bar{a}_n$ is in the direction of $\bar{E}$. Now let us determine the distance of point P (2, 3, –7) from the plane $x-2y+5z = 2$. Let point $Q\ (x, y, z)$ be the point which is the foot of perpendicular dropped from point P (2, 3, –7) on the given plane. Then equations of the line passing through the points P and Q will be

$$\frac{x-2}{1} = \frac{y-3}{-2} = \frac{z+7}{5} \quad \text{or} \quad y = -2x+4 \text{ and } z = 5x-17$$

As $Q\ (x, y, z)$ also lies on the plane $x-2y+5z = 2.$, we have $x-2(-2x+4)+5(5x-17) = 2$ i.e. $x = 3.367$. Hence $y = 0.267$ and $z = -0.167$. Hence the coordinates of foot of perpendicular is (3.367, 0.267, –0.167). Hence the distance of point P from the given plane

$$= \sqrt{(2-3.367)^2 + (3-0.267)^2 + (-7+0.167)^2} = 7.486 \text{ m}$$

Change of potential = – 360 – 50 = – 410 V. Rate of change of potential w.r.t distance = $dV/dL = -410/7.486 = -54.769$ V/m. Similarly the coordinates of foot of perpendicular dropped from point A (–1, 4, 6) on the given plane will be (– 1.633, 5.267, 2.833). Hence distance of point A from the plane $x-2y+5z = 2$ is $\sqrt{(-1+1.633)^2 + (4-5.267)^2 + (6-2.833)^2} =$ 3.469. Let d indicates the distance of any point from the plan $x-2y+5z = 2$. Then potential at that point is given by $V = \mathbf{-54.769\,d + 50}$. When $d = 0$, that is, the point lies on the plane $x-2y+5z = 2$, the potential will be 50 V which is already given and when $d = 7.486$, that is, the point lies on another parallel conducting plane, the potential will be – 360 V. These potential values are obtained from the above expression for the potential.

To determine the potential at point A whose distance from the plane $x-2y+5z = 2$ is 3.469 m, substitute $d = -3.469$ in the expresssion for the potential. Negative value of d is used because point A lies on another side of the plane $x-2y+5z = 2$ as shown in Fig.P6.18(b). It is determined with the help of f evaluated at point A. At point A (-1, 4, 6), $f = -1-2(4)+5(6)-2 = -19$. Fig.P6.18(b) shows also the region corresponding to positive, zero, and negative values of f

$\therefore$ V at $A = -54.769(-3.469)+50 = 239.993$ V and $\bar{E} = dV/dL\ \bar{a}_n$

where $\bar{a}_n$ is a unit vector in the direction of $\bar{E}$. Hence

$\bar{E} = 54.769(-\ 0.183\,\bar{a}_x + 0.365\,\bar{a}_y - 0.913\,\bar{a}_z)$

$= -\ 10.022\,\bar{a}_x + 19.991\,\bar{a}_y - 50.004\,\bar{a}_z$ V/m.

[**Note** : Electric field intensitv remains constant in between two parallel conducting plates.]

Problem 6.19 : Conducting cylinders at $\rho = 1.6$ cm and $\rho = 5$ cm in free space are held at potentials of 80 V and – 40 V, respectively. (a) Find V and $\bar{E}$ at $\rho = 2$ cm. (b) Find the surface at which $V = 0$. (c) Find the surface charge density on each cylinder. (d) Determine the capacitance between the conducting cylinders per meter length.

Solution :

At $\rho = 1.6$ cm $= 0.016$ m, $V = 80$ V

At $\rho = 5$ cm $= 0.05$ m, $V = -\ 40$ V

(a) Potential varies as ρ varies. Hence change of potential w.r.t. ϕ and z is equal to zero. That is $\dfrac{\partial V}{\partial \phi} = \dfrac{\partial V}{\partial z} = 0$ and

$$\therefore\ \bar{\nabla}^2 V = \frac{1}{\rho}\frac{\partial}{\partial \rho}\left(\rho\frac{\partial V}{\partial \rho}\right) + \frac{1}{\rho^2}\frac{\partial^2 V}{\partial \phi^2} + \frac{\partial^2 V}{\partial z^2} = \frac{1}{\rho}\frac{\partial}{\partial \rho}\left(\rho\frac{\partial V}{\partial \rho}\right) = 0$$

Assuming $\rho \neq 0$, we have $\dfrac{d}{d\rho}\left(\rho\dfrac{dV}{d\rho}\right) = 0$. Integrating this equation twice, we get $\rho\dfrac{dV}{d\rho} = A$, $dV = \dfrac{A\,d\rho}{\rho}$, $V = A\ln\rho + B$. Substituting the boundary conditions in the above expression of potential, we have

$$80 = A\ln(0.016) + B \quad \ldots\ldots (1)$$

$$-\ 40 = A\ln(0.05) + B \quad \ldots\ldots (2)$$

From Eq. (1) and (2) we get $A = -\ 105.315$, $B = -\ 355.497$. Hence

$$V = -\ 105.315\ln\rho - 355.497 \quad \ldots\ldots (3)$$

$\therefore\ V(\text{at } \rho = 2\text{ cm} = 0.02\text{ m}) = -\ 105.315\ln(0.02) - 355.497 = 56.498$ V

$$\bar{E} = -\ \bar{\nabla}V = -\left[\frac{\partial V}{\partial \rho}\,\bar{a}_\rho\right] = -\left[\frac{A}{\rho}\,\bar{a}_\rho\right] = \frac{105.315}{\rho}\,\bar{a}_\rho \quad \ldots\ldots (4)$$

$$\therefore\ \bar{E}\ (\text{at } \rho = 2\text{ cm} = 0.02\text{ m}) = \frac{105.315}{0.02}\,\bar{a}_\rho = 5265.75\,\bar{a}_\rho \text{ V/m}$$

(b) From Eq. (3) $V = 0 = -\ 105.315\ln\rho - 355.497$

or $\rho = 0.034199$ m $= 3.420$ cm. On the cylinder $\rho = 3.420$ cm , $V = 0$
(c) From Eq. (4), $\overline{D} = \varepsilon_o \overline{E} = (105.315\,\varepsilon_o)/\rho\ \overline{a}_\rho$
Conductors are in the form of cylinder to which $\pm\ \overline{a}_\rho$ is perpendicular. Hence
$\overline{D} = \overline{D}_N$. Now $\rho_s = |\overline{D}_N| = (105.315\,\varepsilon_o)/\rho$ .....(5)

On the cylinder $\rho = 1.6$ cm , $\rho_s = (105.315\,\varepsilon_o)/0.016 = 58.279\ nC/m^2$

On the cylinder $\rho = 5$ cm , $\rho_s = (-\ 105.315\,\varepsilon_o)/0.05 = -\ 18.649\ nC/m^2$

Note : Negative sign is considered for ρ_s on the cylinder $\rho = 5$ cm due to the fact that the potential on this cylinder is less than that on the cylinder $\rho = 1.6$ cm .

(d) Consider a cylinder of radius ρ and length h . Surface area of the cylinder $= 2\pi\rho h\ m^2$. Using equation (5), the total charge on this cylinder will be
$Q = \rho_s \times$ Area $= (105.315\,\varepsilon_o)/\rho \times 2\pi\rho h = 105.315\,(2\pi)\,\varepsilon_o h$
Potential difference between the cylinders is $80 - (-40) = 120$ V.

$$\text{Hence capacitance} = \frac{105.315\,(2\pi)\,\varepsilon_o h}{120} = 48.823\,h\ \text{pF.} \quad \text{or}$$

$$\text{Capacitance per unit length} = \frac{\text{Capacitance}}{h} = 48.823\ \text{pF/m}$$

Problem 6 .20 : Conducting cylinders at $\rho = 2$ m and $\rho = 6$ cm are at potentials of 100 V and 0, respectively . The region between the cylinders is filled with an inhomogeneous perfect dielectric for which $\varepsilon_r = 0.3/(\rho + 0.04)$. Find : (a) $\overline{D}(\rho)$; (b) $\overline{E}(\rho)$; (c) $V(\rho)$; (d) C per meter length.

Solution :

At $\rho = 0.02$ m, $V = 100$ V and at $\rho = 0.06$ m, $V = 0$ V
For perfect dielectric $\rho_v = 0$
(c) Potential varies with ρ coordinate as seen from given data. Hence change of potential with coordinates ϕ and z is zero. That is $\dfrac{\partial V}{\partial \phi} = \dfrac{\partial V}{\partial z} = 0$. Now

$$\overline{\nabla} \cdot \overline{D} = \rho_v\ ,\ \overline{\nabla} \cdot (\varepsilon_o \varepsilon_r \overline{E}) = \rho_v\ ,\ \overline{\nabla} \cdot (-\ \varepsilon_r \overline{\nabla} V) = \frac{\rho_v}{\varepsilon_o}$$

Since $\rho_v = 0$, $\overline{\nabla} \cdot (\varepsilon_r \overline{\nabla} V) = 0$. Also, we have

$$\bar{\nabla} V = \frac{\partial V}{\partial \rho}\bar{a}_\rho + \frac{1}{\rho}\frac{\partial V}{\partial \rho}\bar{a}_\phi + \frac{\partial V}{\partial z}\bar{a}_z = \frac{\partial V}{\partial \rho}\bar{a}_\rho \text{. Hence}$$

$$\bar{\nabla}\cdot(\varepsilon_r \bar{\nabla} V) = \bar{\nabla}\cdot\left[\frac{0.3}{\rho + 0.04}\frac{\partial V}{\partial \rho}\bar{a}_\rho\right] = 0$$

Let $\bar{A} = A_\rho \bar{a}_\rho = \frac{0.3}{\rho + 0.04}\frac{\partial V}{\partial \rho}\bar{a}_\rho$, i.e. $A_\rho = \frac{0.3}{\rho + 0.04}\frac{\partial V}{\partial \rho}$

Now $\bar{\nabla}\cdot\bar{A} = \frac{1}{\rho}\frac{\partial}{\partial \rho}(\rho A_\rho) + \frac{1}{\rho}\frac{\partial A_\phi}{\partial \phi} + \frac{\partial A_z}{\partial z}$

$$\therefore \bar{\nabla}\cdot\bar{A} = \frac{1}{\rho}\frac{\partial}{\partial \rho}\left[\rho\left(\frac{0.3}{\rho + 0.04}\frac{\partial V}{\partial \rho}\right)\right] = 0$$

Assuming $\rho \neq 0$, we have $\frac{d}{d\rho}\left[\frac{0.3\rho}{\rho + 0.04}\frac{dV}{d\rho}\right] = 0$

Integrating this equation twice, we get

$$\frac{0.3\rho}{\rho + 0.04}\frac{dV}{d\rho} = A\ ,\ dV = A\left(\frac{\rho + 0.04}{0.3\rho}\right)d\rho \quad \text{or,}$$

$V = \frac{A}{0.3}(\rho + 0.04 \ln \rho) + B$ where A and B are constants of integration.

Substituting the boundary conditions in the expression of potential, we have

$$100 = \frac{A}{0.3}(0.02 + 0.04 \ln 0.02) + B \quad \text{.....(1)}$$

$$0 = \frac{A}{0.3}(0.06 + 0.04 \ln 0.06) + B \quad \text{..... (2)}$$

From Eq. (1) and (2) we get, $A = -357.379$ and $B = -62.585$. Hence

$$V(\rho) = \frac{-357.379}{0.3}(\rho + 0.04 \ln \rho) - 62.585$$

$$= -62.585 - 1191.263\rho - 47.651 \ln \rho$$

(b) $\bar{E} = -\bar{\nabla} V = -\frac{\partial V}{\partial \rho}\bar{a}_\rho = -\frac{A(\rho + 0.04)}{0.3\rho}\bar{a}_\rho$

$$\bar{E}(\rho) = \frac{-(-357.379)(\rho + 0.04)}{0.3\rho}\bar{a}_\rho = 1191.263\left(1 + \frac{0.04}{\rho}\right)\bar{a}_\rho \text{ V/m}$$

(a) $\bar{D}(\rho) = \varepsilon_o \varepsilon_r \bar{E}(\rho) = \frac{\varepsilon_o \times 0.3}{\rho + 0.04} \times 1191.263\left(\frac{\rho + 0.04}{\rho}\right)\bar{a}_\rho$

$\therefore \bar{D}(\rho) = 3.16423/\rho\ \bar{a}_\rho \text{ nC/m}^2$

(d) Consider a cylinder of radius ρ and length h. Surface area of the cylinder

$= 2\pi\rho h\, m^2$. Now $\rho_s = |\overline{D}_N| = |\overline{D}(\rho)| = \dfrac{3.16423}{\rho}\ nC/m^2$

Charge on this cylinder = $\rho_s \times$ Area $= \dfrac{3.16423}{\rho} \times 2\pi\rho h = 19.881\, h$ nC

Potential difference between the cylinders = $100 - 0 = 100$ V

Capacitance = $\dfrac{19.881\, h \times 10^{-9}}{100} = 198.814\, h$ pF .

Hence Capacitance / length = Capacitance / h = 198.814 pF/m.

Problem 6.21 : Two semi – infinite planes are located at $\phi = \alpha$ and $\phi = \beta$. A narrow insulating strip separates them along the z axis. Points P_α (5, 2, 0) and P_β (1, 4, 0) lie on the planes. Let V_α = 100 V, while E_ϕ = 20 V/m at P_β. (a) Find V_β. (b) Find E_ϕ at P_α.

Solution :

(a) $\overline{E} = -\overline{\nabla} V = \left[\dfrac{\partial V}{\partial \rho}\, \overline{a}_\rho + \dfrac{1}{\rho}\dfrac{\partial V}{\partial \phi}\, \overline{a}_\rho + \dfrac{\partial V}{\partial z}\, \overline{a}_z \right] = E_\rho\, \overline{a}_\rho + E_\phi\, \overline{a}_\phi + E_z\, \overline{a}_z$

$\therefore\ E_\phi = -\dfrac{1}{\rho}\dfrac{\partial V}{\partial \phi}$

Since E_ϕ is given to us, it means that the potential varies with ϕ and do not varies with ρ and z. That is $\dfrac{\partial V}{\partial \rho} = \dfrac{\partial V}{\partial z} = 0$.

$$\overline{\nabla}^2 V = \frac{1}{\rho}\frac{\partial}{\partial \rho}\left(\rho \frac{\partial V}{\partial \rho}\right) + \frac{1}{\rho}\frac{\partial^2 V}{\partial \phi^2} + \frac{\partial V}{\partial z} = 0 \quad \text{or} \quad \frac{1}{\rho}\frac{\partial^2 V}{\partial \phi^2} = 0$$

Assuming $\rho \neq 0$, we have $\dfrac{d^2 V}{d\phi^2} = 0$

Integrating this equation twice, we get

$\dfrac{dV}{d\phi} = A$, $dV = A\, d\phi$, $V = A\phi + B$ and $E_\phi = -\dfrac{1}{\rho}\dfrac{\partial V}{\partial \phi} = -\dfrac{A}{\rho}$

For point P_α (5, 2, 0) , $\phi_\alpha = \tan^{-1}(2/5) = 21.801^\circ$, $\rho = \sqrt{29}$

and for point P_β (1, 4, 0) , $\phi_\beta = \tan^{-1}(4/1) = 75.964^\circ$, $\rho = \sqrt{17}$

Now $V_\alpha = 100 = A\left(21.801 \times \dfrac{\pi}{180}\right) + B$ .....(1)

and E_ϕ (at P_β) $= 20 = -\dfrac{A}{\sqrt{17}}$ .....(2)

From equations (1) and (2) we have, $A = -82.462$ and $B = 131.377$

$\therefore V = -82.462\,\phi + 131.377$ .....(3)

and $E_\phi = \dfrac{82.462}{\rho}$ ....(4)

$\therefore V_\beta = -82.462\left(75.964 \times \dfrac{\pi}{180}\right) + 131.377 = 22.047$ V

(b) E_ϕ (at P_α) $= 82.462/\sqrt{29} = 15.313$ V/m

Problem 6.22 : The two conducting planes shown in Fig. P 6.22 are defined by $0.001 < \rho < 0.120$ m , $0 < z < 0.1$ m, $\phi = 0.179$ and 0.188 rad. The medium surrounding the planes is air. For region 1, $0.179 < \phi < 0.188$, neglect fringing and find : (a) $V(\phi)$; (b) $\overline{\mathbf{E}}(\rho)$; (c) $\overline{\mathbf{D}}(\rho)$; (d) ρ_s on the upper surface of the lower plane ; (e) Q on the upper surface of the lower plane. (f) Repeat (a) to (c) for region 2 by letting the location of the upper plane be $\phi = 0.188 - 2\pi$, and then find ρ_s and Q on the lower surface of the lower plane. (g) Find the total charge on the lower plane and the capacitance between the planes.

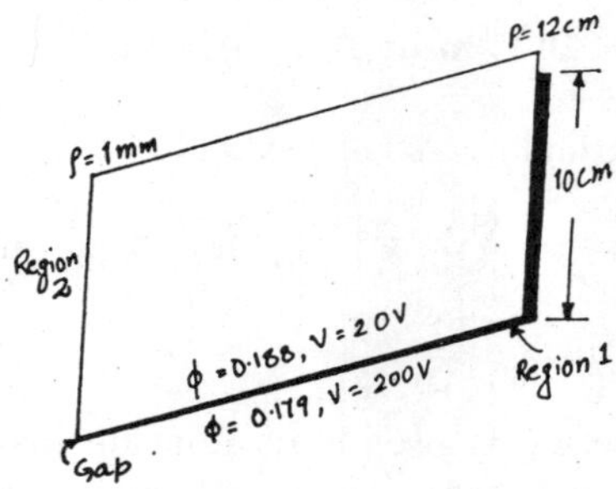

FIG.P6.22

Solution :

At $\phi = 0.188$ $V = 20$ V and at $\phi = 0.179$ $V = 200$ V

(a) The potential changes with ϕ coordinate. Hence change of potential with coordinates ρ and z is zero. That is $\dfrac{\partial V}{\partial \rho} = \dfrac{\partial V}{\partial z} = 0$

$$\overline{\nabla}^2 V = \frac{1}{\rho}\frac{\partial}{\partial \rho}\left(\rho \frac{\partial V}{\partial \rho}\right) + \frac{1}{\rho}\frac{\partial^2 V}{\partial \phi^2} + \frac{\partial^2 V}{\partial z^2} = \frac{1}{\rho}\frac{\partial^2 V}{\partial \phi^2} = 0$$

Assuming $\rho \neq 0$, we have $\dfrac{d^2 V}{d\phi^2} = 0$. Integrating this equation twice,we get $\dfrac{dV}{d\phi} = A$ or $dV = A\,d\phi$ or $V = A\,\phi + B$ (1)

Substituting the boundary conditions in the expression of potential,we have

$$20 = A\,(0.188) + B \quad (2)$$

and $$200 = A\,(0.179) + B \qquad (3)$$

From Eq.(2),and (3),we get $A = -20000$ and $B = 3780$. Hence

$$V(\phi) = -20000\,\phi + 3780$$

(b) $$\overline{\mathrm{E}} = -\overline{\nabla} V = -\left[\frac{\partial V}{\partial \rho}\bar{a}_\rho + \frac{1}{\rho}\frac{\partial V}{\partial \rho}\bar{a}_\phi + \frac{\partial V}{\partial z}\bar{a}_z\right] = -\frac{1}{\rho}\frac{\partial V}{\partial \phi}\bar{a}_\phi = -\frac{A}{\rho}\bar{a}_\phi .$$

$$\therefore\ \overline{\mathrm{E}}(\rho) = -\frac{A}{\rho}\bar{a}_\phi = \frac{20000}{\rho}\bar{a}_\phi \ \mathrm{V/m}$$

(c) $$\overline{\mathrm{D}}(\rho) = \varepsilon_o \overline{\mathrm{E}}(\rho) = \frac{20000\,\varepsilon_o}{\rho}\bar{a}_\phi = \frac{177.08}{\rho}\bar{a}_\phi \ \mathrm{nC/m^2}.$$

(d) ρ_s on the upper surface of the lower plane

$$= |\overline{\mathrm{D}}_{\mathrm{N}}| = |\overline{D}(\rho)| = \frac{177.08}{\rho}\ \mathrm{nC/m^2}$$

(e) To the lower plane,$\pm\ \bar{a}_\phi$ are perpendicular.Hence

$\overline{dS} = \pm\ d\rho\ dz\ \bar{a}_\phi$ or $dS = d\rho\ dz$.

$$\therefore\ Q = \int dQ = \int \rho_s\ dS = \int_{0.001}^{0.12}\int_0^{0.1} \frac{177.08}{\rho}\ d\rho\ dz$$

$$= 177.08\left[\ln\rho\right]_{0.001}^{0.12}\left[z\right]_0^{0.1} = 84.777\ \mathrm{nC}$$

(f) At $\phi = -6.095$ $V = 20\,\mathrm{V}$ and at $\phi = 0.179$ $V = 200\,\mathrm{V}$

Substotuting these boundary conditions in Eq.(1), we have

$$20 = A\,(-6.095) + B \qquad (4)$$

$$200 = A\,(0.179) + B \qquad (5)$$

From these equations, we have $A = 28.690$ and $B = 194.8645$

Hence $V(\phi) = 28.69\,\phi + 194.8645$

$$\overline{\mathrm{E}}(\rho) = -\frac{A}{\rho}\bar{a}_\phi = \frac{-28.690}{\rho}\bar{a}_\phi$$

$$\overline{\mathrm{D}}(\rho) = \varepsilon_o\overline{\mathrm{E}}(\rho) = \frac{-28.690\,\varepsilon_o}{\rho}\bar{a}_\phi = \frac{-254.021}{\rho}\bar{a}_\phi\ \mathrm{pC/m^2}$$

$$\rho_s = \frac{254.021}{\rho}\ \mathrm{pC/m^2}$$

$$Q = \int_{0.001}^{0.12}\int_0^{0.1}\frac{254.021}{\rho}\,d\rho\,dz = 254.021\left[\ln\rho\right]_{0.001}^{0.12}\left[z\right]_0^{0.1} = 0.1216\ \mathrm{nC}$$

(g) Total charge on lower plane = $84.777 + 0.1216 = 84.8986$ nC

Capacitance = $84.8986 \times 10^{-9} / 180 = 471.659$ pF

Problem 6 .23 : (a) Solve Laplace's equation for the potential field in the homogeneous region between two concentric conducting spheres with radii a and $b, b > a$, if $V = 0$ at $r = b$, and $V = V_o$ at $r = a$. (b) Find the capacitance between them.

Solution : $V = 0$ at $r = b$ and $V = V_o$ at $r = a$, $b > a$

(a) The potential varies as the r coordinates varies. Hence change of potential with θ and ϕ coordinates is zero. That is $\frac{\partial V}{\partial \theta} = \frac{\partial V}{\partial \phi} = 0$. Hence

$$\bar{\nabla}^2 V = \frac{1}{r^2}\frac{\partial}{\partial r}\left(r^2 \frac{\partial V}{\partial r}\right) + \frac{1}{r^2 \sin\theta}\frac{\partial}{\partial \theta}\left(\sin\theta \frac{\partial V}{\partial \theta}\right) + \frac{1}{r^2 \sin^2\theta}\frac{\partial^2 V}{\partial \phi^2}$$

$$= \frac{1}{r^2}\frac{\partial}{\partial r}\left(r^2 \frac{\partial V}{\partial r}\right) = 0$$

Assuming r $\neq$ 0 ,we have $\frac{d}{dr}\left(r^2 \frac{dV}{dr}\right) = 0$

Integrating this equation twice, we get

$$r^2 \frac{dV}{dr} = A \text{ or } dV = \frac{A\,dr}{r^2} \text{ or } V = -\frac{A}{r} + B$$

Substituting the boundary conditions in the expressin of potential, we have

$$0 = -\frac{A}{b} + B \quad \ldots. (1)$$

$$V_o = -\frac{A}{a} + B \quad \ldots.. (2)$$

From Eq.(1) and (2) we get, $A = \left[\frac{-V_o}{\frac{1}{a} - \frac{1}{b}}\right]$, and $B = \frac{1}{b}\left[\frac{-V_o}{\frac{1}{a} - \frac{1}{b}}\right]$

$$\therefore V = \left[\frac{-V_o}{\left(\frac{1}{a} - \frac{1}{b}\right) r}\right] + \frac{1}{b}\left[\frac{-V_o}{\frac{1}{a} - \frac{1}{b}}\right] \quad \text{or} \quad V(r) = V_o \frac{\frac{1}{r} - \frac{1}{b}}{\frac{1}{a} - \frac{1}{b}}$$

(b) $$\bar{E} = -\bar{\nabla} V = -\frac{\partial V}{\partial r}\bar{a}_r = -\left(\frac{A}{r^2}\right)\bar{a}_r = \frac{V_o}{\left(\frac{1}{a} - \frac{1}{b}\right) r^2}\bar{a}_r$$

$$\bar{D} = \varepsilon \bar{E} = \frac{V_o \varepsilon}{\left(\frac{1}{a} - \frac{1}{b}\right) r^2}\bar{a}_r \text{ C/m}^2$$

$$\therefore \ \rho_s = |\overline{D}_N| = |\overline{D}| = \frac{V_o \, \varepsilon}{\left(\frac{1}{a} - \frac{1}{b}\right) r^2} \ \text{C/m}^2$$

Consider a sphere of radius r. Surface area of the sphere is $4\pi r^2$.

Hence total charge on the sphere $= \dfrac{V_o \, \varepsilon}{\left(\frac{1}{a} - \frac{1}{b}\right) r^2} \, 4\pi r^2 = \dfrac{4\pi V_o \varepsilon}{\frac{1}{a} - \frac{1}{b}}$

Potential difference between the spheres $= V_o - 0 = V_o$

$$\therefore \ \text{Capacitance} = \frac{\dfrac{4\pi V_o \varepsilon}{1/a - 1/b}}{V_o} = \frac{4\pi\varepsilon}{\frac{1}{a} - \frac{1}{b}}$$

Problem 6 . 24 : Two concentric conducting spheres have radii of 3 and 5 cm. The region between them is filled with a homogeneous dielectric for which $\varepsilon_r = 5$. If the potential of the inner sphere is 100 V while that of the outer is –100 V, find : (a) $V(r)$; (b) $E_r(r)$; (c) the potential midway between the conducting spheres ; (d) the value of r at which $V = 0$; (e) Q on the inner sphere ; (f) the capacitance between the two spheres.

Solution :

At $r = 3$ cm $= 0.03$ m $\quad V = 100$ V

At $r = 5$ cm $= 0.05$ m $\quad V = -100$ V

(a) The potential changes w.r.t. r. Hence change of potential with θ and ϕ is zero. That is $\dfrac{\partial V}{\partial \theta} = \dfrac{\partial V}{\partial \phi} = 0$.

$$\overline{\nabla}^2 V = \frac{1}{r^2}\frac{\partial}{\partial r}\left(r^2 \frac{\partial V}{\partial r}\right) + \frac{1}{r^2 \sin\theta}\frac{\partial}{\partial \theta}\left(\sin\theta \frac{\partial V}{\partial \theta}\right) + \frac{1}{r^2 \sin^2\theta}\frac{\partial^2 V}{\partial \phi^2}$$

$$= \frac{1}{r^2}\frac{\partial}{\partial r}\left(r^2 \frac{\partial V}{\partial r}\right) = 0$$

Assuming $r \neq 0$, we have $\dfrac{d}{dr}\left(r^2 \dfrac{dV}{dr}\right) = 0$

Integrating this equation twice, we get

$$r^2 \frac{dV}{dr} = A \quad \text{or} \quad dV = \frac{A\,dr}{r^2} \quad \text{or} \quad V = -\frac{A}{r} + B$$

Substituting the boundary conditions in the expression of potential, we have

$$100 = \frac{-A}{0.03} + B \qquad .....(1)$$

$$-100 = \frac{-A}{0.05} + B \qquad .....(2)$$

From Eq. (1) and (2), $A = -15$ and $B = -400$. Hence

$$V(r) = \frac{15}{r} - 400$$

(b) $\overline{E} = -\overline{\nabla} V = -\frac{\partial V}{\partial r}\,\bar{a}_r = -\left(\frac{A}{r^2}\right)\bar{a}_r = -\left(\frac{-15}{r^2}\right)\bar{a}_r = \frac{15}{r^2}\,\bar{a}_r$ V/m

Hence $E_r(r) = \frac{15}{r^2}$ V/m^2

(c) $r = (0.03 + 0.05)/2 = 0.04$ m

$V(r = 0.04\text{ m}) = 15/0.04 - 400 = -25V$

(d) $V = 0 = \frac{15}{r} - 400$. Hence $r = 3.75$ cm

(e) $\overline{D} = \varepsilon_o \varepsilon \overline{E} = \varepsilon_o \times 5 \times \frac{15}{r^2}\,\bar{a}_r = \frac{75\,\varepsilon_o}{r^2}\,\bar{a}_r$ C/m^2

$\rho_s = |\overline{D}_N| = |\overline{D}| = \frac{75\,\varepsilon_o}{r^2}$ C/m^2

$Q = \rho_s \times \text{area} = \frac{75\,\varepsilon_o}{r^2} \times 4\pi r^2 = 300\pi\varepsilon_o = 8.345$ nC

(f) Potential difference between the spheres $= 100 - (-100) = 200$ V

$\therefore\; C = \frac{300\,\pi\,\varepsilon_o}{200} = 41.725$ pF

Problem 6.25 : Conducting spheres at $r = 2$ and 6 cm are at potentials of 100 V and 0, respectively. The region between the spheres is filled with an inhomogeneous dielectric for which $\varepsilon_r = 0.3/(r + 0.04)$. Find (a) $\overline{D}(r)$; (b) $\overline{E}(r)$; (c) $V(r)$; (d) C.

Solution :

At $r = 2$ cm $= 0.02$ m $V = 100$ V

At $r = 6$ cm $= 0.06$ m $V = 0$ V

$\varepsilon = 0.3/r + 0.04$

(c) Let us assume that the dielectric is a perfect one, so that $\rho_v = 0$

Potential varies with r coordinate as seen from given data. Hence change of potential with coordinates θ and ϕ is zero. That is $\frac{\partial V}{\partial \theta} = \frac{\partial V}{\partial \phi} = 0$. Now

$$\overline{\nabla} \cdot \overline{D} = \rho_v \ , \ \overline{\nabla} \cdot (\varepsilon_o \varepsilon_r \overline{E}) = \rho_v \ , \ \overline{\nabla} \cdot (- \varepsilon_r \overline{\nabla} V) = \frac{\rho_v}{\varepsilon_o}$$

Since $\rho_v = 0$, $\overline{\nabla} \cdot (\varepsilon_r \overline{\nabla} V) = 0$. Also from Eq.(4.***),we have

$$\overline{\nabla} V = \frac{\partial V}{\partial r} \overline{a} + \frac{1}{r} \frac{\partial V}{\partial \theta} \overline{a}_\theta + \frac{1}{r \sin\theta} \frac{\partial V}{\partial \phi} \overline{a}_\phi = \frac{\partial V}{\partial r} \overline{a}_r \text{ . Hence}$$

$$\overline{\nabla} \cdot (\varepsilon_r \overline{\nabla} V) = \overline{\nabla} \cdot \left[\frac{0.3}{r + 0.04} \frac{\partial V}{\partial r} \overline{a}_r \right] = 0$$

$$\text{Let } \overline{A} = A_r \overline{a}_r = \frac{0.3}{r + 0.04} \frac{\partial V}{\partial r} \overline{a}_r \text{, i.e. } A_r = \frac{0.3}{r + 0.04} \frac{\partial V}{\partial r}$$

$$\text{Now } \overline{\nabla} \cdot \overline{A} = \frac{1}{r^2} \frac{\partial}{\partial r} (r^2 A_r) + \frac{1}{r \sin\theta} \frac{\partial}{\partial \theta} (A_\theta \sin\theta) + \frac{1}{r \sin\theta} \frac{\partial A_\phi}{\partial \phi}$$

$$\therefore \ \overline{\nabla} \cdot \overline{A} = \frac{1}{r^2} \frac{\partial}{\partial r} \left[r^2 \left(\frac{0.3}{r + 0.04} \frac{\partial V}{\partial r} \right) \right] = 0$$

Assuming $r \neq 0$, we have $\frac{d}{dr} \left(\frac{0.3 r^2}{r + 0.04} \frac{dV}{dr} \right) = 0$

Integrating this equation twice, we get

$$\frac{0.3 r^2}{r + 0.04} \frac{dV}{dr} = A \ , \ dV = \frac{A (r + 0.04)}{0.3 r^2} dr \text{, or}$$

$V = \frac{A}{0.3} \left(\ln r - \frac{0.04}{r} \right) + B$ where A and B are constants of integration.

Substituting the boundary conditions in the expression of potential , we have

$$100 = \frac{A}{0.3} \left(\ln 0.02 - \frac{0.04}{0.02} \right) + B \qquad \text{.....(1)}$$

$$0 = \frac{A}{0.3} \left(\ln 0.06 - \frac{0.04}{0.06} \right) + B \qquad \text{.....(2)}$$

From Eq. (1) and (2) we have ~ ~ $A = -12.336$ and $B = -143.098$

$$\therefore \ V = \frac{-12.336}{0.3} \left(\ln r - \frac{0.04}{r} \right) - 143.098$$

$$V(r) = -41.12 \ln r + \frac{1.6448}{r} - 143.098$$

$$(b)\ \overline{E} = -\overline{\nabla} V = -\frac{\partial V}{\partial r}\,\overline{a}_r = -\left[\frac{A\,(r+0.04)}{0.3\,r^2}\right]\overline{a}_r$$

$$\overline{E}\,(r) = \frac{12.336\,(r+0.04)}{0.3r^2}\,\overline{a}_r = 41.12\left(\frac{1}{r}+\frac{0.04}{r^2}\right)\overline{a}_r \text{ V/m}$$

$$(a)\overline{D}\,(r) = \varepsilon_o\varepsilon_r\overline{E}\,(r) = \frac{0.3\varepsilon_o}{r+0.04}\times\frac{12.336\,(r+0.04)}{0.3\,r^2}\,\overline{a}_r = \frac{12.336\,\varepsilon_o}{r^2}\,\overline{a}_r \text{ C/m}^2$$

$$(d)\ \rho_s = |\,\overline{D}\,(r)\,| = \frac{12.336\,\varepsilon_o}{r^2}$$

Area of a sphere = $4\pi r^2$

Charge on the sphere = $\rho_s \times \text{Area} = \frac{12.336\,\varepsilon_o}{r^2}\times 4\pi r^2 = 1.3725 \text{ nC}$

Potential difference between the spheres = $100 - 0 = 100$

$$\therefore\ C = \frac{1.3725\times 10^{-9}}{100} = 13.725 \text{ pF}$$

Problem 6.26 : Two coaxial conducting cones have their vertices at the origin and are defined by $\theta = 40^o$ and $\theta = 55^o$. The inner cone is at a potential of 2000 V, while the outer is at 500 V. A small bit of insulation keeps the cones from touching. (a) Find $V(\theta)$. (b) Find V at $P_1\,(r = 0.6, \theta = 45^o, \phi = 60^o)$. (c) Find $\overline{E}\,(r, \theta)$. (d) Find $\overline{E}$ at P_1. (e) Find the total charge on the surface $\theta = 40^o, 0.1 < r < 0.5, 100^o < \phi < 170^o$, assuming $\varepsilon = \varepsilon_o$ for $40^o < \theta < 55^o$.

Solution :

At $\theta = 40^o \quad V = 2000$ V

At $\theta = 55^o \quad V = 5000$ V

(a) The potential is a function of θ. Hence change of potential w.r.t. r and ϕ coordinate is zero.That is $\frac{\partial V}{\partial r} = 0$ and $\frac{\partial V}{\partial \phi} = 0$.

$$\overline{\nabla}^2 V = \frac{1}{r^2}\frac{\partial}{\partial r}\left(r^2\frac{\partial V}{\partial r}\right) + \frac{1}{r^2\sin\theta}\frac{\partial}{\partial\theta}\left(\sin\theta\frac{\partial}{\partial\theta}\right) + \frac{1}{r^2\sin^2\theta}\frac{\partial^2 V}{\partial\phi^2}$$

$$= \frac{1}{r^2\sin\theta}\frac{\partial}{\partial\theta}\left(\sin\theta\frac{\partial V}{\partial\theta}\right) = 0$$

Assuming $r^2 \sin\theta \neq 0$, we have $\frac{d}{d\theta}\left(\sin\theta \frac{d\theta}{d\theta} \right) = 0$

Integrating this equation twice, we get

$\sin\theta \frac{dV}{d\theta} = A$, $dV = A \operatorname{cosec}\theta \, d\theta$, $V = A \ln\left(\tan\frac{\theta}{2} \right) + B$

where A and B are constants of integration. Hence

$$2000 = A \ln\left(\tan\frac{40^o}{2} \right) + B \qquad \ldots\ldots(1)$$

$$500 = A \ln\left(\tan\frac{55^o}{2} \right) + B \qquad \ldots\ldots(2)$$

From Eq. (1) and (2) we have, $A = -4191.740$ and $B = 3525.669$. Hence

$V(\theta) = -4191.74 \ln\left(\tan\frac{\theta}{2} \right) - 2236.521$

(b) V at $P_1 = = -4191.74 \ln\left(\tan\frac{45^o}{2} \right) - 2236.521 = 1457.968$ V

(c) $\bar{E} = -\bar{\nabla} V = -\frac{1}{r}\frac{\partial V}{\partial\theta}\bar{a}_\theta = -\frac{1}{r}\left[A \operatorname{cosec}\theta \, \bar{a}_\theta \right] = \frac{4191.74}{r\sin\theta}\bar{a}_\theta$ V/m

(d) $\bar{E}$ at $P_1 = \frac{4191.74 \, \bar{a}_\theta}{0.6 \sin 45^o} = 9880.026 \, \bar{a}_\theta$ V/m

(e) $\bar{D} = \varepsilon_o \bar{E} = \frac{4191.74 \, \varepsilon_o}{r\sin\theta} \bar{a}_\theta$ C/m^2

$\therefore \rho_s = |\bar{D}_N| = \frac{4191.74 \, \varepsilon_o}{r\sin\theta}$ C/m^2 . To the surface $\theta = 40^o$, $\pm \, \bar{a}_\theta$ is perpendicular. Hence $\overline{dS} = \pm \left(r\sin\theta \, dr \, d\phi \right) \bar{a}_\theta$

$$Q = \int_{0.1}^{0.5} \int_{100^o}^{170^o} \frac{4191.74 \, \varepsilon_o}{r\sin\theta} \times r\sin\theta \, dr \, d\phi$$

$$= 4191.74 \, \varepsilon_o \left[r \right]_{0.1}^{0.5} \left[\phi \right]_{100}^{70^o} = 18.137 \text{ nC}$$

Problem 6 .27 : In the potential field $V = A \ln[\tan(\theta/2)] + B$ it is specified that $V = 400$ at $P(3, \theta = 60^o, \phi = 0^o)$, while $E_\theta = 2000$ V/m at $Q(4, 40^o, 90^o)$. (a) Find the surface at which $E_\theta = 1000$ V/m . (b) Find the surface at which $|\bar{E}_z| = 500$ V/m .

Solution :

(a) $V = A \ln [\tan(\theta/2)] + B$. Hence

$$400 = A \ln [\tan(60^o/2)] + B \qquad \ldots\ldots(1)$$

$$\therefore \bar{E} = -\bar{\nabla} V = -\frac{1}{r}\frac{\partial V}{\partial \theta}\bar{a}_\theta = -\frac{A \operatorname{cosec}\theta}{r}\bar{a}_\theta \text{ . Hence}$$

$$2000 = -\frac{A \operatorname{cosec}(40^o)}{4} \qquad \ldots\ldots(2)$$

From Eq.(1) and (2), we have, $A = -5142.301$ and $b = -2424.697$. Hence

$$E_\theta = \frac{5142.301 \operatorname{cosec}\theta}{r}$$

$$\therefore 1000 = \frac{5142.301}{r \sin\theta} \quad \text{or} \quad r\sin\theta = 5.142$$

Hence on the cylinder $\rho = 5.142$, $E_\theta = 1000$ V/m

Note :

$$\rho = \sqrt{x^2 + y^2} = \sqrt{(r\sin\theta\cos\phi)^2 + (r\sin\theta\sin\phi)^2} = r\sin\theta$$

(b) $E_z = \bar{E}\cdot\bar{a}_z = E_\theta\,\bar{a}_\theta\cdot\bar{a}_z = E_\theta(-\sin\theta) = -\sin\theta\, E_\theta$

$$= -\sin\theta\left(\frac{5142.301 \operatorname{cosec}\theta}{r}\right) = -\frac{5142.301}{r}$$

$$|E_z| = \frac{5142.301}{r} = 500 \quad \text{or} \quad r = 10.285$$

Hence on the sphere $r = 10.285$, $|E_z| = 500$ V/m

Problem 6 . 28 : : Volume charge density in free space varies as $\rho_v = 100\,\varepsilon_o/r^{2.5}$. (a) Use Poisson's equation to find $V(r)$ if it is assumed that $r^2 E_r \to 0$ when $r \to 0$, while $V \to 0$ as $r \to \infty$. (b) Now find $V(r)$ by using Gauss's law and a line integral.

Solution :

$$\text{(a) } \bar{\nabla}^2 V = -\frac{\rho_v}{\varepsilon_o} = -\frac{100\,\varepsilon_o}{r^{2.5}\,\varepsilon_o} = -100/r^{2.5}$$

$$\therefore \frac{1}{r^2}\frac{\partial}{\partial r}\left(r^2\frac{\partial V}{\partial r}\right) = \frac{-100}{r^{2.5}} \quad \text{or}$$

$$\therefore \frac{d}{dr}\left(r^2\frac{dV}{dr}\right) = \frac{-100}{\sqrt{r}}$$

Integrating this equation, we get

$$r^2 \frac{dV}{dr} = -200\sqrt{r} + A \quad \text{or} \quad \frac{dV}{dr} = -\frac{200}{r^{1.5}} + \frac{A}{r^2}$$

where A is a constant of integration.

$$E_r = -\frac{\partial V}{\partial r} = -\left(\frac{-200}{r^{1.5}} + \frac{A}{r^2}\right)$$

Now $r^2 E_r \to 0$ when $r \to 0$

$$\text{Hence } r^2 E_r = \frac{200\, r^2}{r^{1.5}} - \frac{A\, r^2}{r^2} = 200\sqrt{r} - A$$

$$\lim_{r \to 0} r^2 E_r = \lim_{r \to 0} 200\sqrt{r} - A = -A = 0$$

$$A = 0, \quad \frac{dV}{dr} = -200/r^{1.5}$$

Integrating this equation we get,

$$V = \frac{400}{\sqrt{r}} + B$$

$$\lim_{r \to \infty} V = \lim_{r \to \infty} \left(\frac{400}{\sqrt{r}} + B\right) = B = 0 \qquad \therefore B = 0$$

$$\text{Hence } V(r) = \frac{400}{\sqrt{r}}$$

(b) Consider a sphere of radius r. According to Gauss's law

$$Q = \oint \bar{D} \cdot \overline{dS} = \int_0^{\pi} \int_0^{2\pi} (D_r\, \bar{a}_r) \cdot (r^2 \sin\theta \, d\theta \, d\phi \, \bar{a}_r)$$

$$= \int_0^{\pi} \int_0^{2\pi} D_r\, r^2 \sin\theta \, d\theta \, d\phi = 4\pi r^2 D_r$$

$$Q = \int dQ = \int \rho_v \, dv$$

$$= \int_0^{r} \int_0^{\pi} \int_0^{2\pi} \frac{100\, \varepsilon_0}{r^{2.5}} \times r^2 \sin\theta \, dr \, d\theta \, d\phi$$

$$= 100\, \varepsilon_0 \left[\frac{r^{0.5}}{0.5}\right]_0^r \left[-\cos\theta\right]_0^{\pi} \left[\phi\right]_0^{2\pi} = 800\, \pi\, \varepsilon_0 \sqrt{r}$$

$$\therefore\ 800\,\pi\,\varepsilon_o\,\sqrt{r} = 4\,\pi\,r^2 D_r \quad \text{or} \quad D_r = \frac{200\,\varepsilon_o}{r^{1.5}} \quad \text{or} \quad E_r = \frac{D_r}{\varepsilon_o} = \frac{200}{r^{1.5}}$$

$$\text{Now } V_{PQ} = -\int_Q^P \overline{E}\cdot\overline{dL} = -\int_Q^P (E_r\,\bar{a}_r)\cdot(dr\,\bar{a}_r) = -\int_Q^P E_r\,dr$$

$$V_{PQ} = -\int_\infty^r \frac{200}{r^{1.5}} = -200\left[-\frac{2}{\sqrt{r}}\right]_\infty^r = \frac{400}{\sqrt{r}}$$

Note : We are assuming that point Q lies at infinity and it is given that $V \to 0$ as $r \to \infty$. Hence $V_Q = 0$

Now $V_{PQ} = V_P - V_Q = V_P$. Therefore $V(r) = 400/\sqrt{r}$

Problem 6.29 : The charge distribution near a semiconductor junction is represented as $\rho_v = \rho_{vo}\,a^3 x/(x^2 + a^2)^2$. (a) Determine $\rho_{v,\max}$ and $\rho_{v,\min}$ and their locations. (b) Find E_x and $V(x)$, assuming that $E_x \to 0$ as $x \to \infty$, $V(0) = 0$. (c) Use a development similar to that in Problem 6.3 to show that $C = dQ/dV_o = \varepsilon S/(2\pi a)$.

Solution :

(a) $\rho_v = \rho_{vo}\,a^3 x/(x^2 + a^2)^2$

$$\frac{d\rho_v}{dx} = \rho_{vo}\,a^3\left[\frac{(x^2+a^2)^2(1) - (x)\,2\,(x^2+a^2)\,(2x)}{(x^2+a^2)^4}\right]$$

$$= \rho_{vo}\,a^3\left[\frac{x^2+a^2-4x^2}{(x^2+a^2)^3}\right] = \rho_{vo}\,a^3\,\frac{(a^2-3x^2)}{(x^2+a^2)^3}$$

Equating $\frac{d\rho_v}{dx} = 0$ we get

$$\rho_{vo}\,a^3\,\frac{(a^2-3x^2)}{(x^2+a^2)^3} = 0 \quad \text{or} \quad 3x^2 = a^2 \quad \text{or } x = \pm\,(a/\sqrt{3})$$

Now

$$\frac{d^2\rho_v}{dx^2} = \rho_{vo}\,a^3\left[\frac{(x^2+a^2)^3\,(-6x) - (a^2-3x^2)\,3\,(x^2+a^2)^2\,(2x)}{(x^2+a^2)^6}\right]$$

$$= \frac{\rho_{vo}\,a^3\,(-12x)\,(a^2-x^2)}{(x^2+a^2)^4}$$

When $x = -a/\sqrt{3}$, $\dfrac{d^2\rho_v}{dx^2} = \dfrac{81\,\rho_{vo}}{32\sqrt{3}\,a^2} > 0$

Hence at section $x = -a/\sqrt{3}$, $\rho_{v,\,min}$ occur.

$$\rho_{v,\,min} = \rho_{vo}\, a^3 \,(-a/\sqrt{3})\,/\,(a^2/3 + a^2)^2 = \frac{-9\,\rho_{vo}}{16\sqrt{3}}$$

When $x = a/\sqrt{3}$, $\dfrac{d^2\rho_v}{dx^2} = -\dfrac{81\,\rho_{vo}}{32\sqrt{3}\,a^2} < 0$

Hence at section $x = a/\sqrt{3}$, $\rho_{v,\,max}$ occur.

$$\rho_{v,\,min} = \rho_{vo}\, a^3 \,(a/\sqrt{3})\,/\,(a^2/3 + a^2)^2 = \frac{9\,\rho_{vo}}{16\sqrt{3}}$$

(b) $\overline{\nabla}V \cdot \overline{D} = \overline{\nabla}\cdot(\varepsilon\overline{E}) = \overline{\nabla}\cdot(\varepsilon E_x\,\bar{a}_x) = \dfrac{d}{dx}(\varepsilon E_x) = \rho_v$

$$\therefore\ \varepsilon E_x = \int \rho_v\, dx = \int \frac{\rho_{vo}\,a^3\,x\,dx}{(x^2 + a^2)^2} = \frac{-\rho_{vo}\,a^3}{2(x^2 + a^2)} + C$$

where C is a constant of integration. Now as $x \to \infty$, $E_x \to 0$. Hence $C = 0$. Hence

$$E_x = -\frac{\rho_{vo}\,a^3}{2\varepsilon(x^2 + a^2)} \ . \quad \text{Now } E_x = -\frac{dV}{dx} = -\frac{\rho_{vo}\,a^3}{2\varepsilon(x^2 + a^2)}$$

$\therefore\ V = \dfrac{\rho_{vo}\,a^2}{2\varepsilon}\tan^{-1}\left(\dfrac{x}{a}\right) + D$ where D is a constant of integration.

Now $V(0) = 0$. Hence $D = 0$. Hence $V(x) = \dfrac{\rho_{vo}\,a^2}{2\varepsilon}\tan^{-1}\left(\dfrac{x}{a}\right)$

(c) The total potential difference V_o across the junction is given by

$$V_o = V_{x\to\infty} - V_{x\to-\infty} = \frac{\rho_{vo}\,a^2}{2\varepsilon}\left[\left(\frac{\pi}{2}\right) - \left(-\frac{\pi}{2}\right)\right] = \frac{\pi\,\rho_{vo}\,a^2}{2\varepsilon}$$

$$Q = S\int_0^{\infty} \rho_v\,dx = S\int_0^{\infty} \frac{\rho_{vo}\,a^3\,x}{(x^2 + a^2)^2} = \frac{-S\,\rho_{vo}\,a^3}{2}\left[\frac{1}{x^2 + a^2}\right]_0^{\infty}$$

$= S\,\rho_{vo}\,a/2$ where S is the junction cross section area.

$$I = \frac{dQ}{dt} = C\frac{dV_o}{dt} \quad \text{or} \quad C = \frac{dQ}{dV_o}$$

$$\text{Now } Q = \frac{S\,\rho_{vo}\,a}{2} = \frac{S\,\rho_{vo}}{2}\sqrt{\frac{2\,\varepsilon\,V_o}{\pi\,\rho_{vo}}} = S\sqrt{\frac{\varepsilon\,\rho_{vo}\,V_o}{2\pi}}$$

$$\therefore\ C = \frac{dQ}{dV_o} = \frac{S}{2}\sqrt{\frac{\varepsilon\,\rho_{vo}}{2\pi\,V_o}} = \frac{S}{2}\sqrt{\frac{\varepsilon\,\rho_{vo}}{2\pi}\times\frac{2\varepsilon}{\pi\,\rho_{vo}\,a^2}} = \frac{S\,\varepsilon}{2\pi\,a}$$

Problem 6 .30 : A rectangular trough is formed by four conducting planes located in air $x = 0$ and 8 cm and $y = 0$ and 6 cm. The surface at $y = 6$ cm is at a potential of 100 V, the other three are at zero potential, and the necessary gaps are placed at two corners. Find the potential at $x = 3$ cm, $y = 4$ cm.

Solution :

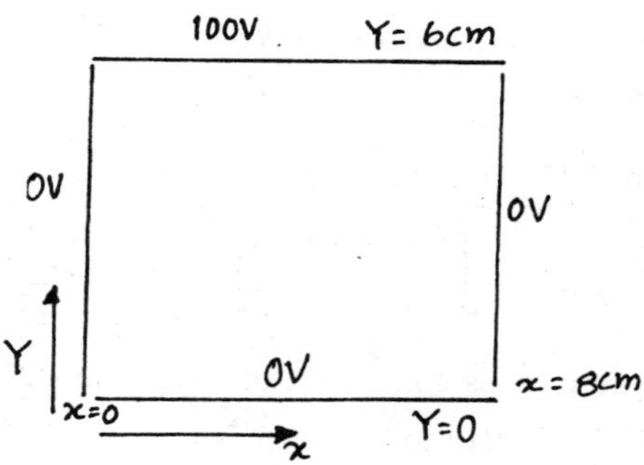

FIG.P6 . 30

When the potential varies with x and y coordinate, it is given by

$$V = (A\cos\alpha x + B\sin\alpha x)\times(C\cosh\alpha y + D\sinh\alpha y)$$

When $x = 0$, $V = 0$. Hence $A = 0$.

When $y = 0$, $V = 0$. Hence $C = 0$

$$\therefore\ V = B\sin\alpha x \times D\sinh\alpha y = E\sin\alpha x\,\sinh\alpha y$$

Now $V = 0$ at $x = 8$ cm $= 0.08$ m

$\therefore\ 0 = E\sin 0.08\,\alpha\ \sinh\alpha\cdot y$ where $E = BD$ or

$$\sin(0.08\,\alpha) = 0 \quad\text{or}\quad 0.08\,\alpha = m\pi \quad (m = 1, 2, 3\ldots.) \quad\text{or}\quad \alpha = \frac{m\pi}{0.08}$$

$$V = E\sin\frac{m\pi x}{0.08}\,\sinh\frac{m\pi y}{0.08}$$

Now $V = 100$ V when $y = 0.06$ m. It is impossible to choose m or E in such a way that $V = 100$ V when $y = 0.06$ m for each and every value of x between $x = 0$ and $x = 0.08$ m. We must combine an infinite numbers of these fields, each with a different value of m and a corresponding value of E. Hence

$$V = \sum_{m=0}^{\infty} E_m \sin\left(\frac{m\pi x}{0.08}\right)\sinh\left(\frac{m\pi y}{0.08}\right)$$

where E_m indicates that this amplitude factor will have a different value for each different value of m. Applying the last boundary condition now,

$$100 = \sum_{m=0}^{\infty} E_m \sin\left(\frac{m\pi x}{0.08}\right) \sinh\left(\frac{0.06\, m\pi}{0.08}\right) = \sum_{m=0}^{\infty} C_m \sin\left(\frac{m\pi x}{0.08}\right)$$

where $C_m = E_m \sinh\left(\dfrac{0.06\, m\pi}{0.08}\right)$

The above equation represents a fourier series with sine terms only. C_m is the fourier coefficient . Let us select a function V which is a function of x and defined as

$$V = 100 \qquad 0 < x < 0.08$$
$$= -100 \qquad 0.08 < x < 0.16$$

$$C_m = \frac{2}{T}\int_0^T V \sin\left(\frac{m\pi x}{0.08}\right) dx$$

$$= \frac{2}{0.16}\left\{\int_0^{0.08} 100 \sin\left(\frac{m\pi x}{0.08}\right) dx - \int_{0.08}^{0.16} 100 \sin\left(\frac{m\pi x}{0.08}\right) dx\right\}$$

$$= \frac{200}{0.16}\left\{\frac{-0.08}{m\pi}\left[\cos\left(\frac{m\pi x}{0.08}\right)\right]_0^{0.08} + \frac{0.08}{m\pi}\left[\cos\left(\frac{m\pi x}{0.08}\right)\right]_{0.08}^{0.16}\right\}$$

$$= \frac{200}{0.16} \times \frac{0.08}{m\pi}\left[-\cos m\pi + 1 + \cos(2m\pi) - \cos(m\pi)\right]$$

$$= \frac{100}{m\pi}(2 - 2\cos m\pi) \qquad (\because \cos 2m\pi = 1 \text{ for } m = 0, 1, 2, \ldots)$$

When m is even such as $m = 0, 2, 4, 6, \ldots\ldots \cos m\pi = 1$ so that $C_m = 0$. When m is an odd number such as $m = 1, 3, 5, 7, \ldots\ldots, \cos m\pi = -1$ so that $C_m = 400/m\pi$. Hence

$$C_m = E_m \sinh\left(\frac{0.06\, m\pi}{0.08}\right) = \frac{400}{m\pi} \quad \text{or} \quad E_m = \frac{400}{m\pi \sinh(0.06\, m\pi/0.08)}$$

$$V = \sum_{m=1,\,odd}^{\infty} \frac{400 \sinh(m\pi y/0.08) \sin(m\pi x/0.08)}{m\pi \sinh(0.06\, m\pi/0.08)}$$

$$V(x = 3\text{ cm}, y = 4\text{ cm}) = \sum_{m=1,\,odd}^{\infty} \frac{400 \sinh\left(\dfrac{m\pi\, 0.04}{0.08}\right) \sin\left(\dfrac{m\pi\, 0.08}{0.08}\right)}{m\pi \sinh(0.06\, m\pi/0.08)}$$

$$V = \frac{400}{\pi}\left[\frac{\sinh\left(\frac{0.04\,\pi}{0.08}\right)\sin\left(\frac{0.03\,\pi}{0.08}\right)}{1\sinh(0.06\,\pi/0.08)} + \frac{\sinh\left(\frac{0.12\,\pi}{0.08}\right)\sin\left(\frac{0.09\,\pi}{0.08}\right)}{3\sinh(0.18\,\pi/0.08)}\right.$$

$$\left. + \sinh\frac{\left(\frac{0.2\,\pi}{0.08}\right)\sin\left(\frac{0.15\,\pi}{0.08}\right)}{5\sinh(0.3\,\pi/0.08)} + \ldots\ldots\right]$$

$$= \frac{400}{\pi}\,(0.40668 - 0.01209 - 0.001508 + \ldots\ldots) = 50.049\ \text{V}$$

Problem 6.31 : The four sides of a square trough are held at potentials of 0, – 20, 40, and 50 V. Find the potential at the center of the trough.

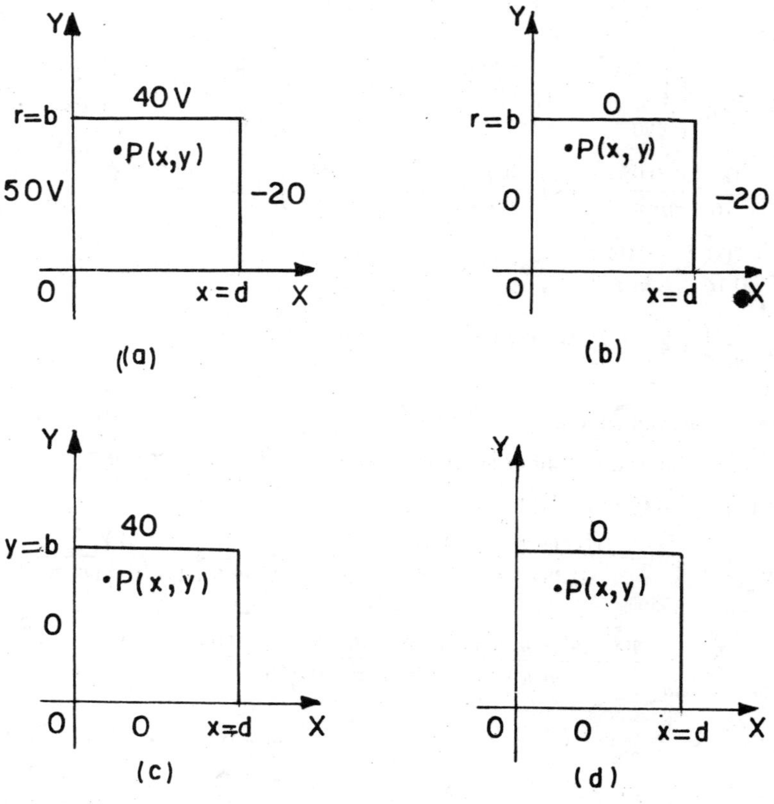

FIG. P 6.31

Solution :

Fig. P 6.31 (a) shows a square trough with potentials of 0, –20, 40, and 50 V, where $b = d$ for a square trough. We will use the principle of super- position. The potential at point $P(x, y)$ can be obtained by adding the potentials of Fig. P6.31 (b), (c), and (d) at point P.

For Fig. P6.31 (b), use the Eq.(6.5)

$$V(x,y) = \frac{4(-20)}{\pi} \sum_{1,\,odd}^{\infty} \frac{1}{m} \times \frac{\sinh(m\pi x/b)}{\sinh(m\pi d/b)} \sin\left(\frac{m\pi y}{b}\right) \quad \ldots\ldots(1)$$

For Fig. P6.31 (c) use the Eq.(6.8)

$$V(x,y) = \frac{4(40)}{\pi} \sum_{1,\,odd}^{\infty} \frac{1}{m} \times \frac{\sinh(m\pi y/d)}{\sinh(m\pi b/d)} \sin\left(\frac{m\pi x}{d}\right) \quad \ldots\ldots(2)$$

For Fig. P6.31 (d), use the the Eq.(6.7)

$$V(x,y) = \frac{4(50)}{\pi} \sum_{1,\,odd}^{\infty} \frac{1}{m} \times \frac{\sinh[m\pi(d-x)/b]}{\sinh(m\pi d/b)} \sin\left(\frac{m\pi y}{b}\right) \ldots(3)$$

Now, the coordinates of the center are $x = d/2$, $y = d/2$ with $b = d$

Hence at the center,

$$V = -\frac{80}{\pi} \sum_{1,\,odd}^{\infty} \frac{1}{m} \times \frac{\sinh(m\pi/2)}{\sinh(m\pi)} \sin\left(\frac{m\pi}{2}\right)$$

$$+ \frac{160}{\pi} \sum_{1,\,odd}^{\infty} \frac{1}{m} \times \frac{\sinh(m\pi/2)}{\sinh(m\pi)} \sin\left(\frac{m\pi}{2}\right)$$

$$+ \frac{200}{\pi} \sum_{1,\,odd}^{\infty} \frac{1}{m} \times \frac{\sinh(m\pi/2)}{\sinh(m\pi)} \sin\left(\frac{m\pi}{2}\right)$$

$$\therefore\ V = \frac{280}{\pi} \sum_{1,\,odd}^{\infty} \frac{1}{m} \times \frac{\sinh(m\pi/2)}{\sinh(m\pi)} \sin\frac{m\pi}{2}$$

Substitute $m = 1, 3, 5, 7, 9, 11, 13, \ldots\ldots$ we get,

$$V = \frac{280}{\pi}\left[\frac{\sinh(\pi/2)}{\sinh(\pi)} \sin\frac{\pi}{2} + \frac{\sinh(3\pi/2)}{3\sinh(3\pi)} \sin\frac{3\pi}{2} + \ldots\ldots\ldots\right]$$

$$= \frac{280}{\pi}(0.199268 - 0.00299 + 0.00007764\ldots\ldots) = 17.5 \text{ V}$$

Problem 6.32 : The potential on the side of the rectangular trough shown in Fig. P 6 . 32 varies sinusoidally as

$V_{(x=d)} = 200 \sin(2\pi y/b)$ **V**

(a) Find $V(x,y)$. (b) Find V at $x = 0.7d$, $y = 0.2b$, if $b = d$.

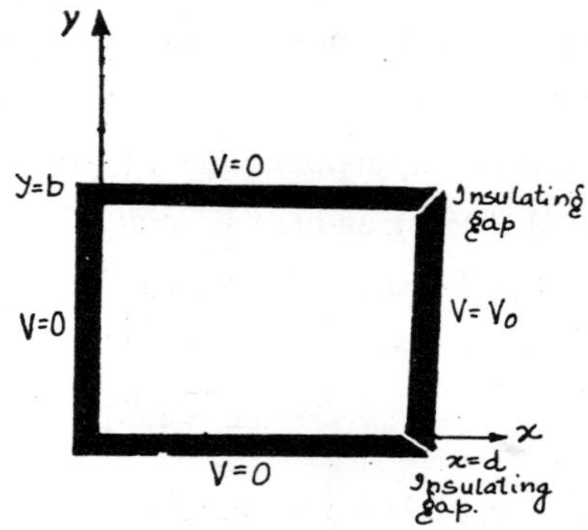

FIG.P6.32

Solution :

The boundary conditions are

$V = 0$ for $y = 0$, $0 < x < d$

$V = 0$ for $x = 0$, $0 < y < b$

$V = 0$ for $y = b$, $0 < x < d$

$$V = 200 \sin\left(\frac{2\pi y}{b}\right) \quad \text{for } x = d, 0 < y < d$$

$$V = (A \cosh \alpha x + B \sinh \alpha x)(C \cos \alpha y + D \sin \alpha y)$$

$V = 0$ for $y = 0$ indicates that C must be equal to 0. $V = 0$ for $x = 0$ indicates that A must be equal to zero.Hence

$$V = BD \sinh \alpha x \sin \alpha y = V_1 \sinh \alpha x \sin \alpha y$$

Now $V = 0$ for $y = b$ indicates that $\sin \alpha y = \sin \alpha b = 0$ or $\alpha b = m\pi$ or

$\alpha = \dfrac{m\pi}{b}$ where $m = 1, 2, 3, ..$

$$\text{Hence} V = V_1 \sinh\left(\frac{m\pi x}{b}\right) \sin\left(\frac{m\pi y}{b}\right) \qquad ......(1)$$

Applying the last boundary condition, we get

$$200 \sin\left(\frac{2\pi y}{b}\right) = V_1 \sinh\left(\frac{m\pi d}{b}\right) \sin\left(\frac{m\pi y}{b}\right)$$

When $m = 2$, we get

$$200 \sin\left(\frac{2\pi y}{b}\right) = V_1 \sin\left(\frac{2\pi d}{b}\right) \sin\left(\frac{2\pi y}{b}\right)$$

$$\therefore\ 200 = V_1 \sinh\left(\frac{2\pi d}{b}\right) \quad \text{or} \quad V_1 = \frac{200}{\sinh\left(\frac{2\pi d}{b}\right)}$$

Hence, from Eq.(1), $V = \dfrac{200 \sinh\left(\dfrac{2\pi x}{b}\right) \sin\left(\dfrac{2\pi y}{b}\right)}{\sinh\left(\dfrac{2\pi d}{b}\right)}$

(b) $x = 0.7\,d$, $y = 0.2\,b = 0.2\,d$ (as $b = d$)

$$\therefore\ V = \frac{200 \sinh\left(\dfrac{2\pi \times 0.7\,d}{d}\right) \sin\left(\dfrac{2\pi \times 0.2\,d}{d}\right)}{\sinh\left(\dfrac{2\pi d}{d}\right)}$$

$$V = \frac{200 \sinh(1.4\pi) \sin(0.4\pi)}{\sinh(2\pi)} = 28.8766 \text{ V}$$

Problem 6.33 : If X is a function only of x and $X'' + xX' - 2X = 0$, assume a solution in the form of an infinite power series and determine numerical values for a_2 to a_8 if $a_o = 1$ and $a_1 = -1$.

Solution :

$$\text{Let } X = \sum_{n=0}^{\infty} a_n x^n = a_o + a_1 x + a_2 x^2 + a_3 x^3 + a_4 x^4 + a_5 x^5$$
$$+ a_6 x^6 + a_7 x^7 + a_8 x^8 + \ldots\ldots$$

$$\therefore\ X' = \frac{dX}{dx} = a_1 + 2a_2 x + 3a_3 x^2 + 4a_4 x^3 + 5a_5 x^4$$
$$+ 6a_6 x^5 + 7a_7 x^6 + 8a_8 x^7 + \ldots\ldots$$

$$xX' = a_1 x + 2a_2 x^2 + 3a_3 x^3 + 4a_4 x^4 + 5a_5 x^5$$
$$+ 6a_6 x^6 + 7a_7 x^7 + 8a_8 x^8 + \ldots\ldots$$

$$X'' = \frac{d^2X}{dx^2} = 2a_2 + 6a_3 x + 12a_4 x^2 + 20a_5 x^3$$
$$+ 30a_6 x^4 + 42a_7 x^5 + 56a_8 x^6 + \ldots\ldots$$

$\therefore\ X'' + xX' - 2X = 0$ gives

$2a_2 + 6a_3 x + 12a_4 x^2 + 20a_5 x^3 + 30a_6 x^4 + 42a_7 x^5 +$
$56a_8 x^6 + \ldots\ldots, + a_1 x + 2a_2 x^2 + 3a_3 x^3 + 4a_4 x^4 + 5a_5 x^5 +$

$6 a_6 x^6 + 7 a_7 x^7 + 8 a_8 x^8 + \ldots\ldots, - 2 a_o - 2 a_1 x - 2 a_2 x^2 -$
$2 a_3 x^3 - 2 a_4 x^4 - 2 a_5 x^5 - 2 a_6 x^6 - 2 a_7 x^7 - 2 a_8 x^8 + \ldots\ldots = 0$

$$\therefore (2 a_2 - 2 a_o) + x(6 a_3 + a_1 - 2 a_1) + x^2 (12 a_4 + 2 a_2 - 2 a_2)$$
$$+ x^3 (20 a_5 + 3 a_3 - 2 a_3) + x^4 (30 a_6 + 4 a_4 - 2 a_4)$$
$$+ x^5 (42 a_7 + 5 a_5 - 2 a_5) + x^6 (56 a_8 + 6 a_6 - 2 a_6)$$
$$+ x^7 (72 a_9 + 7 a_7 - 2 a_7) + x^8 (90 a_{10} + 8 a_8 - 2 a_8) \ldots . = 0$$

Now $a_o = 1$ and $a_1 = -1$. Then $2 a_2 - 2 a_o = 0$ or $a_2 = 1$

$$6 a_3 + a_1 - 2 a_1 = 0 \quad \text{or} \quad a_3 = \frac{a_1}{6} = \frac{-1}{6}$$
$$12 a_4 + 2 a_2 - 2 a_2 = 0 \quad \text{or} \quad a_4 = 0$$
$$20 a_5 + 3 a_3 - 2 a_3 = 0 \quad \text{or} \quad a_5 = \frac{-a_3}{20} = \frac{1}{120}$$
$$30 a_6 + 4 a_4 - 2 a_4 = 0 \quad \text{or} \quad a_6 = \frac{-2 a_4}{30} = 0$$
$$42 a_7 + 5 a_5 - 2 a_5 = 0 \quad \text{or} \quad a_7 = \frac{-3 a_3}{42} = \frac{1}{84}$$
$$56 a_8 + 6 a_6 - 2 a_6 = 0 \quad \text{or} \quad a_8 = \frac{-4 a_6}{56} = 0$$

Problem 6 .34 : It is known that $V = XY$ is a solution of Laplace's equation, where X is a function of x alone and Y is function of y alone. Determine which of the following functions are also solutions of Laplace's equation : (a) $V_a = 100 XY$; (b) $V_b = 100 XY + 2x$; (c) $V_c = X + 3Y$; (d) $V_d = 2XY + y^2 - x^2$; (e) $V_e = (XY)^2$.

Solution :

(a) As $V = XY$ is a solution of Laplace's equation we have

$$Y \frac{\partial^2 X}{\partial x^2} + X \frac{\partial^2 Y}{\partial y^2} = 0$$

Now $V_a = 100 XY$

$$\therefore \quad \bar{\nabla}^2 V_a = 100 \left[Y \frac{\partial^2 X}{\partial x^2} + X \frac{\partial^2 Y}{\partial y^2} \right] = 100(0) = 0$$

Hence the answer is Yes.

(b) $V_b = 100\, X\, Y + 2x$

$$\frac{\partial V_b}{\partial x} = 100\,\frac{\partial X}{\partial x}\,Y + 2\ ; \quad \frac{\partial^2 V_b}{\partial x^2} = 100\,\frac{\partial^2 X}{\partial x^2}\,Y$$

$$\frac{\partial V_b}{\partial y} = 100\,X\,\frac{\partial Y}{\partial y}\ ; \quad \frac{\partial^2 V_b}{\partial y^2} = 100\,X\,\frac{\partial^2 Y}{\partial y^2}$$

$$\therefore\ \bar{\nabla}^2 V_b = \frac{\partial^2 V_b}{\partial x^2} + \frac{\partial^2 V_b}{\partial y^2}$$

$$= 100\left(Y\frac{\partial^2 X}{\partial x^2} + X\frac{\partial^2 Y}{\partial y^2}\right) = 100\,(0) = 0$$

Hence the answer is Yes.

(c) $V_c = X + 3\,Y$

$$\frac{\partial V_c}{\partial x} = \frac{\partial X}{\partial x}\ ; \quad \frac{\partial^2 V_c}{\partial x^2} = \frac{\partial^2 X}{\partial x^2}$$

$$\frac{\partial V_c}{\partial y} = 3\,\frac{\partial Y}{\partial y}\ ; \quad \frac{\partial^2 V_c}{\partial y^2} = 3\,\frac{\partial^2 Y}{\partial y^2}$$

$$\therefore\ \bar{\nabla}^2 V_c = \frac{\partial^2 X}{\partial x^2} + 3\,\frac{\partial^2 Y}{\partial y^2} \neq 0$$

Hence the answer is no.

(d) $V_d = 2\,X\,Y + y^2 - x^2$

$$\frac{\partial V_d}{\partial x} = 2\,\frac{\partial X}{\partial x}\,Y - 2x\ ; \quad \frac{\partial^2 V_d}{\partial x^2} = 2\,\frac{\partial^2 X}{\partial x^2}\,Y - 2$$

$$\frac{\partial V_d}{\partial y} = 2\,\frac{\partial Y}{\partial y}\,X + 2y\ ; \quad \frac{\partial^2 V_d}{\partial y^2} = 2\,\frac{\partial^2 Y}{\partial y^2}\,X + 2$$

$$\therefore\ \bar{\nabla}^2 V_d = 2\,\frac{\partial^2 X}{\partial x^2}\,Y - 2 + 2\,\frac{\partial^2 Y}{\partial y^2}\,X + 2$$

$$= 2\left(Y\frac{\partial^2 X}{\partial x^2} + X\frac{\partial^2 Y}{\partial y^2}\right) = 2\,(0) = 0$$

Hence the answer is Yes.

(e) $V_e = (XY)^2$

$$\frac{\partial V_e}{\partial x} = 2XY\frac{\partial X}{\partial x} \quad ; \quad \frac{\partial^2 V_e}{\partial x^2} = 2Y\frac{\partial X}{\partial x} + 2XY\frac{\partial^2 X}{\partial x^2}$$

$$\frac{\partial V_e}{\partial y} = 2XY\frac{\partial Y}{\partial y} \quad ; \quad \frac{\partial^2 V_e}{\partial y^2} = 2X\frac{\partial Y}{\partial y} + 2XY\frac{\partial^2 Y}{\partial y^2}$$

$$\therefore \ \bar{\nabla}^2 V_e = 2\left(Y\frac{\partial X}{\partial x} + X\frac{\partial Y}{\partial y}\right) + 2XY\left(\frac{\partial^2 X}{\partial x^2} + \frac{\partial^2 Y}{\partial y^2}\right) \neq 0$$

Hence the answer is no.

Problem 6 . 35 : Assume a product solution of Laplace's equation in cylindrical coordinates, $V = PF$, where V is not a function of z , P is a function only of ρ , and F is a function only of ϕ . (a) Obtain the two separated equations if the separation constant in n^2 . Select the sign of n^2 so that the solution of the ϕ equation leads to trigonometric functions. (b) Show that $P = A\rho^n + B\rho^{-n}$ satisfies the ρ equation . (c) Construct the solution $V(\rho, \phi)$. Functions of this form are called circular harmonics.

Solution :

$$\frac{\partial V}{\partial z} = 0 \quad \text{as V is not a function z.}$$

$$V = PF = P(\rho)\, F(\phi)$$

$$\bar{\nabla}^2 V = \frac{1}{\rho}\frac{\partial}{\partial \rho}\left(\rho\frac{\partial V}{\partial \rho}\right) + \frac{1}{\rho^2}\frac{\partial^2 V}{\partial \phi^2} = 0$$

$$\text{or } \rho\frac{\partial}{\partial \rho}\left(\rho\frac{\partial V}{\partial \rho}\right) + \frac{\partial^2 V}{\partial \phi^2} = 0$$

$$\text{or } \rho\frac{\partial V}{\partial \rho} + \rho^2\frac{\partial^2 V}{\partial \rho^2} + \frac{\partial^2 V}{\partial \phi^2} = 0$$

$$F\rho\frac{\partial P}{\partial \rho} + \rho^2 F\frac{\partial^2 P}{\partial \rho^2} + P\frac{\partial^2 F}{\partial \phi^2} = 0$$

$$\frac{\rho^2}{P}\frac{\partial^2 P}{\partial \rho^2} + \frac{\rho}{P}\frac{\partial P}{\partial \rho} = -\frac{1}{F}\frac{\partial^2 F}{\partial \phi^2}$$

The left hand side is a function of ρ only where as the right hand side is a function of ϕ only, both sides are equal to a constant n^2

$$\therefore \quad \frac{\rho^2}{P}\frac{\partial^2 P}{\partial \rho^2} + \frac{\rho}{P}\frac{\partial P}{\partial \rho} = n^2$$

or $$\frac{d^2 P}{d\rho^2} + \frac{1}{\rho}\frac{dP}{d\rho} - \frac{n^2 P}{\rho^2} = 0 \qquad \ldots\ldots (1)$$

and $$-\frac{1}{F}\frac{\partial^2 F}{\partial \phi^2} = n^2$$

or $$\frac{d^2 F}{d\phi^2} + n^2 F = 0 \qquad \ldots\ldots (2)$$

(b) $P = A\rho^n + B\rho^{-n}$

$$\frac{dP}{d\rho} = nA\rho^{n-1} - nB\rho^{-n-1}$$

$$\frac{d^2 P}{d\rho^2} = n(n-1)A\rho^{n-2} - n(-n-1)B\rho^{(size6- n - 1) - 1}$$

$$= n(n-1)A\rho^{n-2} - n(n+1)\rho^{-n-2}$$

The LHS of ρ equation i.e. equation (1) is

$$\text{LHS} = \frac{d^2 P}{d\rho^2} + \frac{1}{\rho}\frac{dP}{d\rho} - n^2\frac{P}{\rho^2}$$

$$= n(n-1)A\rho^{n-2} + n(n+1)B\rho^{-n-2} +$$

$$\frac{1}{\rho}(nA\rho^{n-1} - nB\rho^{-n-1}) - \frac{n^2}{\rho^2}(A\rho^n + B\rho^{-n})$$

$$= \rho^{n-2}(An^2 - nA + nA - n^2 A) +$$

$$\rho^{-n-2}(n^2 B + nB - nB - n^2 B)$$

$$= 0$$

Hence $P = A\rho^n + B\rho^{-n}$ satisfies the ρ equation.

(c) The solution of equation (2) may be written as

$F = C \cos n\phi + D \sin n\phi$

Hence the solution is

$V(\rho, \phi) = (A\rho^{n} + B\rho^{-n})(C \cos n\phi + D \sin n\phi)$

—X—

CHAPTER 7

The Steady Magnetic Field

7.1 Different types of current distributions : Fig. 7.1(a) shows a conductor circular in cross section of radius a. It carries a total current I. If the radius of the conductors is very small than it is known as a filamentary conductor.

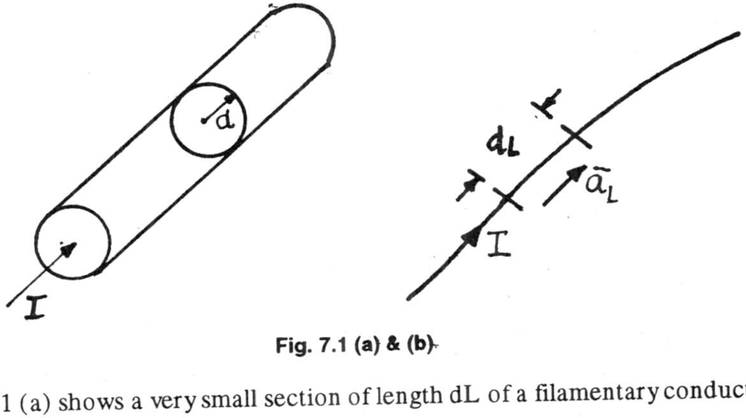

Fig. 7.1 (a) & (b)

Fig. 7.1 (a) shows a very small section of length dL of a filamentary conductor carrying a current I in $\bar{a}_L$ directions. Then a differential current element for a filamentary conductor is given by $I\,\overline{dL}$ where $\overline{dL} = dL\ \bar{a}_L$.

Fig. 7.2 (a) shows a rectangular slab of height h and width b. It carries a total current of I in $\bar{a}$ direction which is perpendicular to the cross sectional area (bh). Then current density $\bar{J}$ is given by

$$\bar{J} = \frac{I}{(bH)} \bar{a} \ A/m^2$$

and the total current is given by

$$I = \int \bar{J} \cdot \overline{dS}$$

The different current element for current, if the height h is very small, then the current I will flow on a surface crossing a width b perpendicular in $\bar{a}$ direction as shown in Fig. 7.2 (b).

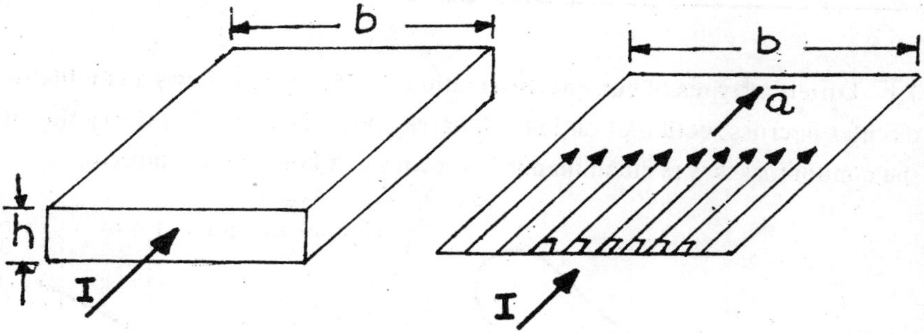

Fig. 7.2 (a) & (b)

Then surface current density $\bar{K}$ is given by

$$\bar{K} = \frac{I}{b} \bar{a} \ A/m$$

and the total current is given by

$$I = \int K \, dL$$

Density $\bar{J}$ is given by $\bar{J}$ dV where dV is the differential volume.
Current density $\bar{K}$ is given by $\bar{K}$ dL, where dL is the differential element of the path across which the current is flowing.

7.1 MAGNETIC FIELD INTENSITY

7.1.1 Biot – Savart Law

Consider a differential current element of a filamentary conductor carrying a current I as shown in Fig. 7.3. The law of Biot–Savart then states that at any point P the magnitude of the magnetic field intensity produced by the differential element is proportional to the product of the current, the magnitude of the differential length, and the sine of the angle lying between the filament and a line connecting the filament to the point P where the field is desired. The magnitude of the magnetic field intensity is inversely proportional to the square of the distance from the differential element to the point P. The direction of the magnetic field intensity is normal to the plane containing the differential filament and the line drawn from the filament to the point P. Of the two possible normals, that one is to be chosen which is in the direction of progress of a right handed screw turned from dL through the smaller angle to the line from the filament to P.

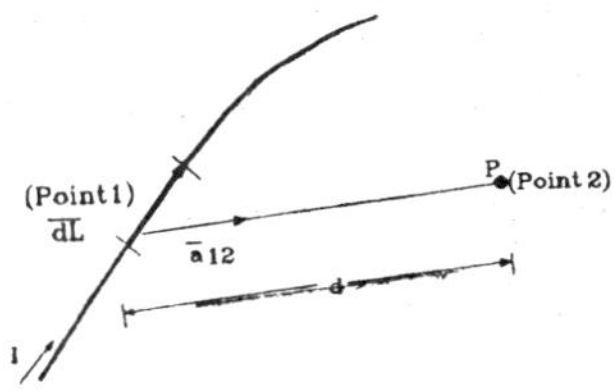

FIG.7.3

The Biot – Savart law may be written using vector notation as

$$\overline{\mathbf{dH}} = \frac{\mathbf{I}\,\overline{\mathbf{dL}} \times \bar{\mathbf{a}}_{12}}{4\pi d^2} \qquad \ldots\ldots (7.1)$$

The unit of $\overline{\mathrm{H}}$ is ampere / m. In terms of surface current density ($\overline{\mathrm{K}}$) and volume current density ($\overline{\mathrm{J}}$) the expression for magnetic field intensity are

$$\overline{\mathbf{dH}} = \frac{\overline{\mathbf{K}}\,dS \times \bar{\mathbf{a}}_{12}}{4\pi d^2} \qquad \ldots\ldots (7.2)$$

and

$$\overline{\mathbf{dH}} = \frac{\overline{\mathbf{J}}\,dv \times \bar{\mathbf{a}}_{12}}{4\pi d^2} \qquad \ldots\ldots (7.3)$$

where point 2 is that point at which magnetic field intensity is desired, point 1 is the point at which differential current element is located and d is the distance between points 1 and 2.

Problem 7.1 : Find the incremental field $\Delta \overline{H}_2$ at P_2 caused by a source at P_1 of $I_1 \Delta \overline{L}_1$ = : (a) $2\pi \bar{a}_z$ μ A · m, given P_1 (4, 0, 0) and P_2 (0, 3, 0) ; (b) $2\pi \bar{a}_z$ μ A · m, given P_1 (4, −2, 3) and P_2 (0, 3, 0) ;(c) $2\pi(0.6\bar{a}_x - 0.8\bar{a}_y)$ μ A · m, given P_1 (4, −2, 3) and P_2 (1, 3, 2).

Solution :

(a) $$d\overline{H}_2 = \frac{I\, d\overline{L} \times \bar{a}_{12}}{4\pi d^2}$$

Point 2 is that point at which magnetic field intensity is desired = (0, 3, 0)

Point1 is that point at which differential current element is located= (4, 0, 0)

$$\overline{R}_{12} = (0-4)\,\bar{a}_x + (3-0)\,\bar{a}_y = -4\,\bar{a}_x + 3\,\bar{a}_y$$

$$d = |\overline{R}_{12}| = \sqrt{(-4)^2 + (3)^2} = 5$$

$$\therefore\ d\overline{H}_2 = \frac{(2\pi\, 10^{-6}\bar{a}_z)}{4\pi(5)^2} \times \left[\frac{-4\,\bar{a}_x + 3\,\bar{a}_y}{5}\right] = -12\,\bar{a}_x - 16\,\bar{a}_y \ \text{nA/m}$$

(b) $$\overline{R}_{12} = (0-4)\,\bar{a}_x + (3+2)\,\bar{a}_y + (0-3)\,\bar{a}_z$$
$$= -4\,\bar{a}_x + 5\,\bar{a}_y - 3\,\bar{a}_z$$

$$d = |\overline{R}_{12}| = \sqrt{(-4)^2 + (5)^2 + (-3)^2} = \sqrt{50} = 7.071$$

$$\therefore\ d\overline{H}_2 = \frac{(2\pi\, 10^{-6}\bar{a}_z)}{4\pi(7.071)^2} \times \left[\frac{-4\,\bar{a}_x + 5\,\bar{a}_y - 3\,\bar{a}_z}{7.071}\right]$$
$$= -7.071\,\bar{a}_x - 5.657\,\bar{a}_y \ \text{n A/m}$$

(c) $$\overline{R}_{12} = (1-4)\,\bar{a}_x + (3+2)\,\bar{a}_y + (2-3)\,\bar{a}_z$$
$$= -3\,\bar{a}_x + 5\,\bar{a}_y - \bar{a}_z$$

$$d = |\overline{R}_{12}| = \sqrt{(-3)^2 + (5)^2 + (-1)^2} = \sqrt{35} = 5.916$$

$$\therefore\ d\overline{H}_2 = \left[\frac{2\pi(0.6\,\bar{a}_x - 0.8\,\bar{a}_y)\,10^{-6}}{4\pi(5.916)^2}\right] \times \left[\frac{-3\,\bar{a}_x + 5\,\bar{a}_y - \bar{a}_z}{5.916}\right]$$
$$= 1.932\,\bar{a}_x + 1.44\,\bar{a}_y + 1.449\,\bar{a}_z \ \text{nA/m}$$

Problem 7.2 : Find the vector magnetic field intensity in cartesian coordinates at P_2 (1.5, 2, 3) caused by a current filament of 24 A in the $\bar{a}_z$ direction

on the z axis and extending from : (a) $z = 0$ to $z = 6$; (b) $z = 6$ to $z = \infty$; (c) $z = -\infty$ to $z = \infty$.

Solution :

Point 2 is that point at which magnetic field intensity is desired = (1.5, 2, 3)

Point 1 is the general point on the current (i.e. z axis)= (0, 0, z)

$$\overline{R}_{12} = (1.5 - 0)\,\overline{a}_x + (2 - 0)\,\overline{a}_y + (3 - z)\,\overline{a}_z$$

$$= 1.5\,\overline{a}_x + 2\,\overline{a}_y + (3 - z)\,\overline{a}_z$$

$$d = |\overline{R}_{12}| = \sqrt{(1.5)^2 + (2)^2 + (3 - z)^2} = \sqrt{6.25 + (3 - z)^2}$$

$\overline{dL} = dz\,\overline{a}_z$ as the current lies on the z axis . Hence

$$d\overline{H}_2 = \frac{(24\,dz\,\overline{a}_z)}{4\pi} \times \left\{ \frac{1.5\,\overline{a}_x + 2\,\overline{a}_y + (3 - z)\,\overline{a}_z}{[6.25 + (3 - z)^2]^{3/2}} \right\}$$

$$= \frac{6(-2\,\overline{a}_x + 1.5\,\overline{a}_y)\,dz}{\pi\,[6.25 + (3 - z)^2]^{3/2}}$$

$$\therefore\ \overline{H}_2 = \int d\overline{H}_2 = \int_0^6 \frac{6(-2\,\overline{a}_x + 1.5\,\overline{a}_y)\,dz}{\pi\,[6.25 + (3 - z)^2]^{3/2}}$$

$$= \frac{6(-2\,\overline{a}_x + 1.5\,\overline{a}_y)}{\pi} \times \left[\frac{1}{-6.25} \times \frac{3 - z}{\sqrt{6.25 + (3 - z)^2}} \right]_0^6$$

$$\overline{H}_2 = -0.939\,\overline{a}_x + 0.704\,\overline{a}_y \quad \text{A/m}$$

(b) $$\overline{H}_2 = \int_6^{\infty} \frac{6}{\pi} \frac{(-2\,\overline{a}_x + 1.5\,\overline{a}_y)\,dz}{[6.25 + (3 - z)^2]^{3/2}}$$

$$= \frac{6(-2\,\overline{a}_x + 1.5\,\overline{a}_y)}{\pi} \times \left[\frac{1}{-6.25} \times \frac{3 - z}{\sqrt{6.25 + (3 - z)^2}} \right]_6^{\infty}$$

$$= -0.1417\,\overline{a}_x + 0.1062\,\overline{a}_y \quad \text{A/m} .$$

(c) $$\overline{H}_2 = \int_{-\infty}^{\infty} \frac{6}{\pi} \frac{(-2\,\overline{a}_x + 1.5\,\overline{a}_y)\,dz}{[6.25 + (3 - z)^2]^{3/2}}$$

$$= \frac{6(-2\bar{a}_x + 1.5\bar{a}_y)}{\pi} \times \left[\frac{1}{-6.25} \times \frac{3-z}{\sqrt{6.25 + (3-z)^2}} \right]_{-\infty}^{\infty}$$

$$= -1.222\,\bar{a}_x + 0.917\,\bar{a}_y \text{ A/m}$$

7.1.2 Expression for $\overline{\mathbf{H}}$ at any point in cylindrical system due to a filamentary conductor carrying a current I on the z axis from $-\infty < z < \infty$.

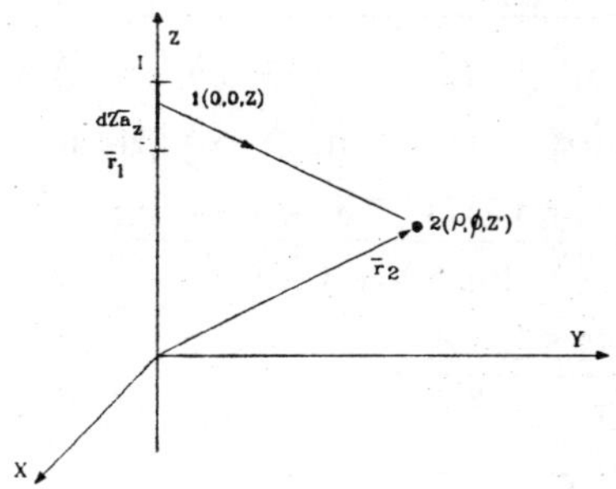

Fig. 7.4.

Fig. 7.4 shows a filamentary conductor on the z axis carrying a current I. Consider any point 1 (0, 0, z) on the z axis at which a differential length $\overline{dL} = dz\,\bar{a}_z$ is located. The point at which $\overline{\mathbf{H}}$ has to be calculated is 2 (ρ, ϕ, Z).

$$\overline{R}_{12} = \bar{r}_2 - \bar{r}_1$$

$$= (\rho\,\bar{a}_\rho + Z\,\bar{a}_z) - (z\,\bar{a}_z)$$

$$= \rho\,\bar{a}_\rho + (Z - z)\,\bar{a}_z$$

$$d = |\overline{R}_{12}|$$

$$= \sqrt{\rho^2 + (Z-z)^2}$$

$$\therefore \overline{dH} = \frac{I\,dz\,\bar{a}_z \times [\rho\,\bar{a}_\rho + (Z - z)\,\bar{a}_z]}{4\pi[\rho^2 + (Z - z)^2]^{3/2}}$$

$$\therefore \overline{H} = \int_{-\infty}^{\infty} \frac{I\,dz\,[\rho\,\bar{a}_\phi]}{4\pi[\rho^2 + (Z - z)^2]^{3/2}}$$

$$= \frac{I \rho \bar{a}_\phi}{4\pi} \left[-\frac{1}{\rho^2} \frac{Z - z}{\sqrt{\rho + (Z - z)^2}} \right]_{-\infty}^{\infty} = \frac{I}{2\pi\rho} \bar{a}_\phi$$

Note :

(1) *In general the magnetic field intensity due to infinitely long filamentary conductor is given by*

$$\boxed{\bar{\mathbf{H}} = \frac{I}{2\pi d} (\bar{\mathbf{a}}_I \times \bar{\mathbf{a}}_{12})} \qquad \ldots\ldots (7.4)$$

where $\bar{a}_I$ is a unit vector in the direction of current, point 2 is that point at which magnetic field intensity is desired, point 1 is the foot of perpendicular dropped from point 2 on the filamentary conductor and d is the distance between the points 1 and 2.

(2) *Whenever the current flows in $\bar{a}_z$ direction, H_ϕ component of magnetic field intensity is produced.*

7.2 AMPERE'S CIRCUITAL LAW

7.2.1 Statement : Ampere's circuital law states that the line integral of $\bar{H}$ about any closed path is exactly equal to the direct current enclosed by that path. Referring to the Fig.7.5 which shows a circular wire carrying a direct current I, the line integral of $\bar{H}$ about the closed paths lettered a and b results in an answer of I ; the integral about the closed path c which passes through the conductor gives an answer less than I and is exactly that portion of the total current which is enclosed by the path c. Mathematically

$$\boxed{\oint \bar{\mathbf{H}} \cdot \bar{\mathbf{dL}} = I_{\text{encl}}} \qquad \ldots\ldots\ldots (7.5)$$

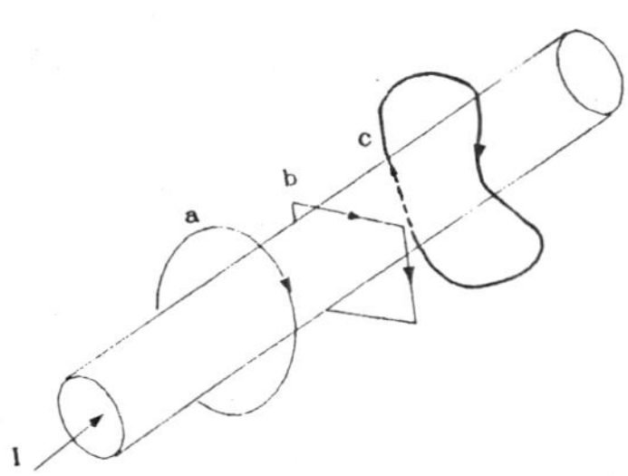

Fig. 7.5.

7.2.2 Expression for $\overline{H}$ at any point due touniform surface current density $\overline{K}$ on an infinite plane

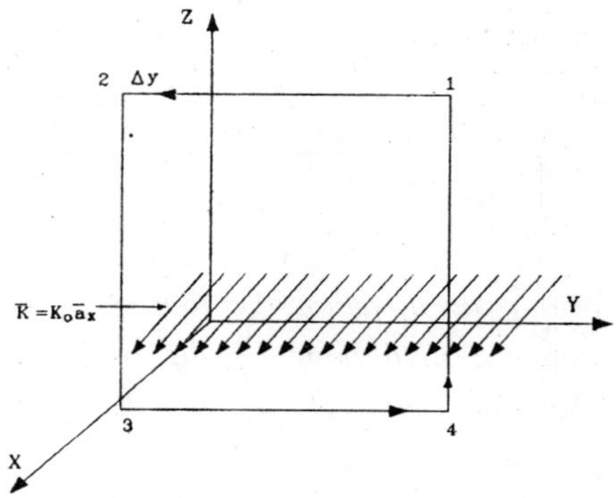

Fig. 7.6.

Let us assume $\overline{K} = K_o \, \overline{a}_x$ A/m in the $z = 0$ plane. The point at which $\overline{H}$ has to be calculated lies above the $z = 0$ plane. The following conclusions have been drawn. :

(i) Since the current is in $\overline{a}_x$ direction, x component of $\overline{H}$ is absent.

(ii) Because of the symmetrical distribution of current with respect to the point at which $\overline{H}$ is desired, z component of $\overline{H}$ will get cancel out and only y component (H_y) will be present.

(iii) $\overline{H}$ above the plane at a point is equal and opposite at a symmetrical point located below the plane. Fig. 7.6 also shows a rectangular path (sides parallel to the axis) and apply Ampere's Circuital law $\oint \overline{H} \cdot \overline{dL} = I$ to this closed path 1 - 2 - 3 - 4 - 1. Hence we have,

$$- H_{y,\,1-2} \, \Delta Y + H_{y,\,3-4} \, \Delta Y = K_o \, \Delta Y$$

Note that $\overline{H} \cdot \overline{dL}$ for path 2 - 3 and path 4 - 1 is equal to zero as $\overline{H}$ has $\pm \, \overline{a}_y$ and $\overline{dL}$ has $\pm \, \overline{a}_z$. Since $- H_{y,\,3-4} = H_{y,\,1-2}$, we have

$$- H_{y,\,1-2} - H_{y,\,1-2} = K_o$$

and

$$H_{y,\,1-2} = - \frac{K_o}{2} \quad \text{or} \quad \overline{H} = - \frac{K_o}{2} \, \overline{a}_y$$

As $- \overline{a}_y = \overline{a}_x \times \overline{a}_z$,we have in general

$$\boxed{\overline{\mathbf{H}} = \frac{\overline{\mathbf{K}} \times \overline{\mathbf{a}}_N}{2}} \qquad \ldots\ldots (7.6)$$

where $\overline{a}_N$ is a unit vector perpendicular to the plane in which surface current density lies and directed towards the point at which $\overline{H}$ is desired.

7.2.3 Expression for $\overline{\mathbf{H}}$ at any point due to a coaxial cable

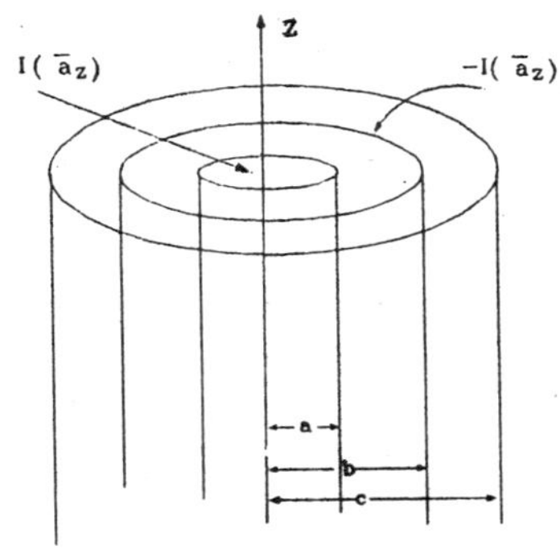

Fig. 7.7.

Fig. 7.7 shows a co–axial cable with inner conductor of radius a and outer conductor consists of two cylinders of radii b and c with $b < c$. The inner conductor carries a total current of I ampere in $\overline{a}_z$ direction where as the outer conductor carries the same current in $-\overline{a}_z$ direction.

(a) Consider a circular path of radius ρ such that $\rho < a$ and apply Ampere's Circutial law to the closed path $\oint \overline{H} \cdot \overline{dL} = I$

Let us assume that the solid conductor is made up of infinitely large number of filamentary conductors, each carrying a current in z direction so that ϕ component of $\overline{H}$ is produced. Hence $\overline{H} = H_\phi \, \overline{a}_\phi$; $\overline{dL} = \rho \, d\phi \, \overline{a}_\phi$.

$$\therefore \text{ LHS} = \int_0^{2\pi} \rho \, H_\phi \, d\phi = 2\pi \rho \, H_\phi$$

Current enclosed by the circular path = RHS = $\dfrac{\rho^2 I}{a^2}$

Equating the two sides $H_\phi = \dfrac{\rho I}{2\pi a^2}$ or

$$\bar{\mathbf{H}} = \frac{\rho I}{2\pi a^2}\,\bar{\mathbf{a}}_\phi \qquad (\rho < \mathbf{a}) \qquad \ldots\ldots (7.7\,a)$$

(b) Consider a circular path of radius ρ such that $a < \rho < b$. Current enclosed by the circular path = R.H.S. = I.

Equating the two sides we get $H_\phi = \dfrac{I}{2\pi\rho}$, or

$$\bar{\mathbf{H}} = \frac{I}{2\pi\rho}\,\bar{\mathbf{a}}_\phi \qquad (a < \rho < b) \qquad \ldots\ldots (7.7\,b)$$

(c) Consider the circular path of radius ρ such that $b < \rho < c$. In this case two currents are enclosed, one in the positive z direction and another in the negative z direction. Hence the current enclosed by the path is subtraction of these two currents.

Current in positive z direction $= I$

Current in negative I direction $= \dfrac{I(\rho^2 - b^2)}{(c^2 - b^2)}$

Hence RHS $= \dfrac{I(c^2 - \rho^2)}{(c^2 - b^2)}$. Equating the two sides, we get

$$H_\phi = \frac{I(c^2 - \rho^2)}{2\pi\rho(c^2 - b^2)}$$

or

$$\bar{\mathbf{H}} = \frac{I(c^2 - \rho^2)}{2\pi\rho(c^2 - b^2)}\,\bar{\mathbf{a}}_\phi \qquad (b < \rho < c) \qquad \ldots (7.7\,c)$$

(d)Consider a circular path of radius ρ such that $\rho > c$.
Current enclosed by the circular path = $I - I = 0$. Therefore

$$\bar{\mathbf{H}} = \mathbf{0} \qquad (\rho > c) \qquad \ldots\ldots\ldots (7.7\,d)$$

7.2.4 Expression for $\overline{H}$ at any point due to an ideal solenoid of infinite length

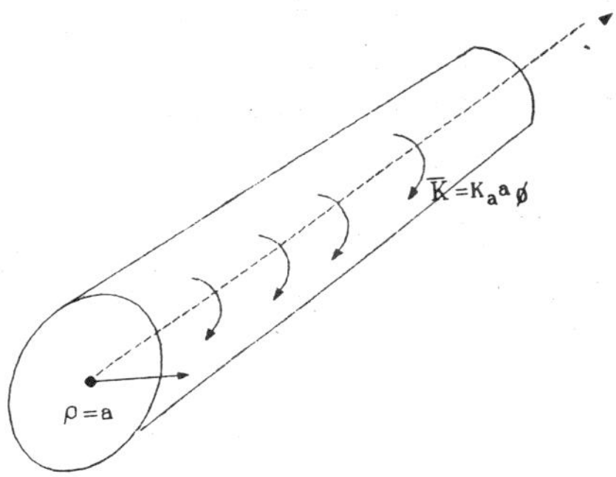

Fig. 7.8 (A)

For the solenoid as shown in Fig. 7.8 (b), which carries a surface current density $\overline{K} = K_a \, \overline{a}_\phi$, the magnetic field intensity is given by

$$\boxed{\overline{\mathbf{H}} = \mathbf{K_a} \, \overline{\mathbf{a}}_z} \qquad \ldots\ldots (7.8)$$

Fig. 7.8(b) shows a solenoid with N as the number of turns, d as the length of solenoid, a as the radius of solenoid and I is the current in $\overline{a}_z$ direction. For this solenoid, it is assumed that the turns are place very close to each other so that it will form a surface current density $\overline{K} = \frac{NI}{d} \overline{a}_\phi$ A/m .

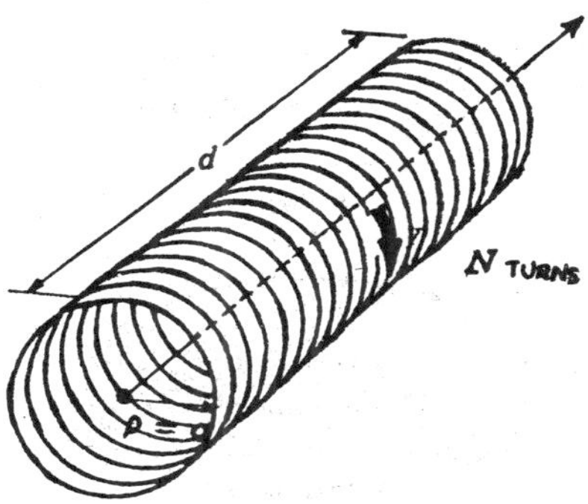

Fig. 7.8 (B)

Hence using equation 7.8, we get

$$\boxed{\overline{H} = \frac{NI}{d}\,\overline{a}_z} \quad \ldots\ldots (7.9)$$

7.2.5 Expression for $\overline{H}$ at any point due to an ideal toroid

Fig. 7.9 (a) shows a toroid with z axis as its axis. The cross section of the toroid is a circle of radius a. The mean radius of the toroid is c ρ_o so that inner and outer radii will be $\rho_o - a$ and $\rho_o + a$. A surface current density $\overline{K} = K_a\,\overline{a}_z$ at $\rho = \rho_o - a$ (i.e. on inner surface) is assumed

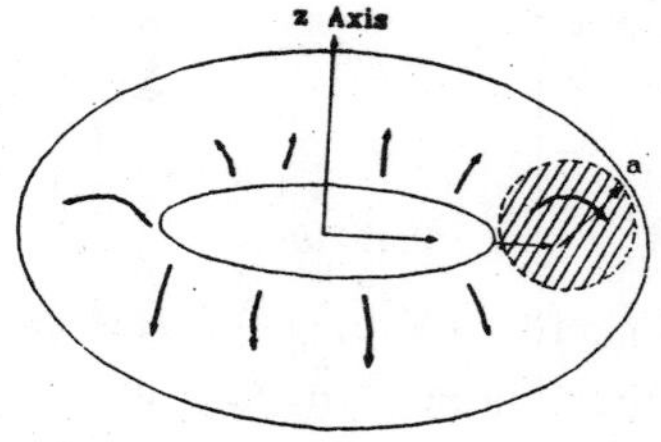

FIG. 7.9 (A)

Consider a circular path of radius ρ such that $\rho_0 - a < \rho < \rho_0 + a$

Using Ampere's circuital law, we get

$$\boxed{\overline{H} = K_a \frac{\rho_o - a}{\rho}\,\overline{a}_\phi \text{ (inside toroid)}} \quad \ldots\ldots (7.10\text{ a})$$

Fig. 7.9 (b) shows a toroid with N turns. It carries a current I in $\overline{a}_z$ direction on the inner surface $\rho = \rho_o - a$. Using Ampere's circuital law, we get

$$\boxed{\overline{H} = \frac{NI}{2\pi\rho}\,\overline{a}_\phi \text{ (inside toroid)}} \quad \ldots.. (7.10\text{b})$$

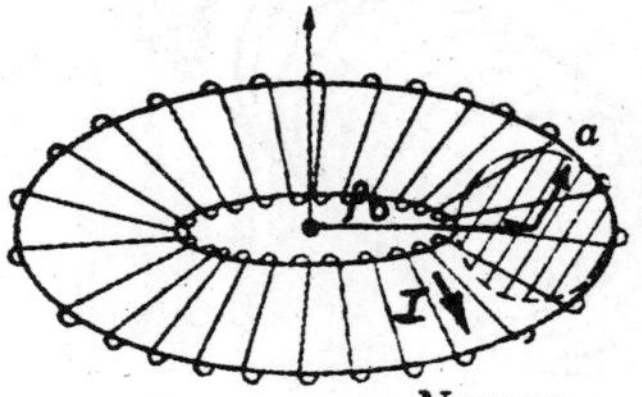

FIG. 7.9 (B)

Problem 7.3 : Express the value of $\overline{H}$ incCartesian components at P (0.01, 0, 0) in the field of : (a) two current filaments, 0.08 A in the $\overline{a}_z$ direction on the z–axis, and 0.08 A in the $-\overline{a}_z$ direction at x = 0.015, y = 0 ; (b) coax, a = 3 mm, b = 9mm, c = 12 mm, I = 0.8 A, centered on the z–axis, $\overline{a}_z$ direction in center conductor ; (c) three current sheets, 1.5 π $\overline{a}_y$ A/m at x = 6 mm, – 3 π $\overline{a}_y$ A/m at x = 9 mm, and 1.5 π $\overline{a}_y$ A/m at x = 12 mm ; (d) infinite solenoid, K = 3 π $\overline{a}_\phi$ A/m at ρ = 12 mm ; (e) toroid, centered at origin, axis on y axis, ρ_o = 12 mm , α = 3 mm , N = 250 turns, I = 2 mA in $\overline{a}_y$ direction at outer radius.

Solution :

(a) Due to infinitely long filamentary conductor

$$\overline{H} = \frac{I}{2\pi d}\,\overline{a} = \frac{I}{2\pi d}\,[\,\overline{a}_I \times \overline{a}_{12}\,]$$

Point 2 is that point at which $\overline{H}$ is desired = (0.01, 0, 0)

Point 1 is the foot of perpendicular dropped from point 2 on the filamentary conductor = (0, 0, 0)

$\therefore\ \overline{R}_{12} = (0.01 - 0)\,\overline{a}_x = 0.01\,\overline{a}_x$; $d = |\overline{R}_{12}| = 0.01$; $\overline{a}_{12} = \overline{a}_x$

$\overline{a}_I$ = unit vector in the direction of current = $\overline{a}_z$

$$\therefore\ \overline{H}_1 = \frac{0.08}{2\pi(0.01)}\,[\,\overline{a}_z \times \overline{a}_x\,] = 1.2732\,\overline{a}_y \quad \text{A/m}$$

Point 2 is that point at which $\overline{H}$ is desired = (0.01, 0, 0)

Point 1 is the floor of perpendicular dropped from point 2 on the filamentary conductor = (0.015, 0, 0)

$|\overline{R}_{12}| = -0.005\,\overline{a}_x$, $d = |\overline{R}_{12}| = 0.005$, $\overline{a}_{12} = -\overline{a}_x$

$\overline{a}_I$ = unit vector in the direction of current = $-\overline{a}_z$

$$\therefore\ \overline{H}_2 = \frac{0.08}{2\pi(0.005)}\,[-\overline{a}_z \times (-\overline{a}_x)] = 2.546\,\overline{a}_y \quad \text{A/m}$$

$$\therefore\ \overline{H} = \overline{H}_1 + \overline{H}_2 = 1.2732\,\overline{a}_y + 2.546\,\overline{a}_y = 3.8196\,\overline{a}_y \quad \text{A/m}$$

(b) Let us determine the location of point P within the coax. If a perpendicular is dropped from point P on the axis of coax. (z axis), the foot of perpendicular is given by (0, 0, 0). hence

$$\rho = \sqrt{(0.01 - 0)^2} = 0.01 = 10 \text{ mm}$$

Hence point P lies in between b = 9 mm and C = 12 mm.

using equation 7.7 (c), we get

$$\overline{H} = \frac{I}{2\pi\rho} \times \left(\frac{c^2 - \rho^2}{c^2 - b^2}\right) \overline{a}_\phi$$

$$= \frac{0.8}{2\pi(0.01)} \times \left[\frac{(0.012)^2 - (0.01)^2}{(0.012)^2 - (0.009)^2}\right] \overline{a}_\phi$$

$$\overline{H} = 8.892\ \overline{a}_\phi$$

Transforming this vector in Cartesian coordinate system, we get

$$\overline{H} = 8.892\ (-\sin\phi\ \overline{a}_x + \cos\phi\ \overline{a}_y)$$

At point P (0.01, 0, 0) , $\phi = \tan^{-1}\left(\frac{0}{0.01}\right) = 0$

$$\therefore\ \overline{H} = 8.892\,(-\sin 0\ \overline{a}_x + \cos 0\ \overline{a}_y) = 8.892\ \overline{a}_y \quad \text{A/m}$$

(c) $\overline{H} = \overline{H}_1 + \overline{H}_2 + \overline{H}_3$

$$\overline{H}_1 = \frac{1.5\pi\ \overline{a}_y \times \overline{a}_x}{2} = -2.356\ \overline{a}_z \quad [\because \text{ x coordinate of point P} > 0.006]$$

$$\overline{H}_2 = -\frac{3\pi\ \overline{a}_y \times \overline{a}_x}{2} = +4.712\ \overline{a}_z \quad [\because \text{ x coordinate of point P} > 0.009]$$

$$\overline{H}_3 = \frac{1.5\pi\ \overline{a}_y \times (-\overline{a}_x)}{2} = 2.356\ \overline{a}_z \quad [\because \text{ x coordinate of point P} < 0.012]$$

$$\therefore\ \overline{H} = -2.356\ \overline{a}_z + 4.712\ \overline{a}_z + 2.356\ \overline{a}_z = 4.712\ \overline{a}_z \quad \text{A/m}$$

(d) The axis of the solenoid is the z axis . As determined in part (b), the distance of point P from the Z axis is 0.01 = 10 mm < 12 mm. Hence point P lies inside the solenoid.

$$\therefore\ \overline{H} = K_o\ \overline{a}_z = 3\pi\ \overline{a}_z = 9.425\ \overline{a}_z \quad \text{A/m}$$

(e) Again point P lies inside the totoid because

$\rho_o - a < 0.01 < \rho_o + a$

$$\therefore\ \overline{H} = \frac{NI}{2\pi\rho}\ (\overline{a}_I \times \overline{a}_{12})$$

$\overline{a}_I$ = unit vector in the direction of curren on the inner surface .

The direction of current on the outer surface is $\overline{a}_y$. Hence on the inner surface

$\bar{a}_I = -\bar{a}_y$. Also $\bar{a}_{12} = \bar{a}_x$

$$\therefore \quad \bar{H} = \frac{250 \times 2 \times 10^{-3}}{2\pi(0.01)} [-\bar{a}_y \times \bar{a}_x] = 7.958\, \bar{a}_z \quad A/m$$

7.3. EXPRESSION FOR CURL OF H OR DERIVE POINT FORM OF AMPERE'S CIRCUITAL LAW

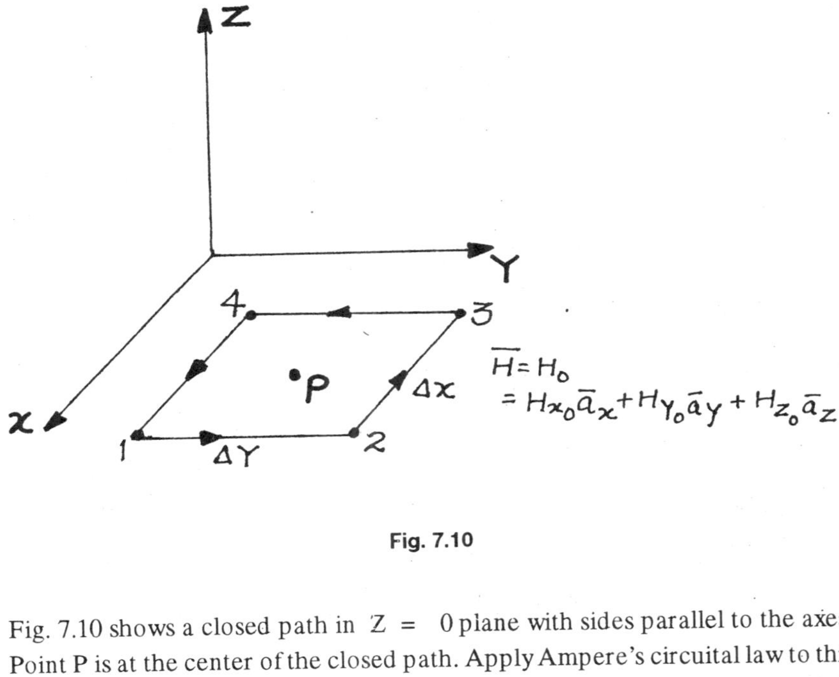

Fig. 7.10

Fig. 7.10 shows a closed path in $Z = 0$ plane with sides parallel to the axes. Point P is at the center of the closed path. Apply Ampere's circuital law to this closed path and let us assume that the current enclosed by the closed path is in Z direction given by I_Z . At point P magnetic field intensity is

$$\bar{H} = \bar{H}_o = H_{xo}\,\bar{a}_x + H_{yo}\,\bar{a}_y + H_{zo}\,\bar{a}_z\,.$$

$$\oint \bar{H}\cdot\overline{dL} = \int_1^2 \bar{H}\cdot\overline{dL} + \int_2^3 \bar{H}\cdot\overline{dL} + \int_3^4 \bar{H}\cdot\overline{dL} + \int_4^1 \bar{H}\;\overline{dL}$$

$$\int_1^2 \overline{H} \cdot \overline{dL} = H_{y\,1-2}\,\Delta y;\; H_{y\,1-2} = H_{yo} + \frac{\partial H_y}{\partial x} \times \frac{\Delta x}{2}$$

$$= \left(H_{yo} + \frac{\partial H_y}{\partial x} \times \frac{\Delta x}{2} \right) \Delta y$$

$$\int_2^3 \overline{H} \cdot \overline{dL} = H_{x\,2-3}\,(-\Delta x)\;;\; H_{x\,2-3} = H_{xo} + \frac{\partial H_x}{\partial y} \times \frac{\Delta y}{2}$$

$$= \left(H_{xo} + \frac{\partial H_x}{\partial y} \times \frac{\Delta y}{2} \right) (-\Delta x)$$

$$\int_3^4 \overline{H} \cdot \overline{dL} = H_{y\,3-4}\,(-\Delta y)\;;\; H_{y\,3-4} = H_{yo} - \frac{\partial H_y}{\partial x} \times \frac{\Delta x}{2}$$

$$= \left(H_{yo} - \frac{\partial H_y}{\partial x} \times \frac{\Delta x}{2} \right) (-\Delta y)$$

$$\int_4^1 \overline{H} \cdot \overline{dL} = H_{x\,4-1}\,(\Delta x)\;;\; H_{x\,4-1} = H_{xo} - \frac{\partial H_x}{\partial y} \times \frac{\Delta y}{2}$$

$$= \left(H_{xo} - \frac{\partial H_x}{\partial y} \frac{\Delta y}{2} \right) (\Delta x)$$

$$\therefore \oint \overline{H} \cdot \overline{dL} = I_z = \left(\frac{\partial H_y}{\partial x} - \frac{\partial H_x}{\partial y} \right) \Delta x\, \Delta y$$

Dividing the above expression by $\Delta x \Delta y$ and taking the limit

$$\lim_{\Delta x\, \Delta y \to 0} \frac{\oint \overline{H} \cdot \overline{dL}}{\Delta x\, \Delta y} = \frac{\partial H_y}{\partial x} - \frac{\partial H_x}{\partial y} = J_z$$

Similarly
$$\lim_{\Delta y \Delta z \to 0} \frac{\oint \overline{H} \cdot \overline{dL}}{\Delta y\, \Delta z} = \frac{\partial H_z}{\partial y} - \frac{\partial H_y}{\partial z} = J_x$$

$$\lim_{\Delta z \Delta x \to 0} \frac{\oint \overline{H} \cdot \overline{dL}}{\Delta z\, \Delta x} = \frac{\partial H_x}{\partial z} - \frac{\partial H_z}{\partial x} = J_y$$

Now $\bar{J} = J_x \bar{a}_x + J_y \bar{a}_y + J_z \bar{a}_z$

$$\therefore \quad \bar{J} = \left(\frac{\partial H_z}{\partial y} - \frac{\partial H_y}{\partial z} \right) \bar{a}_x + \left(\frac{\partial H_x}{\partial z} - \frac{\partial H_z}{\partial x} \right) \bar{a}_y + \left(\frac{\partial H_y}{\partial x} - \frac{\partial H_x}{\partial y} \right) \bar{a}_z$$

$$\nabla = \frac{\partial}{\partial x} \bar{a}_x + \frac{\partial}{\partial y} \bar{a}_y + \frac{\partial}{\partial z} \bar{a}_z$$

Let $\bar{H} = H_x \bar{a}_x + H_y \bar{a}_y + H_z \bar{a}_z$

Then $$\bar{\nabla} \times \bar{H} = \begin{vmatrix} \bar{a}_x & \bar{a}_y & \bar{a}_z \\ \frac{\partial}{\partial x} & \frac{\partial}{\partial y} & \frac{\partial}{\partial z} \\ H_x & H_y & H_z \end{vmatrix}$$

If we expand this determinant we will get the same expression as that of $\bar{J}$

Therefore

$$\boxed{\bar{\mathbf{J}} = \nabla \times \bar{\mathbf{H}}} \qquad \ldots\ldots (7.11)$$

which is the point form of Ampere's Circuital Law.

Definition of Curl : The curl of any vector is a vector, and any component of curl is given by the limit of the quotient of the closed line integral of the vector about a small path in a plane normal to that component desired and the area enclosed, as the path shrinks to zero.

$$(\text{Curl } \bar{H})_N = \lim_{\Delta S_N \to 0} \frac{\oint \bar{H} \cdot \bar{dL}}{\Delta S_N}$$

$$\boxed{\bar{\nabla} \times \bar{\mathbf{H}} = \begin{vmatrix} \bar{\mathbf{a}}_x & \bar{\mathbf{a}}_y & \bar{\mathbf{a}}_z \\ \frac{\partial}{\partial \mathbf{x}} & \frac{\partial}{\partial \mathbf{y}} & \frac{\partial}{\partial z} \\ \mathbf{H}_x & \mathbf{H}_y & \mathbf{H}_z \end{vmatrix} \quad \text{Cartesian}} \qquad \ldots (7.12.\ a)$$

$$\boxed{\bar{\nabla} \times \bar{\mathbf{H}} = \begin{vmatrix} \frac{\bar{\mathbf{a}}_\rho}{\rho} & \bar{\mathbf{a}}_\phi & \frac{\bar{\mathbf{a}}_z}{\rho} \\ \frac{\partial}{\partial \rho} & \frac{\partial}{\partial \phi} & \frac{\partial}{\partial z} \\ \mathbf{H}_\rho & \rho\, \mathbf{H}_\phi & \mathbf{H}_z \end{vmatrix} \quad \text{Cylindrical}} \qquad \ldots\ldots (7.12\ b)$$

$$\bar{\nabla} \times \bar{H} = \begin{vmatrix} \dfrac{\bar{a}_r}{r^2 \sin\theta} & \dfrac{\bar{a}_\phi}{r \sin\theta} & \dfrac{\bar{a}_\phi}{r} \\ \dfrac{\partial}{\partial r} & \dfrac{\partial}{\partial \theta} & \dfrac{\partial}{\partial \phi} \\ H_r & r H_\theta & r \sin\theta \, H_\phi \end{vmatrix} \quad \text{spherical} \qquad \ldots . (7.12 \text{ c})$$

Problem 7.4 : (a) Evaluate the closed line integral of $\bar{H}$ from P_1 (5, 4, 1) to P_2 (5, 6, 1) to P_3 (0, 6, 1) to P_4 (0, 4, 1) to P_1, using straight line segments, if $\bar{H} = 0.1 y^3 \bar{a}_x + 0.4x \bar{a}_z$ A/m. (b) Determine the quotient of the closed line integral and the are enclosed by the path as an approximation to $(\bar{\nabla} \times \bar{H})_z$. (c) Determine $(\bar{\nabla} \times \bar{H})_z$ at the center of the area.

Solution :

(a) $$\int_{P_1}^{P_2} \bar{H} \cdot \overline{dL} = \int_{P_1}^{P_2} (\bar{H} \cdot dy \, \bar{a}_y) = \int_4^6 0 = 0$$

$$\int_{P_2}^{P_3} \bar{H} \cdot \overline{dL} = \int_{P_2}^{P_3} (\bar{H} \cdot dx \, \bar{a}_x) = \int_5^0 0.1 \, y^3 dx \Big|_{y=6}$$

$$= 0.1 (6)^3 \Big[x \Big]_5^0 = - 108$$

$$\int_{P_3}^{P_4} \bar{H} \cdot \overline{dL} = \int_{P_3}^{P_4} (\bar{H} \cdot dy \, \bar{a}_y) = \int_6^4 0 = 0$$

$$\int_{P_4}^{P_1} \bar{H} \cdot \overline{dL} = \int_{P_4}^{P_1} (\bar{H} \cdot dx \, \bar{a}_x) = \int_0^5 0.1 \, y^3 dx \Big|_{y=4}$$

$$= 0.1 (4)^3 \Big[x \Big]_0^5 = 32$$

$$\therefore \oint \bar{H} \cdot \overline{dL} = 0 - 108 + 0 + 32 = - 76 \text{ A}$$

(b) Area enclosed by the path

$P_1 \, P_2 \, P_3 \, P_4 \, P_1 = (5 - 0) (6 - 4) = 10 \text{ m}^2$

$$\therefore (\bar{\nabla} \times \bar{H})_z = \frac{-76}{10} = - 7.6 \text{ A/m}^2$$

(c) $\bar{\nabla} \times \bar{H} = \begin{vmatrix} \bar{a}_x & \bar{a}_y & \bar{a}_z \\ \partial/\partial x & \partial/\partial y & \partial/\partial z \\ 0.1\, y^3 & 0 & 0.4\, x \end{vmatrix}$

$$(\bar{\nabla} \times \bar{H})_z = \bar{a}_z (0.1 \times 3y^2) = 0.3\, y^2\, \bar{a}_z$$

The center is $\left(\frac{5+0}{2}, \frac{4+6}{2}, 1 \right) = (2.5, 5, 1)$

$\therefore \; (\bar{\nabla} \times \bar{H})_z$ at the center $= 0.3\,(5)^2\, \bar{a}_z = 7.5\, \bar{a}_z \quad A/m^2$

Problem 7.5 : Find $\bar{\nabla} \times \bar{G}$: (a) in cartesian coordinates at P_A (3, 2, 1) if $\bar{G} = xyz(\bar{a}_x + \bar{a}_y)$; (b) in cylindrical coordinates at P_B (2, 30°, 3) if $\bar{G} = \rho\, \phi\, z\, \bar{a}_z$; (c) in spherical coordinates at P_c (4, 30°, 45°) if $\bar{G} = \sin\theta\,(\bar{a}_r + \bar{a}_\theta + \bar{a}_\phi)$.

Solution :

(a) $\bar{\nabla} \times \bar{G} = \begin{vmatrix} \bar{a}_x & \bar{a}_y & \bar{a}_z \\ \partial/\partial x & \partial/\partial y & \partial/\partial z \\ xyz & xyz & 0 \end{vmatrix}$

$$= -xy\,\bar{a}_x + xy\,\bar{a}_y + (yz - xz)\,\bar{a}_z$$

$$\bar{\nabla} \times \bar{G} \text{ at } P_A(3, 2, 1) = -(3)(2)\,\bar{a}_x + (3)(2)\,\bar{a}_y + (2 \times 1 - 3 \times 1)\,\bar{a}_z$$

$$= -6\,\bar{a}_x + 6\,\bar{a}_y - \bar{a}_z$$

(b) $\bar{\nabla} \times \bar{G} = \begin{vmatrix} \bar{a}_\rho/\rho & \bar{a}_\phi & \bar{a}_z/\rho \\ \partial/\partial \rho & \partial/\partial \phi & \partial/\partial z \\ 0 & 0 & \rho\,\phi\,z \end{vmatrix}$

$$= z\,\bar{a}_\rho - \phi\, z\, \bar{a}_\phi$$

$$\bar{\nabla} \times \bar{G} \text{ at } P_B\,(2, 30^\circ, 3) = 3\,\bar{a}_\rho - 30 \times \frac{\pi}{180} \times 3\,\bar{a}_\phi$$

$$= 3\,\bar{a}_\rho - 1.571\,\bar{a}_\phi$$

(c) $\bar{\nabla} \times \bar{G} = \begin{vmatrix} \dfrac{\bar{a}_r}{r^2 \sin\theta} & \dfrac{\bar{a}_\theta}{r \sin\theta} & \dfrac{\bar{a}_\phi}{r} \\ \dfrac{\partial}{\partial r} & \dfrac{\partial}{\partial \theta} & \dfrac{\partial}{\partial \phi} \\ \sin\theta & r\sin\theta & r\sin^2\theta \end{vmatrix}$

$$= \frac{2}{r}\cos\theta\,\bar{a}_r - \frac{\sin\theta}{r}\bar{a}_\theta + \frac{(\sin\theta - \cos\theta)}{r}\bar{a}_\phi$$

$$\bar{\nabla}\times\bar{G} \text{ at } P_C\ (4, 30^\circ, 45^\circ) = \frac{2}{4}\cos 30^\circ\,\bar{a}_r - \frac{\sin 30}{4}\bar{a}_\theta + \frac{(\sin 30^\circ - \cos 30^\circ)}{4}\bar{a}_\phi$$

$$= 0.433\,a_r - 0.125\,\bar{a}_\theta - 0.0915\,\bar{a}_\phi$$

7.4 STOKE'S THEOREM

Fig. 7.11 shows an area S which is broken up into incremental surfaces of area Δ S.

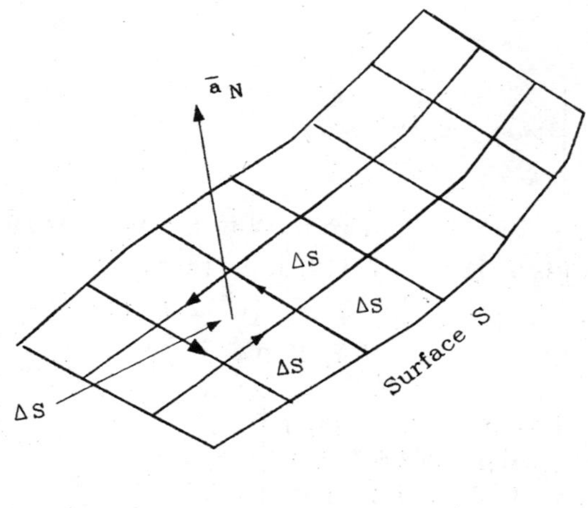

Fig. 7.11

Apply the definition of curl to one of these incremental surfaces then,

$$\frac{\oint \bar{H}\cdot\overline{dL}_{\Delta S}}{\Delta S} \doteq (\bar{\nabla}\times\bar{H})_N$$

where the N subscript indicates the right hand normal to the surface. The subscript on $\overline{dL}_{\Delta S}$ indciates that the closed path is the perimeter of an incremental area ΔS

This result may be writte as $\dfrac{\oint \overline{H} \cdot \overline{dL}_{\Delta S}}{\Delta S} = \text{doe } (\overline{\nabla} \times \overline{H}) \cdot \overline{a}_N$

or $\oint \overline{H} \cdot \overline{dL}_{\Delta S} \doteq (\overline{\nabla} \times \overline{H}) \cdot \overline{a}_N \Delta S = (\overline{\nabla} \times \overline{H}) \cdot \Delta\overline{S}$

where $\overline{a}_N$ is a unit vector in the direction of the right–hand normal to ΔS. Apply the definition to each ΔS and adding the result, some cancellation will occur because everyinterior wall is convered onece in each direction. The only boundaries on which cancellation cannot occur form the outside boundary, the path enclosing S. Therefore

$$\oint \overline{\mathbf{H}} \cdot \overline{\mathbf{dL}} \equiv \int_S (\overline{\nabla} \times \overline{\mathbf{H}}) \cdot \overline{\mathbf{dS}} \qquad \ldots\ldots (7.13)$$

where $\overline{dL}$ is taken only on the perimeter of S. This identity is the mathemacial expression of Stoke's theorem which states that, line integral of magnetic field intensity around any closed path is equal to the integral of the normal component of curl of magnetic field intensity over the surface which is enclosed by that closed path.

Problem 7.6 : Given the field $\overline{H} = 6r \sin\phi\, \overline{a}_r + 18r \sin\theta \cos\phi\, \overline{a}_\phi$, evaluate both sides of Stoke's theorem for the portion of the cone $\theta = 0.1\pi$ bounded by $r = 2, r = 4, \phi = 0$ and $\phi = 0.3\pi$. Let the direction of $\overline{dS}$ be $+\overline{a}_\theta$.

Solution :

The Stoke's theorem is

$$\oint \overline{H} \cdot \overline{dL} = \int (\overline{\nabla} \times \overline{H}) \cdot \overline{dS}$$

$\text{LHS} = \oint \overline{H} \cdot \overline{dL}$ where $\overline{dL} = dr\, \overline{a}_r + r\, d\theta\, \overline{a}_\theta + r \sin\theta\, d\phi\, \overline{a}_\phi$

The closed path is shown in Fig.P. 7.6.

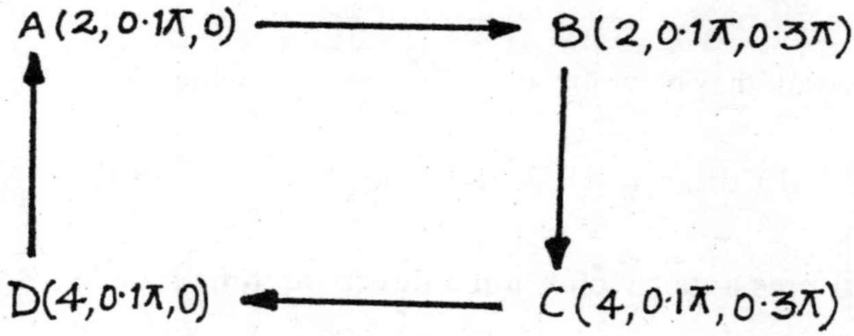

Fig. P. 7.6.

$$\int_A^B \overline{H} \cdot \overline{dL}$$

$$= \int_A^B \overline{H} \cdot (r \sin\theta \, d\phi \, \overline{a}_\phi)$$

$$= \int_0^{0.3\pi} 18 r^2 \sin^2\theta \cos\phi \, d\phi \Big|_{\substack{r = 2 \\ \theta = 0.1\pi}}$$

$$= 18(2)^2 \sin^2(0.1\pi) \Big[\sin\phi \Big]_0^{0.3\pi} = 5.5623 \text{ A}$$

$$\int_B^C \overline{H} \cdot \overline{dL} = \int_B^C \overline{H} \cdot (dr \, \overline{a}_r) = \int_2^4 6 r \sin\phi \, dr \Big|_{\phi = 0.3\pi}$$

$$= 6 \Big[\frac{r^2}{2} \Big]_2^4 \sin(0.3\pi) = 29.1246 \text{ A}$$

$$\int_C^D \overline{H} \cdot \overline{dL} = \int_C^D \overline{H} \cdot (r \sin\theta \, d\phi \, \overline{a}_\phi)$$

$$= \int_{0.3\pi}^{0} 18 r^2 \sin^2\theta \cos\phi \, d\phi \Big|_{\substack{r = 4 \\ \theta = 0.1\pi}}$$

$$= 18(4)^2 \sin^2(0.1\pi) \Big[\sin\phi \Big]_{0.3\pi}^{0} = -22.2492 \text{ A}$$

$$\int_D^A \overline{H} \cdot \overline{dL} = \int_D^A \overline{H} \cdot (dr\,\overline{a}_r) = \int_4^2 6\,r \sin\phi \, dr \Big|_{\phi = 0}$$

$$= 6\left[\frac{r^2}{2}\right]_4^2 \sin 0 = 0$$

$$\therefore \oint \overline{H} \cdot \overline{dL} = 5.5623 + 29.1246 - 22.2492 + 0$$

$$= 12.4379 \text{ A}$$

$$\overline{dS} = r \sin\theta \, dr \, d\phi \, \overline{a}_\theta$$

$$(\overline{\nabla} \times \overline{H})_\theta = \begin{vmatrix} \dfrac{\overline{a}_r}{r^2 \sin\theta} & \dfrac{\overline{a}_\theta}{r \sin\theta} & \dfrac{\overline{a}_\phi}{r} \\ \dfrac{\partial}{\partial r} & \dfrac{\partial}{\partial \theta} & \dfrac{\partial}{\partial \phi} \\ 6\,r \sin\phi & 0 & 18\,r^2 \sin^2\theta \cos\phi \end{vmatrix}$$

$$= \frac{-\overline{a}_\theta}{r \sin\theta}(36\,r \sin^2\theta \cos\phi - 6\,r \cos\phi)$$

$$= \overline{a}_\theta(-36 \sin\theta \cos\phi + 6 \cos\phi \operatorname{cosec}\theta)$$

$$\therefore \text{RHS} = \int (\overline{\nabla} \times \overline{H}) \cdot \overline{dS}$$

$$= \int_2^4 \int_0^{0.3\pi} (-36 \sin\theta \cos\phi + 6 \cos\phi \operatorname{cosec}\theta)\, r \sin\theta \, dr \, d\phi \quad (\theta = 0.1\pi)$$

$$= -36\left[\frac{r^2}{2}\right]_2^4 \left[\sin\phi\right]_0^{0.3\pi} \sin^2(0.1\pi) + 6\left[\sin\phi\right]_o^{0.3\pi}\left[\frac{r^2}{2}\right]_2^4$$

$$= -16.6869 + 29.1246 = 12.4377 \text{ A}$$

Hence Stoke's theorem is verified.

7.5 MAGNETIC FLUX AND MAGNETIC FLUX DENSITY

Magnetic flux density is denoted by the letter $\overline{B}$ and is related to magnetic field intensity $\overline{H}$ by the equation.

$$\boxed{\overline{\mathbf{B}} = \mu_o \mu_r \overline{\mathbf{H}}} \qquad \ldots\ldots (7.14)$$

where μ_r is relation permeability of free space and μ_o is permeability of free space. Its value is $\mu_o = 4\pi \times 10^{-7}$ H/m. The unit of magnetic flux density is Wb/m^2 or T (Tesla). The magnetic flux is denoted by the letter φ and the unit is Wb. It can be determined with the help of creation.

$$\Phi = \int \overline{B} \cdot \overline{dS} \qquad \ldots\ldots (7.15)$$

Problem 7.7 : A solid nonmagnetic conductor of circular cross section has its axis on the z axis and carries a uniformly distributed total current of 60 A in the $\bar{a}_z$ direction. If the radius is 4 mm, find : (a) B_ϕ at ρ = 5 mm ; (b) B_ϕ at ρ = 3 mm ; (c) the total magnetic flux per unit length inside the conductor ; (d) the total flux per unit length outside the conductor.

Solution :

(a) Consider a circle of radius ρ such that $\rho > 0.004$ as shown in Fig.P. 7.7(a). Apply Ampere's circuital law to this closed path

$$\oint \overline{H} \cdot \overline{dL} = I_{encl.}$$

$$\text{LHS} = \oint \overline{H} \cdot \overline{dL}$$

As the current is in $\bar{a}_z$ direction, $H\,\phi$ component of magnetic field intensity is produced. Hence $\overline{H} = H\,\phi\,\bar{a}_\phi$

$$\therefore \quad \text{LHS} = \int (H_\phi\,\bar{a}_\phi) \cdot (\rho\,d\,\phi\,\bar{a}_\phi)$$

$$= \int_0^{2\pi} \rho\,H_\phi\,d\,\phi = 2\,\pi\,\rho\,H_\phi$$

$$\text{RHS} = 60$$

$$\therefore \quad 2\,\pi\,\rho\,H_\phi = 60\ ,\ H_\phi = \frac{60}{2\,\pi\,\rho}$$

$$\text{or} \quad \overline{H} = \frac{30}{\pi\,\rho}\,\bar{a}_\phi \quad \text{A/m}$$

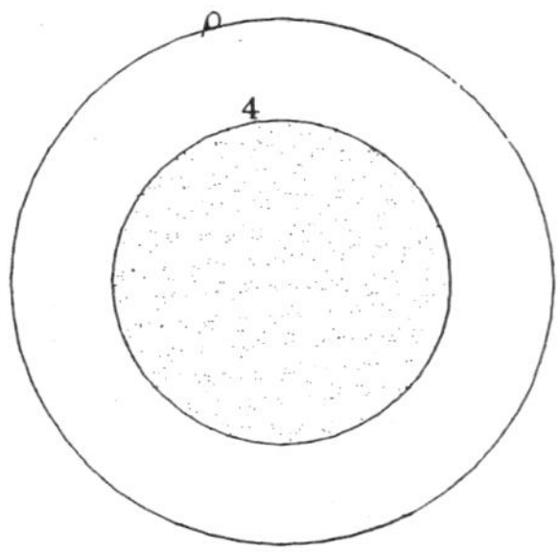

Fig.P. 7.7 (a)

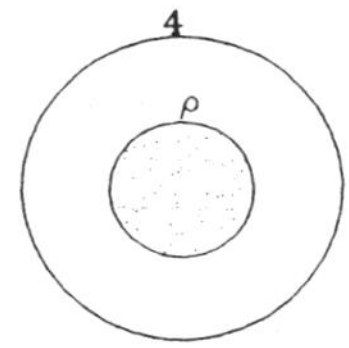

Fig.P. 7.7 (b)

F.∴ $\overline{B} = \mu_o \overline{H} = \dfrac{30 \mu_o}{\pi \rho} \overline{a}_\phi$ Wb/m^2

∴ $B_\phi \text{ (at } \rho = 5 \text{ mm) } = \dfrac{30 \mu_o}{\pi (0.005)} = 2.4 \text{ m T}$

(b) Consider a circle of radius ρ such that $\rho < 0.004$ as shown in Fig.7.7(b). Apply Ampere's Circuital law to this closed path.

$$\oint \overline{H} \cdot \overline{dL} = I_{encl.}$$

$$\text{LHS} = \oint \overline{H} \cdot \overline{dL} = 2 \pi \rho H_\phi$$

$$\text{RHS} = \frac{60}{\pi (0.004)^2} \times \pi \rho^2$$

$$= \frac{60 \rho^2}{1.6 \times 10^{-5}}$$

$$\therefore 2 \pi \rho H_\phi = \frac{60 \rho^2}{1.6 \times 10^{-5}} \; ; \; H_\phi = \frac{1875000 \rho}{\pi} \; ; \; B_\phi = \frac{1875000 \mu_o \rho}{\pi}$$

$$\therefore B_\phi \text{ (at } \rho = 3 \text{ mm) } = \frac{1875000 \mu_o (0.003)}{\pi} = 2.25 \text{ m T}$$

(c) From part b , inside the conductor, $B_\phi = \dfrac{1875000 \mu_o \rho}{\pi}$

$$\therefore \phi = \int \overline{B} \cdot \overline{dS} = \int (B_\phi \overline{a}_\phi) \cdot (d \rho \, dz \, \overline{a}_\phi) = \int B_\phi \, d \rho \, dz$$

$$= \int_0^{0.004} \int_0^1 \frac{1875000\,\mu_o}{\pi}\,\rho\, d\rho\, dz$$

$$= \frac{1875000}{\pi}\,\mu_o \left[\frac{\rho^2}{2}\right]_0^1 \left[z\right]_0^1 = 6\,\mu\text{Wb}$$

(d) From part a, outside the conductor, $B_\phi = \dfrac{30\,\mu_o}{\pi\rho}$

$$\therefore\ \phi = \int_{0.004}^{\infty} \int_0^1 \frac{30\,\mu_o}{\pi\rho}\, d\rho\, dz$$

$$= \frac{30\,\mu_o}{\pi}\left[\ln\rho\right]_{0.004}^{\infty}\left[z\right]_0^1 = \infty$$

7.6 SCALAR MAGNETIC POTENTIAL

Electric potential is denoted by V, the negative gradient of which gives Electric field intensity. On similar line, scalar magnetic potential is denoted by V_m, the negative gradient of which gives magnetic field intensity.

Therefore

$$\boxed{\overline{\mathbf{H}} = -\overline{\nabla}\mathbf{V_m}} \qquad \ldots\ldots (7.16)$$

Unit of V_m is ampere. Since $\overline{\nabla} \times \overline{H} = \overline{J}$, taking the curl of the above expression, $\overline{\nabla} \times \overline{H} = -\overline{\nabla} \times \overline{\nabla} V_m = 0$, because curl of gradient is zero. Therefore if $\overline{H}$ is to be defined as the gradient of a scalar magnetic potential, then current density must be zero throughout the region in which the scalar magnetic potential is defined. The difference between V and V_M is that V_M is not a single valued function while V is a single valued.

$V_{ab} = -\int \overline{E} \cdot \overline{dL}$ is independent of path.

For a specified path

$$\boxed{\mathbf{V_{m,\,ab}} = -\int \overline{\mathbf{H}} \cdot \overline{\mathbf{dL}}} \qquad \ldots\ldots (7.17)$$

Problem 7.8 : The region, $3.5 < \rho < 4.5$ cm, $0 \le \phi < 2\pi$, $0 < z < 2$ cm, is current–free and contains the field $H_\phi = 70/\rho$ A/m . Find V_m at $\rho = 3.8$ cm, $\phi = 110^\circ$, $z = 0.5$ cm if : (a) $V_m = 0$ at $\phi = 0$, and there is a barrier at $\phi = 180^\circ$; (b) $V_m = 0$ at $\phi = 0$, and there is a barrier at $\phi = 90^\circ$; (c) $V_m = 0$ at $\phi = 180^\circ$, and there is a barrier at $\phi = 0^\circ$; (d) $V_m = 0$ at $\phi = 180^\circ$, and there is a barrier at $\phi = 150^\circ$.

Solution :

(a) $\therefore \quad V_{m,PQ} = -\int_Q^P \overline{H} \cdot \overline{dL}$

$$= -\int_Q^P (H_\phi \overline{a}_\phi) \cdot (\rho \, d\phi \, \overline{a}_\phi)$$

$$= -\int_Q^P \rho H_\phi \, d\phi = -\int_0^{110^0} \rho \times \frac{70}{\rho} \, d\phi = -70\Big[\phi\Big]_0^{110^\circ}$$

$$= -70\left[110 \times \frac{\pi}{180}\right] = -134.390$$

Now $V_{m,PQ} = V_{m,P} - V_{m,Q} = V_{m,P}$ as $V_{m,Q} = 0$ (given)

$\therefore \quad V_m$ at P $= V_{m,PQ} = -134.39$ A

(b) $\therefore \quad V_{m,PQ} = -\int_0^{-250^\circ} \rho \times \frac{70}{\rho} \, d\phi = -70\left[-250 \times \frac{\pi}{180}\right]$

$$= 305.433 \text{ A}$$

$\therefore \quad V_{m,P} = 305.433$ A as $V_{m,Q} = 0$ (Given)

(c) $\therefore \quad V_{m,PQ} = -\int_{-180^\circ}^{-250^\circ} \rho \times \frac{70}{\rho} \, d\phi$

$$= -70\Big[-250 - (-180)\Big] \frac{\pi}{180} = 85.521 \text{ A}$$

$\therefore \quad V_{m,P} = 85.521$ A as $V_{m,Q} = 0$ (Given)

$$(d) \therefore \; V_{m,PQ} = -\int_{180^\circ}^{470^\circ} \rho \times \frac{70}{\rho} \, d\phi$$

$$= -70\left[470 - (180)\right] \frac{\pi}{180} = -354.302 \text{ A}$$

7.7 VECTOR MAGNETIC POTENTIAL

We know that $\bar{\nabla} \cdot \bar{B} = 0$ and divergence of the curl of any vector field is zero. Therefore we select

$$\boxed{\bar{\mathbf{B}} = \bar{\nabla} \times \bar{\mathbf{A}}} \quad \ldots\ldots (7.18)$$

where $\bar{A}$ signifies a vector magnetic potential. Vector magnetic potential is defined in a region where current density may be zero or may not be zero. This vector field is useful in studying radiation from antennas, from apertures, and radiation leakage from transmission lines, waveguides, and microwave ovens. Unit is Wb / m.

Problem 7.9 : The value of $\bar{A}$ within a solid nonmagnetic conductor of radius a carrying a total current I in the $\bar{a}_z$ direction may be found easily. Using the known value of $\bar{H}$ or $\bar{B}$ for $\rho < a$, then may be solved for $\bar{A}$. Select $\bar{A} = (\mu_o I \ln 5) / 2\pi$ at $\rho = a$ (to correspond with an example in the next section) and find A at $\rho =$: (a) 0, (b) , 0.3a ; (c) 0.6 a ; (d) a.

Solution :

Fig.P. 7.9 shows the cross section of a conductor of radius a. The axis of the conductor is z axis. The total current through the conductor is I ampere in $\bar{a}_z$ direction. Consider a circular path of radius ρ such that $\rho < a$. Apply Ampere's Circuital law to this closed path.

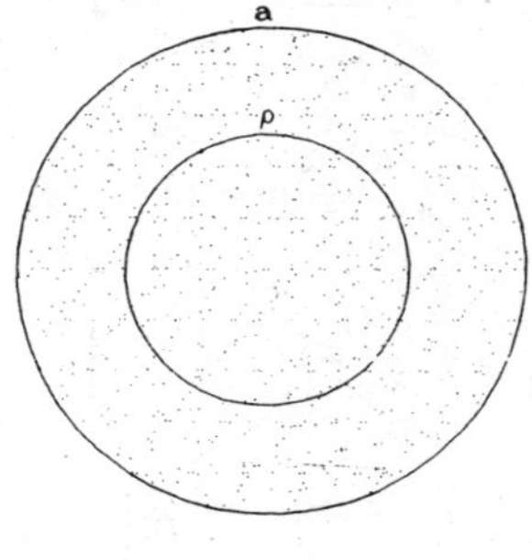

Fig. P. 7.9.

$$\oint \overline{H} \cdot \overline{dL} = I_{encl.}$$

$$LHS = \oint \overline{H} \cdot \overline{dL}$$

$$= \int (H\phi\, \overline{a}_\phi) \cdot (\rho\, d\phi\, \overline{a}_\phi)$$

$$= \int_0^{2\pi} \rho H\phi\, d\phi = 2\pi\rho H\phi$$

$$RHS = I_{encl} = \frac{I}{\pi a^2} \times \pi\rho^2 = \frac{I\rho^2}{a^2}$$

$$\therefore\ 2\pi\rho H_\phi = \frac{I\rho^2}{a^2} \quad \text{or} \quad H_\phi = \frac{I\rho}{2\pi a^2} \quad \text{or} \quad \overline{H} = \frac{I\rho}{2\pi a^2}\,\overline{a}_\phi$$

$$\therefore\ \overline{B} = \mu_o \overline{H} = \frac{I\rho\mu_o}{2\pi a^2}\,\overline{a}_\phi$$

Now $\overline{B} = \overline{\nabla} \times \overline{A}$, As $\overline{B}$ has $\overline{a}_\phi$, we will consider ϕ component of $\overline{\nabla} \times \overline{A}$.

$$(\overline{\nabla} \times \overline{A})_\phi = \left(\frac{\partial A_\rho}{\partial z} - \frac{\partial A_z}{\partial \rho}\right)\overline{a}_\phi$$

$$\therefore\ \frac{I\rho\mu_o}{2\pi a^2}\,\overline{a}_\phi = \left(\frac{\partial A_\rho}{\partial z} - \frac{\partial A_z}{\partial \rho}\right) a_\phi$$

As current is in $+\ \overline{a}_z$ direction, Az is present and A_ρ is absent.

$$\therefore\ \frac{I\rho\mu_o}{2\pi a^2} = -\frac{dA_z}{d\rho}$$

$$\therefore\ A_z = -\frac{I\mu_o}{2\pi a^2} \times \frac{\rho^2}{2} + C$$

where C is a constant of integration.

At $\rho = A$, $A = A_z = \dfrac{\mu_o I \ln 5}{2\pi}$

$$\therefore\ \frac{\mu_o I \ln 5}{2\pi} = -\frac{I\mu_o}{2\pi a^2} \times \frac{a^2}{2} + C$$

$$\therefore \quad C = \frac{\mu_o I}{2\pi}\left[\ln 5 + \frac{1}{2}\right]$$

$$\therefore \quad \bar{A} = A_z \bar{a}_z = \left[-\frac{I\mu_o}{2\pi a^2} \times \frac{\rho^2}{2} + \frac{\mu_o I}{2\pi}\left(\ln 5 + \frac{1}{2}\right)\right]\bar{a}_z$$

(a) At $\rho = 0$, $\bar{A} = \dfrac{\mu_o I}{2\pi}\left[\ln 5 + \dfrac{1}{2}\right]\bar{a}_z = 0.422\, I\, \bar{a}_z\ \mu\, \text{Wb/m}$

(b) At $\rho = 0.3\, a$, $\bar{A} = \left[\dfrac{-I\mu_o}{2\pi a^2} \times \dfrac{(0.3\, a)^2}{2} + \dfrac{\mu_o I}{2\pi}\left(\ln 5 + \dfrac{1}{2}\right)\right]\bar{a}_z$

$$= 0.413\, I\, \bar{a}_z\ \mu\text{Wb/m}$$

(c) At $\rho = 0.6\, a$, $\bar{A} = \left[\dfrac{-I\mu_o}{2\pi a^2} \times \dfrac{(0.6\, a)^2}{2} + \dfrac{\mu_o I}{2\pi}\left(\ln 5 + \dfrac{1}{2}\right)\right]\bar{a}_z$

$$= 0.386\, I\, \bar{a}_z\ \mu\, \text{Wb/m}$$

(d) At $\rho = a$, $\bar{A} = \left[\dfrac{-I\mu_o}{2\pi a^2} \times \dfrac{a^2}{2} + \dfrac{\mu_o I}{2\pi}\left(\ln 5 + \dfrac{1}{2}\right)\right]\bar{a}_z$

$$= 0.322\, I\, \bar{a}_z\ \mu\, \text{Wb/m}$$

Problem 7.10 : Consider a conductor of circular cross section carrying a current I in the $\bar{a}_z$ direction in free space. The zero reference is arbitrarily set at ρ = b. Now consider three conductors, each of 1–cm radius, parallel to the z axis with their axes lying in the x = 0 plane. Two have their axes at y = 4 cm and y = – 4 cm, each carrying a current of 12A in the $\bar{a}_z$ direction. The third is located at the z–axis and carries – 24 A in the $\bar{a}_z$ direction. If each current has its zero reference for A located 2 cm from its axis, find the combined $\bar{A}$ field at : (a) (0, 2, 0) ; (b) (0, –2, 0) ; (c) (2, 0, 0) ; (d) (3, 4, 0) ; (e) (∞ , 0, 0).

Solution :

If a conductor located on the z axis carries a current I in + $\bar{a}_z$ direction, then vector magnetic potential outside the conductor is given by

$$A_z = \frac{\mu_o I}{2\pi}\ln\left(\frac{b}{\rho}\right) = \frac{\mu_o I}{2\pi}\ln\left[\frac{b}{\sqrt{x^2 + y^2}}\right]$$

where at $\rho = b$, A_z is zero. In this problem $b = 0.02$.

For the conductor whose axis is located by $x = 0, y = -0.04$,

we have $$A_z = \frac{\mu_o (12)}{2\pi} \ln\left[\frac{0.02}{\sqrt{x^2 + (y + 0.04)^2}}\right] \quad \ldots\ldots (1)$$

Similarly for the conductor whose axis is located by $x = 0, y = 0.04$,

we have $$A_z = \frac{\mu_o (12)}{2\pi} \ln\left[\frac{0.02}{\sqrt{x^2 + (y - 0.04)^2}}\right] \quad \ldots\ldots (2)$$

For the conductor located on the z axis, we have

$$A_z = \frac{\mu_o (-24)}{2\pi} \ln\left[\frac{0.02}{\sqrt{x^2 + y^2}}\right] \quad \ldots\ldots (3)$$

(a) Substitute $x = 0$ and $y = 0.02$ in equations 1, 2, and 3.

We have $A_z = -2.6367 \times 10^{-6} + 0 + 0 = -2.6367 \times 10^{-6}$

$\therefore \bar{A} = -2.637\, \bar{a}_z\ \mu$ Wb/m.

(b) Substitute $x = 0$ and $y = -0.02$ in equations 1, 2, and 3.

We have $A_z = -2.6367 \times 10^{-6} + 0 = -2.6367 \times 10^{-6}$

$\therefore \bar{A} = -2.63\, \bar{a}_z\ \mu$ Wb/m.

(c) Substitute $x = 0.02$ and $y = 0$ in equations 1, 2, and 3

We have

$A_z = -1.9313 \times 10^{-6} - 1.9313 \times 10^{-6} + 0 = -3.8626 \times 10^{-6}$

$\therefore \bar{A} = -3.863\, \bar{a}_z\ \mu$ Wb/m

(d) Substitute $x = 0.03$ and $y = 0.04$ in equations 1, 2, and 3

We have $A_z = -3.485 \times 10^{-6} - 0.973 \times 10^{-6} + 4.398 \times 10^{-6}$

$= -0.0598 \times 10^{-6}$

$\therefore \bar{A} = -0.0598\, \bar{a}_z\ \mu$ Wb/m

(e) As A_z decreases away from the conductor, $A_z = 0$ when $x = \infty$ and $y = 0$

Problem 7.11 : A filamentary current of 10 A is directed in from infinity to the origin on the postive x axis, and then back out to infinity along the positive y axis Use the Biot – Savart law to find $\overline{H}$ at P (0, 0, 1) .

Solution :

Consider current which lies on the x axis.

$$d\,\overline{H} = \frac{I\,\overline{dL} \times \bar{a}_{12}}{4\pi d^2}$$

$$\overline{dL} = dx\,\bar{a}_x \;,\; 2(0,0,1) \;,\; 1(x,0,0)$$

$$\overline{R}_{12} = -x\,\bar{a}_x + \bar{a}_z \;,\; d = \left(x^2+1\right)^{1/2}$$

$$\therefore\; d\,\overline{H} = \frac{(10\,dx\,\bar{a}_x)}{4\pi} \times \left[\frac{-x\,\bar{a}_x + \bar{a}_z}{\left(x^2+1\right)^{3/2}}\right]^2$$

$$= \frac{-10\,dx\,\bar{a}_y}{4\pi(x^2+1)^{3/2}}$$

Fig.P. 7.11

$$\therefore\; \overline{H} = \frac{-10}{4\pi}\int_{\infty}^{0} \frac{dx\,\bar{a}_y}{(x^2+1)^{3/2}} = \frac{-10}{4\pi}\left[\frac{x}{\sqrt{x^2+1}}\right]_{\infty}^{0} \bar{a}_y$$

$$= \frac{-10}{4\pi}[0-1]\,\bar{a}_y = 0.796\,\bar{a}_y$$

Similarly for the current which lies on the y axis, we have

$$d\overline{H} = \frac{10\, dy\, \bar{a}_y}{\left[4\pi \sqrt{y^2 + 1^2}\right]^2} \times \left[\frac{-y\bar{a}_y + \bar{a}_z}{\sqrt{y^2 + 1}}\right]$$

$$= \frac{10\, dy\, \bar{a}_x}{4\pi (y^2 + 1)^{3/2}}$$

$$\therefore \quad \overline{H} = \frac{10}{4\pi} \int_0^{\infty} \frac{dy\, \bar{a}_x}{(y^2 + 1)^{3/2}} = \frac{10}{4\pi} \left[\frac{y}{\sqrt{y^2 + 1}}\right]_0^{\infty} \bar{a}_x$$

$$= \frac{10}{4\pi} \left[1 - 0\right] \bar{a}_x = 0.796\, \bar{a}_x$$

$$\overline{H}_{total} = 0.796\, \bar{a}_x + 0.796\, \bar{a}_y \text{ A/m}$$

Problem 7.12 : The parallel filamentary conductors shown in Fig.P. 7.12 lie in free space. Plot $|\overline{H}|$ vs. y, $-5 < y < 5$, along the line x = 0, z = 1.

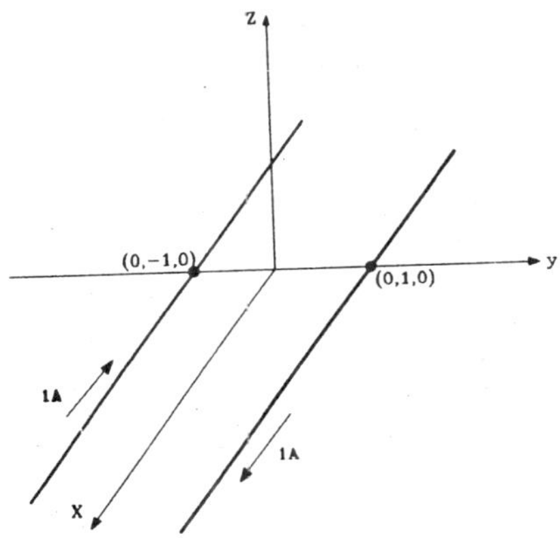

Fig.P. 7.12

Solution :

Consider current carrying conductor located at y = 1, z = 0. As the current is infinite in extent, we will use

$$\overline{H} = \frac{I}{2\pi d}\, \bar{a} = \frac{I}{2\pi d} (\bar{a}_I \times \bar{a}_{12})$$

$$\bar{a}_I = \bar{a}_x \; ; \; 2(0, y, 1) \; ; \; 1(0, 1, 0)$$

$$\overline{R}_{12} = (y - 1)\bar{a}_y + \bar{a}_z \; ; \; d = |\overline{R}_{12}| = \sqrt{(y - 1)^2 + 1}$$

$$\therefore \ \overline{H} = \frac{1}{2\pi\sqrt{(y-1)^2+1}} \ \overline{a}_x \times \frac{(y-1)\,\overline{a}_y + \overline{a}_z}{\sqrt{(y-1)^2+1}}$$

$$= \frac{-\overline{a}_y + (y-1)\,\overline{a}_z}{2\pi\,[(y-1)^2+1]^{3/2}}$$

Another conductor is located at $y = -1$, $z = 0$

$\overline{a}_I = -\overline{a}_x$; $2(0, y, 1)$; $1(0, -1, 0)$

$\overline{R}_{12} = (y+1)\,\overline{a}_y + \overline{a}_z$; $d = |\overline{R}_{12}| = \sqrt{(y+1)^2+1^2}$

$$\therefore \ \overline{H} = \frac{1}{2\pi\sqrt{(y+1)^2+1}} \ \overline{a}_x \times \left[\frac{(y+1)\,\overline{a}_y + \overline{a}_z}{\sqrt{(y+1)^2+1}}\right]$$

$$= \frac{\overline{a}_y - (y+1)\,\overline{a}_z}{2\pi\,[(y+1)^2+1]^{3/2}}$$

$$\therefore \ \overline{H}_{total} = \frac{-\overline{a}_y + (y-1)\,\overline{a}_z}{2\pi\,[(y-1)^2+1]^{3/2}} + \frac{\overline{a}_y - (y+1)\,\overline{a}_z}{2\pi\,[(y+1)^2+1]^{3/2}}$$

$$= Hy\,\overline{a}_y + Hz\,\overline{a}_z \quad \text{where}$$

$$Hy = \frac{1}{2\pi}\left[\frac{-1}{[(y-1)^2+1]^{3/2}} + \frac{1}{[(y+1)^2+1]^{3/2}}\right] \quad \text{and}$$

$$Hz = \frac{1}{2\pi}\left[\frac{(y-1)}{[(y-1)^2+1]^{3/2}} - \frac{(y+1)}{[(y+1)^2+1]^{3/2}}\right]$$

$$\therefore \ |\overline{H}| = \sqrt{(Hy)^2 + (Hz)^2}$$

Plot $|\overline{H}|$ on the y–axis and y on the x–axis plot is not given.

Problem 7.13 : Find $\overline{H}$ at P (2, 3, 5) in cartesian coordinates if there is an infinitely long current filament passing through the origin and point C. The current of 50 A is directed from the origin to C, where the location of C is : (a) C (0, 0, 1) ; (b) C (0, 1, 0).

Solution :

(a) I = 50 A

$$\overline{a}_I = \frac{\overline{R}_{oc}}{|\overline{R}_{oc}|} = \overline{a}_z \ ; \ 2(2, 3, 5) \ ; \ 1(0, 0, 5)$$

$$\overline{R}_{12} = 2\,\overline{a}_x + 3\,\overline{a}_y \;;\; d = |\,\overline{R}_{12}\,|$$

$$= \sqrt{2^2 + (3)^2} = 3.606$$

$$\therefore\; \overline{H} = \frac{I}{2\pi d}\,[\,\overline{a}_I \times \overline{a}_{12}$$

$$= \frac{50}{2\pi\,(3.603\,)}\left[\,\overline{a}_z \times \left(\frac{2\,\overline{a}_x + 3\,\overline{a}_y}{3.606}\right)\right]$$

$$= 1.224\,\overline{a}_x + 1.836\,\overline{a}_y \;\; \text{A/m}$$

(b) $I = 50$ A

$$\overline{a}_I = \frac{\overline{R}_{oc}}{|\,R_{oc}\,|} = \overline{a}_z \;;\; 2\,(2\,,3\,,5\,)\;;\;1\,(0\,,3\,,0\,)$$

$$\overline{R}_{12} = 2\,\overline{a}_x + 5\,\overline{a}_z \;;\; d = |\,\overline{R}_{12}\,| = \sqrt{(2)^2 + (5)^2} = 5.385$$

$$\therefore\; \overline{H} = \frac{50}{2\pi\,(5.385\,)}\left[\,\overline{a}_y \times \left(\frac{2\,\overline{a}_x + 5\,\overline{a}_z}{5.385}\right)\right]$$

$$= 1.372\,\overline{a}_x - 0.549\,\overline{a}_z \;\; \text{A/m}$$

Problem 7.14 : Two current filaments are parallel to the z axis. One passes through (0, 0.5, 0) and carries 100 A in the $\overline{a}_z$ direction; the other passes through (0, – 0.5, 0) and carries 100 A in the – $\overline{a}_z$ direction. Plot H_x vs. y, 0 < y < 2., for points on the y axis.

Solution :

Consider the current filament which passes through (0, 0.5, 0).

$$I = 100\text{ A} \;;\; \overline{a}_I = \overline{a}_z \;;\; 2\,(0\,,y\,,0) \;;\; 1\,(0\,,0.5\,,0)$$

$$\overline{R}_{12} = (y - 0.5\,)\,\overline{a}_y \;;\; d = |\,\overline{R}_{12}\,| = |\,y - 0.5\,|$$

$$\therefore\; \overline{H} = \frac{100}{2\pi\,|\,y - 0.5\,|}\;\overline{a}_z \times \left[\frac{(y - 0.5\,)\,\overline{a}_y}{|\,y - 0.5\,|}\right]$$

$$= \frac{100}{2\pi\,(y - 0.5)^2}\,[\,(y - 0.5)\,\overline{a}_x\,]$$

$$= \frac{-\,50\,\overline{a}_x}{\pi\,(y - 0.5)}$$

Similarly for the second current filament which passes through (0, –0.5, 0) we have

$$\overline{H} = \frac{100}{2\pi \mid y + 0.5 \mid} \left\{ -\overline{a}_z \times \left[\frac{(y + 0.5)\,\overline{a}_y}{\mid y + 0.5 \mid} \right] \right\} = \frac{50\,\overline{a}_x}{\pi\,(y + 0.5)}$$

$$\therefore \quad \overline{H}_{total} = \frac{-50\,\overline{a}_x}{\pi\,(y - 0.5)} + \frac{50\,\overline{a}_x}{\pi\,(y + 0.5)} = \frac{-50\,\overline{a}_x}{\pi\,(y^2 - 0.25)}$$

$$\therefore \quad H_x = \frac{-50}{\pi\,(y^2 - 0.25)} \text{ A/m}$$

Plot Hx on the y axis and y on the x axis plot is not given.

Problem 7.15 : (a) A current filament I is formed into a circle, $\rho = a$, in the $z = z'$ plane. Find H_z at P (0, 0, z) if I flows in the $\overline{a}_\phi$ direction. (b) Find H_z at P caused by a uniform surface current density, $\overline{K} = K_0\,\overline{a}_\phi$, flowing on the cylindrical surface, $\rho = a$, $0 \le z \le h$. The results of part (a) should help.

Solution :

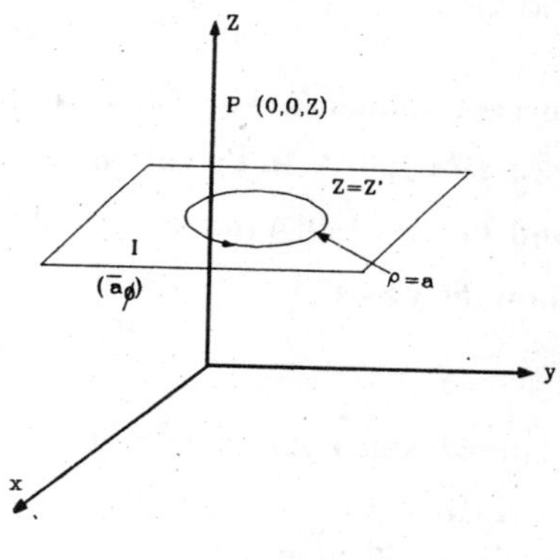

Fig.P. 7.15 (a)

(a) $d\overline{H} = \dfrac{I\,\overline{dL} \times \overline{a}_{12}}{4\pi d^2}$

For the circular current, $\rho = a$, $z = z'$ are constants. Hence $d\rho = dz = 0$. Now $\overline{dL} = d\rho\,\overline{a}_\rho + \rho\,d\phi\,\overline{a}_\phi + dz\,\overline{a}_z$.

Hence $\overline{dL} = \rho\,d\phi\,\overline{a}_\phi = a\,d\phi\,\overline{a}_\phi \quad 2\,(0, 0, z),\ 1\,(a, \phi, z')$

$$\overline{R}_{12} = \overline{r}_2 - \overline{r}_1 = z\,\overline{a}_z - (a\,\overline{a}_\rho + z'\,\overline{a}_z)$$

$$= (z - z')\,\overline{a}_z - a\,\overline{a}_\rho$$

$$d = \mid \overline{R}_{12} \mid = [(z - z')^2 + a^2]^{1/2}$$

$$\therefore \quad d\overline{H} = \frac{I}{4\pi\sqrt{[(z-z')^2+a^2]^2}} \left\{ a d\phi\, \overline{a}_\phi \times 0 \left[\frac{(z-z')\,\overline{a}_z - a\,\overline{a}_\rho}{\sqrt{(z-z')^2+a^2}} \right] \right\}$$

$$= \frac{Ia\, d\phi\,[(z-z')\,\overline{a}_\rho + a\,\overline{a}_z]}{4\pi[(z-z')^2+a^2]^{3/2}}$$

Due to symmetry of the problem, only z component is present. Hence

$$\overline{H} = \int d\overline{H} = \int_0^{2\pi} \frac{I a^2\,\overline{a}_z\, d\phi}{4\pi[(z-z')^2+a^2]^{3/2}}$$

$$= \frac{Ia^2\,\overline{a}_z}{2[(z-z')^2+a^2]^{3/2}}$$

(b) Consider a circular path of thickness dz which lies on the circular surface $\rho = a$ as shown in Fig.P. 7.15 (b)

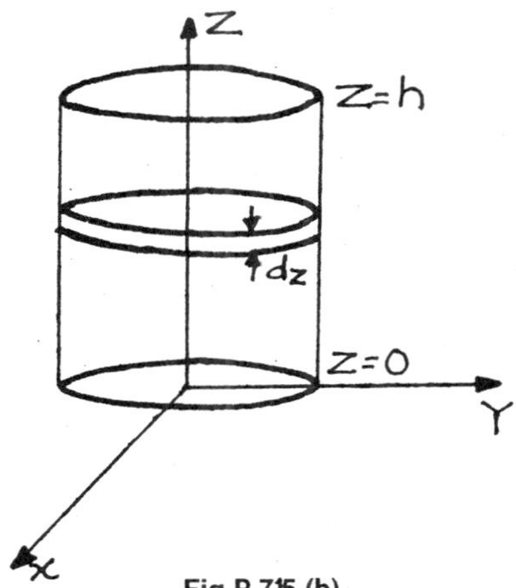

Fig.P 7.15 (b)

Then $I = K_o\, dz$. Because of this current, the magnetic field intensity at point P (0, 0, z) is given by the result of part (a) as

$$d\overline{H} = \frac{K_o\, dz\, a^2}{2[(z-z')^2+a^2]^{3/2}}$$

$$\therefore \quad \overline{H} = \int d\overline{H} = \int_0^h \frac{K_o \, dz \, a^2}{2[(z - z')^2 + a^2]^{3/2}}$$

$$= K_o a^2 \overline{a}_z \left[\frac{z}{\sqrt{(z - z')^2 + a^2}} \right]_o^h$$

$$= K_o a^2 \overline{a}_z \left[\frac{h}{\sqrt{(h - z')^2 + a^2}} \right] \text{A/m}.$$

Problem 7.16 : A current element $I\,d\overline{L} = 10^{-3}(2\,\overline{a}_x + 4\,\overline{a}_y - \overline{a}_z)$ A . m, located at A (–5, 3, –2), produces a field $d\overline{H}$ at B (3, –4, 3). (a) Give a unit vector in the direction of $d\overline{H}$ at B. (b) Find $|d\overline{H}|$ at B. (c) What direction should $I\,d\overline{L}$ have at A so that $d\overline{H} = 0$ at B.

Solution :

$$\text{(a)} \quad d\overline{H} = \frac{I\,d\overline{L} \times \overline{a}_{AB}}{4\pi d^2}$$

$$\overline{R}_{AB} = (3 + 5)\,\overline{a}_x + (-4 - 3)\,\overline{a}_y + (3 + 2)\,\overline{a}_z$$

$$= 8\,\overline{a}_x - 7\,\overline{a}_y + 5\,\overline{a}_z$$

$$d = |\overline{R}_{AB}| = \sqrt{(8)^2 + (-7)^2 + (5)^2} = 11.7473$$

$$\therefore \quad \overline{a}_{AB} = \frac{\overline{R}_{AB}}{|\overline{R}_{AB}|} = \frac{8\,\overline{a}_x - 7\,\overline{a}_y + 5\,\overline{a}_z}{11.7473}$$

$$= 0.681\,\overline{a}_x - 0.5959\,\overline{a}_y + 0.4256\,\overline{a}_z$$

$$I\,d\overline{L} \times \overline{a}_{AB} = 10^{-3} \begin{vmatrix} \overline{a}_x & \overline{a}_y & \overline{a}_z \\ 2 & 4 & -1 \\ 0.681 & -0.5959 & 0.4256 \end{vmatrix}$$

$$= 10^{-3}(1.1065\,\overline{a}_x - 1.5322\,\overline{a}_y - 3.9158\,\overline{a}_z)$$

$$\therefore \quad d\overline{H} = \frac{10^{-3}(1.1065\,\overline{a}_x - 1.5322\,\overline{a}_y - 3.9158\,\overline{a}_z)}{4\pi(11.7473)^2}$$

$$= 0.638\,\overline{a}_x - 0.884\,\overline{a}_y - 2.258\,\overline{a}_z \;\mu\text{A/m}.$$

$$|d\overline{H}| = \sqrt{(0.638)^2 + (-0.884)^2 + (-2.258)^2}$$

$= 2.5074 \, \mu \, A/m$

$\therefore$ Unit vector in the direction of $\overline{dH} = \dfrac{\overline{dH}}{|\, dH \,|}$

$$= \frac{0.638 \, \bar{a}_x - 0.884 \, \bar{a}_y - 2.258 \, \bar{a}_z}{2.5074}$$

$$= 0.2544 \, \bar{a}_x - 0.3526 \, \bar{a}_y - 0.901 \, \bar{a}_z$$

(b) From part (a), $|\, d\overline{H} \,|$ at B $= 2.5074 \, \mu \, A/m$

(c) If the direction of $I \, d\overline{L}$ is same as $\overline{R}_{AB}$ or $- \overline{R}_{AB}$, then, $d\overline{H} = 0$ at B.

$$|\, I \, d\overline{L} \,| = 10^{-3} \sqrt{(2)^2 + (4)^2 + (-1)^2} = (\sqrt{21}) \, 10^{-3} \, A/m$$

$\therefore$ direction of $I \, d\overline{L} = \pm \, \bar{a}_{AB} = \pm \, (0.681 \, \bar{a}_x - 0.5959 \, \bar{a}_y + 0.4256 \, \bar{a}_z)$

and $I \, d\overline{L} = \pm \, \bar{a}_{AB} \, (\sqrt{21}) \, 10^{-3}$

$$= \pm \, 10^{-3} \, (3.1207 \, \bar{a}_x - 2.7308 \, \bar{a}_y + 1.9503 \, \bar{a}_z) \, A/m.$$

Problem 7.17 : Given points A (1, 2, 4), B (–2, –1, 3), and C (3, 1, –2), let a differential current element with I = 6A and $|\, \overline{dL} \,| = 10^{-4}$ m be located at A. The direction of $\overline{dL}$ is from A to B. Find $\overline{dH}$ at C.

Solution :

$$\overline{R}_{AB} = (-2 - 1) \, \bar{a}_x + (-1 - 2) \, \bar{a}_y + (3 - 4) \, \bar{a}_z$$

$$= -3 \, \bar{a}_x - 3 \, \bar{a}_y - \bar{a}_z$$

$$\bar{a}_{AB} = \bar{a}_L = \frac{\overline{R}_{AB}}{|\, R_{AB} \,|} = \frac{-3 \, \bar{a}_x - 3 \, \bar{a}_z}{\sqrt{(-3)^2 + (-3)^2 + (-1)^2}}$$

$$= -0.688 \, \bar{a}_x - 0.688 \, \bar{a}_y - 0.229 \, \bar{a}_z$$

$\therefore \; I \, \overline{dL} = I \, |\, \overline{dL} \,| \, \bar{a}_L$

$$= 6 \times 10^{-4} \, (-0.088 \, \bar{a}_x - 0.688 \, \bar{a}_y - 0.229 \, \bar{a}_z) \, A - m$$

$$\overline{dH} = \frac{I \, \overline{dL} \times \bar{a}_{AC}}{4 \pi d^2}$$

$$\overline{R}_{AC} = (3 - 1) \, \bar{a}_x + (1 - 2) \, \bar{a}_y + (-2 - 4) \, \bar{a}_z$$

$$= 2 \, \bar{a}_x - \bar{a}_y - 6 \, \bar{a}_z$$

$$\bar{a}_{AC} = \frac{\bar{R}_{AC}}{|\bar{R}_{AC}|} = \frac{2\,\bar{a}_x - \bar{a}_y - 6\,\bar{a}_z}{\sqrt{(2)^2 + 9- 1)^2 + (-6)^2}}$$

$$= 0.3123\,\bar{a}_x - 0.1562\,\bar{a}_y - 0.937\,\bar{a}_z$$

$$I\,\overline{dL} \times \bar{a}_{AC} = 6 \times 10^{-4} \begin{vmatrix} \bar{a}_x & \bar{a}_y & \bar{a}_z \\ -0.688 & -0.688 & -0.229 \\ 0.3123 & -0.1562 & -0.937 \end{vmatrix}$$

$$= 6 \times 10^{-4}\,(0.60888\,\bar{a}_x - 0.71617\,\bar{a}_y + 0.32233\,\bar{a}_z)$$

$$\therefore \quad d\,\bar{H} = \frac{6 \times 10^{-4}\,(0.60888\,\bar{a}_x - 0.71617\,\bar{a}_y + 0.32222\,\bar{a}_z)}{4\,\pi\,(\sqrt{41})^2}$$

$$= 0.709\,\bar{a}_x - 0.834\,\bar{a}_y + 0.375\,\bar{a}_z \ \mu A/m .$$

Problem 7.18 : For the finite–length current element on the z axis, as shown in Fig. P. 7.18, using Biot–Savart law, show that

$$\bar{H} = \frac{I}{4\,\pi\,\rho}\,[\,\sin\alpha_2 - \sin\alpha_1\,]\,\bar{a}_\phi .$$

Solution :

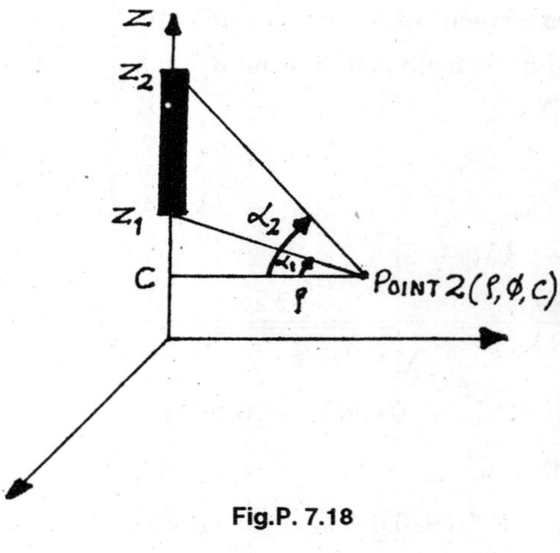

Fig.P. 7.18

$$\overline{dH} = \frac{I\,\overline{dL} \times \bar{a}_{12}}{4\,\pi\,d^2}$$

$\overline{dL} = dz\,\bar{a}_z$ as the current lies on the z axis.

Point 2 is that point at which $\bar{H}$ is desired $= (\rho, \phi, c)$

Point 1 is the general point on the current filament = (0, 0, z)

$$\overline{R}_{12} = \overline{r}_2 - \overline{r}_1 = (\rho\, \overline{a}_\rho + c\, \overline{a}_z) - z\, \overline{a}_z)$$

$$= \rho\, \overline{a}_\rho + (c - z)\, \overline{a}_z$$

$$d = |\overline{R}_{12}| = \sqrt{\rho^2 + (c - z)^2}$$

$$\overline{a}_{12} = \frac{\overline{R}_{12}}{|\overline{R}_{12}|} = \frac{\rho\, \overline{a}_\rho + (c - z)\, \overline{a}_z}{\sqrt{\rho^2 + (c - z)^2}}$$

$$\overline{dH} = \frac{(I\, dz\, \overline{a}_z) \times [\rho\, \overline{a}_\rho + (c \quad z)\, \overline{a}_z)\,]}{4\pi\, [\rho^2 + (c - z)^2\,]^{3/2}}$$

$$\overline{dH} = \frac{I dz\, \rho\, \overline{a}_\phi}{4\pi\, [\rho^2 + (c - z)^2\,]^{3/2}}$$

$$\overline{H} = \int_{z_1}^{z_2} \frac{I\, dz\, \rho\, \overline{a}_\phi}{4\pi\, [\rho^2 + (c - z)^2\,]^{3/2}}$$

Substitute $c - z = \rho \tan\theta$

Than $dz = -\rho \sec^2\theta\, d\theta$

when $z = z_1,\ \theta = \tan^{-1}\dfrac{c - z_1}{\rho} = -\tan^{-1}\dfrac{z_1 - c}{\rho}$

From Fig.P.7.18 $\alpha_1 = \tan^{-1}\dfrac{z_1 - c}{\rho}$ Hence $\theta = -\alpha_1$

when $z = z_2,\ \theta = \tan^{-1}\dfrac{c - z_2}{\rho} = -\tan^{-1}\dfrac{z_2 - c}{\rho}$

From Fig.P.7.18, $\alpha_2 = \tan^{-1}\dfrac{z_2 - c}{\rho}$ Hence $\theta = -\alpha_2$

$$\therefore\ \overline{H} = \int_{-\alpha_1}^{-\alpha_2} \frac{I\, \rho\, \overline{a}_\phi}{4\pi} \times \frac{-\rho \sec^2\theta\, d\theta}{\rho^3 \sec^3\theta}$$

$$= -\frac{I\, \overline{a}_\phi}{4\pi\rho} \Big[\sin\theta\Big]_{-\alpha_1}^{-\alpha_2}$$

$$= \frac{I}{4\pi\rho}\, \overline{a}_\phi \Big[\sin(-\alpha_1) - \sin(-\alpha_2)\Big]$$

$$\overline{H} = \frac{I}{4\pi\rho}\, [\sin\alpha_2 - \sin\alpha_1]\, \overline{a}_\phi \text{ A/m.}$$ Hence proved.

Problem 7.19 : A surface current density, K = 20 $\bar{a}_x$ A/m, flows in the y = 0 plane throughout the region − 5 < z < 5m , − ∞ < x < ∞ . Find $\overline{H}$ at P (0, 10, 0) in free space.

Solution :

Surface current density do not lie on infinite plane.

Hence we will use equation (7.2)

$$d\overline{H} = \frac{K\, ds \times \bar{a}_{12}}{4\pi d^2}$$

$\overline{K}$ lies on the plane y = 0 plane.

$\bar{a}_y$ is perpendicular to the plane y = 0.

Hence $\overline{ds} = dx\,dz\,\bar{a}_y$ or $ds = dx\,dz$

Point 2 is that point at which $\overline{H}$ is desired = (0, 10, 0)

Point 1 is the general point on the surface-current density = (x, 0, z)

$$\therefore \quad \overline{R}_{12} = (0 - x)\,\bar{a}_x + (10 - 0)\,\bar{a}_y + (0 - z\,\bar{a}_z)$$

$$= -x\,\bar{a}_x + 10\,\bar{a}_y - z\,\bar{a}_z$$

$$d = |\overline{R}_{12}| = \sqrt{(-x)^2 + (10)^2 + (-z)^2}$$

$$= \sqrt{x^2 + z^2 + 100}$$

$$\bar{a}_{12} = \frac{\overline{R}_{12}}{|\overline{R}_{12}|} = \frac{-x\,\bar{a}_x + 10\,\bar{a}_y - z\,\bar{a}_z}{\sqrt{x^2 + z^2 + 100}}$$

$$\therefore \quad d\overline{H} = \frac{(20\,\bar{a}_x\,dx\,dz) \times (-x\,\bar{a}_x + 10\,\bar{a}_y - z\,\bar{a}_z)}{4\pi\,(x^2 + z^2 + 100)^{3/2}}$$

$$= \frac{5\,(10\,\bar{a}_z + z\,\bar{a}_y)\,dx\,dz}{\pi\,(x^2 + z^2 + 100)^{3/2}}$$

$$\therefore \quad \overline{H} = \int d\overline{H}$$

$$= \frac{5}{\pi}\left[\int_{-\infty}^{5}\int_{-5}^{5} \frac{10\,\bar{a}_z\,dx\,dz}{(x^2 + z^2 + 100)^{3/2}} + \int_{-\infty}^{\infty}\int_{-5}^{5} \frac{z\,\bar{a}_y\,dx\,dz}{(x^2 + z^2 + 100)^{3/2}}\right.$$

$$\overline{H} = \frac{5}{\pi}\left[\int_{-5}^{5} \frac{20\,\overline{a}_z\,dz}{z^2 + 100} + \int_{-5}^{5} \frac{2z\,\overline{a}_y\,dz}{z^2 + 100}\right]$$

$$= \frac{5}{\pi}\left[\frac{20\,\overline{a}_z}{10}\left[\tan^{-1}\left(\frac{z}{10}\right)\right]_{-5}^{5} + \overline{a}_y\left[\ln(z^2 + 100)\right]_{-5}^{5}\right]$$

$$= 2.952\,\overline{a}_z$$

Problem 7.20. : (a) A filamentary current I flows in the $\overline{a}_\phi$ direction in the circular path $\rho = a$, $z = 0$, in free space. Find $\overline{H}$ at P (0, 0, z) .
(b) Now let identical filamentary currents of 5A flow in the $\overline{a}_\phi$ direction at $\rho = 2m$, $z = \pm 1m$. Plot | H (0, 0, z) | vs. z.

Solution :

(a) Let us consider the circular path $\rho = a$, $z = b$ as shown in Fig.P. 7.20

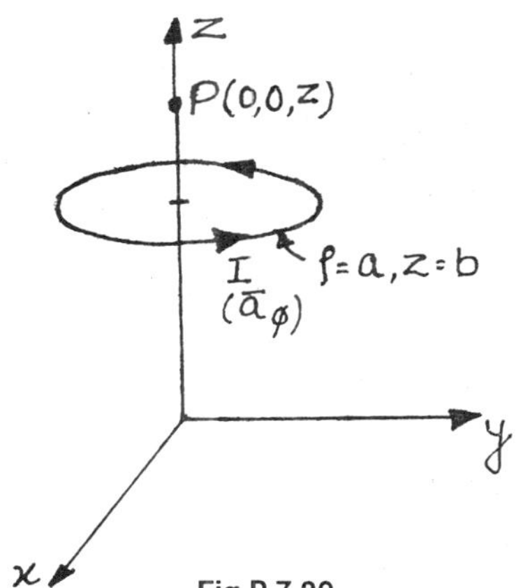

Fig.P.7.20

$$d\overline{H} = \frac{I\,\overline{dL} \times \overline{a}_{12}}{4\pi d^2}$$

$$\overline{dL} = d\rho\,\overline{a}_\rho + \rho\,d\phi\,\overline{a}_\phi + dz\,\overline{a}_z$$

For the circular path $\rho = a$ and $z = b$ are constants.

Hence $d\rho = dz = 0$

$$\overline{dL} = \rho\,d\phi\,\overline{a}_\phi = a\,d\phi\,\overline{a}_\phi$$

Point 2 is that point at which $\overline{H}$ is desired $= (0, 0, z)$

Point 1 is the general point on the current $= (a, \phi, b)$

$$\overline{R}_{12} = \overline{r}_2 - \overline{r}_1 = (z\,\overline{a}_z) - (a\,\overline{a}_\rho + b\,\overline{a}_z)$$

$$= -a\,\overline{a}_\rho + (z - b)\,\overline{a}_z$$

$$d = |\overline{R}_{12}| = \sqrt{(-a)^2 + (z-b)^2}$$

$$= \sqrt{a^2 + (z-b)^2}$$

$$\overline{a}_{12} = \frac{\overline{R}_{12}}{|\overline{R}_{12}|} = \frac{-a\,\overline{a}_\rho + (z-b)\,\overline{a}_z}{\sqrt{a^2 + (z-b)^2}}$$

$$\therefore \quad d\overline{H} = \frac{(I\,a\,d\phi\,\overline{a}_\phi) \times [-a\,\overline{a}_\rho + (z-b)\,\overline{a}_z]}{4\pi\left[a^2 + (z-b)^2\right]^{3/2}}$$

$$= \frac{I\,a\,d\phi\,[a\,\overline{a}_z + (z-b)\,\overline{a}_\rho]}{4\pi\left[a^2 + (z-b)^2\right]^{3/2}}$$

As the current I is symmetrically located w.r.t. point P, ρ component of $\overline{H}$ gets cancelled out.

Hence

$$\overline{H} = \int d\overline{H} = \int_0^{2\pi} \frac{I\,a^2\,d\phi\,\overline{a}_z}{4\pi\left[a^2 + (z-b)^2\right]^{3/2}}$$

$$= \frac{I a^2\,\overline{a}_z}{2\left[a^2 + (z-b)^2\right]^{3/2}}$$

If the circular path is located by $\rho = a$, $b = 0$

$$\text{then } \overline{H} = \frac{I a^2\,\overline{a}_z}{2\left[a^2 + z^2\right]^{3/2}} \text{ A/m}$$

[If the current is in $-\overline{a}_\phi$ direction i.e. clockwise direction, then the limits of integration will be 2π to 0 or 0 to -2π].

(b) For the current of 5A in $\overline{a}_\phi$ direction at $\rho = 2\text{m}$, $z = 1\text{m}$,

$$\overline{H}_1 = \frac{5\,(2)^2\,\overline{a}_z}{2\left[(2)^2 + (z-1)^2\right]^{3/2}}$$

For the current of 5A in $\overline{a}_\phi$ direction at $\rho = 2\text{m}$, $z = -1\text{m}$,

$$\overline{H}_2 = \frac{5\,(2)^2\,\overline{a}_z}{2\left[\,(2)^2 + (z+1)^2\right]^{3/2}}$$

$$\therefore\ \overline{H} = \overline{H}_1 + \overline{H}_2$$

$$= 10\,\overline{a}_z\left\{\frac{1}{\left[4 + (z-1)^2\right]^{3/2}} + \frac{1}{\left[4 + (z+1)^2\right]^{3/2}}\right\}$$

Plot of $|\overline{H}|$ vs. z is not shown.

Problem 7.21 : A surface current, $\overline{K} = 10\,\overline{a}_z$ A/m, flows on that portion of the plane y = 4 in free space bounded by x = ± ∞ , z = ± k. Find $\overline{H}$ at the origin if k = : (a) 3 ; (b) 30 ; (c) ∞ .

Solution :

$$d\overline{H} = \frac{\overline{K}\,ds \times \overline{a}_{12}}{4\pi d^2}$$

$\overline{K} = 10\,\overline{a}_z$ A/m lies on y = 4 plane. To the plane $\overline{a}_y$ is normal.

Hence $\overline{ds} = dx\,dz\,\overline{a}_y$ or $ds = dx\,dz$.

Point 2 is that point at which $\overline{H}$ is desired = (0, 0, 0

Point 1 is the general point on the surface current density = (x, 4, z)

$$\overline{R}_{12} = -x\,\overline{a}_x - 4\,\overline{a}_y - z\,\overline{a}_z$$

$$d = |\overline{R}_{12}| = \sqrt{[x^2 + z^2 + 16}$$

$$\overline{a}_{12} = \frac{\overline{R}_{12}}{|\overline{R}_{12}|} = \frac{-x\,\overline{a}_x - 4\,\overline{a}_y - z\,\overline{a}_z}{\sqrt{x^2 + z^2 + 16}}$$

$$\therefore\ \overline{dH} = \frac{(10\,\overline{a}_z\,dx\,dy) \times (-x\,\overline{a}_x - 4\,\overline{a}_y - z\,\overline{a}_z)}{4\pi\,(x^2 + z^2 + 16)^{3/2}}$$

$$= \frac{10\,dx\,dy\,(-x\,\overline{a}_y + 4\,\overline{a}_x)}{4\pi\,(x^2 + z^2 + 16)^{3/2}}$$

$$\frac{10}{4\pi}\left[\int_{-\infty}^{\infty}\int_{-k}^{k}\frac{-x\,\overline{a}_y\,dx\,dy}{(x^2 + z^2 + 16)^{3/2}} + \int_{-\infty}^{\infty}\int_{-k}^{k}\frac{4\,\overline{a}_x\,dx\,dy}{(x^2 + z^2 + 16)^{3/2}}\right.$$

$$\overline{H} = \frac{10}{4\pi}\left[0 + \int_{-k}^{k} \frac{8\,dy\,\overline{a}_x}{z^2+16}\right]$$

Using equation

$$\overline{H} = \frac{5}{2\pi}\left[\frac{8\,\overline{a}_x}{4}\left[\tan^{-1}\frac{z}{4}\right]_{-k}^{k}\right]$$

$$= \frac{10}{\pi}\tan^{-1}\left(\frac{k}{4}\right)\overline{a}_x$$

(a) If k = 3, then

$$\overline{H} = \frac{10}{\pi}\tan^{-1}\left(\frac{3}{4}\right)\overline{a}_x$$

$$= 2.048\,\overline{a}_x \text{ A/m}$$

(b) If k = 30, then

$$\overline{H} = \frac{10}{\pi}\tan^{-1}\left(\frac{30}{4}\right)\overline{a}_x = 4.578\,\overline{a}_x$$

(c) If k = ∞, then

$$\overline{H} = \frac{10}{\pi}\tan^{-1}(\infty)\,\overline{a}_x = 5\,\overline{a}_x$$

Problem 7.22 : Let 20A flow in a filament in the $\overline{a}_z$ direction on the z axis, let $3\,\overline{a}_z$ A/m flow on the infinite cylinder ρ = 2m, and let $-2.5\,\overline{a}_z$ A/m be at ρ = 3. Sketch H_ϕ vs. ρ, $0 < \rho < 5$.

Solution :

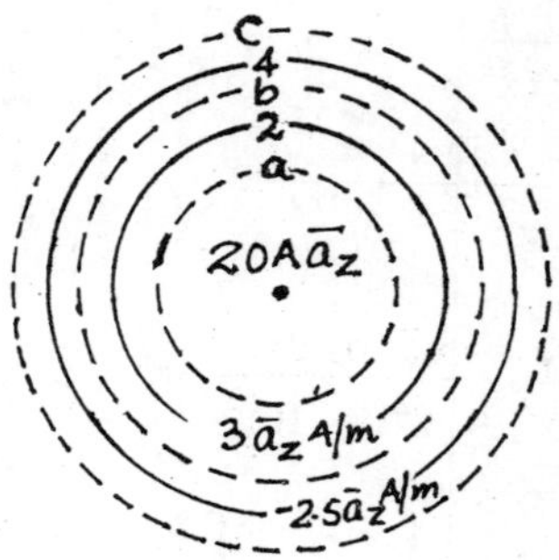

Fig.P. 7.22

Consider a circle of radius a which encloses the current of 20A only. Apply Ampere's circuital law to this closed path $\oint \overline{H} \cdot \overline{dL} = I_{encl}$

$$\text{LHS} = \oint \overline{H} \cdot \overline{dL}$$

As the current is directed in $\overline{a}_z$ direction, H_ϕ component of $\overline{H}$ is produced.

$$\therefore \text{ LHS} = \int_0^{2\pi} (H_\phi \ \overline{a}_\phi) \cdot (a \, d\phi \ \overline{a}_\phi) = \int_0^{2\pi} a \, H_\phi \, d\phi = 2\pi a H_\phi$$

$$\text{RHS} = I_{encl} = 20$$

$$\therefore \ 2\pi a H_\phi = 20 \text{ or } H_\phi = \frac{10}{\pi a}$$

$$\therefore \ \overline{H} = \frac{10}{\pi a} \overline{a}_\phi$$

Replacing a by ρ, $\overline{H} = \left(\frac{10}{\pi \rho} \right) \overline{a}_\phi$ A/m for $\rho < 2$m.

Consider a circle of radius b as shown in Fig.P.7.22 which encloses 20 A $\overline{a}_z$ and 3 $\overline{a}_z$ A/m. Apply Ampere's circuital law.

$$\oint \overline{H} \cdot \overline{dL} = I_{encl}$$

$$\therefore \text{ LHS} = \oint \overline{H} \cdot \overline{dL} = \int_0^{2\pi} (H_\phi \ \overline{a}_\phi) \cdot (b \, d\phi \ \overline{a}_\phi)$$

$$= \int_0^{2\pi} b \, H_\phi \, d\phi = 2\pi b H_\phi$$

$$\text{RHS} = I_{end.} = 20 + 3 \times \text{circumference of } \rho = 2\text{m circle}.$$

$$= 20 + 3 \times 2\pi \times 2 = 57.699 \text{ A}$$

$$\therefore \ 2\pi b H_\phi = 57.669,$$

$$\therefore \ H_\phi = \frac{9.183}{b} \text{ or } \overline{H} = \frac{9.183}{b} \overline{a}_\phi$$

Replacing b by ρ, $\overline{H} = \frac{9.183}{\rho} \overline{a}_\phi$ A/m for $2 < \rho < 4$

Consider a circle of radius C as shown in fig. which encloses 20 A $\overline{a}_z$,

$3\,\bar{a}_z$ A/m and $-\ 2.5\,\bar{a}_z$ A/m. Apply Ampere's circuital law to this circular path, $\oint \overline{H} \cdot \overline{dL} = I_{encl}$

$$\text{LHS} = \int_0^{2\pi} (H_\phi\,\bar{a}_\phi) \cdot (c\,d\phi\,\bar{a}_\phi) = \int_0^{2\pi} c\,H_\phi\,d\phi = 2\pi\,C\,H_\phi$$

$$\text{RHS} = I_{end} = 20 + 3 \times 2\pi \times 2 - 2.5 \times 2\pi \times 4 = -\ 5.13274\ \text{A}$$

$$\therefore\ 2\pi\,C\,H_\phi = -\ 5.13274\ ,\ H_\phi = -\frac{0.8169}{C}$$

or $\overline{H} = -\dfrac{0.8169}{C}\,\bar{a}_\phi$

Replacing C by ρ , $\overline{H} = -\dfrac{0.8169}{\rho}\,\bar{a}_\phi$ A/m for $\rho > 4$m .

Hence

$$\overline{H} = \begin{cases} \dfrac{10}{\pi\rho}\,\bar{a}_\phi & \rho < 2 \\ \dfrac{9.183}{\rho}\,\bar{a}_\phi & 2 < \rho < 4 \\ \dfrac{-\ 0.8169}{\rho}\,\bar{a}_\phi & \rho > 4 \end{cases}$$

Problem 7.23. : The regions, $0 < z < 0.1$m and $0.3 < z < 0.4$ m, are conducting slabs carrying uniform current densities of 10 A / m^2 in opposite directions, as shown in Fig.P.7.23(a). Find H_x at z = $-\ 0.04$, 0.06, 0.26, 0.36, and 0.46 m.

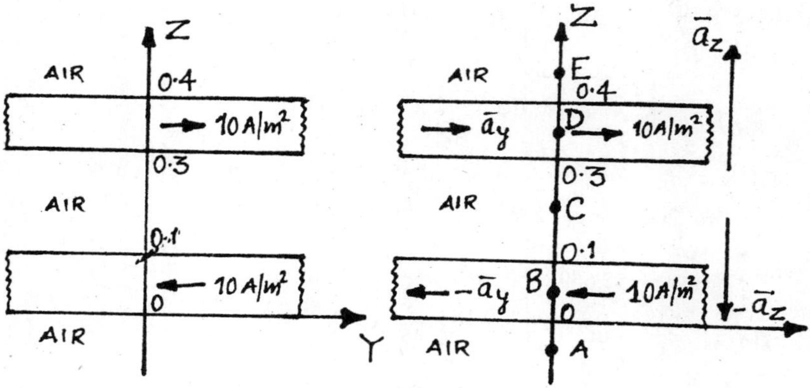

Fig.P. 7.23 (a) Fig.P. 7.23 (b)

Solution :

We will consider a plane of thickness dz located at any of value z, between $0 < z < 0.1$ and $0.3 < z < 0.4$. On this plane the surface current density will be J dz where J is the magnitude of current density. Using the formula

$$\overline{H} = \frac{\overline{K} \times \overline{a}_n}{2}$$ we will determine magnetic field intensity at any point such as A, B, C, D, and E. If a point lies below the surface current density then

$\overline{a}_N = -\overline{a}_z$ and $\overline{a}_N = \overline{a}_z$

if the point lies above the surface current density.

Consider a point such as A where z = – 0.06 then

$$d\overline{H} = \frac{(-10\,dz\,\overline{a}_y) \times (-\overline{a}_z)}{2} + \frac{(10\,dz\,\overline{a}_y) \times (-\overline{a}_z)}{2}$$

$$\therefore \quad \overline{H} = \int d\overline{H} = \int_0^{0.1} 5\,\overline{a}_x\,dz - \int_{0.3}^{0.4} 5\,\overline{a}_x\,dz$$

$$= 5\,\overline{a}_x \Big[z \Big]_0^{0.1} - 5\,\overline{a}_x \Big[z \Big]_{0.3}^{0.4} = 0\,\overline{a}_x$$

Consider a point such as B where z = 0.6 then

$$\overline{H} = \int_0^{0.06} \frac{(-10\,dz\,\overline{a}_y) \times (\overline{a}_z)}{2} + \int_{0.06}^{0.1} \frac{(-10\,dz\,\overline{a}_z) \times (-\overline{a}_z)}{2} + \int_{0.3}^{0.4} \frac{(10\,dz\,\overline{a}_y) \times (-\overline{a}_z)}{2}$$

$$= 5\,\overline{a}_x \Big[z \Big]_0^{0.06} + 5\,\overline{a}_x \Big[z \Big]_{0.06}^{0.1} - 5\,\overline{a}_x \Big[z \Big]_{0.3}^{0.4} = -0.6\,\overline{a}_x$$

Consider a point such as a D where z = 0.36 then

$$\overline{H} = \int_0^{0.1} \frac{(-10\,dz\,\overline{a}_y) \times (\overline{a}_z)}{2} + \int_{0.3}^{0.36} \frac{(10\,dz\,\overline{a}_y) \times (+\overline{a}_z)}{2} + \int_{0.36}^{0.4} \frac{(10\,dz\,\overline{a}_y) \times (-\overline{a}_z)}{2}$$

$$= 5\,\overline{a}_x \Big[z \Big]_0^{0.1} + 5\,\overline{a}_x \Big[z \Big]_{0.3}^{0.36} - 5\,\overline{a}_x \Big[z \Big]_{0.36}^{0.4} = -0.4\,\overline{a}_x \text{A/m}$$

Consider a point such as $\overline{E}$ where z = 0.46 m then

$$\overline{H} = \int_0^{0.1} \frac{(-\ 10\,dz\,\overline{a}_y) \times (\overline{a}_z)}{2} + \int_{0.3}^{0.4} \frac{(10\,dz\,\overline{a}_z) \times (-\ \overline{a}_z)}{2}$$

$$= -\ 5\,\overline{a}_x \Big[\ z\ \Big]_0^{0.1} + 5\,\overline{a}_x \Big[\ z\ \Big]_{0.3}^{0.4} = 0$$

Problem 7.23 : A hollow cylindrical shell of radius a is centered on the z axis and carries a uniform surface current density of $K_a\,\overline{a}_\phi$. (a) Show that H is not a function of ϕ or z. (b) Show that H_ϕ and H_ρ are everywhere zero. (c) Show that $H_z = 0$ for $\rho > a$. (d) Whow that $H_z = K_a$ for $\rho < a$. (e) A second shell, $\rho = b$, carries a current $K_b\,\overline{a}_\phi$. Find $\overline{H}$ everywhere.

Solution :

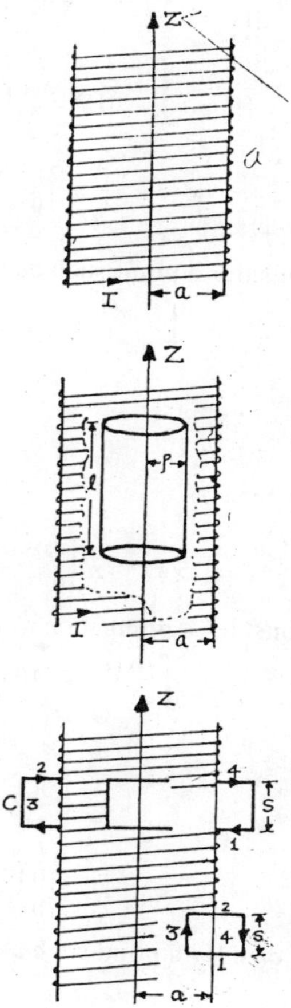

As the current is directed in $\overline{a}_\phi$ direction, ϕ component of magnetic field intensity is absent.Hence $\overline{H} = H_\rho\,\overline{a}_\rho + H_z\,\overline{a}_z$. Further H_ρ and H_z are independent of z as the solenoid is infinitely long in the z direction. If we move on a circle of radius ρ (which may greater or smaller than radius of solenoid), the solenoid appears to be same. Hence H_ρ and H_z are also independent of ϕ . Now we will prove that H_ρ component of magnetic field intensity is zero every where. Consider an axial cylinder of length l and radius ρ, either larger than the solenoid radius (outside to the solenoid) or smaller (inside to the solenoid) as shown in fig. 7.6.

Now $\oint \overline{B} \cdot dS = 0$, i.e. surface integral of magnetic flux density over a closed surface is always zero. As there are three surfaces to the cylinder, we have

$$\oint \overline{B} \cdot \overline{dS} = \int_{\text{circular surface}} \overline{B} \text{ cdod } \overline{dS} + \int_{\text{bottom surface}} \overline{B} \cdot \overline{dS} + \int_{\text{top surface}} \overline{B} \cdot \overline{dS} = 0$$

For the circular surface of radius ρ , $\overline{dS} = \rho \, d\phi \, dz \, \overline{a}_\rho$ for the bottom surface, $\overline{dS} = -\rho \, d\rho \, d\phi \, \overline{a}_z$ and for the top surface, $\overline{dS} = \rho \, d\rho \, d\phi \, \overline{a}_z$. Hence

$$\oint \overline{B} \cdot \overline{dS} = \int_{\text{circular}} (B_\rho \, \overline{a}_\rho) \cdot (\rho \, d\phi \, dz \, \overline{a}_\rho) + \int_{\text{bottom}} (B_z \, \overline{a}_z)(-\rho \, d\rho \, d\phi \, \overline{a}_z)$$

$$+ \int_{\text{top}} (B_z \, \overline{a}_z) \cdot (\rho \, d\phi \, \overline{a}_z) = 0$$

$$= \int_{\text{circular}} B_\rho \, \rho \, d\phi \, dz - \int_{\text{bottom}} B_z \, \rho \, d\rho \, d\phi + \int_{\text{top}} B = \rho \, d\rho \, d\phi = 0$$

As H_ρ and H_z are independent of z and ϕ , B_ρ and B_z will be also independent of z and φ . Further the area of top and bottom surface is same. Hence the last two integrals they get cancelled out. Hence

$$\int_{\text{circular}} B_\rho \text{ rgo } d\phi \, dz = 0 \quad \text{or} \quad B_\rho = 0$$

As $B_\rho = 0$, H_ρ must be equal to zero. Hence

$$\overline{H} = H_z \, \overline{a}_z$$

Now we will prove that $\overline{H}$ is independent of ρ . $\overline{\nabla} \times \overline{H} = \overline{J}$. As the current density is present inside the wire and not outside the wire, we must have $\nabla \times \overline{H}$ equal to zero everywhere except inside the wire.

$$\overline{\nabla} \times \overline{H} = \begin{vmatrix} \overline{a}_\rho/\rho & \overline{a}_\phi & \overline{a}_z/\rho \\ \partial/\partial_\rho & \partial/\partial_\phi & \partial/\partial_z \\ o & 0 & H_z \end{vmatrix} = 0$$

$$= \frac{\overline{a}_\rho}{\rho} \left[\frac{\partial}{\partial_\phi} (H_z) \right] - \overline{a}_\phi \left[\frac{\partial}{\partial_\rho} H_z \right] = 0$$

As H_z is independent of φ , $\dfrac{\partial H_z}{\partial \phi} = 0$

$$\therefore \quad \overline{\nabla} \times \overline{H} = -\overline{a}_\phi \frac{\partial H_z}{\partial_\rho} = 0$$

Hence $\dfrac{dH_z}{\partial_\rho} = 0$, i.e. H_z is independent of ρ also

In short, $\overline{H} = H_z \overline{a}_z$ and H_z is not a function of ρ , ϕ and z. That is H_z is a constant. Now we will prove that H_z is zero outside the solenoid. For this purpose consider path C as shown in fig.7.6. Apply Ampere's circuital law to this closed path C. As the current coming out of this closed path is zero, the line integral of $\overline{H}$ around this closed path must be equal to zero. As the current coming out of this closed path is zero, the line integral of $\overline{H}$ around this closed path must be equal to zero.

$$\oint_C \overline{H} \cdot \overline{dL} = \int_1 \overline{H} \cdot \overline{dL} + \int_2 \overline{H} \cdot \overline{dL} + \int_3 \overline{} \cdot \overline{dL} + \int_4 \overline{H} \cdot \overline{dL}$$

As H_ρ is equal to zero $\int \overline{H} \cdot \overline{dL}$ along the sides 1 and 2 must be equal to zero. Hence $\int \overline{H} \cdot \overline{dL}$ for sides 3 and 4 must cancel. But sides 3 and 4 can each be situated at any distance from the sloenoid, and $\overline{H}_z$ must therefore be either zero or constant along them.

Since the lines of $\overline{B}$ have neither beginning nor end, and since the total return flux in the infinite space outside must equal the finite flux inside the solenoid, B_z or H_z must be zero outside.

To determine H_z inside the solenoid, consider a closed path such as d.

$$\oint_d \overline{H} \cdot \overline{dL} = \int_1 \overline{H} \cdot \overline{dL} + \int_2 \overline{H} \cdot \overline{dL}$$

$$+ \int_3 \overline{} \cdot \overline{dL} + \int_4 \overline{H} \cdot \overline{dL} = I_{end}$$

Now $\int \overline{H} \cdot \overline{dL}$ along side 1 and 2 is zero as $H_\rho = 0$.

Also $\int \overline{H} \cdot \overline{dL}$ along side 4 is zero as $H_z = 0$ outside the solenoid. Hence

$$\oint_d \overline{I} \cdot \overline{dL} = \int_3 \overline{H} \cdot \overline{dL} = \int_3 (H_z \overline{a}_z) \cdot (dz\, \overline{a}_z)$$

$$= \int_3 H_z\, dz = H_z S$$

$I_{end.} = K_a S$. Hence

$H_z S = K_a S$ or $H_z = K_a$ or

Fig shows both the shells. Using the property of solenoid that it produces $\overline{H}$ inside the solenoid we have

$$\overline{H} = \begin{cases} K_a + K_b & \text{for } \rho < a \\ K_b & \text{for } a < \rho < b \\ 0 & \text{for } \rho > b \end{cases}$$

Problem 7.24 : One toroid of rectangular cross section is defined by the cylinders $\rho = 5.5$ cm and $\rho = 10.5$ cm , and the planes $z = \pm 5$ cm . A second toroid is defined by the surfaces $\rho = 6.5$ cm and 9.5 cm , and $z = \pm 1$ cm . Each toroid carries a surface current density of $-20\bar{a}_z$ A/m on its outer surface ($\rho = 9.5$ and 10.5 cm). Find $\overline{H}$ at : (a) $P_A(0, 5\text{cm}, 0)$; (b) $P_B(0, 6, 0)$; (c) $P_C(0, 7, 0)$; (d) $P_D(0, 7, 2)$; (e) $P_E(0, 10, 0)$.

Solution :

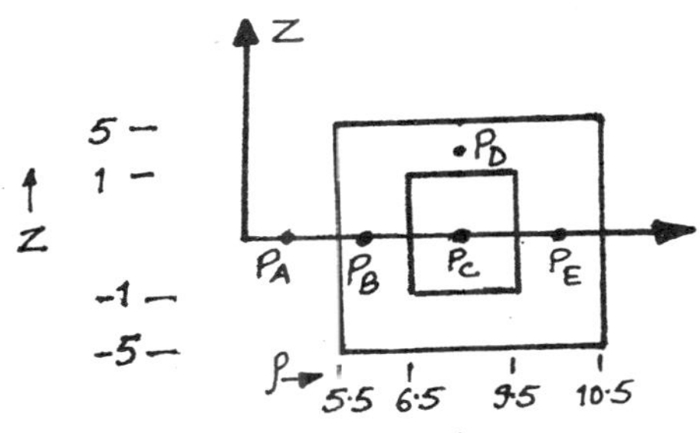

Fig. P 7.24.

Fig. 7.24. shows both the toroids in cross section. Let us first determine the surface current density on the inner surface of both the toroids. Consider the toroid defined by the cylinders $\rho = 5.5$ cm and $\rho = 10.5$ cm . As the current on inner surface is equal to curretn on outer surface, we have

$$K_i \times 2\pi(5.5) = K_o \times 2\pi(10.5)$$

where K_i on d K_o are surface current densities on the inner and outer surface.

$$\therefore \; K_i = \frac{20 \times 2\pi \times 10.5}{2\pi \times 5.5} = 38.18\dot{2} \text{ A/m}$$

As the direction of K_i and K_o must be opposite to each other, we have $\overline{K}$ (on ρ = 5.5 cm) = 38.182 $\overline{a}_z$ A/m.

Consider the toroid formed by the cylinders ρ = 6.5 cm and ρ = 9.5. On the similar basis, we have

$$K_i \times 2\pi \times 6.5 = K_o \times 2\pi \times 9.5$$

$$\therefore \quad K_i = \frac{20 \times 2\pi \times 9.5}{2\pi \times 6.5} = 29.231$$

Hence $\overline{K}$ (on ρ = 6.5 cm) = 29.231 $\overline{a}_z$ A/m.

(a) P_A (0, 5cm, 0). This point lies outside to both the toroids. Hence $\overline{H}$ = 0.

(b) P_B (0, 6cm, 0). This point lies outside to second toroid. Hence

$$\overline{H} = 38.182 \times \frac{5.5}{6}\,\overline{a}_\phi = 35\,\overline{a}_\phi \text{ A/m}.$$

(c) P_C (0, 7cm, 0). This point lies inside to both the toroids. Hence

$$\overline{H} = \left(38.182 \times \frac{5.5}{7} + 29.231 \times \frac{6.5}{7}\right)\overline{a}_\phi$$

$$= 57.143\,\overline{a}_\phi \text{ A/m}.$$

(d) P_D (0, 7cm, 2cm). This point lies outside to the second toroid. Hence

$$\overline{H} = 38.182 \times \frac{5.5}{7}\,\overline{a}_\phi = 30\,\overline{a}_\phi \text{ A/m}.$$

(e) P_E (0, 10cm, 0). This point lies outside to the second toroid. Hence

$$\overline{H} = 38.182 \times \frac{5.5}{10}\,\overline{a}_\phi = 21\,\overline{a}_\phi \text{ A/m}.$$

Problem 7.25 : The planes z = 0 and z = 0.1 m are perfect conductors A voltage of 2 V is maintained between them such that $\overline{E}$ is in the $\overline{a}_z$ direction. For $0 < z < 0.1$ m there is a conducting material for which $\sigma = 1000\,e^{-100\rho}$ mho/m. (a) Find $\overline{E}$ and $\overline{J}$. (b) Find I_{total} in a cylinder of radius ρ. (c) Use Ampere's circuital law to find $\overline{H}$. (d) Sketch $|\overline{H}|$ vs ρ, $0 < z < 0.05$ m.

Solution :

(a) $$\overline{E} = \frac{2}{\text{distance between planes}}\,\overline{a}_z$$

$$= \frac{2}{0.1}\,\overline{a}_z = 20\,\overline{a}_z \text{ V/m}.$$

$$\overline{J} = \sigma \overline{E} = 1000\,e^{-100\rho}\,(20\,\overline{a}_z)$$

$$= 20000\,e^{-100\rho}\,\overline{a}_z \text{ A/m}^2$$

(b) $\overline{dS} = \rho\,d\rho\,d\phi\,\overline{a}_z$.

$$\therefore \quad I_{total} = \int \overline{J}\cdot\overline{dS} = \int_0^{\rho}\int_0^{2\pi} 20000\,e^{-100\rho}\,\rho\,d\rho\,d\phi$$

$$= 20000\left[\frac{-\rho\,e^{-100\rho}}{100} - \frac{e^{-100\rho}}{10000}\right]_0^{\rho}\left[\,\phi\,\right]_0^{2\pi}$$

$$= 4\pi\left[-100\,\rho\,e^{-100\rho} - e^{-100\rho} + 1\right]$$

(c) $$\oint \overline{H}\cdot\overline{dL} = I_{encl}$$

As the current is in $\overline{a}_z$ direction, H_ϕ component of magnetic field intensity is produced. Hence

$$\text{LHS} = \oint \overline{H}\cdot\overline{dL} = \int_0^{2\pi} H_\phi\,\overline{a}_\phi\cdot\rho\,d\phi\,\overline{a}_\phi$$

$$= \int_0^{2\pi} \rho\,H_\phi\,d\phi = 2\pi\rho H_\phi$$

$$\text{RHS} = I_{total} = 4\pi\left[-100\,\rho\,e^{-100\rho} - e^{-100\rho} + 1\right]$$

$$\therefore \quad 2\pi\rho H_\phi = 4\pi\left[-100\,\rho\,e^{-100\rho} - e^{-100\rho} + 1\right]$$

$$\therefore \quad H_\phi = 2\left[-100\,e^{-100\rho} - \frac{e^{-100\rho}}{\rho} + \frac{1}{\rho}\right]$$

$$\therefore \quad \overline{H} = 2\left[-100\,e^{-100\rho} - \frac{e^{-100\rho}}{\rho} + \frac{1}{\rho}\right]\overline{a}_\phi$$

(d) $|\overline{H}| = 2\left[-100\,e^{-100\rho} - \frac{e^{-100\rho}}{\rho} + \frac{1}{\rho}\right]$

Plot $|\overline{H}|$ on the y axis and ρ on the x axis. Plot is not given.

Problem 7.26 : A solid conductor has a circular cross section, radius 1 mm, centered on the z axis, and carries a uniformly distributed total current of 10π A in the $\overline{a}_z$ direction. A current sheet $\overline{K} = -1000\,\overline{a}$ A/m lies at $\rho = 2$ mm . Find $\overline{H}$ in the region : (a) $0 \le \rho < 1$ mm ; (b) $1 < \rho < 2$ mm ; (c) $2 < \rho < 4$ mm . (d) What value of current sheet should be located at $\rho = 4$ mm so that $\overline{H} = 0$ for $\rho > 4$ mm ?

Solution :

(a) As the currents are directed in $\pm\,\overline{a}_z$ direction, H_ϕ component of magnetic field intensity is present. Consider a circular path of radius 'a' where $a < 1$mm. Apply Ampere's circuital law to this circular path.

$$\oint \overline{H} \cdot \overline{dL} = I_{encl}$$

$$\text{L.H.S.} = \oint \overline{H} \cdot \overline{dL}$$

$$= \int_0^{2\pi} (H_\phi\,\overline{a}_\phi) \cdot (\rho\, d\phi\,\overline{a}_\phi)\,\Big|_{\rho = a} = \int_0^{2\pi} \rho\, H_\phi\, d\phi\,\Big|_{\rho = a}$$

$$= 2\pi\, a\, H_\phi$$

$$\text{R.H.S.} = I_{encl}.$$

The current of 10π A is uniformly distributed through a circle of radius 1mm. Hence

$$J = \frac{I}{\text{Area}} = \frac{10\pi}{\pi\,(10^{-3})^2} = 10^7\ \text{A/m}^2$$

$$\therefore\ \text{RHS} = I_{encl} = J \times \text{area end}.$$

$$= 10^7 \times \pi a^2 = \pi a^2\, 10^7$$

$\therefore \quad 2\pi a H_\phi = \pi a^2 10^7 \qquad \therefore \quad H_\phi = 0.5 \times 10^7 a$

$\therefore \quad \overline{H} = 0.5 \times 10^7 a \, \overline{a}_\phi$ where $a < 1mm$

Replacing a by ρ, we have

$$\overline{H} = 0.5 \times 10^7 \rho \, \overline{a}_\phi \text{ A/m for } \rho < 1mm.$$

(b) Consider a circle of radius 'b' where $1mm < b < 2mm$.

Apply Ampere's circuital law to this circular path.

$$\oint \overline{H} \cdot \overline{dL} = I_{encl}$$

$$\text{L.H.S.} = \oint \overline{H} \cdot \overline{dL}$$

$$= \int_0^{2\pi} (H_\phi \, \overline{a}_\phi) \cdot (\rho \, d\phi \, \overline{a}_\phi) \Big|_{\rho = b} = \int_0^{2\pi} \rho H_\phi \, d\phi \Big|_{\rho = b}$$

$$= 2\pi b H_\phi$$

$$\text{R.H.S.} = I_{end.} = 10\pi \text{ A}$$

$$\therefore \quad 2\pi b H_\phi = 10\pi$$

or $H_\phi = \dfrac{5}{b}$

or $\overline{H} = \dfrac{5}{b} \overline{a}_\phi$

where $1mm < b < 2mm$.

Replacing b by ρ, we have

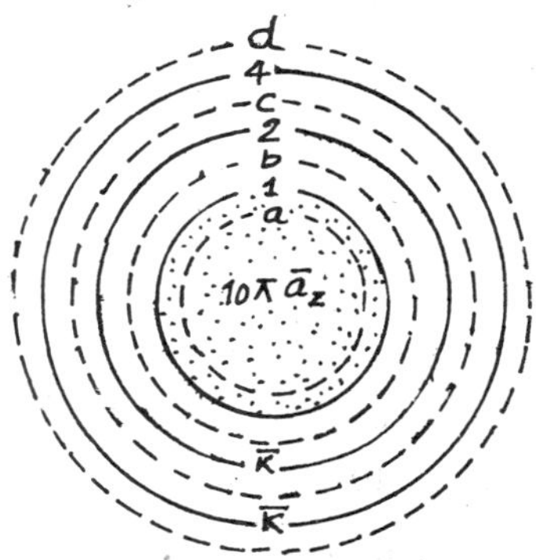

Fig. P. 7.26.

$$\overline{H} = \frac{5}{\rho} \overline{a}_\phi \text{ A/m} \quad 1mm < \rho < 2mm.$$

(c) Considers a circle of radius 4m where $2mm < c < 4$ mm.

Apply Ampere's circuital law to this circular path.

$$\oint \overline{H} \cdot \overline{dL} = I_{encl}$$

$$\text{L.H.S.} = \oint \overline{H} \cdot \overline{dL}$$

$$= \int_0^{2\pi} (H_\phi \, \bar{a}_\phi) \cdot (\rho \, d\phi \, \bar{a}_\phi) \Big|_{\rho = c} = \int_0^{2\pi} \rho \, H_\phi \, d\phi \Big|_{\rho = c}$$

$$= 2\pi c H_\phi$$

R.H.S. = $I_{end.}$

The current of 10π A is in $\bar{a}_z$ direction where asd $\bar{K} = -1000\,\bar{a}_z$ at $\rho = 2$mm is in $-\bar{a}_z$ direction. The current at $\rho = 2$mm is equal to $1000 \times$ circumference of $\rho = 2$mm circle

$$= 1000 \times 2\pi(0.002) = 4\pi \text{ A directed in } -\bar{a}_z \text{ direction}$$

$\therefore$ RHS $= I_{encl} = 10\pi - 4\pi = 6\pi$ A.

$\therefore \quad 2\pi c H_\phi = 6\pi$

$\therefore \quad H_\phi = \frac{3}{c}$

$\therefore \quad \bar{H} = \frac{3}{c}\,\bar{a}_\phi$ for $2\text{mm} < c < 4\text{mm}$

Replacing c by ρ we have

$$\bar{H} = \frac{3}{\rho}\,\bar{a}_\phi \text{ for } 2\text{mm} < \rho < 4\text{mm}$$

(d) If $\bar{H} = 0$ for $\rho > 4$mm , then I_{end} must be equal to zero for $\rho > 4$mm .

Let $\bar{K}$ be the current sheet located at $\rho = 4$mm .

$I_{end} = 10\pi - 4\pi + K \times$ circumference of $\rho = 4$mm circle

$\therefore \quad 0 = 10\pi - 4\pi + K \times 2\pi(0.004)$

$\therefore \quad K = -750$

Hence $\bar{K} = -750\,\bar{a}_z$ A/m must be present at $\rho = 4$mm so that $\bar{H} = 0$ for $\rho > 4$mm .

Problem 7.27 : Let $\bar{J} = 0$ for $|z| > 0.2$m, and $\bar{J} = 20 \sin 5\pi z\,\bar{a}_y$ A/m^2 for $-0.2 < z < 0.2$ m. Use Ampere's law and symmetry to find $\bar{H}$ everywhere.

Solution :

The current is flowing in $\bar{a}_y$ direction as $\bar{J}$ as $\bar{a}_y$.

Consider a small cross sectional area as shown in fig.P. 7.27 (a) of dimensions dx and dz. The area will be dx dz. The current through this area is $\overline{I} = \overline{J}\,dx\,dz = 20 \sin(5\pi z)\,dx\,dz\,\overline{a}_z$. If we assume that dz is very small so that the entire current will flow on a surface (plane) located by z coordinate, then it will be a surface current density.

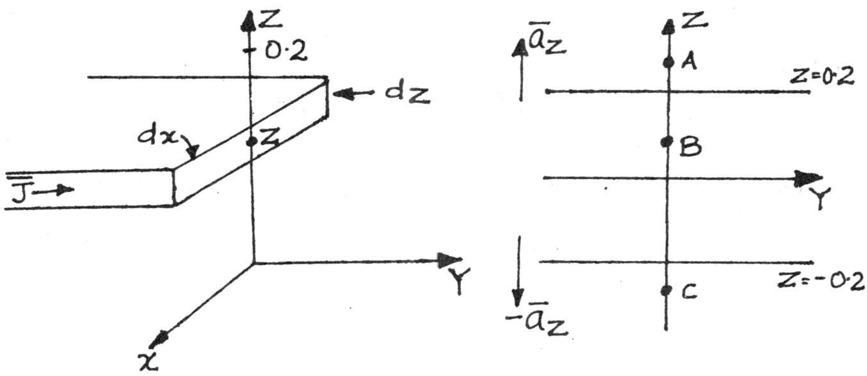

(a) (b)

Fig. P. 7.27

Hence

$$\overline{K} = \frac{\overline{I}}{dx} = 20 \sin(5\pi z)\,dz\,\overline{a}_y \text{ A/m}.$$

Consider point A for which z > 0.2 as shown in fig.P. 7.26 (b)

$$d\overline{H} = \frac{\overline{K} \times \overline{a}_n}{2}$$

$$= \frac{20 \sin(5\pi z)\,dz\,\overline{a}_y \times \overline{a}_z}{2}$$

$$= 10 \sin(5\pi z)\,dz\,\overline{a}_x$$

$$\therefore \quad \overline{H} = \int_{-0.2}^{0.2} 10 \sin(5\pi z)\,dz\,\overline{a}_x$$

$$= \frac{-10}{5\pi}\Big[-\cos(5\pi z)\Big]_{-0.2}^{0.2}\,\overline{a}_x$$

$$= \frac{-10}{5\pi}\left[\cos(\pi) - \cos(-\pi)\right]\bar{a}_x = 0$$

$\therefore$ $\bar{H} = 0$ for $z > 0.2$

Consider point B for which $-0.2 < z < 0.2$

$$\bar{H} = \int_{-0.2}^{z} 10 \sin(5\pi z)\, dz\, \bar{a}_x - \int_{z}^{0.2} 10\sin(5\pi z)\, dz\, \bar{a}_x$$

There will be two differential magnetic field intensity at point B. For the surface current density between –0.2, and z, $\bar{a}_N = \bar{a}_z$ so that $d\bar{H}$ will have $\bar{a}_x$ where as for the surface current density between z and 0.2, $\bar{a}_N = -\bar{a}_z$ so that $d\bar{H}$ will have $-\bar{a}_x$.

$$\therefore \bar{H} = \frac{-10}{5\pi}\left[\cos(5\pi z)\right]_{-0.2}^{z}\bar{a}_x + \frac{10}{5\pi}\left[\cos(5\pi z)\right]_{z}^{0.2}\bar{a}_x$$

$$= \frac{-10}{5\pi}\left[\cos 5\pi z - \cos(-\pi)\right]\bar{a}_x + \frac{10}{5\pi}\left[\cos(\pi) - \cos(5\pi z)\right]\bar{a}_x$$

$$= \frac{10}{5\pi}\bar{a}_x\left[-\cos 5\pi z - 1 - 1 - \cos 5\pi z\right]$$

$$= \frac{4}{\pi}(1 + \cos 5\pi z)\,\bar{a}_x \quad \text{for } -0.2 < z < 0.2$$

Consider point c for which $z < -0.2$

$$\bar{H} = \int_{-0.2}^{0.2} -10\sin(5\pi z)\, dz\, \bar{a}_x$$

$$= \frac{+10}{5\pi}\left[\cos(5\pi z)\right]_{-0.2}^{0.2}\bar{a}_x = 0 \text{ for } z < -0.2$$

Hence

$$\bar{H} = \begin{cases} 0 & z > 0.2 \\ \frac{4}{\pi}(1 + \cos 5\pi z)\,\bar{a}_x & -0.2 < z < 0.2 \\ 0 & z < -0.2 \end{cases}$$

Problem 7.28 : The cylindrical surface $\rho = $ 20mm carries the current $\overline{K} = 100\,\bar{a}_z$ A/m , while the surface $\rho = $ 40 mm has the solenoidal current $\overline{K} = 80\,\bar{a}_\phi$ A/m . Calculate $|\overline{H}|$ at $\rho = $:(a) 10 mm ; (b) 30 mm ; (c) 50 mm.

Solution :

(a) The cylinder $\rho = $ 10mm do not encloses the cylinder of radius 20 mm. Hence $\overline{K} = 100\,\bar{a}_z$ will not contribute in $\overline{H}$

But due to solenoid

$(\rho = 40\text{mm}, \overline{K} = 80\,\bar{a}_\phi)$

there will be magnetic field intensity.

Using equation, we have

$$\overline{H} = 80\,\bar{a}_z \text{ A/m}$$

$$|\overline{H}| = 80 \text{ A/m}$$

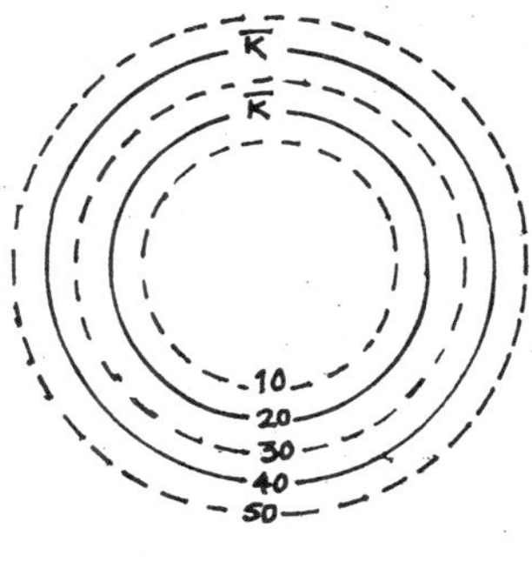

Fig. P. 7.28.

(b) The cylinder $\rho = $ 30 mm encloses the cylinder of radius 20 mm. Hence $\overline{K} = 100\,\bar{a}_z$ will contribute in magnetic field intensity. Apply Ampere circuital law to the circular path $\rho = $ 30mm . As $\overline{K}$ is directed in $\bar{a}_7$ direction, H_ϕ component of magnetic field intensity is produced.

$$\oint \overline{H} \cdot \overline{dL} = I_{encl}$$

$$\text{LHS} = \oint \overline{H} \cdot \overline{dL} = \int_0^{2\pi} (H_\phi\,\bar{a}_\phi) \cdot (\rho\, d\phi\,\bar{a}_\phi)$$

$$= 2\pi\rho H_\phi$$

RHS = I_{encl} = K × circumference of circle with radius 20mm.

$$= 100 \times 2\pi(20 \times 10^{-3}) = 4\pi \text{ A}$$

$$\therefore \quad 2\pi\rho H_\phi = 4\pi \quad \text{or} \quad H_\phi = \frac{2}{\rho}$$

Hence at $\rho = 30mm$,

$$\overline{H} = \frac{2}{30 \times 10^{-3}} \bar{a}_\phi = 66.667\, \bar{a}_\phi$$

Also due to solenoid, $\overline{H} = 80\, \bar{a}_z$ A/m .

Hence $\overline{H}_{total}\ (\rho = 30\ mm) = 66.667\, \bar{a}_\phi + 80\, \bar{a}_z$ A/m

$$|\,H\,| = \sqrt{(66.667)^2 + (80)^2} = 104.137\ A/m\ .$$

(c) The cylinder of radius 50mm lies outside to the solenoid. The property of the solenoid is to produce $\overline{H}$ inside the solenoid and not outside the solenoid. Hence the contribution from the solenoid to $\overline{H}$ will be zero. Using the previous result of part (b), we have

$$\overline{H} = \frac{2}{\rho} \bar{a}_\phi = \frac{2}{50 \times 10^{-3}} \bar{a}_\phi = 40\, \bar{a}_\phi \ \ A/m$$

$$|\,\overline{H}\,| = 40$$

Problem 7.29 : If $\overline{F} = x^2 y\, \bar{a}_x - 2z\, \bar{a}_y + (3z^2 + xy)\, \bar{a}_z$, find $\overline{\nabla} \times [\overline{\nabla}(\overline{\nabla} \cdot \overline{F})]$.

Solution :

$$\overline{\nabla} \cdot \overline{F} = \frac{\partial F_x}{\partial x} + \frac{\partial F_y}{\partial y} + \frac{\partial F_z}{\partial z}$$

$$= \frac{\partial}{\partial x}(x^2 y) + \frac{\partial}{\partial y}(-2z) + \frac{\partial}{\partial z}(3z^2 + xy)$$

$$= 2xy + 6z$$

$$\overline{\nabla}(\overline{\nabla} \cdot \overline{F}) = \frac{\partial}{\partial x}(2xy + 6x)\, \bar{a}_x + \frac{\partial}{\partial y}(2xy + 6z)\, \bar{a}_y + \frac{\partial}{\partial z}(2xy + 6z)\, \bar{a}_z$$

$$= 2y\, \bar{a}_x + 2x\, \bar{a}_y + 6\, \bar{a}_z$$

$$\overline{\nabla} \times [\overline{\nabla}(\overline{\nabla} \cdot \overline{F})] = \begin{vmatrix} \bar{a}_x & \bar{a}_y & \bar{a}_z \\ \partial/\partial x & \partial/\partial y & \partial/\partial z \\ 2y & 2x & 6 \end{vmatrix}$$

$$= \bar{a}_z\, [2 - 2] = 0$$

Problem 7.30 : Given $\overline{H} = y^2 z \,\overline{a}_x + 2(x+1) yz \,\overline{a}_y - (x+1) z^2 \,\overline{a}_z$: (a) find $\oint \overline{H} \cdot \overline{dL}$ around the square path going from P (0, 2, 0) to A (0, 2, + b, 0) to B (0, 2 + b, b) to C (0, 2, b) to P ; (b) evaluate $\oint \overline{H} \cdot \overline{dL}$ for b = 0.1 ; (c) Find $\overline{\nabla} \times \overline{H}$; (d) evaluate $(\overline{\nabla} \times \overline{H})_x$ at P ; (e) show that

$$(\overline{\nabla} \times \overline{H})_x = \frac{\oint \overline{H} \cdot \overline{dL}}{\Delta S}.$$

Solution :

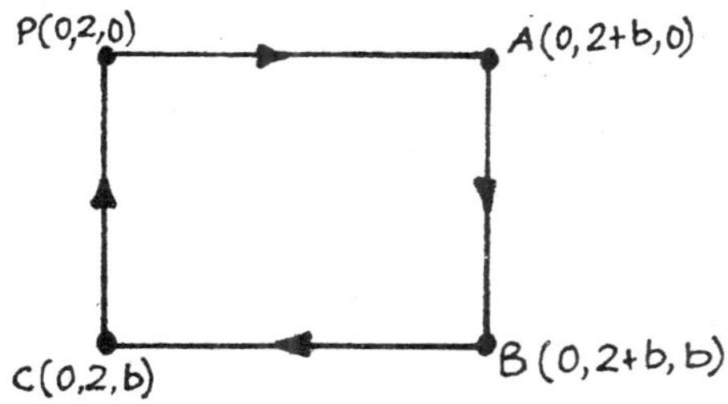

Fig.P. 7.30.

(a) The path P–A–B–C–P lies in x = 0 plane as the x coordinate of all these points is zero.

$$\oint \overline{H} \cdot \overline{dL} = \int_P^A \overline{H} \cdot \overline{dL} + \int_A^B \overline{H} \cdot \overline{dL} + \int_B^C \overline{H} \cdot \overline{dL} + \int_C^P \overline{H} \cdot \overline{dL}$$

Between the points P – A and B –C, only y coordinate changes. Hence $\overline{dL} = dy\,\overline{a}_y$ where as between the points A – B and C – P only z coordinates. Hence $\overline{dL} = dz\,\overline{a}_z$

$$\oint \overline{H} \cdot \overline{dL} = \int_P^A \overline{H} \cdot (dy\,\overline{a}_y) + \int_A^B \overline{H} \cdot (dz\,\overline{a}_z)$$

$$+ \int_{B}^{C} \overline{H} \cdot (dy\, \bar{a}_y) + \int_{C}^{P} \overline{H} \cdot (dz\, \bar{a}_z)$$

$$= \int_{2}^{2+b} 2(x+1)\, yz\, dy \Bigg|_{\substack{x=0\\ z=2}} + \int_{0}^{b} -(x+1)\, z^2\, dz \Bigg|_{\substack{x=0\\ y=2+b}}$$

$$+ \int_{2+b}^{2} 2(x+1)\, yz\, dy \Bigg|_{\substack{x=0\\ z=b}} + \int_{b}^{0} -(x+1)\, z^2\, dz \Bigg|_{\substack{x=0\\ y=2}}$$

$$= 0 - \left[\frac{z^3}{3}\right]_0^b + 2b\left[\frac{y^2}{2}\right]_{2+b}^{2} - \left[\frac{z^3}{3}\right]_b^0$$

$$= -\frac{b^3}{3} + b\,[4 - (2+b)^2] + \frac{b^3}{3}$$

$$= b\,[-4b - b^2] = -4b^2 - b^3$$

(b) $\oint \overline{H} \cdot \overline{dL} \Big|_{b=0.1} = -4(0.1)^2 - (0.1)^3 = -0.041 \text{ A}$

(c) $$\overline{\nabla} \times \overline{H} = \begin{vmatrix} \bar{a}_x & \bar{a}_y & \bar{a}_z \\ \partial/\partial x & \partial/\partial y & \partial/\partial z \\ y^2 z & 2(x+1)\, yz & -(x+1)\, z^2 \end{vmatrix}$$

$$= -2(x+1)\, y\, \bar{a}_x + (y^2 + z^2)\, \bar{a}_y \quad \text{A/m}^2$$

(d) $(\overline{\nabla} \times \overline{H})_x = -2(x+1)\, y$

A at point P (0, 2, 0), $(\overline{\nabla} \times \overline{H})_x = -2(0+1)\,2 = -4 \text{ A/m}^2$

(e) Area enclosed by the closed path

P – A – B – C – P = d (PA) × d (AB) = b × b = b^2

$$\therefore \frac{\oint \overline{H} \cdot \overline{dL}}{\Delta S} = \frac{-4b^2 - b^3}{b^2} = -4 - b$$

If b = 0.1 , then $\dfrac{\oint \overline{H} \cdot \overline{dL}}{\Delta S} = -4 - 0.1 = -4.1$

Hence $-4 = -4.1$

Problem 7.31 : A nonmagnetic wire of circular cross section has a radius of 10 mm and a length of 100 m. It carries a total current of 1000 A in the $\overline{a}_z$ direction and has a potential difference of 10 V between ends. Let $\overline{H} = k\rho^2 \overline{a}_\phi$, where k is an unknown constant. (a) Find k. (b) Give σ as a function of the radius ρ .

Solution :

(a) Consider a circular path of radius 10 mm. Apply Ampere's Circuital law to this closed path.

$$\oint \overline{H} \cdot \overline{dL} = I_{end}$$

$$L\,H\,S = \oint \overline{H} \cdot \overline{dL} = \int_0^{2\pi} K\rho^2 \overline{a}_\phi \cdot \rho\, d\phi\, \overline{a}_\phi \Big|_{\rho = 10\text{ mm}}$$

$$= 2\pi\rho^3 K \Big|_{\rho = 10\text{ mm}} = 2\pi K\, 10^{-6} \text{ A}$$

R H S = 1000 A.

$\therefore\ 2\pi K\, 10^{-6} = 1000$

or $K = 159154943.1$ A/m^3

$$\text{(b) } \nabla \times \overline{H} = \begin{vmatrix} \overline{a}_\rho/\rho & \overline{a}_\phi & \overline{a}_z/\rho \\ \partial/\partial\rho & \partial/\partial\phi & \partial/\partial z \\ 0 & \rho(K\rho^2) & 0 \end{vmatrix} = \overline{a}_z/\rho\,(3K\rho^2) = 3K\rho\,\overline{a}_z \text{ A/m}^2$$

$$= \frac{\overline{a}_z}{\rho}(3K\rho^2)$$

$\therefore\ \overline{J} =, 3K\rho\,\overline{a}_z$ A/m^2

Also $\overline{J} = \sigma\overline{E} = \sigma\dfrac{V}{L}\overline{a}_z = \dfrac{\sigma \times 10}{100}\overline{a}_z = 0.1\,\sigma\,\overline{a}_z$

$$\therefore \quad 0.1\,\sigma = 3\,K\,\rho \quad \text{or} \quad \sigma = \frac{3 \times 159154943.1\,\rho}{0.1}\ \Omega/\text{m}$$

$$= 4.774648293 \times \rho \times 10^9 \text{ ohms/m}$$

Problem 7.32 : Let $\overline{H} = (2/\pi\rho)\,[\,1 + (10^7\rho^3/6)\,]\ \overline{a}_\phi + 8\,\overline{a}_z$ A/m for $0 \le \rho \le 0.01$ m , $\overline{H} = (16/3\pi\rho)\,\overline{a}_\phi + 8\,\overline{a}_z$ A/m for $\rho > 0.01$ m . (a) Find $\overline{J}$ everywhere . (b) What is $\overline{J}$ at $\rho = 0$? (c) Is there a filamentary current at $\rho = 0$? If so, what is its value ? (d) What is $|\,\overline{J}\,|_{max}$ for $\rho > 0$?

Solution :

(a) $$\overline{J} = \overline{\nabla} \times \overline{H} = \begin{vmatrix} \overline{a}_\rho/\rho & \overline{a}_\phi & \overline{a}_z/\rho \\ \partial/\partial\rho & \partial/\partial\phi & \partial/\partial z \\ 0 & \rho\left[\dfrac{2}{\pi\rho}\left(1 + \dfrac{10^7\rho^3}{6}\right)\right] & 8 \end{vmatrix}$$

$$= \frac{\overline{a}_z}{\rho}\left[\frac{2}{\pi}\left(\frac{3 \times 10^7\rho^2}{6}\right)\right] = \frac{10^7}{\pi}\rho\ \overline{a}_z \quad \text{A/m}^2 \text{ for } 0 \le \rho \le 0.01$$

$$\overline{J} = \overline{\nabla} \times \overline{H} = \begin{vmatrix} \overline{a}_\rho/\rho & \overline{a}_\phi & \overline{a}_z/\rho \\ \partial/\partial\rho & \partial/\partial\phi & \partial/\partial z \\ 0 & \rho\left[\dfrac{16}{3\pi\rho}\right] & 8 \end{vmatrix}$$

$$= 0 \quad \text{for} \quad \rho > 0.01 \text{ m}$$

(b) $$\overline{J}\,\Big|_{\rho = 0} = \frac{10^7}{\pi}\,(0)\ \overline{a}_z = 0$$

(c) $$\oint \overline{H} \cdot \overline{dL} = I_{end}$$

$$\text{LHS} = \int_0^{2\pi} \overline{H} \cdot (\rho\, d\phi\, \overline{a}_\phi) = \int_0^{2\pi} \frac{2}{\pi\rho}\left[1 + \frac{10^7\rho^3}{6}\right]\rho\, d\phi$$

$$= \frac{2}{\pi}\left[1 + \frac{10^7\rho^3}{6}\right]\Big[\phi\Big]_0^{2\pi} = 4\left[1 + \frac{10^7\rho^3}{6}\right]$$

Hence $I_{end} = 4\left[1 + \frac{10^7 \rho^3}{6}\right]$

At $\rho = 0$, $I_{end} = 4$ A

(d) $|\bar{J}| = \frac{10^7 \rho}{\pi}$

The maximum value $|\bar{J}|$ will be at $\rho = 0.01$ m

$|\bar{J}| = \frac{10^7}{\pi}(0.01) = 31830.98862 \quad A/m^2$

Note that for $\rho > 0.01$, $\bar{J} = 0$

Problem 7.33 : Let $\bar{H} = -y(x^2 + y^2)\,\bar{a}_x + x(x^2 + y^2)\,\bar{a}_y$ A/m in the $z = 0$ plane for $-5 \leq x \leq 5$ m and $-5 =< y \leq 5$ m. Find the total current passing through the $z = 0$ plane in the $\bar{a}_z$ direction inside the rectangle $-1 < x < 1$ and $-2 < y < 2$ by two different methods.

Solution :

The Stoke's theorem is

$$\oint \bar{H} \cdot \overline{dL} = \int (\bar{\nabla} \times \bar{H}) \cdot \overline{ds}$$

$$LHS = \oint \bar{H} \cdot \overline{dL}$$

The closed path which encloses the given surface as

$$A(-1, -2, 0) \xrightarrow{x\,dir^n} B(1, -2, 0)$$

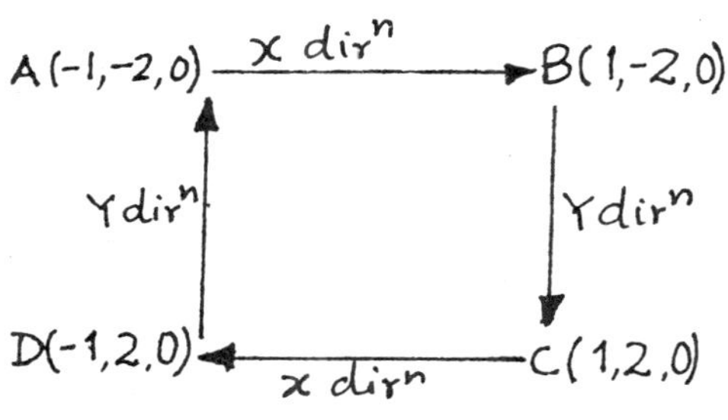

Fig.P. 7.33.

Between the points A — B and C — D, x coordinate changes.
Hence $\overline{dL} = dx\,\bar{a}_x$ where as between the points B — C and D — A, y coordinate changes. Hence $\overline{dL} = dy\,\bar{a}_y$

$$\therefore \text{ LHS } = \oint \overline{H}\cdot\overline{dL} = \int_A^B \overline{H}\cdot\overline{dL} + \int_B^C \overline{H}\cdot\overline{dL} + \int_C^D \overline{H}\cdot\overline{dL} + \int_D^A \overline{H}\cdot\overline{dL}$$

$$= \int_A^B \overline{H}\cdot(dx\,\bar{a}_x) + \int_B^C \overline{H}\cdot(dy\,\bar{a}_y) + \int_C^D \overline{H}\cdot(dx\,\bar{a}_x) + \int_D^A \overline{H}\cdot(dy\,\bar{a}_y)$$

$$= \int_{-1}^{1} -y(x^2+y^2)\,dx\Big|_{\substack{y=-2\\z=0}} + \int_{-2}^{2} x(x^2+y^2)\,dy\Big|_{\substack{x=1\\z=0}}$$

$$+ \int_{1}^{-1} -y(x^2+y^2)\,dx\Big|_{\substack{y=2\\z=0}} + \int_{2}^{-2} x(x^2+y^2)\,dy\Big|_{\substack{x=-1\\z=0}}$$

$$= -(-2)\left[\frac{x^3}{3} + (-2)^2 x\right]_{-1}^{1} + (1)\left[(1)^2 y + \frac{y^3}{3}\right]_{-2}^{2}$$

$$- (2)\left[\frac{x^3}{3} + (2)^2 x\right]_{1}^{-1} + (-1)\left[(-1)^2 y + \frac{y^3}{3}\right]_{2}^{-2}$$

$$= 17.333 + 9.333 + 17.333 + 9.333 = 53.333 \text{ A}$$

$$\text{RHS} = \int (\bar{\nabla}\times\overline{H})\cdot\overline{ds}$$

To the plane $z = 0$, $\bar{a}_z$ is perpendicular. Hence we will determine only $(\bar{\nabla}\times\overline{H})_z$.

$$\bar{\nabla}\times\overline{H} = \begin{vmatrix} \bar{a}_x & \bar{a}_y & \bar{a}_z \\ \partial/\partial x & \partial/\partial y & \partial/\partial z \\ -y(x^2+y^2) & x(x^2+y^2) & 0 \end{vmatrix}$$

$\therefore \quad (\bar{\nabla} \times \bar{H})_z = \bar{a}_z [3x^2 + y^2 + x^2 + 3y^2] = 4(y^2 + x^2)\,\bar{a}_z$

$\overline{ds} = dx\, dy\, \bar{a}_z$

$\therefore \quad (\bar{\nabla} \times \bar{H})_z \cdot \overline{ds} = [4(y^2 + x^2)\,\bar{a}_z] \cdot [dx\,dy\,\bar{a}_z]$

$$= 4[y^2 + x^2)\, dx\, dy$$

$$\therefore \quad RHS = \int_{-1}^{1} \int_{-2}^{2} 4(y^2 + x^2)\, dx\, dy$$

$$= 4\left\{\left[\frac{y^3}{3}\right]_{-2}^{2}\left[x\right]_{-1}^{1} + \left[\frac{x^3}{3}\right]_{-1}^{1}\left[y\right]_{-2}^{2}\right\}$$

$$= 53.333 \text{ A}$$

$$\oint \bar{H} \cdot \overline{dL} = \int (\bar{\nabla} \times \bar{H}) \cdot \overline{ds}$$

Problem 7.34 : Given the field $\bar{G} = \rho^2 \sin^2\phi\, \bar{a}_\rho + \rho^2 \cos^2\phi\, \bar{a}_\phi + 2z^2\, \bar{a}_z$ evaluate both sides of Stoke's theorem for the path formed by the intersection of the cylinder $\rho = 2$ and the plane $z = 1$, and for the surface defined by $\rho = 2, 1 \le z \le 3$, and $z = 3, 0 \le \rho \le 2$.

Solution :

The Stoke's theorem is

$$\oint \bar{G} \cdot \overline{dL} = \int (\bar{\nabla} \times \bar{G}) \cdot \overline{dS}$$

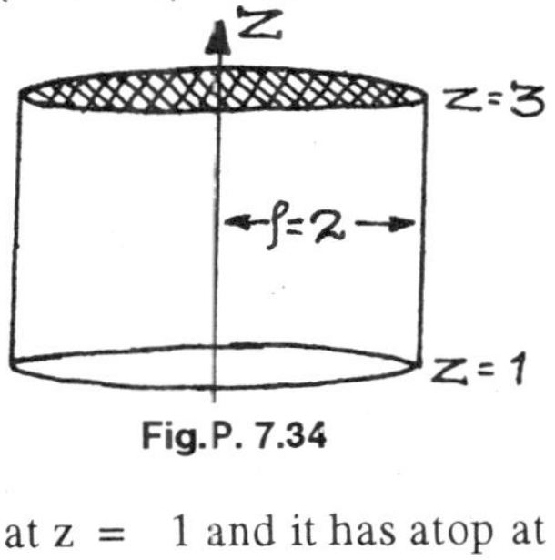

Fig.P. 7.34

The given cylinder is open at z = 1 and it has atop at z = 3 i.e. it is closed at the upper end. The circle formed by the intersection of the cylinder $\rho = 2$, $z = 1$ encloses the given surface.

Hence $\oint \overline{G} \cdot \overline{dL}$ has to be evaluated for the circle represented by $\rho = 2$, $z = 1$. For this circle, $\overline{dL} = \rho\, d\phi\, \bar{a}_\phi$.

$$\overline{G} \cdot \overline{dL} = (\rho^2 \cos^2\phi\, \bar{a}_\phi) \cdot (\rho\, d\phi\, \bar{a}_\phi) = \rho^3 \cos^2\phi\, d\phi$$

$$\therefore \text{ LHS} = \oint \overline{G} \cdot \overline{dL} = \int_0^{2\pi} \rho^3 \cos^2\phi\, d\phi \Big|_{\substack{\rho = 2 \\ z = 1}}$$

$$= (2)^3 \left[\frac{\phi}{2} + \frac{\sin 2\phi}{4} \right]_0^{2\pi} = 8\left[\frac{2\pi}{2}\right] = 8\pi$$

$$\bar{\nabla} \times \overline{G} = \begin{vmatrix} \dfrac{\bar{a}_\rho}{\rho} & \bar{a}_\phi & \dfrac{\bar{a}_z}{\rho} \\ \dfrac{\partial}{\partial\rho} & \dfrac{\partial}{\partial\phi} & \dfrac{\partial}{\partial z} \\ \rho^2 \sin^2\phi & \rho\,(\rho^2\cos^2\phi) & 2z^2 \end{vmatrix}$$

$$= \frac{\bar{a}_\rho}{\rho}\left[0\right] - \bar{a}_\phi\left[0\right] + \frac{\bar{a}_z}{\rho}\left[3\rho^2\cos^2\phi - 2\rho^2 \sin\phi\cos\phi\right]$$

$$= [3\rho\cos^2\phi - \rho\sin 2\phi]\ \bar{a}_z$$

$$\text{RHS} = \int (\bar{\nabla} \times \overline{G}) \cdot \overline{dS}$$

For the cylindrical surface, $\rho = 2$, $\bar{a}_\rho$ is perpendicular.

Hence $\overline{ds} = \rho\, d\phi\, dz\, \bar{a}_\rho$ and

$$(\bar{\nabla} \times \overline{G}) \cdot (\rho\, d\phi\, dz\, \bar{a}_\rho) = 0$$

For the top surface, $z = 3$, $\bar{a}_z$ is perpendicular.

Hence $\overline{ds} = \rho\, d\rho\, d\phi\, \bar{a}_z$ and

$$(\bar{\nabla} \times \overline{G}) \cdot (\rho\, d\rho\, d\phi\, dz\, \bar{a}_z) = (3\rho^2\cos^2\phi - \rho^2 \sin 2\phi)\, d\rho\, d\phi$$

$$\therefore \text{ RHS} = \int_0^2 \int_0^{2\pi} (3\rho^2\cos^2\phi - \rho^2\sin 2\phi)\, d\rho\, d\phi \Big|_{z = 3}$$

$$= 3\left[\frac{\rho^3}{3}\right]_0^2 \left[\frac{\phi}{2} + \frac{\sin 2\phi}{4}\right]_0^{2\phi} - \left[\frac{\rho^3}{3}\right]_0^2 \left[\frac{-\cos 2\phi}{2}\right]_0^{2\pi}$$

$$= 3 \times \frac{8}{3} \times \frac{2\pi}{2} - \frac{8}{3} \times 0 = 8\pi$$

Problem 7.35 : In cylindrical coordinates, a magnetic field is given as $\overline{H} = (2\rho - \rho^2)\,\overline{a}_\phi$ A/m, $0 \le \rho \le 1$. (a) Determine the current density as a function of ρ within this cylinder. (b) What total current passes through the surface $z = 0$, $0 \le \rho \le 1$, in the $\overline{a}_z$ direction ? (c) Check (b) by another method.

Solution :

$$\text{(a)}\ \overline{J} \times \overline{H} = \begin{vmatrix} \frac{\overline{a}_\rho}{\rho} & \overline{a}_\phi & \frac{\overline{a}_z}{\rho} \\ \frac{\partial}{\partial \rho} & \frac{\partial}{\partial \phi} & \frac{\partial}{\partial z} \\ 0 & \rho(2\rho - \rho^2) & 0 \end{vmatrix}$$

$$= \frac{\overline{a}_z}{\rho}\left[4\rho - 3\rho^2\right] = (4 - 3\rho)\,\overline{a}_z \text{ A/m}^2$$

(b) To the surface $z = 0$, $\overline{a}_z$ is perpendicular.

Hence $\overline{ds} = \rho\, d\rho\, d\phi\, \overline{a}_z$.

Hence $\overline{J} \cdot \overline{ds} = (\overline{\nabla} \times \overline{H}) \cdot \overline{ds} = (4\rho - 3\rho^2)\, d\rho\, d\phi$

$$\therefore\ I = \int \overline{J} \cdot \overline{dS} = \int (\overline{\nabla} \times \overline{H}) \cdot \overline{dS}$$

$$= \int_0^1 \int_0^{2\pi} (4\rho - 3\rho^2)\, d\rho\, d\phi = \left[\frac{4\rho^2}{2} - \frac{3\rho^3}{3}\right]_0^1 \left[\phi\right]_0^{2\pi}$$

$$= 2\pi \text{ A}$$

$$\text{(c)}\ I = \oint \overline{H} \cdot \overline{dL} = \int \overline{H} \cdot (\rho\, d\phi\, \overline{a}_\phi)$$

$$= \int_0^{2\pi} (2\rho - \rho^2)\, \rho\, d\phi \Big|_{\rho = 1}$$

$$= \int_0^{2\pi} d\phi = \left[\phi\right]_0^{2\pi} = 2\pi \text{ A}$$

Problem 7.36 : The magnetic field intensity is given as $\overline{H} = (3/\rho)\,\overline{a}_\rho + 12(z - 5)\,\overline{a}_\phi + 2\rho^2 \sin\phi\, \overline{a}_z$ in free space. Determine

the value of the surface integral $\int_s (\bar{\nabla} \times \bar{H}) \cdot \overline{dS}$ over the paraboloidal surface, $z = \rho^2$; $0 \leq z \leq 3$.

Solution :

$$f = z - \rho^2 = 0$$

$$\bar{\nabla} f = \frac{\partial f}{\partial \rho} \bar{a}_\rho + \frac{1}{\rho} \frac{\partial f}{\partial \phi} \bar{a}_\phi + \frac{\partial f}{\partial z} \bar{a}_z$$

$$\bar{a}_n = \frac{\bar{\nabla} f}{|\bar{\nabla} f|} = \frac{-2\rho\, \bar{a}_\rho + \bar{a}_z}{\sqrt{(-2\rho)^2 + (1)^2}}$$

$$= \frac{-2\rho\, \bar{a}_\rho + \bar{a}_z}{\sqrt{4\rho^2 + 1}}$$

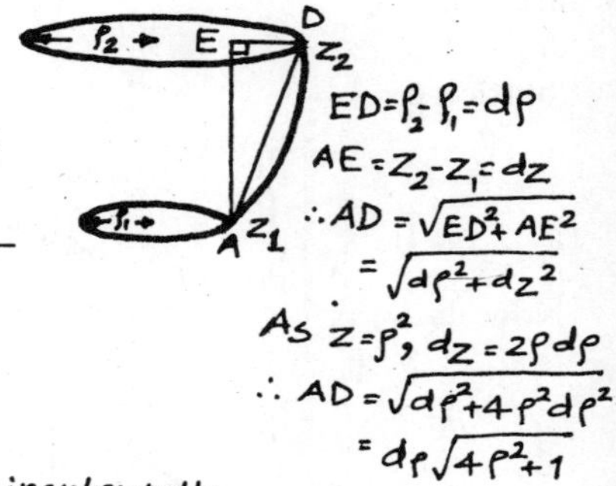

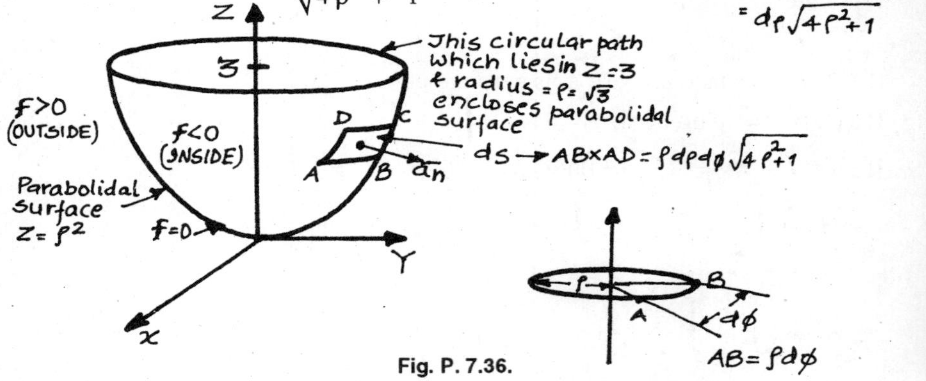

Fig. P. 7.36.

Fig. 7.36 shows the paraboliodal surface whose equation is $z = \rho^2$

let $f = z - \rho^2$

on the paraboliodal surface we have $f = z - \rho^2 = 0$. Inside the paraboliodal surface $f = z - \rho^2 < 0$ and outside the paraboliodal surface $f = z - \rho^2 > 0$ we need $\bar{a}_n$ which is perpendicular to the parabolidal surface and is directed outside as shown in fig. That is we need a unit vector directed from lower value of f towards higher value f which is given by

$$\bar{a}_n = \frac{\bar{\nabla} f}{|\bar{\nabla} f|}$$

$$\bar{\nabla} \times \bar{H} = \begin{vmatrix} \dfrac{\bar{a}_\rho}{\rho} & \bar{a}_\phi & \dfrac{\bar{a}_z}{\rho} \\ \dfrac{\partial}{\partial \rho} & \dfrac{\partial}{\partial \phi} & \dfrac{\partial}{\partial z} \\ 3/\rho & \rho\,[12\,(z-5)\,] & 2\rho^2 \sin\phi \end{vmatrix}$$

$$= \frac{\bar{a}_\rho}{\rho}\left[\,2\rho^2\cos\phi - 12\rho\,\right] - \bar{a}_\phi\left[\,4\rho\sin\phi\,\right] + \frac{\bar{a}_z}{\rho}\,[\,12\,(z-5)\,]$$

$$= (2\rho\cos\phi - 12)\,\bar{a}_\rho - 4\sin\phi\,\bar{a}_\phi + \frac{12\,(z-5)}{\rho}\,\bar{a}_z$$

Now $\overline{ds} = ds\,\bar{a}_n = ds\left[\dfrac{-2\rho\,\bar{a}_\rho + \bar{a}_z}{[4\rho^2+1]^{1/2}}\right]$

$$\therefore\ (\bar{\nabla} \times \bar{H}) \cdot \overline{ds} = \frac{-2\rho(2\rho\cos\phi - 12)\,ds}{\sqrt{4\rho^2+1}} + \frac{12\,(z-5)\,ds}{\rho\sqrt{4\rho^2+1}}$$

$$\therefore\ \int (\bar{\nabla} \times \bar{H}) \cdot \overline{ds}$$

$$= \int_0^{\sqrt{3}}\int_0^{2\pi} \frac{-2\,(2\rho^2\cos\phi - 12\rho)}{\sqrt{4\rho^2+1}}\,\rho\sqrt{4\rho^2+1}\ d\rho\,d\phi$$

$$+ \int_0^{\sqrt{3}}\int_0^{2\pi} \frac{12\,(\rho^2-5)}{\rho\sqrt{4\rho^2+1}}\,\rho\sqrt{4\rho^2+1}\ d\rho\,d\phi$$

$$= -4\left[\frac{\rho^4}{4}\right]_0^{\sqrt{3}}\left[\,\sin\phi\,\right]_0^{2\pi} + 24\left[\frac{\rho^3}{3}\right]_0^{\sqrt{3}}\left[\,\phi\,\right]_0^{2\pi}$$

$$+ 12\left[\frac{\rho^3}{3}\right]_0^{\sqrt{3}}\left[\,\phi\,\right]_0^{2\pi} - 60\left[\,\rho\,\right]_0^{\sqrt{3}}\left[\,\phi\,\right]_0^{2\pi}$$

$$= -261.187\ \text{A}$$

To verify this answer, let us determine $\oint \bar{H} \cdot \overline{dL}$, the close path which encloses the paraboloidal surface is $z = 3$, $\rho = \sqrt{3}$.

For this path $\overline{dL} = \rho\,d\phi\,\bar{a}_\phi$.

$$\therefore\ \oint \bar{H} \cdot \overline{dL} = \int 12\,(z-5)\,\bar{a}_\phi \cdot \rho\,d\phi\,\bar{a}_\phi$$

$$= \int_{0}^{2\pi} 12\,(z - 5)\,\rho\, d\phi \,\Big|_{\substack{z = 3 \\ \rho = \sqrt{3}}}$$

$$= 12\,(3 - 5)\,\sqrt{3}\,\Big[\,\phi\,\Big]_{0}^{2\pi} = -\ 261.187 \text{ A}$$

Problem 7.37 : Given $\overline{H} = (10\,r^2/\sin\theta)\,\overline{a}_\theta + 180\,r\cos\theta\,\overline{a}_\phi$ A/m in free space : (a) find the current in the $\overline{a}_\theta$ direction through the conical surface $\theta = 30^o$, $0 \leq \phi \leq 2\pi$, $0 \leq r \leq 2$, by using whichever side of Stoke's theorem you like the best. (b) Check the result by using the other side of Stoke's theorem.

Solution :

The Stokes theorem is

$$\oint \overline{H} \cdot \overline{dL} = \int (\overline{\nabla} \times \overline{H}) \cdot \overline{ds}$$

$$\text{LHS} = \oint \overline{H} \cdot \overline{dL}$$

The closed path which enclosed the given surface is given by

$$\text{A}\,(0\,,30^o\,,0) \xrightarrow{\phi\ \text{dir}^n} \text{B}\,(0\,,30^o\,,2\pi) \xrightarrow{r\ \text{dir}^n}$$

$$\text{C}\,(2\,,30^o\,,2\pi) \xrightarrow{\phi\ \text{dir}^n} \text{D}\,(2\,,30^o\,,0) \xrightarrow{r\ \text{dir}^n} \text{A}\,(0\,,30^o\,,0)$$

$$\overline{dL} = dr\,\overline{a}_r + r\,d\theta\,\overline{a}_\theta + r\sin\theta\,d\phi\,\overline{a}_\phi$$

$$\text{LHS} = \oint \overline{H} \cdot \overline{dL} \int_{A}^{B} \overline{H} \cdot \overline{dL} + \int_{B}^{C} \overline{H} \cdot \overline{dL}$$

$$+ \int_{C}^{D} \overline{H} \cdot \overline{dL} + \int_{D}^{A} \overline{H} \cdot \overline{dL}$$

Between the points A and B, C and D, only φ coordinate changes.

Hence $\overline{dL} = r\sin\theta\,d\phi\,\overline{a}_\phi$

Hence $\overline{H} \cdot \overline{dL} = 180\,r\cos\theta \times r = 90\,r^2 \sin 2\theta$

Between the points B and C, D and A, only r coordinate changes.

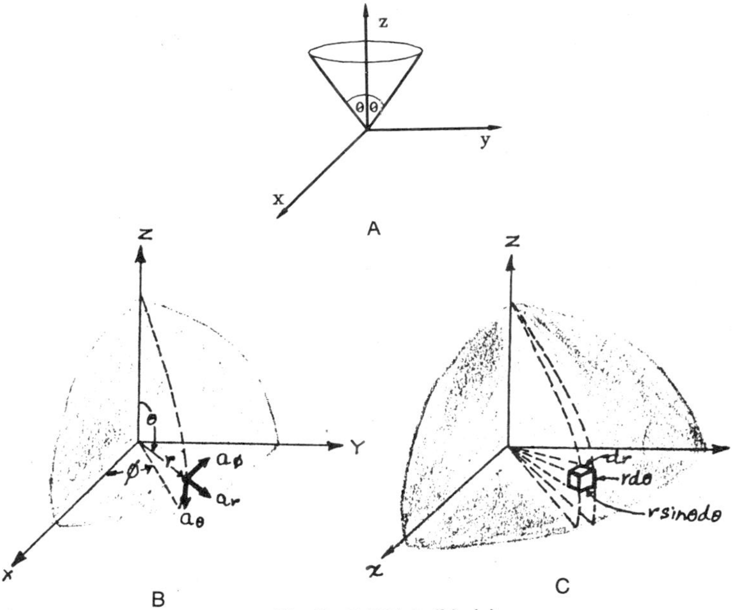

Fig. P. 7.37 (a), (b), (c).

Hence $\overline{dL} = dr\, \bar{a}_r$

Hence $\overline{H} \cdot \overline{dL} = 10\, r^2 \operatorname{cosec}\theta\, \bar{a}_\theta \cdot dr\, \bar{a}_r$

$$\therefore \text{ LHS} = \oint \overline{H} \cdot \overline{dL} = \int_0^{2\pi} 90 r^2 \sin 2\theta\, d\phi \Big|_{\substack{\theta = 90^\circ \\ r = 0}} + \int_{2\pi}^{0} 90\, r^2 \sin 2\,\theta\, d\phi \Big|_{\substack{\theta = 30^\circ \\ r = 2}}$$

$$= 90\,(2)^2 \sin 60^0 \Big[\, \phi \,\Big]_{2\pi}^{0} = -\ 1958.903 \text{ A}$$

To the surface $\theta = 30^\circ$, $\bar{a}_\theta$ is perpendicular.

Hence $\overline{ds} = r \sin\theta\, dr\, d\phi\ \bar{a}_\theta$

$$\bar{\nabla} \times \overline{H} = \begin{vmatrix} \dfrac{\bar{a}_r}{r^2 \sin\theta} & \dfrac{\bar{a}_\theta}{r \sin\theta} & \dfrac{\bar{a}_\phi}{r} \\ \dfrac{\partial}{\partial r} & \dfrac{\partial}{\partial\,\theta} & \dfrac{\partial}{\partial \phi} \\ 0 & r\,(10 r^2 \operatorname{cosec}\theta) & r \sin\theta\,(180\, r \cos\theta) \end{vmatrix}$$

As $\bar{a}_\theta$ is present is $\overline{ds}$, we will also determine $\bar{\nabla} \times \overline{H}$ in which $\bar{a}_\theta$ is present.

$$[\nabla \times \overline{H}]_\theta = \frac{-\bar{a}_\theta}{r \sin\theta}\left[360\, r \sin\theta \cos\theta\right]$$

$$= -360 \cos\theta\, \bar{a}_\theta$$

$$\therefore \text{RHS} = \int_0^2 \int_0^{2\pi} (-360 \cos\theta\, \bar{a}_\theta) \cdot (r\sin\theta\, dr\, d\phi\, \bar{a}_\theta) \Big|_{\theta = 30^\circ}$$

$$= \int_0^2 \int_0^{2\pi} -360 \cos\theta \sin\theta\, r\, dr\, d\phi \Big|_{\theta = 30^\circ}$$

$$= -360 \cos 30 \sin 30 \left[\frac{r^2}{2}\right]_0^2 \left[\phi\right]_0^{2\pi}$$

$$= -1958.903 \text{ A}$$

Problem 7.38 : (a) Use Ampere's circuital law to find $\overline{H}$ and $\overline{B}$ inside a long, straight, nonmagnetic conductor of radius 8 mm carrying a uniform current density of 100 kA / m^2 . (b) Show that $\bar{\nabla} \times \overline{H} = \overline{J}$.

Find the total magnetic flux crossing the surface :

(c) $z = 0$, for $0.2 < \rho < 8$ mm , $\pi < \phi < 2\pi$;

(d) $\phi = 0$, $0 < \rho < 8$ mm , $0 < z < 10$ mm ;

(e) $\rho = 0.2$ mm , $0 < \phi < \pi/2$, $0 < z < 10$ mm .

Solution :

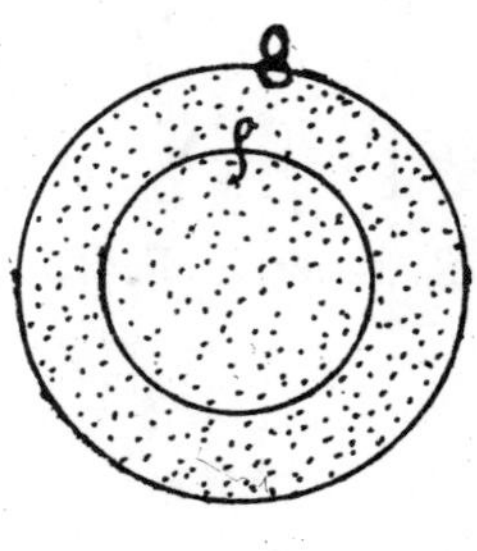

Fig.P. 7.38

Fig.P. 7.38 shows the cross sectional view of the conductor. Let us assume that the current is flowing in + $\bar{a}_z$ direction so that $\bar{J} = J_z \bar{a}_z = 100000\, \bar{a}_z$ A/m^2 and $\bar{H} = H_\phi \bar{a}_\phi$. (whenever current is flowing in + $\bar{a}_z$ direction, H_ϕ component of magnetic field intensity is produced). Consider a circular path of radius ρ as shown in Fig.P. 7.38 and apply Ampere's circuital law.

$$\oint \bar{H} \cdot \overline{dL} = I_{encl}$$

$$\text{LHS} = \oint \bar{H} \cdot \overline{dL} = \int_0^{2\pi} H_\phi \bar{a}_\phi \cdot (\rho\, d\phi\, \bar{a}_\phi)$$

$$= \int_0^{2\pi} \rho H_\phi\, d\phi = 2\pi \rho H_\phi$$

$$\text{RHS} = J \times \text{area enclosed}$$

$$= 10^5 \times \pi \rho^2 = \pi \rho^2 10^5$$

$$\therefore\ 2\pi\rho H_\phi = \pi \rho^2 10^5 \quad \text{or} \quad H_\phi = 0.5 \rho 10^5$$

$$\left.\begin{aligned} \therefore\ \bar{H} &= 0.5\rho 10^5 \bar{a}_\rho \text{ A/m} \\ \bar{B} &= 0.5 \rho 10^5 \mu_o \bar{a}_\phi \text{ Wb/m}^2 \end{aligned}\right] \text{ inside the conductor i.e. } \rho \le 0.008$$

(b) Now $\bar{J} = 10^5 \bar{a}_z$

$$\nabla \times \bar{H} = \begin{vmatrix} \bar{a}_\rho/\rho & \bar{a}\phi & \bar{a}_z/\rho \\ \partial/\partial \rho & \partial/\partial \phi & \partial/\partial z \\ 0 & \rho\,(0.5\rho 10^5) & 0 \end{vmatrix}$$

$$= \frac{\bar{a}_z}{\rho}\left[0.5 \times 2\rho \times 10^5\right] = 10^5 \bar{a}_z = \bar{J}$$

Hence $\bar{J} = \nabla \times \bar{H}$ is proved

(c) For the plane z = 0, $\bar{a}_z$ is perpendicular.

$$\overline{ds} = \rho\, d\rho\, d\phi\, \bar{a}_z$$

$$\therefore\ \bar{B} \cdot \overline{ds} = (0.5\rho 10^5 \mu_o \bar{a}_\phi) \quad (\rho\, d\rho\, d\phi\, \bar{a}_z) = 0$$

$$\therefore\ \phi = \int \bar{B} \cdot \overline{ds} = 0$$

(d) For the plane $\phi = 0$, $\bar{a}_\phi$ is perpendicular.

$\overline{ds} = d\rho\, dz\, \bar{a}_\phi$

$$\therefore \quad \overline{B} \cdot \overline{ds} = (0.5\,\rho\, 10^5\, \mu_o\, \bar{a}_\phi) \cdot (d\rho\, dz\, \bar{a}_\phi)$$

$$= 0.5\,\rho\, 10^5\, \mu_o\, d\rho\, dz$$

$$\therefore \quad \Phi = \int_0^{0.008} \int_0^{0.01} 0.5\,\rho\, 10^5\, \mu_o\, d\rho\, dz$$

$$= 0.5 \times 10^5 \left[\frac{\rho^2}{2} \right]_0^{0.008} \left[z \right]_0^{0.01} \mu_o$$

$$= 20.106 \text{ nWb}$$

(e) For the surface $\rho = 0.2$ mm , $\bar{a}_\rho$ is perpendicular

$\overline{ds} = \rho\, d\phi\, dz\, \bar{a}_\rho$

$$\therefore \quad \overline{B} \cdot \overline{ds} = (0.5\,\rho\, 10^5\, \mu_o\, \bar{a}_\phi) \cdot (\rho\, d\phi\, dz\, \bar{a}_\rho) = 0$$

$$\therefore \quad \phi = \int \overline{B} \cdot \overline{ds} = 0$$

Problem 7.39 : A solid nonmagnetic conductor of circular cross section, $\rho = 2$ cm, carries a total current, I = 60 A, in the $\bar{a}_z$ direction. The conductor is inhomogeneous , having a conductivity that varies with ρ as $\sigma = 10^5 (1 + 2.5 \times 10^5 \rho^2)$ ohms/m . Find the total flux crossing the radial plane defined by $\phi = 0$, $0 < z < 1$ m , and : (a) $0 < \rho < 1$ cm ; (b) $1 < \rho < 2$ cm .

Solution : Suppose V is the potential difference applied to the conductor and l is the length of the conductor.

Then $E = \dfrac{V}{l}$.

Now $\overline{J} = \sigma E$.

As the current is in $\bar{a}_z$ direction, $\overline{J} = J_z\, \bar{a}_z$. Hence

$$\overline{J} = 10^5 (1 + 2.5 \times 10^5 \rho^2) \frac{V}{l}\, \bar{a}_z \text{ A/m}^2$$

As the current density is not constant, we must use,

$$\therefore \quad I = \int \bar{J} \cdot \overline{ds} = \int \bar{J} \cdot (\rho \, d\rho \, d\phi \, \bar{a}_z)$$

$$= \int_0^{0.02} \int_0^{2\pi} 10^5 \, (1 + 2.5 \times 10^5 \rho^2) \, \frac{V}{1} \, \rho \, d\rho \, d\phi$$

$$\text{Hence } I = 60 = 10^5 \frac{V}{1} \left[\frac{\rho^2}{2} + 2.5 \times 10^5 \frac{\rho^4}{4} \right]_0^{0.02} \left[\phi \right]_0^{2\pi}$$

$$\therefore \quad \frac{V}{1} = E = \frac{60}{2040\,\pi} = \frac{1}{34\,\pi} \text{ V/m}$$

$$\therefore \quad \bar{J} = \frac{10^5}{34\,\pi} (1 + 2.5 \times 10^5 \rho^2) \, \bar{a}_z \text{ A/m}^2$$

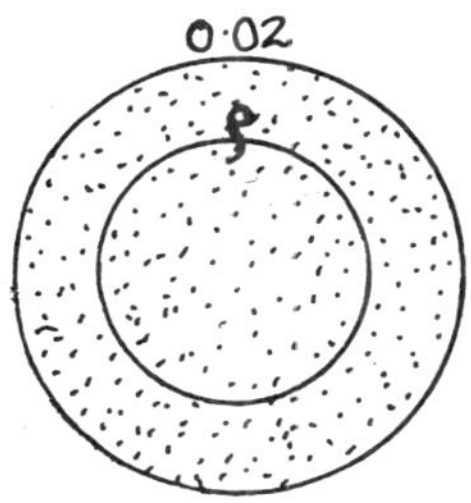

Fig.P. 7.39

Fig.P. 7.39 shows the cross sectional view of the conductor. Consider a circle of radius ρ . Apply Ampere's circuital law to this circular path.

$$\oint \bar{H} \cdot \overline{dL} = I_{encl}$$

$$\text{LHS} = \oint \int \bar{H} \cdot \overline{dL} = \int_0^{2\pi} (H_\phi \, \bar{a}_\phi) \cdot (\rho \, d\phi \, \bar{a}_\phi)$$

$$= \int_0^{2\pi} \rho \, H_\phi \, d\phi = 2\pi \rho \, H_\phi$$

$$\text{RHS} = I_{encl} = \int \bar{J} \cdot \overline{ds} = \int \bar{J} \cdot (\rho \, d\rho \, d\phi \, \bar{a}_z)$$

$$= \int_0^{\rho} \int_0^{2\pi} \frac{10^5}{34\,\pi} \left[1 + 2.5 \times 10^5 \rho^2 \right] \rho \, d\rho \, d\phi$$

$$= \frac{10^5}{34\pi}\left[\frac{\rho^2}{2} + \frac{2.5 \times 10^5 \rho^4}{4}\right]_0^\rho \left[\phi\right]_0^{2\pi}$$

$$= \frac{10^5}{17}\left[\frac{\rho^2}{2} + \frac{2.5 \times 10^5 \rho^4}{4}\right]$$

$$\therefore \quad 2\pi\rho H_\phi = \frac{10^5}{17}\left[\frac{\rho^2}{2} + \frac{2.5 \times 10^5 \rho^4}{4}\right]$$

$$H_\phi = \frac{10^5}{34\pi}\left[\frac{\rho}{2} + \frac{2.5 \times 10^5 \rho^3}{4}\right]$$

$$\therefore \quad \overline{H} = \frac{10^5}{34\pi}\left[\frac{\rho}{2} + \frac{2.5 \times 10^5 \rho^3}{4}\right] \overline{a}_\phi \text{ A/m}$$

$$\overline{B} = \mu_o \overline{H} = \frac{10^5 \mu_o}{34\pi}\left[\frac{\rho}{2} + \frac{2.5 \times 10^5 \rho^3}{4}\right] \overline{a}_\phi \text{ Wb/m}^2$$

(a) To the plane $\phi = 0$, $\overline{a}_\phi$ is perpendicular.

$$\therefore \quad \overline{ds} = d\rho\, dz\, \overline{a}_\phi$$

$$\therefore \quad \Phi = \int \overline{B} \cdot \overline{ds} = \int_0^{0.01}\int_0^1 \frac{10^5 \mu_o}{34\pi}\left[\frac{\rho}{2} + \frac{2.5 \times 10^5 \rho^3}{4}\right] d\rho\, dz$$

$$= \frac{10^5 \mu_o}{34\pi}\left[\frac{\rho^2}{4} + \frac{2.5 \times 10^5 \rho^4}{16}\right]_0^{0.01} \left[z\right]_0^1$$

$$= 0.213 \ \mu \text{Wb}.$$

$$\text{(b)} \quad \phi = \int \overline{B} \cdot \overline{ds} = \int_{0.01}^{0.02}\int_0^1 \frac{10^5 \mu_o}{34\pi}\left[\frac{\rho}{2} + \frac{2.5 \times 10^5 \rho^3}{4}\right] d\rho\, dz$$

$$= \frac{10^5 \mu_o}{34\pi}\left[\frac{\rho^2}{4} + \frac{2.5 \times 10^5 \rho^4}{16}\right]_{0.01}^{0.02} \left[z\right]_0^1$$

$$= 2.846 \ \mu \text{Wb}.$$

Problem 7.40 : Two solid nonmagnetic cylinders of 0.1 – m radius have their axes located in the x = 0 plane at y = ± 0.3 m. The one at y = 0.3 carries 18 A in the $\overline{a}_z$ direction; the other current is 18 A in the – $\overline{a}_z$ direction. Find the total magnetic flux crossing the planar surface in the x = 0 plane

extending from z = 0 to z = 1 m and : (a) − 0.2 < y < 0.2 m ; (b) 0.2 < y < 0.4 m ; (c) 0.4 < y < ∞ .

Solution :

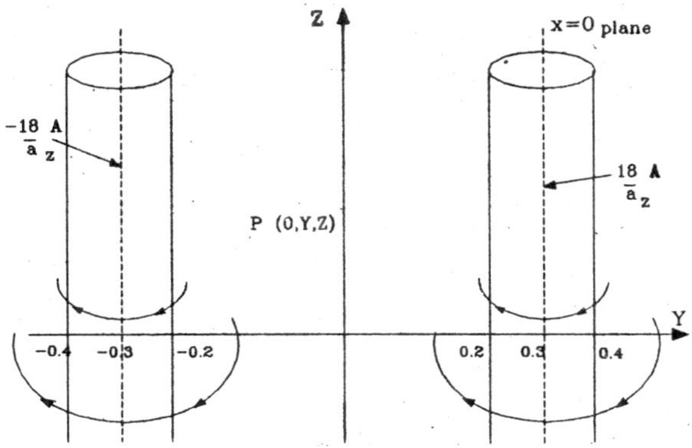

Fig.P. 7. 40 (a).

Fig. 7.40. shows the location two nonmagnetic cylinders. The current at y = 0.3 is 18 A in + $\bar{a}_z$ direction. Hence the flux lines which are circular are directed in anticlockwise direction i.e. + $\bar{a}_\phi$ direction as shown in fig. The current at y = − 0.3 is 18 A in − $\bar{a}_z$ direction. Hence the flux lines are clockwise as shown in fig. Note that when current is in $\bar{a}_z$ direction then $\overline{H} = H_\phi \, \bar{a}_\phi$ and when it is in − $\bar{a}_z$ direction then $\overline{H} = - H_\phi \, \bar{a}_\phi$.

(a) − 0.2 < y < 0.2 m .

Within this region, both the flux lines are coming out through x = 0 i.e. yz plane. Further the total flux is two times that of one conductor flux. Consider the conducting nonmagnetic cylinder located at y = − 0.3 the region − 0.2 < y < 0.2 lies outside to this conductor. Hence using the equation 7.

$$|H| = \frac{I}{2\pi\rho}$$

where ρ is the radius of the circle which is greater than radius of the cylinder. Now consider the point P (0, y, z) which lies in x = 0 plane.

The distance of this point from the axis of the cylinder y = 0.3, x = 0 is calculated as shown below.

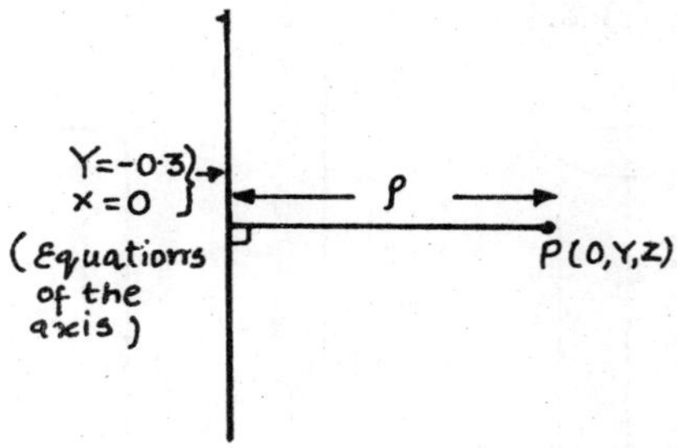

Fig. P. 7.40 (b).

Point 1 is the foot of the perpendicular dropped from point P (0, y, z) on the axis of the conductor. The coordinates of point 1 are (0, –0.3, z).

Hence

$$\rho = d\,(P1) = \sqrt{(y + 0.3)^2} = y + 0.3$$

$$\text{Hence } |\,H\,| = \frac{18}{2\pi\,(y + 0.3)} \quad \text{or} \quad |\,\overline{B}\,| = \frac{18\,\mu_o}{2\pi\,(y + 0.3)}$$

$$\therefore\ \phi = \int_{-0.2}^{0.2}\int_{0}^{1} \frac{18\,\mu_o}{2\pi\,(y + 0.3)}\,dydz$$

$$= \frac{18\,\mu_o}{2\pi}\Big[\ln(y + 0.3)\Big]_{-0.2}^{0.2}\Big[z\Big]_0^1$$

(Note that $\rho = y + 0.3$ is always positive as it is a distance)

$$= 5.7939\ \mu\,\text{Wb}$$

Hence total flux $= 2 \times 5.7939 = 11.5879\ \mu\,\text{Wb}$

(If we use the conducting cylinder located at $y = 0.3$, then

$$\Phi = \frac{18\,\mu_o}{2\pi}\Big[\ln\big|(y - 0.3)\big|\Big]_{0.2}^{-0.2}\Big[z\Big]_0^1$$

(Note that $\rho = y - 0.3$ is always positive as it is a distance).

(b) $0.2 < y < 0.4$ m

First consider the cylinder located at $y = 0.3$. as shown in fig given below.

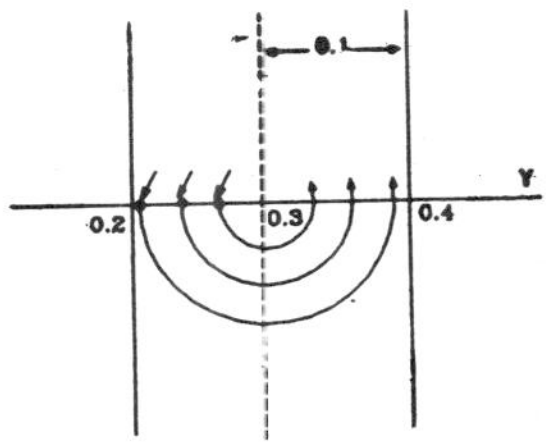

Fig. P. 7.40 (c)

The flux lines which are anticlockwise are shown for $0.2 < y < 0.4$. The region $0.2 < y < 0.4$ is the interior of the conductor. From this fig. it is seen that for $0.2 < y < 0.3$ the flux lines are coming out of the $x = 0$ plane where as for $0.3 < y < 0.4$, the flux lines are going inside of the $x = 0$ plane. Hence the contribution to the total flux will be zero due to this conductor.

If we consider the second cylinder located at $y = -0.3$, then

$$\phi = \int_{0.2}^{0.4} \int_{0}^{1} \frac{18 \mu_o}{2\pi (y + 0.3)} dy\, dz$$

$$= \frac{9 \mu_o}{\pi} \Big[\ln (y + 0.3) \Big]_{0.2}^{0.4} \Big[z \Big]_0^1$$

$$= 1.2113 \ \mu \text{Wb which is the total flux itself.}$$

(c) $0.4 < y < \infty$

$\overline{H}$ or $\overline{B}$ is a function of ρ which represents the distance of a point from the conductor axis. As seen from fig. for the region $0.4 < y < \infty$, the flux lines will be stronger for the cylinder carrying the current of 18 A in $\bar{a}_z$ direction as this region is closer to this cylinder. Further the flux lines are going inside the plane $x = 0$. The flux lines for the second cylinder carrying a current of + 18 A in $-\bar{a}_z$ direction will be weaker as this region is at a longer distance from the axis of this conductor. Further the flux lines are coming out from the $x = 0$ plane.

Hence the total flux will be

$$\Psi_{total} = \Psi_1 - \Psi_2$$

where Ψ_1 is due to 18 A in $+\bar{a}_z$ direction and Ψ_2 is due to 18A in $-\bar{a}_z$ direction.

$$\Psi_{total} = \int_{0.4}^{C}\int_{0}^{1} \frac{18\,\mu_o\,dy\,dz}{2\pi\,(y-0.3)} - \int_{0.4}^{C}\int_{0}^{1} \frac{18\,\mu_o\,dy\,dz}{2\pi\,(y+0.3)}$$

where $C = \infty$

$$\Psi_{total} = \frac{9\,\mu_o}{\pi}\left\{\left[\ln(y-0.3) - \ln(y+0.3)\right]_{0.4}^{C}\right\}$$

$$= \frac{9\,\mu_o}{\pi}\left[\ln\left(\frac{y-0.3}{y+0.3}\right)\right]_{0.4}^{\infty}$$

$$= \frac{9\,\mu_o}{\pi}\left[\ln\frac{1-\frac{0.3}{\infty}}{1+\frac{0.3}{\infty}} - \ln\frac{0.4-0.3}{0.4+0.3}\right]$$

$$= \frac{9\,\mu_o}{\pi}\left[\ln 1 - \ln\left(\frac{0.1}{0.7}\right)\right] = 7.005\,\mu\,\text{Wb}$$

See that the answer is positive and in the expression for Ψ_{total}, the total flux is going inside the $x = 0$ plane. Meaning there by that the total flux is going (directed) inside the plane $x = 0$.

Problem 7.41 : The region in free space enclosed by planes, $z = 0$ and 3 cm, and by cylinders, $\rho = 5$ and 7 cm, forms a toroid with a rectangular cross section. A surface current, $\bar{K} = 600\,\bar{a}_z$ A/m, flows on the inner surface. (a) Determine the current on the other three surfaces. (b) Find $\bar{H}$ within the toroid. (c) Find the total flux within the toroid.

Solution :

(a) On all the surfaces current must be same where as the value of $\bar{K}$ is different. Hence current = circumference of inner circle × K (scalar).

$$= 2\pi\,(0.05) \times 600 = 188.496 \text{ Amp.}$$

The directions of the current on all the three surfaces are as shown in fig (b). Consider a circle of radius ρ such that $0.05 < \rho < 0.07$ and lying on the surface $z = 0.03$ m. Circumference of this circle is $2\pi\rho$.

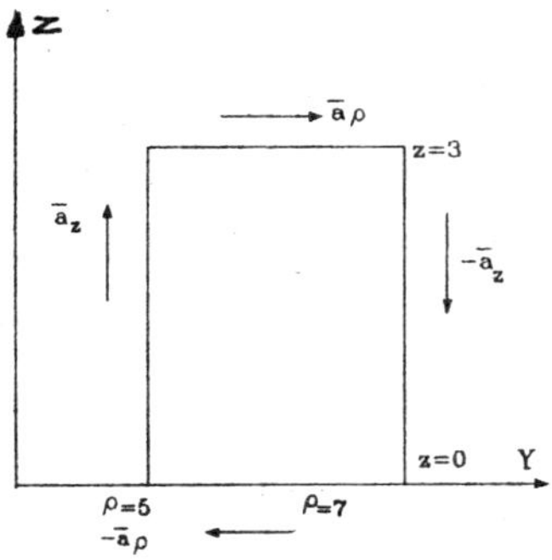

Fig. P. 7.41.

Hence $\bar{K} = \dfrac{\text{current}}{\text{circumference}}\ \bar{a}_\rho$

$$= \frac{188.496}{2\pi\rho}\ \bar{a}_\rho = \frac{30}{\rho}\ \bar{a}_\rho \ \text{A/m}$$

Similarly on the surface $z = 0$,

$$\bar{K} = -\frac{30}{\rho}\ \bar{a}_\rho \ \text{A/m}.$$

On the surface $\rho = 7\text{cm}$,

$$\bar{K} = \frac{188.496}{2\pi(0.07)}\ (-\bar{a}_z)$$

$$= -428.5724\ \bar{a}_z \ \text{A/m}.$$

(b) $\bar{H} = \dfrac{I}{2\pi\rho}\ \bar{a}_\phi$

where I is the current flowing on the inner surface in $\bar{a}_z$ direction.

$$\therefore\ \bar{H} = \frac{188.496}{2\pi\rho}\ \bar{a}_\phi$$

$$= \frac{30}{\rho}\ \bar{a}_\phi \ \text{for } 0.05 < \rho < 0.07 \text{ m}$$

(c) $\bar{B} = \mu_o \bar{H} = \dfrac{30\,\mu_0}{\rho}\ \bar{a}_\phi$

for $0.05 < \rho < 0.07$ m

$$\phi = \int \bar{B} \cdot \overline{dS} = \int \bar{B} \cdot (d\rho\, dz\, \bar{a}_\phi)$$

$$= \int_{0.05}^{0.07} \int_{0}^{0.03} \frac{30\,\mu_o}{\rho}\, d\rho\, dz$$

$$= 30\,\mu_o \Big[\ln \rho \Big]_{0.05}^{0.07} \Big[z \Big]_{0}^{0.03} = 0.38054\ \mu\,\text{Wb}$$

Problem 7.42 : A filamentary conductor on the z axis carries a current of 10 A in the $\bar{a}_z$ direction, and a coaxial conducting shell at $\rho = 8$ mm carries a total current of 5 A in the $-\bar{a}_z$ direction. The region, $3 < \rho < 5$ mm is a solid conducting , material carrying a uniformly distributed total current of 5 A in the $-\bar{a}_z$ direction. All media are nonmagnetic. (a) Find $\overline{H}$, $0 < \rho < 10$ mm . (b) Plot $\overline{H}_\phi$ vs . ρ . (c) Find the total flux Φ in a 1 – m length of the conducting material, $3 < \rho < 5$ mm .

Solution : (a) Consider a circle of radius ρ_a such that $\rho_a < 0.003$m . Apply Ampere circuital law to this closed path. Since in the problem current is in $\pm\,\bar{a}_z$, direction always H_ϕ component of magnetic field intensity which is impedent of ϕ and z is produced. The current in + $\bar{a}_z$ direction is considered as positive and the current in the $-\,\bar{a}_z$ direction as negative.

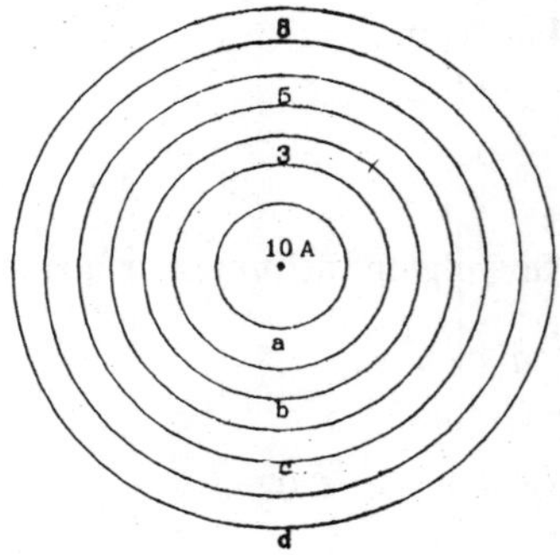

Fig. P 7. 42.

$$\oint \overline{H} \cdot \overline{dL} = I_{encl}\,.$$

$$\oint\!\!\int H_\phi\, \bar{a}_\phi \cdot \rho_a\, d\phi\, \bar{a}_\phi = 10$$

$$\int_0^{2\pi} \rho_a H_\phi \, d\phi = 10$$

$$2\pi \rho_a H_\phi = 10$$

$$\therefore \quad H_\phi = \frac{10}{2\pi \rho_a}$$

or $\overline{H} = \dfrac{5}{\pi \rho} \, \overline{a}_\phi$ when $\rho < 0.003$

(b) Consider a circle of radius ρ_b such that

$$0.003 < \rho_b < 0.005$$

$$\oint \overline{H} \cdot \overline{dL} = I_{encl}$$

$$LHS = \oint \overline{H} \cdot \overline{dL} = 2\pi \rho_b H_\phi$$

$$RHS = 10 - I$$

where I is the current between $\rho = 0.003$ m and $\rho = \rho_b$. and –ve sign is used as the current is in $-\overline{a}_z$ direction. Now 5 A of current is flowing through an area of $\pi [(0.005)^2] - (0.003)^2]$ and I A of current is flowing through an area of $\pi [(\rho_b)^2 - (0.003)^2]$. Hence

$$I = \frac{5\pi [(\rho_b)^2 - (0.003)^2]}{\pi [(0.005)^2 - (0.003)^2]} = 312500 [(\rho_b)^2 - (0.003)^2]$$

$$= 312500 (\rho_b)^2 - 2.8125$$

$$\therefore \quad R.H.S = 10 - [312500 (\rho_b)^2 - 2.8125]$$

$$= 12.8125 - 312500 (\rho_b)^2 .$$

$$\therefore \quad 2\pi \rho_b \overline{H}_\phi = 12.8125 - 312500 (\rho_b)^2$$

$$\overline{H}_\phi = \frac{12.8125 - 312500 (\rho_b)^2}{2\pi \rho_b} \qquad \text{or}$$

$$\overline{H} = \frac{2.03917 - 49735.91972 \rho^2}{\rho} \, \overline{a}_\phi \quad \text{where } 0.003 < \rho < 0.005$$

(c) Consider a circle of radius ρ_c such that $0.005 < \rho_c < 0.008$.

$$\oint \overline{H} \cdot \overline{dL} = I_{encl}$$

$$2\pi \rho_c H_\phi = 10 - 5 = 5$$

$$\therefore \quad H_\phi = \frac{5}{2\pi \rho_c} \quad \text{or}$$

$$\overline{H} = \frac{2.5}{\pi \rho} \overline{a}_\phi \quad \text{where } 0.005 < \rho < 0.008$$

(d) Consider a circle of radius ρ_d such that $\rho_d > 0.008$

$$\oint \overline{H} \cdot \overline{dL} = I_{end}$$

$$2\pi \rho_d H_\phi = 10 - 5 - 5 = 0$$

$$\therefore \quad H_\phi = 0 \quad \text{or}$$

$\overline{H} = 0$ when $\rho > 0.008$ m .

Hence

$$\overline{H} = \begin{cases} 5/\pi\rho \ \overline{a}_\phi & \text{when } \rho < 0.003 \\ \dfrac{2.03917 - 49735.91972\,\rho^2}{\rho}\ \overline{a}_\phi & \text{when } 0.003 < \rho < 0.005 \\ \dfrac{2.5}{\pi\rho}\ \overline{a}_\phi & \text{when } 0.005 < \rho < 0.008 \\ 0 & \rho > 0.008 \end{cases}$$

(b) Plot of H_ϕ is ρ is not given .

(c) $\overline{B}$ (in the conducting region) = $\left(\dfrac{2.03917 - 49735.91972\,\rho^2}{\rho} \right) \mu_o \overline{a}_\phi$

where $0.003 < \rho < 0.005$.

$$\phi = \int \overline{B} \cdot \overline{ds} = \int (B_\phi \overline{a}_\phi) \cdot (d\rho\, dz)\, \overline{a}_\phi$$

$$= \int_{0.003}^{0.005} \int_0^1 B_\phi \, d\rho\, dz$$

$$= \int_{0.003}^{0.005} \int_{0}^{1} \left(\frac{2.03917}{\rho} - 49735.91972\,\rho \right) \mu_o \, d\rho \, dz$$

$$= \mu_o \left[2.03917 \ln \rho - 49735.91972 \frac{\rho^2}{2} \right]_{0.003}^{0.005} [z]_0^1$$

$$= 0.8089889 \ \mu \ \text{Wb}$$

Problem 7.43 : A current sheet, $\overline{K} = 40\,\overline{a}_z$ A/m , is located in free space at x = 0.25 m, and a second sheet, $\overline{K} = -40\,\overline{a}_z$ A/m , is at x = – 0.25 m. (a) Let $V_m = 0$ at P (0.1 , 0.2 , 0.3) and find $V_m (x, y, z)$ for $-0.25 < x, y, z < 0.25$. (b) Let $\overline{A} = 0$ at P and determine $\overline{A}(x, y, z)$ for $-0.25 < x, y, z < 0.25$.

Solution :

(a) $\overline{H} = \overline{H}_1 + \overline{H}_2$ where $\overline{H}_1$ is due to $\overline{K} = 40\,\overline{a}_z$ A/m and $\overline{H}_2$ is due to $\overline{K} = -40\,\overline{a}_z$ A/m . Consider a point Q (x, y, z) such that $-0.25 < x, y, z < 0.25$ as shown in Fig.P. 7.43. To the plane x = – 0.25, $\overline{a}_x$ is perpendicular and directed towards the point Q (x, y, z). Therefore

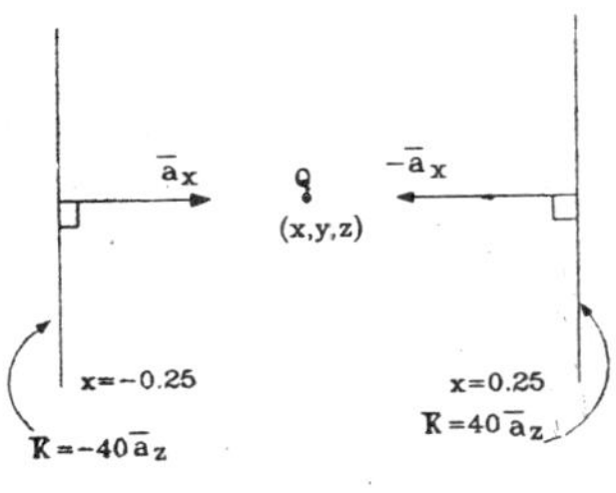

Fig.P. 7.43.

$$\overline{H}_1 = \frac{\overline{K} \times \overline{a}_N}{2} = \frac{(-40\,\overline{a}_z) \times (\overline{a}_x)}{2} = -20\,\overline{a}_y \ \text{A/m}$$

To the plane x = 0.25 , $-\overline{a}_x$ is perpendicular and directed towards the points Q (x, y, z). Therefore

$$\overline{H}_2 = \frac{(40\,\overline{a}_z) \times (-a_x)}{2} = -20\,\overline{a}_y \quad \text{A/m}$$

$$\therefore \quad \overline{H} = \overline{H}_1 + \overline{H}_2 = -20\,\bar{a}_y - 20\,\bar{a}_y = -40\,\bar{a}_y \quad A/m$$

$$V_{mQP} = -\int_P^Q \overline{H}\cdot\overline{dL} = -\int_P^Q (-40\,\bar{a}_y)\cdot(dy\,\bar{a}_y)$$

$$= \int_{0.2}^{y} 40\,dy = 40\Big[\,y\,\Big]_{0.2}^{y} = 40\,(y - 0.2)$$

Now $V_{mQP} = V_{mQ} - V_{mP} = V_{mQ}$ as $V_{mP} = V_m$ at point P is given as zero.

$$\therefore \quad V_{mQ} = V_m\,(x, y, z) = 40\,y - 8 \quad A$$

(b) $\overline{B} = \mu_o\,\overline{H} = -40\,\mu_o\,\bar{a}_y \quad Wb/m^2$

$$\overline{B} = \nabla\times\overline{A} = \left(\frac{\partial A_x}{\partial z} - \frac{\partial \overline{A}_z}{\partial x}\right)\bar{a}_y$$

$$\therefore \quad -40\,\mu_o\,\bar{a}_y = \left(\frac{\partial A_x}{\partial z} - \frac{\partial A_z}{\partial x}\right)\bar{a}_y$$

$$\therefore \quad -40\,\mu_o = \frac{\partial A_x}{\partial z} - \frac{\partial A_z}{\partial x}$$

As the current on the current sheet is in z direction, A_z is present where as A_x is zero.

$$\therefore \quad \frac{d\,A_z}{dx} = 40\,\mu_o \qquad \text{or} \qquad A_z = 40\,\mu_o\,x + C$$

where C is a constant of integration.

At point P (0.1 , 0.2 , 0.3) , $\overline{A} = 0$. Hence

$$0 = 40\,\mu_o\,(0.1) + C \qquad \text{or} \qquad C = -4\,\mu_o$$

$$\therefore \quad A_z = 40\,\mu_o\,x - 4\,\mu_o \quad \text{or}$$

$$\overline{A} = 40\,\mu_o\,(x - 0.1)\,\bar{a}_z \quad Wb/m$$

Problem 7.44 : Given the scalar magnetic potential field, $V_m = 30 - 12\ln\rho$ A : (a) sketch the traces (intersections) of the equipotential surfaces $V_m = 0, 2.5, 5, 7.5,$ and 10 A with the z = 0 plane ; (b) find $\overline{H}$ and note the direction of $\overline{H}$ in the z = 0 plane. The direction lines for $\overline{H}$ are given by ϕ = constant (not ϕ = constant). Draw the line for $\phi = 0$. (c) Determine ϕ_1 such that the direction line $\phi = \phi_1$ forms curvilinear squares with the equipotentials. Estimate ϕ_1 from your sketch.

Solution :

$V_m = 30 - 12 \ln \rho$.

(a) $V_m = 0$, $0 = 30 - 12 \ln \rho$, $\rho = 12.182$ m

The intersection of $\rho = 12.182$ m and $z = 0$ is a circle of radius $\rho = 12.182$ m which lies in $z = 0$ plane.

$V_m = 2.5$, $2.5 = 30 - 12 \ln \rho$, $\rho = 9.891$ m .

$V_m = 5$, $5 = 30 - 12 \ln \rho$, $\rho = 8.031$ m .

$V_m = 7.5$, $7.5 = 30 - 12 \ln \rho$, $\rho = 6.521$ m .

Hence there are four circles of above radii in $z = 0$ plane which gives required values of V_m .

(b) $\overline{H} = -\bar{\nabla} V_m = -\left[\frac{\partial V_m}{\partial \rho} \bar{a}_\rho \right]$

$$= -\left[-\frac{12}{\rho} \bar{a}_\rho \right] = \frac{12}{\rho} \bar{a}_\rho \quad \text{A/m}$$

The direction of $\overline{H}$ in $z = 0$ plane is given by $\bar{a}_\rho$.

The equations of the direction lines are given by

$$\frac{H_\rho}{d\rho} = \frac{H_\phi}{\rho\, d\phi} = \frac{H_z}{dz}$$

In $z = 0$ plane, equation of the direction line is given by

$$\frac{H_\rho}{d\rho} = \frac{H_\phi}{\rho\, d\phi}$$

$$\therefore \quad \frac{d\phi}{d\rho} = \frac{H_\phi}{\rho H_\rho} = \frac{0}{\rho \times \frac{12}{\rho}} = 0$$

$\therefore \quad \phi = C$

where C is the constant of integration.

As seen from Fig. P. 7.44 (b), for curvilinear square, we have

$$\frac{d\,(ab)}{d\,(aa')} = \frac{d\,(bc)}{d\,(bb')} = \frac{d\,(cd)}{d\,(cc')} = 1$$

i.e. $\frac{6.521\,\phi_1}{1.51} = \frac{8.031\,\phi_1}{1.86} = \frac{9.891\,\phi_1}{2.291} = 1$

or $\phi_1 = 0.231623642 = 13.271^\circ$

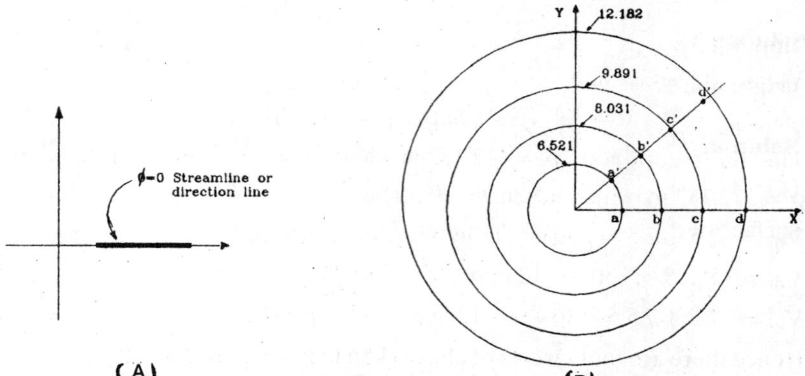

Fig.P. 7.44.

Problem 7.45 : Find V_m for the region within the toroid of Prob. 35, where $\overline{H} = (30/\rho)\,\overline{a}_\phi$ A/m and $V_m = 0$ at $\rho = 6$ cm, $\Phi = 0.6$ rad, and z= 2 cm. Keep Φ within the range $0 \leq \Phi \leq 2\pi$.

Solution :

$\overline{H} = (30/\rho)\,\overline{a}_\phi$

$$V_{mPQ} = -\int_Q^P \overline{H} \; \overline{dL} = -\int_Q^P \left(\frac{30}{\rho}\,\overline{a}_\phi\right) \cdot (\rho\, d\phi\, \overline{a}_\rho$$

$$= -\int_Q^P 30\, d\phi$$

The coordinates of points P and Q are (ρ, ϕ, z) and $(0.06, 0.6, 0.02)$ respectively.

$$\therefore \quad V_{mPQ} = -\int_{0.6}^{\phi} 30\, d\phi = -30\phi - 0.6) = -30(\phi + 18$$

Now $V_{mPQ} = V_{mP} - V_{mQ}$

Now $V_m = V_{mQ} = 0$ (given)

$\therefore \quad V_m = -30\phi + 18$ A

Problem 7.46 : A solenoid in free space is centred on the z axis and has a current of $24\,\overline{a}_\phi$ A/m at $\rho = a = 4$ cm . (a) Let $V_m = 0$ at the origin and

find $V_m(\rho, \phi, z)$ for all points within the solenoid. (b) Let $\bar{A} = 0$ at the origin and find $\bar{A}(\rho, \phi, z)$ for all points within the solenoid.

Solution :

(a) $\bar{H} = K_a\,\bar{a}_z$ where $\bar{K} = K_a\,\bar{a}_\phi$ is the surface current density on the surface of the solenoid $\rho = a$.

$$\therefore \quad \bar{H} = 24\,\bar{a}_z \ \text{A/m}$$

$$V_m = -\int \bar{H}\cdot\overline{dL} = -\int (24\,\bar{a}_z)\cdot(dz\,\bar{a}_z)$$

$$= -\int 24\,dz = -24\,z \ \text{A}$$

$$\bar{B} = \bar{\nabla}\times\bar{A} \quad \text{and} \quad \bar{B} = \mu_o\bar{H} = 24\,\mu_o\,\bar{a}_z \ \text{Wb/m}^2$$

$$24\,\mu_o\,\bar{a}_z = \frac{1}{\rho}\left[\frac{\partial(\rho A_\phi)}{\partial\rho} - \frac{\partial A_\rho}{\partial\phi}\right]\bar{a}_z$$

$$\therefore \quad 24\,\mu_o = \frac{1}{\rho}\left[\frac{\partial}{\partial\rho}(\rho A_\phi) - \frac{\partial A_\rho}{\partial\phi}\right]$$

As the surface current density is directed in $\bar{a}_\phi$ direction, A_ϕ component if $\bar{A}$ is present.

$$\therefore \quad 24\,\mu_o = \frac{1}{\rho}\frac{\partial}{\partial\rho}(\rho A_\phi) = \frac{1}{\rho}\frac{d}{d\rho}(\rho A_\phi)$$

$$24\,\mu_o\,\rho = \frac{d}{d\rho}(\rho A_\phi) \quad \text{or} \quad \frac{24\,\mu_o\,\rho^2}{2} = \rho A_\phi$$

$$\therefore \quad A_\phi = 12\,\mu_o\,\rho\ ; \quad \bar{A} = 12\,\mu_o\,\rho\,\bar{a}_\phi \ \text{Wb/m}$$

Problem 7.47 : Let $\bar{J} = 8\,\bar{a}_y$ A/m^2 for $-3 < x < 3$ m, and $\bar{J} = 0$ elsewhere. (a) Obtain expressions for $\bar{H}(x, y, z)$ for $|x| < 3$ and $|x| > 3$m . (b) Obtain expressions for $\bar{B}(x, y, z)$ for $|x| < 3$ and $|x| > 3$m if $\mu = \mu_o$. (c) Find $V_m(x, y, z)$ for $x > 3$ if $V_m = 0$ at P (4, –2, 5). (d) Find $\bar{A}(x, y, z)$ for $-3 < x < 3$ if $\bar{A} = 0$ at Q (2, –2, 4).

Solution :

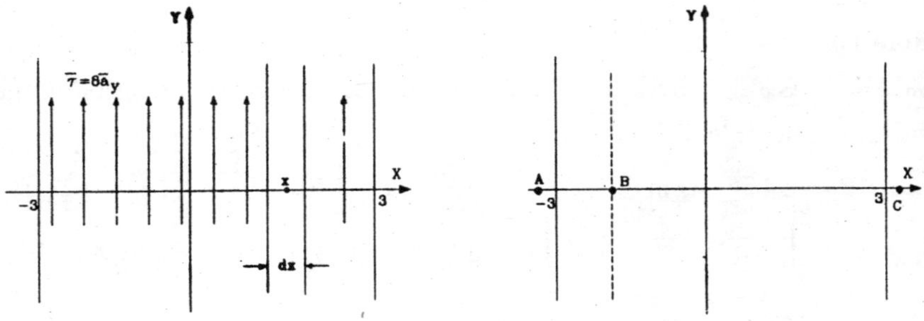

Fig.P. 7.47 (a) (b).

Consider a plane located at x of thickness dx. The surface current density on this plane is $\overline{K} = \overline{J}\, dx = 8\, dx\, \overline{a}_y$ A/m . This plane is infinite in y and z directions. Because of infinite surface current density.

$\overline{H} = \dfrac{\overline{K} \times \overline{a}_N}{2}$ where $\overline{a}_N$ is a unit vector perpendicular to the surface current density and directed towards the point at which $\overline{H}$ is desired.

Consider a point located at A for which $x < -3$

i.e. $|x| > 3$. At this point.

$$\overline{dH} = \frac{8\, dx\, \overline{a}_y \times (-\overline{a}_x)}{2} = 4\, dx\, \overline{a}_z$$

$$\therefore \quad \overline{H} = \int_{-3}^{3} 4\, dx\, \overline{a}_z = 4\Big[\, x \,\Big]_{-3}^{3} \overline{a}_z = 24\, \overline{a}_z \text{ A/m}$$

Consider a point at C fro which $x > 3$

i.e. $|x| > 3$. At this point

$$\overline{dH} = \frac{8\, dx\, \overline{a}_y \times (\overline{a}_x)}{2} = -4\, dx\, \overline{a}_z$$

$$\therefore \quad \overline{H} = \int_{-3}^{3} -4\, dx\, \overline{a}_z = -4\Big[\, x \,\Big]_{-3}^{3} \overline{a}_z$$

$$= -24\, \overline{a}_z \text{ A/m}$$

Consider a point such as B for which $-3 < x < 3$,
i.e. $| x | < 3$. At this point.

$$\overline{H} = \int_{-3}^{x} \frac{(8\,dx\,\overline{a}_y) \times (\overline{a}_x)}{2} + \int_{x}^{3} \frac{(8\,dx\,\overline{a}_y) \times (-\overline{a}_x)}{2}$$

$$= -4\,\overline{a}_z \left\{ \Big[x \Big]_{-3}^{x} - \Big[x \Big]_{x}^{3} \right\}$$

$$= -4\,\overline{a}_z\,[x - (-3)] - [3 - x]$$

$$= -4\,\overline{a}_z\,[2x] = -8\,x\,\overline{a}_z$$

Hence

$$\overline{H} = \begin{cases} 24\,\overline{a}_z \ \text{A/m} & x < -3 \\ -8\,x\,\overline{a}_z \ \text{A/m} & -3 < x < 3 \\ -24\,\overline{a}_z \ \text{A/m} & x > 3 \end{cases}$$

(b) $$\overline{B} = \mu_o \overline{H} = \begin{cases} 24\,\mu_o\,\overline{a}_z \ \text{Wb/m}^2 & x < -3 \\ -8\,\mu_o\,x\,\overline{a}_z \ \text{Wb/m}^2 & -3 < x < 3 \\ -24\,\mu_o\,\overline{a}_z \ \text{Wb/m}^2 & x > 3 \end{cases}$$

(c) For $x > 3$,

$$\overline{H} = -24\,\overline{a}_z \ .$$

$$V_{mQP} = -\int_{P}^{Q} \overline{H} \cdot \overline{dL} = -\int_{P}^{Q} (-24\,\overline{a}_z) \cdot (dz\,\overline{a}_z)$$

$$= \int^{z} 24\,dz = 24\,(z - 5)$$

$$V_{mQP} = V_{mQ} - V_{mP} \quad \text{and} \quad V_{mP} = 0 \ (\text{given})$$

$$\therefore \ V_m = 24\,(z - 5) = 24\,z - 120 \ \text{A}$$

(d) For $-3 < x < 3$,

$$\overline{B} = -8\,\mu_o\,x\,\overline{a}_z$$

$$\overline{B} = \overline{\nabla} \times \overline{A}$$

$$-\,8\,\mu_o\,x\,\bar{a}_z = \left(\frac{\partial\,Ay}{\partial\,x} - \frac{\partial\,Ax}{\partial\,y}\right)\bar{a}_z$$

$$-\,8\,\mu\,x = \frac{\partial\,Ay}{\partial\,x} - \frac{\partial\,A\,x}{\partial\,y}$$

As the current density ($\bar{J}$) or surface current density ($\bar{K}$) is directed in $\bar{a}_y$ direction, Ay is present.

$$\therefore\ -\,8\,\mu_o\,x = \frac{\partial\,Ay}{\partial\,x} = \frac{d\,Ay}{dx}$$

$$Ay = \frac{-\,8\,\mu_o\,x^2}{2} + C$$ where C is a constant of integration.

Ay = 0 at point Q (2, –2, 4)

Hence $0 = -\,4\,\mu_o\,(2)^2 + C$ or $C = 16\,\mu_o$

$$\therefore\ Ay = -\,4\,\mu_o\,x^2 + 16\,\mu_o \quad \text{or}$$

$$\bar{A} = 4\,\mu_o\,(4 - x^2)\,\bar{a}_y\ \text{Wb/m}$$

Problem 7.48 : A certain magnetic field intensity is given in free space as $\bar{H} = \dfrac{20\,(x\,\bar{a}_x + y\,\bar{a}_y)}{(x^2 + y^2)}$ A/m . (a) Show that $\bar{\nabla}\cdot\bar{B} = 0$. (b) Find the current density $\bar{J}$. (c) Find $V_m\,(x\,,y\,,z)$ if $V_m = 0$ at P (1, 1, 1). (d) Find $\bar{A}$ (x, y, z) if $A_x = A_y = 0$ and $A_z = 0$ at P (1, 1, 1).

Solution :

(a) $\bar{\nabla}\cdot\bar{B} = \bar{\nabla}\,.\,(\mu_o\,\bar{H}) = \mu_o\,(\bar{\nabla}\cdot\bar{H})$

$$= \mu_o\left[\frac{\partial\,Hx}{\partial\,x} + \frac{\partial\,Hy}{\partial\,y} + \frac{\partial\,Hz}{\partial\,z}\right]$$

$$= \mu_o\left[\frac{\partial}{\partial\,x}\left(\frac{20x}{x^2 + y^2}\right) + \frac{\partial}{\partial\,y}\left(\frac{20y}{x^2 + y^2}\right)\right]$$

$$= 20\,\mu_o\left[\frac{y^2 - x^2}{(x^2 + y^2)^2} + \frac{x^2 - y^2}{(x^2 + y^2)^2}\right] = 0$$

(b) $$\bar{J} = \bar{\nabla}\times\bar{H} = \begin{vmatrix} \bar{a}_x & \bar{a}_y & \bar{a}_z \\ \partial/\partial\,x & \partial/\partial\,y & \partial/\partial\,z \\ \dfrac{20\,x}{x^2 + y^2} & \dfrac{20\,y}{x^2 + y^2} & 0 \end{vmatrix}$$

$$= \bar{a}_z \left[\frac{-40\,xy}{(x^2 + y^2)^2} + \frac{40\,xy}{(x^2 + y^2)^2} \right] = 0$$

(c) $$V_{mQP} = -\int_P^Q \bar{H} \cdot \overline{dL} = -\left[\int_1^x \frac{20\,x\,dx}{x^2 + y^2} + \int_1^y \frac{20\,y\,dy}{x^2 + y^2} \right]$$

The path from P (1, 1, 1) to Q (x, y, z) is splitted as given below.

P (1 , 1 , 1) --> A (x , 1 , 1) --> B (x , y , 1) --> Q (x y , z)

Between points P and A, x coordinate changes from 1 to x where as y and z are constants equal to 1.

Between points A to B, y coordinate changes from 1 to y, where as z and z are constants, equal to x and 1 respectively.

$$\therefore \quad V_{mQP} = -\left[\int_1^x \frac{20\,x\,dx}{x^2 + (1)^2} + \int_1^y \frac{20\,y\,dy}{x^2 + y^2} \right]$$

$$= -\left\{ 10\Big[\ln (x^2 + 1) \Big]_1^x + 10\Big[\ln (x^2 + y^2) \Big]_1^y \right\}$$

$$= -10\Big[\ln (x^2 + 1) - \ln (1 + 1) + \ln (x^2 + y^2) - \ln (x^2 + 1) \Big]$$

$$= -10 \ln \left(\frac{x^2 + y^2}{2} \right) = 10 \ln \left(\frac{2}{x^2 + y^2} \right) \text{ A}$$

Now $V_{mQP} = V_{mQ} - V_{mP}$ and $V_{mP} = 0$ (given)

$$\therefore \quad V_m = 10 \ln \left(\frac{2}{x^2 + y^2} \right) \text{ A}$$

(d) $$\bar{B} = B_x \bar{a}_x + B_y \bar{a}_y = \bar{\nabla} \times \bar{A} = \left(\frac{\partial Az}{\partial y} - \frac{\partial Ay}{\partial z} \right) \bar{a}_x + \left(\frac{\partial Ax}{\partial z} - \frac{\partial Az}{\partial x} \right) \bar{a}_y$$

$$\therefore \quad Bx = \frac{\partial Az}{\partial z} - \frac{\partial Ay}{\partial z} \quad \text{and} \quad By = \frac{\partial Ax}{\partial z} - \frac{\partial Az}{\partial x}$$

As $Ax = Ay = 0$, we have

$$Bx = \frac{20\,\mu_o\,x}{x^2 + y^2} = \frac{d\,Az}{dy} \quad \text{and} \quad \frac{20\,\mu_o\,y}{x^2 + y^2} = -\frac{d\,Az}{dx}$$

$$A_z = 20\,\mu_o \tan^{-1}\left(\frac{y}{x}\right) + C_1 \quad \text{and} \quad A_z = -\,20\,\mu_o \tan^{-1}\left(\frac{x}{y}\right) + C_2$$

At point P (1, 1, 1), Az = 0

$$\therefore\ 0 = 20\,\mu_o \tan^{-1}(1) + C_1 \quad \text{or} \quad C_1 = -\,5\,\pi\,\mu_o$$

$$\text{and}\ \ 0 = -\,20\,\mu_o \tan^{-1}(1) + C_2 \quad \text{or} \quad C_2 = 5\,\pi\,\mu_o$$

$$\therefore\ A_z = 20\,\mu_o \tan^{-1}\left(\frac{y}{x}\right) - 5\,\pi\,\mu_o \quad \text{and}$$

$$A_z = -\,20\,\mu_o \tan^{-1}\left(\frac{x}{y}\right) + 5\,\pi\,\mu_o$$

Both the answers are correct.

Problem 7.49 : If $\overline{A} = 10\,\rho^{1.5}\,\overline{a}_z$ Wb/m in free space, find : (a) $\overline{H}$; (b) $\overline{J}$; (c) Show that $\oint \overline{H} \cdot \overline{dL} =$ I for a circular path with $\rho = 1$.

Solution :

$$\text{(a)}\quad \overline{B} = \overline{\nabla} \times \overline{A} = \begin{vmatrix} \overline{a}_\rho/\rho & \overline{a}_\rho & \overline{a}_z/\rho \\ \partial/\partial\rho & \partial/\partial\phi & \partial/\partial z \\ 0 & 0 & 10\,\rho^{1.5} \end{vmatrix}$$

$$= -\,\overline{a}_\phi \left[\frac{\partial}{\partial\rho}\,(10\,\rho^{1.5})\right]$$

$$= -\,15\,\rho^{1.5}\,\overline{a}_\phi$$

$$\overline{H} = \overline{B}_{\mu_o} = \frac{0\!-\,16\,\rho^{1.5}}{\mu_o}\,\overline{a}_\phi \ \text{A/m}$$

$$\text{(b)}\quad \overline{J} = \overline{\nabla} \times \overline{H} = \begin{vmatrix} \overline{a}_\rho/\rho & \overline{a}_\phi & \overline{a}_z/\rho \\ \partial/\partial\rho & \partial/\partial\phi & \partial/\partial z \\ 0 & \rho\left(\frac{-15}{\mu_o}\rho^{1.5}\right) & 0 \end{vmatrix}$$

$$= \frac{\overline{a}_z}{\rho}\left[\frac{\partial}{\partial\rho}\left(\frac{-15}{\mu_o}\,\rho^{2.5}\right)\right] = \frac{-\,15 \times 2.5}{\rho\,\mu_o}\,\rho^{1.5}\,\overline{a}_z$$

$$= -\,\frac{37.5}{\rho\,\mu_o}\,\rho^{1.5}\,\overline{a}_z \ \text{A/m}^2 = \frac{-\,37.5\,\rho^{0.5}}{\mu_o}\,\overline{a}_z \ \text{A/m}^2$$

(c) $\oint \overline{H} \quad \overline{dL} = \oint \overline{H} \cdot (\rho\, d\phi\, \overline{A}_\phi)$

$$= \int_0^{2\pi} \left(\frac{-15\rho^{1.5}}{\mu_o} \bar{a}_\phi \right) \cdot (\rho\, d\phi\, \bar{a}_\phi) \Big|_{\rho = 1}$$

$$= \int_0^{2\pi} \frac{-15\rho^{2.5}}{\mu_o} d\phi = \frac{-15(1)^{2.5}}{\mu_o} \left[\phi \right]_0^{2\pi} = -75 \text{ MA}$$

$$I = \int \overline{J} \cdot \overline{ds} = \int \overline{J} \cdot (\rho\, d\rho\, d\phi\, \bar{a}_z)$$

$$= \int_0^1 \int_0^{2\pi} \frac{-37.5\rho^{0.5}}{\mu_o} \rho\, d\rho\, d\phi$$

$$= -\frac{37.5}{\mu_o} \left[\frac{\rho^{2.5}}{2.5} \right]_0^1 \left[\phi \right]_0^{2\pi} = -75 \text{ MA}$$

Problem 7.50 : Show that $\bar{\nabla}_2 \left(\frac{1}{R_{12}} \right) = -\bar{\nabla}_{12} \left(\frac{1}{R_{12}} \right) = \frac{\overline{R}_{21}}{R_{12}^3}$.

Solution :

$$\overline{R}_{12} = (x_2 - x_1)\,\bar{a}_x + (y_2 - y_1)\,\bar{a}_y + (z_2 - z_1)\,\bar{a}_z$$

$$|\overline{R}_{12}| = R_{12} = \left[(x_2 - x_1)^2 + (y_2 - y_1)^2 + (z_2 - z_1)^2 \right]^{1/2}$$

$$\bar{\nabla}_2 (1/R_{12}) = \frac{\partial}{\partial x_2} \left(1/R_{12} \right)^{\bar{a}_x} + \frac{\partial}{\partial y_2} \left(1/R_{12} \right)^{\bar{a}_y} + \frac{\partial}{\partial z_2} \left(1/R_{12} \right)^{\bar{a}_z}$$

$$\frac{\partial}{\partial x_2} \left[\frac{1}{\sqrt{(x_2 - x_1)^2 + (y_2 - y_1)^2 + (z_2 - z_1)^2}} \right]$$

$$= \frac{-(x_2 - x_1)}{\left[(x_2 - x_1)^2 + (y_2 - y_1)^2 + (z_2 - z_1)^2 \right]^{3/2}}$$

$$\frac{\partial}{\partial y_2} \left[\frac{1}{\sqrt{(x_2 - x_1)^2 + (y_2 - y_1)^2 + (z_2 - z_1)^2}} \right]$$

$$= \frac{-(y_2 - y_1)}{\left[(x_2 - x_1)^2 + (y_2 - y_1)^2 + (z_2 - z_1)^2 \right]^{3/2}}$$

$$\frac{\partial}{\partial z_2} \left[\frac{1}{\sqrt{(x_2 - x_1)^2 + (y_2 - y_1)^2 + (z_2 - z_1)^2}} \right]$$

$$= \frac{-(z_2 - z_1)}{\left[(x_2 - x_1)^2 + (y_2 - y_1)^2 + (z_2 - z_1)^2\right]^{3/2}}$$

$$\therefore \; \nabla_2 (1/R_{12}) = -\left\{\frac{(x_2 - x_1)\bar{a}_x + (y_2 - y_1)\bar{a}_y + (z_2 - z_1)\bar{a}_z}{\sqrt{(x_2 - x_1)^2 + (y_2 - y_1)^2 + (z_2 - z_1)}}\right\}$$

$$= \frac{\bar{R}_{21}}{|\bar{R}_{12}|^3}$$

Similarly we can prove that $-\nabla_1 (1/R_{12}) = \dfrac{\bar{R}_{21}}{|\bar{R}_{12}|^3}$

Problem 7.51 : Compute the vector magnetic potential within the outer conductor for the coaxial line whose vector magnetic potential is shown in Fig.P. 7.51 if the outer radius of the outer conductor is 7a. Select the proper zero reference and sketch the results on the figure.

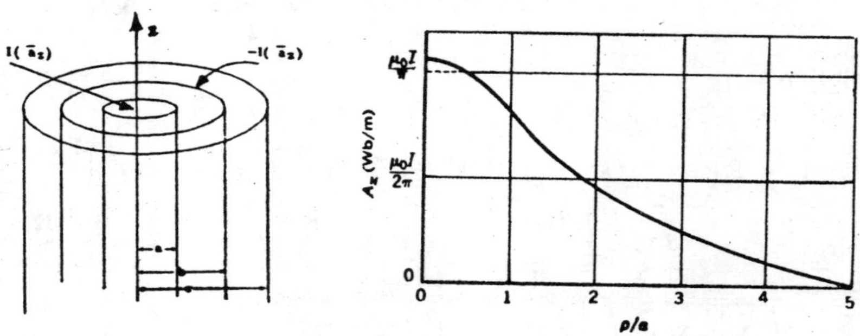

A B

Fig.P. 7.51

Solution :

Within the outer conductor of the coaxial line $\bar{H} = \dfrac{I}{2\pi\rho} \times \dfrac{c^2 - \rho^2}{c^2 - b^2}\,\bar{a}_\phi$

$b = 5a$ and $c = 7a$ is given

$$\therefore \; \bar{H} = \frac{I}{2\pi\rho} \times \left(\frac{49\,a^2 - \rho^2}{49\,a^2 - 25\,a^2}\right) \bar{a}_\phi \;\; \text{A/m}$$

$$\bar{B} = \mu_o \bar{H} = \frac{I \mu_o}{2 \pi \rho} \times \left(\frac{49 a^2 - \rho^2}{24 a^2} \right) \bar{a}_\phi \quad Wb/m^2$$

$$= \bar{\nabla} \times \bar{A} = \left(\frac{\partial A_\rho}{\partial z} - \frac{\partial A_z}{\partial \rho} \right) \bar{a}_\phi$$

As the current is flowing in $\bar{a}_z$ direction, A_ρ component is zero and A_z component is present.

$$\therefore \frac{- d A_z}{d \rho} = \frac{I \mu_o}{2 \pi \rho} \left(\frac{49 a^2 - \rho^2}{24 a^2} \right)$$

$$\therefore A_z = \frac{I \mu_o}{2 \pi} \left(\frac{\rho^2}{48 a^2} - \frac{49}{24} \times \ln \rho \right) + C$$

where C is a constant of integration .

Assuming $A_z = 0$ at $\rho = b = 5a$ we have

$$O = \frac{I \mu_o}{2 \pi} \left(\frac{25 a^2}{48 a^2} - \frac{49}{24} \times \ln 5a \right) + C$$

$$\therefore C = \left(49 a^2 \ln 5a - \frac{25}{48} \right) \frac{I \mu_o}{2 \pi}$$

$$\therefore A_z = \frac{I \times 4 \pi \times 10^{-7}}{2 \pi} \left(\frac{\rho^2}{48 a^2} - 49 \times \ln \rho + 49 \times \ln 5a - \frac{25}{48} \right)$$

$$= \frac{10^{-7} I}{24} \left[\frac{\rho^2}{a^2} - 25 - 98 \ln \left(\frac{\rho}{5a} \right) \right] \quad Wb/m$$

Problem 7.52 : By expanding equation $\bar{\nabla} \times \bar{\nabla} \times \bar{A} = \bar{\nabla}(\bar{\nabla} \cdot \bar{A}) - \bar{\nabla}^2 \bar{A}$ in cartesian coordinates, show that $\bar{\nabla}^2 \bar{A} = \bar{\nabla}^2 A_x \bar{a}_x + \bar{\nabla}^2 A_y \bar{a}_y + \bar{\nabla}^2 A_z \bar{a}_z$ is correct.

Solution :

$$\bar{\nabla}^2 A = - \bar{\nabla} \times \bar{\nabla} \times \bar{A} + \bar{\nabla}(\bar{\nabla} . \bar{A})$$

$$R H S = - \bar{\nabla} \times \bar{\nabla} \times \bar{A} + \bar{\nabla}(\bar{\nabla} . \bar{A})$$

Let $\bar{A} = A_x \bar{a}_x + A_y \bar{a}_y + A_z \bar{a}_z$.

$$\text{Then } \bar{\nabla} \times \bar{A} = \begin{vmatrix} \bar{a}_x & \bar{a}_y & \bar{a}_z \\ \partial/\partial x & \partial/\partial y & \partial/\partial z \\ A_x & A_y & A_z \end{vmatrix}$$

$$= \left(\frac{\partial A_z}{\partial y} - \frac{\partial A_y}{\partial z} \right) \bar{a}_x + \left(\frac{\partial A_x}{\partial z} - \frac{\partial A_z}{\partial x} \right) \bar{a}_y + \left(\frac{\partial A_y}{\partial x} - \frac{\partial A_x}{\partial y} \right) \bar{a}_z$$

$$\therefore \quad \bar{\nabla} \times (\bar{\nabla} \times \bar{A}) = \begin{vmatrix} \bar{a}_x & \bar{a}_y & \bar{a}_z \\ \partial/\partial x & \partial/\partial y & \partial/\partial z \\ \dfrac{\partial A_z}{\partial y} - \dfrac{\partial A_y}{\partial z} & \dfrac{\partial A_x}{\partial z} - \dfrac{\partial A_z}{\partial x} & \dfrac{\partial A_y}{\partial x} - \dfrac{\partial A_x}{\partial y} \end{vmatrix}$$

$$= \left(\frac{\partial^2 A_y}{\partial y \partial x} - \frac{\partial^2 A_x}{\partial y^2} - \frac{\partial^2 A_x}{\partial z^2} + \frac{\partial^2 A_z}{\partial z \partial x} \right) \bar{a}_x$$

$$- \left(\frac{\partial^2 A_y}{\partial x^2} - \frac{\partial^2 A_x}{\partial x \partial y} - \frac{\partial^2 A_z}{\partial z \partial y} + \frac{\partial^2 A_y}{\partial z^2} \right) \bar{a}_y$$

$$+ \left(\frac{\partial^2 A_x}{\partial x \partial z} - \frac{\partial^2 A_z}{\partial x^2} - \frac{\partial^2 A_z}{\partial y^2} + \frac{\partial^2 A_y}{\partial y \partial z} \right) \bar{a}_z$$

Also $\bar{\nabla} \cdot \bar{A} = \dfrac{\partial A_x}{\partial x} + \dfrac{\partial A_y}{\partial y} + \dfrac{\partial A_z}{\partial z}$

$$\therefore \quad \bar{\nabla} (\bar{\nabla} \cdot \bar{A}) = \frac{\partial}{\partial x} \left(\frac{\partial A_x}{\partial x} + \frac{\partial A_y}{\partial y} + \frac{\partial A_z}{\partial z} \right) \bar{a}_x$$

$$+ \frac{\partial}{\partial y} \left(\frac{\partial A_x}{\partial x} + \frac{\partial A_y}{\partial y} + \frac{\partial A_z}{\partial z} \right) \bar{a}_y + \frac{\partial}{\partial z} \left(\frac{\partial A_x}{\partial x} + \frac{\partial A_y}{\partial y} + \frac{\partial A_z}{\partial z} \right) \bar{a}_z$$

$$= \left(\frac{\partial^2 A_x}{\partial x^2} + \frac{\partial^2 A_y}{\partial x \partial y} + \frac{\partial^2 A_z}{\partial x \partial z} \right) \bar{a}_x$$

$$+ \left(\frac{\partial^2 A_x}{\partial y \partial x} + \frac{\partial^2 A_y}{\partial y^2} + \frac{\partial^2 A_z}{\partial y \partial z} \right) \bar{a}_y + \left(\frac{\partial^2 A_x}{\partial z \partial x} + \frac{\partial^2 A_y}{\partial z \partial y} + \frac{\partial^2 A_z}{\partial z^2} \right) \bar{a}_z$$

Hence $\text{RHS} = -\bar{\nabla} \times \bar{\nabla} \times \bar{A} + \bar{\nabla} (\bar{\nabla} \cdot \bar{A})$

$$= \left(\frac{\partial^2 A_x}{\partial x^2} + \frac{\partial^2 A_x}{\partial y^2} + \frac{\partial^2 A_x}{\partial z^2} \right) \bar{a}_x$$

$$+ \left(\frac{\partial^2 A_y}{\partial x^2} + \frac{\partial^2 A_y}{\partial y^2} + \frac{\partial^2 A_y}{\partial z^2} \right) \bar{a}_y + \left(\frac{\partial^2 A_z}{\partial x^2} + \frac{\partial^2 A_z}{\partial y^2} + \frac{\partial^2 A_z}{\partial z^2} \right) \bar{a}_z$$

$$= \bar{\nabla}^2 A_x \bar{a}_x + \bar{\nabla}^2 A_y \bar{a}_y + \bar{\nabla}^2 A_z \bar{a}_z$$

Hence $\bar{\nabla}^2 A = \bar{\nabla}^2 A_x \bar{a}_x + \bar{\nabla}^2 A_y \bar{a}_y + \bar{\nabla}^2 A_z \bar{a}_z$

—X—

CHAPTER 8

Magnetic Forces, Materials & Inductance

8.1. FORCE ON A MOVING CHARGE

The force acting on a charge Q if it is placed in an electric field $\overline{E}$ is given by

$$\overline{F} = Q\overline{E} \qquad \ldots\ldots (8.1)$$

The force acting on a charge Q if it is moving at a velocity $\overline{v}$ in the presence of magnetic flux density $\overline{B}$ is given by

$$\overline{F} = Q\overline{V} \times \overline{B} \qquad \ldots\ldots (8.2)$$

The force on charge Q due to combined electric and magnetic field is given by

$$\overline{F} = Q(\overline{E} + \overline{V} \times \overline{B}) \qquad \ldots\ldots (8.3)$$

This equation is known as the Lorentz force equation.

Problem 8.1 : A negative point charge, Q = −40 nC, is moving with a velocity of 6×10^6 m/s in a direction specified by the unit vector $\overline{a}_v = -0.48\,\overline{a}_x - 0.6\,\overline{a}_y + 0.64\,\overline{a}_z$. Find the magnitude of the vector force exerted on the moving particle by the field : (a) $\overline{B} = 2\overline{a}_x - 3\overline{a}_y + 5\overline{a}_z$ mT ; (b) $\overline{E} = 2\overline{a}_x - 3\overline{a}_y + 5\overline{a}_z$ kV/m ; (c) $\overline{B}$ and $\overline{E}$ acting together.

Solution :

(a) $\bar{V} = 6 \times 10^6 (-0.48\,\bar{a}_x - 0.6\,\bar{a}_y + 0.64\,\bar{a}_z)\ m/sec.$

$$\bar{F} = Q\,\bar{V} \times \bar{B} = -40 \times 10^{-9} \times 6 \times 10^6 \times 10^{-3} \begin{vmatrix} \bar{a}_x & \bar{a}_y & \bar{a}_z \\ -0.48 & -0.6 & 0.64 \\ 2 & -3 & 5 \end{vmatrix}$$

$$= 259.2\,\bar{a}_x + 883.2\,\bar{a}_y - 633.6\,\bar{a}_z\ \mu N$$

$$|\bar{F}| = \sqrt{(259.2)^2 + (883.2)^2 + (-633.6)^2} = 1117.442\ \mu N$$

(b) $\bar{F} = Q\,\bar{E} = -40 \times 10^{-9} (2\,\bar{a}_x - 3\,\bar{a}_y + 5\,\bar{a}_z)\, 10^3$

$$= -80\,\bar{a}_x + 120\,\bar{a}_y - 200\,\bar{a}_z\ \mu N$$

$$|\bar{F}| = \sqrt{(-80)^2 + (120)^2 + (-200)^2} = 246.577\ \mu N$$

(c) $\bar{F} = Q\,\bar{E} + Q\,\bar{V} \times \bar{B}$

$$= (-80\,\bar{a}_x + 120\,\bar{a}_y - 200\,\bar{a}_z) + (259.2\,\bar{a}_x + 883.2\,\bar{a}_y - 633.6\,\bar{a}_z)$$

$$= 179.2\,\bar{a}_x + 1003.2\,\bar{a}_y - 833.6\,\bar{a}_z\ \mu N$$

$$\therefore\ |\bar{F}| = \sqrt{(179.2)^2 + (1003.2)^2 + (-833.6)^2}$$

$$= 1316.591\ \mu N$$

8.2. FORCE ON A DIFFERENTIAL CURRENT ELEMENT

Using Eq. 8.2 , the differential amount of force acting on a differential charge dQ is given by

$$d\bar{F} = dQ\,\bar{V} \times \bar{B}$$

Now

$$\bar{J} = \rho_v \bar{V}$$

and

$$dQ = \rho_v\, dV$$

Thus

$$d\bar{F} = \rho_v\, dV\ \bar{V} \times \bar{B}$$

or

$$\boxed{\mathbf{d\bar{F} = \bar{J} \times \bar{B}\, dV}} \qquad \ldots\ldots (8.4)$$

As $\bar{J}\,dV = \bar{K}\,dS = I\,\overline{dL}$ we have

$$\boxed{\mathbf{d\bar{F} = \bar{K} \times \bar{B}\ dS}} \qquad \ldots\ldots (8.5)$$

and

$$\boxed{\mathbf{d\bar{F} = I\ \overline{dL} \times \bar{B}}} \qquad \ldots\ldots (8.6)$$

Problem 8.2 : Let $V_m = 2x^2 + 4x - 2y^2$ A in a certain region of free space. Find the vector force exerted on a straight wire segment in this region if it extends from the origin to ; (a) $P_A(1, 0, 0)$ and carries 5 A in the $\bar{a}_x$ direction ; (b) $P_B(0, 0, 1)$ and carries 5A in the $\bar{a}_z$ direction ; (c) $P_C(0.6, 0.8, 0)$ and carries 5A away from the origin.

Solution :

$$\bar{H} = -\bar{\nabla} V_m = -\left[\frac{\partial}{\partial x}(2x^2 + 4x - 2y^2)\,\bar{a}_x + \frac{\partial}{\partial y}(2x^2 + 4x - 2y^2)\,\bar{a}_y + \frac{\partial}{\partial z}(2x^2 + 4x - 2y^2)\,\bar{a}_z\right]$$

$$= -(4x + 4)\,\bar{a}_x + 4y\,\bar{a}_y \text{ A/m}$$

$$\therefore\ \bar{B} = 4\mu_o[-(x + 1)\,\bar{a}_x + y\,\bar{a}_y) \text{ Wb/m}^2$$

(a) $d\bar{F} = I\,\overline{dL} \times \bar{B}$ & $\overline{dL} = dx\,\bar{a}_x$ as the straight wire segment lies on the x axis.

$$\therefore\ d\bar{F} = (5\,dx\,\bar{a}_x) \times \left\{4\mu_o[-(x + 1)\,\bar{a}_x + y\,\bar{a}_y]\right\}$$

$$= 20\,\mu_o\,dx\,y\,\bar{a}_z$$

$$\therefore\ \bar{F} = \int d\bar{F} = \int_0^1 20\,\mu_o\,y\,\bar{a}_z\,dx\,\Big|_{y=0} = 0$$

(On the x–axis y coordinate is zero. Also the current is directed from $x = 0$ to $x = 1$).

(b) $\overline{dL} = dz\,\bar{a}_z$ as the straight wire segment lies on the z–axis.

$$\therefore\ d\bar{F} = (5\,dz\,\bar{a}_z) \times \left\{4\mu_o[-(x + 1)\,\bar{a}_x + y\,\bar{a}_y]\right\}$$

$$= (20\,\mu_o\,dz)\left[-(x+1)\,\bar{a}_y - y\,\bar{a}_x\right]$$

$$\therefore\ \bar{F} = -20\,\mu_o\left\{\int_0^1 (x+1)\,\bar{a}_y\,dz\,\Big|_{x=0} + \int_0^1 y\,\bar{a}_x\,dz\,\Big|_{y=0}\right\}$$

$$= -20\,\mu_o\left\{\left[Z\right]_0^1 \bar{a}_y\right\}$$

$$= -20\,\mu_o\,\bar{a}_y = -25.133\,\bar{a}_y\,\mu N$$

(c) $\overline{dL} = dx\,\bar{a}_x + dy\,\bar{a}_y + dz\,\bar{a}_z$

$$\overline{dL} \times \bar{B} = 4\,\mu_o \times \begin{vmatrix} \bar{a}_x & \bar{a}_y & \bar{a}_z \\ dx & dy & dz \\ -(x+1) & y & 0 \end{vmatrix}$$

$$= 4\,\mu_o\left\{(-y\,dz)\,\bar{a}_x - (x+1)\,dz\,\bar{a}_y + (y\,dx + (x+1)\,dy)\,\bar{a}_z\right\}$$

The path from origin (0, 0, 0) to point P_c (0.6 , 0.8 , 0) is splitted as given below

0 (0 , 0 , 0) – – – – > (In x dirn.) A (0.6 , 0 , 0) – – – – > (In y dirn.) B (0.6 , 0.8 , 0) ↓ In z dirn.

P_c (0.6 , 0.8 , 0)

In the X direction, the integration is w.r.t. x. The limits are x = 0 to x = 0.6 where as y = z = 0 are constants.

In the Y direction, the integration is w.r.t. y. The limits are y = 0 to y = 0.8 where as x = 0.6 and z = 0 are contants.

In the Z direction, the integration is w.r.t. Z. The limits are z = 0 to z = 0. the answer of integration is zero.

$$d\bar{F} = I\,\overline{dL} \times \bar{B}\ .\ \text{Hence } \bar{F} = \int \overline{dF} = I \int \overline{dL} \times \bar{B}$$

$$\therefore\ \bar{F} = 5 \times 4\,\mu_o\left\{\int_0^0 -y\,dz\,\bar{a}_x - \int_0^0 (x+1)\,dz\,\bar{a}_y + \int_0^{0.6} y\,dx\,\bar{a}_z + \int_0^{0.8} (x+1)\,dy\,\bar{a}_z\,\Big|_{x=0.6}\right.$$

$$= 20\,\mu_o\left\{\left[0.6y + y\right]_0^{0.8} \bar{a}_z\right\}$$

$$= 20\,\mu_o\,(1.28)\,\bar{a}_z = 32.170\,\bar{a}_z\,\mu N$$

Problem 8.3 : The semiconductor sample shown in Fig.P. 8.3 is p–type silicon, having a rectangular cross section of 0.8 mm by 0.9 cm and a length of 1.25 cm. Assume the electron and hole mobilities are 0.14 and 0.05 m^2 / V.s, respectively, at the operating temperature. With B = 0.05 T and an electric field intensity in the direction of the current flow of 750 V/m, find the magnitude of : (a) the voltage across the sample length ; (b) the drift velocity ; (c) the transverse force per coulomb of moving charge caused by B ; (d) the transverse electric field intensity ; (e) the Hall voltage.

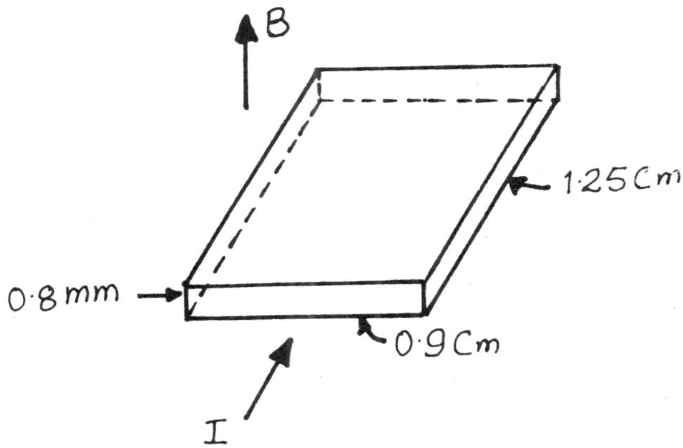

Fig P. 8.3

Solution :

$\mu_e = 0.14 \text{ m}^2/\text{V.s}$

$\mu_h = 0.05 \text{ m}^2/\text{V.s}$

(a) Voltage across the sample length = Electric field intensity in the direction of current flow × length of the sample

$$= 750 \times (1.25 \times 10^{-2}) = 9.375 \text{ V}$$

(b) $V_d = \mu E$

where $\mu = \mu_n$ as the material is p–type silicion.

$\therefore \quad V_d = 0.05 \times 750 = 37.5$ m/s

(d) $\bar{F} = Q(\bar{V} \times \bar{B})$ or $\left|\frac{\bar{F}}{Q}\right| = V B \sin\theta$

where $\theta = 90^{o}$ as $\bar{V}$ and $\bar{B}$ are perpendicular to each other. Hence

$$\left|\frac{\bar{F}}{Q}\right| = 37.5 \times 0.05 \sin 90^{o} = 1.875 \text{ N/C}$$

(d) Transverse electric field intensity $= \left|\frac{\bar{F}}{Q}\right| = 1.875$ V/m .

(e) Hall voltage = Transverse electric field intensity × (0.9×10^{-2})

$$= 1.875 \times (0.9 \times 10^{-2}) = 16.875 \text{ mV.}$$

8.3. FORCE BETWEEN DIFFERENTIAL CURRENT ELEMENTS

Suppose two differential current elements $I_1 \overline{dL_1}$ and $I_2 \overline{dL_2}$ are located at points 1 and 2 respectively. The magnetic flux density at point 2 due to $I_1 \overline{dL_1}$ is given by

$$d\bar{B}_2 = \mu_o \frac{I_1 \overline{dL_1} \times \bar{a}_{12}}{4\pi R_{12}^2}$$

Hence the force on second current element due to first is given by

$$d(d\bar{F}_2) = I_2 \overline{dL_2} \times d\bar{B}_2$$

$$= \frac{\mu_o I_1 I_2}{4\pi R_{12}^2} \overline{dL_2} \times (\overline{dL_1} \times \bar{a}_{R12})$$

Problem 8.4 : A current element, $I_1 \Delta L_1 = 10^{-5} \bar{a}_z$ A · m, is located at $P_1(1, 0, 0)$, while a second element, $I_2 \Delta L_2 = 10^{-5}$ $(0.6\bar{a}_x - 2\bar{a}_y + 3\bar{a}_z)$ A · m is at $P_2(-1, 0, 0)$, both in free space. (a) Find the vector force exerted on $L_2 \Delta L_2$ by $L_1 \Delta L_1$. (b) Find the vector force exerted on $L_1 \Delta L_1$ by $L_2 \Delta I_2$.

Solution :

(a) Magnetic field intensity at point 2 due to $I_1 \Delta\bar{L}_1$ is equal to

$$d\bar{H} = \frac{I_1 \overline{\Delta L_1} \times \bar{a}_{P_1 P_2}}{4\pi [d(P_1 P_2)]^2}$$

$$\bar{R}_{P_1P_2} = (-1 - 1)\,\bar{a}_x = -2\,\bar{a}_x\ ;$$

$$d\,(P_1P_2) = |\,\bar{R}_{P_1P_2}\,| = 2$$

$$\bar{a}_{P_1P_2} = \frac{\bar{R}_{P_1P_2}}{|\,\bar{R}_{P_1P_2}\,|} = \frac{-2\,\bar{a}_x}{2} = -\,\bar{a}_x\text{. Hence}$$

$$d\bar{H} = \frac{10^{-5}\,\bar{a}_z) \times (-\,\bar{a}_x)}{4\,\pi\,(2)^2} = -\,\frac{10^{-5}}{16\,\pi}\,\bar{a}_y\ \text{A/m}$$

$$\therefore\ d\bar{B} = \mu_o\,d\bar{H} = \frac{-\,10^{-5} \times 4\,\pi \times 10^{-7}}{16\,\pi}\,\bar{a}_y$$

$$= -\,0.25 \times 10^{-12}\,\bar{a}_y\ \text{Wb/m}^2$$

Force on $I_2\,\overline{\Delta L_2}$ due to $I_1\overline{\Delta L_1}$ is given by

$$\bar{F} = I_2\,\overline{\Delta L_2} \times d\bar{B}$$

$$= 10^{-5}\,(0.6\,\bar{a}_x - 2\,\bar{a}_y + 3\,\bar{a}_z) \times (-\,0.25 \times 10^{-12}\,\bar{a}_y)$$

$$= 10^{-17}\,(-\,0.15\,\bar{a}_z + 0.75\,\bar{a}_x) = (7.5\,\bar{a}_x - 1.5\,\bar{a}_z)\ 10^{-18}\,\text{N}.$$

(b) Magnetic field intensity at point 2 due to $I_2\,\overline{\Delta L_2}$ is equal to

$$d\bar{H} = \frac{L_2\,\overline{\Delta L_2} \times \bar{a}_{P_2P_1}}{4\,\pi\,[\,d\,(P_2P_1)\,]^2}$$

Now $\bar{a}_{P_2P_1} = -\,\bar{a}_{P_1P_2} = \bar{a}_x$

$$\therefore\ d\bar{H} = \frac{10^{-5}\,(0.6\,\bar{a}_x - 2\,\bar{a}_y + 3\,\bar{a}_z) \times \bar{a}_x}{4\,\pi\,(2)^2}$$

$$= \frac{10^{-5}}{16\,\pi}\,(2\,\bar{a}_z + 3\,\bar{a}_y)\ \text{A/m}$$

$$\therefore\ d\bar{B} = \mu_o\,d\bar{H} = \frac{4\,\pi \times 10^{-7} \times 10^{-5}}{16\,\pi}\,(2\,\bar{a}_z + 3\,\bar{a}_y)$$

$$= 0.25 \times 10^{-12}\,(2\,\bar{a}_z + 3\,\bar{a}_y)\ \text{Wb/m}^2$$

Force on $L_1\,\overline{\Delta L_1}$ due to $I_2\,\overline{\Delta I_2}$ is given by

$$\bar{F} = L_1\,\overline{\Delta L_1} \times d\bar{B}$$

$$= 10^{-5}\,\bar{a}_z \times \left[\,0.25 \times 10^{-12}\,(2\,\bar{a}_z + 3\,\bar{a}_y)\,\right]$$

$$= 10^{-17}\,(-\,0{,}75\,\bar{a}_x) = -\,7.5 \times 10^{-18}\,\bar{a}_x\,\text{N}$$

8.4. FORCE BETWEEN CURRENT CARRYING CONDUCTORS

If two infinitely long current filaments carrying a current I and seperated by a distance d are present, then it can be proved that

(i) if the currents are in the same direction, then the forces are attractive

(ii) if the currents are in the opposite direction, then the forces are repulsive.

(iii) the magnitude of the force is

$$\boxed{F = \frac{\mu_o I_1 I_2}{2 \pi d}} \qquad \ldots\ldots (8.7)$$

8.5. FORCE AND TORQUE ON A CLOSED CIRCUIT

8.5.1. When ever a closed circuit is subjected to a uniform magnetic flux density, the total force on the closed circuit is zero. For a closed circuit, we have

$$\overline{F} = -I \oint \overline{B} \times \overline{dL}$$

If uniform magnetic flux density is assumed, then $\overline{B}$ may be removed from the integral :

$$\overline{F} = \overline{B} \times \oint \overline{dL}$$

However for a closed circuit, $\oint \overline{dL} = 0$. Hence the above statement.

8.5.2. Expression for the torque on a differential current loop in a magnetic field $\overline{B}$.

Fig. 8.1. shows a differential current loop in the xy plane; the sides of the loop are parallel to the x and y axes and are of length dx and dy. The value of the magnetic field at the center of the loop is taken as

$\overline{B}_o = B_{ox}\,\overline{a}_x + B_{oy}\,\overline{a}_y + B_{oz}\,\overline{a}_z$.

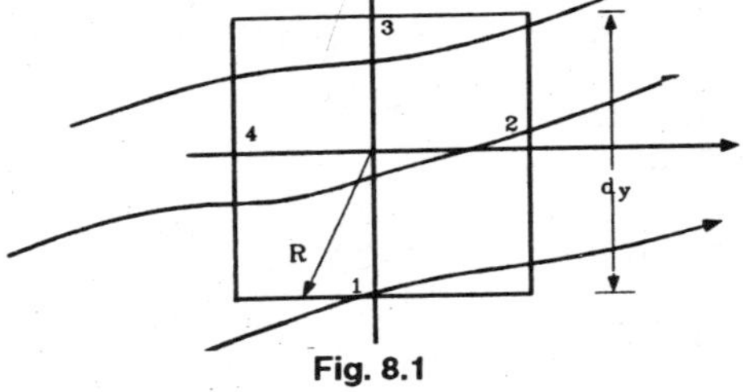

Fig. 8.1

The center of the loop is selected as the origin for the torque.

The vector force on side 1 is

$$d\overline{F}_1 = I\,dx\,\overline{a}_x \times \overline{B}_o = I\,dx(B_{oy}\,\overline{a}_z - B_{oz}\,\overline{a}_y)$$

For this side of the loop, the lever arm $\overline{R}$ extends from the origin to the midpoint of the side, $\overline{R}_1 = -\frac{1}{2}\,dy\,\overline{a}_y$.

Hence the total torque is

$$d\overline{T}_1 = \overline{R}_1 \times d\overline{F}_1$$

$$= -\frac{1}{2}\,dy\,\overline{a}_y \times I\,dx\,(B_{oy}\,\overline{a}_z - B_{oz}\,\overline{a}_y)$$

$$= -\frac{1}{2}\,dx\,dy\,I\,B_{oy}\,\overline{a}_x$$

On the similar way, the torque on side 3 is

$$d\overline{T}_3 = \overline{R}_3 \times d\overline{F}_3 = \frac{1}{2}\,dy\,\overline{a}_y \times (-I\,dx\,\overline{a}_x \times \overline{B}_o)$$

$$= -\frac{1}{2}\,dx\,dy\,I\,B_{oy}\,\overline{a}_x = d\overline{T}_1$$

$$\therefore\; d\overline{T}_1 + d\overline{T}_3 = -dx\,dy\,I\,B_{oy}\,\overline{a}_x$$

Similarly

$$d\overline{T}_2 + d\overline{T}_4 = dx\,dy\,I\,B_{ox}\,\overline{a}_y$$

and hence the total torque is

$$d\overline{T} = I\,dx\,dy\,(B_{ox}\,\overline{a}_y - B_{oy}\,\overline{a}_x)$$

or $d\overline{T} = I\,dx\,dy\,(\overline{a}_z \times \overline{B}_o)$

or

$$\boxed{\mathbf{d\overline{T} = I\,\overline{dS} \times \overline{B}}} \qquad \ldots\ldots\ (8.8)$$

where $\overline{dS}$ is the vector area of the differential current loop.

The quantity $I\,\overline{dS}$ is defined as the differential magnetic dipole moment $\overline{dm}$ with units of $A - m^2$. Hence

$$\boxed{\mathbf{\overline{dm} = I\,\overline{dS}}} \qquad \ldots\ldots\ (8.9)$$

and

$$\boxed{\mathbf{\overline{dT} = \overline{dm} \times \overline{B}}} \qquad \ldots\ldots\ (8.10)$$

Problem 8.5 : A loop of wire is constructed of three straight segments connecting (0, 0, 0) to (0.6, 0, 0) to (0.4, 1, 0.7) to (0, 0, 0). A current of 8 mA is in the $\bar{a}_x$ direction in the first segment. Given a uniform magnetic field, $\bar{B} = 0.2\bar{a}_x - 0.1\bar{a}_y + 0.2\bar{a}_z$ T, find : (a) the force on the segment extending from (0, 0, 0) to (0.6, 0, 0) ; (b) the total force on the loop ; (c) the torque on the loop about an origin at (0, 0, 0) ; (d) the torque on the loop about an orign at (0.45, 0.6, 0.3).

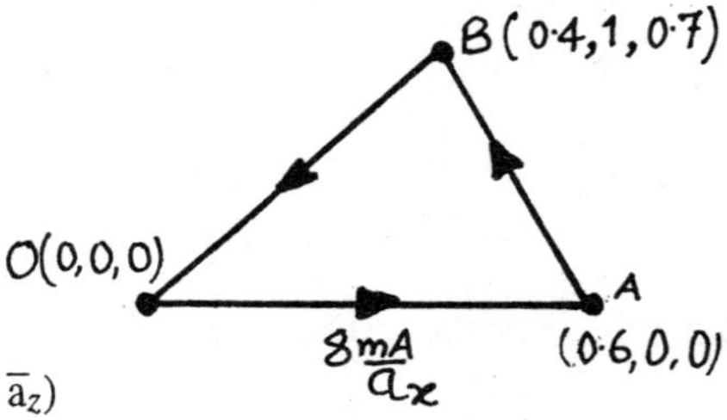

Fig P. 8.5

Solution :

(a) $d\bar{F} = I\,\overline{dL} \times \bar{B}$

$= (8 \times 10^{-3}\, dx\, \bar{a}_x)$

$\times\ (0.2\,\bar{a}_x - 0.1\,\bar{a}_y + 0.2\,\bar{a}_z)$

$= (-\ 0.8\,\bar{a}_z - 1.6\,\bar{a}_y)\, dx$ mN .

$$\therefore\quad \bar{F} = \int d\bar{F} = \int_0^{0.6} (-\ 0.8\,\bar{a}_z - 1.6\,\bar{a}_y)\, dx \text{ mN}$$

$= -\ 0.96\,\bar{a}_y - 0.48\,\bar{a}_z$ mN

(b) As a closed path is subjected to uniform magnetic flux density, the total force on the loop is zero.

(c) $\bar{R}_{OA} = 0.6\,\bar{a}_x$ and $\bar{R}_{OB} = 0.4\,\bar{a}_x + \bar{a}_y + 0.7\,\bar{a}_z$

$$\bar{R}_{OA} \times \bar{R}_{OB} = \begin{vmatrix} \bar{a}_x & \bar{a}_y & \bar{a}_z \\ 0.6 & 0 & 0 \\ 0.4 & 1 & 0.7 \end{vmatrix}$$

$= -\ 0.42\,\bar{a}_y + 0.6\,\bar{a}_z$

$\bar{S}$ = area of triangle OAB $= \frac{1}{2}\left(\bar{R}_{OA} \times \bar{R}_{OB}\right)$

$= -\ 0.21\,\bar{a}_y + 0.3\,\bar{a}_z$ m^2

As the current is in anticlockwise direction, the cross product $\bar{R}_{OA} \times \bar{R}_{OB}$ is considered.

Now, $\overline{T} = I\overline{S} \times \overline{B}$

$$= 8 \times 10^{-3} (-0.21\,\overline{a}_y + 0.3\,\overline{a}_z) \times (0.2\,\overline{a}_x - 0.1\,\overline{a}_y + 0.2\,\overline{a}_z)$$

$$= 8 \times 10^{-3} (-0.012\,\overline{a}_x + 0.06\,\overline{a}_y + 0.042\,\overline{a}_z)$$

$$= -0.096\,\overline{a}_x + 0.48\,\overline{a}_y + 0.336\,\overline{a}_z \text{ mN} - \text{m}$$

(d) Same as c part.

8.6. MAGNETIZATION AND PERMEABILITY

$$\overline{\mathbf{M}} = \mathbf{N}\,\overline{\mathbf{m}} \qquad \ldots\ldots (8.12a)$$

where $\overline{M}$ is the magnetization, i.e. magnetic dipole moment per unit volume, N is the number of atoms per unit volume and m is the magnetic dipole moment per unit volume.

$$\mathbf{I_b} = \oint \overline{\mathbf{M}} \cdot \overline{\mathbf{dL}} = \oint_S \overline{\mathbf{J_b}} \cdot \overline{\mathbf{dS}} \qquad \ldots\ldots (8.12b)$$

$$\mathbf{I_T} = \oint \frac{\overline{\mathbf{B}}}{\mu_o} \cdot \overline{\mathbf{dL}} = \oint_S \overline{\mathbf{J_T}} \cdot \overline{\mathbf{dS}} \qquad \ldots\ldots (8.12c)$$

$$\mathbf{I} = \mathbf{I_T} - \mathbf{I_b} = \oint \left(\frac{\overline{\mathbf{B}}}{\mu_o} - \overline{\mathbf{M}} \right) \cdot \overline{\mathbf{dL}} = \oint_S \overline{\mathbf{J}} \cdot \overline{\mathbf{dS}} \qquad \ldots\ldots (8.12d)$$

$$\overline{\mathbf{J_b}} = \overline{\nabla} \times \overline{\mathbf{M}} \qquad \ldots\ldots (8.12e)$$

$$\overline{\mathbf{J_T}} = \overline{\nabla} \times \frac{\overline{\mathbf{B}}}{\mu_o} \qquad \ldots\ldots (8.12f)$$

$$\overline{\mathbf{J}} = \overline{\nabla} \times \overline{\mathbf{H}} \qquad \ldots\ldots (8.12g)$$

$$\overline{\mathbf{B}} = \mu_o (\overline{\mathbf{H}} + \overline{\mathbf{M}}) \qquad \ldots\ldots (8.12h)$$

$$\overline{\mathbf{M}} = \chi_m \overline{\mathbf{H}} \qquad \ldots\ldots (8.12i)$$

$$\mu_R = 1 + \chi_m \qquad \ldots\ldots (8.12j)$$

where I_b is the bound current, $\bar{J}_b$ is the bound current density, $\bar{M}$ is the magnetization, I_T is the total current, $\bar{J}_T$ is the total current density, $\bar{B}$ is magnetic flux denstiy, I is the free current, $\bar{J}$ is the free current density, and χ_m is magnetic susceptibility.

Problem 8.6 : Find the magnetic field intensity within a magnetic material where : (a) $M = 150$ A/m & $\mu = 1.5 \times 10^{-5}$ H/m ; (b) $\bar{B} = 300\ \mu$T and $\chi_m = 15$; (c) there are 8.2×10^{28} atoms/m^3, each atom has a dipole moment of 5×10^{-27} A · m^2 , and $\mu_R = 30$.

Solution :

(a) $m = \chi_m H = (\mu_R - 1)\ H$

$$\therefore\ H = \frac{M}{\mu_R - 1} = \frac{150}{(1.5 \times 10^{-5} / 4\pi \times 10^{-7}) - 1}$$

$$= 13.715 \text{ A/m}$$

(b) $B = \mu_o \mu_r H = \mu_o (1 + \chi_m) H$

$$\therefore\ H = \frac{B}{\mu_o (1 + \chi_m)} = \frac{300 \times 10^{-6}}{4\pi \times 10^{-7} (1 + 15)} = 14.921 \text{ A/m}$$

(c) $M = 8.2 \times 10^{28} \times 5 \times 10^{-27} = 410$ A/m

$$H = \frac{M}{\mu_R - 1} = \frac{410}{30 - 1} = 14.138 \text{ A/m} .$$

Problem 8.7 : The magnetic flux density in a magnetic material with $\chi_m = 6$ is given in a certain region as $\bar{B} = 0.005\, y^2\, \bar{a}_x$ T. At y = 0.4 m, find the magnitude of : (a) $\bar{J}$; (b) $\bar{J}_b$; (c) $\bar{J}_T$.

Solution :

(a) $\bar{J} = \bar{\nabla} \times \bar{H} = \bar{\nabla} \times \left(\frac{\bar{B}}{\mu_o \mu_r}\right)$;

$\mu_r = 1 + \chi_m = 1 + 6 = 7$

$$\bar{\nabla} \times \bar{B} = \begin{vmatrix} \bar{a}_x & \bar{a}_y & \bar{a}_z \\ \partial/\partial x & \partial/\partial y & \partial/\partial z \\ 0.005\, y^2 & 0 & 0 \end{vmatrix} = -0.01\, y\, \bar{a}_z$$

$$\therefore \quad \bar{J} = \frac{-0.01\, y\, \bar{a}_z}{\mu_o \times 7} ;$$

At $y = 0.4$ m, $\bar{J} = -454.728\, \bar{a}_z$ A/m^2

(b) $\bar{J}_b = \bar{\nabla} \times \bar{m} = \bar{\nabla} \times (\chi_m \bar{H}) = \bar{\nabla} \times \left[\dfrac{\chi_m \bar{B}}{\mu_o \mu_r} \right]$

$$= -0.01\, y\, \bar{a}_z \left(\frac{\chi_m}{\mu_o \mu_r} \right)$$

At $y = 0.4$ m, $\bar{J}_b = -0.01\,(0.4)\, \bar{a}_z \left(\dfrac{6}{4\pi \times 10^{-7} \times 7} \right)$

$$= -2728.37\, \bar{a}_z \text{ A/m}^2$$

(c) $\bar{J}_T = \bar{J}_b + \bar{J} = -2728.37\, \bar{a}_z - 454.728\, \bar{a}_z$

$$= -3183.098\, \bar{a}_z \text{ A/m}^2$$

8.7. MAGNETIC BOUNDARY CONDITIONS

Fig. 8.2. shows a boundary between two isotropic homogeneous linear materials with permeabilities μ_1 and μ_2 .

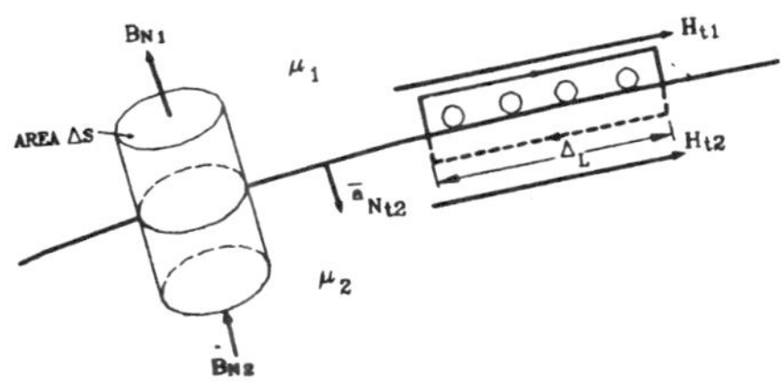

Fig. 8.2

Boundary comditions for normal components :

These conditions are determined by applying Gauss's law for the magnetic field to a small cylindrical surface.

$$\oint_S \overline{B} \cdot \overline{dS} = 0$$

$\therefore \quad B_{N1}\, \Delta S - B_{N2}\, \Delta S = 0$

or

$$\boxed{B_{N1} = B_{N2}} \qquad \text{..... (8.13a)}$$

or

$$\boxed{H_{N2} = \frac{\mu_1}{\mu_2} H_{N1}} \qquad \text{..... (8.13b)}$$

Thus the normal component of magnetic flux density is contineous but the normal component of magnetic field intensity is discontinous by the ratio μ_1 / μ_2 .

Also

$$\boxed{M_{N2} = \frac{\chi_{m2}\, \mu_1}{\chi_{m1}\, \mu_2} M_{N1}} \qquad \text{..... (8.13c)}$$

Boundary conditions for tangential components :

These conditions are determined by applying Ampere's circuital law about a small closed path in a plane normal to the boundary surface as shown in Fig. 8.2. Hence, $H_{t1}\, \Delta L - H_{t2}\, \Delta L = k\, \Delta L$

where it is assumed that the boundary may carry a surfacecurrent $\overline{K}$ whose component normal to the plane of the closed path is K.

$$\boxed{H_{t1} = H_{t2} = K} \qquad \text{..... (8.13d)}$$

This is expressed by the vector formula

$$\boxed{(\overline{H}_1 - \overline{H}_2) \times \overline{a}_{N12} = \overline{K}} \qquad \text{..... (8.13e)}$$

where $\overline{a}_{N12}$ is the unit normal at the boundary directed from region 1 to region 2. In terms of magnetic flux density, we have

$$\boxed{\frac{B_{t1}}{\mu_1} - \frac{B_{t2}}{\mu_2} = K} \qquad \text{..... (8.13f)}$$

Also

$$\boxed{M_{t2} = \frac{\chi_{m2}}{\chi_{m1}} M_{t1} - \chi_{m2}\, K} \qquad \text{..... (8.13g)}$$

Problem 8.8 : Region 1 is the semi–infinite space in which 2x – 5y > 0, while region 2 is defined by 2x – 5y < 0. Let $\mu_{R1} = 3$, $\mu_{R2} = 4$, and $\overline{H}_1 = 30\,\overline{a}_x$ A/m. Find : (a) $|\overline{B}_1|$; (b) $|\overline{B}_{N1}|$; (c) $|H_{t1}|$; (d) $|\overline{H}_2|$.

Solution :

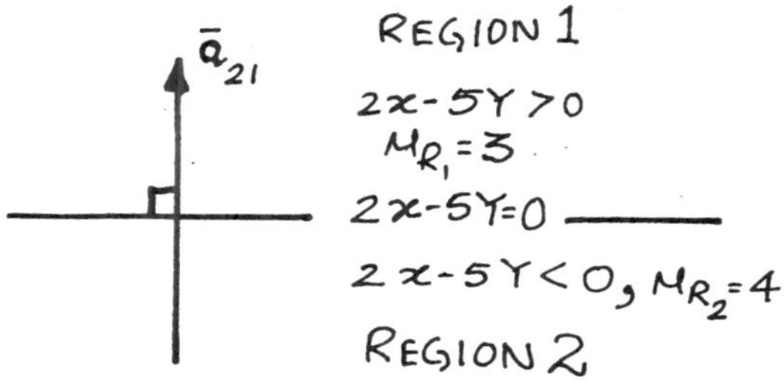

Fig P. 8.8

Let f = 2x – 5y

then f increases from region 2 to region 1.

$\overline{\nabla} f$ gives a vector in the direction of increasing value of f.

Hence, $\overline{\nabla} f = \frac{\partial f}{\partial x}\,\overline{a}_x + \frac{\partial f}{\partial y}\,\overline{a}_y = 2\,\overline{a}_x - 5\,\overline{a}_y$

$$\overline{a}_{21} = \frac{\overline{\nabla} f}{|\overline{\nabla} f|} = \frac{2\,\overline{a}_x - 5\,\overline{a}_y}{\sqrt{(2)^2 + (-5)^2}} = 0.371\,\overline{a}_x - 0.928\,\overline{a}_y$$

(a) $|\overline{B}_1| = \mu_o \mu_1 |\overline{H}_1| = 4\pi \times 10^{-7} \times 3 \times 30 = 113.097\,\mu T$

(b) $\overline{B}_{N1} = (\overline{B}_1 \cdot \overline{a}_{21})\,\overline{a}_{21} = (\mu_o \mu_1 \overline{H}_1 \cdot \overline{a}_{21})\,\overline{a}_{21}$

$= 3\mu_o\,[(30\,\overline{a}_x) \cdot (0.371\,\overline{a}_x - 0.928\,\overline{a}_y)]\,(0.371\,\overline{a}_x - 0.928\,\overline{a}_y)$

$= 15.567\,\overline{a}_x - 38.938\,\overline{a}_y\,\mu T$

$\therefore\ |\overline{B}_{N1}| = \sqrt{(15.567)^2 + (-38.938)^2} = 41.934\,\mu T$

(c) $\overline{B}_1 = \overline{B}_{t1} + \overline{B}_{N1}$; $\overline{B}_{t1} = \overline{B}_1 - \overline{B}_{N1}$

$\therefore\ \overline{B}_{t1} = \mu_o \mu_1 \overline{H}_1 - \overline{B}_{N1} = 113.097\,\overline{a}_x - (15.567\,\overline{a}_x - 38.938\,\overline{a}_y)$

$= 97.530\,\overline{a}_x + 38.938\,\overline{a}_y\,\mu T$

$\therefore\ \overline{H}_{t1} = \frac{\overline{B}_{t1}}{\mu_o \mu_1}$ or $|\overline{H}_{t1}| = \frac{|\overline{B}_{t1}|}{\mu_o \mu_1}$

$$= \frac{\sqrt{(97.530)^2 + (38.938)^2} \times 10^{-6}}{3 \times 4\pi \times 10^{-7}} = 27.856 \text{ A/m}$$

(d) $\overline{B}_{N2} = \overline{B}_{N1} = 15.567\,\overline{a}_x - 38.938\,\overline{a}_y\,\mu\,T$

$$\overline{B}_{t2} = \frac{\mu_2}{\mu_1}\overline{B}_{t1} = \frac{4}{3}(97.530\,\overline{a}_x + 38.938\,\overline{a}_y)$$

$$= 130.04\,\overline{a}_x + 51.917\,\overline{a}_y\,\mu\,T$$

$$\overline{B}_2 = \overline{B}_{N2} + \overline{B}_{t2} = 145.607\,\overline{a}_x + 12.979\,\overline{a}_y\,\mu\,T$$

$$|\overline{H}_2| = \frac{|\overline{B}_2|}{\mu_o\,\mu_{r2}} = \frac{\sqrt{(145.607)^2 + (12.979)^2} \times 10^{-6}}{4\pi \times 10^{-7} \times 4}$$

$$= 29.082 \text{ A/m}$$

8.8. THE MAGNETIC CIRCUIT

Use the following equations for the analysis of magnetic circuits.

$$\boxed{\overline{\mathbf{H}} = -\overline{\nabla}\,\mathbf{V_m}} \qquad \ldots\ldots (8.14a)$$

where $\overline{H}$ and V_m are the magnetic field intensity (A / m) and scalar magnetic potential (A) repectively. V_m is also known as magnetomotive force or mmf. The unit of mmf is ampere or ampere turns. V_m is responsible for the flow of magnetic flux (ϕ) in the magnetic circuit.

$$\boxed{\mathbf{V_{mAB}} = \int_A^B \overline{\mathbf{H}} \cdot \overline{\mathbf{dL}}} \qquad \ldots\ldots (8.14b)$$

$$\boxed{\overline{\mathbf{B}} = \mu_o\,\mu_r\,\overline{\mathbf{H}}} \qquad \ldots\ldots (8.14c)$$

where $\overline{B}$ is the magnetic flux density and the unit is Wb / m^2 or Tesla (T)

$$\boxed{\phi = \int \overline{\mathbf{B}}\ \ \overline{\mathbf{dS}}} \qquad \ldots\ldots (8.14d)$$

$$\boxed{\mathbf{V_m} = \phi\,\mathbf{R}} \qquad \ldots\ldots (8.14e)$$

where R is the reluctance of the magnetic circuit and the unit is $A - t / Wb$. The reluctance opposes the flow of magnetic flux on the application of magnetomotive force. For a linear isotropic homogeneous magnetic material of length d and uniform cross section, the reluctance is given by

$$\boxed{R = \frac{d}{\mu_o \, \mu_r \, S}} \qquad \ldots\ldots (8.14f)$$

Finally Amperes Circuital law gives

$$\boxed{\oint \overline{H} \cdot \overline{dL} = I_{total} = NI} \qquad \ldots\ldots (8.14g)$$

The B – H curve for cast iron, cast steel, silicon steel, nickel–iron alloy is shown in Fig. 8.3.

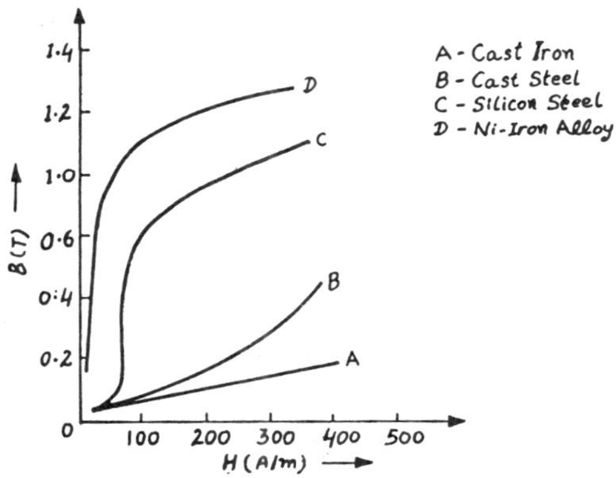

B-H Curve

Problem 8.9 : Given the magnetic circuit of fig.P. 8.9, assume B = 0.6 T at the midpoint of the left leg and find : (a) $V_{m,air}$; (b) $V_{m,steel}$; (c) the current required in a 1250–turn coil linking the left leg.

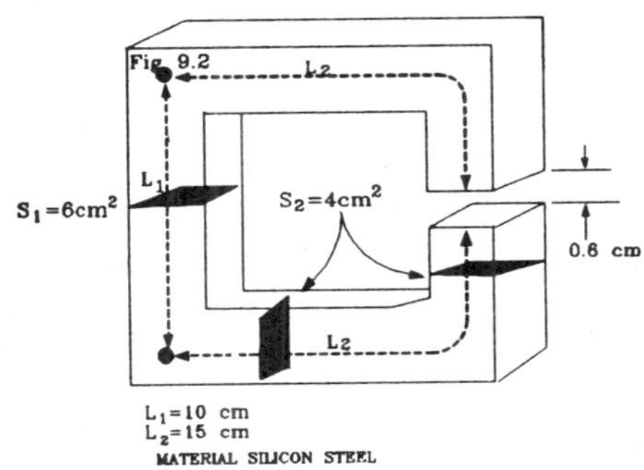

Fig.P. 8.9

Solution :

(a) $R_{air} = \dfrac{d_{air}}{\mu S}$

$$= \frac{0.6 \times 10^{-2}}{4\pi \times 10^{-7} \times 4 \times 10^{-4}} = 11936620.73 \text{ A-t/Wb}$$

Total flux $\phi = BS = 0.6 \times 6 \times 10^{-4} = 3.6 \times 10^{-4}$ Wb

$$V_{m,\,air} = 11936620.73 \times 3.6 \times 10^{-4} = 4297.183 \text{ A-t}$$

(b) Referring to B–H curve for silicon steel, (Fig. 8.3) a magnetic field strength of 100 A-t / m is required to produce a flux density of 0.6 T in the steel.

Thus $H_{steel} = 100$ A– t

$V_{m,\,steel}$ (only for left left) $= H_{steel}\, d_{steel} = 100 \times 0.1 = 10$ A– t

The cross sectional area of right, top and bottom legs is 4 cm^2. Hence the flux density will be

$\dfrac{3.6 \times 10^{-4}}{4 \times 10^{-4}} = 0.9$ T . Corresponding to this value of flux density, from the B–H curve, we have a magnetic field strength of 160 A-t / m.

Thus $V_{m,\,steel}$ (top, bottom and right leg)

$$= 160.0\,(0.15 + 0.15) = 48 \text{ A-t}$$

Hence $V_{m,\,steel} = 10 + 48 = 58$ A– t

(c) Total $V_m = V_{m,\,air} + V_{m,\,steel} = 4297.183 + 48$

$$= 4345.183$$

Now $V_m = NI = 4345.183$ where $N = 1250$

$\therefore\ I = 3.476$ A

Problem 8.10 : For values of B below the knee on the magnetization curve for silicon steel, approximate the curve by a straight line with $\mu = 0.005$ H/m. The core shown in fig.P. 8.10 (a) has areas of 1.5 cm^2 and lengths of 12 cm in each outer leg, and an area of 2.5 cm^2 and a length of 4 cm in the central leg. A coil of 1250 turns carrying 10 mA is placed around the central leg. Find B in the : (a) center leg ; (b) left leg : (c) right leg , if a 0.25 mm air gap is now installed in the center and left legs.

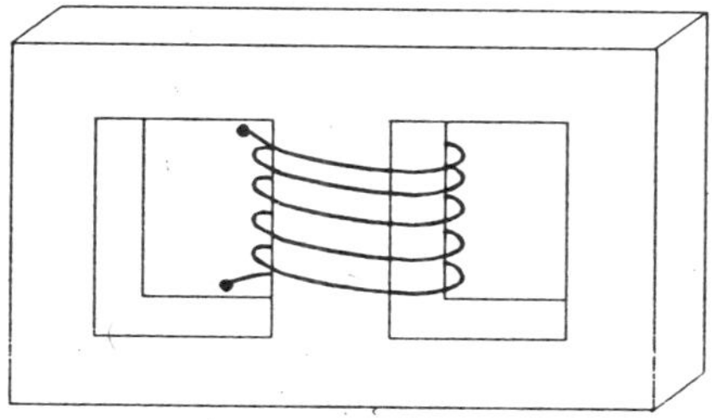

Fig. P. 8.10 (a)

Solution :

(a) $V_m = NI = 1250 \times 10 \times 10^{-3} = 12.5$ A– t

The equivalent electrical circuit for the given problem is shown in Fig.P 8.10(b).

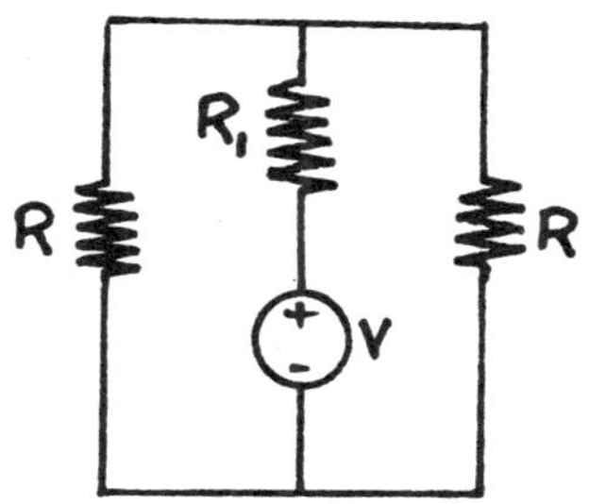

Fig.P. 8.10 (b)

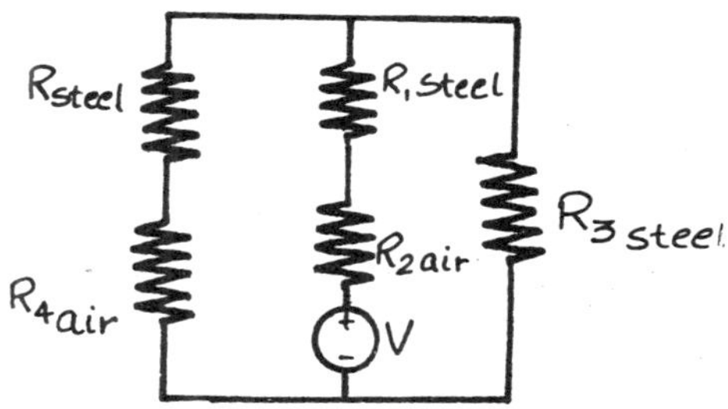

Fig. P. 8.10 (c)

$$\text{Total Resistance} = \frac{R \times R}{R + R} = \frac{R}{2} + R_1$$

$$\text{Hence total reluctance} = R = \frac{1}{2} \times \frac{12 \times 10^{-2}}{0.005 \times 1.5 \times 10^{-4}}$$

$$+ \frac{4 \times 10^{-2}}{0.005 \times 2.5 \times 10^{-4}} = 112000 \text{ A– t/Wb}$$

$$\phi \text{ (total)} = \frac{V_m}{R} = \frac{12.5}{112000} = 1.116071429 \times 10^{-4} \text{ Wb}$$

$$\therefore \text{ B (in center leg)} = \frac{\phi}{S} = \frac{1.116071429 \times 10^{-4}}{2.5 \times 10^{-4}} = 0.4464 \text{ T}$$

(b) The total flux ϕ lies in the center leg. It divides equally in left and right leg.

$$\text{Hence the flux in left leg} = \frac{\phi}{2} = \frac{1.116071429 \times 10^{-4}}{2}$$

$$= 5.580357145 \times 10^{-5}\ \text{Wb}$$

$$\therefore\ B\ (\text{in left leg}) = \frac{\phi}{S} = \frac{5.580357145 \times 10^{-5}}{1.5 \times 10^{-4}} = 0.37202\ \text{T}$$

(c) The equivalent electrical circuit will be as shown in fig. P. 8.10 (c).

$$R_{steel} = \frac{(0.12 - 0.00025)}{0.005 \times 1.5 \times 10^{-4}} = 159666.6667\ \text{A-t/Wb}$$

$$R_{4,\,steel} = \frac{0.00025}{4\pi \times 10^{-7} \times 1.5 \times 10^{-4}} = 1326291.192\text{A-t/Wb}$$

$$R_{1,\,steel} = \frac{(0.04 - 0.00025)}{0.005 \times 2.5 \times 10^{-4}} = 31800\ \text{A-t/Wb}$$

$$R_{2,\,air} = \frac{0.00025}{4\pi \times 10^{-7} \times 2.5 \times 10^{-4}} = 795774.7155\ \text{A-t/Wb}$$

$$R_{3,\,steel} = \frac{0.12}{0.005 \times 1.5 \times 10^{-4}} = 160000\ \text{A-t / Wb}$$

$\therefore$ Total Reluctance

$$= R_{1,\,steel} + R_{2,\,air} + \frac{(R_{steel} + R_{4\,air}) \times R_{3,\,steel}}{R_{steel} + R_{4\,air} + R_{3,\,steel}}$$

$$= 31800 + 795774.7155 + \frac{(159666.6667 + 1326291.192)\ (160000)}{159666.6667 + 1326291.192 + 160000}$$

$$= 972021.4619\ \text{A-t/Wb}$$

$$\text{Hence } \phi\ (\text{total}) = \frac{12.5}{972021.4619} = 1.285979836\ \text{Wb}$$

$$\text{Flux in the right leg} = \frac{\phi\ (R_{steel} + R_{4\,air})}{R_{steel} + R_{4,\,air} + R_{3,\,steel}}$$

$$= \frac{1.285979836 \times 10^{-5}\ (159666.6667 + 1326291.192)}{159666.6667 + 1326291.192 + 160000}$$

$= 1.16097252 \times 10^{-5}$ Wb.

Hence flux density in right leg $= \dfrac{1.16097252 \times 10^{-5}}{1.5 \times 10^{-4}}$

$= 0.077398$ Wb/m^2 or T

8.9 ENERGY DENSITY IN THE MAGNETOSTATIC FIELD

The energy stored in the electric field is given by

$$W_E = \frac{1}{2}\int_{vol} \overline{D} \cdot \overline{E}\, dV$$

Similarly, the energy stored in the magnetic field is given by,

$$\boxed{W_H = \frac{1}{2}\int_{vol} \overline{B}\ \ \overline{H}\, dV} \qquad \ldots\ldots (8.15a)$$

Substituting $\overline{B} = \mu\overline{H}$, we get

$$\boxed{W_H = \frac{1}{2}\int_{vol} \mu\,|\,\overline{H}\,|^2\, dV} \qquad \ldots\ldots (8.15b)$$

or

$$\boxed{W_H = \frac{1}{2}\int_{vol} \frac{B^2\, dV}{\mu}} \qquad \ldots\ldots (8.15c)$$

Problem 8.11 : (a) What force is being exerted on the pole faces of the circuit described in Problem 8.9 and fig.P. 8.9. (b) Is the force trying to open or close the air gap ?

Solution :

$$F = \frac{B_{st}^2\, S}{2\,\mu_o} = \frac{(0.9)^2 \times (4 \times 10^{-4})}{2\,\mu_o} = 128.916\ \text{N}$$

8.10. INDUCTANCE AND MUTUAL INDUCTANCE

8.10.1. Definition :

Let us consider a coil of N turns in which a current I produced a total flux ϕ . Let us assume that this flux links or encircles each of the N turns. The flux linkage N ϕ is defined as the product of thenumber of turns N and the flux

ϕ linking each of them.

Inductance (or self – inductance) is defined as the ratio of the total flux linkages to the current which they link.

$$\mathbf{L = \frac{N\phi}{I}} \qquad \ldots\ldots (8.16)$$

The unit of inductance is the henry (H).

The mutual inductance between circuits 1 and 2, M_{12} , in terms of mutual flux linkages is defined as

$$\mathbf{M_{12} = \frac{N_2 \phi_{12}}{I_1}} \qquad \ldots\ldots (8.17)$$

where ϕ_{12} is the flux produced by I_1 which links the path of the filamentary current I_2 , and N_2 is the number of turns in circuit 2.

8.10.2. Important formulae for inductance :

(1) For a coaxial cable of inner radius a and outer radius b,

$$\mathbf{L = \frac{\mu_o \mu_r}{2\pi} \ln\left(\frac{b}{a}\right) \;\; H/m} \qquad \ldots\ldots (8.18)$$

(2) For a solenoid, with N turns, S as the cross sectional area and d as the length,

$$\mathbf{L = \frac{\mu_o \mu_r N^2 S}{d}} \qquad \ldots\ldots (8.19)$$

(3) For a toroid, with N as the number of turns, h as the height of toroid, pouter and pinner are outer and inner radii respectively,

$$\mathbf{L = \frac{\mu_o \mu_r N^2 h}{2\pi} \ln\left(\frac{pouter}{pinner}\right)} \qquad \ldots\ldots (8.20)$$

(4) For a toroid, with N as the number of turns, S as the cross sectional area, ρ_o as the mean radius,

$$\mathbf{L = \frac{\mu_o \mu_r N^2 S}{2\pi \rho_o}} \qquad \ldots\ldots (8.21)$$

Problem 8.12 : Calculate the inductance of : (a) a 10–m length of coaxial cable filled with a material for which $\varepsilon_R = 18$, $\sigma = 0$, and $\mu_R = 80$, and

having dimensions of a = 1 mm, b = 4 mm ; (b) a toroid formed by surfaces ρ = 3 cm , ρ = 5 cm , z = 0, and z = 1.5 cm, wrapped with 5000 turns of wire, and filled with a magnetic material for which $\mu_R = 6$; (c) a solenoid, 8 cm in length, 2 cm in radius, having $\mu_R = 100$, and carrying 900 turns of wire.

Soluiton :

(a) $$L = \frac{\mu_o \mu_r}{2\pi} \ln\left(\frac{b}{a}\right) \text{ H/m}$$

$$= \frac{4\pi \times 10^{-7} \times 80}{2\pi} \ln\left(\frac{4}{1}\right) = 22.1807 \,\mu\text{ H/m}.$$

For 10 m length, L = 22.1807 × 10 = 221.807 μ H

(b) Using the formula

$$L = \frac{\mu_o \mu_r N^2}{2\pi} \ln\left(\frac{p_o + a}{p_o - a}\right) (Z_2 - Z_1)$$

$$L = \frac{4\pi \times 10^{-7} \times 6\,(5000)^2}{2\pi} \ln\left(\frac{5}{3}\right) (1.5 \times 10^{-2})$$

$$= 0.22987 \text{ H}$$

(c) Using the formula ......(8.19)

$$L = \frac{\mu_o \mu_r N^2 S}{d} = \frac{4\pi \times 10^{-7} \times 100 \times (900)^2 \,[\pi\,(2 \times 10^{-2})^2]}{8 \times 10^{-2}}$$

$$= 1.5988 \text{ H}$$

Problem 8.13 : Each of the square coils shown in fig.P. 8.13 has 1200 turns and a length of 25 cm. The dimensions of the coils are 1.2 × 1.2 cm and 3 × 3cm . (a) With the inner coil and core removed, and a current of 1 mA established in the outer coil, neglect fringing and determine the magnetic field intensity within the outer coil. (b) Calculate the self–inductance of this coil. (c) With the inner coil and core centered inside the larger coil, as shown in fig.P. 8.13, determine the mutual inductance between the coils.

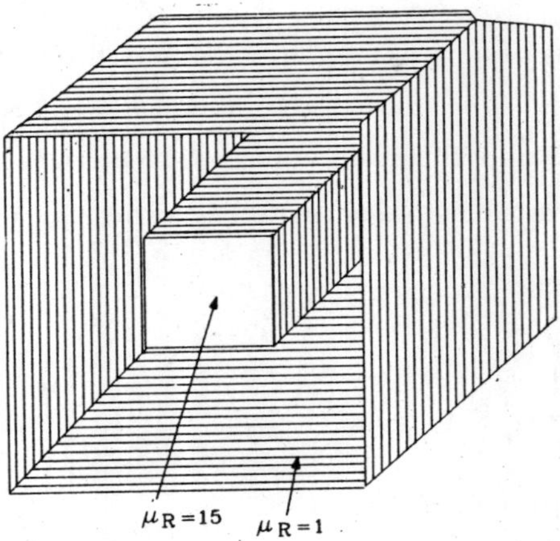

Fig.P. 8.13

Solution :

(a) $H = \dfrac{NI}{d} = \dfrac{1200 \times 1 \times 10^{-3}}{25 \times 10^{-2}} = 4.80 \text{ A/m}$

(b) $L = \dfrac{\mu_o \cdot \mu_r N^2 S}{d} = \dfrac{4\pi \times 10^{-7} \times 1 \times (1200)^2 (3 \times 3 \times 10^{-4})}{25 \times 10^{-2}}$

$= 6.514 \text{ mH}$

(c) $M = \dfrac{\mu_o \mu_r N_1 N_2 S}{d}$

$= \dfrac{4\pi \times 10^{-7} \times 15 \times 1200 \times 1200 \times (1.2 \times 1.2 \times 10^{-4})}{25 \times 10^{-2}}$

$= 15.635 \text{ mH}$

Problem 8.14 : A 0.2 – μC point charge, moving through the field $\overline{E} = 20\,\overline{a}_z$ V/m, is located at the origin at t = 0 with a velocity $v = 10^5\,\overline{a}_y$ m/s. Let m = 2×10^{-16} kg. Make use of Eq. (8.1) and Newton's laws to develop the appropriate differential equations and solve them to evaluate at t = 2 μs: (a) the position P(x,y,z) of the charge; (b) its velocity $\overline{v}$, (c) and its kinetic energy.

Solution :

(a) $\overline{F} = Q\overline{E} = 0.2 \times 10^{-6} \times 20\,\overline{a}_z = 4 \times 10^{-6}\,\overline{a}_z$ N

According to Newoton's Law,

$$\overline{F} = m\,\overline{a} = m\frac{d\,\overline{V}}{dt} = m\frac{d^2(\overline{r})}{dt^2}$$

where $\overline{r}$ is the position vector of the charge

Hence $4 \times 10^{-6}\,\overline{a}_z = 2 \times 10^{-16}\frac{d\,\overline{V}}{dt}$ or $\frac{d\,\overline{v}}{dt} = 2 \times 10^{10}\,\overline{a}_z$

Integrating this equation, we have

$$\overline{V} = 2 \times 10^{10}\,t\,\overline{a}_z + C$$

where C is a constant of integration.

At $t = 0$, $\overline{V} = 10^5\,\overline{a}_y$ m/sec.

$\therefore\quad 10^5\,\overline{a}_y = 2 \times 10^{10}(0)\,\overline{a}_z + C$ or $C = 10^5\,\overline{a}_y$

Hence $\overline{V} = \frac{d(\overline{r})}{dt} = 2 \times 10^{10}\,t\,\overline{a}_z + 10^5\,\overline{a}_y$ m/sec.

Integrating this equation, we have

$$\overline{r} = \frac{2 \times 10^{10}\,t^2}{2}\,\overline{a}_z + 10^5\,t\,\overline{a}_y + C_1$$

where C_1 is a constant of integration.

At $t = 0$, the charge is located at (0, 0, 0).

Hence $\overline{r} = 0\,\overline{a}_x + 0\,\overline{a}_y + 0\,\overline{a}_z = C_1$

Hence $\overline{r} = 10^{10}\,t^2\,\overline{a}_z + 10^5\,t\,\overline{a}_y$

At $t = 2\,\mu S$, $\overline{r} = 10^{10}(2 \times 10^{-6})\,\overline{a}_z + 10^5(2 \times 10^{-6})\,\overline{a}_y$

$$\overline{r} = 0.04\,\overline{a}_z + 0.2\,\overline{a}_y.$$

Hence the position of the charge at $2\,\mu$ S is (0, 0.2, 0.04).

(b) $\overline{V} = 2 \times 10^{10}\,t\,\overline{a}_z + 10^5\,\overline{a}_y$

At $t = 2 \times 10^{-6}$ sec., $\overline{V} = 2 \times 10^{10} \times 2 \times 10^{-6}\,\overline{a}_z + 10^5\,\overline{a}_y$

or $\overline{V} = 10^5\,\overline{a}_y + 4 \times 10^4\,\overline{a}_z$ m/sec.

(c) K.E. $= \frac{1}{2} \times mV^2 = \frac{1}{2} \times 2 \times 10^{-16}(10^{10} + 16 \times 10^8)$

$$= 1.16 \times 10^{-6}\,J = 1.16\,\mu\,J$$

Problem 8.15 : A 0.2 – μC point charge, moving through the field $\overline{B} = 20\,\bar{a}_z$ mT , is located at the origin at t = 0 with a velocity v = $10^5\,\bar{a}_y$ m/s . Let m = 2×10^{-16} kg. Make use of Eq. (8.2) and Newton's laws to develop the appropriate differential equations and solve them to evaluate at t = 2 μs : (a) the position P(x,y,z) of the charge; (b) its velocity $\overline{v}$; (c) and its kinetic energy.

Solution :

Let $\overline{V} = V_x\,\bar{a}_x + V_y\,\bar{a}_y + V_z\,\bar{a}_z$ m/sec.

(a) $\overline{F} = Q\,\overline{V} \times \overline{B}$

$$= 0.2 \times 10^{-6}\left[(V_x\,\bar{a}_x + V_y\,\bar{a}_y + V_z\,\bar{a}_z) \times (20 \times 10^{-3}\,\bar{a}_z)\right]$$

$$= 0.2 \times 10^{-6}(-0.02\,V_x\,\bar{a}_y + 0.02\,V_y\,\bar{a}_x)$$

$$\overline{F} = m\,\bar{a} = m\frac{d\overline{V}}{dt} = 0.2 \times 10^{-6}(-0.02\,V_x\,\bar{a}_y + 0.02\,V_y\,\bar{a}_x)$$

$$(2 \times 10^{-6})\frac{d}{dt}[V_x\,\bar{a}_x + V_y\,\bar{a}_y + V_z\,\bar{a}_z)$$

$$= 0.2 \times 10^{-6}(-0.02\,V_x + 0.02\,V_y\,\bar{a}_x)$$

$$\therefore\ 2 \times 10^{-16}\frac{dV_x}{dt} = 0.2 \times 10^{-6}(0.02\,V_y) \qquad \ldots\ldots (1)$$

$$2 \times 10^{-16}\frac{dV_y}{dt} = 0.2 \times 10^{-6}(-0.02\,V_x) \qquad \ldots\ldots (2)$$

$$2 \times 10^{-16}\frac{dV_z}{dt} = 0 \qquad \ldots\ldots (3)$$

From equation (1) $$\frac{dV_x}{dt} + 2 \times 10^7\,V_y = 0 \qquad \ldots\ldots (4)$$

From equation (2) $$2 \times 10^7\,V_x + \frac{dV_y}{dt} = 0 \qquad \ldots\ldots (5)$$

At t = 0 , $\overline{V} = 10^5\,\bar{a}_y = V_y\,\bar{a}_y$

Hence at t = 0 , $V_y = 10^5$ m/sec.

Solving equations (4) and (5) which are actually simultaneous differential equations with the initial condition that

$$t = 0,\ V_y = 10^5,\ V_x = V_z = 0,$$

we get $V_x(t) = 10^5 \sin(2 \times 10^7 t)$ and

$V_y(t) = 10^5 \cos(2 \times 10^7 t)$.

From equation (3), $V_z = 0$. Hence

$$\overline{V} = 10^5 \sin(2 \times 10^7 t)\,\overline{a}_x + 10^5 \cos(2 \times 10^7 t)\,\overline{a}_y$$

$$\text{Now } \overline{V} = \frac{d(\overline{r})}{dt} = 10^5 \sin(2 \times 10^7 t)\,\overline{a}_x + 10^5 \cos(2 \times 10^7 t)\,\overline{a}_y$$

Integrating above equation, we get

$$\overline{r} = \frac{-10^5 \cos(2 \times 10^7 t)\,\overline{a}_x}{2 \times 10^7} + \frac{10^5 \sin(2 \times 10^7 t)\,\overline{a}_y}{2 \times 10^7} + C$$

$$= -0.005 \cos(2 \times 10^7 t)\,\overline{a}_x + 0.005 \sin(2 \times 10^7 t)\,\overline{a}_y + C$$

At $t = 0$, the charge is located at origin. That is at

$$t = 0,\ \overline{r} = 0\,\overline{a}_x + 0\,\overline{a}_y + 0\,\overline{a}_z$$

$$= -0.005\,\overline{a}_x + 0\,\overline{a}_y + C \text{ or } C = 0.005\,\overline{a}_x$$

Hence $\overline{r} = 0.005[1 - \cos(2 \times 10^7 t]\,\overline{a}_x + 0.005 \sin(2 \times 10^7 t)\,\overline{a}_y$

At $t = 2\,\mu S$, $\overline{r} = (8.335\,\overline{a}_x + 3.726\,\overline{a}_y)\,10^{-3}$

Hence the position of the charge is (0.008335, 0.003726, 0).

(b) $\overline{V} = 10^5 \sin(2 \times 10^7 t)\,\overline{a}_x + 10^5 \cos(2 \times 10^7 t)\,\overline{a}_y$

At $t = 2\,\mu S$, $\overline{V} = 10^5 (0.745\,\overline{a}_x - 0.667\,\overline{a}_y)$ m/sec..

(c) $$\text{K.E.} = \frac{1}{2} mV^2$$

$$= \frac{1}{2} \times 2 \times 10^{-16} \times 10^{10} \left[(0.745)^2 + (-0.667)^2 \right]$$

$$= 1\,\mu J$$

Problem 8.16 : A charge, $Q = 5 \times 10^{-18}$ C, is moving through a uniform magnetic field, $\overline{B} = -0.4\,\overline{a}_x + 0.2\,\overline{a}_y - 0.1\,\overline{a}_z$ T, with a velocity, $\overline{V} = (2\,\overline{a}_x - 3\overline{a}_y + 6\,\overline{a}_z)\,10^5$ m/s, at $t = 0$. (a) What electric field is present at $t = 0$ if the net force on the electron is zero ? (b) If the electric field intensity is entirely in the $\overline{a}_x$ direction and $|\overline{F}_{net}| = 2$ pN at $t = 0$, find E_x.

Solution :

(a) $\overline{F} = Q(\overline{E} + \overline{V} \times \overline{B}) = 0$

Hence $\overline{E} + \overline{V} \times \overline{B} = 0$

i.e. $\overline{E} = -\overline{V} \times \overline{B}$

$$\therefore \quad \overline{E} = -10^5 \times \begin{vmatrix} \overline{a}_x & \overline{a}_y & \overline{a}_z \\ 2 & -3 & 6 \\ -0.4 & 0.2 & -0.1 \end{vmatrix}$$

$$= 90\,\overline{a}_x + 220\,\overline{a}_y + 80\,\overline{a}_z \text{ kV/m}$$

(b) $\overline{E} = E_x\,\overline{a}_x$

$$\overline{F} = Q(\overline{E} + \overline{V} \times \overline{E})$$

$$= 5 \times 10^{-18}\left[E_x\,\overline{a}_x - 10^3(90\,\overline{a}_x + 220\,\overline{a}_y + 80\,\overline{a}_z) \right]$$

$$= 5 \times 10^{-18}\left[(E_x - 90000)\,\overline{a}_x - 220000\,\overline{a}_y - 80000\,\overline{a}_z \right]$$

$$|\overline{F}| = 5 \times 10^{-18}\sqrt{(E_x - 90000)^2 + (-220000)^2 + (-80000)^2}$$

$$= 2 \times 10^{-12}$$

$$\therefore \quad (E_x - 90000)^2 + (220000)^2 + (80000)^2 = (400000)^2$$

$$\therefore \quad E_x - 90000 = \pm\ 324345.4948$$

$$\therefore \quad E_x = -234.345 \text{ KV/m} \quad \text{or} \quad 414.345 \text{ KV/m}.$$

Problem 8.17 : An electron ($q_e = -1.602\,18 \times 10^{-19}$ C, m = 9.109 39 $\times\ 10^{-31}$ kg) is moving at a constant velocity, $\overline{v} = 4 \times 10^7\,\overline{a}_x$ m/s, along the negative x axis. At the origin, it encounters the uniform magnetic field. $\overline{B} = -2 \times 10^{-3}\,\overline{a}_y\ \overline{T}$, and remains in it up to x = 3 cm. If we assume (with good accuracy) that the electron remains on the x axis while it is in the magnetic field, find its y– and z– coordinate values when x = 0.4 m.

Solution :

$$\overline{F} = Q\,\overline{V} \times \overline{B}$$

$$= -1.60218 \times 10^{-19}\left[(4 \times 10^7\,\overline{a}_x) \times (-2 \times 10^{-3}\,\overline{a}_y) \right]$$

$$= 1.281744 \times 10^{-14}\,\overline{a}_z \text{ N}$$

Now, $\overline{F} = m\,\overline{a}$

$$\therefore \ \bar{a} = \frac{\bar{F}}{m} = \frac{1.281744 \times 10^{-14}\,\bar{a}_z}{9.10939 \times 10^{-31}}$$

$$= 1.407 \times 10^{16}\,\bar{a}_z$$

Problem 8.18 : In a certain region of space, $\bar{B}$ is given by $0.1\,x\,\bar{a}_x + 0.2\,y\,\bar{a}_y - 0.3\,z\,\bar{a}_z$ T. Find the total force on the rectangular loop shown in Fig.P 8.18 if it lies in the z = 0 plane and is bounded by x = 1, x = 3, y = 2, and y = 5 m.

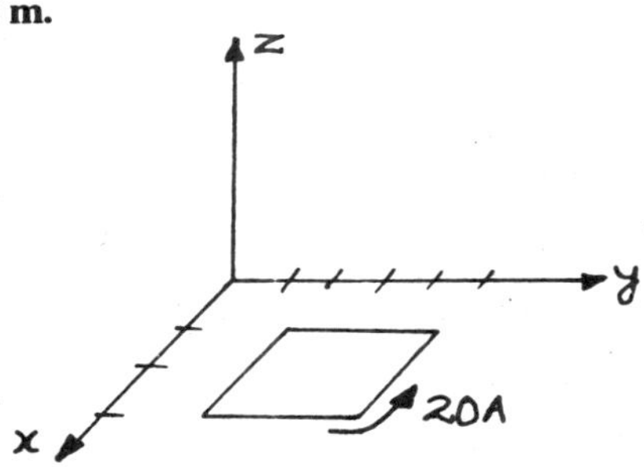

Fig. P. 8.18

Solution :

Force on AB : x and z coordinates are constant.

Hence $\overline{dL} = dy\,\bar{a}_y$

$$d\bar{F} = (20\,dy\,\bar{a}_y) \times (0.1x\,\bar{a}_x + 0.2y\,\bar{a}_y - 0.3z\,\bar{a}_z)$$

$$= 20dy\,(-0.1x\,\bar{a}_z - 0.3z\,\bar{a}_x)$$

$$\therefore \ \bar{F}_{AB} = \int_2^5 20\,(-0.1x\,\bar{a}_z - 0.3z\,\bar{a}_x)\,dy \,\Big|_{\substack{x=3\\ z=0}}$$

$$= -2 \times 3\,\Big[\,y\,\Big]_2^5\,\bar{a}_z = -18\,\bar{a}_z \text{ N}$$

$$\text{Similarly } \bar{F}_{CD} = \int_5^2 20\,(-0.1x\,\bar{a}_z - 0.3z\,\bar{a}_x)\,dy \,\Big|_{\substack{x=1\\ z=0}}$$

$$= -2 \times 1\Big[\,y\,\Big]_5^2\,\bar{a}_z = 6\,\bar{a}_z \text{ N}.$$

Force on BC : y and z coordinates are constant.

Hence $\overline{dL} = dx\,\bar{a}_x$

$$d\bar{F} = (20\,dx\,\bar{a}_x) \times (0.1x\,\bar{a}_x + 0.2y\,\bar{a}_y - 0.3z\,\bar{a}_z)$$

$$= 20dx\,(0.2y\,\bar{a}_z + 0.3z\,\bar{a}_y)$$

$$\therefore \quad \overline{F}_{BC} = \int_3^1 20\,(0.2y\,\overline{a}_z + 0.3z\,\overline{a}_y)\,dx \,\Big|_{\substack{y=5\\z=0}}$$

$$= 4 \times 5 \left[x \right]_3^1 \overline{a}_z = -40\,\overline{a}_z \text{ N}$$

Similarly $\overline{F}_{DA} = \int_1^3 20\,(0.2y\,\overline{a}_z + 0.3z\,\overline{a}_y)\,dx \,\Big|_{\substack{y=2\\z=0}}$

$$= 4 \times 2 \left[x \right]_1^3 \overline{a}_z = 16\,\overline{a}_z \text{ N}$$

$$\overline{F}_{total} = -18\,\overline{a}_z + 6\,\overline{a}_z - 40\,\overline{a}_z + 16\,\overline{a}_z, \quad = -36\,\overline{a}_z \text{ N}$$

Problem 8.19 : A current fliament on the z axis produces the fielo $\overline{H} = -2\overline{a}_x / Y$ A/m at P (0, y, z) in free space. (a) Find the force exerted on a current strip lying between y = 1 and 3 m in the x = 0 plane carrying a current density $\overline{K} = -2\pi\,\overline{a}_z$ A/m . (b) What force is exerted on the current filament by the current strip ?

Solution :

(a) $\overline{a}_x$ is perpendicular to x = 0 plane.

Hence $\overline{dS} = dy\,dz\,\overline{a}_x$ or $dS = dy\,dz$.

$$\therefore \quad \overline{dF} = \overline{K} \times \overline{B}\,dS = (-2\pi\,\overline{a}_z) \times \frac{(-2\mu_o\,\overline{a}_x)}{y}\,dy\,dz$$

$$= \frac{4\pi\mu_o\,\overline{a}_y}{y}\,dy\,dz$$

$$\therefore \quad \overline{F} = \int_1^3 \int_0^L \frac{4\pi\mu_o}{y}\,dy\,dz\,\overline{a}_y$$

$$= 4\pi\mu_o \left[\ln y \right]_1^3 \left[z \right]_0^L \overline{a}_y$$

$$\therefore \quad \frac{\overline{F}}{L} = 4\pi\mu_o \ln 3\,\overline{a}_y = 17.3485\,\overline{a}_y \text{ uN/m}.$$

(b) $\overline{dH} = \dfrac{\overline{K}\,dS \times \overline{a}_{12}}{4\pi d^2}$

Consider any point (0, 0, a) on the z axis as point 2. Point 1 is (0, y, z) which is the general point on the current strip.

$$\therefore \quad \overline{dH} = \frac{(-2\pi\,\overline{a}_z\,dy\,dz) \times [-y\,\overline{a}_y + (a-z)\,\overline{a}_z]}{4\pi\,[y^2 + (a-z)^2]^{3/2}}$$

$$= \frac{-\, y\, dy\, dz\, \bar{a}_x}{2\,[\,y^2 + (a - z)^2\,]^{3/2}}$$

$$\therefore \quad \bar{H} = \int d\bar{H} = \int_1^3 \int_{-\infty}^{\infty} \frac{-\, y\, dy\, dz\, \bar{a}_z}{2\,[\,y^2 + (a - z)^2\,]^{3/2}}$$

Substitute $a - z = y \tan \theta$. So that $dz = -\, y \sec^2 \theta \, d\theta$ and $[\,y^2 + [a - z)^2\,]^{3/2} = y^3 \sec^3 \theta$. Further, when $z = -\infty$, $\theta = 90^o$ and when $z = \infty$, $\theta = -90^o$. Hence

$$\bar{H} = \int_1^3 \int_{90^o}^{-90^o} \frac{y^2\, dy\, \sec^2 \theta\, d\theta\, \bar{a}_x}{2\, y^3 \sec^3 \theta}$$

$$= \int_1^3 \frac{dy}{2y} \Big[\sin \theta \Big]_{90^o}^{-90^o} \bar{a}_x$$

$$= \Big[\ln y \Big]_1^3 \bar{a}_x = 1.0986\, \bar{a}_x \text{ A/m}$$

when the current filament is located in the axis then, at point (0, y, z) the magnetic field intensity is given by $\dfrac{I}{2\pi y}$

aS $\bar{H} = -\, 2\,\bar{a}_x / y$ is given, we have

$$|\,\bar{H}\,| = \frac{2}{y} = \frac{I}{2\pi y} \quad \text{or} \quad I = 4\pi \text{ A}$$

Consider a differential current element located at point (0, 0, a). Hence force on the current filament by the current strip is given by

$$d\bar{F} = I\, d\bar{L} \times \bar{B} = 4\pi\, dz\, \bar{a}_z \times (1.0986\, \mu_o\, \bar{a}_x)$$

$$= -\, 13.8054\, \mu_o\, dz\, \bar{a}_y$$

$$\therefore \quad \frac{d\bar{F}}{dz} = -\, 13.8054\, \mu_o\, \bar{a}_y = -\, 17.348\, \bar{a}_y \;\mu\text{ N/m}.$$

Problem 8.20. : A straight solid wire segment carrying a current $2\,\bar{a}_y$ A extends from A (0, 1, 2) to B (0, 4, 2) in free spae. This wire is subjected to the magnetic field of an infinite current filament lying along the z axis and carrying 25 A in the $\bar{a}_z$ direction. Find the vector torque on the wire segment about an origin at : (a) P_A (0, 0, 2) ; (b) P_B (0, 0, 0) ; (c) P_C (0, 2, 0).

Solution :

The wire segment extends from A (0, 1, 2) to B (0, 4, 2). In points A and B, x and z coordinates are constant. Hence the general point on the wire segment is given by (0, y, 2). At this point we will consider a differential current element $I\,\overline{dL} = 2\,,dy\,\bar{a}_y$ A . x . At point (0, y, 2), the magnetic fielld intensity due to infinite current filament lying along the z axis is given by

$$\overline{H} = \frac{-25}{2\pi y}\,\bar{a}_x \text{ A/m}$$

$$\text{or } \overline{B} = \frac{-25\mu_o}{2\pi y}\,\bar{a}_x \text{ Wb/m}^2$$

$$\therefore\quad d\overline{F} = 2\,dy\,\bar{a}_y \times \left(\frac{-25\mu_o}{2\pi y}\,\bar{a}_x\right) = \frac{25\mu_o\,dy\,\bar{a}_z}{\pi y}\text{ N}$$

Now $d\overline{T} = \bar{r} \times d\overline{F}$

(a) $\bar{r}$ = a vector directed from (0, 0, 2) towards (0, y, 2).

$$= y\,\bar{a}_y$$

$$\therefore\quad d\overline{T} = y\,\bar{a}_y \times \left[\frac{25\mu_o\,dy}{\pi y}\,\bar{a}_z\right] = \frac{25\mu_o\,dy}{\pi}\,\bar{a}_x$$

$$\overline{T} = \int d\overline{T} = \int_1^4 \frac{25\mu_o\,dy}{\pi}\,\bar{a}_x = \frac{25\mu_o}{\pi}\,\bar{a}_x\,(4 - 1)$$

$$= 30\,\bar{a}_x\ \mu\text{N} - \text{m}.$$

(b) $\bar{r}$ = a vector directed from (0, 0, 0) towards (0, y, 2).

$$= y\,\bar{a}_y + 2\,\bar{a}_z$$

$$\therefore\quad d\overline{T} = (y\,\bar{a}_y + 2\,\bar{a}_z) \times \left[\frac{25\mu_o\,dy}{\pi y}\,\bar{a}_z\right]> \ , = \frac{25\mu_o\,dy}{\pi}\,\bar{a}_x$$

$$\therefore\quad \overline{T} = \int dT = \int_1^4 \frac{25\mu_8}{\pi}\left[\frac{1-2}{y}\right] dy\,\bar{a}_x$$

$$= \frac{25\mu_o}{\pi}\Big[y - 2\ln y\Big]_1^4\,\bar{a}_x\ ,\ = 2.274\,\bar{a}_x\ \mu\text{N} - \text{m}$$

Problem 8.21 : In An infinite filament carrying 12 $\bar{a}_z$ A lies on the z axis, and a straight solid conductor segment extends from P_A (1, –1, 0) to P_B (1, 1, 0), carrying 2 $\bar{a}_y$ A. Find the torque on the conductor segment about an origin at P_0 (1, 0, 0) as a function of I.

Solution :

The solid conductor extends from P_A (1, –1, 0) to P_B (1, 1, 0). In these points x and z coordinates are constant. Hence general point on solid conductor is given by (1, y, 0). At this point $\overline{H}$ due infinite current of $12\,\overline{a}_z$ A on the z axis is given by

$$\overline{H} = \frac{12}{2\pi\sqrt{1+y^2}}\left\{\overline{a}_z \times \left[\frac{\overline{a}_x + y\,\overline{a}_y}{\sqrt{1+y^2}}\right]\right\}$$

$$= \frac{6(-y\,\overline{a}_x + \overline{a}_y)}{\pi(1+y^2)}\ \text{A/m}$$

$$\therefore\quad \overline{B} = \mu_o\overline{H} = \frac{6\mu_o(-y\,\overline{a}_x + \overline{a}_y)}{\pi(1+y^2)}\ \text{Wb/m}^2$$

Consider a differential current element $I\,\overline{dL} = 2\,dy\,\overline{a}_y$ A – m located at point (1, y, 0). Then

$$d\overline{F} = I\,\overline{dL} \times \overline{B} = 2\,dy\,\overline{a}_y \times \left[\frac{6\mu_o(-y\,\overline{a}_x + \overline{a}_y)}{\pi(1+y^2)}\right]$$

$$= \frac{12\mu_o\,y\,\overline{a}_z\,dy}{\pi(1+y^2)}\ \text{N}$$

$$d\overline{T} = \overline{r} \times d\overline{F}$$

where $\overline{r}$ = a vector directed from (1, 0, 0) towards (1, y, 0)

$$= y\,\overline{a}_y$$

$$\therefore\quad d\overline{T} = y\,\overline{a}_y \times \frac{12\mu_o\,y\,\overline{a}_z}{\pi(1+y^2)}\,dy$$

$$= \frac{12\mu_o\,y^2\,\overline{a}_x}{\pi(1+y^2)}\,dy\ \text{N – m}$$

$$\text{M}\quad \overline{T} = \int d\overline{T} = \int_{-1}^{1} \frac{12\mu_o\,y^2\,\overline{a}_x}{\pi(1+y^2)}$$

$$= \frac{12\mu_o\,\overline{a}_x}{\pi}\left[y - \tan^{-1}y\right]_{-1}^{1}$$

$$= 9.6\,[1 - \tan^{-1}1]\,\overline{a}_x\ \mu\text{N – m}.$$

Problem 8.22 : Find the vector torque on the square wire loop shown in Fig. 9.18 (a) about the origin by B = $0.8\,\overline{a}_x$ T ; (b) about the origin by

$B = 0.6\,\bar{a}_x - 0.8\,\bar{a}_y$ T ; (c) about P (2, 3, 4) by $B = 0.4\,\bar{a}_x - 0.6\,\bar{a}_y - 0.7\,\bar{a}_z$ T ; (d) about P (2, 3, 4) by the field $B = 0.4\,\bar{a}_x + 0.6\,\bar{a}_y - 0.7\,\bar{a}_z$ T acting only on that side of the loop which y = 1.

Solution :

(a) $\bar{B}$ is uniform. Hence

$$\bar{T} = I\bar{S} \times \bar{B} = 0.6\,(2 \times 2\,\bar{a}_z) \times (0.8\,\bar{a}_x) = 1.92\,\bar{a}_y \text{ N} - \text{m}$$

(b) $\bar{T} = 0.6\,(2 \times 2\,\bar{a}_z) \times (0.6\,\bar{a}_x - 0.8\,\bar{a}_y) = 1.92\,\bar{a}_x + 1.44\,\bar{a}_y$ N − m

(c) $\bar{T} = 0.6\,(2 \times 2\,\bar{a}_z) \times (0.4\,\bar{a}_x + 0.6\,\bar{a}_y - 0.7\,\bar{a}_z)$

$= -1.44\,\bar{a}_x + 0.96\,\bar{a}_y$ N − m.

(d) Torque remains the same as that of part (c).

(e) Any point on the side y = 1 is given by (x, 1, 0).

At this point consider a differential current elemtn $I\,d\bar{L} = -0.6\,dx\,\bar{a}_x$ A.m. On this differential current element, the force due to $\bar{B}$ is

$$d\bar{F} = I\,d\bar{L} \times \bar{B} = (-0.6\,dx\,\bar{a}_x) \times (0.4\,\bar{a}_x + 0.6\,\bar{a}_y - 0.7\,\bar{a}_z)$$
$$= (-0.42\,\bar{a}_y - 0.36\,\bar{a}_z)\,dx$$

$\bar{r}$ = a vector directed from (2, 3, 4) towards (x, 1, 0)

$= (x - 2)\,\bar{a}_x - 2\,\bar{a}_y - 4\,\bar{a}_z$

$$\therefore \quad d\bar{T} = \bar{r} \times \bar{F} = dx \begin{vmatrix} \bar{a}_x & \bar{a}_y & \bar{a}_z \\ x-2 & -2 & -4 \\ 0 & -0.42 & -0.36 \end{vmatrix}$$
$$= \left[-0.96\,\bar{a}_x + 0.36\,(x - 2)\,\bar{a}_y - 0.42\,(x - 2)\,\bar{a}_z \right] dx$$

$$\therefore \quad \bar{T} = \int d\bar{T} = \int_1^{-1} -0.96\,dx\,\bar{a}_x + \int_1^{-1} 0.36\,(x - 2)\,dx\,\bar{a}_y - \int_1^{-1} 0.42\,(x - 2)\,dx\,\bar{a}_z$$

$$= -0.96\Big[\,x\,\Big]_1^{-1}\bar{a}_x + 0.36\Big[\frac{x^2}{2} - 2x\Big]_1^{-1}\bar{a}_y - 0.42\Big[\frac{x^2}{2} - x\Big]_1^{-1}\bar{a}_z$$
$$= 1.92\,\bar{a}_x + 1.44\,\bar{a}_y - 0.84\,\bar{a}_z \text{ N} - \text{m}.$$

Problem 8.23 : Find the magnitude of the magnetic flux density in a material for which : (a) the magnetization is 2.8 A/m and the magnetic susceptibility is 0.0025; (b) the magnetic field intensity is 1300 A/m and the relative permeability is 1.006 ; (c) there are 8.2×10^{28} atoms per cubic meter, each

having a dipole moment of d3 × 10^{-30} A − m^2 in the same direction, and the magnetic susceptibility is 2 × 10^{-4}.

Solution :

(a) $B = \mu_o \left(M + \dfrac{M}{\chi_m} \right)$

$= 4\pi \times 10^{-7} \left(2.8 + \dfrac{2.8}{0.0025} \right)$, $= 1.411$ mT

(b) $B = \mu_o \mu_r H$

$= 4\pi \times 10^{7} \times 1.006 \times 1300$, $= 1.643$ mT

(c) $M = 8.2 \times 10^{28} \times 3 \times 10^{-30} = 0.246$ A − m

$B = \mu_o \left(M + \dfrac{M}{\chi_m} \right)$

$= 4\pi \times 10^{-7} \left(0.246 + \dfrac{0.246}{2 \times 10^{-4}} \right)$, $= 1.546$ mT

Problem 8.24 : If $\bar{B} = 0.05\, x\, \bar{a}_y$ T in a material for which $\chi_m = 2.5$, find: (a) μ_R; (b) μ; (c) $\bar{H}$; (d) $\bar{M}$; (e) $\bar{J}$; (f) $\bar{J}_b$.

Solution :

(a) $\mu_r = \chi_m + 1 = 2.5 + 1 = 3.5$

(b) $\mu = \mu_o \mu_r = 4\pi \times 10^{-7} \times 3.5$

$= 4.3982297 \times 10^{-6}$ H/m

(c) $\bar{H} = \dfrac{\bar{B}}{\mu_o \mu_r}$

$= \dfrac{0.05\, x\, \bar{a}_y}{4.3982297 \times 10^{-6}} = 11368.210\, x\, \bar{a}_y$ A/m

(d) $\bar{M} = \chi_m \bar{H} = 2.5 \times 11368.210\, x\, \bar{a}_y$

$= 28420.525\, x\, \bar{a}_y$ A/m

(e) $\bar{J} = \bar{\nabla} \times \bar{H} = \left(\dfrac{\partial H_y}{\partial x} \right) \bar{a}_z = 11368.210\, \bar{a}_z$ A/m^2

(f) $\bar{J}_b = \bar{\nabla} \times \bar{M} = \left(\dfrac{\partial M_y}{\partial x} \right) \bar{a}_z = 28420.525\, \bar{a}_z$ A/m^2

Problem 8.25 : In Prob. 8.10 (a), the linear approximation suggested in the statement of the problem leads to a flux density of 0.446 T in the central leg. Using this value of B and the magnetization curve for silicon steel, what current is required in the 1250–turn coil ?

Solution :

$B = 0.446$ (in the central leg)

From the B – H curve for silicon steel, corresponding to a value of 0.446 T, the magnetic field strength is 88.889 A – t / m. Hence

V_m (in the central leg) $= 0.04 \times 88.889 = 3.5556$ A · t .

$B = 0.37202$ T (in the left leg)

From the B – H curve for silicon steel, corresponding to a value of 0.37202 T , the magnetic field strength is 77.778 A . t / m. Hence

V_m (in the left leg) $= 0.12 \times 77.778 = 9.33336$ A. t

Total $V_m = 3.5556 + 9.33336 = 12.88896$ A . t

Now $V_m = NI$ where $N = 1250$

$$\therefore \quad I = \frac{V_m}{N} = \frac{12.88896}{1250} = 10.311 \text{ mA} \quad \text{(Approximate answer)}$$

Problem 8.26 : Determine the energy denisty stored in free space by the field: (a) $\overline{H}_A = 10^3\,\bar{a}_x$ A/m ; (b) $\overline{H}_B = 10^3\,\bar{a}_x + 10^3\,\bar{a}_y$ A/m ; (c) $\overline{H}_C = 10^3\,\bar{a}_x - 10^3\,\bar{a}_y$ A/m ; (d) $\overline{H}_B + \overline{H}_C$; (e) $10^3\,\bar{a}_x$ A/m + $10^{-3}\,\bar{a}_y$ T .

Solution : (a) Energy density $= \frac{1}{2}\mu_o H^2$

$$= \frac{1}{2}\mu_o \times 10^6 = 0.628318 \text{ J/m}^3$$

(b) Energy density $= \frac{1}{2}\mu_o H^2$

$$= \frac{1}{2}\mu_o \left[\sqrt{(10^3)^2 + (10^3)^2}\,\right]^2 = 1.25663 \text{ J/m}^3$$

(c) Energy density $= \frac{1}{2}\mu_o \left[\sqrt{(10^3)^2 + (-10^3)^2}\,\right]^2$

$$= 1.25663 \text{ J/m}^3$$

(d) $\overline{H}_B + \overline{H}_C = 10^3\,\bar{a}_x + 10^3\,\bar{a}_y + {}'10^3\,\bar{a}_x - 10^3\,\bar{a}_y$

$= 2000\,\bar{a}_x$ A/m

Energy density $= \frac{1}{2}\mu_o\,[2000]^2 = 2.513274$ J/m^3

(e) Energy density $= \frac{B^2}{2\mu_o}$

$$= \frac{\left[\sqrt{((10^{-3})^2 + (10^{-3})^2}\right]^2}{2\mu_o} = 0.79577 \text{ J/m}^3$$

Problem 8.27 : A toroid having a square cross section is defined by the surfaces z = 0, z = 20 mm, ρ = 30 mm, and ρ = 50 mm. The surface at ρ = 30 mm carries a current $\overline{K} = 10\,\bar{a}_z$ k A/m . (a) Find the surface currents on the other three surfaces. (b) Find $\overline{H}$ within the toroid. (c) If $\mu_R = 20$ within the toroid, find the total stored energy.

Solution :

(a) Current on the ρ = 30 mm surface

$= (10 \times 10^3)\,(2\pi \times 30 \times 10^{-3})$

$= 1884.955592$ A

Current on all the surfaces remains the same

Surface current on Z = 20 mm

Surface $= \frac{1884.955592}{2\pi\rho}\,\bar{a}_\rho = \frac{300}{\rho}\,\bar{a}_\rho$ A/m .

surface current on ρ = 50 mm surface $= \frac{1884.955592}{2\pi\,(50 \times 10^{-3})}\,(-\bar{a}_z)$

$= -6000\,\bar{a}_z$ A/m

Surface current on z = 0 surface $= \frac{1884.955592}{2\pi\rho}\,(-\bar{a}_\rho)$

$= \frac{-300}{\rho}\,\bar{a}_\rho$ A/m

For the direction of $\overline{K}$, refer to the fig.P. 8.32

(b) Consider a circular path of radius ρ such that it lies between ρ = 30 mm and ρ = 50 mm. Apply Ampere's circuital law to this closed path.

$\oint \overline{H} \quad \overline{dL} = I_{encl}$. This circular path encloses the current $(\overline{K})$ on the surface $\rho = 30$ mm which is directed in $\bar{a}_z$ direction. Hence H_ϕ component of $\overline{H}$ is produced.

$$\text{LHS} = \oint \overline{H} \cdot \overline{dL} = \int_0^{2\pi} H_\phi \, \bar{a}_\phi \cdot \rho \, d\phi \, \bar{a}_\phi$$

$$= \int_0^{2\pi} \rho \, H_\phi \, d\phi = 2\pi\rho H_\phi$$

RHS = 1884.955592

$$\therefore \quad 2\pi\rho H_\phi = 1884.955592 \quad \text{or} \quad H_\phi = \frac{300}{\rho}$$

$$\therefore \quad \overline{H} = \frac{300}{\rho} \bar{a}_\phi \quad \text{where } 30 \text{ mm} < \rho < 50 \text{ mm}$$

$$\text{(c)} \quad W_H = \frac{\mu_o \mu_r}{2} \int H^2 \, dV$$

$$= \frac{\mu_o (20)}{2} \int_{0.03}^{0.05} \int_0^{2\pi} \int_0^{0.02} \left(\frac{300}{\rho}\right)^2 \rho \, d\rho \, d\phi \, dz$$

$$= 9 \times 10^5 \, \mu_o \left[\ln \rho \right]_{0.03}^{0.05} \left[\phi \right]_0^{2\pi} \left[z \right]_0^{0.02}$$

$$= 0.0725997 \text{ J}$$

Problem 8.28 : A toroidal core is composed of a material with $\mu_R = 25$. The boundary surfaces are $z = 0$, $z = 0.025$, $\rho = 0.025$, and $\rho = 0.04$ m. The core is wound symmetrically with 12000 turns of wire such that H is in the $\bar{a}_\phi$ direction. Find the inductance.

Solution :

Using the Equation 8.20, we have

$$L = \frac{\mu_o \mu_r N^2 h}{2\pi} \ln\left(\frac{\rho_{outer}}{\rho_{inner}}\right)$$

where $\mu_r = 25$

h = height of toroidal = $(z = 0.025) - (z = 0)$

$= 0.025$ m

$\rho_{outer} = 0.04$ m

$\rho_{inner} = 0.025$ m

N = no. of turns = 12000.

$\therefore$ L = 8.46006 .

Problem 8.29 : Conducting planes in air at z = 0 and z = d carry surface currents of $\pm K_o \bar{a}_x$ A/m . (a) Find the energy stored in the magnetic field per unit length (0 < x < 1) in a width w (0 < y < w) and separation d (0 < z < d). (b) Calculate the inductance per unit length of this transmission line from $W_H = \frac{1}{2} LI^2$, where I is the total current in a width w in either conductor. (c) Calculate the total flux passing through the rectangle 0 < x < 1, 0 < z < d, in the plane y = 0, and from this result again find the inductance per unit length.

Solution :

Let $\bar{K} = K_o \bar{a}_x$ at $z = 0$ and

$\bar{K} = -K_o \bar{a}_x$ at $z = d$.

In between the conducting planes,

$$\bar{H} = \bar{H}_1 + \bar{H}_2$$

$$= \frac{K_o \bar{a}_x \times \bar{a}_z}{2} + \frac{-K_o \bar{a}_x \times (-\bar{a}_z)}{2}$$

$$= -0.5 K_o \bar{a}_y - 0.5 K_o \bar{a}_y = K_o \bar{a}_y \text{ A / m}$$

$$\therefore W_H = \frac{\mu_o}{2} \int H^2 dV = \frac{\mu_o}{2} \int_0^1 \int_0^w \int_0^d K_o^2 \, dx \, dy \, dz$$

$$= 0.5\ \mu_o K_o^2\, w\, d \quad J$$

(b) Current I in a width w in one conductor only =

surface current density × width

$$\therefore\ I = K_o w$$

$$\frac{1}{2} LI^2 = 0.5\, \mu_o K_0^2\ wd$$

$$\frac{1}{2} L\,(K_o w)^2 = 0.5\ \mu_o K_o^2\ w\,d$$

$$\text{or}\quad L = \frac{\mu_o d}{w} \quad H/m\ .$$

(c) $\overline{B} = \mu_o \overline{H} = K_o \mu_o \overline{a}_y$

$$\overline{dS} = dx\ dz\ \overline{a}_y$$

$$\phi = \int \overline{B} \cdot \overline{dS} = \int K_0 \mu_o \overline{a}_y \cdot dx\, dz\, \overline{a}_y$$

$$= \int_0^1 \int_0^d K_o \mu_o\, dx\, dz = K_0 \mu_o d$$

$$L = \frac{\phi}{I} = \frac{K_o \mu_o d}{K_o w} = \frac{\mu_o d}{w} \quad H/m$$

–X–

CHAPTER 9

Time Varying Fields & Maxwell's Equation

9.1 FARADAY'S LAW

It states that a time - varying magnetic field produces an electromotive force. (emf). An electromotive force is the voltage that arises from conductors moving in a magnetic field or from changing magnetic fields. Faraday's law is customarily stated as

$$\textbf{emf} = -\frac{d\Phi}{dt} \qquad \ldots\ldots (9.1)$$

If the conductor circuit is closed, this emf will circulate a current which produces a flux such that it opposes the cause producing it. emf is also defined by the equation

$$\textbf{emf} = \oint \overline{\mathbf{E}} \cdot \overline{\mathbf{dL}} \qquad \ldots\ldots (9.2)$$

Hence

$$\oint \overline{\mathrm{E}} \cdot \overline{\mathrm{dL}} = -\frac{d\Phi}{dt}$$

where Φ indicates the magnetic flux.

A nonzero value of $d\Phi/dt$ may result from any of the following situations:

1. A time - changing flux linking a stationary closed path.
2. Relative motion between a steady flux and a closed path.
3. A combination of the two.

9.2 DERIVATION OF MAXWELL'S EQUATIONS

9.2.1 From Faraday's law

According to Faraday's Law

$$\text{emf} = \oint \overline{E} \cdot \overline{dL} = -\frac{d\Phi}{dt}$$

But

$$\Phi = \int \overline{B} \cdot \overline{dS}$$

Hence

$$\oint \overline{E} \cdot \overline{dL} = -\frac{d}{dt}\int \overline{B} \cdot \overline{dS}$$

$$\boxed{\oint \overline{\mathbf{E}} \cdot \overline{\mathbf{dL}} = -\int \frac{\partial \overline{\mathbf{B}}}{\partial t} \cdot \overline{\mathbf{dS}}} \quad \text{.... (9.3)}$$

This is Maxwell's first equation in integral form derived from Faraday's law which states that

The electro motive force around a closed path is equal to the time derivative of the magnetic displacement through any surface bounded by the path.

Now the Stoke's theorem is $\oint \overline{A} \cdot \overline{dL} = \int (\overline{\nabla} \times \overline{A}) \cdot \overline{dS}$, where $\overline{A}$ is any vector field. Apply Stoke's theorem to the left handside of the Eq.(9.3), we have

$$\oint \overline{E} \cdot \overline{dL} = \int (\overline{\nabla} \times \overline{E}) \cdot \overline{dS}$$

Hence

$$\int (\overline{\nabla} \times \overline{E}) \cdot \overline{dS} = -\int \frac{\partial \overline{B}}{\partial t} \cdot \overline{dS}$$

where the surface intergrals may be taken over identical surfaces.

$$(\bar{\nabla} \times \bar{E}) \cdot \overline{dS} = -\frac{\partial \bar{B}}{\partial t} \cdot \overline{dS}$$

and

$$\boxed{\bar{\nabla} \times \bar{\mathbf{E}} = -\frac{\partial \bar{\mathbf{B}}}{\partial t}} \qquad \ldots\ldots (9.4)$$

This is Maxwell's first equation for time varying fields derived from Faraday's law in point form or the differential form..

Problem 9.1 : Given the magnetic flux density, $\bar{B} = 6 \cos 10^6 t \sin 0.01 x\ \bar{a}_z$ mT, find : (a) the magnetic flux passing through the surface $z = 0, 0 < x < 20$ m, $0 < y < 3$ m, at $t = 1\ \mu s$; (b) the value of the closed line integral of $\bar{E}$ around the perimeter of the surface specified above, at $t = 1\ \mu s$.

Solution :

(a) $\Phi = \int \bar{B} \cdot \overline{dS}$. To the surface $z = 0$, $\pm\ \bar{a}_z$ is perpendicula· Hence $\overline{dS} = \pm\ (dx\ dy)\ \bar{a}_z$. As $+\ \bar{a}_z$ is present in $\bar{B}$, $\overline{dS} = +\ (dx\ dy)\ \bar{a}_z$. Hence

$$\Phi = \int (6 \cos 10^6 t\ \sin\ 0.01\, x\ \bar{a}_z) \cdot (dx\ dy\ \bar{a}_z)\ 10^{-3}$$

$$= \int_0^{20} \int_0^{3} 6 \cos 10^6 t\ \sin 0.01\, x\ dx\ dy\ 10^{-3}$$

$$= 6 \cos 10^6 t \left[\frac{-\cos 0.01\, x}{0.01} \right]_0^{20} \Big[y \Big]_0^3\ 10^{-3}$$

$$= 0.03588 \cos 10^6 t$$

At $1\ \mu$ sec , $\Phi = 0.03588 \cos (10^6 \times 10^{-6}) = 19.386$ mWb

(b) $$\text{emf} = \oint \bar{E} \cdot \overline{dL} = -\int \frac{\partial \bar{B}}{\partial t} \cdot \overline{dS}$$

$$= -\int (-6 \times 10^6 \sin 10^6 t\ \sin 0.01\, x\ \bar{a}_z) \cdot (dx\ dy\ \bar{a}_z)\ 10^{-3}$$

$$\text{emf} = \int_0^{20}\int_0^{3} 6 \times 10^3 \sin 10^6 t \sin 0.01\, x\, dx\, dy$$

$$= 6 \times 10^3 \sin 10^6 t \left[\frac{-\cos 0.01\, x}{0.01} \right]_0^{20} \left[y \right]_0^3$$

$$= 35880.15988 \sin 10^6 t$$

At 1 μ sec , $\oint \bar{E} \cdot \overline{dL} = 35880.15988 \sin (10^6 \times 10^{-6}) = 30.192$ kV

Problem 9.2 : Parameters for the device shown in Fig. P 9.2 are $d = 5$ cm, $\bar{B} = 0.25\, \bar{a}_z$ T, and $\bar{v} = 20\sqrt{y}\ \bar{a}_y$ m / s. Let $y = 4$ cm at $t = 0$ and find value at $t = 0.06$ s for : (a) v_y ; (b) y ; (c) V_{12} ; (d) the current entering terminal 1 of the voltmeter if the matter resistance is 200 kΩ .

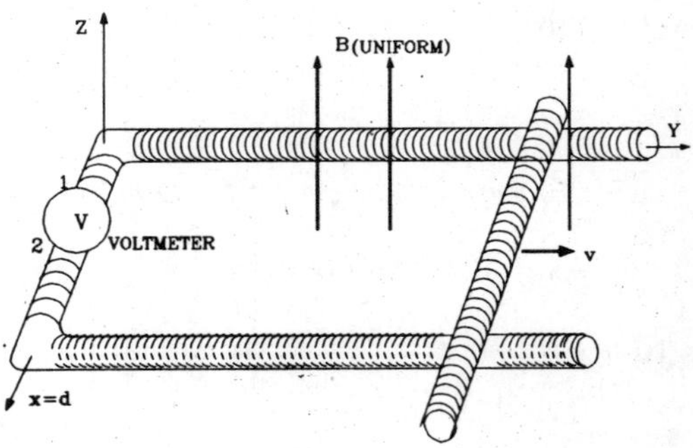

FIG.P 9.2

Solution :

(a) $\bar{v} = v_y \bar{a}_y = 20\sqrt{y}\ \bar{a}_y$ m/sec.

Also $v_y = \dfrac{dy}{dt} = 20\sqrt{y}$.

Hence $2\sqrt{y} = 20t + C$ where C is a constant of integration.

At $t = 0$, $y = 4$ cm $= 0.04$ m. Hence $2\sqrt{0.04} = C$ or $C = 0.4$.

Therefore $2\sqrt{y} = 20t + 0.4$ or $\sqrt{y} = 10t + 0.2$

At $t = 0.06$ s, $\sqrt{y} = 10(0.06) + 0.2$ or $y = 0.64$ m $= 64$ cm

Hence v_y (at 0.06 s) $= 20\sqrt{0.64} = 16$ m/sec.

(b) $y = 64$ cm at $t = 0.06$ s

(c) $V_{12} = -\frac{d\Phi}{dt} = -\frac{d}{dt}(B \times \text{Area}) \quad \left[\because \text{ As } \overline{B} \text{ is constant}\right]$

$$= -\frac{d}{dt}[0.25 \times d \times y] = -0.25\, d\, \frac{dy}{dt} = -0.25\, d \times V_y$$

$$= -0.25\, d \times 20\sqrt{y} = -5\, d\sqrt{y}$$

At 0.06 s, $V_{12} = -5(0.05)\sqrt{0.64} = -0.2$ V

(d) $I = \frac{V_{12}}{R} = \frac{-0.2}{200 \times 10^3} = -1\ \mu\text{A}$

9.2.2 From Ampere's circuital law

Point from of Ampere's circuital law is

$$\overline{\nabla} \times \overline{H} = \overline{J} \qquad \ldots\ldots(1)$$

Taking the divergence of each side.

$$\overline{\nabla} \cdot (\overline{\nabla} \times \overline{H}) = \overline{\nabla} \cdot \overline{J} = 0$$

because left hand side is the divergence of curl which is zero. But the continuity equation is

$$\overline{\nabla} \cdot \overline{J} = -\frac{\partial \rho_v}{\partial t}$$

which shows that Eq. (1) is true if

$$\frac{\partial \rho_v}{\partial t} = 0$$

It is necessary to have a generalized form of Ampere's circuital law is point form which is valid for both d.c. and a.c. fields.

Suppose an unknown term $\overline{G}$ is added to Eq. (1). Then

$$\overline{\nabla} \times \overline{H} = \overline{J} + \overline{G}$$

Again taking the divergence we have,

$$\overline{\nabla} \cdot (\overline{\nabla} \times \overline{H}) = \overline{\nabla} \cdot \overline{J} + \overline{\nabla} \cdot \overline{G}$$

$$0 = -\frac{\partial \rho_v}{\partial t} + \overline{\nabla} \cdot \overline{G}$$

$$\bar{\nabla} \cdot \bar{G} = + \frac{\partial \rho_v}{\partial t}$$

Now $\rho_v = \bar{\nabla} \cdot \bar{D}$. Hence we have,

$$\bar{\nabla} \cdot \bar{G} = \frac{\partial}{\partial t} (\bar{\nabla} \cdot \bar{D}) = \bar{\nabla} \cdot \frac{\partial \bar{D}}{\partial t}$$

$$\bar{G} = \frac{\partial \bar{D}}{\partial t}$$

$$\boxed{\bar{\nabla} \times \bar{\mathbf{H}} = \bar{\mathbf{J}} + \frac{\partial \bar{\mathbf{D}}}{\partial t}} \quad \ldots\ldots (9.5)$$

This is Maxwell's second equation in point form or differential form derived from Ampere's law. The term $\frac{\partial \bar{D}}{\partial t}$ is knowns as displacement current density and is denoted by $\bar{J}_d$.

Hence

$$\boxed{\bar{\mathbf{J}}_d = \frac{\partial \bar{\mathbf{D}}}{\partial t}} \quad \ldots\ldots (9.6)$$

Integrating Eq.(9.5) over a surface we get

$$\int (\bar{\nabla} \times \bar{H}) \cdot \overline{dS} = \int \left(\bar{J} + \frac{\partial \bar{D}}{\partial t} \right) \cdot \overline{dS}$$

and applying Stoke's theorem we obtain

$$\oint \bar{H} \quad \overline{dL} = \int \bar{J} \cdot \overline{dS} + \int \frac{\partial \bar{D}}{\partial t} \cdot \overline{dS}$$

$$\boxed{\oint \bar{\mathbf{H}} \cdot \overline{\mathbf{dL}} = I + \int \frac{\partial \bar{\mathbf{D}}}{\partial t} \cdot \overline{\mathbf{dS}}} \quad \ldots\ldots (9.7)$$

This is Maxwell's second equation in integral form which states that :

The magnetomotive force around a closed path is equal to the conduction current plus the time derivative of the electric displacement through any surface bounded by the path.

Problem 9.3 : FInd the amplitude of the displacement current density : (a) in the air near a car antenna, where the field strength of an FM signal is $\overline{E} = 80 \cos(6.277 \times 10^8 t - 2.092 y)\, \bar{a}_z$ V/m ; (b) in an air space within a large power transformer where $\overline{H} = 10^6 \cos(377 t + 1.2566 \times 10^{-6} z)\, \bar{a}_y$ A / m ; (c) inside a capacitor where $\varepsilon_R = 600$ and $\overline{D} = 3 \times 10^{-6} \sin(6 \times 10^6 t - 0.3464 x)\, \bar{a}_z$ C/m^2 ; (d) inside a typical metallic conductor where $f = 1$ kHz , $\sigma = 5 \times 10^7$ mho/m , $\varepsilon_R = 1$; and the conduction current density is $\overline{J} = 10^7 \sin(6283 t - 444 z)\, \bar{a}_x$ A/m^2 .

Solution :

(a) $\overline{E} = 80 \cos(6.277 \times 10^8 t - 2.092 y)\, \bar{a}_z$ V/m

$$\overline{J}_d = \frac{\partial \overline{D}}{\partial t} = \frac{\partial}{\partial t}(\varepsilon_o \overline{E}) = \varepsilon_o \frac{\partial \overline{E}}{\partial t}$$

$$= \varepsilon_o \times 80 \times (-6.277 \times 10^8) \sin(6.277 \times 10^8 t - 2.092 y)\, \bar{a}_z$$

$$= -0.445 \sin(6.277 \times 10^8 t - 2.092 y)\, \bar{a}_y \text{ A/m}^2$$

Hence the amplitude of displacement current density is 0.445 A / m^2.

(b) $\overline{H} = 10^6 \cos(377 t + 1.2566 \times 10^{-6} z)\, \bar{a}_y$ A/m .

$$\overline{\nabla} \times \overline{E} = -\frac{\partial \overline{B}}{\partial t} = -\frac{\partial}{\partial t}(\mu_o \overline{H}) = -\mu_o \frac{\partial \overline{H}}{\partial t}$$

$$= -\mu_o \times 10^6 (-377) \sin(377 t + 1.2566 \times 10^{-6} z)\, \bar{a}_y$$

$$= 473.752 \sin(377 t + 1.2566 \times 10^{-6} z)\, \bar{a}_y$$

As $\bar{a}_y$ is present on RHS, we will consider y component of $\overline{\nabla} \times \overline{E}$. Hence

$$\left(\frac{\partial E_x}{\partial z} - \frac{\partial E_z}{\partial x}\right) \bar{a}_y = 473.752 \sin(377 t + 1.2566 \times 10^{-6} z)\, \bar{a}_y$$

As z is present on RHS, $\frac{\partial E_z}{\partial x} = 0$, so that

$$\frac{d E_x}{dz} = 473.752 \sin(377 t + 1.2566 \times 10^{-6} z)$$

or $$E_x = 473.752 \left[\frac{-\cos(377 t + 1.2566 \times 10^{-6} z)}{1.2566 \times 10^{-6}}\right]$$

or $\overline{E} = -377.011 \times 10^6 \cos(377t + 1.2566 \times 10^{-6} z)\, \overline{a}_x$

$$\therefore \overline{J}_d = \varepsilon_o \frac{\partial \overline{E}}{\partial t}$$

$$= \varepsilon_o(-377.011 \times 10^6)(-377) \sin(377t + 1.2566 \times 10^{-6} z)\, \overline{a}_x$$

$$= 1.258 \sin(377t + 1.2566 \times 10^{-6} z)\, \overline{a}_x \text{ A/m}^2$$

Hence amplitude of displacement current density is 1.258 A / m^2.

(c) $\overline{D} = 3 \times 10^{-6} \sin(6 \times 10^6 t - 0.3464x)\, \overline{a}_z$

$$\overline{J}_d = \frac{\partial \overline{D}}{\partial t} = 3 \times 10^{-6} \times 6 \times 10^6 \cos(6 \times 10^6 t - 0.3464x)\, \overline{a}_z$$

$$= 18 \cos(6 \times 10^6 t - 0.3464x)\, \overline{a}_z \text{ A/m}^2$$

Hence amplitude of displacement current density is 18 A / m^2.

(d) $\overline{J} = \sigma \overline{E}$ or $\overline{E} = \dfrac{\overline{J}}{\sigma}$.

Hence $\overline{J}_d = \varepsilon_o \varepsilon_R \dfrac{\partial \overline{E}}{\partial t}$ or $\overline{J}_d = \dfrac{\varepsilon_o \varepsilon_R}{\sigma} \dfrac{\partial \overline{J}}{\partial t}$

$$= \frac{\varepsilon_o}{5 \times 10^7} \times 10^7 \times 6283 \cos(6283t - 444z)\, \overline{a}_x$$

$$= 11.126 \cos(6283\text{t} - 444\text{z})\, \overline{a}_x \text{ nA/m}^2$$

Hence amplitude of displacement current density is 11.126 nA/m^2.

9.2.3 From Gauss's law for electric field

Gauss's law states that

$$Q = \oint \overline{D} \cdot \overline{dS}$$

and

$$Q = \int \rho_v \, dv$$

Hence

$$\boxed{\oint \overline{\mathbf{D}} \cdot \overline{\mathbf{dS}} = \int \rho_v \, dv} \qquad \text{..... (9.8)}$$

This is the integral form of Maxwell's third equation for time varying field which states that :

The total electric displacement through the surface enclosing a volume is equal to the total charge within the volume.

The Divergence theorem is $\int (\overline{\nabla} \cdot \overline{A})\, dv = \oint \overline{A} \cdot \overline{dS}$, where $\overline{A}$ is any vector flux density. Applying divergence theorem to LHS of Eq.(9.8), we get,

$$\oint \overline{D} \cdot \overline{dS} = \int (\overline{\nabla} \cdot \overline{D})\, dv$$

$$\int (\overline{\nabla} \cdot \overline{D})\, dv = \int \rho_v\, dv$$

$$\boxed{\overline{\nabla} \cdot \overline{\mathbf{D}} = \rho_v} \qquad \ldots\ldots (9.9)$$

This is Maxwell's third equation in point form or differential form derived from Gauss's law for electric field.

9.2.4 From Gauss's law for magnetic field

Maxwell's first equation in point form or differential form is

$$\overline{\nabla} \times \overline{E} = -\frac{\partial \overline{B}}{\partial t}$$

Taking divergence of both sides,we have

$$\overline{\nabla} \cdot (\overline{\nabla} \times \overline{E}) = -\operatorname{div} \frac{\partial \overline{B}}{\partial t}$$

and since divergence of curl of $\overline{E}$ is identically equal to zero, we have

$$\operatorname{div} \frac{\partial \overline{B}}{\partial t} = 0$$

Intercharging the div operator and $\frac{\partial}{\partial t}$, we have

$$\frac{\partial}{\partial t} (\operatorname{div} \overline{B}) = 0$$

Integrating this equation shows that div $\overline{B}$ is a constant i.e. invariant with time. This means that the process of changing $\overline{B}$ from zero to some finite magnitude will not change div $\overline{B}$. Hence

$$\text{div } \overline{\text{B}} = 0$$

or

$$\boxed{\overline{\nabla} \cdot \overline{\mathbf{B}} = \mathbf{0}} \qquad \ldots\ldots. (9.10)$$

This is Maxwell's fourth equation in point or differential form . Integrating above equation over a volume we get

$$\int (\overline{\nabla} \cdot \overline{\text{B}})\, dv = 0$$

and applying divergence theorem ,we get,

$$\boxed{\oint \overline{\mathbf{B}} \cdot \overline{\mathbf{dS}} = \mathbf{0}} \qquad \ldots\ldots (9.11)$$

This is integral form of Maxwell's fourth equation derived from Gauss's law for magnetic field which states that :

The net magnetic flux emerging through any closed surface is zero.

9.2.5 Table 9.1 gives all the Maxwells's equations for time varying field in tabular form

TABLE 9.1. Maxwell's Equations, General Set

	Maxwells's Equation in	
Law from which derived	Point form or differential form	Integral form
Faraday's law	$\overline{\nabla} \times \overline{\text{E}} = -\dfrac{\partial \overline{\text{B}}}{\partial t}$	$\oint \overline{\text{E}} \cdot \overline{\text{dL}} = \int \dfrac{\partial \overline{\text{B}}}{\partial t} \cdot \overline{\text{dS}}$
Ampere's circuital law	$\overline{\nabla} \times \overline{\text{H}} = \overline{\text{J}} + \dfrac{\partial \overline{\text{D}}}{\partial t}$	$\oint \overline{\text{H}} \cdot \overline{\text{dL}} = I + \int \dfrac{\partial \overline{\text{D}}}{\partial t} \cdot \overline{\text{dS}}$
Gauss's law (Electric)	$\overline{\nabla} \cdot \overline{\text{D}} = \rho_v$	$\oint_S \overline{\text{D}} \cdot \overline{\text{dS}} = \int \rho_v \, dv$
Gauss's law (Magnetic)	$\overline{\nabla} \cdot \overline{\text{B}} = 0$	$\oint_S \overline{\text{B}} \cdot \overline{\text{dS}} = 0$

9.2.6 Table 9.2 gives all the Maxwell's equations for static conditions in tabular form

TABLE 9.2 Maxwell's Equations , Static conditions

	Maxwells's Equation in	
Law from which derived	Point form or differential form	Integral form
Faraday's law	$\bar{\nabla} \times \bar{E} = 0$	$\oint \bar{E} \cdot \overline{dL} = 0$
Ampere's circuital law	$\bar{\nabla} \times \bar{H} = \bar{J}$	$\oint \bar{H} \cdot \overline{dL} = I$
Gauss's law (Electric)	$\bar{\nabla} \cdot \bar{D} = \rho_v$	$\oint_S \bar{D} \cdot \overline{dS} = \int \rho_v \, dv$
Gauss's law (Magnetic)	$\bar{\nabla} \cdot \bar{B} = 0$	$\oint_S \bar{B} \cdot \overline{dS} = 0$

9.2.7 Table 9.3 gives all the Maxwell's equations for free space conditions in tabular form

TABLE 9.3 Maxwell's Equations , Free - Space set

	Maxwells's Equation in	
Law from which derived	Point form or differential form	Integral form
Faraday's law	$\bar{\nabla} \times \bar{E} = -\frac{\partial \bar{B}}{\partial t}$	$\oint \bar{E} \cdot \overline{dL} = -\int \frac{\partial \bar{B}}{\partial t} \cdot \overline{dS}$
Ampere's circuital law	$\bar{\nabla} \times \bar{H} = \frac{\partial \bar{D}}{\partial t}$	$\oint \bar{H} \cdot \overline{dL} = \int \frac{\partial \bar{D}}{\partial t} \cdot \overline{dS}$
Gauss's law (Electric)	$\bar{\nabla} \cdot \bar{D} = 0$	$\oint_S \bar{D} \cdot \overline{dS} = 0$
Gauss's law (Magnetic)	$\bar{\nabla} \cdot \bar{B} = 0$	$\oint_S \bar{B} \cdot \overline{dS} = 0$

Problem 9.4 : Select the value of K so that each of the following pairs of fields satisfies Maxwell's equations in a region where $\sigma = 0$ and $\rho_v = 0$: (a) $\overline{E} = (Kx - 100t)\,\overline{a}_y$ V/m, $\overline{H} = (x + 20t)\,\overline{a}_z$ A/m, if $\mu = 0.25$ H/m and $\varepsilon = 0.01$ F/m ; (b) $\overline{D} = 5x\,\overline{a}_x - 2y\,\overline{a}_y + Kz\,\overline{a}_z\ \mu C/m^2$, $\overline{B} = 2\,\overline{a}_y$ mT , $\mu = \mu_o$, $\varepsilon = \varepsilon_o$; (c) $\overline{E} = 60 \sin 10^6 t \sin 0.01 z\ \overline{a}_x$ V / m, $\overline{H} = 0.6 \cos 10^6 t \cos 0.01 z\ \overline{a}_y$ A/m , $\mu = K$, $\varepsilon = C_1$.

Solution :

$$\text{(a) } \overline{\nabla} \times \overline{E} = -\frac{\partial}{\partial z}(Kx - 100t)\,\overline{a}_x + \frac{\partial}{\partial x}(Kx - 100t)\,\overline{a}_z = K\,\overline{a}_z$$

$$\frac{\partial \overline{B}}{\partial t} = \mu \frac{\partial \overline{H}}{\partial t} = 0.25\ (20)\,\overline{a}_z = 5\,\overline{a}_z$$

As $\overline{\nabla} \times \overline{E} = -\dfrac{\partial \overline{B}}{\partial t}$, we have $K = -5\ \text{V/m}^2$

(b) $\overline{\nabla}\ \ \overline{D} = \rho_v = 0$.

$$\text{Hence } \frac{\partial D_x}{\partial x} + \frac{\partial D_y}{\partial y} + \frac{\partial D_z}{\partial z} = 0$$

or $(5 - 2 + K)\ 10^{-6} = 0$ or $K = -3\ \mu C/m^3$

(c)

$$\overline{\nabla} \times \overline{E} = \frac{\partial}{\partial z}(60 \sin 10^6 t \sin 0.01 z)\,\overline{a}_y - \frac{\partial}{\partial y}(60 \sin 10^6 t \sin 0.01 z)\,\overline{a}_z$$

$$= 0.6 \sin 10^6 t \cos 0.01 z\ \overline{a}_y$$

$$\frac{\partial \overline{B}}{\partial t} = \mu \frac{\partial \overline{H}}{\partial t} = K \times 0.6\,(-10^6) \sin 10^6 t \cos(0.01 z)\,\overline{a}_y$$

$$= -K \times 6 \times 10^5 \sin 10^6 t \cos(0.01 z)\,\overline{a}_y$$

Using the equation $\overline{\nabla} \times \overline{E} = -\dfrac{\partial \overline{B}}{\partial t}$,we get

$0.6 = K \times 6 \times 10^5$ or $K = 10^{-6}$ H/m

Problem 9.5 : The unit vector $0.48\,\overline{a}_x - 0.6\,\overline{a}_y + 0.64\,\overline{a}_z$ is directed from region 2 ($\varepsilon_{R2} = 2.5$, $\mu_{R2} = 2$, $\sigma_2 = 0$) toward region 1 ($\varepsilon_{R1} = 4$, $\mu_{R1} = 10$, $\sigma_1 = 0$) . If $\overline{H}_1 = (-100\,\overline{a}_x - 50\,\overline{a}_y + 200\,\overline{a}_z) \sin 400t$ A / m

at point P in region 1 adjacent to the boundary, find the amplitude at P of : (a) $\overline{H}_{N1}$; (b) $\overline{H}_{t1}$; (c) $\overline{H}_{N2}$; (d) $\overline{H}_2$.

Solution :

(a) $H_{N1} = \overline{H}_1 \cdot \overline{a}$

$= (-100\,\overline{a}_x - 50\,\overline{a}_y + 200\,\overline{a}_z)\sin 400t \cdot (0.48\,\overline{a}_x - 0.6\,\overline{a}_y + 0.64\,\overline{a}_z)$

$= (-48 + 30 + 128)\sin 400t = 110\sin 400t$

$\overline{H}_{N1} = H_{N1}\,\overline{a} = 110\,(0.48\,\overline{a}_x - 0.6\,\overline{a}_y + 0.64\,\overline{a}_z)\sin 400t$

$= (52.8\,\overline{a}_x - 66\,\overline{a}_y + 70.4\,\overline{a}_z)\sin 400t$

Amplitude of $\overline{H}_{N1} = \sqrt{(52.8)^2 + (-66)^2 + (70.4)^2} = 110$ A/m

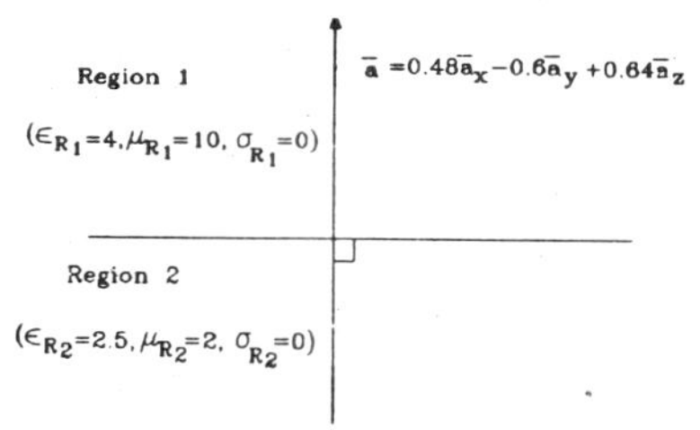

FIG.P 9.5

(b) $\overline{H}_1 = \overline{H}_{N1} + \overline{H}_{t1}$ or $\overline{H}_{t1} = \overline{H}_1 - \overline{H}_{N1}$

$\overline{H}_{t1} = [(-100\,\overline{a}_x - 50\,\overline{a}_y + 200\,\overline{a}_z) - (52.8\,\overline{a}_x - 66\,\overline{a}_y + 70.4\,\overline{a}_z)]\sin 400t$

$= (-152.8\,\overline{a}_x + 16\,\overline{a}_y + 129.6\,\overline{a}_z)\sin 400t$

Amplitude of $\overline{H}_{t1} = \sqrt{(-152.8)^2 + (16)^2 + (129.6)^2} = 200.988$ A/m

(c) $\overline{H}_{N2} = \dfrac{\mu_1}{\mu_2}\,\overline{H}_{N1} = \dfrac{10}{2}\,(52.8\,\overline{a}_x - 66\,\overline{a}_y + 70.4\,\overline{a}_z)\sin 400t$

$= (264\,\overline{a}_x - 330\,\overline{a}_y + 352\,\overline{a}_z)\sin 400t$

Amplitude of $\overline{H}_{N2} = \sqrt{(264)^2 + (-330)^2 + (352)^2} = 550$ A/m

(d) $\overline{H}_2 = \overline{H}_{N2} + \overline{H}_{t2} = \overline{H}_{N2} + \overline{H}_{t1}$

$= [(264\,\overline{a}_x - 330\,\overline{a}_y + 352\,\overline{a}_z) + (-152.8\,\overline{a}_x + 16\,\overline{a}_y + 129.6\,\overline{a}_z)]\sin 400t$

$= (111.2\,\overline{a}_x - 314\,\overline{a}_y + 481.6\,\overline{a}_z)\sin 400t$

Amplitude of $\overline{H}_2 = \sqrt{(111.2)^2 + (-314)^2 + (481.6)^2} = 585.577$ A/m

Problem 9.6 : The surface $x = 0$ is a perfectly conducting plane, while the region $x > 0$ has $\mu_R = 2, \varepsilon_R = 8$, and $\sigma = 0$. Let $\overline{H} = 2\cos(3 \times 10^8 t - 4z)\,\overline{a}_y$ A/m for $x > 0$, and find : (a) $\overline{K}$ at P (0, 2, 0) at $t = 5$ ns ; (b) ρ_S at P at $t = 5$ ns ; (c) $\overline{E}$ at Q (1, 2, 3) at $t = 4$ ns.

Solution :

(a) Let $\overline{K}$ be the surface current density on the conducting surface $x = 0$. Then $\overline{H}$ above the $x = 0$ plane $(x > 0)$ is $\overline{H} = \overline{K} \times \overline{a}_N$ or $\overline{K} \times \overline{a}_x$

$$= 2\cos(3 \times 10^8 t - 4z)\,\overline{a}_y$$

On the RHS, $\overline{a}_y$ is present. Hence $\overline{K}$ must have $\overline{a}_z$ so that $\overline{a}_z \times \overline{a}_x = \overline{a}_y$. The magnitude of $\overline{K}$ is equal to $2\cos(3 \times 10^8 t - 4z)$.

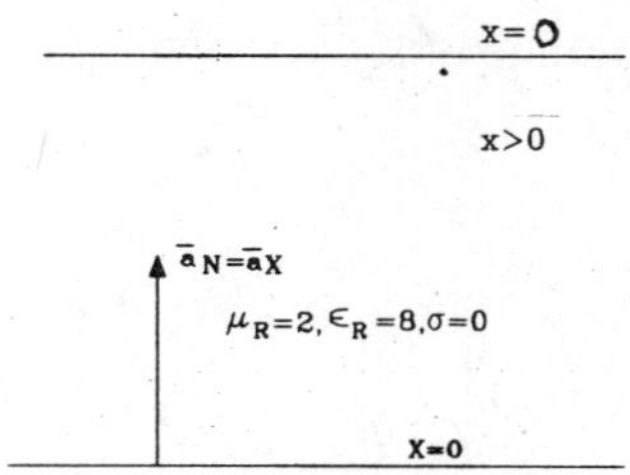

FIG. P 9.6

Hence

$$\overline{K} = 2\cos(3 \times 10^8 t - 4z)\,\overline{a}_z$$

$$\overline{K}\left[\text{ at } P\,(0, 2.0, 0.1) \text{ at } t = 5\text{ ns}\right] = 2\cos(3 \times 10^8 \times 5 \times 10^{-9} - 4 \times 0.1)\,\overline{a}_z$$

$$= 0.907\,\overline{a}_z \text{ A/m}$$

Note : We are assuming that another perfectly conducting surface is located at $x = \infty$ on which $\overline{K} = -2\cos(3 \times 10^8 t - 4z)\,\overline{a}_z$. Hence magnetic field intensity between two conducting planes carrying surface current density in opposite directions is given by $\overline{H} = \overline{K} \times \overline{a}_N$. If only one surface current density is present then $\overline{H} = 0.5\,\overline{K} \times \overline{a}_N$

(b) $$\overline{\nabla} \times \overline{E} = -\frac{\partial \overline{B}}{\partial t} = -\mu \frac{\partial \overline{H}}{\partial t}$$

$$= -2\mu_o \left[-2 \times 3 \times 10^8 \sin(3 \times 10^8 t - 4z)\,\overline{a}_y\right]$$

$$\overline{\nabla} \times \overline{E} = 1507.964 \sin(3 \times 10^8 t - 4z)\,\overline{a}_y$$

As $\overline{a}_y$ is present on RHS, we will consider y component of $\overline{\nabla} \times \overline{E}$.

Hence $$\left(\frac{\partial E_x}{\partial z} - \frac{\partial E_z}{\partial x}\right)\overline{a}_y = 1507.964 \sin(3 \times 10^8 t - 4z)\,\overline{a}_y$$

As z is present on RHS, $\frac{\partial E_z}{\partial x} = 0$.

Hence $\frac{dE_x}{dz} = 1507.964 \sin(3 \times 10^8 t - 4z)$

or $E_x = 1507.964 \left[\frac{-\cos(3 \times 10^8 t - 4z)}{-4} \right]$

$= 376.991 \cos(3 \times 10^8 t - 4z)$

$\therefore \overline{E} = 376.991 \cos(3 \times 10^8 t - 4z)\, \overline{a}_x$

$\overline{D} = 376.991\, \varepsilon_o \varepsilon_R \cos(3 \times 10^8 t - 4z)\, \overline{a}_x$

As $\overline{E}$ or $\overline{D}$ is always perpendicular to a perfectly conducting surface, we have

$\rho_s = |\overline{D}_N| = |\overline{D}| = \left| 376.991\, \varepsilon_o \varepsilon_R \cos(3 \times 10^8 t - 4z) \right|$

$\therefore \rho_s$ [at $P(0, 2.0, 0.1)$ at $t = 5$ ns]

$= \left| 376.991 \times \varepsilon_o \times 8 \cos(3 \times 10^8 \times 5 \times 10^{-9} - 4 \times 0.1) \right|$

$= 12.112$ nC/m^2

(c) $\overline{E} = 376.991 \cos(3 \times 10^5 t - 4z)\, \overline{a}_x$

$\overline{E}$ [at $Q(1, 2, 3)$ at $t = 4$ ns]

$= 376.991 \cos(3 \times 10^8 \times 4 \times 10^{-9} - 4 \times 3)\, \overline{a}_x$

$= -73.261\, \overline{a}_x$ V/m

9.3 CONCEPT OF DISPLACEMENT CURRENT OR DISPLACEMENT CURRENT DENSITY

Maxwell's equation derived from Ampere's circuital law is

$$\overline{\nabla} \times \overline{H} = \overline{J} + \frac{\partial \overline{D}}{\partial t}$$

The term $\frac{\partial \overline{D}}{\partial t}$ has the dimensions of current density (A/m^2) . Since it results from a time varying electric flux density (or displacement current density) Maxwell termed it a displacement current density. It is denoted by $\overline{J}_d = \frac{\partial \overline{D}}{\partial t}$. The total displacement current crossing any given surface is given by the surface integral

$$I_d = \int_S \overline{J}_d \cdot \overline{dS} = \int_S \frac{\partial \overline{D}}{\partial t} \overline{dS}$$

Hence

$$\oint \overline{H} \cdot \overline{dL} = I + I_d$$

where I = conduction current and I_d = displacement current.

Consider a circuit containing a filamentary loop and a parallel plate capacitor as shown in Fig. 9.1. Within the loop a magnetic field varying sinusoidally with time is applied to produce an emf about the closed path (the filament plus the dashed portion between the capacitor plates). Let emf = $V_o \cos \omega t$. Let for the loop $R = 0, L = 0$

$$I = C\frac{dV}{dt} = -\omega C V_o \sin \omega t = -\frac{\omega \varepsilon S}{d} V_o \sin \omega t$$

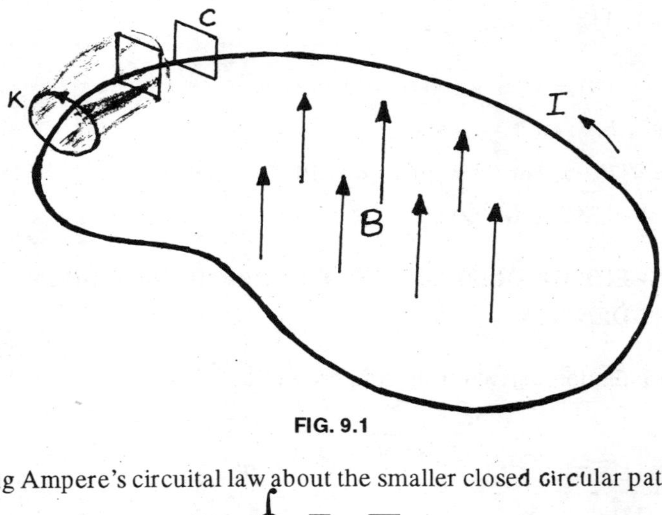

FIG. 9.1

Applying Ampere's circuital law about the smaller closed circular path K and neglect I_d for the time being. $\oint_K \overline{H} \cdot \overline{dL} = I_K$. The current I_K is that current through every surface whose perimeter is the path K. If we consider a circular path K punctured by the filament then the current is evidently the conduction current. Now consider a closed path K between the capacitor plates. The path is not punctured by the filament and the conduction current is zero.

$$D = \varepsilon E = \frac{\varepsilon V_o}{d} \cos \omega t$$

and therefore

$$I_d = -\frac{\omega \varepsilon S}{d} V_o \sin \omega t$$

which is equal to conduction current in the filamentary loop. For some surface the current is almost entirely conduction current byt for those. Surfaces passing bewteen the capacitor plates the conduction current is zero and it is the displacement current.

9.4 THE RETARDED POTENTIALS

The time – varying potentials, are usually called as retarded potentials, find their greatest application in radiation problems in which the distribution of the source is known approximately. It can be shown that any electromagentic disturbance is found to travel at a velocity given by

$$v = \frac{1}{\sqrt{\mu \varepsilon}}$$

through any homogeneous medium described by μ and ε. In the case of free space this velocity turns out to be velocity of light, i.e. 3×10^8 m/sec . Then the potential at any point is due not to the value of the charge density at some distant point at the same instant, but to its value at the some previous time, since the effect propogates at a finite velocity. In case of static field.

$$V = \int_{vol} \frac{\rho_v \, dv}{4 \pi \varepsilon R}$$

which then becomes

$$V = \int_{vol} \frac{[\rho_v]}{4 \pi \varepsilon R} \, dv$$

where $[\rho_v]$ indicates that every t appearing in the expression for ρ_v has been replaced by a retarded time t' given by

$$t' = t - \frac{R}{v}$$

Problem 9.7 : A point charge of $5 \cos 10^7 \pi t$ μC is located at P_+ (0, 0, 15), while $-5 \cos 10^7 \pi t$ μC is at P_- (0, 0, –15), both in free space. Find V at $P(r = 3000, \theta, \phi = 0)$ at $t = 150$ ns, where $\theta =$: (a) 0° ; (b) 90° ; (c) 60°.

Solution :

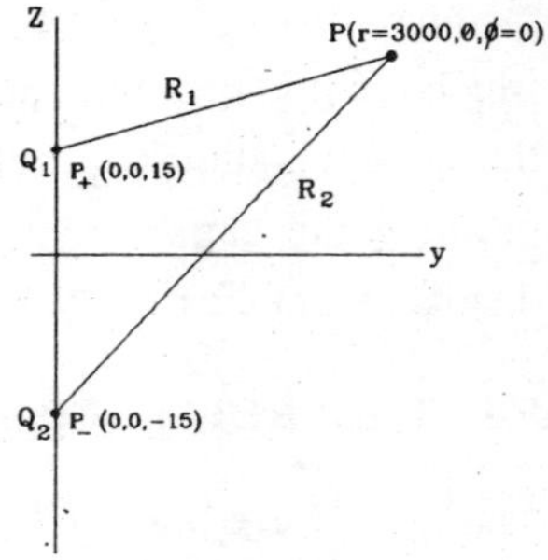

FIG. P 9.7

(a) $V = \dfrac{Q_1}{4\pi\varepsilon_o R_1} + \dfrac{Q_2}{4\pi\varepsilon_o R_2}$

Cartesian coordinates of point P with $\theta = 0^\circ$ are ($x = 0, y = 0, z = 3000$)

Hence

$R_1 = d(PP_+) = \sqrt{(3000 - 15)^2} = 2985$

$R_2 = d(PP_-) = \sqrt{(3000 + 15)^2} = 3015I$

In charge Q_1, t is replaced by a retarded time $t' = t - \dfrac{R_1}{v}$ and in charge Q_2 by $t' = t - \dfrac{R_2}{v}$, where $v =$ velocity of light $= 3 \times 10^8$ m/sec.

$$V = \frac{5 \cos 10^7 \pi \left(t - \dfrac{R_1}{3 \times 10^8} \right)}{4\pi\varepsilon_o R_1} + \frac{-5 \cos 10^7 \pi \left(t - \dfrac{R_2}{3 \times 10^8} \right)}{4\pi\varepsilon_o R_2} \ \mu V \qquad \ldots..(1)$$

Substitute $t = 150$ ns, $R_1 = 2985$ and $R_2 = 3015$ in Eq.(1), we get

$V = 30$ V

(b) Cartesian coordinates of point P with $\theta = 90^\circ$ are ($x = 3000, y = 0, 0$). Hence

$R_1 = d(PP_+) = \sqrt{(3000 - 0)^2 + (0 - 15)^2} = 3000.0375$

$R_2 = d(PP_-) = \sqrt{(3000 - 0)^2 + (0 + 15)^2} = 3000.0375$

Substitute $t = 150$ ns along with above values of R_1 and R_2 in equation (1), we get $V = 0$ V

(c) Cartesian coordinates of point P with $\theta = 60^\circ$ are (2598.0762, 0, 1500). Hence

$R_1 = d(PP_+) = \sqrt{(2598.0762 - 0)^2 + (1500 - 15)^2} = 2992.528$

$R_2 = d(PP_-) = \sqrt{(2598.0762 - 0)^2 + (1500 + 15)^2} = 3007.528$

Substitute $t = 150$ ns along with $R_1 = 2992.528$ and $R_2 = 3007.528$ in Eq. (1) we get $V = 21.106$ V

Problem 9.8 : In Fig P. 9.2, replace the uniform $\overline{B}$ field by $\overline{B} = 2e^{-50y}\,\overline{a}_z$ T. Find $V_{12}(t)$ if $d = 4$ cm and $\overline{v} = 65\,\overline{a}_y$ mi/h with $y = 0$ at $t = 0$.

Solution :

$$\overline{v} = 65\,\overline{a}_y \text{ mi/h} = 29.06\,\overline{a}_y \text{ m/sec} \quad \text{and} \quad \overline{v} = v_y\,\overline{a}_y = \frac{dy}{dt}\overline{a}_y = 29.06\,\overline{a}_y$$

$$\therefore \frac{dy}{dt} = 29.06 \quad \text{or} \quad y = 29.06\,t + C$$

where C is an integration constnat. As at $t = 0$, $y = 0$ (given), C must be equal to zero. Hence $y = 29.06\,t$

$$\Phi = \int \overline{B} \cdot \overline{dS} = \int_0^{0.04} \int_0^{29.06t} 2\,e^{-50y}\,dx\,dy$$

$$= 2\left[x\right]_0^{0.04} \left[\frac{e^{-50y}}{-50}\right]_0^{29.06t}$$

$$\Phi = -0.0016\,(e^{-1453t} - 1)$$

$$V_{12}(t) = -\frac{d\Phi}{dt} = 0.0016\,(-1453)\,e^{-1453t} = -2.3248\,e^{-1453t} \text{ V}$$

Problem 9.9 : In Fig. P 9.9, let the separation of the rails be 8 cm and let $\overline{B} = 0.5\,e^{0.1x}\,\overline{a}_z$ T.

(a) Find $V_{ab}(t)$ if the bar moves with a velocity $\overline{v} = 5\,\overline{a}_x$ m/s and $x = 0$ at $t = 0$.

(b) Find $\overline{v}(t)$ if $V_{ab}(t) = -0.1\,t$ V and $x = 0$ at $t = 0$.

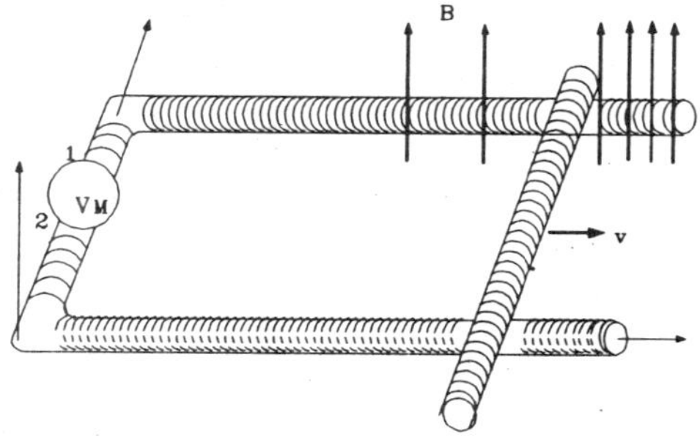

FIG. P 9.9

Solution :

(a) $\overline{v} = 5\,\overline{a}_x$ m/s $= v_x\,\overline{a}_x$ m/s .

Hence $v_x = \dfrac{dx}{dt} = 5$ or $x = 5t + C$, where C is constant of integration

$C = 0$ as $x = 0$ at $t = 0$. Hence $x = 5t$. The rails lies in $z = 0$ plane to which $\overline{a}_z$ is perpendicular. Hence $\overline{dS} = dx\,dy\,\overline{a}_z$

$$\therefore\ \Phi = \int \overline{B}\cdot\overline{dS} = \int_0^{5t}\int_0^{0.08} 0.5\,e^{0.1x}\,dx\,dy$$

$$= 0.5\left[\frac{e^{0.1x}}{0.1}\right]_0^{5t}\left[y\right]_0^{0.08} = 0.4\left[e^{0.5t} - 1\right]$$

$$\text{emf} = V_{ab}(t) = -\frac{d\Phi}{dt} = -0.4\left[0.5\,e^{0.5t}\right] = -0.2\,e^{0.5t}$$

(b) $\overline{v} = v(t)\,\overline{a}_x$ m/s $= v_x\,\overline{a}_x$ m/s .

Hence $v_x = \dfrac{dx}{dt} = v(t)$ or $x = \displaystyle\int v(t)\,dt + C$

$C = 0$ as $x = 0$ when $t = 0$.

Hence $x = \displaystyle\int v(t)\,dt$ or $\dfrac{dx}{dt} = v(t)$

$$\therefore\ \Phi = \int_0^{x}\int_0^{0.08} 0.5\,e^{0.1x}\,dx\,dy$$

$$= 0.5\left[\frac{e^{0.1x}}{0.1}\right]_0^{x}\left[y\right]_0^{0.08} = 0.4\left[e^{0.1x} - 1\right]$$

$$V_{ab}(t) = -0.1t = -\frac{d\Phi}{dt} = -\frac{d\Phi}{dx}\times\frac{dx}{dt} = -v(t)\times 0.4\left[0.1\,e^{0.1x}\right]$$

$$\therefore\ v(t) = \frac{0.1t}{0.4\times 0.1\,e^{0.1x}} = \frac{2.5t}{e^{0.1x}} = \frac{dx}{dt}$$

$$\therefore\ \int_0^{x} e^{0.1x}\,dx = \int_0^{t} 2.5t\,dt$$

$$\therefore\ \frac{e^{0.1x} - 1}{0.1} = \frac{2.5\,t^2}{2} \text{ or } e^{0.1x} = 0.125\,t^2 + 1$$

Hence $v(t) = \dfrac{2.5t}{0.125\,t^2 + 1}$

or $\bar{v}(t) = \dfrac{2.5t}{0.125\,t^2 + 1}\,\bar{a}_x = \dfrac{20t}{t^2 + 8}\,\bar{a}_x$

Problem 9.10 : Let us assume that the sliding bar of Fig. P 9.9, is fixed in position at $x = 1$ m with a rail separation of 8 cm. Find $V_{ab}(t)$ if $B_z =$: (a) $2\cos 10^8 t\ \mu$T ; (b) $2\cos\left(10^8 t - x/3\right)\ \mu$T .

Solution :

(a) $\bar{B} = B_z\,\bar{a}_z = 2\cos 10^8 t\ \bar{a}_z\ \mu\text{T}$ and $\overline{dS} = dx\,dy\,\bar{a}_z$

$$\therefore\ \Phi = \int \bar{B}\cdot\overline{dS} = \int_0^1\int_0^{0.08} 2\cos 10^8 t \times 10^{-6}\,dx\,dy$$

$$= 2\times 10^{-6}\cos 10^8 t\,\Big[x\Big]_0^1\Big[y\Big]_0^{0.08} = 0.16\times 10^{-6}\cos 10^8 t\ \text{Wb}$$

$$V_{ab}(t) = -\frac{d\Phi}{dt} = -\,0.16\times 10^{-6}\times(-\,10^8)\sin 10^8 t = 16\sin 10^8 t\ \text{V}.$$

(b) $\bar{B} = 2\cos\left(10^8 t - x/3\right)\bar{a}_z\ \mu\text{T}$

$$\therefore\ \Phi = \int \bar{B}\cdot\overline{dS} = \int_0^1\int_0^{0.08} 2\times 10^{-6}\cos\left(10^8 t - x/3\right)dx\,dy$$

$$= 2\times 10^{-6}\left[\frac{\sin(10^8 t - x/3)}{-\,1/3}\right]_0^1\Big[y\Big]_0^{0.08}$$

$$= -\,0.48\times 10^{-6}\left[\sin(10^8 t - 1/3) - \sin(10^8 t)\right]$$

$$= -\,0.48\times 10^{-6}\left[-\,0.05504\sin(10^8 t) - 0.32719\cos(10^8 t)\right]$$

Let $A = \sqrt{(0.05504)^2 + (0.32719)^2} = 0.331787$

$$\therefore\ \Phi = 0.48\times 10^6 A\left[\frac{0.05504}{A}\sin(10^8 t) + \frac{0.32719}{A}\cos(10^8 t)\right]$$

$$= 0.48\times 10^6\ \text{A}\left[\cos\theta\sin(10^8\,\text{t}) + \sin\theta\cos(10^8\,\text{t})\right]$$

where $\theta = \cos^{-1}\left(\dfrac{0.05504}{\text{A}}\right)$ or $\theta = \sin^{-1}\left(\dfrac{0.32719}{\text{A}}\right)$

$$\therefore\ \phi = 1.5925776\times 10^{-7}\sin(10^8\,\text{t} + 80.45^\circ)$$

$$V_{ab}(t) = -\frac{d\phi}{dt} = -1.5925776 \times 10^{-7} \times 10^{8} \cos(10^{8} t + 80.45^{\circ})$$

$$= -15.926 \cos(10^{8} t + 80.45^{\circ})$$

$$= 15.926 \cos(10^{8} t + 80.45^{\circ} - 180^{\circ})$$

$$= 15.926 \cos(10^{8} t - 99.55^{\circ})$$

Problem 9.11 : A coaxial transmission line has inner and outer radii of a and b, respectively, and has the magnetic field intensity, $H_\phi = (A/\rho) \sin \omega t \cos \beta z$. The conductors are perfect and the dielectric has $\mu_R = 1$, $\varepsilon_R = 2.25$, and $\sigma_{diel} = 0$. The line extends from z = 0 to z = d. (a) Determine the flux $\phi(t)$ passing through that portion of the plane $\phi = 0$ bounded by $\rho = a$, $\rho = b$, z = 0 and z = d. (b) Find $-d\Phi/dt$. (c) If the plane z = 0 is perfectly conducting, find $V_{ab}(t)$ at z = d.

Solution :

(a) $\overline{H} = H_\phi \, \overline{a}_\phi = \left(\frac{A}{\rho}\right) \sin \omega t \cos \beta z \, \overline{a}_\phi$

$\therefore \quad \overline{B} = \mu_o \mu_r \overline{H} = \left(\frac{A}{\rho}\right) \mu_o \sin \omega t \cos \beta z \, \overline{a}_\phi$

To the plane $\phi = 0$, $\overline{a}_\phi$ is perpendicular. Hence $\overline{dS} = d\rho \, dz \, \overline{a}_\phi$.

$$\therefore \quad \phi = \int \overline{B} \cdot \overline{dS} = \int_a^b \int_O^d \frac{A}{\rho} \mu_o \sin \omega t \cos \beta z \, d\rho \, dz$$

$$= A \mu_o \sin \omega t \left[\ln \rho\right]_a^b \left[\frac{\sin \beta z}{\beta}\right]_0^d$$

$$= \frac{A \mu_o}{\beta} \sin \omega t \sin \beta d \ln\left(\frac{b}{a}\right)$$

(b) $$-\frac{d\phi}{dt} = \frac{-A \mu \omega}{\beta} \cos \omega t \sin \beta d \ln\left(\frac{b}{a}\right)$$

(c) $$\overline{\nabla} \times \overline{H} = \overline{J}_C + \overline{J}_D = \sigma \overline{E} + \frac{\partial \overline{D}}{\partial t} = \frac{\partial \overline{D}}{\partial t} \quad (\because \sigma_{dielec.} = 0)$$

$$\overline{\nabla} \times \overline{H} = -\frac{\partial}{\partial z}\left(\frac{A}{\rho} \sin \omega t \cos \beta z\right) \overline{a}_\rho + \frac{1}{\rho} \frac{\partial}{\partial \rho}\left[\rho \left(\frac{A}{\rho} \sin \omega t \cos \beta z\right)\right] \overline{a}_z$$

$$\frac{\partial \overline{D}}{\partial t} = \frac{\beta A}{\rho} \sin \omega t \sin \beta z \, \overline{a}_\rho$$

$$\therefore \quad \overline{D} = \int \frac{\beta A}{\rho} \sin \omega t \sin \beta z \, dt \, \overline{a}_\rho = -\frac{\beta A}{\rho \omega} \cos \omega t \sin \beta z \, \overline{a}_\rho$$

$$\text{or} \quad \bar{E} = \frac{\bar{D}}{\varepsilon_o \varepsilon_r} = \frac{-\beta A \cos \omega t \sin \beta z}{\rho \omega \varepsilon_o \varepsilon_r} \bar{a}_\rho$$

$$V_{ab}(t) = \int_b^a \bar{E} \quad \overline{dL} = \int_b^a \bar{E} \cdot (d\rho \; \bar{a}_\rho)$$

$$= \int_b^a \frac{-\beta A \cos \omega t \; \sin \beta z}{\rho \omega \varepsilon_o \varepsilon_r} d\rho$$

$$= -\frac{\beta A \cos \omega t \; \sin \beta z}{\omega \varepsilon_o \varepsilon_r} \ln\left(\frac{a}{b}\right)$$

$$\text{At } z = d, \; V_{ab}(t) = \frac{-\beta A \; \cos \omega t \; \sin \beta d}{\omega \varepsilon_o \varepsilon_r} \ln\left(\frac{a}{b}\right)$$

Problem 9.12 : Let $\bar{B} = 0.5 \, x \, \bar{a}_z$ T in fig.P 9.9, The position of the sliding bar is given by $x = 4t - 2t^2$. If the separation of the rails is 10 cm, find the voltmeter reading V_{ab} at : (a) t = 0.5 s ; (b) x = 1 m.

Solution :

(a) To the z = 0 plane, $\bar{a}_z$ is perpendicular. $\therefore$ $\overline{dS} = dx\,dy\,\bar{a}_z$.

Hence flux leaving through loop is given by

$$\Phi = \int \bar{B} \cdot \overline{dS} = \int (0.5x\,\bar{a}_z) \cdot (dx\,dy\,\bar{a}_z)$$

$$= \int_0^x \int_0^{0.1} 0.5x \; dx\,dy = 0.5\left[\frac{x^2}{2}\right]_0^x \left[y\right]_0^{0.1}$$

$$\Phi = 0.025x^2 = 0.025\,(4t - 2t^2)^2$$

$$\text{emf} = -\frac{d\phi}{dt} = -0.025 \times 2\,(4t - 2t^2)(4 - 4t)$$

Substitute t = 0.5 s in the above expression.

$\therefore$ emf (t = 0.5 s) = − 0.15 V

(b) $x = 1 = 4t - 2t^2$

or $2t^2 - 4t + 1 = 0$

$$\text{or} \quad t = \frac{4 \pm \sqrt{(-4)^2 - 4 \times 2 \times 1}}{2 \times 2}$$

$$= 1.7071 \;,\; 0.29289$$

t = 0.29289 is considered, because at this time, the sliding Bar is at x = 1m for the first time.

$\therefore$ emf (t = 0.29289) = – 0.1414 V

Problem 9.13 : A perfectly conducting filament containing a small 500 –Ω resistor is formed into a square, as shown in fig.P 9.13. Find I (t) if $\bar{B}$ = : (a) 0.2 cos 120 π t $\bar{a}_z$ T (b) 2 cos $\left[3\pi\, 10^8 \left(t - x/c \right) \right]$ $\bar{a}_z$ μ T, where c = 3 × 10^8 m/s.

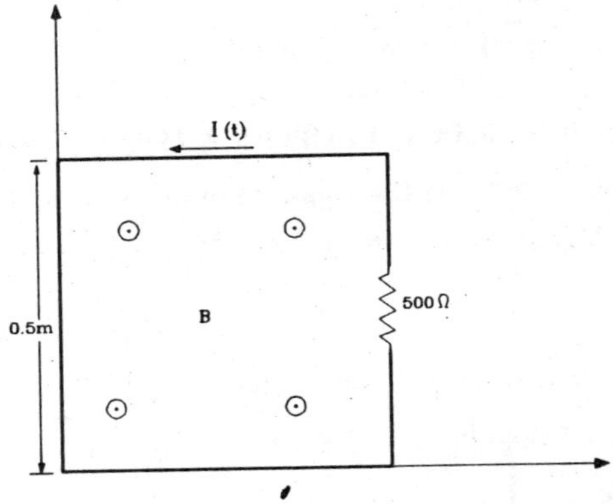

Fig. P 9..13

Solution :

Area of the loop = $(0.5)^2 = 0.25\ m^2$

(a) $\therefore\ \phi = 0.2 \cos(120\pi t) \times 0.25$

$= 0.05 \cos(120\pi t)$ T

$$\text{emf} = -\frac{d\phi}{dt} = -0.05(-120\pi)\sin(120\pi t)$$

$$= 6\pi \sin(120\pi t)\ V$$

$$\therefore\ I(t) = \frac{\text{emf}}{R} = \frac{6\pi\sin(120\pi t)}{500}$$

$$= 37.699 \sin(120\pi t)\ mA$$

(b) $\bar{B} = 2\cos[3\pi\, 10^8 t - \pi x]\ \bar{a}_z\ \mu T\ ;\ \overline{dS} = dx\, dy\, \bar{a}_z$

$$\therefore\ \phi = \int \bar{B}\cdot\overline{dS} = \int_0^{0.05}\int_0^{0.05} 2\cos(3\pi\, 10^8 t - \pi x)\, dx\, dy\, 10^{-6}$$

$$= 2\left[\frac{\sin(3\pi 10^8 t - \pi x)}{-\pi}\right]_0^{0.5} \left[y\right]_0^{0.5} 10^{-6}$$

$$= -\frac{1}{\pi}\left[-\cos(3\pi 10^8 t) - \sin(3\pi 10^8 t)\right] 10^{-6}$$

$$= \frac{\sqrt{2}}{\pi}\left[\frac{1}{\sqrt{2}}\cos(3\pi 10^8 t) + \frac{1}{\sqrt{2}}\sin(3\pi 10^8 t)\right] 10^{-6}$$

$$= \frac{\sqrt{2}}{\pi}\left[\cos\left(\frac{\pi}{4}\right)\cos(3\pi 10^8 t) + \sin\left(\frac{\pi}{4}\right)\sin(3\pi 10^8 t)\right] 10^{-6}$$

$$\Phi = \frac{\sqrt{2}}{\pi}\cos\left(3\pi 10^8 t - \frac{\pi}{4}\right) 10^{-6}$$

$$\therefore \quad \text{emf} = -\frac{d\phi}{dt} = +\frac{\sqrt{2}}{\pi} \times 3\pi 10^8 \sin\left(3\pi 10^8 t - \frac{\pi}{4}\right) 10^{-6}$$

$$\therefore \quad I(t) = \frac{300\sqrt{2}}{500}\sin\left(3\pi 10^8 t - \frac{\pi}{4}\right)$$

$$= 0.8485 \sin\left(3\pi 10^8 t - \frac{\pi}{4}\right) \text{ Amp.}$$

Problem 9.14 : The rails in fig.P 9.14(a) each have a resistance of 2.5 Ω/m. The Bar moves at a constant speed of 8 m/s in a uniform field of 0.9 T. Find I as a function of time, 0 < t < 2s, if the bar leaves the left end at t = 0 and : (a) 0.2 –Ω resistor is present across the left end with the right end open–circuited ; (b) a 0.2–Ω resistor is present across each end.

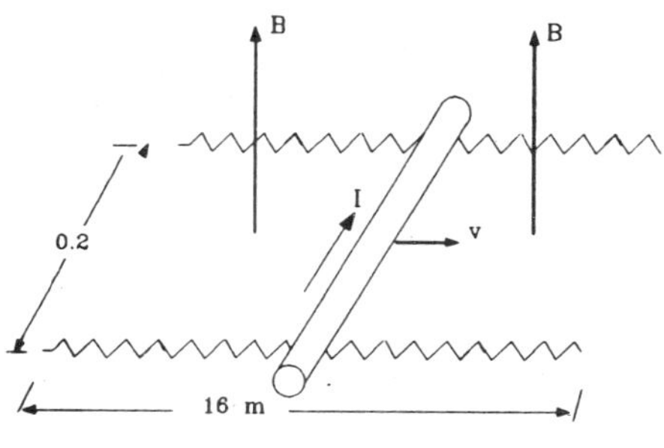

Fig.P 9.14(a)

Solution :

(a) At any time, the position of the sliding Bar is given by y = 8t m.

Hence area of the loop $= xy = 0.2 \times 8t = 1.6t \text{ m}^2$

$\therefore \quad \Phi = 0.9 \times 1.6t = 1.44t \text{ Wb}$

$\therefore \quad \text{emf} = -\dfrac{d\phi}{dt} = -1.44 \text{ V}$

Total resistance in the circuit $= 0.2 + 2.5y + 2.5y$

$= 0.2 + 5y = 0.2 + 5 \times 8t$

$= 0.2 + 40t\ \Omega$

$\therefore \quad I(t) = \dfrac{-1.44}{R} = \dfrac{-1.44}{0.2 + 40t} \text{ A}$

(b)

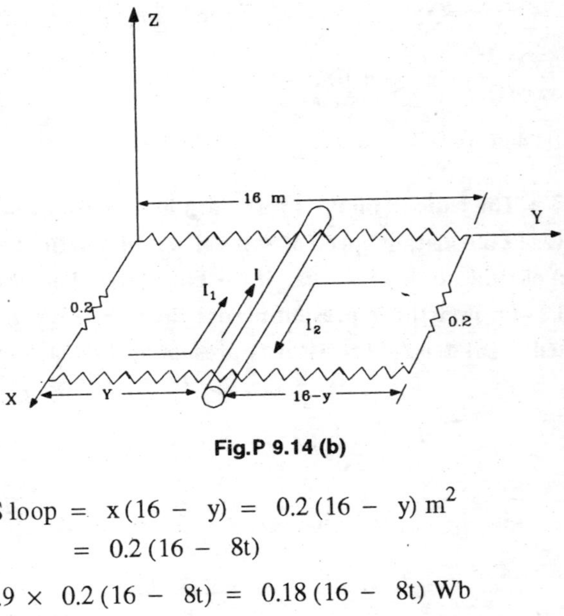

Fig.P 9.14 (b)

Area of RHS loop $= x(16 - y) = 0.2(16 - y) \text{ m}^2$

$= 0.2(16 - 8t)$

$\therefore \quad \Phi = 0.9 \times 0.2(16 - 8t) = 0.18(16 - 8t) \text{ Wb}$

$\therefore \quad \text{emf} = \dfrac{-d\phi}{dt} = -0.18(-8) = 1.44$

Total resistance in the RHS circuit

$= 2.5(16 - y) + 2.5(16 - y) + 0.2$

$= 80.2 - 5y = 80.2 - 40t$

Hence current $I_2 = \dfrac{1.44}{80.2 - 40t}$

Direction of this current will be as shown in fig.P 9.14 (b).

$$\therefore \quad I = I_1 - I_2 = -\frac{1.44}{0.2 + 40t} - \frac{1.44}{80.2 - 40t}$$

$$= \frac{-115.776}{16.04 + 3200t - 1600t^2}$$

$$= \frac{-0.07236}{0.010025 + 2t - t^2}$$

Problem 9.15 : The nearer rail in fig.P 9.1 is replaced by a nonstraight conductor such that the separation is given by d = f (y). The sliding bar is sufficiently long to make contact with both rails. Assuming a uniform field $\overline{B} = B_o \, \overline{a}_z$, a constant velocity $\overline{v} = v_o \, \overline{a}_y$, and a location y = 0 at t = 0, find $V_{12}(t)$.

Solution :

$\overline{V} = v_o \, \overline{a}_y = v_y \, \overline{a}_y$.

Hence $v_y = v_o$ or $\frac{dy}{dt} = v_o$ or $y = v_o t + C$

where C is a constant of integration.

At t = 0, y = 0. So that C = 0.

Hence $y = v_o t$

To the z = 0 plane $+ \overline{a}_z$ is perpendicular.

Hence $\overline{dS} = dx \, dy \, \overline{a}_z$.

$$\text{Now } \Phi = \int \overline{B} \cdot \overline{dS} = \int (B_o \, \overline{a}_z) \cdot (dx \, dy \, \overline{a}_z)$$

$$= \int_0^{f(y)} \int_0^{y} B_o \, dx \, dy$$

$$= B_o \int_0^{y} \Big[x \Big]_0^{f(y)} dy = B_o \int_0^{y} f(y) \, dy$$

$$= B_o \left[\int f(y) \, dy \Big|_{y=y} - \int f(y) \, dy \Big|_{y=0} \right]$$

$$V_{12}(t) = \frac{-d\Phi}{dt} = \frac{-d\Phi}{dy} \times \frac{dy}{dt}$$

$$= -V_o \frac{d\Phi}{dy} = -V_o \frac{d}{dy}\left\{B_o \int_{(y=y)} f(y)\, dy - \int_{(y=0)} f(y)\, dy\right\}$$

$$= -V_o B_o f(y)$$

Problem 9.16 : The square loop of wire has corners at (0, 0, 0), (0.2, 0, 0), (0.2, 0.2, 0), and (0, 0.2, 0) at t = 0. The loop is perfectly conducting expect for a small 100–Ω resistor in one side. It is moving through the field $\overline{B} = 5\left[\cos(6 \times 10^8 t - 2x)\right] \overline{a}_z$ μ T with a constant velocity of 40 $\overline{a}_y$ m/s . Calculate the power being delivered to the resistor as a function of time.

Solution :

The loop lies in z = 0 plane to which $\overline{a}_z$ is perpendicular.

Therefore $\overline{dS} = dx\,dy\,\overline{a}_z$

$$\therefore \quad \Phi = \int \overline{B}\; \overline{dS}$$

$$= \int_0^{0.2}\int_0^{0.2} 5 \times 10^{-6}\left[\cos(6 \times 10^8 t - 2x)\right] dx\,dy$$

$$= 5 \times 10^{-6}\left[\frac{\sin(6 \times 10^8 t - 2x)}{-2}\right]_0^{0.2} \left[y\right]_0^{0.2}$$

$$= -5 \times 10^{-7}\left[\sin(6 \times 10^8 t - 0.4) - \sin(6 \times 10^8 t)\right]$$

$$= -5 \times 10^{-7}\left[-0.078939 \sin(6 \times 10^8 t) - 0.389418 \cos(6 \times 10^8 t)\right]$$

Let $A = \left[(0.078939)^2 + (0.389418)^2\right]^{1/2} = 0.39733866$

$$\therefore \quad \Phi = A \times 5 \times 10^{-7}\left[\frac{0.078939}{A} \sin(6 \times 10^8 t)\right.$$

$$+ \frac{0.389418}{A} \cos (6 \times 10^8 t) \Big]$$

$$= A \times 5 \times 10^{-7} \Big[\cos\theta \sin (6 \times 10^8 t) + \sin\theta \cos (6 \times 10^8 t) \Big]$$

$$= A \times 5 \times 10^{-7} \sin (6 \times 10^8 t + \theta)$$

where $\theta = \cos^{-1}\left(\frac{0.078939}{A} \right)$ or $\theta = \sin^{-1}\left(\frac{0.389418}{A} \right)$

$$\therefore \quad \Phi = 1.98669 \times 10^{-7} \sin (6 \times 10^8 t + 78.541^\circ)$$

$$\text{emf} = -\frac{d\Phi}{dt} = -1.98669 \times 10^{-7} \times 6 \times 10^8 \cos (6 \times 10^8 t + 78.541^\circ)$$

$$= -119.2014 \cos (6 \times 10^8 t + 78.541^\circ)$$

Power consumed by the resistor

$$P(t) = I^2(t) R = \left[\frac{V(t)}{R} \right]^2 R = \frac{[V(t)]^2}{R}$$

$$= \frac{[-119.2014 \cos (6 \times 10^8 t + 78.541^\circ)]^2}{100}$$

$$= 142.09 \cos^2 (6 \times 10^8 t + 78.541^\circ) \text{ watts}$$

Problem 9.17 : (a) Show that the ratio of the amplitudes of the conduction current density and displacement current density is $\frac{\sigma}{\omega\varepsilon}$ for the applied field $E = E_m \cos \omega t$. Assume $\mu = \mu_o$. (b) What is this amplitude ratio if the applied fiel is $E = E_m e^{-t/\tau}$, where τ is real ?

Solution :

(a) $J_C = \sigma E = \sigma E_m \cos \omega t$

$$J_D = \frac{\partial D}{\partial t} = \varepsilon \frac{\partial E}{\partial t} = -\omega \varepsilon E_m \sin \omega t$$

$$\therefore \quad \frac{\text{Amplitude of } J_C}{\text{Amplitude of } J_D} = \frac{\sigma E_m}{\omega \varepsilon E_m} = \frac{\sigma}{\omega \varepsilon}$$

(b) $J_C = \sigma E = \sigma E_m e^{-t/\tau}$

$$J_D = \frac{\partial \overline{D}}{\partial t} = \varepsilon \frac{\partial \overline{E}}{\partial t} = \varepsilon E_m \left(\frac{-1}{\tau} \right) e^{-t/\tau}$$

$$= -\frac{\varepsilon E_m}{\tau} e^{-t/\tau}$$

$$\therefore \quad \frac{\text{Amplitude of } J_C}{\text{Amplitude of } J_D} = \frac{\sigma E_m}{\varepsilon E_m / \tau} = \frac{\sigma \tau}{\varepsilon}$$

Problem 9.18 : A coaxial transmission line with $\alpha = 1$cm and $b = 2.718$ cm has a voltage $V_{ab} = 100 \cos(3 \times 10^8 t - z)$ V between conductors $\varepsilon_R = \mu_R = 1$. (a) If the spatial relationship between the time–varying $\overline{E}$ and V fields is the same as it is for electrostatics, find $\overline{E}(\rho, z, t)$. (b) Find the displacement current density $\overline{J}_d(\rho, z, t)$.

Solution :

(a) $V_{ab} = V_a - V_b = 100 \cos(3 \times 10^8 t - z)$

Suppose $V = 100 \cos(3 \times 10^8 t - z)$ at $\rho = a = 0.01$ m

$V = 0$ at $\rho = b = 0.02718$ m

The potential varies w.r.t. ρ coordinate.

Hence $V = A \ln \rho + B$, which is a solution of Laplace's equation.

$$\therefore \quad 100 \cos(3 \times 10^8 t - z) = A \ln(0.01) + B \text{ and}$$

$$0 = A \ln(0.02718) + B$$

$$\text{Hence } A = \frac{100 \cos(3 \times 10^8 t - z)}{\ln(0.01) - \ln(0.02718)}$$

$$= -100.01 \cos(3 \times 10^8 t - z)$$

$$\text{Now } \overline{E} = -\overline{\nabla} V = -\frac{\partial V}{\partial \rho} \bar{a}_\rho = -\frac{A}{\rho} \bar{a}_\rho$$

$$\therefore \quad \overline{E} = -\frac{[-100.01 \cos(3 \times 10^8 t - z)]}{\rho} \bar{a}_\rho$$

$$= \left(\frac{100.01}{\rho} \right) \cos(3 \times 10^8 t - z) \bar{a}_\rho \text{ V/m}$$

(b) $\overline{D} = \varepsilon_o \varepsilon_r \overline{E}$

$$\overline{J}_d = \frac{\partial \overline{D}}{\partial t} = \varepsilon_o \varepsilon_r \frac{\partial \overline{E}}{\partial t}$$

$$\overline{J}_D = \varepsilon_o \left(\frac{100.01}{\rho} \right) (-3 \times 10^8) \sin (3 \times 10^8 t - z) \overline{a}_\rho$$

$$= \left(-\frac{0.2656}{\rho} \right) \sin (3 \times 10^8 t - z) \overline{a}_\rho \ A/m^2$$

Problem 9.19 : A coaxial capacitor has the parameters a = 5 mm, b = 30 mm, l = 20 cm, ε_R = 8 , and $\sigma = 10^{-6}$ mho/m . If the conduction current density in the capacitor is $\left(\frac{2}{\rho}\right) \sin 10^6 t \, \overline{a}_\rho$ A/m^2 , find : (a) the total conduction current through the capacitor, (b) the maximum instantaneous value of the displacement current density ; (c) the total displacement current ; (d) the Q (= ω CR = I_d / I_σ) of the capacitor.

Solution :

(a) $\overline{J}_C = \left(\frac{2}{\rho}\right) \sin 10^6 t \, \overline{a}_\rho$

Consider a cylinder of radius ρ such that $a < \rho < b$. $\overline{a}_\rho$ is perpendicular to this cylinder.

Hence $\overline{dS} = \rho \, d\phi \, dz \, \overline{a}_\rho$

$$\therefore \quad I_C = \int \overline{J}_C \cdot \overline{dS}$$

$$= \int \left(\frac{2}{\rho} \sin 10^6 t \, \overline{a}_\rho \right) \cdot \left(\rho \, d\phi \, dz \, \overline{a}_\rho \right)$$

$$= \int_0^{2\pi} \int_0^{0.2} 2 \sin 10^6 t \, d\phi \, dz = 2 \sin 10^6 t \left[\phi \right]_0^{2\pi} \left[z \right]_0^{0.2}$$

$$= 2.513 \sin 10^6 t \ A$$

(b) $\overline{J}_C = \sigma \overline{E}$

$$\therefore \quad \overline{E} = \frac{\overline{J}_C}{\sigma}$$

$$\therefore \quad \bar{D} = \varepsilon_o \varepsilon_r \bar{E} = \frac{\varepsilon_o \varepsilon_r}{\sigma} \bar{J}_C$$

$$\bar{J}_D = \frac{\partial \bar{D}}{\partial t} = \frac{\varepsilon_o \varepsilon_r}{\sigma} \frac{\partial \bar{J}_C}{\partial t}$$

$$= \frac{8 \varepsilon_o}{10^{-6}} \left(\frac{2}{\rho} \right) 10^6 \cos 10^6 t \, \bar{a}_\rho$$

$$= \frac{141.664}{\rho} \cos 10^6 t \, \bar{a}_\rho \ \text{A/m}^2$$

At $\rho = 5\text{mm} = 0.005$ m

$$(\bar{J}_D)_{max} = \frac{141.664}{0.005} \cos 10^6 t \, \bar{a}_\rho$$

$$= 28332.8 \cos 10^6 t \, \bar{a}_\rho \ \text{A/m}^2$$

(c) $I_D = \int \bar{J}_D \cdot \overline{dS}$

$$= \int \left(\frac{141.664}{\rho} \cos 10^6 t \, \bar{a}_\rho \right) \cdot (\rho \, d\phi \, dz \, \bar{a}_\rho)$$

$$= \int_0^{2\pi} \int_0^{0.2} 141.664 \cos 10^6 t \ d\phi \, dz$$

$$= 141.664 \cos 10^6 t \left[\phi \right]_0^{2\pi} \left[z \right]_0^{0.2}$$

$$= 178.020 \cos 10^6 t \ \text{A/m}^2$$

(d) $Q = \dfrac{I_d}{I_\sigma} = \dfrac{178.020}{2.513} = 70.840$

Problem 9.20 : A voltage source $V_0 \sin \omega t$ is connected between two concentric conducting spheres, $r = a$ and $r = b$, $b > a$, where the region between them is a material for which $\varepsilon = \varepsilon_R \varepsilon_o$, $\mu = \mu_o$, and $\sigma = 0$. Find the total displacement current through the dielectric and compare it with the source current as determined from the capacitance (sec. 9.3) and circuit–analysis methods.

Solution :

Let $V = V_o \sin \omega t$ at $r = a$ and

$V = 0$ at $r = b$

The potential varies w.r.t. r coordinate.

Hence $\bar{\nabla}^2 V = 0$ gives $\dfrac{1}{r^2}\dfrac{\partial}{\partial r}\left(r^2 \dfrac{\partial V}{\partial r}\right) = 0$

or $\dfrac{d}{dr}\left(r^2 \dfrac{dV}{dr}\right) = 0$ or $r^2 \dfrac{dV}{dr} = A$ or $V = -\dfrac{A}{r} + B$

where A and B are constants of integration.

$$\therefore \quad V_o \sin \omega t = -\frac{A}{a} + B$$

$$\text{and } 0 = -\frac{A}{b} + B$$

$$\therefore \quad A = \frac{V_o \sin \omega t}{1/b - 1/a}$$

$$\text{Now } \bar{E} = -\bar{\nabla} V = -\frac{\partial V}{\partial r}\bar{a}_r = -\left(\frac{A}{r^2}\right)\bar{a}_r$$

$$= -\left(\frac{V_o \sin \omega t}{\frac{1}{b} - \frac{1}{a}}\right)\frac{1}{r^2}\bar{a}_r$$

$$\bar{D} = \varepsilon_o \varepsilon_r \bar{E} = -\varepsilon_o \varepsilon_r \left(\frac{V_o \sin \omega t}{\frac{1}{b} - \frac{1}{a}}\right)\frac{1}{r^2}\bar{a}_r \ C/m^2$$

$$\bar{J}_D = \frac{\partial \bar{D}}{\partial t} = -\frac{\varepsilon_o \varepsilon_r V_o \omega \cos \omega t}{\left(\frac{1}{b} - \frac{1}{a}\right) r^2}\bar{a}_r \ A/m^2$$

Consider a sphere of radius r such that $a < r < b$.

Hence $\overline{dS} = r^2 \sin\theta \, d\theta \, d\phi \, \bar{a}_r$

$$\therefore \quad I_D = \int \bar{J}_D \cdot \overline{dS}$$

$$= \int \left[\frac{-\varepsilon_o \varepsilon_r V_o \omega \cos \omega t \, \bar{a}_r}{\left(\frac{1}{b} - \frac{1}{a}\right) r^2}\right] \cdot \left[r^2 \sin\theta \, d\theta \, d\phi \, \bar{a}_r\right]$$

$$= \int_0^{\pi} \int_0^{2\pi} - \frac{\varepsilon_o \varepsilon_r V_o \omega \cos \omega t}{\left(\frac{1}{b} - \frac{1}{a}\right)} \sin \theta \, d\theta \, d\phi$$

$$= - \frac{4 \pi \varepsilon_o \varepsilon_r V_o \omega \cos \omega t}{\left(\frac{1}{b} - \frac{1}{a}\right)}$$

Now for a spherical capacitor,

$$C = \frac{4 \pi \varepsilon_o \varepsilon_r}{1/a - 1/b}$$

$$I = C \frac{dV}{dt} = \frac{4 \pi \varepsilon_o \varepsilon_r}{1/a - 1/b} \frac{d}{dt} (V_o \sin \omega t)$$

$$= \frac{- 4 \pi \varepsilon_o \varepsilon_r V_o \omega \cos \omega t}{1/b - 1/a}$$

which is same as that of I_d.

Problem 9.21 : The displacement current density is $5 \cos (2 \times 10^8 t - kz) \bar{a}_x \mu A/m^2$ in a material for which $\sigma = 0$, $\varepsilon = 5 \varepsilon_0$, and $\mu = 4 \mu_0$. (a) Use the defnintion of displacement current density to find $\bar{D}$ and $\bar{E}$. (b) Now use the point from of Faraday's law and a time integration to find $\bar{B}$ and $\bar{H}$. (c) Finally, use the point form of Ampere's circuital law to find the displacement current density. What must k be ?

Solution :

(a) $$\bar{J}_D = \frac{\partial \bar{D}}{\partial t}$$

$$\therefore \partial \bar{D} = \bar{J}_D \, \partial t \quad \text{or} \quad \bar{D} = \int \bar{J}_D \, dt$$

$$\therefore \bar{D} = \int 5 \cos (2 \times 10^8 t - kz) \bar{a}_x \times 10^{-6} dt$$

$$= \frac{5 \sin (2 \times 10^8 t - kz) 10^{-6}}{2 \times 10^8} \bar{a}_x$$

$$\bar{D} = 0.025 \sin (2 \times 10^8 t - kz) \bar{a}_x \text{ pC/m}^2$$

Now $\overline{D} = \varepsilon \overline{E} = \varepsilon_o \varepsilon_r \overline{E}$

Hence $\overline{E} = \dfrac{\overline{D}}{\varepsilon} = \dfrac{\overline{D}}{5\varepsilon_o}$

$\therefore \quad \overline{E} = 564.717 \sin(2 \times 10^8 t - kz)\, \overline{a}_x \, \mu V/m$

(b) $\overline{\nabla} \times \overline{E} = -\dfrac{\partial \overline{B}}{\partial t}$

Now

$$\overline{\nabla} \times \overline{E} = \frac{\partial}{\partial z}\left[564.717 \times 10^{-6} \sin(2 \times 10^8 t - kz)\right] \overline{a}_y - \frac{\partial}{\partial y}\left[564.717 \times 10^{-6} \sin(2 \times 10^8 t - kz)\right] \overline{a}_z$$

$$= -\overline{a}_y\left[-564.717 \times 10^{-6}(-k)\cos(2 \times 10^8 t - kz)\right]$$

$$-\frac{\partial \overline{B}}{\partial t} = -564.717 \times 10^{-6} k \cos(2 \times 10^8 t - kz)\, \overline{a}_y$$

$$\therefore \quad \overline{B} = \int + 564.717 \times 10^{-6} K \cos(2 \times 10^8 t - kz)\, \overline{a}_y \, dt$$

$$\overline{B} = \frac{+564.717 \times 10^{-6} k \sin(2 \times 10^8 t - kz)\, \overline{a}_y}{2 \times 10^8}$$

$$= +2.823585\, k \sin(2 \times 10^8 t - kz)\, \overline{a}_y \ pWb/m^2$$

$\overline{B} = \mu \overline{H}$ or $\therefore \quad \overline{H} = \dfrac{\overline{B}}{\mu} = \dfrac{\overline{B}}{4\mu_o}$

or $\overline{H} = +0.561734\, k \sin(2 \times 10^8 t - kz)\, \overline{a}_y \ \mu A/m$

(c) $\overline{\nabla} \times \overline{H} = \overline{J}_C + \overline{J}_D = \sigma \overline{E} + \overline{J}_D = \overline{J}_D \quad [\because \sigma = 0]$

$$\overline{\nabla} \times \overline{H} = -\frac{\partial}{\partial z}\left[0.561734k\, 10^{-6} \sin(2 \times 10^8 t - kz)\right] \overline{a}_x + \frac{\partial}{\partial x}\left[0.561734k\, 10^{-6} \sin(2 \times 10^8 t - kz)\right] \overline{a}_z$$

$\overline{J}_D = +\overline{a}_x\, 0.561734\, k^2\, 10^{-6} \cos(2 \times 10^8 t - kz)$

Comparing this value of $\overline{J}_D$ with the given value of $\overline{J}_D$,

we have $5 \times 10^{-6} = +0.561734\, k^2\, 10^{-6}$

or $k = \pm 2.9834$ rad/m.

Problem 9.22 : Assume a homogeneous material of inifinite extent with $\varepsilon = 2 \times 10^{-10}$ F/m , $\mu = 1.25 \times 10^{-5}$ H/m, and $\sigma = 0$. Let $\overline{E} = 400 \cos(10^9 t - kz)\, \overline{a}_y$ V/m . If all the fields vary sinusoidally (or cosinusoidaly), use Maxwell's equations to find $\overline{D}$, $\overline{B}$, $\overline{H}$, and k.

Solution :

$$\overline{D} = \varepsilon \overline{E} = 2 \times 10^{-10} \times 400 \cos(10^9 t - kz)\, \overline{a}_x$$

$$= 80 \cos(10^9 t - kz)\, \overline{a}_x \text{ nC/m}^2 \qquad \ldots\ldots (1)$$

$$\overline{\nabla} \times \overline{E} = -\frac{\partial \overline{B}}{\partial t}$$

$$\overline{\nabla} \times \overline{E} = \frac{\partial}{\partial z}\left[400 \cos(10^9 t - kz)\right] \overline{a}_y - \frac{\partial}{\partial y}\left[400 \cos(10^9 t - kz)\right] \overline{a}_z$$

$$= -\overline{a}_y \left[-400(-1) \sin(10^9 t - kz)(-k)\right]$$

$$-\frac{\partial \overline{B}}{\partial t} = 400k \sin(10^9 t - kz)\, \overline{a}_y$$

$$\therefore \overline{B} = \int -400k \sin(10^9 t - kz)\, \overline{a}_y\, dt$$

$$= \frac{-400k(-1)\cos(10^9 t - kz)}{10^9}\, \overline{a}_y$$

$$= 0.4\, k \cos(10^9 t - kz)\, \overline{a}_y\ \mu\text{Wb/m}^2$$

$$\overline{H} = \frac{\overline{B}}{\mu} = \frac{k\, 0.4 \cos(10^9 t - kz)\, \overline{a}_y\, 10^{-6}}{1.25 \times 10^{-5}}$$

$$\overline{H} = 0.032\, k \cos(10^9 t - kz)\, \overline{a}_y \text{ A/m}$$

$$\overline{\nabla} \times \overline{H} = \overline{J}_C + \overline{J}_D = \sigma \overline{E} + \frac{\partial \overline{D}}{\partial t} = \frac{\partial \overline{D}}{\partial t} \qquad [\because \sigma = 0]$$

$$\overline{\nabla} \times \overline{H} = -\frac{\partial}{\partial z}\left[0.032k \cos(10^9 t - kz)\right] \overline{a}_x + \frac{\partial}{\partial x}\left[0.032k \cos(10^9 t - kz)\right] \overline{a}_z$$

$$= \overline{a}_x\left[-0.032(-1)\, k \sin(10^9 t - kz)(-k)\right]$$

$$\frac{\partial \overline{D}}{\partial t} = -0.032\, k^2 \sin(10^9 t - kz)\, \overline{a}_x$$

$$\therefore \overline{D} = \int -0.032\, k^2 \sin(10^9 t - kz)\, \overline{a}_x$$

$$= \frac{-0.032\, k^2 (-1) \cos(10^9 t - kz)\, \overline{a}_x}{10^9}$$

$$= +\ 0.032\ k^2 \cos(10^9 t - kz)\ \bar{a}_x\ nC/m^2$$

Comparing with equation(1), we have

$$0.032\ k^2 = 80 \quad \text{or} \quad k = \pm\ 50\ rad/m$$

Problem 9.23 : (a) A certain material has $\sigma = 0$ and $\mu_R = 1$. If $\bar{E} = 800 \sin(10^6 t - 0.01 z)\ \bar{a}_y$ V/m, make use of Maxwells's equations to find ε_R. (b) find $\bar{H}(z, t)$.

Solution :

(a) $\bar{\nabla} \times \bar{E} = -\dfrac{\partial \bar{B}}{\partial t}$

$$\bar{\nabla} \times \bar{E} = \begin{vmatrix} \bar{a}_x & \bar{a}_y & \bar{a}_z \\ \dfrac{\partial}{\partial x} & \dfrac{\partial}{\partial y} & \dfrac{\partial}{\partial z} \\ 0 & 800 \sin(10^6 t - 0.01z) & 0 \end{vmatrix}$$

$$= \bar{a}_x \left[-800 \cos(10^6 t - 0.01 z)\ (-0.01) \right]$$

$$-\frac{\partial \bar{B}}{\partial t} = 8 \cos(10^6 t - 0.01 z)\ \bar{a}_x$$

$$\therefore \quad \bar{B} = \int -8 \cos(10^6 t - 0.01 z)\ \bar{a}_x\, dt$$

$$= -8 \sin(10^6 t - 0.01z)\ \bar{a}_x \times 10^{-6}\ Wb/m^2$$

$$\therefore \quad \bar{H} = \frac{\bar{B}}{\mu_o \mu_r} = \frac{-8 \times 10^{-6}}{\mu_o} \sin(10^6 t - 0.01z)\ \bar{a}_x \qquad \ldots.\ (1)$$

$$\bar{\nabla} \times \bar{H} = \bar{J}_C + \bar{J}_D = \bar{J}_D = \frac{\partial \bar{D}}{\partial t}$$

$$\left[\because \ \sigma = 0\ ,\ \bar{J}_C = 0 \right]$$

$$\bar{\nabla} \times \bar{H} = \begin{vmatrix} \bar{a}_x & \bar{a}_y & \bar{a}_z \\ \dfrac{\partial}{\partial x} & \dfrac{\partial}{\partial y} & \dfrac{\partial}{\partial z} \\ \dfrac{-8 \times 10^{-6}}{\mu_o} \sin(10^6 t - 0.01z) & 0 & 0 \end{vmatrix}$$

$$= -\ \bar{a}_y \left[\frac{8 \times 10^{-6}}{\mu_o} \cos(10^6 t - 0.01 z)\ (-0.01) \right]$$

$$\frac{\partial \overline{D}}{\partial t} = \frac{8 \times 10^{-8}}{\mu_o} \cos(10^6 t - 0.01z)\, \overline{a}_y$$

$$\text{or} \quad \overline{D} = \int \frac{8 \times 10^{-8}}{\mu_o} \cos(10^6 t - 0.01\, z)\, \overline{a}_y\, dt$$

$$= \frac{8 \times 10^{-8}}{10^6\, \mu_o} \sin(10^6 t - 0.01\, z)\, \overline{a}_y$$

$$\text{or} \quad \overline{E} = \frac{\overline{D}}{\varepsilon_o\, \varepsilon_r} = \frac{8 \times 10^{-8}}{10^6\, \mu_o\, \varepsilon_o\, \varepsilon_r} \sin(10^6 t - 0.01\, z)\, \overline{a}_y$$

Comparing this value of $\overline{E}$ with the given value of $\overline{E}$,

we have $800 = \dfrac{8 \times 10^{-8}}{10^6\, \mu_o\, \varepsilon_o\, \varepsilon_r}$

$\therefore \quad \varepsilon_r = 8.9877$

Problem 9.24 : The parallel–strip transmission line shown in fig.P 9.24 has dimensions b = 5 cm and d = 1mm, while the medium between the strips is characterized by $\mu_R = 1$, $\varepsilon_R = 25$, and $\sigma = 0$. Neglect fields outside the dielectric. Given the field $\overline{E} = 10^4 \cos(10^9 t - \beta z)\, \overline{a}_x$ V/m, use Maxwell's equations to be help find : (a) β, if $\beta > 0$; (b) the displacement current density at z = 0; (c) the total displacement current crossing the surface x = 0.5 d, 0 < y < b, 0 < z < 0.1 m in the $\overline{a}_x$ direction.

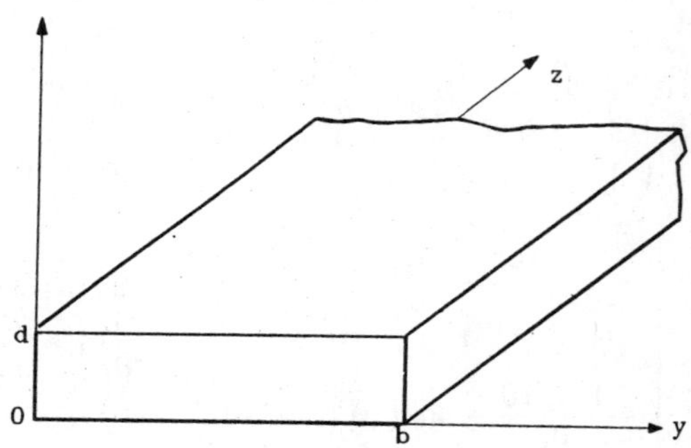

Fig.P 9.24

Solution :

(a) $$\bar{\nabla} \times \bar{E} = \begin{vmatrix} \bar{a}_x & \bar{a}_y & \bar{a}_z \\ \frac{\partial}{\partial x} & \frac{\partial}{\partial y} & \frac{\partial}{\partial z} \\ 10^4 \cos(10^9 t - \beta z) & 0 & 0 \end{vmatrix}$$

$$= -\bar{a}_y \left[-10^4 (-1) \sin(10^9 t - \beta z) \times (-\beta) \right]$$

$$= \bar{a}_y \left[\beta\, 10^4 \sin(10^9 t - \beta z) \right]$$

$$\bar{\nabla} \times \bar{E} = -\frac{\partial \bar{B}}{\partial t}$$

Hence $$\frac{d\bar{B}}{dt} = -\bar{a}_y \left[\beta\, 10^4 \sin(10^9 t - \beta z) \right]$$

$$\therefore \quad \bar{B} = -\bar{a}_y \left\{ \beta\, 10^4 (-1) \frac{\cos(10^9 t - \beta z)}{10^9} \right\}$$

$$= \beta\, 10^{-5} \cos(10^9 t - \beta z)\, \bar{a}_y$$

$$\therefore \quad \bar{H} = \frac{\bar{B}}{\mu_o \mu_r} = \frac{\beta\, 10^{-5} \cos(10^9 t - \beta z)\, \bar{a}_y}{\mu_o}$$

$$\bar{\nabla} \times \bar{H} = \begin{vmatrix} \bar{a}_x & \bar{a}_y & \bar{a}_z \\ \frac{\partial}{\partial x} & \frac{\partial}{\partial y} & \frac{\partial}{\partial z} \\ 0 & \frac{\beta\, 10^{-5} \cos(10^9 t - \beta z)}{\mu_o} & 0 \end{vmatrix}$$

$$= \bar{a}_x \left[-\frac{\beta\, 10^{-5}}{\mu_o} (-1) \sin(10^9 t - \beta z)\, (-\beta) \right]$$

$$= -\frac{\beta^2\, 10^{-5}}{\mu_o} \sin(10^9 t - \beta z)\, \bar{a}_x$$

$$\bar{\nabla} \times \bar{H} = \sigma \bar{E} + \bar{J}_D = \bar{J}_D \text{ as } \sigma = 0$$

$$\therefore \quad \bar{\nabla} \times \bar{H} = \frac{\partial \bar{D}}{\partial t} = \varepsilon_o \varepsilon_r \frac{\partial \bar{E}}{\partial t}$$

$$-\frac{\beta^2 10^{-5}}{\mu_o} \sin(10^9 t - \beta z)\ \bar{a}_x = \varepsilon_o (25)\ 10^4 (-1) \sin(10^9 t - \beta z) \times \bar{a}_x \times (10^9)$$

$$\therefore\ \frac{\beta^2 10^{-5}}{\mu_o} = 25 \times 10^{13} \times \varepsilon_o$$

or $\beta = 16.678$ rad/m .

(b) $\bar{J}_D = -25 \times 10^{13} \varepsilon_o \sin(10^9 t - \beta z)\ \bar{a}_x$

when $z = 0$

$$\bar{J}_D = -2213.5 \sin(10^9 t)\ \bar{a}_x \text{ A/m}^2$$

(c) $I_D = \int \bar{J}_D \cdot \overline{dS}$

$$= \int \left[-2213.5 \sin(10^9 t - \beta z)\ \bar{a}_x\right] \cdot \left[dy\,dz\,\bar{a}_x\right]$$

$$= \int_0^b \int_0^{0.1} -2213.5 \sin(10^9 t - \beta z)\ dy\,dz$$

$$= -2213.5 \left[\frac{-\cos(10^9 t - \beta z)}{-\beta}\right]_0^{0.1} \left[y\right]_0^b$$

$$= -\frac{2213.5\,b}{\beta}\left[\cos(10^9 t - 0.1 \times 16.678) - \cos(10^9 t)\right]$$

$$= -\frac{2213.5\,b}{\beta}\left[\cos(10^9 t - 1.6678) - \cos(10^9 t)\right]$$

$$= -\frac{2213.5\,b}{\beta}\left[-0.09685 \cos 10^9 t + 0.9953 \sin 10^9 t - \cos 10^9 t\right]$$

$$= \frac{2213.5\,b}{\beta}\left[1.09685 \cos 10^9 t - 0.9953 \sin 10^9 t\right]$$

$$= \frac{2213.5\,b}{\beta} M \left[\frac{1.09685}{M} \cos 10^9 t - \frac{0.9953}{M} \sin 10^9 t \right]$$

where $M = \left[(1.09685)^2 + (0.9953)^2 \right]^{1/2} = 1.481115$

$$I_D = \frac{2213.5\,bM}{\beta} \left[\cos\theta \cos 10^9 t - \sin\theta \sin 10^9 t \right]$$

$$= \frac{2213.5\,bM}{\beta} \cos(10^9 t + \theta)$$

where $\tan\theta = \dfrac{0.9953}{1.09685}$ or $\theta = 42.221^\circ$

Hence

$$I_D = \frac{2213.5 \times (0.05)}{16.678} \times (1.481115) \cos(10^9 t + 42.221^\circ)$$

$$= 9.829 \cos(10^9 t + 42.221^\circ)$$

Problem 9.25 : The electric field intensity throughout a certain enclosed region of free space is $E_y = A \sin 20x \sin bz \sin(12 \times 10^9 t)$ V/m . Beginning with the $\bar{\nabla} \times \bar{E}$ relationship, use Maxwell's equations to find a numerical value of b, assuming $b > 0$.

Solution :

$\bar{E}_y = A \sin 20x \sin bz \sin(12 \times 10^9 t)$ V/m

or $\bar{E} = A \sin 20x \sin bz \sin(12 \times 10^9 t)\,\bar{a}_y$ V/m

$$\bar{\nabla} \times \bar{E} = \begin{vmatrix} \bar{a}_x & \bar{a}_y & \bar{a}_z \\ \dfrac{\partial}{\partial x} & \dfrac{\partial}{\partial y} & \dfrac{\partial}{\partial z} \\ 0 & A \sin 20x \sin bz \sin(12 \times 10^9 t) & 0 \end{vmatrix}$$

$$= -bA \sin 20x \cos bz \sin(12 \times 10^9 t)\,\bar{a}_x +$$

$$20A \cos 20x \sin bz \sin(12 \times 10^9 t)\,\bar{a}_z$$

Now $\overline{\nabla} \times \overline{E} = -\dfrac{\partial \overline{B}}{\partial t}$. Hence

$$\frac{\partial \overline{B}}{\partial t} = bA \sin 20x \cos bz \sin(12 \times 10^9 t)\, \bar{a}_x - 20 A \cos 20x \sin bz \sin(12 \times 10^9 t)\, \bar{a}_z$$

or $$\overline{B} = \frac{-bA \sin 20x \cos bz \cos(12 \times 10^9 t)\, \bar{a}_x}{12 \times 10^9} + \frac{20 A \cos 20x \sin bz \cos(12 \times 10^9 t)\, \bar{a}_z}{12 \times 10^9}$$

$\overline{B} = \mu_o \overline{H}$, or $\overline{H} = \dfrac{\overline{B}}{\mu_o}$. Hence,

$$\overline{H} = \frac{-bA \sin 20x \cos bz \cos(12 \times 10^9 t)\, \bar{a}_x}{12 \times 10^9 \mu_o} + \frac{20 A \cos 20x \sin bz \cos(12 \times 10^9 t)\, \bar{a}_z}{12 \times 10^9 \mu_o}$$

$$\overline{\nabla} \times \overline{H} = \begin{vmatrix} \bar{a}_x & \bar{a}_y & \bar{a}_z \\ \dfrac{\partial}{\partial x} & \dfrac{\partial}{\partial y} & \dfrac{\partial}{\partial z} \\ \dfrac{-bA \sin 20x \cos bz \cos(12 \times 10^9 t)}{12 \times 10^9 \mu_o} & 0 & \dfrac{20 A \cos 20x \sin bz \cos(12 \times 10^9 t)}{12 \times 10^9 \mu_o} \end{vmatrix}$$

$$= -\bar{a}_y \left[\frac{-400 A \sin 20x \sin bz \cos(12 \times 10^9 t)}{12 \times 10^9 \mu_o} - \frac{b^2 A \sin 20x \sin bz \cos(12 \times 10^9 t)}{12 \times 10^9 \mu_o} \right] \quad \ldots. (1)$$

For free space $\sigma = 0$.

Hence $\overline{\nabla} \times \overline{H} = \dfrac{\partial \overline{D}}{\partial t} = \dfrac{\varepsilon_o \partial \overline{E}}{\partial t}$

$\therefore \quad \overline{J}_D = \varepsilon_o\, 12 \times 10^9 A \sin 20x \sin bz \cos(12 \times 10^9 t)\, \bar{a}_y \quad \ldots. (2)$

Equating the coefficient of $A \sin 20x \sin bz \cos(12 \times 10^9 t)\, \bar{a}_y$ from equation (1) and (2), we have

$$\frac{400}{12 \times 10^9 \mu_o} + \frac{b^2}{12 \times 10^9 \mu_o} = \varepsilon_o \times 12 \times 10^9,$$

or $b = 34.672$ rad/m

Problem 9.26 : Let $\overleftarrow{E} = (-20\,\bar{a}_x + 30\,\bar{a}_y + 70\,\bar{a}_z)\cos 10^6 t$ V/m at point P (3, –4, 1) which lies on the surface of a perfect conductor. If the material adjacent to the conductor has $\varepsilon_R = 5$, $\mu_R = 2$, $\sigma = 0$, find : (a) a unit vector normal to the concutor surface at P ; (b) the surface charge density on the concutor surface at P.

Solution :

(a) $\overline{E}$ is always normal to the surface of a perfect conductor.

$$\text{Hence } \bar{a}_N = \frac{\overline{E}}{|\overline{E}|} = \frac{-20\,\bar{a}_x + 30\,\bar{a}_y + 70\,\bar{a}_z}{\sqrt{(-20)^2 + (30)^2 + (70)^2}}$$

$$= \pm(-0.254\,\bar{a}_x + 0.381\,\bar{a}_y + 0.889\,\bar{a}_z)$$

(b) $\overline{D} = \varepsilon_o \varepsilon_r \overline{E} = 5\varepsilon_o(-20\,\bar{a}_x + 30\,\bar{a}_y + 70\,\bar{a}_z)\cos 10^6 t\ \text{C/m}^2$

$$\rho_S = |\overline{D}_N| = |\overline{D}|$$

$$= 5\varepsilon_o\sqrt{(-20)^2 + (30)^2 + (70)^2}\ \cos 10^6 t$$

$$= \pm\ 3.486\ \cos 10^6 t\ \text{nC/m}^2$$

Surface charge density may be positive or negative.

Problem 9.27 : In the first section of this chapter, Faraday's law was used to show that the field $\overline{E} = -\frac{1}{2} k B_o e^{kt}\,\bar{a}_\phi$ results from the changing magnetic field $\overline{B} = B_o e^{kt}\,\bar{a}_z$. (a) Show that these fields do not satisfy Maxwell's other curl equation. (b) If we let $B_o = 1$T and $k = 10^6\ s^{-1}$, we are establishing a farily large magnetic flux density in 1 μs. Use the $\overline{\nabla} \times \overline{H}$ equation to show that the rate at which B_z should (but does not) change with ρ is only about 5×10^{-6} T per meter in free space at $t = 0$.

Solution :

(a) $\overline{B} = \mu H$

$$\therefore \overline{H} = \frac{\overline{B}}{\mu} = \frac{B_o e^{kt} \overline{a}_z}{\mu}$$

$$\overline{\nabla} \times \overline{H} = \begin{vmatrix} \frac{\overline{a}_\rho}{\rho} & \overline{a}_\phi & \frac{\overline{a}_z}{\rho} \\ \frac{\partial}{\partial \rho} & \frac{\partial}{\partial \phi} & \frac{\partial}{\partial z} \\ 0 & 0 & \frac{B_o e^{kt}}{\mu} \end{vmatrix} = 0$$

Now $\overline{\nabla} \times \overline{H} = \sigma \overline{E} + \overline{J}_D$

$$\text{RHS} = \sigma \overline{E} + \overline{J}_D = \frac{-\sigma k B_o e^{kt}}{2} \overline{a}_\phi + \varepsilon \left(-\frac{1}{2} k B_o k e^{kt} \overline{a}_\phi \right)$$

$$= \left[\frac{-\sigma k B_o}{2} - \frac{k^2 B_o \varepsilon}{2} \right] e^{kt} \overline{a}_\phi$$

The value of σ is not given. Even if it is zero, RHS will be

$\frac{-k^2 B_o \varepsilon}{2} e^{kt} \overline{a}_\phi$ which is not equal to LHS.

Hence the given fields does not satisfy Maxwell's equations.

(b) $\overline{H} = \frac{\overline{B}}{\mu} = \frac{B_z \overline{a}_z}{\mu}$ where $B_z = B_o e^{kt}$

$$\overline{\nabla} \times \overline{H} = \begin{vmatrix} \frac{\overline{a}_\rho}{\rho} & \overline{a}_\phi & \frac{\overline{a}_z}{\rho} \\ \frac{\partial}{\partial \rho} & \frac{\partial}{\partial \phi} & \frac{\partial}{\partial z} \\ 0 & 0 & \frac{B_z}{\mu} \end{vmatrix} = -\overline{a}_\phi \frac{\partial}{\partial \rho} \left(\frac{B_z}{\mu} \right)$$

For free space $\sigma = 0$, $\mu = \mu_o$, $\varepsilon = \varepsilon_o$

Now $\overline{\nabla} \times \overline{H} = \sigma \overline{E} + \overline{J}_D = \overline{J}_D$

$$= \frac{-K^2 B_o \varepsilon}{2} e^{kt} \overline{a}_\phi$$

$$\therefore \quad -\frac{\bar{a}_\phi}{\mu}\frac{\partial}{\partial \rho}(B_z) = \frac{-K^2 B_o \varepsilon}{2} e^{kt}\,\bar{a}_\phi$$

Substitute $B_o = 1T$, $K = 10^6 s^{-1}$, $t = 0$, $\mu = \mu_o$, $\varepsilon = \varepsilon_o$

We have $\frac{1}{\mu_o}\frac{\partial}{\partial \rho}(B_z) = \frac{(10^6)^2 \times \varepsilon_o}{2}$

or $\frac{\partial}{\partial \rho}(B_z) = \frac{10^{12}\mu_o \varepsilon_o}{2} = 5.563 \times 10^{-6}\,T/m$

Problem 9.28 : A section of coaxial cable is bounded by conducting surfaces at z = 0 and z = 0.2 m. The conductor radii are a = 2mm and b = 8 mm, and in the region a < ρ < b we have $\varepsilon_R = 6.25$, $\mu_R = 1$, and σ = 0. If $\bar{E} = \left(\frac{100}{\rho}\right)\sin 5\pi z \sin \omega t\,\bar{a}_\rho$ V/m, find : (a) ω with the help of Maxwell's equations ; (b) the total charge on the inner conductor at $t = \frac{\pi}{(2\omega)}$; (c) the surface current density at ρ = 3mm , φ = 0, z = 0, t = 0.

Solution :

$$\text{(a)}\ \bar{\nabla} \times \bar{E} = \begin{vmatrix} \frac{\bar{a}_\rho}{\rho} & \bar{a}_\phi & \frac{\bar{a}_z}{\rho} \\ \frac{\partial}{\partial \rho} & \frac{\partial}{\partial \phi} & \frac{\partial}{\partial z} \\ (100/\rho)\sin 5\pi z \sin \omega t & 0 & 0 \end{vmatrix}$$

$$= -\bar{a}_\phi \left[(-1)\frac{100}{\rho}\right] 5\pi \cos 5\pi z \sin \omega t$$

$$= \left[\frac{500\pi}{\rho}\right] \cos 5\pi z \sin \omega t\ \bar{a}_\phi$$

Now $\bar{\nabla} \times \bar{E} = -\frac{\partial \bar{B}}{\partial t}$. Hence

$$\frac{\partial \bar{B}}{\partial t} = \left(-\frac{500\pi}{\rho}\right)\cos 5\pi z \sin \omega t\ \bar{a}_\phi$$

$$\therefore \quad \bar{B} = \left(\frac{500\pi}{\rho\,\omega}\right)\cos 5\pi z \cos \omega t\ \bar{a}_\phi$$

$$\bar{H} = \frac{\bar{B}}{\mu_o \mu_r} = \frac{\bar{B}}{\mu_o}$$

$$\overline{\nabla} \times \overline{H} = \begin{vmatrix} \frac{\bar{a}_\rho}{\rho} & \bar{a}_\phi & \frac{\bar{a}_z}{\rho} \\ \frac{\partial}{\partial \rho} & \frac{\partial}{\partial \phi} & \frac{\partial}{\partial z} \\ 0 & \rho\left(\frac{500\pi}{\mu_o \rho \omega}\right)\cos 5\pi z \cos \omega t & 0 \end{vmatrix} = 0$$

$$= \frac{\bar{a}_\rho}{\rho}\left[-\left(\frac{500\pi}{\mu_o \omega}\right)(-5\pi \sin 5\pi z)\cos \omega t\right]$$

$$= \frac{2500\pi^2}{\rho \mu_o \omega} \sin 5\pi z \cos \omega t \ \bar{a}_\rho$$

$$\overline{\nabla} \times \overline{H} = \sigma\overline{E} + \overline{J}_D = \overline{J}_D = \varepsilon_o \varepsilon_r \frac{\partial \overline{E}}{\partial t}$$

$$= 6.25\,\varepsilon_o\,\omega \left(\frac{100}{\rho}\right) \sin(5\pi z)\cos(\omega t)\ \bar{a}_\rho$$

$$\therefore \frac{2500\pi^2}{\rho \mu_o \omega} \sin 5\pi z \cos \omega t \ \bar{a}_\rho$$

$$= \frac{625\,\omega\,\varepsilon_o}{\rho} \sin 5\pi z \cos \omega t \ \bar{a}_\rho$$

or $$\frac{2500\pi^2}{\mu_o \omega} = 625\,\omega\,\varepsilon_o$$

$$\therefore \omega = 1.884 \times 10^9 \text{ rad/sec}.$$
$$= 1.884 \text{ Grad/sec.}$$

(b) $\rho_S = |\overline{D}_N| = |\varepsilon_o \varepsilon_r \overline{E}|$

$$= \left| 6.25 \times \left(\frac{100}{\rho}\right) \varepsilon_o \sin 5\pi z \sin \omega t \right|$$

$$= \left(\frac{625\,\varepsilon_o}{\rho}\right) \sin 5\pi z \sin \omega t$$

To the inner conductor $\rho = 2$, $\pm\ \bar{a}_\rho$ is perpendicular.
Hence $\overline{dS} = \pm\ (\rho\, d\phi\, dz)\ \bar{a}_\rho$ or $dS = \rho\, d\phi\, dz$

$$\therefore Q = \int dQ = \int \rho_S\, dS$$

$$= \int_0^{2\pi} \int_0^{0.2} \frac{625\,\varepsilon_o}{\rho} \sin 5\pi z \sin \omega t \ \rho\, d\phi\, dz$$

$$= 625\,\varepsilon_o \left[\frac{-\cos 5\pi z}{5\pi} \right]_0^{0.2} \left[\phi \right]_0^{2\pi} \sin \omega t$$

$$= 4.427 \sin \omega t \ \text{nC}$$

at $t = \left(\frac{\pi}{2\omega} \right)$, $Q = 4.427 \sin\left[\omega \left(\frac{\pi}{2\omega} \right) \right] = 4.427\ \text{nC}$

(c) $\overline{H} = \left(\frac{500\pi}{\rho\,\omega\,\mu_o} \right) \cos 5\pi z \cos \omega t\, \overline{a}_\phi$

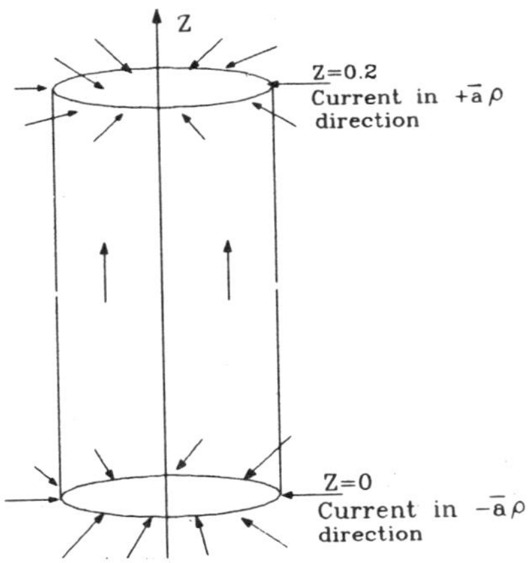

Fig.P 9.28(a)

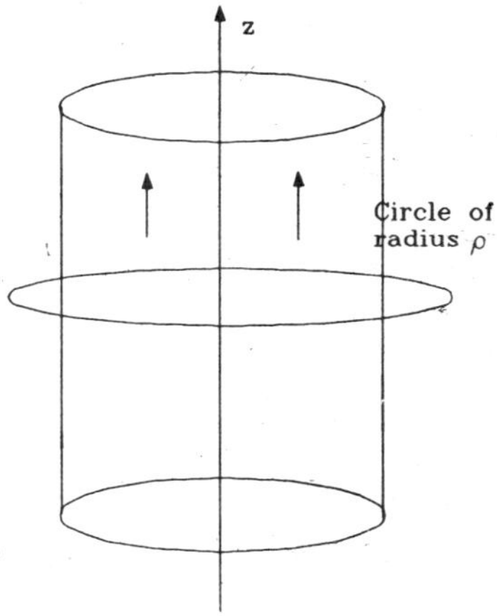

Fig.P 9.28(b)

Consider that current I flows in the + $\overline{a}_z$ direction on the inner conductor $\rho = a = 2$mm. This current produces Φ component of $\overline{H}$.

Hence $H = H\,\Phi\,\overline{a}_\Phi$. Consider a circle of radius ρ such that $\rho > 2$mm as shown in Fig. P. 9.28(a). Apply Ampere circuital law to this closed path, we get

$$\oint \overline{H} \cdot \overline{dL} = I_{encl.}$$

LHS

$$= \oint \overline{H} \cdot \overline{dL} = \int (H_\phi\, \overline{a}_\phi) \cdot (d\rho\, \overline{a}_\rho + \rho\, d\phi\, \overline{a}_\phi + dz\, \overline{a}_z)$$

$$= \int_0^{2\pi} \rho\, H_\phi\, d\phi = 2\pi\rho H_\phi = I$$

From the expression for $\overline{H}$, we have

$$H_\phi = \left(\frac{500\,\pi}{\rho\,\omega\,\mu_o}\right) \cos 5\pi z \cos \omega t$$

$$\therefore \quad I = 2\pi\rho H_\phi = \frac{1000\,\pi^2}{\omega\,\mu_o} \cos 5\pi z \cos \omega t$$

Circumference of a circle with radius 3 mm is equal to

$$2\pi\,(0.003) = 0.006\,\pi \text{ m}.$$

Hence surface current density $= \dfrac{1000\,\pi^2 \cos 5\pi z \cos \omega t}{0.006\,\pi\,\omega\,\mu_o}$

Substitute $\omega = 1.884 \times 10^9$, $z = 0$, $t = 0$, we get

Surface current density $= -\ 221.161$ A/sec.

Negative sign is used because the current is directed in $-\ \overline{a}_\rho$ direction as shown in the fig.P 9.28 (b)

Problem 9.29 : Consider a region where $\varepsilon = \varepsilon_o$, $\mu = \mu_o$, $\sigma = 0$, $\overline{J} = 0$, and $\rho_v = 0$. Use Maxwell's equations in cartesian coordinates and assume that $\overline{E} = E_x(z, t)\ \overline{a}_x$ and $\overline{H} = H_y(z, t)\ \overline{a}_y$. (a) Find the second–order partial differential equation that E_x must satisfy. (b) Show that $E_x = (200t - az)^2$ is a solution of this equation for a certain value of a, and find that value.

Solution :

(a) $\overline{\nabla} \times \overline{E} = -\dfrac{\partial \overline{B}}{\partial t}$ or $\nabla \times \overline{E} = -\mu_o \dfrac{\partial \overline{H}}{\partial t}$

Hence $\overline{\nabla} \times \overline{\nabla} \times \overline{E} = -\mu_o \overline{\nabla} \times \dfrac{\partial \overline{H}}{\partial t}$

$$= -\mu_o \frac{\partial}{\partial t} (\nabla \times \overline{H})$$

Now $\overline{\nabla} \times \overline{H} = \sigma \overline{E} + \dfrac{\partial \overline{D}}{\partial t} \qquad = \dfrac{\partial \overline{D}}{\partial t} = \varepsilon_o \dfrac{\partial \overline{E}}{\partial t}$

$$\therefore \quad \bar{\nabla} \times \bar{\nabla} \times \bar{E} = -\mu_o \frac{\partial}{\partial t}\left(\frac{\partial \bar{E}}{\partial t}\right)$$

$$= -\varepsilon_o \mu_o \frac{\partial^2 \bar{E}}{\partial t^2}$$

$$\bar{\nabla} \times \bar{E} = \begin{vmatrix} \bar{a}_x & \bar{a}_y & \bar{a}_z \\ \frac{\partial}{\partial x} & \frac{\partial}{\partial y} & \frac{\partial}{\partial z} \\ E_x(z,t) & 0 & 0 \end{vmatrix}$$

$$= -\bar{a}_y \left[\frac{-\partial E_x(z,t)}{\partial x} \right] = \frac{\partial E_x(z,t)}{\partial z} \bar{a}_y$$

$$\therefore \quad \bar{\nabla} \times \bar{\nabla} \times \bar{E} = \bar{\nabla} \times \left[\frac{\partial}{\partial z} E_x(z,t)\, \bar{a}_y \right]$$

$$= \begin{vmatrix} \bar{a}_x & \bar{a}_y & \bar{a}_z \\ \frac{\partial}{\partial x} & \frac{\partial}{\partial y} & \frac{\partial}{\partial z} \\ 0 & \frac{\partial}{\partial z} E_x(z,t) & 0 \end{vmatrix} = \bar{a}_x \left[-\frac{\partial^2 E_x}{\partial z^2}(z,t) \right] = -\frac{\partial^2 E_x}{\partial z^2} \bar{a}_x$$

Hence $\frac{\partial^2 E_x}{\partial z^2} = \varepsilon_o \mu_o \frac{\partial^2 E_x}{\partial t^2}$ is the required second order partial differential equation.

(b) $E_x = (200t - a z)^2$

$$\therefore \quad \frac{\partial^2 E_x}{\partial z^2} = 2a^2 \quad \text{and} \quad \frac{\partial^2 E_x}{\partial t^2} = 80000$$

Hence $2a^2 = \varepsilon_o \mu_o \times 80000$ or $a = \pm\ 6.6712 \times 10^{-7}$

Problem 9.30 : A current $\bar{I} = 100\, t\, \bar{a}_y$ is present in free space on the y axis in the interval $0 < y < 0.2$ m. Find A_y at P (0, 0, 2) and sketch A_y over the time interval $-\ 0.1 < t < 0.1\, \mu$ s .

Solution :

As the current is a function of time, time t is replaced by the retarded time t′ equal to $t - \frac{R}{V}$, where R is the distance between the differential current filament and point P and V is the velocity of light.

$$\overline{A} = \int \frac{\mu_o [I] \overline{dL}}{4 \pi R}$$

where [I] indicates that every t appearing in the expression for I has been replaced by a retarded time, $t' = t - \frac{R}{v}$.

R is the distance between the differential current element and the point P. As the current is present on the y axis, $\overline{dL} = dy \, \overline{a}_y$ and suppose the differential current filament is located at the general point on the y axis i.e. (0, y, 0). Then we have

$$R = \sqrt{(0 - y)^2 + (2 - 0)^2} = \sqrt{y^2 + 4}$$

$$\therefore \overline{A} = \int_0^{0.2} \frac{\mu_o \times 100 \left(t - \dfrac{\sqrt{y^2 + 4}}{3 \times 10^8} \right) \times dy \, \overline{a}_y}{4 \pi \sqrt{y^2 + 4}}$$

$$= \frac{25 \mu_o}{\pi} \left[t \ln \left(\frac{y + \sqrt{y^2 + 4}}{2} \right) - \frac{y}{3 \times 10^8} \right]_0^{0.2} \overline{a}_y$$

$$= (9.9834 \times 10^{-7} t - 6.6667 \times 10^{-15}) \, \overline{a}_y = A_y \, \overline{a}_y$$

Hence $A_y = 9.9834 \times 10^{-7} t - 6.6667 \times 10^{-15}$ Wb/m

Plot A_y on the y axis and time t on the x axis for

$-0.1 < t < 0.1 \; \mu s$. Plot is not given.

Problem 9.31 : In a region where $\varepsilon_R = \mu_R = 1$ and $\sigma = 0$, let $\overline{A} = 10^{-3} y \cos 3 \times 10^8 t \cos z \, \overline{a}_z$ Wb/m a nd $V = 3 \times 10^5 y \sin 3 \times 10^8 t \sin z$ V. Find $\overline{E}$ and $\overline{H}$.

Solution :

$$\overline{E} = -\overline{\nabla} V - \frac{\partial \overline{A}}{\partial t} = -\left[\frac{\partial V}{\partial x} \overline{a}_x + \frac{\partial V}{\partial y} \overline{a}_y + \frac{\partial V}{\partial z} \overline{a}_z \right] - \frac{\partial \overline{A}}{\partial t}$$

$$= -3 \times 10^5 \sin (3 \times 10^8 t) \left[\sin z \, \overline{a}_y + y \cos z \, \overline{a}_z \right]$$

$$- 10^{-3} y \times (-3 \times 10^8) \sin (3 \times 10^8 t) \cos z \, \overline{a}_z$$

$$= -3 \times 10^5 \sin z \sin (3 \times 10^8) \, \overline{a}_y \text{ V/m}$$

$$\bar{B} = \bar{\nabla} \times \bar{A} = \begin{vmatrix} \bar{a}_x & \bar{a}_y & \bar{a}_z \\ \frac{\partial}{\partial x} & \frac{\partial}{\partial y} & \frac{\partial}{\partial z} \\ 0 & 0 & 10^{-3} y \cos 3 \times 10^8 t \cos z \end{vmatrix}$$

$$= 10^{-3} \cos (3 \times 10^8 t) \cos z \; \bar{a}_x$$

$$\therefore \; \bar{H} = \frac{\bar{B}}{\mu_o} = 795.7747 \cos (3 \times 10^8 t) \cos z \, \bar{a}_x \text{ A/m}$$

Problem 9. 32 : Given the retarded potentials, $V = -y(x + ct)$ V and $\bar{A} = y\left(\frac{x}{c} + t\right) \bar{a}_x$ Wb/m , where $c = \frac{1}{\sqrt{\mu_o \varepsilon_o}}$: (a) show that $\bar{\nabla} \cdot \bar{A} = -\mu \varepsilon \partial V / \partial t$; (b) find $\bar{B}$, $\bar{H}$, $\bar{E}$ and $\bar{D}$. (c) Showthat these results satisfy Maxwell's equations if $\bar{J} = 0$ andc $\rho_v = 0$.

Solution :

(a) $$\bar{\nabla} \cdot \bar{A} = \frac{\partial A_x}{\partial x} + \frac{\partial A_y}{\partial y} + \frac{\partial A_z}{\partial z} = \frac{\partial}{\partial x}\left[y\left(\frac{x}{C} + t\right)\right] = \frac{y}{C}$$

Now $$- \mu \varepsilon \frac{\partial V}{\partial t} = - \mu \varepsilon \frac{\partial}{\partial t}\left[- y(x + Ct) \right] = \mu \varepsilon C y$$

$$= \frac{\mu \varepsilon y}{[\mu \varepsilon]^{1/2}} = [\mu \varepsilon]^{1/2} y = y/C$$

Hence $$\nabla \cdot \bar{A} = - \mu \varepsilon \frac{\partial V}{\partial t}$$

(b) $$\bar{B} = \bar{\nabla} \times \bar{A} = \frac{\partial}{\partial z}\left[y\left(\frac{x}{C} + t\right)\right] \bar{a}_y - \frac{\partial}{\partial y}\left[y\left(\frac{x}{C} + t\right)\right] \bar{a}_z$$

$$= -\left(\frac{x}{C} + t\right) \bar{a}_z \text{ Wb/m}^2$$

$$\bar{H} = \frac{\bar{B}}{\mu} = -\frac{1}{\mu}\left(\frac{x}{C} + t\right) \bar{a}_z \text{ A/m}$$

$$\bar{E} = -\bar{\nabla} V - \frac{\partial \bar{A}}{\partial t} = -\left[- y \bar{a}_x - (x + Ct) \bar{a}_y \right] - y \bar{a}_x$$

$$= (x + Ct) \; \bar{a}_y \text{ V/m}$$

$$\therefore \; \bar{D} = \varepsilon \bar{E} = \varepsilon (x + Ct) \; \bar{a}_y \text{ C/m}^2$$

$$(c)\bar{\nabla} \times \bar{H} = \frac{\partial}{\partial y}\left[-\frac{1}{\mu}\left(\frac{x}{C}+t\right)\right]\bar{a}_x - \frac{\partial}{\partial x}\left[-\frac{1}{\mu}\left(\frac{x}{C}+t\right)\right]\bar{a}_y$$

$$= -\bar{a}_y\left[-\frac{1}{\mu C}\right] = \frac{1}{\mu C}\bar{a}_y$$

$$\bar{J} + \frac{\partial \bar{D}}{\partial t} = \bar{J} + \varepsilon C \bar{a}_y$$

Now $\bar{\nabla} \times \bar{H} = \bar{J} + \dfrac{\partial \bar{D}}{\partial t}$

Hence $\dfrac{1}{\mu C}\bar{a}_y = \bar{J} + \varepsilon C \bar{a}_y$

As $C = \dfrac{1}{\sqrt{\mu \varepsilon}}$, we have $\dfrac{1}{\mu C} = \varepsilon C$

So that $\bar{J}$ must be equal to zero.

Now $\bar{\nabla}^2 V = -\dfrac{\rho_v}{\varepsilon} + \mu \varepsilon \dfrac{\partial^2 V}{\partial t^2}$

$$\nabla^2 V = \frac{\partial^2 V}{\partial x^2} + \frac{\partial^2 V}{\partial y^2} + \frac{\partial^2 V}{\partial z^2} = 0$$

$$\frac{\partial^2 V}{\partial t^2} = 0$$

Hence ρv must be equal to zero.

—X—

CHAPTER 10

The Uniform Plane Wave

Definition :

A uniform plane wave is defined as a wave whose value remains constant throughout a plane which is transverse to the direction of propagation of the wave. For example, if the wave is travelling in $\overline{a}_z$ direction, then the plane $z =$ constant will be perpendicular to the direction of propagation. In the plane $z =$ constant, the variables are x and y. Hence for this uniform plane wave, the electric field intensity and magnetic field intensity are independent of x and y coordinate and are the functions of z coordinates only.

10.1 UNIFORM PLANE WAVE OR TRANSVERSE ELECTRO - MAGNETIC WAVE :

10.1.1 Transformation of a time varying quantity into phasor :

Let E_x be the time varying quantity which is either a function of sine or cosine. Then the phasor of E_x is written as E_{xs} and is obtained by dropping Re and suppressing $e^{j\omega t}$. If

$$E_x = E_o \cos(\omega t + \psi) = E_o \text{ Re}[\cos(\omega t + \psi) + j\sin(\omega t + \psi)]$$

$$= E_o \text{ Re}[e^{j(\omega t + \psi)}]$$

$$= E_o \text{ Re}[e^{j\omega t} e^{j\Psi}]$$

then,

$$E_{xs} = E_o e^{j\Psi}$$

where Re means the real part of the following quantity is to be taken. Further it can be proved that the phasor of $\frac{dE_x}{dt}$ is given by multiplying the phasor of E_x by $j\omega$, i.e.phasor of $\left(\frac{dE_x}{dt}\right) = j\omega E_{xs}$

10.1.2 Transformation of phasor into a time varying quantity.

To transform a phasor into a time varying quantity, multiply that phasor by $e^{j\omega t}$ and then consider the real part (Re) of that product. If $E_{xs} = E_o e^{j\psi}$, then ,

$$E_x = \text{Re}\left[E_{xs}\, e^{j\omega t} \right] = \text{Re}\left[E_o e^{j\omega t}\, e^{j\psi} \right] = \text{Re}\left[E_o e^{j(\omega t + \psi)} \right]$$
$$= E_o \cos(\omega t + \psi)$$

10.1.3 Maxwell's equations in phasor form .

The four Maxwell's equations in point form or differential form are

$$\bar{\nabla} \times \bar{E} = -\frac{\partial \bar{B}}{\partial t} = -\mu \frac{\partial \bar{H}}{\partial t}$$

$$\bar{\nabla} \times \bar{H} = \bar{J} + \frac{\partial \bar{D}}{\partial t} = \sigma \bar{E} + \varepsilon \frac{\partial \bar{E}}{\partial t}$$

$$\bar{\nabla} \cdot \bar{D} = \rho_v$$

$$\bar{\nabla} \cdot \bar{B} = 0$$

These equations in phasor form may be written as

$$\boxed{\bar{\nabla} \times \bar{\mathbf{E}}_s = -j\omega\mu \bar{\mathbf{H}}_s} \quad \text{......(10.1a)}$$

$$\boxed{\bar{\nabla} \times \bar{\mathbf{H}}_s = (\sigma + j\omega\varepsilon) \bar{\mathbf{E}}_s} \quad \text{....(10.1b)}$$

$$\boxed{\bar{\nabla} \cdot \bar{\mathbf{D}}_s = \rho_v} \quad \text{......(10.1c)}$$

$$\boxed{\bar{\nabla} \cdot \bar{\mathbf{B}}_s = 0} \quad \text{....(10.1d)}$$

10.1.4 Wave equation for electric field intensity.

Using the vector identity,

$$\bar{\nabla} \times \bar{\nabla} \times \bar{E}_s = \bar{\nabla}(\bar{\nabla} \cdot \bar{E}_s) - \bar{\nabla}^2 \bar{E}_s$$

$$\text{LHS} = \bar{\nabla} \times \bar{\nabla} \times \bar{E}_s$$

$$= \bar{\nabla} \times (-j\omega\mu\bar{H}_s) \qquad [\because \text{ Using Eq.(10.1a)}]$$

$$= -j\omega\mu\bar{\nabla} \times \bar{H}_s$$

$$= -j\omega\mu(\sigma + j\omega\varepsilon)\bar{E}_s \qquad [\because \text{ Using Eq.(10.1b)}]$$

$$\text{RHS} = \bar{\nabla}(\bar{\nabla} \cdot \bar{E}_s) - \bar{\nabla}^2 \bar{E}_s = \bar{\nabla}\left(\bar{\nabla} \cdot \frac{\bar{D}_s}{\varepsilon}\right) - \bar{\nabla}^2 \bar{E}_s$$

$$= \frac{1}{\varepsilon}\bar{\nabla}(\rho_v) - \bar{\nabla}^2 \bar{E}_s \qquad [\because \text{ Using Eq.(10.1c)}]$$

Assuming ρ_v is equal to zero, we get

$$\text{RHS} = -\bar{\nabla}^2 \bar{E}_s$$

$$\therefore -\bar{\nabla}^2 \bar{E}_s = -j\omega\mu(\sigma + j\omega\varepsilon)\bar{E}_s \quad \textbf{or}$$

$$\bar{\nabla}^2 \bar{E}_s = j\omega\mu(\sigma + j\omega\varepsilon)\bar{E}_s$$

The quantity $j\omega\mu(\sigma + j\omega\varepsilon)$ is denoted by γ^2. Hence

$$\boxed{\bar{\nabla}^2 \bar{E}_s = \gamma^2 \bar{E}_s} \qquad \ldots\ldots(10.2a)$$

This is the wave equation for electric field intensity. Similarly it can be proved that, the wave equation for magnetic field intensity is

$$\boxed{\bar{\nabla}^2 \bar{H}_s = \gamma^2 \bar{H}_s} \qquad \ldots\ldots(10.2b)$$

10.1.5 Propagation constant, Attenuation constant and Phase constant

In the wave equation

$$\bar{\nabla}^2 \bar{E}_s = \gamma^2 \bar{E}_s$$

The quantity γ is known as propogation constant and is defined by the equation

$$\gamma^2 = j\omega\mu(\sigma + j\omega\varepsilon)$$

or

$$\boxed{\gamma = \sqrt{j\omega\mu(\sigma + j\omega\varepsilon)}} \qquad \ldots\ldots(10.3)$$

The units of propogation constant is per m. It is a complex number. The real part is denoted by α and is known as attenuation constant. The unit is Np / m where Np stands for neper. The imaginary part is denoted by β and is known as phase constant. The unit is rad / m. Hence

$$\gamma = \alpha + j\omega \qquad \ldots\ldots (10.4)$$

If Eq.(10.3) and Eq.(10.4) are solved for positive values of α and β, then we have

$$\boxed{\alpha = \omega \sqrt{\frac{\mu\varepsilon}{2}\left[\sqrt{1 + \left(\frac{\sigma}{\omega\varepsilon}\right)^2} - 1\right]}} \qquad \ldots\ldots (10.5)$$

and

$$\boxed{\beta = \omega \sqrt{\frac{\mu\varepsilon}{2}\left[\sqrt{1 + \left(\frac{\sigma}{\omega\varepsilon}\right)^2} + 1\right]}} \qquad \ldots\ldots (10.6)$$

10.1.6 Solution of wave equation

The wave equation for electric field intensity is

$$\bar{\nabla}^2 \bar{E}_s = \gamma^2 \bar{E}_s$$

It is vector differential equation. Let us consider x component of electric field intensity. Then

$$\bar{\nabla}^2 E_{xs} = \gamma^2 E_{xs}$$

or

$$\frac{\partial^2 E_{xs}}{\partial x^2} + \frac{\partial^2 E_{xs}}{\partial y^2} + \frac{\partial^2 E_{xs}}{\partial z^2} = \gamma^2 E_{xs}$$

Let us assume that the wave is travelling in + $\bar{a}_z$ direction so that as per the definition of uniform plane wave, E_{xs} will not change with x and y coordinates. Hence

$$\frac{d^2 E_{xs}}{dz^2} = \gamma^2 E_{xs}$$

By inspection, the solution of above differential equation may be written as

$$E_{xs} = A\, e^{-\gamma z}$$

Multiply both sides by $e^{j\omega t}$ and consider the real part. Then we get

$$E_x = \text{Re}\left[A\, e^{-\gamma z}\, e^{j\omega t} \right]$$

Substitute the value of $\gamma\ (= \alpha + j\beta)$ in the above equation

$$E_x = \text{Re}\left[A\, e^{-(\alpha + j\beta) z}\, e^{j\omega t} \right] = \text{Re}\left[A\, e^{-\alpha z}\, e^{j(\omega t - \beta z)} \right]$$

$$E_x = A\, e^{-\alpha z} \cos(\omega t - \beta z)$$

Let E_{xo} be the value of E_x at $t = z = 0$. Then $A = E_{xo}$ so that

$$\boxed{E_x = E_{xo}\, e^{-\alpha z} \cos(\omega t - \beta z)} \qquad \ldots\ldots(10.7)$$

10.1.7 Frequency, wavelength, velocity, and intrinsic impedance.

Frequency : The number of oscillations carried out by a wave per second is known as frequency of the wave. It is denoted by f and the unit is Hz (hertz). The radian frequency will be given by $\omega = 2\pi f$ and the unit is rad/ sec.

Wavelength : By the term wavelength is meant the distance travelled by the wave in which its phase angle changes by 2π radians. It is denoted be λ and the unit is m. Thus as per definition, $\beta\lambda = 2\pi$, or

$$\boxed{\lambda = \frac{2\pi}{\beta}} \qquad \ldots\ldots.(10.8)$$

Velocity : We have $v = f\lambda$, where v is the phase velocity of the wave .

$$\therefore\ v = \frac{\omega}{2\pi} \times \frac{2\pi}{\beta} \quad \text{or}$$

$$\boxed{v = \frac{\omega}{\beta}} \qquad \ldots\ldots(10.9)$$

The unit of velocity is m / sec.

Intrinsic impedance : The ratio of amplitudes of electric field intensity to the magnetic field intensity has the dimensions of impedance (ohms). This ratio is known as intrinsic impedance of the medium in which the wave is travelling. It is denoted by η. Thus we have

$$\boxed{\eta = \frac{E}{H}} \qquad \ldots\ldots(10.10)$$

In article 10.1.6, the solution which is assumed for electric field intensity is

$E_{xs} = A\,e^{-\gamma z}$ or $\overline{E} = A\,e^{-\gamma z}\,\overline{a}_x$

Now $\overline{\nabla} \times \overline{E}_s = -j\omega\mu\overline{H}_s$

$$\text{LHS} = \overline{\nabla} \times \overline{E}_s = \begin{vmatrix} \overline{a}_x & \overline{a}_y & \overline{a}_z \\ \partial/\partial x & \partial/\partial y & \partial/\partial z \\ A\,e^{-\gamma z} & 0 & 0 \end{vmatrix} = -A\,e^{-\gamma z}\,\overline{a}_y$$

Hence

$$-A\gamma\,e^{-\gamma z}\,\overline{a}_y = -j\omega\mu\overline{H}_s \quad \text{or} \quad \overline{H}_s = \frac{A\gamma\,e^{-\gamma z}}{j\omega\mu}\,\overline{a}_y$$

$$\therefore \eta = \frac{A\,e^{-\gamma z}}{A\gamma\,e^{-\gamma z}/j\omega\mu} = \frac{j\omega\mu}{\gamma} = \frac{j\omega\mu}{\sqrt{j\omega\mu(\sigma + j\omega\varepsilon)}}$$

$$= \sqrt{\frac{j^2\omega^2\mu^2}{j\omega\mu(\sigma + j\omega\varepsilon)}} = \sqrt{\frac{j\omega\mu}{\sigma + j\omega\varepsilon}}$$

Hence

$$\boxed{\eta = \sqrt{\frac{j\omega\mu}{\sigma + j\omega\varepsilon}}} \qquad \text{......(10.11)}$$

10.2 WAVE MOTION IN FREE SPACE

The free space is characterized by

$$\sigma = 0\,,\ \mu = \mu_o\,,\ \varepsilon = \varepsilon_o$$

Substituting these values in Eq.10.5, 10.6 and 10.11, we get

$$\boxed{\alpha = 0} \qquad \text{......(10.12a)}$$

$$\boxed{\beta = \omega\sqrt{\mu_o\varepsilon_o}} \qquad \text{......(10.12b)}$$

$$\boxed{\eta = \sqrt{\frac{\mu_o}{\varepsilon_o}} = 377 \doteq 120\pi} \qquad \text{......(10.12c)}$$

Substituting the value of β in Eq. (10.9) we get

$$v = \frac{\omega}{\omega\sqrt{\mu_o\varepsilon_o}} = \frac{1}{\sqrt{\mu_o\varepsilon_o}} = 3 \times 10^8 \ \text{m/sec}$$

Hence electromagnetic wave in free space travels at the speed of light.

$$\boldsymbol{v = 3 \times 10^8 \ \text{m/sec}} \qquad \text{......(10.12d)}$$

Problem 10.1 : The magnetic field intensity of a uniform plane wave in air is 20 A / m in the $\bar{a}_y$ direction. The wave is propagating in the $\bar{a}_z$ direction at a frequency of 2 Grad / s (2×10^9 rad/s) . Find : (a) the wavelength ; (b) the frequency ; (c) the period ; (d) the amplitude of $\bar{E}$.

Solution :

$$\text{(a) } \beta = \omega\sqrt{\mu_0 \varepsilon_0} = 2\times 10^9 \sqrt{4\pi\times 10^{-7}\times 8.854\times 10^{-12}}$$

$$= 6.6712 \text{ rad/m}$$

$$\therefore \lambda = \frac{2\pi}{\beta} = \frac{2\pi}{6.6712} = 0.942 \text{ m}$$

$$\text{(b) } f = \frac{\omega}{2\pi} = \frac{2\times 10^9}{2\pi} = 318.310 \text{ MHz.}$$

$$\text{(c) } T = \frac{1}{f} = 3.142 \text{ ns}$$

$$\text{(d) } \eta = \frac{E}{H} = \sqrt{\frac{\mu_0}{\varepsilon_0}} = 376.734\ \Omega = 376.734\ \Omega$$

$$\therefore \text{E} = 376.734\times 20 = 7534.68 \text{ V/m}$$

Problem 10.2 : Let $\bar{E}_s = [\,500 \angle -40^\circ\ \bar{a}_y + (200 - j\,600)\ \bar{a}_z\,]\ e^{-j\,0.4\,x}$ V/m in free space. Find : (a) ω ; (b) $\bar{E}$ at (2, 3, 1) $t = 0$; (c) $\bar{E}$ at (2, 3, 1), $t = 10$ ns ; (d) $\bar{E}$ at (3, 4, 2), $t = 20$ ns.

Solution :

$$\text{(a) } \beta = 0.4 = \omega\sqrt{\mu_0 \varepsilon_0}$$

$$\therefore \omega = \frac{0.4}{\sqrt{4\pi\times 10^{-7}\times 8.854\times 10^{-12}}} = 119.918 \text{ rad/s}$$

$$\text{(b) } \bar{E}_s = [\,500 \angle -40^\circ\ \bar{a}_y + (200 - j\,600)\ \bar{a}_z\,]\ e^{-j\,0.4x}$$

$$= [\,500 \angle -40^\circ\ \bar{a}_y + 632.456 \angle -71.565^\circ\ \bar{a}_z\,]\ e^{-j\,0.4x}$$

$$= 500\,e^{-j\,40^\circ}\ e^{-j\,0.4x}\ \bar{a}_y + 632.456\,e^{-j\,71.565^\circ}\ e^{-j\,0.4x}\ \bar{a}_z$$

$$= 500\,e^{-j\,(0.4x + 40^\circ)}\ \bar{a}_y + 632.456\ e^{-j\,(0.4x + 71.565^\circ)}\ \bar{a}_z$$

$$\therefore \bar{E}(t) = 500\,\text{Re}\left[e^{-j\,(0.4x + 40^\circ)}\ e^{j\,\omega t}\right] \bar{a}_y +$$

$$632.456\ \text{Re}\left[e^{-j\,(0.4x + 71.565^\circ)}\ e^{j\,\omega t}\right] \bar{a}_z$$

$$\bar{E}(t) = 500 \, \text{Re}\left[e^{j(\omega t - 0.4x - 40^\circ)} \right] \bar{a}_y + 632.456 \, \text{Re}\left[e^{j(\omega t - 0.4x - 71.565^\circ)} \right] \bar{a}_z$$

$$\therefore \bar{E}(t) = 500 \cos(\omega t - 0.4x - 40^\circ) \, \bar{a}_y + 632.456 \cos(\omega t - 0.4x - 71.565^\circ) \, \bar{a}_z$$

$$\bar{E} \text{ at } (2, 3, 1, t = 0) = 500 \cos(-0.4 \times 2 - 40^\circ) \, \bar{a}_y + 632.456 \cos(-0.4 \times 2 - 71.565^\circ) \, \bar{a}_z$$

$$= 36.297 \, \bar{a}_y - 291.076 \, \bar{a}_z \text{ V/m}$$

(c) $\bar{E}$ at $(2, 3, 1, t = 10 \text{ ns})$

$$= 500 \cos(119.918 \times 10^6 \times 10 \times 10^{-9} - 0.4 \times 2 - 40^\circ) \, \bar{a}_y + 632.456 \cos(119.918 \times 10^6 \times 10 \times 10^{-9} - 0.4 \times 2 - 71.565^\circ) \, \bar{a}_z$$

$$= 477.823 \, \bar{a}_y + 417.473 \, \bar{a}_z \text{ V/m}$$

(d) $\bar{E}$ at $(3, 4, 2, t = 20 \text{ ns})$

$$= 500 \cos(119.918 \times 10^6 \times 20 \times 10^{-9} - 0.4 \times 3 - 40^\circ) \, \bar{a}_y + 632.456 \cos(119.918 \times 10^6 \times 20 \times 10^{-9} - 0.4 \times 3 - 71.565^\circ) \, \bar{a}_z$$

$$= 438.736 \, \bar{a}_y + 631.644 \, \bar{a}_z \text{ V/m}$$

10.3 WAVE MOTION IN PERFECT DIELECTRICS

A perfect dielectric is characterized by

$$\sigma = 0, \quad \mu = \mu_o \mu_r, \quad \varepsilon = \varepsilon_o \varepsilon_r$$

Substituting these values in Eq.(10.5), (10.6) and (10.11), we get

$$\alpha = 0 \qquad \text{...... (10.13a)}$$

$$\beta = \omega \sqrt{\mu_o \mu_r \varepsilon_o \varepsilon_r} \qquad \text{...... (10.13b)}$$

$$\eta = \sqrt{\frac{\mu_o \mu_r}{\varepsilon_o \varepsilon_r}} \qquad \text{..... (10.13c)}$$

Substituting the value of β in Eq.(10.9), we get

$$v = \frac{1}{\sqrt{\mu_o \mu_r \varepsilon_o \varepsilon_r}} \qquad \text{...... (10.13d)}$$

Problem 10.3 : A 9375 – MHz uniform plane wave is propagating in polystyrene . If the amplitude of the electric field intensity is 20 V / m and the material is assumed to be lossless, find : (a) the phase constant ; (b) the wavelength (in the polystyrene) ; (c) the velocity of propagation ; (d) the intrinsic impedance ; (e) the propagation constant ; (f) the amplitude of the magnetic field intensity.

Solution :

(a) For polystyrene we have $\varepsilon_r = 2.56$, $\mu_r = 1$, $\sigma = 0$ (loss less)

$$\beta = \omega\sqrt{\mu\varepsilon} = \omega\sqrt{\mu_o\,\mu_r\,\varepsilon_o\,\varepsilon_r}$$

$$= 2\pi\times 9375\times 10^{6}\sqrt{4\pi\times 10^{-7}\,1\times 8.854\times 10^{-12}\times 2.56}$$

$$= 314.373 \text{ rad/m}$$

(b) $\lambda = \dfrac{2\pi}{\beta} = 0.02$ m

(c) $v = \dfrac{1}{\sqrt{\mu\varepsilon}} = \dfrac{1}{\sqrt{4\pi\times 10^{-7}\times 1\times 8.854\times 10^{-12}\times 2.56}}$

$$= 1.874\times 10^{8} \text{ m/sec}$$

(d) $\eta = \sqrt{\dfrac{\mu}{\varepsilon}} = \sqrt{\dfrac{4\pi\times 10^{-7}\times 1}{8.854\times 10^{-12}\times 2.56}} = 235.459\ \Omega$

(e) $\lambda = \alpha + j\beta = 0 + j\,314.373 = j\,314.373$ per m

($\because$ $\alpha = 0$) as the medium is loss less.

(f) $\eta = \dfrac{E}{H} = 235.459\ \Omega$

Hence $H = \dfrac{20}{235.459} = 0.0849$ A/m

10.4 WAVE MOTION IN LOSSY DIELECTRICS

A lossy dielectric is characterized by

$$\sigma(\text{ very small }),\ \mu = \mu_o\,\mu_r\ ,\ \varepsilon = \varepsilon_o\,\varepsilon_r$$

The Eq.(10.5), (10.6), (10.8), (10.9), and (10.11) gives the values of α , β , λ , v , and η respectively. From Eq.(10.1b), we have

$$\overline{J}_s = \overline{\nabla}\times\overline{H}_s = \overline{J}_{cs} + \overline{J}_{ds} = \sigma\,\overline{E}_s + j\,\omega\,\varepsilon\,\overline{E}_s$$

The time phase relationship between $\overline{J}_{ds}$, $\overline{J}_{\sigma s}$, $\overline{J}_s$ and $\overline{E}_s$ is shown in Fig. 10.1. The angle θ may be identified as the angle by which the displacement current density leads the total current density. From Fig. 10.1, we have

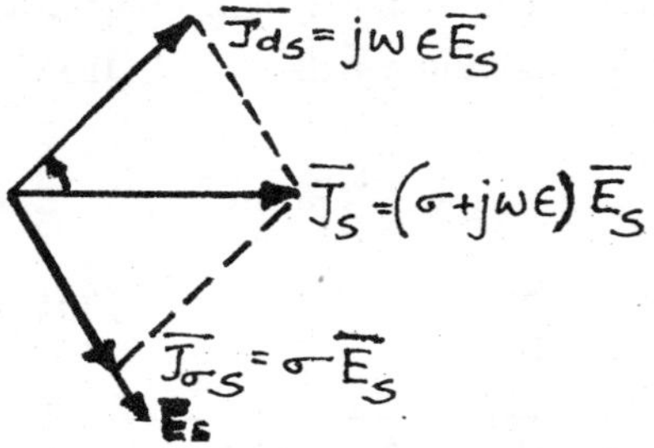

$$\frac{\overline{J}_{\sigma s}}{\overline{J}_{ds}} = \frac{\sigma}{j\omega\varepsilon} \quad \text{.. (10.14)}$$

and

$$\tan\theta = \frac{\sigma}{\omega\varepsilon} \quad \text{.. (10.15)}$$

where tan θ is known as loss tangent. The common power factor angle, or the angle by which $\overline{J}_s$ leads $\overline{E}_s$ is equal to $90^\circ - \theta$. Using the value of tan θ, Eq. (10.5) and (10.6) may be written in a compact form as

$$\alpha = \omega\sqrt{0.5\,\mu\,\varepsilon\,(\sec\theta - 1)} \quad \text{...... (10.16)}$$

and

$$\beta = \omega\sqrt{0.5\,\mu\,\varepsilon\,(\sec\theta + 1)} \quad \text{...... (10.17)}$$

Problem 10.4 : Given a nonmagnetic material having $\varepsilon_r = 2.25$ and $\sigma = 10^{-4}$ mhos/m , find numerical values at 2.5 MHz for : (a) the loss tangent ; (b) the attenuation constant ; (c) the phase constant ; (d) the intrinsic impedance.

Solution :

(a) $\mu_r = 1$ (non magnetic material)

$\varepsilon_r = 2.25$, $\sigma = 10^{-4}$ mhos/m , $f = 2.5$ MHz

$$\tan\theta = \frac{\sigma}{\omega\varepsilon} = \frac{10^{-4}}{2\pi \times 2.5 \times 10^6 \times 2.25 \times 8.854 \times 10^{-12}} = 0.32$$

(b)Now $\tan\theta = 0.32$ .Hence $\theta = \tan^{-1}(0.32) = 17.745^\circ$

$\therefore \sec\theta - 1 = \sec 17.745^\circ - 1 = 0.04995$

Using Eq.(10.16), we have

$$\alpha = 2\pi \times 2.5 \times 10^6\sqrt{0.5 \times 4\pi \times 10^{-7} \times 8.854 \times 10^{-12} \times 2.25 \times 0.04995}$$

$= 0.01242$ Np/m

(c) $\sec\theta + 1 = \sec 17.745^\circ + 1 = 2.04995$

Using Eq.(10.17), we have

$$\beta = 2\pi \times 2.5 \times 10^6 \sqrt{0.5 \times 4\pi \times 10^{-7} \times 8.854 \times 10^{-12} \times 2.25 \times 2.04995}$$
$$= 0.0796 \text{ rad/m}$$

(d) Using Eq.(10.11), we have,

$$\eta = \sqrt{\frac{j2\pi \times 2.5 \times 10^6 \times 4\pi \times 10^{-7}}{10^{-4} + j2\pi \times 2.5 \times 10^6 \times 2.25 \times 8.854 \times 10^{-12}}}$$

$$= \sqrt{\frac{19.7392 \angle 90^\circ}{3.285 \times 10^{-4} \angle 72.278^\circ}} = 245.130 \angle 8.861^\circ \ \Omega$$

Problem 10.5 : Consider a material for which $\mu_r = 1, \varepsilon_r = 4$, and the loss tangent is 0.1. If these three values are constant with frequency for 0.5 MHz $\le f \le$ 80 MHz, calculate : (a) σ at 1 and 50 MHz ; (b) λ at 1 and 50 MHz ; (c) v at 1 and 50 MHz.

Solution :

(a) $\mu_r = 1$, $\varepsilon_r = 4$, $\tan\theta = 0.1$ or $\theta = \tan^{-1}(0.1) = 5.711^\circ$

From Eq(10.15), we have $\sigma = \omega\varepsilon\tan\theta = 2\pi f \varepsilon \tan\theta$ and at 1 MHz,

$$\sigma = 2\pi \times 10^6 \times 4 \times 8.845 \times 10^{-12} \times 0.1 = 22.253 \ \mu\text{ mho/m}$$

Similarly at 50 MHz, we have ,

$$\sigma = 2\pi \times 50 \times 10^6 \times 4 \times 8.854 \times 10^{-12} \times 0.1 = 1112.62 \ \mu\text{ mho/m}$$

(b) Using Eq.(10.17), at 1 MHz, we have,

$$\beta = 2\pi \times 10^6 \sqrt{4\pi \times 10^{-7} \times 4 \times 8.854 \times 10^{-12} (\sec 5.711^\circ + 1)}$$

$$= 41.968696 \times 10^{-3} \text{ rad/m}$$

and at 50 MHz ,

$$\beta = 50 \times 41.968696 \times 10^{-3} = 2.0984348 \text{ rad/m}$$

$$\therefore \lambda \text{ at } 1\text{ MHz} = \frac{2\pi}{41.968696 \times 10^{-3}} = 149.711 \text{ m}$$

$$\text{and } \lambda \text{ at } 50\text{ MHz} = \frac{2\pi}{2.0984348} = 2.994 \text{ m}$$

$$(c)\ v = \frac{\omega}{\beta} = \frac{2\pi \times 10^6}{41.968696 \times 10^{-3}} = 1.497 \times 10^8 \text{ m/sec at } 1\text{ MHz}$$

$$v = \frac{2\pi \times 50 \times 10^6}{2.0984348} = 1.497 \times 10^8 \text{ m/sec at } 50\text{ MHz}$$

10.5 WAVE MOTION IN GOOD CONDUCTORS :

10.5.1

A good conductor is characterized by σ(high) , $\mu = \mu_o$ (nonmagnetic conductor), $\varepsilon = \varepsilon_o$. The Eq.(10.5), (10.6), (10.8), (10.9) and (10.11) gives the values of α , β , λ , v and η respectively.

10.5.2

It can be proved that for a good conductor

$$\alpha = \beta = \sqrt{\pi f \mu \sigma} \qquad \ldots\ldots(10.18)$$

The proof is given below. The propogation constant γ is given by Eq. (10.3) as

$$\gamma = \sqrt{j\omega\mu(\sigma + j\omega\varepsilon)} = j\omega\sqrt{\mu\varepsilon}\sqrt{1 - j\frac{\sigma}{\omega\varepsilon}}$$

As $\dfrac{\sigma}{\omega\varepsilon} >>> 1$, we can write

$$\gamma = j\omega\sqrt{\omega\varepsilon}\sqrt{-j\frac{\sigma}{\omega\varepsilon}} = j\sqrt{-j\omega\mu\sigma}$$

Now $\sqrt{-j} = \sqrt{1\angle -90^\circ} = 1\angle -45^\circ = \dfrac{1}{\sqrt{2}} - j\dfrac{1}{\sqrt{2}}$

Hence we have,

$$\gamma = \left(\frac{1}{\sqrt{2}} - j\frac{1}{\sqrt{2}}\right)\sqrt{\omega\mu\sigma} = (1 + j)\sqrt{\pi f \mu\sigma} = \alpha + j\beta$$

Hence

$$\alpha = \beta = \sqrt{\pi f \mu \sigma}$$

10.5.3 SKIN EFFECT :

In conductor, the primary transmission of energy must take place in the region outside the conductor, because all time varying fields attenuate very quickly within a good conductor. The good conductor has a high conductivity and large conduction currents. The energy represented by the wave travelling through the material threrfore decreases as the wave propagates because ohmic losses are continuously present. As $\alpha = \beta = \sqrt{\pi f \mu \sigma}$, the expression for electric field intensity may be written as

$$E_x = E_{xo}\, e^{-z\sqrt{\pi f \mu \sigma}} \cos(\omega t - z\sqrt{\pi f \mu \sigma})$$

Let us consider the region $z > 0$ be the good conductor and the region $z < 0$ be a perfect dielectric. As displacement current is negligible, conduction current density is given by

$$J_x = \sigma E_x = \sigma E_{xo}\, e^{-z\sqrt{\pi f \mu \sigma}} \cos(\omega t - z\sqrt{\pi f \mu \sigma})$$

The amplitudes of E_x & J_x are $E_{xo}\, e^{-z\sqrt{\pi f \mu \sigma}}$ and $\sigma E_{xo}\, e^{-z\sqrt{\pi f \mu \sigma}}$ respectively. As the wave travels inside the conductor i.e. as z increases, the amplitude of E_x and J_x, both decreases exponentially. At $z = 0$, the exponential factor is unity ($e^0 = 1$) and decreases to $e^{-1} = 0.368$ when $z = \frac{1}{\sqrt{\pi f \mu \sigma}}$. This distance is denoted by δ and is termed as depth of penetration, or the skin depth. Fig. 10.2 shows the graphical illustration of skin depth.

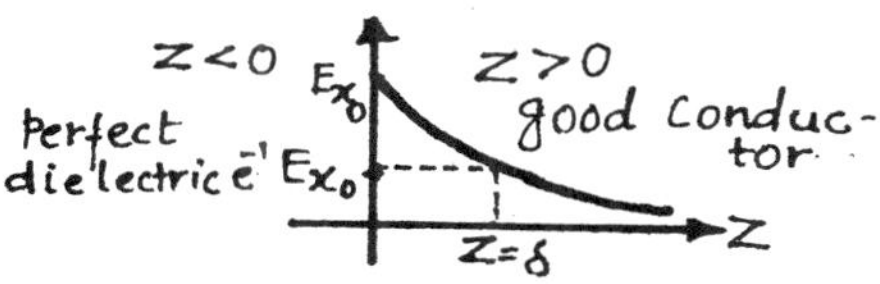

$$\frac{1}{\sqrt{\pi f \mu \sigma}} = \frac{1}{\alpha} = \frac{1}{\beta} \qquad (10.19)$$

Thus, all fields in a good conductor are essentially zero at distances greater than a few skin depths from the surface. Electromagnetic energy is not transmitted in the interior of a conductor; it travels in the region surrounding the conductor, while the conductor merely guides the waves. The price of guiding the waves is paid in terms of ohmic losses. This is known as skin effect. With skin depth δ, the resistance of a conductor is given by

$$R_{ac} = \frac{L}{2\pi a \sigma \delta} \qquad \ldots\ldots (10.20)$$

where L is the length, a is the outer radius of the conductor, σ is the conductivity. Also

$$R_{dc} = \frac{L}{\sigma S} \qquad \text{.....(10.21)}$$

10.6 THE POINTING VECTOR :

The Maxwell's equation derived from Ampere's circuital law is

$$\bar{\nabla} \times \bar{H} = \bar{J} + \frac{\partial \bar{D}}{\partial t}$$

Taking the dot product of each side with $\bar{E}$,

$$\bar{E} \cdot \bar{\nabla} \times \bar{H} = \bar{J} \cdot \bar{E} + \bar{E} \cdot \frac{\partial \bar{D}}{\partial t}$$

using the vector identity,

$$\bar{\nabla} \cdot (\bar{E} \times \bar{H}) = -\bar{E} \cdot (\bar{\nabla} \times \bar{H}) + \bar{H} \quad \bar{\nabla} \times \bar{E}$$

which may be proved by expansion in cartesian coordinates. Thus

$$\bar{H} \cdot \bar{\nabla} \times \bar{E} - \bar{\nabla} \cdot (\bar{E} \times \bar{H}) = \bar{J} \cdot \bar{E} + \bar{E} \cdot \frac{\partial \bar{D}}{\partial t}$$

But

$$\bar{\nabla} \times \bar{E} = -\frac{\partial \bar{B}}{\partial t}$$

and therefore

$$-\bar{H} \cdot \frac{\partial \bar{B}}{\partial t} - \bar{\nabla} \cdot (\bar{E} \times \bar{H}) = \bar{J} \cdot \bar{E} + \bar{E} \cdot \frac{\partial \bar{D}}{\partial t}$$

or

$$-\bar{\nabla} \cdot (\bar{E} \times \bar{H}) = \bar{J} \cdot \bar{E} + \varepsilon \bar{E} \cdot \frac{\partial \bar{E}}{\partial t} + \mu \bar{H} \cdot \frac{\partial \bar{H}}{\partial t}$$

However,

$$\varepsilon \bar{E} \cdot \frac{\partial \bar{E}}{\partial t} = \frac{\varepsilon}{2} \frac{\partial E^2}{\partial t} = \frac{\partial}{\partial t}\left(\frac{\varepsilon E^2}{2}\right)$$

and

$$\mu \bar{H} \cdot \frac{\partial \bar{H}}{\partial t} = \frac{\partial}{\partial t}\left(\frac{\mu H^2}{2}\right)$$

Thus

$$-\bar{\nabla} \cdot (\bar{E} \times \bar{H}) = \bar{J} \cdot \bar{E} + \frac{\partial}{\partial t}\left(\frac{\varepsilon E^2}{2} + \frac{\mu H^2}{2}\right)$$

Finally, integrating throughout a volume,

$$-\int_{vol} \overline{\nabla} \cdot (\overline{E} \times \overline{H})\, dv = \int_{vol} \overline{J} \cdot \overline{E}\, dv + \int_{vol} \left(\frac{\varepsilon E^2}{2} + \frac{\mu H^2}{2} \right) dv$$

and applying the divergence theorem, we obtain

$$-\oint_S (\overline{E} \times \overline{H}) \cdot \overline{dS} = \int_{vol} \overline{J} \cdot \overline{E}\, dv + \frac{\partial}{\partial t} \int_{vol} \left(\frac{\varepsilon E^2}{2} + \frac{\mu H^2}{2} \right) dv$$

Let us assume that there are no sources within the volume, then the first integral on the right is the total (but instantaneous) ohmic power dissipated within the volume. If sources are present within the volume, then the result of integrating over the volume of the source will be positive if power is being delivered to the source, but it will be negative if power is being delivered by the source.

The integral in the second term on the right is total energy stored in the electric and magnetic fields, and the partial derivatives with respect to time cause this term to be the time rate of increase of energy stored within this volume, or the instantaneous power going to increase the stored energy within this volume. The sum of the expressions on the right must therefore be the total power flowing into this volume, and thus the total power flowing out of the volume is

$$\oint_S (\overline{E} \times \overline{H}) \cdot \overline{dS}$$

where the integral is over the closed surface surrounding the volume. The cross product $\overline{E} \times \overline{H}$ is known as the Pointing vector ($\overline{P}$).

$$\overline{\mathbf{P}} = \overline{\mathbf{E}} \times \overline{\mathbf{H}} \qquad \text{......(10.22)}$$

In the case of perfect dielectric, $\alpha = 0$ so that from Eq. (10.7) ,

$$E_x = E_{xo} \cos(\omega t - \beta z)$$

$$H_y = \frac{E_{xo}}{\eta} \cos(\omega t - \beta z)$$

and, thus

$$P_z = \frac{E_{xo}^2}{\eta} \cos^2(\omega t - \beta z)$$

$$P_{z,av} = \frac{1}{T}\int_0^T \frac{E_{xo}^2}{\eta}\cos^2(\omega t - \beta z)\,dt$$

$$= \frac{1}{2T}\frac{E_{xo}^2}{\eta}\left[t + \frac{1}{2\omega}\sin(2\omega t - 2\beta z)\right]_o^T$$

and

$$\boldsymbol{P_{z,av} = \frac{1}{2}\frac{E_{xo}^2}{\eta}} \qquad \ldots..(10.23)$$

For a lossy dielectric,

$$E_x = E_{xo}e^{-\alpha z}\cos(\omega t - \beta z)$$

and $\eta = \eta_m \angle\theta_\eta$ as η is a complex number. Thus,

$$H_y = \frac{E_{xo}}{\eta_m}e^{-\alpha z}\cos(\omega t - \beta z - \theta_\eta)$$

$$\therefore\ P_z = E_x H_y = \frac{E_{xo}^2}{\eta_m}e^{-2\alpha z}\cos(\omega t - \beta z)\cos(\omega t - \beta z - \theta_\eta)$$

$$= \frac{1}{2}\frac{E_{xo}^2}{\eta_m}e^{-2\alpha z}\left[\cos(2\omega t - 2\beta z - \theta_\eta) + \cos\theta_\eta\right]$$

and the time average value of the Poynting vector is

$$\boldsymbol{P_{z,av} = \frac{1}{2}\frac{E_{xo}^2}{\eta_m}e^{-2\alpha z}\cos\theta_\eta} \qquad \ldots.(10.24)$$

Problem 10.6 : At frequencies of 1, 100, and 3000 MHz, the dielectric constant of ice made from pure water has values of 4.15, 3.45, and 3.20, respectively, while the loss tangent is 0.12, 0.035, 0.0009, also respectively. If a uniform plane wave with an amplitude of 100 V / m at $z = 0$ is propagating through such ice, find the time average power density at $z = 0$ and $z = 10$ m for each frequency.

Solution : (a)

Frequency	ε_r	$\tan\theta$

1 MHz	4.15	0.12
100 MHz	3.45	0.035
3000 MHz	3.20	0.0009

At 1 MHz, $\varepsilon_r = 4.15$, $\tan\theta = 0.12$ or $\sec\theta = 7.1743\times 10^{-3}$

Using Eq. (10.6), we have,

$$\alpha = 2\pi\times 10^{6}\sqrt{0.5\times 4\pi\times 10^{-7}\times 4.15\times 8.854\times 10^{-12}\times 7.1743\times 10^{-3}}$$

$$= 2.557\times 10^{-3}\ \text{Np/m}$$

Similarly, α at 100 MHz and 3000 MHz is calculated which is equal to 68.114×10^{-3} and 50.613×10^{-3} respectively. Now,

$$\eta = \sqrt{\frac{j\omega\mu}{\sigma + j\omega\varepsilon}} = \sqrt{\frac{\mu/\varepsilon}{1 - j\tan\theta}}$$

$$\eta \text{ at } 1\text{ MHz} = \sqrt{\frac{4\pi\times 10^{-7}}{4.15\times 8.854\times 10^{-12}(1 - j0.12)}} = 184.272\ \angle 3.421^\circ$$

Similarly at 100 MHz and 3000 MHz, the value of η is $202.765\ \angle 1.002^\circ$ and $210.601\ \angle 0.026^\circ$

Using Eq.(10. 24), we have , at 1 MHz , $z = 0$,

$$P_{z,av} = \frac{1}{2}\times\frac{(100)^2}{184.272}\,e^{-2\times 2.557\times 10^{-3}(0)}\cos(3.421^\circ) = 27.085\ \text{W/m}^2$$

At 1 MHz , $z = 10$,

$$P_{z,av} = \frac{1}{2}\times\frac{(100)^2}{184.272}\,e^{-2\times 2.557\times 10^{-3}(10)}\cos(3.421^\circ) = 24.735\ \text{W/m}^2$$

At 100 MHz , $z = 0$,

$$P_{z,av} = \frac{1}{2}\times\frac{(100)^2}{202.765}\,e^{-2\times 68.114\times 10^{-3}(0)}\cos(1.002^\circ) = 24.655\ \text{W/m}^2$$

At 100 MHz , $z = 10$,

$$P_{z,av} = \frac{1}{2}\times\frac{(100)^2}{202.765}\,e^{-2\times 68.114\times 10^{-3}(10)}\cos(1.002^\circ) = 6.314\ \text{W/m}^2$$

At 3000 MHz , $z = 0$,

$$P_{z,\,av} = \frac{1}{2} \times \frac{(100)^2}{210.601} e^{-2 \times 50.613 \times 10^{-3} \times (0)} \cos(0.026^\circ) = 23.742 \text{ W/m}^2$$

At 3000 MHz , $z = 10$,

$$P_{z,\,av} = \frac{1}{2} \times \frac{(100)^2}{210.601} e^{-2 \times 50.613 \times 10^{-3} \times (10)} \cos(0.026^\circ) = 8.628 \text{ W/m}^2$$

Problem 10.7 : A steel pipe is constructed of a material for which we may assume $\mu_r = 200$ and $\sigma = 5 \times 10^6$ mho/m . The outer and inner radii are 8 and 6 mm, respectively, and the length is 80 m. If the total current carried by the pipe is $2 \cos 10^4 \pi t$ A , find : (a) the skin depth ; (b) the effective resistance ; (c) the dc resistance ; (d) the time average power loss.

Solution :

(a) $\omega = 2\pi f = 10^4 \pi$. Hence $f = 0.5 \times 10^4$ Hz

Usig Eq.(10.19), we have

$$\delta = \frac{1}{\sqrt{\pi \times 0.5 \times 10^4 \times 200 \times 4\pi \times 10^{-7} \times 5 \times 10^6}} = 0.225 \times 10^{-3}$$

(b) a = radius of conductor = 0.008 m.

$\delta = 0.225 \times 10^{-3}$ m, $\sigma = 5 \times 10^6$ mho/m, $L = 80$ m

Using Eq. (10.20), we have

$$R_{eff} = \frac{80}{2\pi \times 0.008 \times 0.225 \times 10^{-3} \times 5 \times 10^6} = 1.4147 \ \Omega$$

c) L = length = 80 m

S = Cross sectional area $= \pi[(0.008)^2 - (0.006)^2] = 2.8\pi \times 10^{-5}$

Usin Eq. (10 . 21), we have

$$R_{dc} = \frac{80}{5 \times 10^6 \times 2.8\pi \times 10^{-5}} = 0.1819 \ \Omega$$

Problem 10.8 : Show that $E_{xs} = A\, e^{\pm j(kz + \alpha)}$ is a solution of the vector Helmholtz equation, for $k = \omega\sqrt{\mu_o \varepsilon_o}$ and α and A.

Solution :

The vector Helmholtz equation in terms of x component of electric field intensity is given by

$$\frac{d^2 E_{xs}}{dz^2} = -\omega^2 \mu_o \varepsilon_o E_{xs}$$

$$\text{L.H.S.} = \frac{d^2 E_{xs}}{dz^2}$$

Now $E_{xs} = A\, e^{\pm j(kz+\alpha)}$

$$\therefore \frac{d E_{xs}}{dz} = \pm jkA\, e^{\pm j(kz+\alpha)}$$

$$\frac{d^2 E_{xs}}{dz^2} = (\pm jk)(\pm jk) A\, e^{\pm j(kz+\alpha)} = j^2 k^2 A\, e^{\pm j(kz+\alpha)}$$

$$= -(\omega\sqrt{\mu_o \varepsilon_o})^2 A\, e^{\pm j(kz+\alpha)} = -\omega^2 \mu_o \varepsilon_o A\, e^{\pm j(kz+\alpha)}$$

$$= -\omega^2 \mu_o \varepsilon_o E_{xs} = \text{RHS}$$

Hence $E_{xs} = A\, e^{\pm j(kz+\alpha)}$ is a solution of vector Helmholtz equation

Problem 10.9 : An $\overline{\mathrm{E}}$ field in free space is given as $\overline{\mathrm{E}} = 800 \cos(10^8 t - \beta y)\, \bar{a}_z$ V/m . Find : (a) β ; (b) λ ; (c) $\overline{\mathrm{H}}$ at P (0.1, 1.5, 0.4) at $t =$ 8 ns.

Solution :

(a) $\omega = 10^8$; $\beta = \omega\sqrt{\mu_o \varepsilon_o} = 10^8 \sqrt{4\pi \times 10^{-7} \times 8.854 \times 10^{-12}}$

$$= 0.33356 \text{ rad/m}$$

(b) $\lambda = 2\pi/\beta = 18.8367$ m

(c) For free space, $\eta = \dfrac{E}{H} = 376.7343\ \Omega$ and $E = 800$ V/m (given)

$$\therefore H = \frac{E}{376.7343} = \frac{800}{376.7343} = 2.1235 \text{ A/m}$$

Since $-y$ is involved in the equation of $\overline{\mathrm{E}}$, the direction of $\overline{\mathrm{P}}$ is $\bar{a}_y$.

Now $\overline{\mathrm{P}} = \overline{\mathrm{E}} \times \overline{\mathrm{H}}$

$\bar{a}_y = \bar{a}_z \times ?$

Hence the direction of $\overline{\mathrm{H}}$ is given by $\bar{a}_x$

$\therefore \; \overline{H} = 2.1235 \; (\cos \; (10^8 \, t - \beta y) \; \bar{a}_x$ A/m

$\overline{H}$ at $P(0.1, 1.5, 0.4)$ at $t = 8$ ns

$$= 2.1235 \; \cos (\; 10^8 \times 8 \times 10^{-9} - 0.33356 \times 1.5 \;) \; \bar{a}_x$$

$$= 2.1235 \; \cos (\; 0.29966 \;) \; \bar{a}_x = 2.1235 \; \cos (\; 17.169^\circ \;) \; \bar{a}_x$$

$$= 2.0288 \; \bar{a}_x \text{ A/m}$$

Problem 10.10 : A uniform plane wave in free space is given by $\overline{E}_s = (\; 200 \angle \; 20^\circ \;) \; e^{-j250z} \; \bar{a}_x$ V/m . Find : (a) β ; (b) ω ; (c) f ; (d) λ ; (e) η ; (f) $\overline{H}_s$; (g) $| \; \overline{E} \; |$ at $z =$ 8 mm, $t =$ 6 ps.

Solution :

(a) The coefficient of z represents the value of β which is equal to 250 rad / m.

(b) $\beta = \omega \sqrt{\mu_o \varepsilon_o} = 250$ rad / m . Hence

$$\omega = \frac{250}{\sqrt{\mu_o \varepsilon_o}} = 74.9489 \text{ G rad/s}$$

(c) $\omega = 2 \pi f$. Hence $f = \dfrac{\omega}{2\pi} = 11.9285$ GHz

(d) $\lambda = \dfrac{2\pi}{\beta} = \dfrac{2\pi}{250} = 25.133$ mm

(e) $\eta = 376.7343$ (a ccnstant for free space)

(f) Since $-z$ is involved in the equation for $\overline{E}_s$, the direction of $\overline{P}$ is $\bar{a}_z$. Now

$$\overline{P} = \overline{E} \times \overline{H}$$

$$\bar{a}_z = \bar{a}_x \times ?$$

As $\bar{a}_x \times \bar{a}_y = \bar{a}_z$, the direction of $\overline{H}$ is given by $\bar{a}_y$

Also $E/H = \eta = 376.7343 \; \Omega$. Hence

$\therefore \; H = E/376.7343 = 200/376.7343 = 0.531$ A/m

$$\therefore \; \overline{H}_s = 0.531 \; \angle 30^\circ \; e^{-j250z} \; \bar{a}_y = 0.531 \; e^{j30^\circ} \; e^{-j250z} \; \bar{a}_y$$

$$= 0.531 \; e^{-j250z + j30^\circ} \; \bar{a}_y \text{ A/m}$$

(g) $\overline{E}_s = 200 \; \angle 30^\circ \; e^{-j250z} \; \bar{a}_x = 200 \, e^{j30^\circ} \; e^{-j250z} \; \bar{a}_x$

$$= 200 \, e^{j(-250z + 30^\circ)} \; \bar{a}_x$$

$\bar{E} = 200\,\text{Re}\left[\, e^{j\omega t} e^{j(-250z+30^\circ)} \right] \bar{a}_x = 200\,\text{Re}\left[\, e^{j(\omega t - 250z + 30^\circ)} \right] \bar{a}_x$

$= 200 \cos(\omega t - 250z + 30^\circ)\, \bar{a}_x$ V/m

$\bar{E}$ at $z = 8$ mm $= 0.008$, $t = 6$ ps $= 6 \times 10^{-12}$ s

$= 200 \cos[74.9489 \times 10^9 \times 6 \times 10^{-12} - 250 \times 0.008 + 30^\circ]\, \bar{a}_x$

$= 200 \cos(-1.550 + 30^\circ)\, \bar{a}_x = 200 \cos(-88.826^\circ + 30^\circ)\, \bar{a}_x$

$= 200 \cos(-58.826^\circ)\, \bar{a}_x = 103.52768\, \bar{a}_x$ V/m

$|\bar{E}| = 103.52768$ V/m

Problem 10.11 : A uniform plane wave in free space is propagating in the $-\bar{a}_y$ direction at a frequency of 10 MHz. If $\bar{E} = 400 \cos \omega t\, \bar{a}_z$ V/m at $y = 0$, write expressions for : (a) $\bar{E}(x, y, z, t)$; (b) $\bar{E}_s(x, y, z)$; (c) $\bar{H}_s(x, y, z)$; (d) $\bar{H}(x, y, z, t)$.

Solution :

(a) Since the wave is propagating in the $-\bar{a}_y$ direction, $+y$ must be involved in the equation for $\bar{E}$.

$\therefore\ \bar{E} = 400 \cos(\omega t + \beta y)\, \bar{a}_z$

$\omega = 2\pi f = 2\pi \times 10 \times 10^6 = 62.832 \times 10^6$ rad/sec.

$\beta = \omega \sqrt{\mu_o \varepsilon_o} = 62.832 \times 10^6 \sqrt{4\pi \times 10^{-6}\, 8.854 \times ^{-12}} = 0.210$ rad/m

$\therefore\ \bar{E} = 400 \cos(62.832 \times 10^6 t + 0.21 y)\, \bar{a}_z$ V/m

(b) $\bar{E} = 400\,\text{Re}\,[\cos(\omega t + \beta y) + j \sin(\omega t + \beta y)]\, \bar{a}_z$

$= 400\,\text{Re}\left[\, e^{j(\omega t + \beta y)} \right] \bar{a}_z = 400\,\text{Re}\left[\, e^{j\omega t} e^{j\beta y} \right] \bar{a}_z$

$\therefore\ \bar{E}_s = 400\, e^{j\beta y}\, \bar{a}_z = 400\, e^{j\,0.21 y}\, \bar{a}_z$ V/m

(c) $\eta = E/H = 376.7343\ \Omega$ (a constant for free space)

$\therefore\ H = 400/376.7343 = 1.06176$ A/m

The direction of $\bar{P}$ is $-\bar{a}_y$ and that of $\bar{E}$ is $\bar{a}_z$

Now $\bar{P} = \bar{E} \times \bar{H}$

$-\bar{a}_y = \bar{a}_z \times\ ?$

Since $\bar{a}_z \times (-\bar{a}_x) = -\bar{a}_y$, the direction of $\bar{H}$ must be $-\bar{a}_x$

$\therefore\ \overline{H}_s = 1.06176\ e^{j\,0.21\,y}\ (-\ \overline{a}_x) = -\ 1.06176\ e^{j\,0.21\,y}\ \overline{a}_x$ A/m

$$(d)\ \overline{H} = -\ 1.06176\ \text{Re}\left[\ e^{j\omega t}\ e^{j\,0.21\,y}\right]\overline{a}_x$$
$$= -\ 1.06176\ \text{Re}\left[\ e^{j(\omega t + 0.21\,y)}\right]\overline{a}_x$$
$$= -\ 1.06176\ \cos(\omega t + 0.21\,y)\ \overline{a}_x\ \ \text{A/m}$$

Problem 10.12 : A 150 – MHz uniform plane wave in free space in is travelling in the $\overline{a}_x$ direction. The electric field intensity has a maximum amplitude of 200 $\overline{a}_y$ + 400 $\overline{a}_z$ V/m at P (10, 30, – 40) at t = 0. Find : (a) ω ; (b) β ; (c) λ ; (d) v ; (e) η ; (f) $\overline{E}(x, y, z)$. (g) Use one or more of Maxwell's equation to find $\overline{H}(x, y, z, t)$.

Solution :

(a) $\omega = 2\pi f = 2\pi \times 150 \times 10^6 = 942.478$ M rad/sec.

(b) $\beta = \omega\sqrt{\mu_o\,\varepsilon_o} = 942.478 \times 10^6\sqrt{4\pi \times 10^{-7}\ 8.854 \times 10^{-12}} = \pi$ rad/m

(c) $\lambda = \dfrac{2\pi}{\beta} = 2$ m

(d) $v = \dfrac{1}{\sqrt{\mu_o\,\varepsilon_o}} = 3 \times 10^8$ m/sec [a constant for free space]

(e) $\eta = \sqrt{\mu_o/\varepsilon_o} = 376.7343\ \Omega$

(f) Maximum amplitude = 200 $\overline{a}_y$ + 400 $\overline{a}_z$ V/m at P (10, 30,– 40) at t = 0

As the wave is travelling in the $\overline{a}_x$ direction, – x must be present in the equation of $\overline{E}$ as well as $\overline{H}$.

$$\therefore\ \overline{E}_s = (200\,\overline{a}_y + 400\,\overline{a}_z)\ e^{-j\beta x}$$

$$\therefore\ \overline{E} = (200\,\overline{a}_y + 400\,\overline{a}_z)\ \text{Re}\left[\ e^{j\omega t}\ e^{-j\beta x}\right]$$
$$= (200\,\overline{a}_y + 400\,\overline{a}_z)\ \text{Re}\left[\ e^{j(\omega t - \beta x)}\right]$$
$$= (200\,\overline{a}_y + 400\,\overline{a}_z)\ \cos(300\pi\,10^6\,t - \pi x)\ \text{V/m}$$

Maxwells's equation in phasor form is $\overline{\nabla} \times \overline{E}_s = -\ j\,\omega\,\mu_o\,\overline{H}_s$

$$\overline{\nabla} \times \overline{E}_s = \begin{vmatrix} \overline{a}_x & \overline{a}_y & \overline{a}_z \\ \partial/\partial x & \partial/\partial y & \partial/\partial z \\ 0 & 200\,e^{-j\beta x} & 400\,e^{-j\beta x} \end{vmatrix}$$

$$= -\bar{a}_y[-j\beta 400 e^{-j\beta x}] + \bar{a}_z[-j\beta 200 e^{-j\beta x}]$$

$$= j\beta e^{-j\beta x}(400\,\bar{a}_y - 200\,\bar{a}_z) = -j\,\omega\mu_o\,\overline{H}_s$$

$$\overline{H}_s = \frac{\beta}{\omega\mu_o} e^{-j\beta x}(-400\,\bar{a}_y + 200\,\bar{a}_z) = (-1.0617\,\bar{a}_y + 0.5309\,\bar{a}_z)\,e^{-j\beta x}$$

or $\overline{H} = (-1.0617\,\bar{a}_y + 0.5309\,\bar{a}_z)\cos(300\pi\,10^6 t - \pi x)$ A/m

Problem 10.13 : The electric field intensity of a 300 – MHz uniform plane wave in free space is given as $\overline{E}_s = (20 + j\,50)(\bar{a}_x + 2\,\bar{a}_y)\,e^{-j\beta z}$ V/m. (a) Find ω, λ, v, and β. (b) Find $\overline{E}$ at $t = 1$ ns, $z = 10$ cm. (c) What is $|\overline{H}|_{max}$?

Solution :

(a) $\omega = 2\pi f = 2\pi \times 300 \times 10^6 = 1.885$ Grad/s

$$\beta = \omega\sqrt{\mu_o\varepsilon_o} = 6.287468 \text{ rad/m}$$

$$\lambda = \frac{2\pi}{\beta} = 0.999318 \text{ m}$$

$$v = \frac{1}{\sqrt{\mu_o\varepsilon_o}} = 3 \times 10^8 \text{ m/sec}$$

(b) $\overline{E}_s = (20 + j\,50)(\bar{a}_x + 2\,\bar{a}_y)\,e^{-j\beta z}$ V/m

$$= 53.852\,e^{j\,68.2^0}(\bar{a}_x + 2\,\bar{a}_y)\,e^{-j\beta z}$$

$$= 53.852(\bar{a}_x + 2\,\bar{a}_y)\,e^{j(68.2^0 - \beta z)}$$

$$\overline{E} = 53.852(\bar{a}_x + 2\,\bar{a}_y)\,\text{Re}[e^{j(68.2^o - \beta z)}\,e^{j(\omega t)}]$$

$$= 53.852(\bar{a}_x + 2\,\bar{a}_y)\cos(\omega t - \beta z + 68.2^o)$$

$\overline{E}$ at $t = 1$ ns , $z = 10$ cm

$$= 53.852(\bar{a}_x + 2\,\bar{a}_y)\cos(1.885 \times 10^9 \times 10^{-9} - 6.287468 \times 0.1 + 68.2^o)$$

$$= 53.852(\bar{a}_x + 2\,\bar{a}_y)\cos(1.25625 + 68.2^o)$$

$$= 53.852(\bar{a}_x + 2\,\bar{a}_y)\cos(71.978^o + 68.2^o)$$

$$= -41.360\,\bar{a}_x - 82.721\,\bar{a}_y \text{ V/m}$$

(c) $|\overline{E}|_{max} = 53.852\sqrt{(1)^2 + (2)^2} = 120.4167$ V/m

$$\therefore |\overline{H}|_{max} = \frac{120.4167}{\eta} = \frac{120.4167}{376.7343} = 0.3196 \text{ A/m}$$

Problem 10.14 : A wave propagating in a lossless dielectric has the components,$\overline{E} = 500 \cos(10^7 t - \beta z)\,\overline{a}_x$ V/m and $\overline{H} = 1.1 \cos(10^7 t - \beta z)\,\overline{a}_y$. If the wave is travelling at $v = 0.5$ c , find : (a) μ_r ; (b) ε_r ; (c) β ; (d) λ ; (e) η .

Solution :

$\omega = 10^7$, amplitude of electric field = 500 V / m and amplitude of magnetic field = 1.1 A / m

$v = 0.5\,c = 0.5 \times 299.7956 \times 10^6 = 149.8978 \times 10^{-6}$ m/sec

(a) $\eta = \dfrac{E}{H} = \sqrt{\dfrac{\mu_o \mu_r}{\varepsilon_o \varepsilon_r}} = \dfrac{500}{1.1}$

$\therefore \dfrac{\mu_r}{\varepsilon_r} = 1.45574$ ....(1)

$v = \dfrac{1}{\sqrt{\mu_o \mu_r \varepsilon_o \varepsilon_r}} = 149.8978 \times 10^6$ m/sec

$\therefore \mu_r \varepsilon_r = 4$ ....(2)

Multiplying equation (1) by equation (2) we have

$\dfrac{\mu_r}{\varepsilon_r} \times \mu_r \varepsilon_r = 1.45574 \times 4 = 5.82296$ or $\mu_r = \sqrt{5.822} = 2.413$

(b) Using equation (2) , we have $\varepsilon_r = 4/2.413 = 1.6576$

(c) Using Eq.(10.13b),we have,

$\beta = 10^7 \sqrt{4\pi \times 10^{-7} \times 2.413 \times 8.854 \times 10^{-12} \times 1.6576} = 0.0667$ rad/m

(d) $\lambda = 2\pi/\beta = 94.201$ m

(e) $\eta = 500/1.1 = 454.545\ \Omega$

Problem 10.15 : Let $\overline{E}_s = (1000\,\overline{a}_x + 400\,\overline{a}_z)\, e^{-j10y}$ V/m for a 250 – MHz uniform plane wave propagating in a perfect dielectric. If the maximum amplitude of $\overline{H}$ is 3 A / m, find $\beta, \eta, \lambda, v, \varepsilon_r, \mu_r$, and $\overline{E}(x, y, z, t)$).

Solution :

Coefficient of y gives the value of $\beta = 10$ rad/m

Amplitude of $\overline{E}_s = \sqrt{(1000)^2 + (400)^2} = 1077.03296$ V/m

$\therefore \eta = E/H = 1077.03296/3 = 359.011\ \Omega$

$\lambda = 2\pi/\beta = 2\pi/10 = 0.6283$ m

$$v = \frac{\omega}{\beta} = 2\pi \times 250 \times 10^6/10 = 1.571 \times 10^8 \text{ m/sec}$$

$$\eta = \sqrt{\frac{\mu_o \mu_r}{\varepsilon_o \varepsilon_r}} = 359.011 \quad \therefore \quad \frac{\mu_r}{\varepsilon_r} = 0.9081 \qquad \ldots\ldots(1)$$

$$v = \frac{1}{\sqrt{\mu_o \mu_r \varepsilon_o \varepsilon_r}} = 1.571 \times 10^8 \quad \therefore \quad \mu_r \varepsilon_r = 3.64165 \qquad \ldots.(2)$$

From equation (1) and (2)

$$\frac{\mu_r}{\varepsilon_r} \times \mu_r \varepsilon_r = 0.9081 \times 3.64165 = 3.30698 \text{ or } \mu_r = \sqrt{3.30698} = 1.8185$$

From equation (2), we have $\varepsilon_r = 3.30698/1.8185 = 1.8185$

$$\overline{E}_s = (1000\,\overline{a}_x + 400\,\overline{a}_z)\, e^{-j10y}$$

$$\overline{E} = (1000\,\overline{a}_x + 400\,\overline{a}_z) \text{ Re } [e^{j\omega t} e^{-j10y}]$$

$$= (1000\,\overline{a}_x + 400\,\overline{a}_z) \text{ Re } [e^{j(\omega t - 10y)}]$$

$$= (1000\,\overline{a}_x + 400\,\overline{a}_z) \cos(\omega t - 10y) \text{ V/m}$$

Problem 10.16 : A uniform plane wave, $\overline{E}_s = (5\,\overline{a}_x + j\,10\,\overline{a}_y)\, e^{-j2z}$ V/m, has a frequency of 50 MHz in a lossless dielectric material for which $\mu_r = 1$. (a) Find $\beta, \omega, v, \lambda, \varepsilon_r$ and η. (b) Calculate $\overline{E}$ at the origin for $\omega t = 0, \pi/4, \pi/2, 3\pi/4$, and π.

Solution :

(a) β = coefficient of z = 2 rad/m

$$\omega = 2\pi f = 2\pi \times 50 \times 10^6 = 314.159 \times 10^6 \text{ rad/sec}$$

$$v = \omega/\beta = 157.080 \times 10^6 \text{ m/sec}$$

$$\lambda = 2\pi/\beta = 3.14159 \text{ m}$$

Using Eq.(10.13b),we have

$$\beta = 314.159 \times 10^6 \sqrt{4\pi \times 10^{-7} \times 1 \times 8.854 \times 10^{-12}\, \varepsilon_r} = 2$$

$$\therefore \varepsilon_r = 3.6426$$

$$\eta = \sqrt{\frac{\mu_o \mu_r}{\varepsilon_o \varepsilon_r}} = \sqrt{\frac{4\pi \times 10^{-7} \times 1}{8.854 \times 10^{-12} \times 3.6426}} = 197.392 \ \Omega$$

(b) $\overline{E}_s = (5\,\overline{a}_x + j\,10\,\overline{a}_y)\, e^{-j2z} = 5\,e^{-j2z}\,\overline{a}_x + j\,10\,e^{-j2z}\,\overline{a}_y$

$$= 5\,e^{-j2z}\,\overline{a}_x + 10\angle 90^\circ\, e^{-j2z}\,\overline{a}_y$$

$$= 5e^{-j2z}\,\bar{a}_x + 10e^{j90^\circ}e^{-j2z}\,\bar{a}_y$$

$$\therefore\ \bar{E} = 5\,\text{Re}\left[e^{-j2z}e^{j\omega t}\right]\bar{a}_x + 10\,\text{Re}\left[e^{j90^\circ}e^{-j2z}e^{j\omega t}\right]\bar{a}_y$$

$$= 5\,\text{Re}\left[e^{j(\omega t-2z)}\right]\bar{a}_x + 10\,\text{Re}\left[e^{j(\omega t-2z+90^\circ)}\right]\bar{a}_y$$

$$= 5\cos(\omega t-2z)\,\bar{a}_x + 10\cos(\omega t-2z+90^\circ)\,\bar{a}_y$$

$$= 5\cos(\omega t-2z)\,\bar{a}_x - 10\sin(\omega t-2z)\,\bar{a}_y$$

$\bar{E}$ at $(0,0,0) = 5\cos\omega t\,\bar{a}_x - 10\sin\omega t\,\bar{a}_y$

$\bar{E}$ at origin (for $\omega t = 0$) $= 5\,\bar{a}_x$ V/m

$\bar{E}$ at origin (for $\omega t = \pi/4$) $= 3.536\,\bar{a}_x - 7.071\,\bar{a}_y$

$\bar{E}$ at origin (for $\omega t = \pi/2$) $= -10\,\bar{a}_y$

$\bar{E}$ at origin (for $\omega t = 3\pi/4$) $= -3.536\,\bar{a}_x - 7.071\,\bar{a}_y$

$\bar{E}$ at origin (for $\omega t =$ pi) $= -5\,\bar{a}_x$

Problem 10.17 : A 1– MHz uniform plane wave with an amplitude of 25 V/m is propagating in a material for which $\varepsilon_r = 4$, $\mu_r = 9$, and $\sigma = 0$. Propagation is in the $\bar{a}_x$ direction. (a) Find v, β, and η. (b) Find $\bar{E}(t)$ if $E_z(x,y,z,t) = 0$ and $E_y = 25$ V/m at P (10, 10, 10) at 10^{-7} s. (c) Find $\bar{H}(t)$.

Solution :

$$\text{(a) } v = \frac{1}{\sqrt{\mu_o\mu_r\varepsilon_o\varepsilon_r}} = \frac{1}{\sqrt{4\pi\times10^{-7}\times9\times8.854\times10^{-12}\times4}}$$

$$= 49.966\times10^6 \text{ m/sec.}$$

Using Eq. (10.13b), we have

$$\beta = 2\pi\times10^6\sqrt{4\pi\times10^{-7}\times9\times8.854\times10^{-12}\times4} = 0.1257 \text{ rad/m}$$

$$\lambda = 2\pi/\beta = 49.966 \text{ m}$$

$$\eta = \sqrt{\frac{\mu_o\mu_r}{\varepsilon_o\varepsilon_r}} = \sqrt{\frac{4\pi\times10^{-7}\times9}{8.854\times10^{-12}\times4}} = 565.101\ \Omega$$

(b) $\bar{E}(t) = E_o\cos(\omega t - \beta x)\,\bar{a}_y$

Since the wave is propagating in + $\bar{a}_x$ direction, $-x$ is present in the equation for $\overline{\text{E}}(t)$. Also y component of electric field (E_y) is given. Therefore $\overline{\text{E}}(t)$ is multiplied by $\overline{\text{a}}_y$.

Now $E_y = 25$ V/m at $P(10, 10, 10)$ at 10^{-7} s

$\therefore 25 = E_o \cos(2\pi \times 10^6 \times 10^7 - 0.1257 \times 10) = E_o \cos(-0.62868)$

$= E_o \cos(-36.021^\circ)$. Hence $E_o = 30.910$

$\therefore \overline{\text{E}}(t) = 30.910 \cos(2\pi \times 10^6 t - 0.1257x)\, \overline{\text{a}}_y$ V/m

(c) Now $\overline{\text{P}} = \overline{\text{E}} \times \overline{\text{H}}$

$\overline{\text{a}}_x = \overline{\text{a}}_y \times ?$

Since $\overline{\text{a}}_y \times \overline{\text{a}}_z = \overline{\text{a}}_x$, z component of $\overline{\text{H}}$ is present.

$$\therefore \overline{\text{H}}(t) = \frac{30.910}{\eta} \cos(2\pi \times 10^6 t - 0.1257x)\, \overline{\text{a}}_z$$

$$= 54.698 \cos(2\pi \times 10^6 t - 0.1257x)\, \overline{\text{a}}_z \quad \text{mA/m}$$

Problem 10.18 : Using data given for Pyranol 1467, determine values of $\alpha, \beta, \lambda, v$ and η at 10^8 Hz for a uniform plane wave.

Solution :

For Pyranol 1467, we have

$\sigma = 2.5 \times 10^{-3}$ mhos/m and $\varepsilon_r = 4$ at 10^8 Hz

$$\tan\theta = \frac{\sigma}{\omega\varepsilon} = \frac{2.5 \times 10^{-3}}{2\pi \times 10^8 \times 8.854 \times 10^{-12} \times 4} = 0.112346 \quad \text{or} \quad \theta = 6.41^\circ$$

Using Eq. (10.16), we have

$$\alpha = 2\pi \times 10^8 \sqrt{0.5 \times 4\pi \times 10^{-7} \times 1 \times 8.854 \times 10^{-12} \times 4\,(\sec 6.41^\circ - 1)}$$

$= 0.235085$ Np/m

Using Eq.(10.17), we have

$$\beta = 2\pi \times 10^8 \sqrt{0.5 \times 4\pi \times 10^{-7} \times 1 \times 8.854 \times 10^{-12} \times 4\,(\sec 6.41^\circ + 1)}$$

$= 4.19823$ rad/m

$\lambda = 2\pi/\beta = 1.4966$ m

$$v = \frac{\omega}{\beta} = 149.663 \times 10^6 \text{ m/sec.}$$

$$\eta = \sqrt{\frac{\mu_o \mu_r}{\varepsilon_o \varepsilon_r}} = \sqrt{\frac{4\pi \times 10^{-7} \times 1}{8.854 \times 10^{-12} \times 4}} = 188.367\ \Omega$$

Problem 10.19 : A wave having an amplitude $E_{xo} =$ 600 V/m at (0, 0, 0, $t = 0$) is propagating in the $\bar{a}_z$ direction in a material where $\sigma = 0.1$ mohs/m, $\varepsilon = 2 \times 10^{-11}$ F/m, and $\mu = 2 \times 10^{-6}$ H/m. If $\omega = 10^{10}$ rad/s, find: (a) E_x at (0, 0, 2 cm, 0.1 ns) ; (b) H_y at (0,0, 2 cm, 0.1 ns).

Solution :

(a) $\tan\theta = \dfrac{\sigma}{\omega\varepsilon} = \dfrac{0.1}{10^{10} \times 2 \times 10^{-11}} = 0.5$ Hence $\theta = 26.565^\circ$

Using Eq.(10.16), wer have

$$\therefore \alpha = 10^{10}\sqrt{0.5 \times 2 \times 10^{-6} \times 2 \times 10^{-11}(\sec 26.565^\circ - 1)}$$

$$= 15.3645 \text{ Np/m}$$

Using Eq.(10.17),we have

$$\beta = 10^{10}\sqrt{0.5 \times 2 \times 10^{-6} \times 2 \times 10^{-11}(\sec 26.565^\circ + 1)}$$

$$= 65.0851 \text{ rad/m}$$

Equation for electric field intensity is $E_x = E_{xo}\, e^{-\alpha z} \cos(\omega t - \beta z)$ where E_{xo} is the value of electric field intensity at $z = 0$ and $t = 0$ which is equal to 600 V / m

$$E_x(0, 0, 0.02\text{ m}, 0.1 \times 10^{-9}\text{ s})$$

$$= 600\, e^{-15.3645 \times 0.02} \cos(10^{10} \times 0.1 \times 10^{-9} - 65.0851 \times 0.02)$$

$$= 600\, e^{-0.30729} \cos(-0.301702) = 600\, e^{-0.30729} \cos(-17.28625^\circ)$$

$$= 421.331 \text{ V/m}$$

(b) $$\eta = \sqrt{\frac{\mu/\varepsilon}{1 - j\tan\theta}} = \sqrt{\frac{2 \times 10^{-6}/2 \times 10^{-11}}{1 - j0.5}} = 299.070 \angle 13.283^\circ$$

Hence equation for H_y is

$$H_y = \frac{E_{xo}\, e^{-\alpha z}}{\eta_m} \cos(\omega t - \beta z - \theta_\eta)$$

$$= \frac{600\, e^{-15.3645 \times 0.02}}{299.07} \cos(10^{10} \times 0.1 \times 10^{-9} - 65.0851 \times 0.02 - 13.283^\circ)$$

$$= 1.475448 \cos(-0.301702 - 13.283^\circ)$$

$$= 1.475448 \cos(-17.286^\circ - 13.283^\circ)$$

$= 1.475448 \cos(-30.569^\circ) = 1.27038 \text{ A/m}$

Problem 10.20 : A lossy dielectric is characterized by $\varepsilon_r = 2.5$, $\mu_r = 4$, and $\sigma = 10^{-3}$ mho/m at 10 MHz. Let $\overline{E}_s = 20\,e^{-\gamma z}\,\bar{a}_x$ V/m and find : (a) α ; (b) β ; (c) v ; (d) λ; (e) η ; (f) $\overline{H}_s$; (g) $\overline{E}$ at (0, 0, 2 cm, 0.1 ns).

Solution :

$\varepsilon_r = 2.5$, $\mu_r = 4$, $\sigma = 10^{-3}$ mho/m, $f = 10$ MHz

(a) $$\tan\theta = \frac{\sigma}{\omega\varepsilon} = \frac{10^{-3}}{2\pi \times 10 \times 10^{6} \times 8.854 \times 10^{-12} \times 2.5} = 0.719$$

$\therefore\ \theta = 35.717^\circ$. Hence $\sec\theta - 1 = 0.23166$ and $\sec\theta + 1 = 2.23166$

Using Eq.(10.16),we have

$$\alpha = 2\pi \times 10 \times 10^{6} \sqrt{0.5 \times 4\pi \times 10^{7} \times 4 \times 8.854 \times 10^{-12} \times 2.5 \times 0.23166}$$
$$= 0.226 \text{ Np/m}$$

(b) Using Eq.(10.17),we have

$$\beta = 2\pi \times 10 \times 10^{6} \sqrt{0.5 \times 4\pi \times 10^{-7} \times 4 \times 8.854 \times 10^{-12} \times 2.5 \times 2.23166}$$
$$= 0.7 \text{ rad/m}$$

(c) $v = \omega/\beta = 89.760 \times 10^{6}$ m/sec

(d) $\lambda = 2\pi/\beta = 8.97598$ m

(e) $$\eta = \sqrt{\frac{\mu/\varepsilon}{1 - j\tan\theta}} = \sqrt{\frac{4\pi \times 10^{-7} \times 4/(8.854 \times 10^{-12} \times 2.5)}{1 - j\,0.719}}$$

$$= 429.390\ \angle 17.858^\circ\ \Omega$$

(f) The direction of $\overline{P}$ is $\bar{a}_z$ and x component of electric field intensity is given

Now $\overline{P} = \overline{E} \times \overline{H}$

$\bar{a}_z = \bar{a}_x \times\ ?$

Since $\bar{a}_x \times \bar{a}_y = \bar{a}_z$, y component of magnetic field intensity is present.

$$\therefore\ \overline{H}_s = \frac{20\,e^{-\gamma z}}{\eta}\,\bar{a}_y = \frac{20\,e^{-\gamma z}}{429.39\,e^{j\,17.858^\circ}}\,\bar{a}_y$$

$$= 46.578\ e^{-(\alpha + j\beta)z}\ e^{-j\,17.858^\circ}\ \bar{a}_y \text{ mA / m.}$$

$$= 46.578\ e^{-\alpha z}\ e^{-j(\beta z + 17.858^\circ)}\ \bar{a}_y$$

$$= 46.578\ e^{-0.226z}\ e^{-j(0.7z + 17.858^\circ)}\ \bar{a}_y \ \text{mA/m}$$

(g) $\bar{E}_s = 20\,e^{-\gamma z}\ \bar{a}_x = 20\ e^{-(\alpha + j\beta)z}\ \bar{a}_x = 20\,e^{-\alpha z}\ e^{-j\beta z}\ \bar{a}_x$

$\bar{E} = 20\,e^{-\alpha z}\ \text{Re}\left[\ e^{j\omega t}\ e^{-j\beta z}\right]\ \bar{a}_x = 20\,e^{-\alpha z}\ \cos(\omega t - \beta z)\ \bar{a}_x$

$\therefore\ \bar{E}\,(2, 3, 4, 10\ \text{ns})$

$$= 20\,e^{-0.226\times 4}\ \cos(2\pi\times 10\times 10^6\times 10\times 10^{-9} - 0.7\times 4)\ \bar{a}_x$$

$$= 8.0989\ \cos(-\ 2.17168)\ \bar{a}_x = 8.0989\ \cos(-\ 124.428^\circ)\ \bar{a}_x$$

$$= -\ 4.5789\ \bar{a}_x\ \text{V/m}$$

Problem 10.21 : Parameters for a 50 – MHz uniform plane wave propagating in the $\bar{a}_z$ direction in a certain lossy material are $\gamma = 0.25 + j\,2\ \text{m}^{-1}$ and $\eta = 600 + j\,75\ \Omega$. (a) Find μ_r, ε_r, and $\sigma = 0$; (b) $\bar{E}_s = 100\ \bar{a}_x$ V/m at $z = 0$, find $\bar{E}_s$ at $z = 2$ m, and also (c) $\bar{H}_s$ at $z = 2$ m.

Solution :

(a) $f = 50\ \text{MHz}$; $\gamma = 0.25 + j\,2\ \text{m}^{-1} = \alpha + j\beta$

$\eta = 600 + j\,75\ \Omega$

$\alpha = \omega\sqrt{0.5\,\mu\,\varepsilon\,(\sec\theta - 1)}$ ......(1)

$\beta = \omega\sqrt{0.5\,\mu\,\varepsilon\,(\sec\theta + 1)}$ ..... (2)

$\tan\theta = \dfrac{\sigma}{\omega\,\varepsilon}$ .....(3)

Divide equation (1) by equation (2)

$$\frac{\alpha}{\beta} = \sqrt{\frac{\sec\theta - 1}{\sec\theta + 1}} = \frac{0.25}{2} = 0.125$$

$\therefore\ \sec\theta - 1 = 0.015625\,(\sec\theta + 1)$ or $\sec\theta = \dfrac{1.015625}{0.984375} = 1.031746$

$\therefore\ \theta = 14.25^\circ$

From equation (1), $0.25 = 2\pi\times 50\times 10^6\sqrt{0.5\,\mu\,\varepsilon\,(\sec 14.25^\circ - 1)}$, or

$\mu\,\varepsilon = 3.9895\times 10^{-17}$ ......(4)

Now $\eta = \sqrt{\dfrac{\mu/\varepsilon}{1 - j\tan\theta}} = \sqrt{\dfrac{\mu/\varepsilon}{1 - j\tan 14.25^\circ}} = 600 + j\,75$

$\therefore\ \sqrt{(\mu/\varepsilon)\times 0.96923}\ \angle 7.125^\circ = 604.669\ \angle 7.125^\circ$

$\therefore\ \mu/\varepsilon = 377232.029$ ......(5)

Multiplying equation (4) by equation (5) we have

$\mu\,\varepsilon \times (\mu/\varepsilon) = 3.9895 \times 10^{-17} \times 377232.029$ or $\mu^2 = 1.504967 \times 10^{-11}$

Hence $\mu = 3.87939 \times 10^{-6}$ or $\mu_r = 3.08712$

From equation (4), we have

$$\varepsilon = \frac{3.9895 \times 10^{-17}}{3.87939 \times 10^{-6}} = 1.02838 \times 10^{-11} \text{ or } \varepsilon_r = 1.16149$$

From equation (3), we have

$$\tan 14.25^\circ = \frac{\sigma}{2\pi \times 50 \times 10^6 \times 1.02838 \times 10^{-11}}$$

$\therefore\ \sigma = 8.20506 \times 10^{-4}$ mho / m

(b) $\overline{E}_s = E_{xo}\, e^{-\gamma z}\, \overline{a}_x$ where $E_{xo} = 100$ V/m

$$\therefore\ \overline{E}_s = 100\, e^{-(0.25 + j2)2}\, \overline{a}_x = 100\, e^{-0.5}\, e^{-j4}\, \overline{a}_x$$

$$= 60.653\ \angle -4\ \overline{a}_x = 60.653\ \angle -229.183^\circ\ \overline{a}_x$$

(c) The direction of $\overline{P}$ is $\overline{a}_z$ and that of $\overline{E}$ is $\overline{a}_x$

$\therefore\ \overline{P} = \overline{E} \times \overline{H}$

$\overline{a}_z = \overline{a}_x \times\ ?$

Hence y component of $\overline{H}$ must be present.

$$\therefore\ \overline{H}_s = \frac{\overline{E}_s}{\eta} = \frac{E_{xo}\, e^{-\gamma z}\, \overline{a}_y}{600 + j75} = \frac{E_{xo}\, e^{-(\alpha + j\beta)z}\, \overline{a}_y}{604.669\ \angle 7.125^\circ}$$

$$= \frac{E_{xo}\, e^{-\alpha z}\, e^{-j\beta z}\, \overline{a}_y}{604.669\ e^{j7.125^\circ}} = \frac{100\, e^{-0.25z}}{604.669}\, e^{-j(2z + 7.125^\circ)}\, \overline{a}_y$$

At $z = 2$,

$$\overline{H}_s = \frac{100\, e^{-0.25 \times 2}}{604.669}\, e^{-j(2 \times 2 + 7.125^\circ)}\, \overline{a}_y = 100.308\ \angle -(4 + 7.125^\circ)\ \overline{a}_y$$

$$= 100.308\ \angle -(229.183^\circ + 7.125^\circ)\ \overline{a}_y = 100.308\ \angle -236.308^\circ\ \overline{a}_y \text{ mA/m}$$

Problem 10.22 : A 30 – GHz radar signal may be represented as a uniform plane wave in a sufficiently small region. Calculate the wavelength in centimeters and the attenuation in decibels per foot if the wave is

propagating in a nonmagnetic material for which : (a) $\varepsilon_r = 1$ and $\sigma = 0$; (b) $\varepsilon_r = 1.01$ and $\sigma = 10^{-3}$ mho/m ; (c) $\varepsilon_r = 2.1$ & $\sigma = 5$ mho/m.

Solution :

(a) $f = 30 \times 10^9$ Hz , $\mu_r = 1$, $\varepsilon_r = 1$, $\sigma = 0$, $\omega = 2\pi f = 60\pi \times 10^9$ rad/m

Using Eq. (10.13b),we have,

$$\beta = 60\pi \times 10^9 \sqrt{4\pi \times 10^{-7} \times 8.854 \times 10^{-12}} = 628.7468 \text{ rad/m}$$

$$\therefore \lambda = 2\pi/\beta = 0.9993 \text{ cm}$$

$$\alpha = 0 \text{ as } \sigma = 0$$

(b) $\varepsilon_r = 1.01$, $\sigma = 10^{-3}$ mho/m ,

$$\tan\theta = \frac{10^{-3}}{2\pi \times 30 \times 10^9 \times 8.854 \times 10^{-12} \times 1.01} = 5.9325 \times 10^{-4}$$

$$\therefore \theta = 0.0339907^\circ , \sec\theta + 1 = 2 , \sec\theta - 1 = 1.7597 \times 10^{-7}$$

Using Eq. (10.17), we have

$$\beta = 60\pi \times 10^9 \sqrt{0.5 \times 4\pi \times 10^{-7} \times 1 \times 8.854 \times 10^{-12} \times 1.01 \times 2}$$

$$= 631.883 \text{ rad/m}$$

$$\therefore \lambda = 2\pi/\beta = 0.994 \text{ cm}$$

Using Eq.(10.16),we have

$$\alpha = 60\pi \times 10^9 \sqrt{0.5 \times 4\pi \times 10^{-7} \times 8.854 \times 10^{-12} \times 1.01 \times 1.7597 \times 10^{-7}}$$

$$= 0.187446716 \text{ Np/m}$$

The amplitude of electric field is given by $E_{xo}\, e^{-\alpha z}$. If $z = 0$ the amplitude is E_{xo} . If $z = 1$ feet $= 0.3048$ m, then the amplitude is

$$E_{xo}\, e^{-0.187446716 \times 0.3048} = 0.94446\, E_{xo}$$

$$\therefore \text{Attenuation in db / feet} = 20 \log\left(\frac{0.9446\, E_{xo}}{E_{xo}}\right) = -0.4962585 \text{ db/feet}$$

(c) $\varepsilon_r = 2.1$, $\sigma = 5$ mho/m

$$\tan\theta = \frac{5}{2\pi \times 30 \times 10^9 \times 8.854 \times 10^{-12} \times 2.1} = 1.4266 \text{ or } \theta = 54.971^\circ$$

$$\sec\theta - 1 = 0.7421789 \text{ and } \sec\theta + 1 = 2.7421789$$

Using Eq. (10.17), we have

$$\beta = 60\pi \times 10^9 \sqrt{0.5 \times 4\pi \times 10^{-7} \times 1 \times 8.854 \times 10^{-12} \times 2.1 \times 2.7421789}$$

$= 1066.891$ rad/

$\therefore\ \lambda = 2\pi/\beta = 0.5889$ cm

Using Eq.(10.16), we have

$$\alpha = 60\pi \times 10^9 \sqrt{0.5 \times 4\pi \times 10^{-7} \times 1 \times 8.854 \times 10^{-12} \times 2.1 \times 0.7421789}$$

$= 555.044$ Np/m

$$\text{Attenuation in decibels / foot} = 20 \log \frac{E_{xo}\, e^{-\alpha z}}{E_{xo}}$$

where 1 foot $= 0.3048$ m and $\alpha = 555.044$ Np/m

$\therefore$ Attenuation in db / feet $=$ 1469.459 db / ft.

—X—